Texts in Applied Mathematics 17

Editors
J.E. Marsden
L. Sirovich
M. Golubitsky
W. Jäger

Advisors
D. Barkley
M. Dellnitz
P. Holmes
G. Iooss
P. Newton

Springer
New York
Berlin
Heidelberg
Barcelona
Hong Kong
London
Milan
Paris
Singapore
Tokyo

Texts in Applied Mathematics

Jerrold E. Marsden   Tudor S. Ratiu

Introduction to Mechanics and Symmetry

A Basic Exposition of Classical Mechanical Systems

Second Edition

With 54 Illustrations

 Springer

Jerrold E. Marsden
Control and Dynamical Systems, 107-81
California Institute of Technology
Pasadena, CA 91125
USA

Tudor S. Ratiu
Ecole polytechnique fédérale
  de Lausanne
Département de mathématiques
CH-1015 Lausanne
Switzerland

Series Editors

J.E. Marsden
Control and Dynamical Systems, 107–81
California Institute of Technology
Pasadena, CA 91125
USA

L. Sirovich
Division of Applied Mathematics
Brown University
Providence, RI 02912
USA

M. Golubitsky
Department of Mathematics
University of Houston
Houston, TX 77204-3476
USA

W. Jäger
Department of Applied Mathematics
Universität Heidelberg
Im Neuenheimer Feld 294
69120 Heidelberg
Germany

ISBN 978-1-4419-3143-6

e-ISBN 978-0-387-21792-5

Mathematics Subject Classification (1991): 58-01, 70-01, 70Hxx

Library of Congress Cataloging-in-Publication Data
Marsden, Jerrold E.
  Introduction to mechanics and symmetry / Jerrold E. Marsden,
Tudor S. Ratiu. – 2nd ed.
    p. cm. – (Texts in applied mathematics ; 17)
  Includes bibliographical references and index.

  1. Mechanics, Analytic.   2. Symmetry (Physics)   I. Ratiu, Tudor S.
II. Title.   III. Series.
    QA808.M33 1999
    531–dc21   98-44773

Printed on acid-free paper.

To Barbara and Lilian for their love and support

Series Preface

Mathematics is playing an ever more important role in the physical and biological sciences, provoking a blurring of boundaries between scientific disciplines and a resurgence of interest in the modern as well as the classical techniques of applied mathematics. This renewal of interest, both in research and teaching, has led to the establishment of the series: *Texts in Applied Mathematics (TAM)*.

The development of new courses is a natural consequence of a high level of excitement on the research frontier as newer techniques, such as numerical and symbolic computer systems, dynamical systems, and chaos, mix with and reinforce the traditional methods of applied mathematics. Thus, the purpose of this textbook series is to meet the current and future needs of these advances and encourage the teaching of new courses.

TAM will publish textbooks suitable for use in advanced undergraduate and beginning graduate courses, and will complement the *Applied Mathematical Sciences (AMS)* series, which will focus on advanced textbooks and research level monographs.

Preface

Symmetry and mechanics have been close partners since the time of the founding masters: Newton, Euler, Lagrange, Laplace, Poisson, Jacobi, Hamilton, Kelvin, Routh, Riemann, Noether, Poincaré, Einstein, Schrödinger, Cartan, Dirac, and to this day, symmetry has continued to play a strong role, especially with the modern work of Kolmogorov, Arnold, Moser, Kirillov, Kostant, Smale, Souriau, Guillemin, Sternberg, and many others. This book is about these developments, with an emphasis on concrete applications that we hope will make it accessible to a wide variety of readers, especially senior undergraduate and graduate students in science and engineering.

The geometric point of view in mechanics combined with solid analysis has been a phenomenal success in linking various diverse areas, both within and across standard disciplinary lines. It has provided both insight into fundamental issues in mechanics (such as variational and Hamiltonian structures in continuum mechanics, fluid mechanics, and plasma physics) and provided useful tools in specific models such as new stability and bifurcation criteria using the energy–Casimir and energy–momentum methods, new numerical codes based on geometrically exact update procedures and variational integrators, and new reorientation techniques in control theory and robotics.

Symmetry was already widely used in mechanics by the founders of the subject, and has been developed considerably in recent times in such diverse phenomena as reduction, stability, bifurcation and solution symmetry breaking relative to a given system symmetry group, methods of finding explicit solutions for integrable systems, and a deeper understanding of special systems, such as the Kowalewski top. We hope this book will provide a reasonable avenue to, and foundation for, these exciting developments.

Because of the extensive and complex set of possible directions in which one can develop the theory, we have provided a fairly lengthy introduction. *It is intended to be read lightly at the beginning and then consulted from time to time as the text itself is read.*

This volume contains much of the basic theory of mechanics and should prove to be a useful foundation for further, as well as more specialized, topics. Due to space limitations we warn the reader that many important topics in mechanics are not treated in this volume. We are preparing a second volume on general reduction theory and its applications. With luck, a little support, and yet more hard work, it will be available in the near future.

Solutions Manual.   A solution manual is available for instructors. It contains complete solutions to many of the exercises, as well as other supplementary comments. For further information, see

> http://www.cds.caltech.edu/~marsden/books/.

Internet Supplements.   To keep the size of the book within reason, we are making some material available (free) on the Internet. These are a collection of sections whose omission does not interfere with the main flow of the text. See http://www.cds.caltech.edu/~marsden/books/. Updates and information about the book can also be found at this website.

What Is New in the Second Edition?   In this second edition, the main structural changes are the creation of a solutions manual (along with many more exercises in the text) and the Internet supplements. The Internet supplements contain, for example, the material on the Maslov index that was not needed for the main flow of the book. As for the substance of the text, much of the book was rewritten throughout to improve the flow of material and to correct inaccuracies. Some examples: The material on the Hamilton–Jacobi theory was completely rewritten, a new section on Routh reduction (§8.9) was added, Chapter 9 on Lie groups was substantially improved and expanded. The presentation of examples of coadjoint orbits (Chapter 14) was improved by stressing matrix methods throughout.

Acknowledgments.   We thank Rudolf Schmid, Rich Spencer, and Alan Weinstein for helping with an early set of notes that helped us on our way. Our many colleagues, students, and readers, especially Henry Abarbanel, Vladimir Arnold, Larry Bates, Michael Berry, Tony Bloch, Dong-Eui Chang, Hans Duistermaat, Marty Golubitsky, Mark Gotay, George Haller, Aaron Hershman, Darryl Holm, Phil Holmes, Sameer Jalnapurkar, Edgar Knobloch, P.S. Krishnaprasad, Naomi Leonard, Debra Lewis, Robert Littlejohn, Richard Montgomery, Phil Morrison, Richard Murray, Peter Olver, Oliver O'Reilly, Juan-Pablo Ortega, George Patrick, Octavian Popp, Mason Porter, Matthias Reinsch, Shankar Sastry, Tanya Schmah, Juan Simo,

Hans Troger, Loc Vu-Quoc, and Steve Wiggins, have our deepest gratitude for their encouragement and suggestions. We also collectively thank all our students and colleagues who have used these notes and have provided valuable advice.

We are also indebted to Carol Cook, Anne Kao, Nawoyuki Gregory Kubota, Sue Knapp, Barbara Marsden, Marnie McElhiney, June Meyermann, Teresa Wild, and Ester Zack for their dedicated and patient work on the typesetting and artwork for this book. We want to single out with special thanks Hendra Adiwidjaja, Nawoyuki Gregory Kubota, and Wendy McKay for their special effort with the typesetting, the scripts for automatic conversion of references, the macros for indexing, and the figures (including the cover illustration). We also thank the staff at Springer-Verlag, especially Achi Dosanjh, Laura Carlson, MaryAnn Cottone, David Kramer, Ken Dreyhaupt, and Rüdiger Gebauer for their skillful editorial work and production of the book.

JERRY MARSDEN
Pasadena, California

TUDOR RATIU
Santa Cruz, California

December, 1998

About the Authors

Jerrold E. Marsden is professor of control and dynamical systems at Caltech. He received his B.Sc. at Toronto in 1965 and his Ph.D. from Princeton University in 1968, both in applied mathematics. He has done extensive research in mechanics, with applications to rigid-body systems, fluid mechanics, elasticity theory, plasma physics, as well as to general field theory. His primary current interests are in the area of dynamical systems and control theory, especially how it relates to mechanical systems with symmetry. He is one of the founders in the early 1970s of reduction theory for mechanical systems with symmetry, which remains an active and much studied area of research today. He was the recipient of the prestigious Norbert Wiener prize of the American Mathematical Society and the Society for Industrial and Applied Mathematics in 1990, and was elected a fellow of the American Academy of Arts and Sciences in 1997. He has been a Carnegie Fellow at Heriot–Watt University (1977), a Killam Fellow at the University of Calgary (1979), recipient of the Jeffrey–Williams prize of the Canadian Mathematical Society in 1981, a Miller Professor at the University of California, Berkeley (1981–1982), a recipient of the Humboldt Prize in Germany (1991), and a Fairchild Fellow at Caltech (1992). He has served in several administrative capacities, such as director of the Research Group in Nonlinear Systems and Dynamics at Berkeley, 1984–86, the Advisory Panel for Mathematics at NSF, the Advisory committee of the Mathematical Sciences Institute at Cornell, and as director of the Fields Institute, 1990–1994. He has served as an editor for Springer-Verlag's *Applied Mathematical Sciences* series since 1982 and serves on the editorial boards of several journals in mechanics, dynamics, and control.

Tudor S. Ratiu is professor of mathematics at the University of California, Santa Cruz and the Swiss Federal Institute of Technology, in Lausanne. He received his B.Sc. in mathematics and M.Sc. in applied mathematics, both at the University of Timişoara, Romania, and his Ph.D. in mathematics at Berkeley in 1980. He has previously taught at the University of Michigan, Ann Arbor, as a T. H. Hildebrandt Research Assistant Professor (1980–1983) and at the University of Arizona, Tucson (1983–1987). His research interests center on geometric mechanics, symplectic geometry, global analysis, and infinite-dimensional Lie theory, with applications to integrable systems, nonlinear dynamics, continuum mechanics, plasma physics, and bifurcation theory. He was a National Science Foundation Postdoctoral Fellow (1983–86), a Sloan Foundation Fellow (1984–87), a Miller Research Professor at Berkeley (1994), and a recipient of of the Humboldt Prize in Germany (1997). Since his arrival at UC Santa Cruz in 1987, he has been on the executive committee of the Nonlinear Sciences Organized Research Unit. He is managing editor of the AMS *Surveys and Monographs* series and on the editorial board of the *Annals of Global Analysis* and the *Annals of the University of Timişoara*. He was also a member of various research institutes such as MSRI in Berkeley, the Center for Nonlinear Studies at Los Alamos, the Max Planck Institute in Bonn, MSI at Cornell, IHES in Bures-sur-Yvette, The Fields Institute in Toronto (Waterloo), the Erwin Schroödinger Institute for Mathematical Physics in Vienna, the Isaac Newton Institute in Cambridge, and RIMS in Kyoto.

Contents

1

Introduction and Overview

1.1 Lagrangian and Hamiltonian Formalisms

Mechanics deals with the dynamics of particles, rigid bodies, continuous media (fluid, plasma, and elastic materials), and field theories such as electromagnetism and gravity. This theory plays a crucial role in quantum mechanics, control theory, and other areas of physics, engineering, and even chemistry and biology. Clearly, mechanics is a large subject that plays a fundamental role in science. Mechanics also played a key part in the development of mathematics. Starting with the creation of calculus stimulated by Newton's mechanics, it continues today with exciting developments in group representations, geometry, and topology; these mathematical developments in turn are being applied to interesting problems in physics and engineering.

Symmetry plays an important role in mechanics, from fundamental formulations of basic principles to concrete applications, such as stability criteria for rotating structures. The theme of this book is to emphasize the role of symmetry in various aspects of mechanics.

This introduction treats a collection of topics fairly rapidly. The student should not expect to understand everything perfectly at this stage. *We will return to many of the topics in subsequent chapters.*

Lagrangian and Hamiltonian Mechanics. Mechanics has two main points of view, ***Lagrangian mechanics*** and ***Hamiltonian mechanics***. In one sense, Lagrangian mechanics is more fundamental, since it is based on variational principles and it is what generalizes most directly to the general relativistic context. In another sense, Hamiltonian mechanics is more fundamental, since it is based directly on the energy concept and it is what is more closely tied to quantum mechanics. Fortunately, in many cases these branches are equivalent, as we shall see in detail in Chapter 7. Needless to say, the merger of quantum mechanics and general relativity remains one of the main outstanding problems of mechanics. In fact, the methods of mechanics and symmetry are important ingredients in the developments of string theory, which has attempted this merger.

Lagrangian Mechanics. The Lagrangian formulation of mechanics is based on the observation that there are variational principles behind the fundamental laws of force balance as given by Newton's law $\mathbf{F} = m\mathbf{a}$. One chooses a configuration space Q with coordinates q^i, $i = 1, \ldots, n$, that describe the *configuration* of the system under study. Then one introduces the *Lagrangian* $L(q^i, \dot{q}^i, t)$, which is shorthand notation for $L(q^1, \ldots, q^n, \dot{q}^1, \ldots, \dot{q}^n, t)$. Usually, L is the kinetic *minus* the potential energy of the system, and one takes $\dot{q}^i = dq^i/dt$ to be the system velocity. The *variational principle of Hamilton* states

$$\delta \int_a^b L(q^i, \dot{q}^i, t)\, dt = 0. \tag{1.1.1}$$

In this principle, we choose curves $q^i(t)$ joining two fixed points in Q over a fixed time interval $[a, b]$ and calculate the integral regarded as a function of this curve. Hamilton's principle states that this function has a critical point at a solution within the space of curves. If we let δq^i be a variation, that is, the derivative of a family of curves with respect to a parameter, then by the chain rule, (1.1.1) is equivalent to

$$\sum_{i=1}^n \int_a^b \left(\frac{\partial L}{\partial q^i} \delta q^i + \frac{\partial L}{\partial \dot{q}^i} \delta \dot{q}^i \right) dt = 0 \tag{1.1.2}$$

for all variations δq^i.

Using equality of mixed partials, one finds that

$$\delta \dot{q}^i = \frac{d}{dt} \delta q^i.$$

Using this, integrating the second term of (1.1.2) by parts, and employing the boundary conditions $\delta q^i = 0$ at $t = a$ and b, (1.1.2) becomes

$$\sum_{i=1}^n \int_a^b \left[\frac{\partial L}{\partial q^i} - \frac{d}{dt} \left(\frac{\partial L}{\partial \dot{q}^i} \right) \right] \delta q^i\, dt = 0. \tag{1.1.3}$$

Since δq^i is arbitrary (apart from being zero at the endpoints), (1.1.2) is equivalent to the *Euler–Lagrange equations*

$$\frac{d}{dt} \frac{\partial L}{\partial \dot{q}^i} - \frac{\partial L}{\partial q^i} = 0, \quad i = 1, \ldots, n. \tag{1.1.4}$$

As Hamilton [1834] realized, one can gain valuable information by *not* imposing the fixed endpoint conditions. We will have a deeper look at such issues in Chapters 7 and 8.

For a system of N particles moving in Euclidean 3-space, we choose the configuration space to be $Q = \mathbb{R}^{3N} = \mathbb{R}^3 \times \cdots \times \mathbb{R}^3$ (N times), and L often has the form of kinetic minus potential energy:

$$L(\mathbf{q}_i, \dot{\mathbf{q}}_i, t) = \frac{1}{2} \sum_{i=1}^{N} m_i \|\dot{\mathbf{q}}_i\|^2 - V(\mathbf{q}_i), \qquad (1.1.5)$$

where we write points in Q as $\mathbf{q}_1, \ldots, \mathbf{q}_N$, where $\mathbf{q}_i \in \mathbb{R}^3$. In this case the Euler–Lagrange equations (1.1.4) reduce to **Newton's second law**

$$\frac{d}{dt}(m_i \dot{\mathbf{q}}_i) = -\frac{\partial V}{\partial \mathbf{q}_i}, \quad i = 1, \ldots, N, \qquad (1.1.6)$$

that is, $\mathbf{F} = ma$ for the motion of particles in the potential field V. As we shall see later, in many examples more general Lagrangians are needed.

Generally, in Lagrangian mechanics, one identifies a configuration space Q (with coordinates $(q^1, \ldots, q^n)$) and then forms the **velocity phase space** TQ, also called the **tangent bundle** of Q. Coordinates on TQ are denoted by

$$(q^1, \ldots, q^n, \dot{q}^1, \ldots, \dot{q}^n),$$

and the Lagrangian is regarded as a function $L : TQ \to \mathbb{R}$.

Already at this stage, interesting links with geometry are possible. If $g_{ij}(q)$ is a given metric tensor or **mass matrix** (for now, just think of this as a q-dependent positive definite symmetric $n \times n$ matrix) and we consider the kinetic energy Lagrangian

$$L(q^i, \dot{q}^i) = \frac{1}{2} \sum_{i,j=1}^{n} g_{ij}(q) \dot{q}^i \dot{q}^j, \qquad (1.1.7)$$

then *the Euler–Lagrange equations are equivalent to the equations of geodesic motion*, as can be directly verified (see §7.5 for details). Conservation laws that are a result of symmetry in a mechanical context can then be applied to yield interesting geometric facts. For instance, theorems about geodesics on surfaces of revolution can be readily proved this way.

The Lagrangian formalism can be extended to the infinite-dimensional case. One view (but not the only one) is to replace the q^i by *fields* $\varphi^1, \ldots, \varphi^m$ that are, for example, functions of spatial points x^i and time. Then L is a function of $\varphi^1, \ldots, \varphi^m, \dot{\varphi}^1, \ldots, \dot{\varphi}^m$ and the spatial derivatives of the fields. We shall deal with various examples of this later, but we emphasize that properly interpreted, the variational principle and the Euler–Lagrange equations remain intact. One replaces the partial derivatives in the Euler–Lagrange equations by *functional derivatives* defined below.

Hamiltonian Mechanics. To pass to the Hamiltonian formalism, introduce the *conjugate momenta*

$$p_i = \frac{\partial L}{\partial \dot{q}^i}, \qquad i = 1, \dots, n, \tag{1.1.8}$$

make the change of variables $(q^i, \dot{q}^i) \mapsto (q^i, p_i)$, and introduce the *Hamiltonian*

$$H(q^i, p_i, t) = \sum_{j=1}^{n} p_j \dot{q}^j - L(q^i, \dot{q}^i, t). \tag{1.1.9}$$

Remembering the change of variables, we make the following computations using the chain rule:

$$\frac{\partial H}{\partial p_i} = \dot{q}^i + \sum_{j=1}^{n} \left(p_j \frac{\partial \dot{q}^j}{\partial p_i} - \frac{\partial L}{\partial \dot{q}^j} \frac{\partial \dot{q}^j}{\partial p_i} \right) = \dot{q}^i \tag{1.1.10}$$

and

$$\frac{\partial H}{\partial q^i} = \sum_{j=1}^{n} p_j \frac{\partial \dot{q}^j}{\partial q^i} - \frac{\partial L}{\partial q^i} - \sum_{j=1}^{n} \frac{\partial L}{\partial \dot{q}^j} \frac{\partial \dot{q}^j}{\partial q^i} = -\frac{\partial L}{\partial q^i}, \tag{1.1.11}$$

where (1.1.8) has been used twice. Using (1.1.4) and (1.1.8), we see that (1.1.11) is equivalent to

$$\frac{\partial H}{\partial q^i} = -\frac{d}{dt} p_i. \tag{1.1.12}$$

Thus, *the Euler–Lagrange equations are equivalent to* **Hamilton's equations**

$$\frac{dq^i}{dt} = \frac{\partial H}{\partial p_i},$$
$$\frac{dp_i}{dt} = -\frac{\partial H}{\partial q^i}, \tag{1.1.13}$$

where $i = 1, \dots, n$. The analogous Hamiltonian partial differential equations for time-dependent *fields* $\varphi^1, \dots, \varphi^m$ and their conjugate momenta $\pi_1, \dots, \pi_m$ are

$$\frac{\partial \varphi^a}{\partial t} = \frac{\delta H}{\delta \pi_a},$$
$$\frac{\partial \pi_a}{\partial t} = -\frac{\delta H}{\delta \varphi^a}, \tag{1.1.14}$$

where $a = 1, \ldots, m$, H is a functional of the fields φ^a and π_a, and the *variational*, or *functional*, *derivatives* are defined by the equation

$$\int_{\mathbb{R}^n} \frac{\delta H}{\delta \varphi^1} \delta \varphi^1 \, d^n x = \lim_{\varepsilon \to 0} \frac{1}{\varepsilon} [H(\varphi^1 + \varepsilon \delta \varphi^1, \varphi^2, \ldots, \varphi^m, \pi_1, \ldots, \pi_m)$$
$$- H(\varphi^1, \varphi^2, \ldots, \varphi^m, \pi_1, \ldots, \pi_m)], \quad (1.1.15)$$

and similarly for $\delta H/\delta \varphi^2, \ldots, \delta H/\delta \pi_m$. Equations (1.1.13) and (1.1.14) can be recast in *Poisson bracket form*:

$$\dot{F} = \{F, H\}, \quad (1.1.16)$$

where the brackets in the respective cases are given by

$$\{F, G\} = \sum_{i=1}^{n} \left(\frac{\partial F}{\partial q^i} \frac{\partial G}{\partial p_i} - \frac{\partial F}{\partial p_i} \frac{\partial G}{\partial q^i} \right) \quad (1.1.17)$$

and

$$\{F, G\} = \sum_{a=1}^{m} \int_{\mathbb{R}^n} \left(\frac{\delta F}{\delta \varphi^a} \frac{\delta G}{\delta \pi_a} - \frac{\delta F}{\delta \pi_a} \frac{\delta G}{\delta \varphi^a} \right) d^n x. \quad (1.1.18)$$

Associated to any configuration space Q (coordinatized by $(q^1, \ldots, q^n)$) is a phase space T^*Q called the *cotangent bundle* of Q, which has coordinates $(q^1, \ldots, q^n, p_1, \ldots, p_n)$. On this space, the canonical bracket (1.1.17) is intrinsically defined in the sense that the value of $\{F, G\}$ is independent of the choice of coordinates. Because the Poisson bracket satisfies $\{F, G\} = -\{G, F\}$ and in particular $\{H, H\} = 0$, we see from (1.1.16) that $\dot{H} = 0$; that is, *energy is conserved*. This is the most elementary of many deep and beautiful *conservation properties* of mechanical systems.

There is also a variational principle on the Hamiltonian side. For the Euler–Lagrange equations, we deal with curves in q-space (configuration space), whereas for Hamilton's equations we deal with curves in (q, p)-space (momentum phase space). The principle is

$$\delta \int_a^b \sum_{i=1}^{n} [p_i \dot{q}^i - H(q^j, p_j)] \, dt = 0, \quad (1.1.19)$$

as is readily verified; one requires $p_i \delta q^i = 0$ at the endpoints.

This formalism is the basis for the analysis of many important systems in particle dynamics and field theory, as described in standard texts such as Whittaker [1927], Goldstein [1980], Arnold [1989], Thirring [1978], and Abraham and Marsden [1978]. The underlying geometric structures that are important for this formalism are those of *symplectic* and *Poisson geometry*. How these structures are related to the Euler–Lagrange equations and variational principles via the Legendre transformation is an essential ingredient of the story. Furthermore, in the infinite-dimensional case it is fairly well understood how to deal rigorously with many of the functional analytic difficulties that arise; see, for example, Chernoff and Marsden [1974] and Marsden and Hughes [1983].

Exercises

◇ **1.1-1.** Show by *direct calculation* that the classical Poisson bracket satisfies the *Jacobi identity*. That is, if F and K are both functions of the $2n$ variables $(q^1, q^2, \ldots, q^n, p_1, p_2, \ldots, p_n)$ and we define

$$\{F, K\} = \sum_{i=1}^{n} \left(\frac{\partial F}{\partial q^i} \frac{\partial K}{\partial p_i} - \frac{\partial K}{\partial q^i} \frac{\partial F}{\partial p_i} \right),$$

then the identity $\{L, \{F, K\}\} + \{K, \{L, F\}\} + \{F, \{K, L\}\} = 0$ holds.

1.2   The Rigid Body

It was already clear in the last century that certain mechanical systems resist the canonical formalism outlined in §1.1. For example, to obtain a Hamiltonian description for fluids, Clebsch [1857, 1859] found it necessary to introduce certain nonphysical potentials.[1] We will discuss fluids in §1.4 below.

Euler's Rigid-Body Equations.   In the absence of external forces, the Euler equations for the rotational dynamics of a rigid body about its center of mass are usually written as follows, as we shall derive in detail in Chapter 15:

$$I_1 \dot{\Omega}_1 = (I_2 - I_3)\Omega_2 \Omega_3,$$
$$I_2 \dot{\Omega}_2 = (I_3 - I_1)\Omega_3 \Omega_1, \qquad (1.2.1)$$
$$I_3 \dot{\Omega}_3 = (I_1 - I_2)\Omega_1 \Omega_2,$$

where $\Omega = (\Omega_1, \Omega_2, \Omega_3)$ is the body angular velocity vector (the angular velocity of the rigid body as seen from a frame fixed in the body) and I_1, I_2, I_3 are constants depending on the shape and mass distribution of the body—the principal moments of inertia of the rigid body.

Are equations (1.2.1) Lagrangian or Hamiltonian in any sense? Since there is an *odd* number of equations, they obviously cannot be put in canonical Hamiltonian form in the sense of equations (1.1.13).

---

[1]For a geometric account of Clebsch potentials and further references, see Marsden and Weinstein [1983], Marsden, Ratiu, and Weinstein [1984a, 1984b], Cendra and Marsden [1987], and Cendra, Ibort, and Marsden [1987].

A classical way to see the Lagrangian (or Hamiltonian) structure of the rigid-body equations is to use a description of the orientation of the body in terms of three Euler angles denoted by θ, φ, ψ and their velocities $\dot{\theta}, \dot{\varphi}, \dot{\psi}$ (or conjugate momenta $p_\theta, p_\varphi, p_\psi$), relative to which the equations *are* in Euler–Lagrange (or canonical Hamiltonian) form. However, this procedure requires using *six equations*, while many questions are easier to study using the *three equations* (1.2.1).

Lagrangian Form. To see the sense in which (1.2.1) are Lagrangian, introduce the Lagrangian

$$L(\Omega) = \frac{1}{2}(I_1\Omega_1^2 + I_2\Omega_2^2 + I_3\Omega_3^2), \tag{1.2.2}$$

which, as we will see in detail in Chapter 15, is the (rotational) kinetic energy of the rigid body. Regarding $I\Omega = (I_1\Omega_1, I_2\Omega_2, I_3\Omega_3)$ as a vector, write (1.2.1) as

$$\frac{d}{dt}\frac{\partial L}{\partial \Omega} = \frac{\partial L}{\partial \Omega} \times \Omega. \tag{1.2.3}$$

These equations appear explicitly in Lagrange [1788, Volume 2, p. 212] and were generalized to arbitrary Lie algebras by Poincaré [1901b]. We will discuss these general **Euler–Poincaré equations** in Chapter 13. We can also write a variational principle for (1.2.3) that is analogous to that for the Euler–Lagrange equations but is written *directly* in terms of Ω. Namely, (1.2.3) is equivalent to

$$\delta \int_a^b L \, dt = 0, \tag{1.2.4}$$

where variations of Ω are restricted to be of the form

$$\delta\Omega = \dot{\Sigma} + \Omega \times \Sigma, \tag{1.2.5}$$

where Σ is a curve in $\mathbb{R}^3$ that vanishes at the endpoints. This may be proved in the same way as we proved that the variational principle (1.1.1) is equivalent to the Euler–Lagrange equations (1.1.4); see Exercise 1.2-2. In fact, later on, in Chapter 13, we shall see how to *derive* this variational principle from the more "primitive" one (1.1.1).

Hamiltonian Form. If instead of variational principles we concentrate on Poisson brackets and drop the requirement that they be in the canonical form (1.1.17), then there is also a simple and beautiful Hamiltonian structure for the rigid-body equations. To state it, introduce the **angular momenta**

$$\Pi_i = I_i\Omega_i = \frac{\partial L}{\partial \Omega_i}, \quad i = 1, 2, 3, \tag{1.2.6}$$

so that the Euler equations become

$$\dot{\Pi}_1 = \frac{I_2 - I_3}{I_2 I_3} \Pi_2 \Pi_3,$$

$$\dot{\Pi}_2 = \frac{I_3 - I_1}{I_3 I_1} \Pi_3 \Pi_1, \qquad (1.2.7)$$

$$\dot{\Pi}_3 = \frac{I_1 - I_2}{I_1 I_2} \Pi_1 \Pi_2,$$

that is,

$$\dot{\Pi} = \Pi \times \Omega. \qquad (1.2.8)$$

Introduce the **rigid-body Poisson bracket** on functions of the Π's,

$$\{F, G\}(\Pi) = -\Pi \cdot (\nabla F \times \nabla G), \qquad (1.2.9)$$

and the Hamiltonian

$$H = \frac{1}{2} \left(\frac{\Pi_1^2}{I_1} + \frac{\Pi_2^2}{I_2} + \frac{\Pi_3^2}{I_3} \right). \qquad (1.2.10)$$

One checks (Exercise 1.2-3) that Euler's equations (1.2.7) are equivalent to[2]

$$\dot{F} = \{F, H\}. \qquad (1.2.11)$$

For *any* equation of the form (1.2.11), conservation of total angular momentum holds regardless of the Hamiltonian; indeed, with

$$C(\Pi) = \frac{1}{2}(\Pi_1^2 + \Pi_2^2 + \Pi_3^2),$$

we have $\nabla C(\Pi) = \Pi$, and so

$$\frac{d}{dt} \frac{1}{2}(\Pi_1^2 + \Pi_2^2 + \Pi_3^2) = \{C, H\}(\Pi) \qquad (1.2.12)$$

$$= -\Pi \cdot (\nabla C \times \nabla H) \qquad (1.2.13)$$

$$= -\Pi \cdot (\Pi \times \nabla H) = 0. \qquad (1.2.14)$$

The same calculation shows that $\{C, F\} = 0$ for any F. Functions such as these that **Poisson commute** with *every* function are called **Casimir functions**; they play an important role in the study of *stability*, as we shall see later.[3]

---

[2]This simple result is implicit in many works, such as Arnold [1966a, 1969], and is given explicitly in this form for the rigid body in Sudarshan and Mukunda [1974]. (Some preliminary versions were given by Pauli [1953], Martin [1959], and Nambu [1973].) On the other hand, the variational form (1.2.4) appears to be due to Poincaré [1901b] and Hamel [1904], at least implicitly. It is given explicitly for fluids in Newcomb [1962] and Bretherton [1970] and in the general case in Marsden and Scheurle [1993a, 1993b].

[3]H. B. G. Casimir was a student of P. Ehrenfest and wrote a brilliant thesis on the quantum mechanics of the rigid body, a problem that has not been adequately

Exercises

◇ **1.2-1.** Show by direct calculation that the rigid-body Poisson bracket satisfies the Jacobi identity. That is, if F and K are both functions of (Π_1, Π_2, Π_3) and we define

$$\{F, K\}(\Pi) = -\Pi \cdot (\nabla F \times \nabla K),$$

then the identity $\{L, \{F, K\}\} + \{K, \{L, F\}\} + \{F, \{K, L\}\} = 0$ holds.

◇ **1.2-2.** Verify directly that the Euler equations for a rigid body are equivalent to

$$\delta \int L \, dt = 0$$

for variations of the form $\delta\Omega = \dot{\Sigma} + \Omega \times \Sigma$, where Σ vanishes at the endpoints.

◇ **1.2-3.** Verify directly that the Euler equations for a rigid body are equivalent to the equations

$$\frac{d}{dt} F = \{F, H\},$$

where $\{\,,\}$ is the rigid-body Poisson bracket and H is the rigid-body Hamiltonian.

◇ **1.2-4.**

(a) Show that the rotation group SO(3) can be identified with the **Poincaré sphere**, that is, the **unit circle bundle** of the two-sphere S^2, defined to be the set of unit tangent vectors to the two-sphere in $\mathbb{R}^3$.

(b) Using the known fact from basic topology that any (continuous) vector field on S^2 must vanish somewhere, show that SO(3) *cannot* be written as $S^2 \times S^1$.

1.3 Lie–Poisson Brackets, Poisson Manifolds, Momentum Maps

The rigid-body variational principle and the rigid-body Poisson bracket are special cases of general constructions associated to any **Lie algebra**

---

addressed in the detail that would be desirable, even today. Ehrenfest in turn wrote his thesis under Boltzmann around 1900 on variational principles in fluid dynamics and was one of the first to study fluids from this point of view in material, rather than Clebsch, representation. Curiously, Ehrenfest used the Gauss–Hertz principle of least curvature rather than the more elementary Hamilton principle. This is a seed for many important ideas in this book.

$\mathfrak{g}$, that is, a vector space together with a bilinear, antisymmetric bracket $[\xi, \eta]$ satisfying **Jacobi's identity**:

$$[[\xi, \eta], \zeta] + [[\zeta, \xi], \eta] + [[\eta, \zeta], \xi] = 0 \tag{1.3.1}$$

for all $\xi, \eta, \zeta \in \mathfrak{g}$. For example, the Lie algebra associated to the rotation group is $\mathfrak{g} = \mathbb{R}^3$ with bracket $[\xi, \eta] = \xi \times \eta$, the ordinary vector cross product.

The Euler–Poincaré Equations. The construction of a variational principle on $\mathfrak{g}$ replaces

$$\delta\Omega = \dot{\Sigma} + \Omega \times \Sigma \quad \text{by} \quad \delta\xi = \dot{\eta} + [\eta, \xi].$$

The resulting general equations on $\mathfrak{g}$, which we will study in detail in Chapter 13, are called the **Euler–Poincaré equations**. These equations are valid for either finite- or infinite-dimensional Lie algebras. To state them in the finite-dimensional case, we use the following notation. Choosing a basis $e_1, \ldots, e_r$ of $\mathfrak{g}$ (so dim $\mathfrak{g} = r$), the **structure constants** C_{ab}^d are defined by the equation

$$[e_a, e_b] = \sum_{d=1}^{r} C_{ab}^d e_d, \tag{1.3.2}$$

where a, b run from 1 to r. If ξ is an element of the Lie algebra, its components relative to this basis are denoted by ξ^a so that $\xi = \sum_{a=1}^{r} \xi^a e_a$. If $e^1, \ldots, e^r$ is the corresponding dual basis, then the components of the differential of the Lagrangian L are the partial derivatives $\partial L / \partial \xi^a$. Then the Euler–Poincaré equations are

$$\frac{d}{dt} \frac{\partial L}{\partial \xi^d} = \sum_{a,b=1}^{r} C_{ad}^b \frac{\partial L}{\partial \xi^b} \xi^a. \tag{1.3.3}$$

The coordinate-free version reads

$$\frac{d}{dt} \frac{\partial L}{\partial \xi} = \operatorname{ad}_\xi^* \frac{\partial L}{\partial \xi},$$

where $\operatorname{ad}_\xi : \mathfrak{g} \to \mathfrak{g}$ is the linear map $\eta \mapsto [\xi, \eta]$, and $\operatorname{ad}_\xi^* : \mathfrak{g}^* \to \mathfrak{g}^*$ is its dual. For example, for $L : \mathbb{R}^3 \to \mathbb{R}$, the Euler–Poincaré equations become

$$\frac{d}{dt} \frac{\partial L}{\partial \Omega} = \frac{\partial L}{\partial \Omega} \times \Omega,$$

which generalize the Euler equations for rigid-body motion. As we mentioned earlier, these equations were written down for a fairly general class of L by Lagrange [1788, Volume 2, equation A, p. 212], while it was Poincaré [1901b] who generalized them to any Lie algebra.

The generalization of the rigid-body variational principle states that the Euler–Poincaré equations are equivalent to

$$\delta \int L \, dt = 0 \qquad (1.3.4)$$

for all variations of the form $\delta \xi = \dot{\eta} + [\xi, \eta]$ for some curve η in $\mathfrak{g}$ that vanishes at the endpoints.

The Lie–Poisson Equations.   We can also generalize the rigid-body Poisson bracket as follows: Let F, G be defined on the dual space $\mathfrak{g}^*$. Denoting elements of $\mathfrak{g}^*$ by μ, let the *functional derivative* of F at μ be the unique element $\delta F / \delta \mu$ of $\mathfrak{g}$ defined by

$$\lim_{\varepsilon \to 0} \frac{1}{\varepsilon} [F(\mu + \varepsilon \delta \mu) - F(\mu)] = \left\langle \delta \mu, \frac{\delta F}{\delta \mu} \right\rangle, \qquad (1.3.5)$$

for all $\delta \mu \in \mathfrak{g}^*$, where $\langle , \rangle$ denotes the pairing between $\mathfrak{g}^*$ and $\mathfrak{g}$. This definition (1.3.5) is consistent with the definition of $\delta F / \delta \varphi$ given in (1.1.15) when $\mathfrak{g}$ and $\mathfrak{g}^*$ are chosen to be appropriate spaces of fields. Define the $(\pm)$ *Lie–Poisson brackets* by

$$\{F, G\}_\pm(\mu) = \pm \left\langle \mu, \left[\frac{\delta F}{\delta \mu}, \frac{\delta G}{\delta \mu} \right] \right\rangle. \qquad (1.3.6)$$

Using the coordinate notation introduced above, the $(\pm)$ Lie–Poisson brackets become

$$\{F, G\}_\pm(\mu) = \pm \sum_{a,b,d=1}^{r} C_{ab}^d \mu_d \frac{\partial F}{\partial \mu_a} \frac{\partial G}{\partial \mu_b}, \qquad (1.3.7)$$

where $\mu = \mu_a e^a$.

Poisson Manifolds.   The Lie–Poisson bracket and the canonical brackets from the last section have four simple but crucial properties:

PB1 $\{F, G\}$ is real bilinear in F and G.

PB2 $\{F, G\} = -\{G, F\}$,   antisymmetry.

PB3 $[F, C], H] + [\{H, F\}, G] + \{\{G, H\}, F\} = 0$,   Jacobi identity.

PB4 $\{FG, H\} = F\{G, H\} + \{F, H\}G$,   Leibniz identity.

A manifold (that is, an n–dimensional "smooth surface") P together with a bracket operation on $\mathcal{F}(P)$, the space of smooth functions on P, and satisfying properties **PB1–PB4**, is called a *Poisson manifold*. In particular, $\mathfrak{g}^*$ *is a Poisson manifold*. In Chapter 10 we will study the general concept of a Poisson manifold.

For example, if we choose $\mathfrak{g} = \mathbb{R}^3$ with the bracket taken to be the cross product $[\mathbf{x}, \mathbf{y}] = \mathbf{x} \times \mathbf{y}$, and identify $\mathfrak{g}^*$ with $\mathfrak{g}$ using the dot product on $\mathbb{R}^3$ (so $\langle \mathbf{\Pi}, \mathbf{x} \rangle = \mathbf{\Pi} \cdot \mathbf{x}$ is the usual dot product), then the $(-)$ Lie–Poisson bracket becomes the rigid-body bracket.

Hamiltonian Vector Fields. On a Poisson manifold $(P, \{\cdot, \cdot\})$, associated to any function H there is a vector field, denoted by X_H, which has the property that for any smooth function $F : P \to \mathbb{R}$ we have the identity

$$\langle \mathbf{d}F, X_H \rangle = \mathbf{d}F \cdot X_H = \{F, H\},$$

where $\mathbf{d}F$ is the differential of F and $\mathbf{d}F \cdot X_H$ denotes the derivative of F in the direction X_H. We say that the vector field X_H is **generated** by the function H, or that X_H is the **Hamiltonian vector field** associated with H. We also define the associated **dynamical system** whose points z in phase space evolve in time by the differential equation

$$\dot{z} = X_H(z). \tag{1.3.8}$$

This definition is consistent with the equations in Poisson bracket form (1.1.16). The function H may have the interpretation of the energy of the system, but of course the definition (1.3.8) makes sense for *any* function. For canonical systems with the Poisson bracket given by (1.1.17), X_H is given by the formula

$$X_H(q^i, p_i) = \left(\frac{\partial H}{\partial p_i}, -\frac{\partial H}{\partial q^i} \right), \tag{1.3.9}$$

whereas for the rigid-body bracket given on $\mathbb{R}^3$ by (1.2.9),

$$X_H(\mathbf{\Pi}) = \mathbf{\Pi} \times \nabla H(\mathbf{\Pi}). \tag{1.3.10}$$

The general Lie–Poisson equations, determined by $\dot{F} = \{F, H\}$, read

$$\dot{\mu}_a = \mp \sum_{b,c=1}^{r} \mu_d C_{ab}^d \frac{\partial H}{\partial \mu_b},$$

or intrinsically,

$$\dot{\mu} = \mp \operatorname{ad}_{\delta H/\delta \mu}^* \mu. \tag{1.3.11}$$

Reduction. There is an important feature of the rigid-body bracket that also carries over to more general Lie algebras, namely, *Lie–Poisson brackets arise from canonical brackets on the cotangent bundle* (phase space) T^*G associated with a Lie group G that has $\mathfrak{g}$ as its associated Lie algebra. (The general theory of Lie groups is presented in Chapter 9.) Specifically, there is a general construction underlying the association

$$(\theta, \varphi, \psi, p_\theta, p_\varphi, p_\psi) \mapsto (\Pi_1, \Pi_2, \Pi_3) \tag{1.3.12}$$

defined by

$$\Pi_1 = \frac{1}{\sin\theta}[(p_\varphi - p_\psi\cos\theta)\sin\psi + p_\theta\sin\theta\cos\psi],$$

$$\Pi_2 = \frac{1}{\sin\theta}[(p_\varphi - p_\psi\cos\theta)\cos\psi - p_\theta\sin\theta\sin\psi], \qquad (1.3.13)$$

$$\Pi_3 = p_\psi.$$

This rigid-body map takes the canonical bracket in the variables (θ,φ,ψ) and their conjugate momenta $(p_\theta, p_\varphi, p_\psi)$ to the $(-)$ Lie–Poisson bracket in the following sense. If F and K are functions of Π_1, Π_2, Π_3, they determine functions of $(\theta,\varphi,\psi,p_\theta,p_\varphi,p_\psi)$ by substituting (1.3.13). Then a (tedious but straightforward) exercise using the chain rule shows that

$$\{F,K\}_{(-)\{\text{Lie-Poisson}\}} = \{F,K\}_{\text{canonical}}. \qquad (1.3.14)$$

We say that the map defined by (1.3.13) is a *canonical map* or a *Poisson map* and that the $(-)$ Lie–Poisson bracket has been obtained from the canonical bracket by *reduction*.

For a rigid body free to rotate about is center of mass, G is the (proper) rotation group SO(3), and the Euler angles and their conjugate momenta are coordinates for T^*G. The choice of T^*G as the primitive phase space is made according to the classical procedures of mechanics: The configuration space SO(3) is chosen, since each element $A \in$ SO(3) describes the orientation of the rigid body relative to a reference configuration, that is, the rotation A maps the reference configuration to the current configuration. For the description using Lagrangian mechanics, one forms the velocity-phase space T SO(3) with coordinates $(\theta,\varphi,\psi,\dot\theta,\dot\varphi,\dot\psi)$. The Hamiltonian description is obtained as in §1.1 by using the Legendre transform that maps TG to T^*G.

The passage from T^*G to the space of Π's (body angular momentum space) given by (1.3.13) turns out to be determined by *left* translation on the group. This mapping is an example of a *momentum map*, that is, a mapping whose components are the "Noether quantities" associated with a symmetry group. That the map (1.3.13) is a Poisson (canonical) map (see equation (1.3.14)) *is a general fact about momentum maps* proved in §12.6. To get to *space coordinates* one would use *right* translations and the $(+)$ bracket. This is what is done to get the standard description of fluid dynamics.

Momentum Maps and Coadjoint Orbits.   From the general rigid-body equations, $\dot\Pi = \Pi \times \nabla H$, we see that

$$\frac{d}{dt}\|\Pi\|^2 = 0.$$

In other words, Lie–Poisson systems on $\mathbb{R}^3$ conserve the total angular momenta, that is, they leave the spheres in Π-space invariant. The generalization of these objects associated to arbitrary Lie algebras are called *coadjoint orbits*.

Coadjoint orbits are submanifolds of $\mathfrak{g}^*$ with the property that any Lie–Poisson system $\dot{F} = \{F, H\}$ leaves them invariant. We shall also see how these spaces are Poisson manifolds in their own right and are related to the right $(+)$ or left $(-)$ invariance of the system regarded on T^*G, and the corresponding conserved Noether quantities.

On a general Poisson manifold $(P, \{\cdot, \cdot\})$, the definition of a momentum map is as follows. We assume that a Lie group G with Lie algebra $\mathfrak{g}$ acts on P by canonical transformations. As we shall review later (see Chapter 9), the infinitesimal way of specifying the action is to associate to each Lie algebra element $\xi \in \mathfrak{g}$ a vector field ξ_P on P. A *momentum map* is a map $\mathbf{J} : P \rightarrow \mathfrak{g}^*$ with the property that for every $\xi \in \mathfrak{g}$, the function $\langle \mathbf{J}, \xi \rangle$ (the pairing of the $\mathfrak{g}^*$-valued function $\mathbf{J}$ with the vector ξ) generates the vector field ξ_P; that is,

$$X_{\langle \mathbf{J}, \xi \rangle} = \xi_P.$$

As we shall see later, this definition generalizes the usual notions of linear and angular momentum. The rigid body shows that the notion has much wider interest. A fundamental fact about momentum maps is that if the Hamiltonian H is invariant under the action of the group G, then the vector-valued function $\mathbf{J}$ is a constant of the motion for the dynamics of the Hamiltonian vector field X_H associated to H.

One of the important notions related to momentum maps is that of *infinitesimal equivariance*, or the *classical commutation* relations, which state that

$$\{\langle \mathbf{J}, \xi \rangle, \langle \mathbf{J}, \eta \rangle\} = \langle \mathbf{J}, [\xi, \eta] \rangle \qquad (1.3.15)$$

for all Lie algebra elements ξ and η. Relations like this are well known for the angular momentum and can be directly checked using the Lie algebra of the rotation group. Later, in Chapter 12, we shall see that the relations (1.3.15) hold for a large important class of momentum maps that are given by computable formulas. Remarkably, it is the condition (1.3.15) that is exactly what is needed to prove that $\mathbf{J}$ *is, in fact, a Poisson map*. It is via this route that one gets an intellectually satisfying generalization of the fact that the map defined by equations (1.3.13) is a Poisson map; that is, equation (1.3.14) holds.

Some History. The Lie–Poisson bracket was discovered by Sophus Lie (Lie [1890, Vol. II, p. 237]). However, Lie's bracket and his related work was not given much attention until the work of Kirillov, Kostant, and Souriau (and others) revived it in the mid-1960s. Meanwhile, it was noticed by Pauli and Martin around 1950 that the rigid-body equations are in Hamiltonian form using the rigid-body bracket, but they were apparently unaware of the underlying Lie theory. Meanwhile, the generalization of the Euler equations to any Lie algebra $\mathfrak{g}$ by Poincaré [1901b] (and picked up by Hamel [1904]) proceeded as well, but without much contact with Lie's work until recently. The symplectic structure on coadjoint orbits also has a complicated history and itself goes back to Lie (Lie [1890, Ch. 20]).

The general notion of a Poisson manifold also goes back to Lie. However, the four defining properties of the Poisson bracket have been isolated by many authors such as Dirac [1964, p. 10]. The term "Poisson manifold" was coined by Lichnerowicz [1977]. We shall give more historical information on Poisson manifolds in §10.3.

The notion of the momentum map (the English translation of the French words "application moment") also has roots going back to the work of Lie.[4]

Momentum maps have found an astounding array of applications beyond those already mentioned. For instance, they are used in the study of the space of all solutions of a relativistic field theory (see Arms, Marsden, and Moncrief [1982]) and in the study of singularities in algebraic geometry (see Atiyah [1983] and Kirwan [1984]). They also enter into convex analysis in many interesting ways, such as the Schur–Horn theorem (Schur [1923], Horn [1954]) and its generalizations (Kostant [1973]) and in the theory of integrable systems (Bloch, Brockett, and Ratiu [1990, 1992] and Bloch, Flaschka, and Ratiu [1990, 1993]). It turns out that the image of the momentum map has remarkable convexity properties: see Atiyah [1982], Guillemin and Sternberg [1982, 1984], Kirwan [1984], Delzant [1988], and Lu and Ratiu [1991].

Exercises

⋄ **1.3-1.** A linear operator D on the space of smooth functions on $\mathbb{R}^n$ is called a *derivation* if it satisfies the Leibniz identity: $D(FG) = (DF)G + F(DG)$. Accept the fact from the theory of manifolds (see Chapter 4) that in local coordinates the expression of DF takes the form

$$(DF)(x) = \sum_{i=1}^{n} a^i(x)\frac{\partial F}{\partial x^i}(x)$$

for some smooth functions $a^1, \ldots, a^n$.

(a) Use the fact just stated to prove that for any bilinear operation $\{,\}$ on $\mathcal{F}(\mathbb{R}^n)$ which is a derivation in each of its arguments, we have

$$\{F, G\} = \sum_{i,j=1}^{n} \{x^i, x^j\}\frac{\partial F}{\partial x^i}\frac{\partial G}{\partial x^j}.$$

---

[4]Many authors use the words "moment map" for what we call the "momentum map." In English, unlike French, one does not use the phrases "linear moment" or "angular moment of a particle," and correspondingly, we prefer to use "momentum map." We shall give some comments on the history of momentum maps in §11.2.

(b) Show that the Jacobi identity holds for any operation $\{,\}$ on $\mathcal{F}(\mathbb{R}^n)$ as in (a), if and only if it holds for the coordinate functions.

$\diamond$ **1.3-2.**   Define, for a fixed function $f : \mathbb{R}^3 \to \mathbb{R}$,

$$\{F, K\}_f = \nabla f \cdot (\nabla F \times \nabla K).$$

(a) Show that this is a Poisson bracket.

(b) Locate the bracket in part (a) in Nambu [1973].

$\diamond$ **1.3-3.**   Verify directly that (1.3.13) defines a Poisson map.

$\diamond$ **1.3-4.**   Show that a bracket satisfying the Leibniz identity also satisfies

$$F\{K, L\} - \{FK, L\} = \{F, K\}L - \{F, KL\}.$$

1.4   The Heavy Top

The equations of motion for a rigid body with a fixed point in a gravitational field provide another interesting example of a system that is Hamiltonian relative to a Lie–Poisson bracket. See Figure 1.4.1.

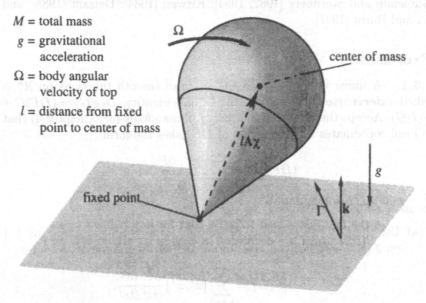

M = total mass

g = gravitational acceleration

Ω = body angular velocity of top

l = distance from fixed point to center of mass

center of mass

FIGURE 1.4.1. Heavy top

The underlying Lie algebra consists of the algebra of infinitesimal Euclidean motions in $\mathbb{R}^3$. (These do *not* arise as Euclidean motions of the body, since the body has a fixed point.) As we shall see, there is a close parallel with the Poisson structure for compressible fluids.

The basic phase space we start with is again $T^* \, \mathrm{SO}(3)$, coordinatized by Euler angles and their conjugate momenta. In these variables, the equations are in canonical Hamiltonian form; however, the presence of gravity breaks the symmetry, and the system is no longer $\mathrm{SO}(3)$ invariant, so it cannot be written entirely in terms of the body angular momentum $\mathbf{\Pi}$. One also needs to keep track of $\mathbf{\Gamma}$, the "direction of gravity" as seen from the body. This is defined by $\mathbf{\Gamma} = \mathbf{A}^{-1}\mathbf{k}$, where $\mathbf{k}$ points upward and $\mathbf{A}$ is the element of $\mathrm{SO}(3)$ describing the current configuration of the body. The equations of motion are

$$\dot{\Pi}_1 = \frac{I_2 - I_3}{I_2 I_3} \Pi_2 \Pi_3 + Mgl(\Gamma^2 \chi^3 - \Gamma^3 \chi^2),$$

$$\dot{\Pi}_2 = \frac{I_3 - I_1}{I_3 I_1} \Pi_3 \Pi_1 + Mgl(\Gamma^3 \chi^1 - \Gamma^1 \chi^3), \qquad (1.4.1)$$

$$\dot{\Pi}_3 = \frac{I_1 - I_2}{I_1 I_2} \Pi_1 \Pi_2 + Mgl(\Gamma^1 \chi^2 - \Gamma^2 \chi^1),$$

and

$$\dot{\mathbf{\Gamma}} = \mathbf{\Gamma} \times \mathbf{\Omega}, \qquad (1.4.2)$$

where M is the body's mass, g is the acceleration of gravity, χ is the body fixed unit vector on the line segment connecting the fixed point with the body's center of mass, and l is the length of this segment. See Figure 1.4.1.

The Lie algebra of the Euclidean group is $\mathfrak{se}(3) = \mathbb{R}^3 \times \mathbb{R}^3$ with the Lie bracket

$$[(\boldsymbol{\xi}, \mathbf{u}), (\boldsymbol{\eta}, \mathbf{v})] = (\boldsymbol{\xi} \times \boldsymbol{\eta}, \boldsymbol{\xi} \times \mathbf{v} - \boldsymbol{\eta} \times \mathbf{u}). \qquad (1.4.3)$$

We identify the dual space with pairs $(\mathbf{\Pi}, \mathbf{\Gamma})$; the corresponding $(-)$ Lie–Poisson bracket, called the **heavy top bracket**, is

$$\{F, G\}(\mathbf{\Pi}, \mathbf{\Gamma}) = -\mathbf{\Pi} \cdot (\nabla_\Pi F \times \nabla_\Pi G)$$
$$- \mathbf{\Gamma} \cdot (\nabla_\Pi F \times \nabla_\Gamma G - \nabla_\Pi G \times \nabla_\Gamma F). \qquad (1.4.4)$$

The above equations for $\mathbf{\Pi}, \mathbf{\Gamma}$ can be checked to be equivalent to

$$\dot{F} = \{F, H\}, \qquad (1.4.5)$$

where the **heavy top Hamiltonian**

$$H(\mathbf{\Pi}, \mathbf{\Gamma}) = \frac{1}{2}\left(\frac{\Pi_1^2}{I_1} + \frac{\Pi_2^2}{I_2} + \frac{\Pi_3^2}{I_3}\right) + Mgl\mathbf{\Gamma} \cdot \chi \qquad (1.4.6)$$

is the total energy of the body (Sudarshan and Mukunda [1974]).

The Lie algebra of the Euclidean group has a structure that is a special case of what is called a *semidirect product*. Here it is the product of the group of rotations with the translation group. It turns out that semidirect products occur under rather general circumstances when the symmetry in T^*G is broken. The general theory for semidirect products was developed by Sudarshan and Mukunda [1974], Ratiu [1980, 1981, 1982], Guillemin and Sternberg [1982], Marsden, Weinstein, Ratiu, Schmid, and Spencer [1983], Marsden, Ratiu, and Weinstein [1984a, 1984b], and Holm and Kuperschmidt [1983]. The Lagrangian approach to this and related problems is given in Holm, Marsden, and Ratiu [1998a].

Exercises

⋄ **1.4-1.**   Verify that $\dot{F} = \{F, H\}$ is equivalent to the heavy top equations using the heavy top Hamiltonian and bracket.

⋄ **1.4-2.**   Work out the Euler–Poincaré equations on $\mathfrak{se}(3)$. Show that with

$$L(\mathbf{\Omega}, \mathbf{\Gamma}) = \frac{1}{2}(I_1\Omega_1^2 + I_2\Omega_2^2 + I_3\Omega_3^2) - Mgl\mathbf{\Gamma} \cdot \boldsymbol{\chi},$$

the Euler–Poincaré equations are *not* the heavy top equations.

1.5   Incompressible Fluids

Arnold [1966a, 1969] showed that the Euler equations for an incompressible fluid could be given a Lagrangian and Hamiltonian description similar to that for the rigid body. His approach[5] has the appealing feature that one sets things up just the way Lagrange and Hamilton would have done: One begins with a configuration space Q and forms a Lagrangian L on the velocity phase space TQ and then H on the momentum phase space T^*Q, just as was outlined in §1.1. Thus, one automatically has variational principles, etc. For ideal fluids, $Q = G$ is the group $\mathrm{Diff}_{\mathrm{vol}}(\Omega)$ of volume-preserving transformations of the fluid container (a region Ω in $\mathbb{R}^2$ or $\mathbb{R}^3$, or a Riemannian manifold in general, possibly with boundary). Group multiplication in G is composition.

Kinematics of a Fluid.   The reason we select $G = \mathrm{Diff}_{\mathrm{vol}}(\Omega)$ as the configuration space is similar to that for the rigid body; namely, each φ in G is a mapping of Ω to Ω that takes a reference point $X \in \Omega$ to a current point $x = \varphi(X) \in \Omega$; thus, knowing φ tells us where each particle

---

[5]Arnold's approach is consistent with what appears in the thesis of Ehrenfest from around 1904; see Klein [1970]. However, Ehrenfest bases his principles on the more sophisticated curvature principles of Gauss and Hertz.

of fluid goes and hence gives us the *fluid configuration*. We ask that φ be a diffeomorphism to exclude discontinuities, cavitation, and fluid inter-penetration, and we ask that φ be volume-preserving to correspond to the assumption of incompressibility.

A *motion* of a fluid is a family of time-dependent elements of G, which we write as $x = \varphi(X, t)$. The *material velocity field* is defined by

$$\mathbf{V}(X, t) = \frac{\partial \varphi(X, t)}{\partial t},$$

and the *spatial velocity field* is defined by $\mathbf{v}(x, t) = \mathbf{V}(X, t)$, where x and X are related by $x = \varphi(X, t)$. If we suppress "t" and write $\dot{\varphi}$ for $\mathbf{V}$, note that

$$\mathbf{v} = \dot{\varphi} \circ \varphi^{-1}, \quad \text{i.e.,} \quad \mathbf{v}_t = \mathbf{V}_t \circ \varphi_t^{-1}, \tag{1.5.1}$$

where $\varphi_t(x) = \varphi(X, t)$. See Figure 1.5.1.

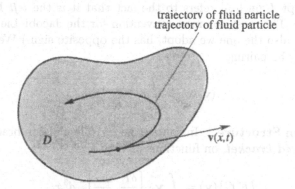

traiectory of fluid particle
trajectory of fluid particle

D

$\mathbf{v}(x, t)$

FIGURE 1.5.1. The trajectory and velocity of a fluid particle.

We can regard (1.5.1) as a map from the space of $(\varphi, \dot{\varphi})$ (material or Lagrangian description) to the space of $\mathbf{v}$'s (spatial or Eulerian description). Like the rigid body, the material to spatial map (1.5.1) takes the canonical bracket to a Lie–Poisson bracket; one of our goals is to understand this reduction. Notice that if we replace φ by $\varphi \circ \eta$ for a fixed (time-independent) $\eta \in \text{Diff}_{\text{vol}}(\Omega)$, then $\dot{\varphi} \circ \varphi^{-1}$ is independent of η; this reflects the *right* invariance of the Eulerian description ($\mathbf{v}$ is invariant under composition of φ by η on the right). This is also called the *particle relabeling symmetry* of fluid dynamics. The spaces TG and T^*G represent the Lagrangian (material) description, and we pass to the Eulerian (spatial) description by right translations and use the $(+)$ Lie–Poisson bracket. One of the things we want to do later is to better understand the reason for the switch between right and left in going from the rigid body to fluids.

Dynamics of a Fluid. The *Euler equations* for an ideal, incompressible, homogeneous fluid moving in the region Ω are

$$\frac{\partial \mathbf{v}}{\partial t} + (\mathbf{v} \cdot \nabla)\mathbf{v} = -\nabla p \qquad (1.5.2)$$

with the constraint div $\mathbf{v} = 0$ and the boundary condition that $\mathbf{v}$ is tangent to the boundary, $\partial\Omega$.

The *pressure* p is determined implicitly by the divergence-free (volume-preserving) constraint div $\mathbf{v} = 0$. (See Chorin and Marsden [1993] for basic information on the derivation of Euler's equations.) The associated Lie algebra $\mathfrak{g}$ is the space of all divergence-free vector fields tangent to the boundary. This Lie algebra is endowed with the *negative Jacobi–Lie bracket* of vector fields given by

$$[v, w]_L^i = \sum_{j=1}^{n} \left(w^j \frac{\partial v^i}{\partial x^j} - v^j \frac{\partial w^i}{\partial x^j} \right). \qquad (1.5.3)$$

(The subscript L on $[\cdot,\cdot]$ refers to the fact that it is the *left* Lie algebra bracket on $\mathfrak{g}$. The most common convention for the Jacobi–Lie bracket of vector fields, also the one we adopt, has the opposite sign.) We identify $\mathfrak{g}$ and $\mathfrak{g}^*$ using the pairing

$$\langle \mathbf{v}, \mathbf{w} \rangle = \int_\Omega \mathbf{v} \cdot \mathbf{w} \, d^3x. \qquad (1.5.4)$$

Hamiltonian Structure. Introduce the $(+)$ Lie–Poisson bracket, called the *ideal fluid bracket*, on functions of $\mathbf{v}$ by

$$\{F, G\}(\mathbf{v}) = \int_\Omega \mathbf{v} \cdot \left[\frac{\delta F}{\delta \mathbf{v}}, \frac{\delta G}{\delta \mathbf{v}} \right]_L d^3x, \qquad (1.5.5)$$

where $\delta F/\delta \mathbf{v}$ is defined by

$$\lim_{\varepsilon \to 0} \frac{1}{\varepsilon}[F(\mathbf{v} + \varepsilon\delta\mathbf{v}) - F(\mathbf{v})] = \int_\Omega \left(\delta\mathbf{v} \cdot \frac{\delta F}{\delta\mathbf{v}} \right) d^3x. \qquad (1.5.6)$$

With the energy function chosen to be the kinetic energy,

$$H(\mathbf{v}) = \frac{1}{2} \int_\Omega \|\mathbf{v}\|^2 \, d^3x, \qquad (1.5.7)$$

one can verify that the Euler equations (1.5.2) are equivalent to the Poisson bracket equations

$$\dot{F} = \{F, H\} \qquad (1.5.8)$$

for all functions F on $\mathfrak{g}$. To see this, it is convenient to use the orthogonal decomposition $\mathbf{w} = \mathbb{P}\mathbf{w} + \nabla p$ of a vector field $\mathbf{w}$ into a divergence-free part $\mathbb{P}\mathbf{w}$ in $\mathfrak{g}$ and a gradient. The Euler equations can be written

$$\frac{\partial \mathbf{v}}{\partial t} + \mathbb{P}(\mathbf{v} \cdot \nabla \mathbf{v}) = 0. \tag{1.5.9}$$

One can express the Hamiltonian structure in terms of the vorticity as a basic dynamic variable and show that the preservation of coadjoint orbits amounts to Kelvin's circulation theorem. Marsden and Weinstein [1983] show that the Hamiltonian structure in terms of Clebsch potentials fits naturally into this Lie–Poisson scheme, and that Kirchhoff's Hamiltonian description of point vortex dynamics, vortex filaments, and vortex patches can be derived in a natural way from the Hamiltonian structure described above.

Lagrangian Structure. The general framework of the Euler–Poincaré and the Lie–Poisson equations gives other insights as well. For example, this general theory shows that the Euler equations are derivable from the "variational principle"

$$\delta \int_a^b \int_\Omega \frac{1}{2} \|\mathbf{v}\|^2 \, d^3x = 0,$$

which is to hold for all variations $\delta \mathbf{v}$ of the form

$$\delta \mathbf{v} = \dot{\mathbf{u}} + [\mathbf{v}, \mathbf{u}]_L$$

(sometimes called **Lin constraints**), where $\mathbf{u}$ is a vector field (representing the infinitesimal particle displacement) vanishing at the temporal endpoints.[6]

There are important functional-analytic differences between working in material representation (that is, on T^*G) and in Eulerian representation (that is, on $\mathfrak{g}^*$) that are important for proving existence and uniqueness theorems, theorems on the limit of zero viscosity, and the convergence of numerical algorithms (see Ebin and Marsden [1970], Marsden, Ebin, and Fischer [1972], and Chorin, Hughes, Marsden, and McCracken [1978]). Finally, we note that for *two-dimensional flow*, a collection of Casimir functions is given by

$$C(\omega) = \int_\Omega \Phi(\omega(x)) \, d^2x \tag{1.5.10}$$

for $\Phi : \mathbb{R} \to \mathbb{R}$ any (smooth) function, where $\omega \mathbf{k} = \nabla \times \mathbf{v}$ is the **vorticity**. For three-dimensional flow, (1.5.10) is no longer a Casimir.

---

[6]As mentioned earlier, this form of the variational (strictly speaking, a Lagrange–d'Alembert type) principle is due to Newcomb [1962]; see also Bretherton [1970]. For the case of general Lie algebras, it is due to Marsden and Scheurle [1993b]; see also Cendra and Marsden [1987].

Exercises

⋄ **1.5-1.**   Show that any divergence-free vector field X on $\mathbb{R}^3$ can be written
globally as a curl of another vector field and, away from equilibrium points,
can *locally* be written as

$$X = \nabla f \times \nabla g,$$

where f and g are real-valued functions on $\mathbb{R}^3$. Assume that this (so-called
Clebsch–Monge) representation also holds globally. Show that the particles
of fluid, which follow trajectories satisfying $\dot{x} = X(x)$, are trajectories of a
Hamiltonian system with a bracket in the form of Exercise 1.3-2.

1.6   The Maxwell–Vlasov System

Plasma physics provides another beautiful application area for the techniques discussed in the preceding sections. We shall briefly indicate these
in this section. The period 1970–1980 saw the development of noncanonical
Hamiltonian structures for the Korteweg–de Vries (KdV) equation (due to
Gardner, Kruskal, Miura, and others; see Gardner [1971]) and other soliton equations. This quickly became entangled with the attempts to understand integrability of Hamiltonian systems and the development of the
algebraic approach; see, for example, Gelfand and Dorfman [1979], Manin
[1979] and references therein. More recently, these approaches have come together again; see, for instance, Reyman and Semenov-Tian-Shansky [1990],
Moser and Veselov [1991]. KdV type models are usually derived from or
are approximations to more fundamental fluid models, and it seems fair to
say that the reasons for their complete integrability are not yet completely
understood.

Some History.   For fluid and plasma systems, some of the key early
works on Poisson bracket structures were Dashen and Sharp [1968], Goldin
[1971], Iwiński and Turski [1976], Dzyaloshinskii and Volovick [1980], Morrison and Greene [1980], and Morrison [1980]. In Sudarshan and Mukunda
[1974], Guillemin and Sternberg [1982], and Ratiu [1980, 1982], a general
theory for Lie–Poisson structures for special kinds of Lie algebras, called
semidirect products, was begun. This was quickly recognized (see, for example, Marsden [1982], Marsden, Weinstein, Ratiu, Schmid, and Spencer
[1983], Holm and Kuperschmidt [1983], and Marsden, Ratiu, and Weinstein
[1984a, 1984b]) to be relevant to the brackets for compressible flow; see §1.7
below.

Derivation of Poisson Structures.   A rational scheme for systematically *deriving* brackets is needed since for one thing, a direct verification
of Jacobi's identity can be inefficient and time–consuming. Here we outline a derivation of the Maxwell–Vlasov bracket by Marsden and Weinstein

[1982]. The method is similar to Arnold's, namely by performing a reduction starting with:

(i) canonical brackets in a material representation for the plasma; and

(ii) a potential representation for the electromagnetic field.

One then identifies the symmetry group and carries out reduction by this group in a manner similar to that we described for Lie–Poisson systems.

For plasmas, the physically correct material description is actually slightly more complicated; we refer to Cendra, Holm, Hoyle, and Marsden [1998] for a full account.

Parallel developments can be given for many other brackets, such as the charged fluid bracket by Spencer and Kaufman [1982]. Another method, based primarily on Clebsch potentials, was developed in a series of papers by Holm and Kupershmidt (for example, Holm and Kuperschmidt [1983]) and applied to a number of interesting systems, including superfluids and superconductors. They also pointed out that semidirect products are appropriate for the MHD bracket of Morrison and Greene [1980].

The Maxwell–Vlasov System. The Maxwell–Vlasov equations for a collisionless plasma are the fundamental equations in plasma physics.[7] In Euclidean space, the basic dynamical variables are

$f(\mathbf{x}, \mathbf{v}, t)$: the plasma particle number density per phase space; volume $d^3x\, d^3v$;

$\mathbf{E}(\mathbf{x}, t)$: the electric field;

$\mathbf{B}(\mathbf{x}, t)$: the magnetic field.

The equations for a collisionless plasma for the case of a single species of particles with mass m and charge e are

$$\frac{\partial f}{\partial t} + \mathbf{v} \cdot \frac{\partial f}{\partial \mathbf{x}} + \frac{e}{m}\left(\mathbf{E} + \frac{1}{c}\mathbf{v} \times \mathbf{B}\right) \cdot \frac{\partial f}{\partial \mathbf{v}} = 0,$$

$$\frac{1}{c}\frac{\partial \mathbf{B}}{\partial t} = -\operatorname{curl}\mathbf{E},$$

$$\frac{1}{c}\frac{\partial \mathbf{E}}{\partial t} = \operatorname{curl}\mathbf{B} - \frac{1}{c}\mathbf{j}_f, \qquad\qquad (1.6.1)$$

$$\operatorname{div}\mathbf{E} = \rho_f$$

$$\operatorname{div}\mathbf{B} = 0.$$

The *current* defined by f is given by

$$\mathbf{j}_f = e\int \mathbf{v}f(\mathbf{x}, \mathbf{v}, t)\, d^3v$$

---

[7]See, for example, Clemmow and Dougherty [1959], van Kampen and Felderhof [1967], Krall and Trivelpiece [1973], Davidson [1972], Ichimaru [1973], and Chen [1974].

and the **charge density** by

$$\rho_f = e \int f(\mathbf{x}, \mathbf{v}, t)\, d^3v.$$

Also, $\partial f/\partial \mathbf{x}$ and $\partial f/\partial \mathbf{v}$ denote the gradients of f with respect to $\mathbf{x}$ and $\mathbf{v}$, respectively, and c is the speed of light. The evolution equation for f results from the Lorentz force law and standard transport assumptions. The remaining equations are the standard Maxwell equations with charge density ρ_f and current $\mathbf{j}_f$ produced by the plasma.

Two limiting cases will aid our discussions. First, if the plasma is constrained to be static, that is, f is concentrated at $\mathbf{v} = 0$ and t-independent, we get the **charge-driven Maxwell equations**:

$$\frac{1}{c}\frac{\partial \mathbf{B}}{\partial t} = -\operatorname{curl}\mathbf{E},$$
$$\frac{1}{c}\frac{\partial \mathbf{E}}{\partial t} = \operatorname{curl}\mathbf{B}, \tag{1.6.2}$$
$$\operatorname{div}\mathbf{E} = \rho, \quad \text{and} \quad \operatorname{div}\mathbf{B} = 0.$$

Second, if we let $c \to \infty$, electrodynamics becomes electrostatics, and we get the **Poisson–Vlasov equation**

$$\frac{\partial f}{\partial t} + \mathbf{v} \cdot \frac{\partial f}{\partial \mathbf{x}} - \frac{e}{m}\frac{\partial \varphi_f}{\partial \mathbf{x}} \cdot \frac{\partial f}{\partial \mathbf{v}} = 0, \tag{1.6.3}$$

where $-\nabla^2 \varphi_f = \rho_f$. In this context, the name "Poisson–Vlasov" seems quite appropriate. The equation is, however, formally the same as the earlier Jeans [1919] equation of stellar dynamics. Henon [1982] has proposed calling it the "collisionless Boltzmann equation."

Maxwell's Equations. For simplicity, we let $m = e = c = 1$. As the basic configuration space we take the space $\mathcal{A}$ of vector potentials $\mathbf{A}$ on $\mathbb{R}^3$ (for the Yang–Mills equations this is generalized to the space of connections on a principal bundle over space). The corresponding phase space $T^*\mathcal{A}$ is identified with the set of pairs $(\mathbf{A}, \mathbf{Y})$, where $\mathbf{Y}$ is also a vector field on $\mathbb{R}^3$. The canonical Poisson bracket is used on $T^*\mathcal{A}$:

$$\{F, G\} = \int \left(\frac{\delta F}{\delta \mathbf{A}}\frac{\delta G}{\delta \mathbf{Y}} - \frac{\delta F}{\delta \mathbf{Y}}\frac{\delta G}{\delta \mathbf{A}} \right) d^3x. \tag{1.6.4}$$

The **electric field** is $\mathbf{E} = -\mathbf{Y}$, and the **magnetic field** is $\mathbf{B} = \operatorname{curl}\mathbf{A}$. With the Hamiltonian

$$H(\mathbf{A}, \mathbf{Y}) = \frac{1}{2}\int \left(\|\mathbf{E}\|^2 + \|\mathbf{B}\|^2 \right) d^3x, \tag{1.6.5}$$

Hamilton's canonical field equations (1.1.14) are checked to give the equations for $\partial \mathbf{E}/\partial t$ and $\partial \mathbf{A}/\partial t$, which imply the vacuum Maxwell's equations.

Alternatively, one can begin with $T\mathcal{A}$ and the Lagrangian

$$L(\mathbf{A}, \dot{\mathbf{A}}) = \frac{1}{2} \int \left(\|\dot{\mathbf{A}}\|^2 - \|\nabla \times \mathbf{A}\|^2 \right) d^3 x \qquad (1.6.6)$$

and use the Euler–Lagrange equations and variational principles.

It is of interest to incorporate the equation div $\mathbf{E} = \rho$ and, correspondingly, to use directly the field strengths $\mathbf{E}$ and $\mathbf{B}$, rather than $\mathbf{E}$ and $\mathbf{A}$. To do this, we introduce the *gauge group* $\mathcal{G}$, the additive group of real-valued functions $\psi : \mathbb{R}^3 \to \mathbb{R}$. Each $\psi \in \mathcal{G}$ transforms the fields according to the rule

$$(\mathbf{A}, \mathbf{E}) \mapsto (\mathbf{A} + \nabla \psi, \mathbf{E}). \qquad (1.6.7)$$

Each such transformation leaves the Hamiltonian H invariant and is a canonical transformation, that is, it leaves Poisson brackets intact. In this situation, as above, there will be a corresponding conserved quantity, or *momentum map*, in the same sense as in §1.3. As mentioned there, some simple general formulas for computing momentum maps will be studied in detail in Chapter 12. For the action (1.6.7) of $\mathcal{G}$ on $T^*\mathcal{A}$, the associated momentum map is

$$\mathbf{J}(\mathbf{A}, \mathbf{Y}) = \operatorname{div} \mathbf{E}, \qquad (1.6.8)$$

so we recover the fact that div $\mathbf{E}$ is preserved by Maxwell's equations (this is easy to verify directly using the identity div curl $= 0$). Thus we see that we can incorporate the equation div $\mathbf{E} = \rho$ by restricting our attention to the set $\mathbf{J}^{-1}(\rho)$. The theory of reduction is a general process whereby one reduces the dimension of a phase space by exploiting conserved quantities and symmetry groups. In the present case, the reduced space is $\mathbf{J}^{-1}(\rho)/\mathcal{G}$, which is identified with Max_ρ, the space of $\mathbf{E}$'s and $\mathbf{B}$'s satisfying div $\mathbf{E} = \rho$ and div $\mathbf{B} = 0$.

The space Max_ρ inherits a Poisson structure as follows. If F and K are functions on Max_ρ, we substitute $\mathbf{E} = -\mathbf{Y}$ and $\mathbf{B} = \nabla \times \mathbf{A}$ to express F and K as functionals of $(\mathbf{A}, \mathbf{Y})$. Then we compute the canonical brackets on $T^*\mathcal{A}$ and express the result in terms of $\mathbf{E}$ and $\mathbf{B}$. Carrying this out using the chain rule gives

$$\{F, K\} = \int \left(\frac{\delta F}{\delta \mathbf{E}} \cdot \operatorname{curl} \frac{\delta K}{\delta \mathbf{B}} - \frac{\delta K}{\delta \mathbf{E}} \cdot \operatorname{curl} \frac{\delta F}{\delta \mathbf{B}} \right) d^3 x, \qquad (1.6.9)$$

where $\delta F/\delta \mathbf{E}$ and $\delta F/\delta \mathbf{B}$ are vector fields, with $\delta F/\delta \mathbf{B}$ *divergence-free*. These are defined in the usual way; for example,

$$\lim_{\varepsilon \to 0} \frac{1}{\varepsilon} [F(\mathbf{E} + \varepsilon \delta \mathbf{E}, \mathbf{B}) - F(\mathbf{E}, \mathbf{B})] = \int \frac{\delta F}{\delta \mathbf{E}} \cdot \delta \mathbf{E} \, d^3 x. \qquad (1.6.10)$$

This bracket makes Max_ρ into a Poisson manifold and the map $(\mathbf{A}, \mathbf{Y}) \mapsto (-\mathbf{Y}, \nabla \times \mathbf{A})$ into a Poisson map. The bracket (1.6.9) was discovered (by a different procedure) by Pauli [1933] and Born and Infeld [1935]. We refer to (1.6.9) as the *Pauli–Born–Infeld bracket* or the *Maxwell–Poisson bracket* for Maxwell's equations.

With the energy H given by (1.6.5) regarded as a function of $\mathbf{E}$ and $\mathbf{B}$, Hamilton's equations in bracket form $\dot{F} = \{F, H\}$ on Max_ρ capture the full set of Maxwell's equations (with external charge density ρ).

The Poisson–Vlasov Equation.   The papers Iwiński and Turski [1976] and Morrison [1980] showed that the Poisson–Vlasov equations form a Hamiltonian system with

$$H(f) = \frac{1}{2} \int \|\mathbf{v}\|^2 f(\mathbf{x}, \mathbf{v}, t) \, d^3x \, d^3v + \frac{1}{2} \int \|\nabla \varphi_f\|^2 \, d^3x \qquad (1.6.11)$$

and the *Poisson–Vlasov bracket*

$$\{F, G\} = \int f \left\{ \frac{\delta F}{\delta f}, \frac{\delta G}{\delta f} \right\}_{\mathbf{xv}} d^3x \, d^3v, \qquad (1.6.12)$$

where $\{ \, , \, \}_{\mathbf{xv}}$ is the canonical bracket on $(\mathbf{x}, \mathbf{v})$-space. As was observed in Gibbons [1981] and Marsden and Weinstein [1982], this is the $(+)$ Lie–Poisson bracket associated with the Lie algebra $\mathfrak{g}$ of functions of $(\mathbf{x}, \mathbf{v})$ with Lie bracket the canonical Poisson bracket.

According to the general theory, this Lie–Poisson structure is obtained by reduction from canonical brackets on the cotangent bundle of the group underlying $\mathfrak{g}$, just as was the case for the rigid body and incompressible fluids. This time, the group $G = \text{Diff}_{\text{can}}$ is the group of canonical transformations of $(\mathbf{x}, \mathbf{v})$-space. The Poisson–Vlasov equations can equally well be written in canonical form on T^*G. This is related to the Lagrangian and Hamiltonian description of a plasma that goes back to Low [1958], Katz [1961], and Lundgren [1963]. Thus, one can start with the particle description with canonical brackets and, through reduction, derive the brackets here. See Cendra, Holm, Hoyle, and Marsden [1998] for exactly how this goes. There are other approaches to the Hamiltonian formulation using analogues of Clebsch potentials; see, for instance, Su [1961], Zakharov [1971], and Gibbons, Holm, and Kuperschmidt [1982].

The Poisson–Vlaslov to Compressible Flow Map.   Before going on to the Maxwell–Vlasov equations, we point out a remarkable connection between the Poisson–Vlasov bracket (1.6.12) and the bracket for compressible flow.

The Euler equations for compressible flow in a region Ω in $\mathbb{R}^3$ are

$$\rho \left(\frac{\partial \mathbf{v}}{\partial t} + (\mathbf{v} \cdot \nabla) \mathbf{v} \right) = -\nabla p \qquad (1.6.13)$$

and

$$\frac{\partial \rho}{\partial t} + \text{div}(\rho \mathbf{v}) = 0, \qquad (1.6.14)$$

with the boundary condition

$$\mathbf{v} \text{ tangent to } \partial\Omega.$$

Here the pressure p is determined from an internal energy function per unit mass given by $p = \rho^2 w'(\rho)$, where $w = w(\rho)$ is the constitutive relation. (We ignore entropy for the present discussion—its inclusion is straightforward to deal with.) The **compressible fluid Hamiltonian** is

$$H = \frac{1}{2}\int_\Omega \rho\|\mathbf{v}\|^2 \, d^3x + \int_\Omega \rho w(\rho) \, d^3x. \tag{1.6.15}$$

The relevant Poisson bracket is most easily expressed if we use the momentum density $\mathbf{M} = \rho\mathbf{v}$ and density ρ as our basic variables. The **compressible fluid bracket** is

$$\{F, G\} = \int_\Omega \mathbf{M} \cdot \left[\left(\frac{\delta G}{\delta \mathbf{M}} \cdot \nabla \right) \frac{\delta F}{\delta \mathbf{M}} - \left(\frac{\delta F}{\delta \mathbf{M}} \cdot \nabla \right) \frac{\delta G}{\delta \mathbf{M}} \right] d^3x$$

$$+ \int_\Omega \rho \left[\left(\frac{\delta G}{\delta \mathbf{M}} \cdot \nabla \right) \frac{\delta F}{\delta \rho} - \left(\frac{\delta F}{\delta \mathbf{M}} \cdot \nabla \right) \frac{\delta G}{\delta \rho} \right] d^3x. \tag{1.6.16}$$

Notice the similarities in structure between the Poisson bracket (1.6.16) for compressible flow and (1.4.4). For compressible flow it is the density that prevents a full $\text{Diff}(\Omega)$ invariance; the Hamiltonian is invariant only under those diffeomorphisms that preserve the density.

The space of $(\mathbf{M}, \rho)$'s can be shown to be the dual of a semidirect product Lie algebra and it can also be shown that the preceding bracket is the associated $(+)$ Lie–Poisson bracket (see Marsden, Weinstein, Ratiu, Schmid, and Spencer [1983], Holm and Kuperschmidt [1983], and Marsden, Ratiu, and Weinstein [1984a, 1984b]).

The relationship with the Poisson–Vlasov bracket is this: Suppressing the time variable, define the map $f \mapsto (\mathbf{M}, \rho)$ by

$$\mathbf{M}(\mathbf{x}) = \int_\Omega \mathbf{v} f(\mathbf{x}, \mathbf{v}) d^3v \quad \text{and} \quad \rho(\mathbf{x}) = \int_\Omega f(\mathbf{x}, \mathbf{v}) \, d^3v. \tag{1.6.17}$$

Remarkably, this plasma to fluid map is a Poisson map taking the Poisson–Vlasov bracket (1.6.12) to the compressible fluid bracket (1.6.16). In fact, this map is a momentum map (Marsden, Weinstein, Ratiu, Schmid, and Spencer [1983]). The Poisson–Vlasov Hamiltonian is *not* invariant under the associated group action, however.

The Maxwell–Vlasov Bracket.   A bracket for the Maxwell–Vlasov equations was given by Iwiński and Turski [1976] and Morrison [1980]. Marsden and Weinstein [1982] used systematic procedures involving reduction and momentum maps to derive (and correct) the bracket starting with a canonical bracket.

The procedure starts with the material description[8] of the plasma as the cotangent bundle of the group Diff_{can} of canonical transformations of $(\mathbf{x}, \mathbf{p})$-space and the space $T^*\mathcal{A}$ for Maxwell's equations. We justify this by noticing that the motion of a charged particle in a fixed (but possibly time-dependent) electromagnetic field via the Lorentz force law defines a (time-dependent) canonical transformation. On $T^* \text{Diff}_{\text{can}} \times T^*\mathcal{A}$ we put the sum of the two canonical brackets, and then we reduce. First we reduce by Diff_{can}, which acts on $T^* \text{Diff}_{\text{can}}$ by right translation but does not act on $T^*\mathcal{A}$. Thus we end up with densities $f_{\text{mom}}(\mathbf{x}, \mathbf{p}, t)$ on position-momentum space and with the space $T^*\mathcal{A}$ used for the Maxwell equations. On this space we get the $(+)$ Lie–Poisson bracket, plus the canonical bracket on $T^*\mathcal{A}$. Recalling that $\mathbf{p}$ is related to $\mathbf{v}$ and $\mathbf{A}$ by $\mathbf{p} = \mathbf{v} + \mathbf{A}$, we let the gauge group $\mathcal{G}$ of electromagnetism act on this space by

$$(f_{\text{mom}}(\mathbf{x}, \mathbf{p}, t), \mathbf{A}(\mathbf{x}, t), \mathbf{Y}(\mathbf{x}, t)) \mapsto$$
$$(f_{\text{mom}}(\mathbf{x}, \mathbf{p} + \nabla\varphi(\mathbf{x}), t), \mathbf{A}(\mathbf{x}, t) + \nabla\varphi(x), \mathbf{Y}(\mathbf{x}, t)). \quad (1.6.18)$$

The momentum map associated with this action is computed to be

$$\mathbf{J}(f_{\text{mom}}, \mathbf{A}, \mathbf{Y}) = \text{div } \mathbf{E} - \int f_{\text{mom}}(\mathbf{x}, \mathbf{p}) \, d^3 p. \quad (1.6.19)$$

This corresponds to $\text{div } \mathbf{E} - \rho_f$ if we write $f(\mathbf{x}, \mathbf{v}, t) = f_{\text{mom}}(\mathbf{x}, \mathbf{p} - \mathbf{A}, t)$. This reduced space $\mathbf{J}^{-1}(0)/\mathcal{G}$ can be identified with the space $\mathcal{MV}$ of triples $(f, \mathbf{E}, \mathbf{B})$ satisfying $\text{div } \mathbf{E} = \rho_f$ and $\text{div } \mathbf{B} = 0$. The bracket on $\mathcal{MV}$ is computed by the same procedure as for Maxwell's equations. These computations yield the following **Maxwell–Vlasov bracket**:

$$\{F, K\}(f, \mathbf{E}, \mathbf{B}) = \int f \left\{ \frac{\delta F}{\delta f}, \frac{\delta K}{\delta f} \right\}_{xv} d^3 x \, d^3 v$$
$$+ \int \left(\frac{\delta F}{\delta \mathbf{E}} \cdot \text{curl } \frac{\delta K}{\delta \mathbf{B}} - \frac{\delta K}{\delta \mathbf{E}} \cdot \text{curl } \frac{\delta F}{\delta \mathbf{B}} \right) d^3 x$$
$$+ \int \left(\frac{\delta F}{\delta \mathbf{E}} \cdot \frac{\delta f}{\delta \mathbf{v}} \frac{\delta K}{\delta f} - \frac{\delta K}{\delta \mathbf{E}} \cdot \frac{\delta f}{\delta \mathbf{v}} \frac{\delta F}{\delta f} \right) d^3 x \, d^3 v \quad (1.6.20)$$
$$+ \int f \mathbf{B} \cdot \left(\frac{\partial}{\partial \mathbf{v}} \frac{\delta F}{\delta f} \times \frac{\partial}{\partial \mathbf{v}} \frac{\delta K}{\delta f} \right) d^3 x \, d^3 v.$$

With the **Maxwell–Vlasov Hamiltonian**

$$H(f, \mathbf{E}, \mathbf{B}) = \frac{1}{2} \int \|\mathbf{v}\|^2 f(\mathbf{x}, \mathbf{v}, t) \, d^3 x \, d^3 v$$
$$+ \frac{1}{2} \int (\|\mathbf{E}(x, t)\|^2 + \|\mathbf{B}(\mathbf{x}, t)\|^2) \, d^3 x, \quad (1.6.21)$$

---

[8] As shown in Cendra, Holm, Hoyle, and Marsden [1998], the correct physical description of the material representation of a plasma is a bit more complicated than simply Diff_{can}; however the end result is the same.

the Maxwell–Vlasov equations take the Hamiltonian form

$$\dot{F} = \{F, H\} \tag{1.6.22}$$

on the Poisson manifold $\mathcal{MV}$.

Exercises

⋄ **1.6-1.** Verify that one obtains the Maxwell equations from the Maxwell–Poisson bracket.

⋄ **1.6-2.** Verify that the action (1.6.7) has the momentum map $\mathbf{J}(\mathbf{A}, \mathbf{Y}) =$ div $\mathbf{E}$ in the sense given in §1.3.

1.7 Nonlinear Stability

There are various meanings that can be given to the word "stability." Intuitively, stability means that small disturbances do not grow large as time passes. Being more precise about this notion is not just capricious mathematical nitpicking; indeed, different interpretations of the word stability can lead to *different* stability criteria. Examples like the double spherical pendulum and stratified shear flows, which are sometimes used to model oceanographic phenomena show that one can get *different* criteria if one uses linearized or nonlinear analyses (see Marsden and Scheurle [1993a] and Abarbanel, Holm, Marsden, and Ratiu [1986]).

Some History. The history of stability theory in mechanics is very complex, but certainly has its roots in the work of Riemann [1860, 1861], Routh [1877], Thomson and Tait [1879], Poincaré [1885, 1892], and Liapunov [1892, 1897].

Since these early references, the literature has become too vast to even survey roughly. We do mention, however, that a guide to the large Soviet literature may be found in Mikhailov and Parton [1990].

The basis of the nonlinear stability method discussed below was originally given by Arnold [1965b, 1966b] and applied to two-dimensional ideal fluid flow, substantially augmenting the pioneering work of Rayleigh [1880]. Related methods were also found in the plasma physics literature, notably by Newcomb [1958], Fowler [1963], and Rosenbluth [1964]. However, these works did not provide a general setting or key convexity estimates needed to deal with the nonlinear nature of the problem. In retrospect, we may view other stability results, such as the stability of solitons in the Korteweg–de Vries (KdV) equation (Benjamin [1972] and Bona [1975]) as being instances of the same method used by Arnold. A crucial part of the method exploits the fact that the basic equations of nondissipative fluid and plasma dynamics are Hamiltonian in character. We shall explain below how the Hamiltonian structures discussed in the previous sections are used in the stability analysis.

Dynamics and Stability.   Stability is a dynamical concept. To explain it, we shall use some fundamental notions from the theory of dynamical systems (see, for example, Hirsch and Smale [1974] and Guckenheimer and Holmes [1983]). The laws of dynamics are usually presented as equations of motion, which we write in the abstract form of a *dynamical system*:

$$\dot{u} = X(u). \tag{1.7.1}$$

Here, u is a variable describing the state of the system under study, X is a system-specific function of u, and $\dot{u} = du/dt$, where t is time. The set of all allowed u's forms the state, or phase space P. We usually view X as a vector field on P. For a classical mechanical system, u is often a $2n$-tuple $(q^1, \ldots, q^n, p_1, \ldots, p_n)$ of positions and momenta, and for fluids, u is a velocity field in physical space.

As time evolves, the state of the system changes; the state follows a curve $u(t)$ in P. The trajectory $u(t)$ is assumed to be uniquely determined if its initial condition $u_0 = u(0)$ is specified. An *equilibrium state* is a state u_e such that $X(u_e) = 0$. The unique trajectory starting at u_e is u_e itself; that is, u_e does not move in time.

The language of dynamics has been an extraordinarily useful tool in the physical and biological sciences, especially during the last few decades. The study of systems that develop spontaneous oscillations through a mechanism called the Poincaré–Andronov–Hopf bifurcation is an example of such a tool (see Marsden and McCracken [1976], Carr [1981], and Chow and Hale [1982], for example). More recently, the concept of "chaotic dynamics" has sparked a resurgence of interest in dynamical systems. This occurs when dynamical systems possess trajectories that are so complex that they behave as if they were, in some sense, random. Some believe that the theory of turbulence will use such notions in its future development. We are not concerned with chaos directly, although it plays a role in some of what follows. In particular, we remark that in the definition of stability below, stability does not preclude chaos. In other words, the trajectories near a stable point can still be temporally very complex; stability just prevents them from moving very far from equilibrium.

To define stability, we choose a measure of nearness in P using a "metric" d. For two points u_1 and u_2 in P, d determines a positive number denoted by $d(u_1, u_2)$, the *distance* from u_1 to u_2. In the course of a stability analysis, it is necessary to specify, or construct, a metric appropriate for the problem at hand. In this setting, one says that an equilibrium state u_e is *stable* when trajectories that start near u_e remain near u_e for all $t \geq 0$. In precise terms, given any number $\epsilon > 0$, there is $\delta > 0$ such that if $d(u_0, u_e) < \delta$, then $d(u(t), u_e) < \epsilon$ for all $t > 0$. Figure 1.7.1 shows examples of stable and unstable equilibria for dynamical systems whose state space is the plane.

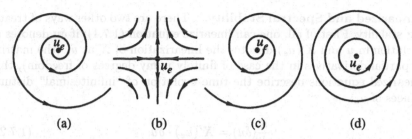

$$(a) \qquad (b) \qquad (c) \qquad (d)$$

FIGURE 1.7.1. The equilibrium point (a) is unstable because the trajectory $u(t)$ does not remain near u_e. Similarly, (b) is unstable, since most trajectories (eventually) move away from u_e. The equilibria in (c) and (d) are stable because all trajectories near u_e stay near u_e.

Fluids can be stable relative to one distance measure and, simultaneously, unstable relative to another. This seeming pathology actually reflects important physical processes; see Wan and Pulvirente [1984].

Rigid-Body Stability. A physical example illustrating the definition of stability is the motion of a free rigid body. This system can be simulated by tossing a book, held shut with a rubber band, into the air. It rotates stably when spun about its longest and shortest axes, but unstably when spun about the middle axis (Figure 1.7.2). One possible choice of a distance measure defining stability in this example is a metric in body angular momentum space. We shall return to this example in detail in Chapter 15 when we study rigid-body stability.

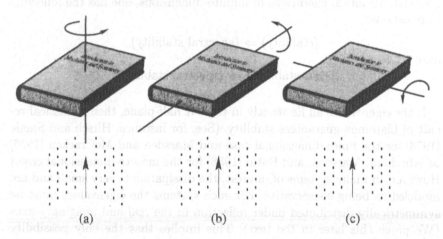

$$(a) \qquad\qquad (b) \qquad\qquad (c)$$

FIGURE 1.7.2. If you toss a book into the air, you can make it spin stably about its shortest axis (a), and its longest axis (b), but it is unstable when it rotates about its middle axis (c).

Linearized and Spectral Stability.   There are two other ways of treating stability. First of all, one can linearize equation (1.7.1); if δu denotes a variation in u and $X'(u_e)$ denotes the linearization of X at u_e (the matrix of partial derivatives in the case of finitely many degrees of freedom), the linearized equations describe the time evolution of "infinitesimal" disturbances of u_e:

$$\frac{d}{dt}(\delta u) = X'(u_e) \cdot \delta u. \tag{1.7.2}$$

Equation (1.7.1), on the other hand, describes the nonlinear evolution of *finite* disturbances $\Delta u = u - u_e$. We say that u_e is **linearly stable** if (1.7.2) is stable at $\delta u = 0$, in the sense defined above. Intuitively, this means that there are no infinitesimal disturbances that are growing in time. If $(\delta u)_0$ is an eigenfunction of $X'(u_e)$, that is, if

$$X'(u_e) \cdot (\delta u)_0 = \lambda(\delta u)_0 \tag{1.7.3}$$

for a complex number λ, then the corresponding solution of (1.7.2) with initial condition $(\delta u)_0$ is

$$\delta u = e^{t\lambda}(\delta u)_0. \tag{1.7.4}$$

The right side of this equation is growing when λ has positive real part. This leads us to the third notion of stability: We say that (1.7.1) or (1.7.2) is **spectrally stable** if the eigenvalues (more precisely, points in the spectrum) all have nonpositive real parts. In finite dimensions and, under appropriate technical conditions in infinite dimensions, one has the following implications:

$$\text{(stability)} \Rightarrow \text{(spectral stability)}$$

and

$$\text{(linear stability)} \Rightarrow \text{(spectral stability)}.$$

If the eigenvalues all lie strictly in the left half-plane, then a classical result of Liapunov guarantees stability. (See, for instance, Hirsch and Smale [1974] for the finite-dimensional case and Marsden and McCracken [1976] or Abraham, Marsden, and Ratiu [1988] for the infinite-dimensional case.) However, in many systems of interest, the dissipation is very small and are modeled as being conservative. For such systems the eigenvalues must be symmetrically distributed under reflection in the real and imaginary axes (We prove this later in the text). This implies that the only possibility for spectral stability occurs when the eigenvalues lie exactly on the imaginary axis. Thus, *this version of the Liapunov theorem is of no help in the Hamiltonian case.*

Spectral stability need not imply stability; instabilities can be generated (even in Hamiltonian systems) through, for example, *resonance*. Thus, to obtain general stability results, one must use other techniques to augment or replace the linearized theory. We give such a technique below.

Here is a planar example of a system that is spectrally stable at the origin but that is unstable there. In polar coordinates (r, θ), consider the evolution of $u = (r, \theta)$ given by

$$\dot{r} = r^3(1 - r^2) \quad \text{and} \quad \dot{\theta} = 1. \tag{1.7.5}$$

In (x, y) coordinates this system takes the form

$$\dot{x} = x(x^2 + y^2)(1 - x^2 - y^2) - y,$$
$$\dot{y} = y(x^2 + y^2)(1 - x^2 - y^2) + x.$$

The eigenvalues of the linearized system at the origin are readily verified to be $\pm\sqrt{-1}$, so the origin is spectrally stable; however, the phase portrait, shown in Figure 1.7.3, shows that the origin is unstable. (We include the factor $1 - r^2$ to give the system an attractive periodic orbit—this is merely to enrich the example and show how a stable periodic orbit can attract the orbits expelled by an unstable equilibrium.) This is not, however, a conservative system; next, we give two examples of Hamiltonian systems with similar features.

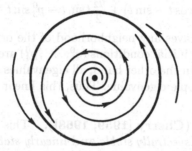

FIGURE 1.7.3. The phase portrait for $\dot{r} = r^3(1 - r^2)$, $\dot{\theta} = 1$.

Resonance Example. The linear system in $\mathbb{R}^2$ whose Hamiltonian is given by

$$H(q, p) = \frac{1}{2}p^2 + \frac{1}{2}q^2 + pq$$

has zero as a double eigenvalue, so it is spectrally stable. On the other hand,

$$q(t) = (q_0 + p_0)t + q_0 \quad \text{and} \quad p(t) = -(q_0 + p_0)t + p_0$$

is the solution of this system with initial condition (q_0, p_0), which clearly leaves any neighborhood of the origin no matter how close to it (q_0, p_0) is. Thus, *spectral stability need not imply even linear stability.* An even simpler example of the same phenomenon is given by the free particle Hamiltonian $H(q,p) = p^2/2$.

Another higher-dimensional example with resonance in $\mathbb{R}^8$ is given by the linear system whose Hamiltonian is

$$H = q_2 p_1 - q_1 p_2 + q_4 p_3 - q_3 p_4 + q_2 q_3.$$

The general solution with initial condition $(q_1^0, \ldots, p_4^0)$ is given by

$$q_1(t) = q_1^0 \cos t + q_2^0 \sin t,$$
$$q_2(t) = -q_1^0 \sin t + q_2^0 \cos t,$$
$$q_3(t) = q_3^0 \cos t + q_4^0 \sin t,$$
$$q_4(t) = -q_3^0 \sin t + q_4^0 \cos t,$$

and

$$p_1(t) = -\frac{q_3^0}{2} t \sin t + \frac{q_4^0}{2}(t \cos t - \sin t) + p_1^0 \cos t + p_2^0 \sin t,$$
$$p_2(t) = -\frac{q_3^0}{2}(t \cos t + \sin t) - \frac{q_4^0}{2} t \sin t - p_1^0 \sin t + p_2^0 \cos t,$$
$$p_3(t) = \frac{q_1^0}{2} t \sin t - \frac{q_2^0}{2}(t \cos t + \sin t) + p_3^0 \cos t + p_4^0 \sin t,$$
$$p_4(t) = \frac{q_1^0}{2}(t \cos t - \sin t) + \frac{q_2^0}{2} t \sin t - p_3^0 \sin t + p_4^0 \cos t.$$

One sees that $p_i(t)$ leaves any neighborhood of the origin, no matter how close to the origin the initial conditions $(q_1^0, \ldots, p_4^0)$ are; that is, the system is linearly unstable. On the other hand, all eigenvalues of this linear system are $\pm i$, each a quadruple eigenvalue. Thus, this linear system is spectrally stable.

Cherry's Example (Cherry [1959, 1968]). This example is a *Hamiltonian system that is spectrally stable **and** linearly stable but is nonlinearly unstable.* Consider the Hamiltonian on $\mathbb{R}^4$ given by

$$H = \frac{1}{2}(q_1^2 + p_1^2) - (q_2^2 + p_2^2) + \frac{1}{2}p_2(p_1^2 - q_1^2) - q_1 q_2 p_1. \qquad (1.7.6)$$

This system has an equilibrium at the origin, which is linearly stable, since the linearized system consists of two uncoupled oscillators in the $(\delta q_2, \delta p_2)$ and $(\delta q_1, \delta p_1)$ variables, respectively, with frequencies in the ratio $2:1$ (the eigenvalues are $\pm i$ and $\pm 2i$, so the frequencies are in resonance). A family of solutions (parametrized by a constant τ) of Hamilton's equations for (1.7.6) is given by

$$q_1 = -\sqrt{2}\frac{\cos(t - \tau)}{t - \tau}, \qquad q_2 = \frac{\cos 2(t - \tau)}{t - \tau},$$

$$p_1 = \sqrt{2}\frac{\sin(t - \tau)}{t - \tau}, \qquad p_2 = \frac{\sin 2(t - \tau)}{t - \tau}.$$

$$(1.7.7)$$

The solutions (1.7.7) clearly blow up in finite time; however, they start at time $t = 0$ at a distance $\sqrt{3}/\tau$ from the origin, so by choosing τ large, we can find solutions starting arbitrarily close to the origin, yet going to infinity in a finite time, so *the origin is nonlinearly unstable*.

Despite the above situation relating the linear and nonlinear theories, there has been much effort devoted to the development of spectral stability methods. When *instabilities* are present, spectral estimates give important information on growth rates. As far as stability goes, spectral stability gives necessary, but not sufficient, conditions for stability. In other words, for the nonlinear problems *spectral instability can predict instability, but not stability*. This is a basic result of Liapunov; see Abraham, Marsden, and Ratiu [1988], for example. Our immediate purpose is the opposite: *to describe sufficient conditions for stability*.

Casimir Functions.   Besides the energy, there are other conserved quantities associated with group symmetries such as linear and angular momentum. Some of these are associated with the group that underlies the passages from material to spatial or body coordinates. These are called *Casimir functions*; such a quantity, denoted by C, is characterized by the fact that it Poisson commutes with every function, that is,

$$\{C, F\} = 0 \tag{1.7.8}$$

for all functions F on phase space P. We shall study such functions and their relation with momentum maps in Chapters 10 and 11. For example, if Φ is any function of one variable, the quantity

$$C(\mathbf{\Pi}) = \Phi(\|\mathbf{\Pi}\|^2) \tag{1.7.9}$$

is a Casimir function for the rigid-body bracket, as is seen by using the chain rule. Likewise,

$$C(\omega) = \int_\Omega \Phi(\omega) \, dx \, dy \tag{1.7.10}$$

is a Casimir function for the two-dimensional ideal fluid bracket. (This calculation ignores boundary terms that arise in an integration by parts—see Lewis, Marsden, Montgomery, and Ratiu [1986] for a treatment of these boundary terms.)

Casimir functions are conserved by the dynamics associated with any Hamiltonian H, since $\dot{C} = \{C, H\} = 0$. Conservation of (1.7.9) corresponds to conservation of total angular momentum for the rigid body, while conservation of (1.7.10) represents Kelvin's circulation theorem for the Euler equations. It provides infinitely many independent constants of the motion that mutually Poisson commute; that is, $\{C_1, C_2\} = 0$, but this does *not* imply that these equations are integrable.

Lagrange–Dirichlet Criterion.   For Hamiltonian systems in canonical
form, an equilibrium point (q_e, p_e) is a point at which the partial derivatives
of H vanish, that is, it is a critical point of H. If the $2n \times 2n$ *matrix $\delta^2 H$
of second partial derivatives evaluated at (q_e, p_e) is positive or negative
definite (that is, all the eigenvalues of $\delta^2 H(q_e, p_e)$ have the same sign), then*
(q_e, p_e) *is stable.* This follows from conservation of energy and the fact from
calculus that the level sets of H near (q_e, p_e) are approximately ellipsoids.
As mentioned earlier, this condition implies, but is not implied by, spectral
stability. The KAM (Kolmogorov, Arnold, Moser) theorem, which gives
stability of periodic solutions for *two*-degree-of-freedom systems, and the
Lagrange–Dirichlet theorem are the most basic *general* stability theorems
for equilibria of Hamiltonian systems.

For example, let us apply the Lagrange–Dirichlet theorem to a classical
mechanical system whose Hamiltonian has the form kinetic plus potential
energy. If (q_e, p_e) is an equilibrium, it follows that p_e is zero. Moreover, the
matrix $\delta^2 H$ of second-order partial derivatives of H evaluated at (q_e, p_e)
block diagonalizes, with one of the blocks being the matrix of the quadratic
form of the kinetic energy, which is always positive definite. Therefore, if
$\delta^2 H$ is definite, it must be positive definite, and this in turn happens if and
only if $\delta^2 V$ is positive definite at q_e, where V is the potential energy of the
system. We conclude that *for a mechanical system* whose Lagrangian is
kinetic minus potential energy, $(q_e, 0)$ *is a stable equilibrium, provided that
the matrix $\delta^2 V(q_e)$ of second-order partial derivatives of the potential V at
q_e is positive definite (or, more generally, q_e is a strict local minimum for
V). If $\delta^2 V$ at q_e has a negative definite direction, then q_e is an unstable
equilibrium.*

The second statement is seen in the following way. The linearized Hamiltonian system at $(q_e, 0)$ is again a Hamiltonian system whose Hamiltonian
is of the form kinetic plus potential energy, the potential energy being given
by the quadratic form $\delta^2 V(q_e)$. From a standard theorem in linear algebra,
which states that two quadratic forms, one of which is positive definite, can
be simultaneously diagonalized, we conclude that the linearized Hamiltonian system decouples into a family of Hamiltonian systems of the form

$$\frac{d}{dt}(\delta p_k) = -c_k \delta q^k, \qquad \frac{d}{dt}(\delta q^k) = \frac{1}{m_k} \delta p_k,$$

where $1/m_k > 0$ are the eigenvalues of the positive definite quadratic form
given by the kinetic energy in the variables δp_j, and c_k are the eigenvalues
of $\delta^2 V(q_e)$. Thus the eigenvalues of the linearized system are given by
$\pm\sqrt{-c_k/m_k}$. Therefore, if some c_k is negative, the linearized system has
at least one positive eigenvalue, and thus $(q_e, 0)$ is spectrally and hence
linearly and nonlinearly unstable. For generalizations of this, see Oh [1987],
Grillakis, Shatah, and Strauss [1987], Chern [1997] and references therein.

The Energy–Casimir Method.   This is a generalization of the classical Lagrange–Dirichlet method. Given an equilibrium u_e for $\dot{u} = X_H(u)$ on a Poisson manifold P, it proceeds in the following steps.

To test an equilibrium (satisfying $X_H(z_e) = 0$) for stability:

Step 1.   *Find a conserved function C (C will typically be a Casimir function plus other conserved quantities) such that the first variation vanishes:*

$$\delta(H + C)(z_e) = 0.$$

Step 2.   *Calculate the second variation*

$$\delta^2(H + C)(z_e).$$

Step 3.   *If $\delta^2(H + C)(z_e)$ is definite (either positive or negative), then z_e is called **formally stable**.*

With regard to Step 3, we point out that an equilibrium solution need not be a critical point of H alone; in general, $\delta H(z_e) \neq 0$. An example where this occurs is a rigid body spinning about one of its principal axes of inertia. In this case, a critical point of H alone would have zero angular velocity; but a critical point of $H + C$ is a (nontrivial) stationary rotation about one of the principal axes.

The argument used to establish the Lagrange–Dirichlet test formally works in infinite dimensions too. Unfortunately, for systems with infinitely many degrees of freedom (like fluids and plasmas), there is a serious technical snag. The calculus argument used before runs into problems; one might think that these are just technical and that we just need to be more careful with the calculus arguments. In fact, there is widespread belief in this "energy criterion" (see, for instance, the discussion and references in Marsden and Hughes [1983, Chapter 6], and Potier-Ferry [1982]). However, Ball and Marsden [1984] have shown using an example from elasticity theory that the difficulty is genuine: They produce a critical point of H at which $\delta^2 H$ is positive definite, yet this point is *not* a local minimum of H. On the other hand, Potier-Ferry [1982] shows that asymptotic stability is restored if suitable dissipation is added. Another way to overcome this difficulty is to modify Step 3 using a convexity argument of Arnold [1966b].

Modified Step 3.   Assume that P is a *linear* space.

(a) *Let $\Delta u = u - u_e$ denote a finite variation in phase space.*

(b) *Find quadratic functions Q_1 and Q_2 such that*

$$Q_1(\Delta u) \leq H(u_e + \Delta u) - H(u_e) - \delta H(u_e) \cdot \Delta u$$

and

$$Q_2(\Delta u) \leq C(u_e + \Delta u) - C(u_e) - \delta C(u_e) \cdot \Delta u,$$

(c) *Require $Q_1(\Delta u) + Q_2(\Delta u) > 0$ for all $\Delta u \neq 0$.*

(d) *Introduce the norm $\|\Delta u\|$ by*

$$\|\Delta u\|^2 = Q_1(\Delta u) + Q_2(\Delta u),$$

so $\|\Delta u\|$ is a measure of the distance from u to u_e; that is, we choose $d(u, u_e) = \|\Delta u\|$.

(e) *Require*

$$|H(u_e + \Delta u) - H(u_e)| \leq C_1 \|\Delta u\|^\alpha$$

and

$$|C(u_e + \Delta u) - C(u_e)| \leq C_2 \|\Delta u\|^\alpha$$

for constants $\alpha, C_1, C_2 > 0$ and $\|\Delta u\|$ sufficiently small.

These conditions guarantee stability of u_e and provide the distance measure relative to which stability is defined. The key part of the proof is simply the observation that if we add the two inequalities in (b), we get

$$\|\Delta u\|^2 \leq H(u_e + \Delta u) + C(u_e + \Delta u) - H(u_e) - C(u_e)$$

using the fact that $\delta H(u_e) \cdot \Delta u$ and $\delta C(u_e) \cdot \Delta u$ add up to zero by Step 1. But H and C are constant in time, so

$$\|(\Delta u)_{\text{time}=t}\|^2 \leq [H(u_e + \Delta u) + C(u_e + \Delta u) - H(u_e) - C(u_e)]|_{\text{time}=0}.$$

Now employ the inequalities in (e) to get

$$\|(\Delta u)_{\text{time}=t}\|^2 \leq (C_1 + C_2)\|(\Delta u)_{\text{time}=0}\|^\alpha.$$

This estimate bounds the temporal growth of finite perturbations in terms of initial perturbations, which is what is needed for stability. For a survey of this method, additional references, and numerous examples, see Holm, Marsden, Ratiu, and Weinstein [1985].

There are some situations (such as the stability of elastic rods) in which the above techniques do not apply. The chief reason is that there may be a lack of sufficiently many Casimir functions to achieve even the first step. For this reason a modified (but more sophisticated) method has been developed called the "energy–momentum method." The key to the method is to avoid the use of Casimir functions by applying the method *before* any reduction has taken place. This method was developed in a series of papers of Simo, Posbergh, and Marsden [1990, 1991] and Simo, Lewis, and Marsden [1991]. A discussion and additional references are found later in this section.

Gyroscopic Systems. The distinctions between "stability by energy methods," that is, *energetics* and "spectral stability" become especially interesting when one adds dissipation. In fact, building on the classical work of Kelvin and Chetaev, one can prove that if $\delta^2 H$ is indefinite, yet the spectrum is on the imaginary axis, then adding dissipation necessarily makes the system *linearly unstable*. That is, at least one pair of eigenvalues of the linearized equations move into the right half-plane. This is a phenomenon called *dissipation-induced instability*. This result, along with related developments, is proved in Bloch, Krishnaprasad, Marsden, and Ratiu [1991, 1994, 1996]. For example, consider the linear *gyroscopic system*

$$M\ddot{q} + S\dot{q} + Vq = 0, \tag{1.7.11}$$

where $q \in \mathbb{R}^n$, M is a positive definite symmetric $n \times n$ matrix, S is skew, and V is symmetric. This system is Hamiltonian (Exercise 1.7-2). If V has negative eigenvalues, then (1.7.11) is *formally unstable*. However, due to S, the system can be spectrally stable. However, if R is positive definite symmetric and $\epsilon > 0$ is small, the system with friction

$$M\ddot{q} + S\dot{q} + \epsilon R\dot{q} + Vq = 0 \tag{1.7.12}$$

is linearly unstable. A specific example is given in Exercise 1.7-4.

Outline of the Energy–Momentum Method. The energy momentum method is an extension of the Arnold (or energy–Casimir) method for the study of stability of relative equilibria, which was developed for Lie–Poisson systems on duals of Lie algebras, especially those of fluid dynamical type. In addition, the method extends and refines the fundamental stability techniques going back to Routh, Liapunov, and, in more recent times, to the work of Smale.

The motivation for these extensions is threefold.

First of all, the energy–momentum method can deal with Lie–Poisson systems for which there are not sufficient Casimir functions available, such as 3-D ideal flow and certain problems in elasticity. In fact, Abarbanel and Holm [1987] use what can be recognized retrospectively as the energy–momentum method to show that 3-D equilibria for ideal flow are generally formally unstable due to vortex stretching. Other fluid and plasma situations, such as those considered by Chern and Marsden [1990] for ABC flows and certain multiple-hump situations in plasma dynamics (see Holm, Marsden, Ratiu, and Weinstein [1985] and Morrison [1987], for example), provided additional motivation in the Lie–Poisson setting.

A second motivation is to extend the method to systems that need not be Lie–Poisson and still make use of the powerful idea of using reduced spaces, as in the original Arnold method. Examples such as rigid bodies with vibrating antennas (Sreenath, Oh, Krishnaprasad, and Marsden [1988], Oh,

Sreenath, Krishnaprasad, and Marsden [1989], Krishnaprasad and Marsden [1987]) and coupled rigid bodies (Patrick [1989]) motivated the need for such an extension of the theory.

Finally, it gives sharper stability conclusions in material representation and links with geometric phases.

The Idea of the Energy–Momentum Method.   The setting of the energy–momentum method is that of a mechanical system with symmetry with a configuration space Q and phase space T^*Q and a symmetry group G acting, with a standard momentum map $\mathbf{J} : T^*Q \to \mathfrak{g}^*$, where $\mathfrak{g}^*$ is the Lie algebra of G. Of course, one gets the Lie–Poisson case when $Q = G$.

The rough idea for the energy momentum method is first to formulate the problem directly on the unreduced space. Here, relative equilibria associated with a Lie algebra element ξ are critical points of the augmented Hamiltonian $H_\xi := H - \langle \mathbf{J}, \xi \rangle$. The idea is now to compute the second variation of H_ξ at a relative equilibrium z_e with momentum value μ_e subject to the constraint $J = \mu_e$ and on a space transverse to the action of G_{μ_e}, the subgroup of G that leaves μ_e fixed. Although the augmented Hamiltonian plays the role of $H + C$ in the Arnold method, notice that Casimir functions are not required to carry out the calculations.

The surprising thing is that the second variation of H_ξ at the relative equilibrium can be arranged to be block diagonal, using splittings that are based on the mechanical connection, while *at the same time*, the symplectic structure also has a simple block structure, so that the linearized equations are put into a useful canonical form. Even in the Lie–Poisson setting, this leads to situations in which one gets much simpler second variations. This block diagonal structure is what gives the method its computational power.

The general theory for carrying out this procedure was developed in Simo, Posbergh, and Marsden [1990, 1991] and Simo, Lewis, and Marsden [1991]. An exposition of the method may be found, along with additional references, in Marsden [1992]. It is of interest to extend this to the singular case, which is the subject of ongoing work; see Ortega and Ratiu [1997, 1998] and references therein.

The energy–momentum method may also be usefully formulated in the Lagrangian setting, which is very convenient for the calculations in many examples. The general theory for this was developed in Lewis [1992] and Wang and Krishnaprasad [1992]. This Lagrangian setting is closely related to the general theory of Lagrangian reduction. In this context one reduces variational principles rather than symplectic and Poisson structures, and for the case of reducing the tangent bundle of a Lie group, this leads to the Euler–Poincaré equations rather than the Lie–Poisson equations.

Effectiveness in Examples.   The energy–momentum method has proven its effectiveness in a number of examples. For instance, Lewis and Simo [1990] were able to deal with the stability problem for pseudo-rigid bodies, which was thought up to that time to be analytically intractable.

The energy–momentum method can sometimes be used in contexts where the reduced space is singular or at nongeneric points in the dual of the Lie algebra. This is done at singular points in Lewis, Ratiu, Simo, and Marsden [1992], who analyze the heavy top in great detail and, in the Lie–Poisson setting for compact groups at nongeneric points in the dual of the Lie algebra, in Patrick [1992, 1995]. One of the key things is to keep track of group drifts, because the isotropy group G_μ can change for nearby points, and these are important for the reconstruction process and for understanding the Hannay–Berry phase in the context of reduction (see Marsden, Montgomery, and Ratiu [1990] and references therein). For noncompact groups and an application to the dynamics of rigid bodies in fluids (underwater vehicles), see Leonard and Marsden [1997]. Additional work in this area is still needed in the context of singular reduction.

The Benjamin–Bona theorem on stability of solitons for the KdV equation can be viewed as an instance of the energy momentum method, see also Maddocks and Sachs [1993], and for example, Oh [1987] and Grillakis, Shatah, and Strauss [1987], although there are many subtleties in the PDE context.

Hamiltonian Bifurcations. The energy–momentum method has also been used in the context of Hamiltonian bifurcation problems. We shall give some simple examples of this in §1.8. One such context is that of free boundary problems building on the work of Lewis, Marsden, Montgomery, and Ratiu [1986], which gives a Hamiltonian structure for dynamic free boundary problems (surface waves, liquid drops, etc.), generalizing Hamiltonian structures found by Zakharov. Along with the Arnold method itself, this is used for a study of the bifurcations of such problems in Lewis, Marsden, and Ratiu [1987], Lewis [1989, 1992], Kruse, Marsden, and Scheurle [1993], and other references cited therein.

Converse to the Energy–Momentum Method. Because of the block structure mentioned, it has also been possible to prove, in a sense, a converse of the energy–momentum method. That is, if the second variation is indefinite, then the system is unstable. One cannot, of course, hope to do this literally as stated, since there are many systems (e.g., gyroscopic system mentioned earlier—an explicit example is given in Exercise 1.7-4) that are formally unstable, and yet their linearizations have eigenvalues lying on the imaginary axis. Most of these are presumably unstable due to "Arnold diffusion," but of course this is a very delicate situation to prove analytically. Instead, the technique is to show that with the addition of dissipation, the system is destabilized. This idea of *dissipation-induced instability* goes back to Thomson and Tait in the last century. In the context of the energy–momentum method, Bloch, Krishnaprasad, Marsden, and Ratiu [1994, 1996] show that with the addition of appropriate dissipation, the indefiniteness of the second variation is sufficient to induce linear instability in the problem.

There are related eigenvalue movement formulas (going back to Krein) that are used to study non-Hamiltonian perturbations of Hamiltonian normal forms in Kirk, Marsden, and Silber [1996]. There are interesting analogues of this for reversible systems in O'Reilly, Malhotra, and Namamchchivaya [1996].

Extension to Nonholonomic Systems. It is possible to partially extend the energy–momentum method to the case of nonholonomic systems. Building on the work on nonholonomic systems in Arnold [1988], Bates and Sniatycki [1993] and Bloch, Krishnaprasad, Marsden, and Murray [1996], on the example of the Routh problem in Zenkov [1995], and on the large Russian literature in this area, Zenkov, Bloch, and Marsden [1998] show that there is a generalization to this setting. The method is effective in the sense that it applies to a wide variety of interesting examples, such as the rolling disk, a three-wheeled vehicle known as the the roller racer and the rattleback.

Exercises

⋄ **1.7-1.** Work out Cherry's example of the Hamiltonian system in $\mathbb{R}^4$ whose energy function is given by (1.7.6). Show explicitly that the origin is a linearly and spectrally stable equilibrium but that it is nonlinearly unstable by proving that (1.7.7) is a solution for every $\tau > 0$ that can be chosen to start arbitrarily close to the origin and that goes to infinity for $t \to \tau$.

⋄ **1.7-2.** Show that (1.7.11) is Hamiltonian with $\mathbf{p} = M\dot{\mathbf{q}}$,

$$H(\mathbf{q}, \mathbf{p}) = \frac{1}{2}\mathbf{p} \cdot M^{-1}\mathbf{p} + \frac{1}{2}\mathbf{q} \cdot V\mathbf{q},$$

and

$$\{F, K\} = \frac{\partial F}{\partial q^i}\frac{\partial K}{\partial p_i} - \frac{\partial K}{\partial q^i}\frac{\partial F}{\partial p_i} - S^{ij}\frac{\partial F}{\partial p_i}\frac{\partial K}{\partial p_j}.$$

⋄ **1.7-3.** Show that (up to an overall factor) the characteristic polynomial for the linear system (1.7.11) is

$$p(\lambda) = \det[\lambda^2 M + \lambda S + V]$$

and that this actually is a polynomial of degree n in λ^2.

⋄ **1.7-4.** Consider the two-degree-of-freedom system

$$\ddot{x} - g\dot{y} + \gamma\dot{x} + \alpha x = 0,$$
$$\ddot{y} + g\dot{x} + \delta\dot{y} + \beta y = 0.$$

(a) Write it in the form (1.7.12).

(b) For $\gamma = \delta = 0$ show:

    (i) it is spectrally stable if $\alpha > 0$, $\beta > 0$;

    (ii) for $\alpha\beta < 0$, it is spectrally unstable;

    (iii) for $\alpha < 0$, $\beta < 0$, it is formally unstable (that is, the energy function, which is a quadratic form, is indefinite); and

        A. if $D := (g^2 + \alpha + \beta)^2 - 4\alpha\beta < 0$, then there are two roots in the right half-plane and two in the left; the system is spectrally unstable;

        B. if $D = 0$ and $g^2 + \alpha + \beta \geq 0$, the system is spectrally stable, but if $g^2 + \alpha + \beta < 0$ then it is spectrally unstable; and

        C. if $D > 0$ and $g^2 + \alpha + \beta \geq 0$, the system is spectrally stable, but if $g^2 + \alpha + \beta < 0$, then it is spectrally unstable.

(c) For a polynomial $p(\lambda) = \lambda^4 + \rho_1\lambda^3 + \rho_2\lambda^2 + \rho_3\lambda + \rho_4$, the *Routh–Hurwitz criterion* (see Gantmacher [1959, Volume 2]) says that the number of right half-plane zeros of p is the number of sign changes of the sequence

$$\left\{ 1, \rho_1, \frac{\rho_1\rho_2 - \rho_3}{\rho_1}, \frac{\rho_3\rho_1\rho_2 - \rho_3^2 - \rho_4\rho_1^2}{\rho_1\rho_2 - \rho_3}, \rho_4 \right\}.$$

Apply this to the case in which $\alpha < 0$, $\beta < 0$, $g^2 + \alpha + \beta > 0$, $\gamma > 0$, and $\delta > 0$ to show that the system is spectrally unstable.

1.8 Bifurcation

When the energy–momentum or energy–Casimir method indicates that an instability might be possible, techniques of bifurcation theory can be brought to bear to determine the emerging dynamical complexities such as the development of multiple equilibria and periodic orbits.

Ball in a Rotating Hoop. For example, consider a particle moving with no friction in a rotating hoop (Figure 1.8.1).

In §2.8 we derive the equations and study the phase portraits for this system. One finds that as ω increases past $\sqrt{g/R}$, the stable equilibrium at $\theta = 0$ becomes unstable through a *Hamiltonian pitchfork bifurcation* and two new solutions are created. These solutions are symmetric in the vertical axis, a reflection of the original $\mathbb{Z}_2$ symmetry of the mechanical system in Figure 1.8.1. Breaking this symmetry by, for example, putting the rotation axis slightly off center is an interesting topic that we shall discuss in §2.8.

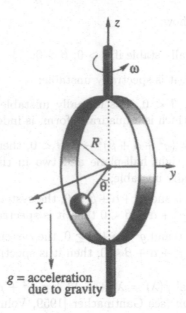

FIGURE 1.8.1. A particle moving in a hoop rotating with angular velocity ω.

Rotating Liquid Drop. The system consists of the two-dimensional Euler equations for an ideal fluid with a free boundary. An equilibrium solution consists of a rigidly rotating circular drop. The energy–Casimir method shows stability, provided that

$$\Omega < 2\sqrt{\frac{3\tau}{R^3}}. \tag{1.8.1}$$

In this formula, Ω is the angular velocity of the circular drop, R is its radius, and τ is the surface tension, a constant. As Ω increases and (1.8.1) is violated, the stability of the circular solution is lost and is picked up by elliptical-like solutions with $\mathbb{Z}_2 \times \mathbb{Z}_2$ symmetry. The bifurcation is actually subcritical relative to the *angular velocity* Ω (that is, the new solutions occur *below* the critical value of Ω) and is supercritical (the new solutions occur *above* criticality) relative to the *angular momentum*. This is proved in Lewis, Marsden, and Ratiu [1987] and Lewis [1989], where other references may also be found (see Figure 1.8.2).

For the ball in the hoop, the eigenvalue evolution for the linearized equations is shown in Figure 1.8.3(a). For the rotating liquid drop, the movement of eigenvalues is the same: They are constrained to *stay* on the imaginary axis because of the symmetry of the problem. Without this symmetry, eigenvalues typically split, as in Figure 1.8.3(b). These are examples of a general theory of the movement of such eigenvalues given in Golubitsky and Stewart [1987], Dellnitz, Melbourne, and Marsden [1992], Knobloch, Mahalov, and Marsden [1994], and Kirk, Marsden, and Silber [1996].

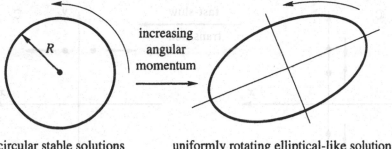

circular stable solutions   uniformly rotating elliptical-like solutions

FIGURE 1.8.2. A circular liquid drop losing its stability and its symmetry.

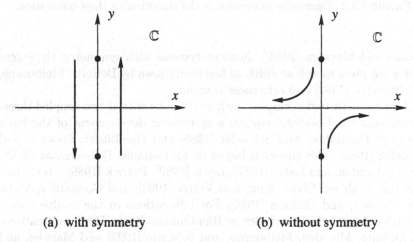

    (a) with symmetry   (b) without symmetry

FIGURE 1.8.3. The movement of eigenvalues in bifurcation of equilibria.

More Examples.   Another example is the heavy top: a rigid body with
one point fixed, moving in a gravitational field. When the top makes the
transition from a fast top to a slow top, the angular velocity ω decreases
past the critical value

$$\omega_c = \frac{2\sqrt{MglI_1}}{I_3}, \tag{1.8.2}$$

stability is lost, and a **resonance bifurcation** occurs. Here, when the
bifurcation occurs, the eigenvalues of the equations linearized at the equilibrium behave as in Figure 1.8.4.

For an extensive study of bifurcations and stability in the dynamics of
a heavy top, see Lewis, Ratiu, Simo, and Marsden [1992]. Behavior of this
sort is sometimes called a **Hamiltonian Krein–Hopf bifurcation**, or a
gyroscopic instability (see van der Meer [1985, 1990]). Here more complex dynamic behavior ensues, including periodic and chaotic motions (see

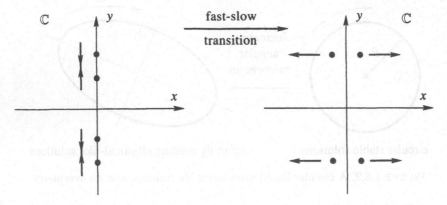

FIGURE 1.8.4. Eigenvalue movement in the Hamiltonian Hopf bifurcation.

Holmes and Marsden [1983]). In some systems with symmetry, the eigenvalues can *pass* as well as *split*, as has been shown by Dellnitz, Melbourne, and Marsden [1992] and references therein.

More sophisticated examples, such as the dynamics of two coupled three-dimensional rigid bodies, requires a systematic development of the basic theory of Golubitsky and Schaeffer [1985] and Golubitsky, Stewart, and Schaeffer [1988]. This theory is begun in, for example, Duistermaat [1983], Lewis, Marsden, and Ratiu [1987], Lewis [1989], Patrick [1989], Meyer and Hall [1992], Broer, Chow, Kim, and Vegter [1993], and Golubitsky, Marsden, Stewart, and Dellnitz [1994]. For bifurcations in the double spherical pendulum (which includes a Hamiltonian–Krein–Hopf bifurcation), see Dellnitz, Marsden, Melbourne, and Scheurle [1992] and Marsden and Scheurle [1993a].

Exercises

⬦ **1.8-1.** Study the bifurcations (changes in the phase portrait) for the equation

$$\ddot{x} + \mu x + x^2 = 0$$

as μ passes through zero. Use the second derivative test on the potential energy.

⬦ **1.8-2.** Repeat Exercise 1.8-1 for

$$\ddot{x} + \mu x + x^3 = 0$$

as μ passes through zero.

1.9   The Poincaré–Melnikov Method

The Forced Pendulum.   To begin with a simple example, consider the equation of a forced pendulum:

$$\ddot{\phi} + \sin\phi = \epsilon\cos\omega t. \tag{1.9.1}$$

Here ω is a constant angular forcing frequency and ϵ is a small parameter. Systems of this or a similar nature arise in many interesting situations. For example, a double planar pendulum and other "executive toys" exhibit chaotic motion that is analogous to the behavior of this equation; see Burov [1986] and Shinbrot, Grebogi, Wisdom, and Yorke [1992].

For $\epsilon = 0$ (1.9.1) has the phase portrait of a simple pendulum (the same as shown later in Figure 2.8.2a). For ϵ small but nonzero, (1.9.1) possesses no analytic integrals of the motion. In fact, it possesses transversal intersecting stable and unstable manifolds (separatrices); that is, the Poincaré map $P_{t_0} : \mathbb{R}^2 \to \mathbb{R}^2$ defined as the map that advance solutions by one period $T = 2\pi/\omega$ starting at time t_0 possess transversal homoclinic points. This type of dynamic behavior has several consequences, besides precluding the existence of analytic integrals, that lead one to use the term "chaotic." For example, (1.9.1) has infinitely many periodic solutions of arbitrarily high period. Also, using the shadowing lemma, one sees that given any bi-infinite sequence of zeros and ones (for example, use the binary expansion of e or π), there exists a corresponding solution of (1.9.1) that successively crosses the plane $\phi = 0$ (the pendulum's vertically downward configuration) with $\phi > 0$ corresponding to a zero and $\phi < 0$ corresponding to a one. The origin of this chaos on an intuitive level lies in the motion of the pendulum near its unperturbed homoclinic orbit, the orbit that does one revolution in infinite time. Near the top of its motion (where $\phi = \pm\pi$) small nudges from the forcing term can cause the pendulum to fall to the left or right in a temporally complex way.

The dynamical systems theory needed to justify the preceding statements is available in Smale [1967], Moser [1973], Guckenheimer and Holmes [1983], and Wiggins [1988, 1990]. Some key people responsible for the development of the basic theory are Poincaré, Birkhoff, Kolmogorov, Melnikov, Arnold, Smale, and Moser. The idea of transversal intersecting separatrices comes from Poincaré's famous paper on the three-body problem (Poincaré [1890]). His goal, not quite achieved for reasons we shall comment on later, was to prove the nonintegrability of the restricted three-body problem and that various series expansions used up to that point diverged (he began the theory of asymptotic expansions and dynamical systems in the course of this work). See Diacu and Holmes [1996] for additional information about Poincaré's work.

Although Poincaré had all the essential tools needed to prove that equations like (1.9.1) are not integrable (in the sense of having no analytic integrals), his interests lay with harder problems, and he did not develop the easier basic theory very much. Important contributions were made by Melnikov [1963] and Arnold [1964] that lead to a simple procedure for proving that (1.9.1) is not integrable. The Poincaré–Melnikov method was revived by Chirikov [1979], Holmes [1980b], and Chow, Hale, and Mallet-Paret [1980]. We shall give the method for Hamiltonian systems. We refer to Guckenheimer and Holmes [1983] and to Wiggins [1988, 1990] for generalizations and further references.

The Poincaré–Melnikov Method.   This method proceeds as follows:

1. Write the dynamical equation to be studied in the form

$$\dot{x} = X_0(x) + \epsilon X_1(x, t), \qquad (1.9.2)$$

where $x \in \mathbb{R}^2$, X_0 is a Hamiltonian vector field with energy H_0, X_1 is periodic with period T and is Hamiltonian with energy a T-periodic function H_1. Assume that X_0 has a homoclinic orbit $\bar{x}(t)$, so $\bar{x}(t) \to x_0$, a hyperbolic saddle point, as $t \to \pm\infty$.

2. Compute the **Poincaré–Melnikov function** defined by

$$M(t_0) = \int_{-\infty}^{\infty} \{H_0, H_1\}(\bar{x}(t - t_0), t)\, dt, \qquad (1.9.3)$$

where $\{\,,\,\}$ denotes the Poisson bracket.

If $M(t_0)$ has simple zeros as a function of t_0, then (1.9.2) has, for sufficiently small ϵ, homoclinic chaos in the sense of transversal intersecting separatrices (in the sense of Poincaré maps as mentioned above).

We shall prove this result in §2.11. To apply it to equation (1.9.1) one proceeds as follows. Let $x = (\phi, \dot{\phi})$, so we get

$$\frac{d}{dt} \begin{bmatrix} \phi \\ \dot{\phi} \end{bmatrix} = \begin{bmatrix} \dot{\phi} \\ -\sin\phi \end{bmatrix} + \epsilon \begin{bmatrix} 0 \\ \cos\omega t \end{bmatrix}.$$

The homoclinic orbits for $\epsilon = 0$ are given by (see Exercise 1.9-1)

$$\bar{x}(t) = \begin{bmatrix} \phi(t) \\ \dot{\phi}(t) \end{bmatrix} = \begin{bmatrix} \pm 2\tan^{-1}(\sinh t) \\ \pm 2\,\text{sech}\,t \end{bmatrix},$$

and one has

$$H_0(\phi, \dot{\phi}) = \tfrac{1}{2}\dot{\phi}^2 - \cos\phi \quad \text{and} \quad H_1(\phi, \dot{\phi}, t) = \phi\cos\omega t. \qquad (1.9.4)$$

Hence (1.9.3) gives

$$M(t_0) = \int_{-\infty}^{\infty} \left(\frac{\partial H_0}{\partial \phi} \frac{\partial H_1}{\partial \dot\phi} - \frac{\partial H_0}{\partial \dot\phi} \frac{\partial H_1}{\partial \phi} \right) (\bar{x}(t - t_0), t) \, dt$$

$$= - \int_{-\infty}^{\infty} \dot\phi(t - t_0) \cos \omega t \, dt$$

$$= \mp \int_{-\infty}^{\infty} [2 \operatorname{sech}(t - t_0) \cos \omega t] \, dt.$$

Changing variables and using the fact that sech is even and sin is odd, we get

$$M(t_0) = \mp 2 \left(\int_{-\infty}^{\infty} \operatorname{sech} t \cos \omega t \, dt \right) \cos(\omega t_0).$$

The integral is evaluated by residues (see Exercise 1.9-2):

$$M(t_0) = \mp 2\pi \operatorname{sech} \left(\frac{\pi \omega}{2} \right) \cos(\omega t_0), \qquad (1.9.5)$$

which clearly has simple zeros. Thus, this equation has chaos for ϵ small enough.

Exercises

⋄ **1.9-1.**   Verify directly that the homoclinic orbits for the simple pendulum equation $\ddot\phi + \sin\phi = 0$ are given by $\phi(t) = \pm 2 \tan^{-1}(\sinh t)$.

⋄ **1.9-2.**   Evaluate the integral $\int_{-\infty}^{\infty} \operatorname{sech} t \cos \omega t \, dt$ to prove (1.9.5) as follows. Write $\operatorname{sech} t = 2/(e^t + e^{-t})$ and note that there is a simple pole of

$$f(z) = \frac{e^{i\omega z} + e^{-i\omega z}}{e^z + e^{-z}}$$

in the complex plane at $z = \pi i/2$. Evaluate the residue there and apply Cauchy's theorem.[9]

1.10   Resonances, Geometric Phases, and Control

The work of Smale [1970] shows that topology plays an important role in mechanics. Smale's work employs Morse theory applied to conserved quantities such as the energy–momentum map. In this section we point out other ways in which geometry and topology enter mechanical problems.

---

[9]Consult a book on complex variables such as Marsden and Hoffman, *Basic Complex Analysis*, Third Edition, Freeman, 1998.

The One-to-One Resonance.   When one considers resonant systems, one often encounters Hamiltonians of the form

$$H = \frac{1}{2}(q_1^2 + p_1^2) + \frac{\lambda}{2}(q_2^2 + p_2^2) + \text{ higher-order terms.} \qquad (1.10.1)$$

The quadratic terms describe two oscillators that have the same frequency when $\lambda = 1$, which is why one speaks of a one-to-one resonance. To analyze the dynamics of H, it is important to utilize a good geometric picture for the critical case

$$H_0 = \frac{1}{2}(q_1^2 + p_1^2 + q_2^2 + p_2^2). \qquad (1.10.2)$$

The energy level $H_0 = \text{constant}$ is the three-sphere $S^3 \subset \mathbb{R}^4$. If we think of H_0 as a function on complex two-space $\mathbb{C}^2$ by letting

$$z_1 = q_1 + ip_1 \quad \text{and} \quad z_2 = q_2 + ip_2,$$

then $H_0 = (|z_1|^2 + |z_2|^2)/2$, so H_0 is left-invariant by the action of SU(2), the group of complex 2×2 unitary matrices of determinant one. The corresponding conserved quantities are

$$\begin{aligned} W_1 &= 2(q_1q_2 + p_1p_2), \\ W_2 &= 2(q_2p_1 - q_1p_2), \\ W_3 &= q_1^2 + p_1^2 - q_2^2 - p_2^2, \end{aligned} \qquad (1.10.3)$$

which comprise the components of a (momentum) map

$$\mathbf{J} : \mathbb{R}^4 \to \mathbb{R}^3. \qquad (1.10.4)$$

From the readily verified relation $4H_0^2 = W_1^2 + W_2^2 + W_3^2$, one finds that $\mathbf{J}$ restricted to S^3 gives a map

$$j : S^3 \to S^2. \qquad (1.10.5)$$

The fibers $j^{-1}(\text{point})$ are circles, and the trajectories for the dynamics of H_0 move along these circles. The map j is the **Hopf fibration**, which describes S^3 as a topologically nontrivial circle bundle over S^2. The role of the Hopf fibration in mechanics was known to Reeb [1949].

One also finds that the study of systems like (1.10.1) that are close to H_0 can, to a good approximation, be reduced to dynamics on S^2. These dynamics are in fact Lie–Poisson and S^2 sits as a coadjoint orbit in $\mathfrak{so}(3)^*$, so the evolution is of rigid-body type, just with a different Hamiltonian. For a computer study of the Hopf fibration in the one-to-one resonance, see Kocak, Bisshopp, Banchoff, and Laidlaw [1986].

The Hopf Fibration in Rigid-Body Mechanics.   When doing reduction for the rigid body, one studies the reduced space

$$\mathbf{J}^{-1}(\mu)/G_\mu = \mathbf{J}^{-1}(\mu)/S^1,$$

which in this case is the sphere S^2. As we shall see in Chapter 15, $\mathbf{J}^{-1}(\mu)$ is topologically the same as the rotation group SO(3), which in turn is the same as $S^3/\mathbb{Z}_2$. Thus, the reduction map is a map of SO(3) to S^2. Such a map is given explicitly by taking an orthogonal matrix A and mapping it to the vector on the sphere given by $A\mathbf{k}$, where $\mathbf{k}$ is the unit vector along the z-axis. This map, which does the projection, is in fact a restriction of a momentum map and is, when composed with the map of $S^3 \cong$ SU(2) to SO(3), just the Hopf fibration again. Thus, not only does the Hopf fibration occur in the one-to-one resonance, *it occurs in the rigid body in a natural way as the reduction map from material to body representation!*

Geometric Phases.   The history of this concept is complex. We refer to Berry [1990] for a discussion of the history, going back to Bortolotti in 1926, Vladimirskii and Rytov in 1938 in the study of polarized light, Kato in 1950, and Longuet–Higgins and others in 1958 in atomic physics. Some additional historical comments regarding phases in rigid-body mechanics are given below.

We pick up the story with the classical example of the Foucault pendulum. The Foucault pendulum gives an interesting phase shift (a shift in the angle of the plane of the pendulum's swing) when the overall system undergoes a cyclic evolution (the pendulum is carried in a circular motion due to the Earth's rotation). This phase shift is geometric in character: If one parallel transports an orthonormal frame along the same line of latitude, it returns with a phase shift equaling that of the Foucault pendulum. This phase shift $\Delta\theta = 2\pi \cos\alpha$ (where α is the co-latitude) has the geometric meaning shown in Figure 1.10.1.

In geometry, when an orthonormal frame returns to its original position after traversing a closed path but is rotated, the rotation is referred to as *holonomy* (or *anholonomy*). This is a unifying mathematical concept that underlies many geometric phases in systems such as fiber optics, MRI (magnetic resonance imaging), amoeba propulsion, molecular dynamics, and micromotors. These applications represent one reason the subject is of such current interest.

In the quantum case a seminal paper on geometric phases is Kato [1950]. It was Berry [1984, 1985], Simon [1983], Hannay [1985], and Berry and Hannay [1988] who realized that holonomy is the crucial geometric unifying thread. On the other hand, Golin, Knauf, and Marmi [1989], Montgomery [1988], and Marsden, Montgomery, and Ratiu [1989, 1990] demonstrated that averaging connections and reduction of mechanical systems with symmetry also plays an important role, both classically and quantum-mechanically. Aharonov and Anandan [1987] have shown that the geomet-

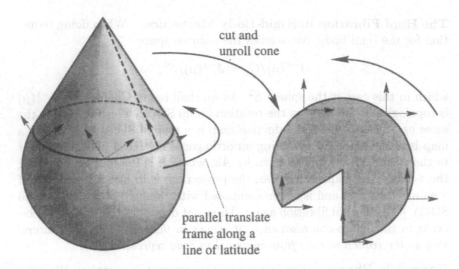

cut and
unroll cone

parallel translate
frame along a
line of latitude

FIGURE 1.10.1. The geometric interpretation of the Foucault pendulum phase shift.

ric phase for a closed loop in projectivized complex Hilbert space occurring in quantum mechanics equals the exponential of the symplectic area of a two–dimensional manifold whose boundary is the given loop. The symplectic form in question is naturally induced on the projective space from the canonical symplectic form of complex Hilbert space (minus the imaginary part of the inner product) via reduction. Marsden, Montgomery, and Ratiu [1990] show that this formula is the holonomy of the closed loop relative to a principal S^1-connection on the unit ball of complex Hilbert space and is a particular case of the holonomy formula in principal bundles with abelian structure group.

Geometric Phases and Locomotion. Geometric phases naturally occur in families of integrable systems depending on parameters. Consider an integrable system with action-angle variables

$$(I_1, I_2, \ldots, I_n, \theta_1, \theta_2, \ldots, \theta_n);$$

assume that the Hamiltonian $H(I_1, I_2, \ldots, I_n; m)$ depends on a parameter $m \in M$. This just means that we have a Hamiltonian independent of the angular variables θ and we can identify the configuration space with an n-torus $\mathbb{T}^n$. Let c be a loop based at a point m_0 in M. We want to compare the angular variables in the torus over m_0, while the system is slowly changed as the parameters traverse the circuit c. Since the dynamics in the fiber vary as we move along c, even if the actions vary by a negligible amount, there will be a shift in the angle variables due to the frequencies $\omega^i = \partial H/\partial I^i$ of the integrable system; correspondingly, one defines

$$dynamic\ phase = \int_0^1 \omega^i\left(I, c(t)\right) dt.$$

Here we assume that the loop is contained in a neighborhood whose standard action coordinates are defined. In completing the circuit c, we return to the same torus, so a comparison between the angles makes sense. The actual shift in the angular variables during the circuit is the *dynamic phase* plus a correction term called the *geometric phase*. One of the key results is that this geometric phase is the holonomy of an appropriately constructed connection (called the *Hannay–Berry connection*) on the torus bundle over M that is constructed from the action–angle variables. The corresponding angular shift, computed by Hannay [1985], is called *Hannay's angles*, so the actual phase shift is given by

$$\Delta\theta = \text{dynamic phases } + \text{ Hannay's angles.}$$

The geometric construction of the Hannay–Berry connection for classical systems is given in terms of momentum maps and averaging in Golin, Knauf, and Marmi [1989] and Montgomery [1988]. Weinstein [1990] makes precise the geometric structures that make possible a definition of the Hannay angles for a cycle in the space of Lagrangian submanifolds, even without the presence of an integrable system. Berry's phase is then seen as a "primitive" for the Hannay angles. A summary of this work is given in Woodhouse [1992].

Another class of examples where geometric phases naturally arise in the dynamics of coupled rigid bodies. The three-dimensional single rigid body is discussed below. For several coupled rigid bodies, the dynamics can be quite complex. For instance, even for three coupled bodies in the plane, the dynamics are known to be chaotic, despite the presence of stable relative equilibria; see Oh, Sreenath, Krishnaprasad, and Marsden [1989]. Geometric phase phenomena for this type of example are quite interesting as they are in some of the work of Wilczek and Shapere on locomotion in microorganisms. (See, for example, Shapere and Wilczek [1987, 1989] and Wilczek and Shapere [1989].) In this problem, control of the system's *internal or shape variables* can lead to phase changes in the *external or group variables*. These choices of variables are related to the variables in the reduced and the unreduced phase spaces. In this setting one can formulate interesting questions of optimal control such as "When a falling cat turns itself over in mid-flight (all the time with zero angular momentum!), does it do so with optimal efficiency in terms of, say, energy expended?" There are interesting answers to these questions that are related to the dynamics of Yang–Mills particles moving in the associated gauge field of the problem. See Montgomery [1984, 1990] and references therein.

We give two simple examples of geometric phases for linked rigid bodies. Additional details can be found in Marsden, Montgomery, and Ratiu [1990].

First, consider three uniform coupled bars (or coupled planar rigid bodies) linked together with pivot (or pin) joints, so the bars are free to rotate relative to each other. Assume that the bars are moving freely in the plane with no external forces and that the angular momentum is zero. However, assume that the joint angles can be controlled with, say, motors in the joints. Figure 1.10.2 shows how the joints can be manipulated, each one going through an angle of 2π and yet the overall assemblage rotates through an angle π.

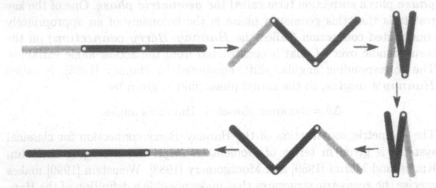

FIGURE 1.10.2. Manipulating the joint angles can lead to an overall rotation of the system.

Here we assume that the moments of inertia of the two outside bars (about an axis through their centers of mass and perpendicular to the page) are each one-half that of the middle bar. The statement is verified by examining the equation for zero angular momentum (see, for example Sreenath, Oh, Krishnaprasad, and Marsden [1988] and Oh, Sreenath, Krishnaprasad, and Marsden [1989]). General formulas for the reconstruction phase applicable to examples of this type are given in Krishnaprasad [1989].

A second example is the dynamics of linkages. This type of example is considered in Krishnaprasad [1989], Yang and Krishnaprasad [1990], including comments on the relation with the three-manifold theory of Thurston. Here one considers a linkage of rods, say four rods linked by pivot joints as in Figure 1.10.3.

The system is free to rotate without external forces or torques, but there are assumed to be torques at the joints. When one turns the small "crank" the whole assemblage turns, even though the angular momentum, as in the previous example, stays zero.

For an overview of how geometric phases are used in robotic locomotion problems, see Marsden and Ostrowski [1998]. (This paper is available at http://www.cds.caltech.edu/~marsden.)

Phases in Rigid-Body Dynamics. As we shall see in Chapter 15, the motion of a rigid body is a geodesic with respect to a left-invariant Rieman-

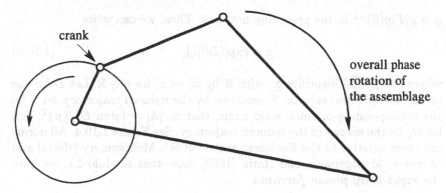

crank

overall phase
rotation of
the assemblage

FIGURE 1.10.3. Turning the crank can lead to an overall phase shift.

nian metric (the inertia tensor) on the rotation group SO(3). The corresponding phase space is $P = T^* \text{SO}(3)$ and the momentum map $\mathbf{J} : P \to \mathbb{R}^3$ for the *left* SO(3) action is *right* translation to the identity. We identify $\mathfrak{so}(3)^*$ with $\mathfrak{so}(3)$ via the standard inner product and identify $\mathbb{R}^3$ with $\mathfrak{so}(3)$ via the map $v \mapsto \hat{v}$, where $\hat{v}(w) = v \times w$, $\times$ being the standard cross product. Points in $\mathfrak{so}(3)^*$ are regarded as the left reduction of $T^* \text{SO}(3)$ by $G = \text{SO}(3)$ and are the angular momenta as seen from a *body-fixed* frame.

The reduced spaces $P_\mu = \mathbf{J}^{-1}(\mu)/G_\mu$ are identified with spheres in $\mathbb{R}^3$ of Euclidean radius $\|\mu\|$, with their symplectic form $\omega_\mu = -dS/\|\mu\|$, where dS is the standard area form on a sphere of radius $\|\mu\|$ and where G_μ consists of rotations about the μ-axis. The trajectories of the reduced dynamics are obtained by intersecting a family of homothetic ellipsoids (the energy ellipsoids) with the angular momentum spheres. In particular, all but at most four of the reduced trajectories are periodic. These four exceptional trajectories are the well-known homoclinic trajectories; we shall determine them explicitly in §15.8.

Suppose a reduced trajectory $\mathbf{\Pi}(t)$ is given on P_μ, with period T. *After time T, by how much has the rigid body rotated in space?* The spatial angular momentum is $\pi = \mu = g\mathbf{\Pi}$, which is the conserved value of $\mathbf{J}$. Here $g \in \text{SO}(3)$ is the attitude of the rigid body and $\mathbf{\Pi}$ is the body angular momentum. If $\mathbf{\Pi}(0) = \mathbf{\Pi}(T)$, then

$$\mu = g(0)\mathbf{\Pi}(0) = g(T)\mathbf{\Pi}(T),$$

and so $g(T)^{-1}\mu = g(0)^{-1}\mu$, that is, $g(T)g(0)^{-1}\mu = \mu$, so $g(T)g(0)^{-1}$ is a rotation about the axis μ. We want to give the angle of this rotation.

To determine this angle, let $c(t)$ be the corresponding trajectory in $\mathbf{J}^{-1}(\mu) \subset P$. Identify $T^* \text{SO}(3)$ with $\text{SO}(3) \times \mathbb{R}^3$ by left trivialization, so $c(t)$ gets identified with $(g(t), \mathbf{\Pi}(t))$. Since the reduced trajectory $\mathbf{\Pi}(t)$ closes after time T, we recover the fact that $c(T) = gc(0)$ for some $g \in G_\mu$. Here,

$g = g(T)g(0)^{-1}$ in the preceding notation. Thus, we can write

$$g = \exp[(\Delta\theta)\zeta], \tag{1.10.6}$$

where $\zeta = \mu/\|\mu\|$ identifies $\mathfrak{g}_\mu$ with $\mathbb{R}$ by $a\zeta \mapsto a$, for $a \in \mathbb{R}$. Let D be one of the two spherical caps on S^2 enclosed by the reduced trajectory, let Λ be the corresponding oriented solid angle, that is, $|\Lambda| = (\text{area } D)/\|\mu\|^2$, and let H_μ be the energy of the reduced trajectory. See Figure 1.10.4. All norms are taken relative to the Euclidean metric of $\mathbb{R}^3$. Montgomery [1991a] and Marsden, Montgomery, and Ratiu [1990] show that modulo 2π, we have the *rigid-body phase formula*

$$\Delta\theta = \frac{1}{\|\mu\|}\left\{\int_D \omega_\mu + 2H_\mu T\right\} = -\Lambda + \frac{2H_\mu T}{\|\mu\|}. \tag{1.10.7}$$

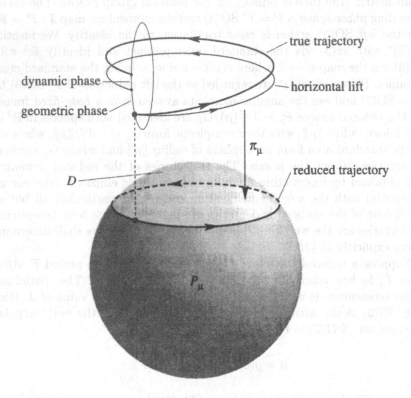

FIGURE 1.10.4. The geometry of the rigid-body phase formula.

More History. The history of the rigid-body phase formula is quite interesting and seems to have proceeded independently of the other developments above.[10] The formula has its roots in work of MacCullagh dating back to 1840 and Thomson and Tait [1867, §§123, 126]. (See Zhuravlev [1996] and O'Reilly [1997] for a discussion and extensions.) A special case of formula (1.10.7) is given in Ishlinskii [1952]; see also Ishlinskii [1963].[11] The formula referred to covers a special case in which only the geometric phase is present. For example, in certain precessional motions in which, up to a certain order in averaging, one can ignore the dynamic phase, and only the geometric phase survives. Even though Ishlinskii found only special cases of the result, he recognized that it is related to the geometric concept of parallel transport. A formula like the one above was found by Goodman and Robinson [1958] in the context of drift in gyroscopes; their proof is based on the Gauss–Bonnet theorem. Another interesting approach to formulas of this sort, also based on averaging and solid angles, is given in Goldreich and Toomre [1969], who applied it to the interesting geophysical problem of polar wander (see also Poincaré [1910]!).

The special case of the above formula for a *symmetric* free rigid body was given by Hannay [1985] and Anandan [1988, formula (20)]. The proof of the general formula based on the theory of connections and the formula for holonomy in terms of curvature was given by Montgomery [1991a] and Marsden, Montgomery, and Ratiu [1990]. The approach using the Gauss–Bonnet theorem and its relation to the Poinsot construction along with additional results is taken up by Levi [1993]. For applications to general resonance problems (such as the three-wave interaction) and nonlinear optics, see Alber, Luther, Marsden and Robbins [1998].

An analogue of the rigid-body phase formula for the heavy top and the Lagrange top (symmetric heavy top) was given in Marsden, Montgomery, and Ratiu [1990]. Links with vortex filament configurations were given in Fukumoto and Miyajima [1996] and Fukumoto [1997].

Satellites with Rotors and Underwater Vehicles.   Another example that naturally gives rise to geometric phases is the rigid body with one or more internal rotors. Figure 1.10.5 illustrates the system considered. To specify the position of this system we need an element of the group of rigid motions of $\mathbb{R}^3$ to place the center of mass and the attitude of the carrier, and an angle (element of S^1) to position each rotor. Thus the configuration space is $Q = \text{SE}(3) \times S^1 \times S^1 \times S^1$. The equations of motion of this system are an extension of Euler's equations of motion for a freely spinning rotor. Just as holding a spinning bicycle wheel while sitting on a swivel chair can affect the carrier's motion, so the spinning rotors can affect the dynamics of the rigid carrier.

---

[10]We thank V. Arnold for valuable help with these comments.

[11]On page 195 of Ishlinskii [1976], a later book on mechanics, it is stated that "the formula was found by the author in 1943 and was published in Ishlinskii [1952]."

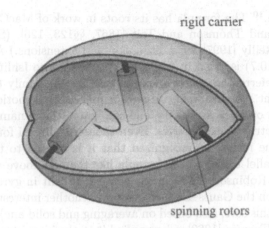

rigid carrier

spinning rotors

FIGURE 1.10.5. The rigid body with internal rotors.

In this example, one can analyze equilibria and their stability in much the same way as one can with the rigid body. However, what one often wants to do is to forcibly spin, or control, the rotors so that one can achieve attitude control of the structure in the same spirit that a falling cat has *control* of its attitude by manipulating its body parts while falling. For example, one can attempt to prescribe a relation between the rotor dynamics and the rigid-body dynamics by means of a *feedback law*. This has the property that the total system angular momentum is still preserved and that the resulting dynamic equations can be expressed entirely in terms of the free rigid-body variable. (A falling cat has zero angular momentum even though it is able to turn over!) In some cases the resulting equations are again Hamiltonian on the invariant momentum sphere. Using this fact, one can compute the geometric phase for the problem generalizing the free rigid-body phase formula. (See Bloch, Krishnaprasad, Marsden, and Sánchez de Alvarez [1992] and Bloch, Leonard, and Marsden [1997, 1998] for details.) This type of analysis is useful in designing and understanding attitude control devices.

Another example that combines some features of the satellite and the heavy top is the *underwater vehicle*. This is in the realm of the dynamics of rigid bodies in fluids, a subject going back to Kirchhoff in the late 1800s. We refer to Leonard and Marsden [1997] and Holmes, Jenkins, and Leonard [1998] for modern accounts and many references.

Miscellaneous Links.   There are many continuum-mechanical examples to which the techniques of geometric mechanics apply. Some of those are free boundary problems (Lewis, Marsden, Montgomery, and Ratiu [1986], Montgomery, Marsden, and Ratiu [1984], Mazer and Ratiu [1989]), spacecraft with flexible attachments (Krishnaprasad and Marsden [1987]), elasticity (Holm and Kuperschmidt [1983], Kuperschmidt and Ratiu [1983],

Marsden, Ratiu, and Weinstein [1984a, 1984b], Simo, Marsden, and Krish-naprasad [1988]), and reduced MHD (Morrison and Hazeltine [1984] and Marsden and Morrison [1984]). We also wish to look at these theories from both the spatial (Eulerian) and body (convective) points of view as reductions of the canonical material picture. These two reductions are, in an appropriate sense, dual to each other.

The geometric-analytic approach to mechanics finds use in a number of other diverse areas as well. We mention just a few samples.

- Integrable systems (Moser [1980], Perelomov [1990], Adams, Harnad, and Previato [1988], Fomenko and Trofimov [1989], Fomenko [1988a, 1988b], Reyman and Semenov-Tian-Shansky [1990], and Moser and Veselov [1991]).

- Applications of integrable systems to numerical analysis (like the QR algorithm and sorting algorithms); see Deift and Li [1989] and Bloch, Brockett, and Ratiu [1990, 1992].

- Numerical integration (Sanz-Serna and Calvo [1994], Marsden, Patrick, and Shadwick [1996], Wendlandt and Marsden [1997], Marsden, Patrick, and Shkoller [1998]).

- Hamiltonian chaos (Arnold [1964], Ziglin [1980a, 1980b, 1981], Holmes and Marsden [1981, 1982a, 1982b, 1983], Wiggins [1988]).

- Averaging (Cushman and Rod [1982], Iwai [1982, 1985], Ercolani, Forest, McLaughlin, and Montgomery [1987]).

- Hamiltonian bifurcations (van der Meer [1985], Golubitsky and Schaeffer [1985], Golubitsky and Stewart [1987], Golubitsky, Stewart, and Schaeffer [1988], Lewis, Marsden, and Ratiu [1987], Lewis, Ratiu, Simo, and Marsden [1992], Montaldi, Roberts, and Stewart [1988], Golubitsky, Marsden, Stewart, and Dellnitz [1994]).

- Algebraic geometry (Atiyah [1982, 1983], Kirwan [1984, 1985 1998]).

- Celestial mechanics (Deprit [1983], Meyer and Hall [1992]).

- Vortex dynamics (Ziglin [1980b], Koiller, Soares, and Melo Neto [1985], Wan and Pulvirente [1984], Wan [1986, 1988a, 1988b, 1988c], Kirwan [1988], Szeri and Holmes [1988], Newton [1994], Pekarsky and Marsden [1998]).

- Solitons (Flaschka, Newell, and Ratiu [1983a, 1983b], Newell [1985], Kovačič and Wiggins [1992], Alber and Marsden [1992]).

- Multisymplectic geometry, PDEs, and nonlinear waves (Gotay, Isenberg, and Marsden [1997], Bridges [1994, 1997], Marsden and Shkoller [1997], and Marsden, Patrick, and Shkoller [1998]).

- Relativity and Yang–Mills theory (Fischer and Marsden [1972, 1979], Arms [1981], Arms, Marsden, and Moncrief [1981, 1982]).

60   1. Introduction and Overview

- Fluid variational principles using Clebsch variables and "Lin constraints" (Seliger and Whitham [1968], Cendra and Marsden [1987], Cendra, Ibort, and Marsden [1987], Holm, Marsden, and Ratiu [1998a]).

- Control, stabilization, satellite and underwater vehicle dynamics (Krishnaprasad [1985], van der Schaft and Crouch [1987], Aeyels and Szafranski [1988], Bloch, Krishnaprasad, Marsden, and Sánchez de Alvarez [1992], Wang, Krishnaprasad, and Maddocks [1991], Leonard [1997], Leonard and Marsden [1997]), Bloch, Leonard, and Marsden [1998], and Holmes, Jenkins, and Leonard [1998]).

- Nonholonomic systems (Naimark and Fufaev [1972], Koiller [1992], Bates and Sniatycki [1993], Bloch, Krishnaprasad, Marsden, and Murray [1996], Koon and Marsden [1997a, 1997b, 1998], Zenkov, Bloch, and Marsden [1998]).

Reduction theory for mechanical systems with symmetry is a natural historical continuation of the works of Liouville (for integrals in involution) and of Jacobi (for angular momentum) for reducing the phase space dimension in the presence of first integrals. It is intimately connected with work on momentum maps, and its forerunners appear already in Jacobi [1866], Lie [1890], Cartan [1922], and Whittaker [1927]. It was developed later in Kirillov [1962], Arnold [1966a], Kostant [1970], Souriau [1970], Smale [1970], Nekhoroshev [1977], Meyer [1973], and Marsden and Weinstein [1974]. See also Guillemin and Sternberg [1984] and Marsden and Ratiu [1986] for the Poisson case and Sjamaar and Lerman [1991] for basic work on the singular symplectic case.

2
Hamiltonian Systems on Linear Symplectic Spaces

A natural arena for Hamiltonian mechanics is a symplectic or Poisson manifold. The next few chapters concentrate on the symplectic case, while Chapter 10 introduces the Poisson case. The symplectic context focuses on the symplectic two-form $\sum dq^i \wedge dp_i$ and its infinite-dimensional analogues, while the Poisson context looks at the Poisson bracket as the fundamental object.

To facilitate an understanding of a number of points, we begin this chapter with the theory in linear spaces in which case the symplectic form becomes a skew-symmetric bilinear form that can be studied by means of linear-algebraic methods. This linear setting is already adequate for a number of interesting examples such as the wave equation and Schrödinger's equation.

Later, in Chapter 4, we make the transition to manifolds, and we generalize symplectic structures to manifolds in Chapters 5 and 6. In Chapters 7 and 8 we study the basics of Lagrangian mechanics, which are based primarily on variational principles rather than on symplectic or Poisson structures. This apparently very different approach is, however, shown to be equivalent to the Hamiltonian one under appropriate hypotheses.

2.1 Introduction

To motivate the introduction of symplectic geometry in mechanics, we briefly recall from §1.1 the classical transition from Newton's second law to the Lagrange and Hamilton equations. *Newton's second law* for a particle moving in Euclidean three-space $\mathbb{R}^3$, under the influence of a *potential energy* $V(\mathbf{q})$, is

$$\mathbf{F} = m\mathbf{a}, \tag{2.1.1}$$

where $\mathbf{q} \in \mathbb{R}^3$, $\mathbf{F}(\mathbf{q}) = -\nabla V(\mathbf{q})$ is the *force*, m is the mass of the particle, and $\mathbf{a} = d^2\mathbf{q}/dt^2$ is the acceleration (assuming that we start in a postulated privileged coordinate frame called an *inertial frame*).[1] The potential energy V is introduced through the notion of work and the assumption that the force field is conservative as shown in most books on vector calculus. The introduction of the *kinetic energy*

$$K = \frac{1}{2}m\left\|\frac{d\mathbf{q}}{dt}\right\|^2$$

is through the *power*, or *rate of work*, *equation*

$$\frac{dK}{dt} = m\langle \dot{\mathbf{q}}, \ddot{\mathbf{q}} \rangle = \langle \dot{\mathbf{q}}, \mathbf{F} \rangle,$$

where $\langle\,,\rangle$ denotes the inner product on $\mathbb{R}^3$.

The *Lagrangian* is defined by

$$L(q^i, \dot{q}^i) = \frac{m}{2}\|\dot{\mathbf{q}}\|^2 - V(\mathbf{q}), \tag{2.1.2}$$

and one checks by direct calculation that Newton's second law is equivalent to the *Euler–Lagrange equations*

$$\frac{d}{dt}\frac{\partial L}{\partial \dot{q}^i} - \frac{\partial L}{\partial q^i} = 0, \tag{2.1.3}$$

which are second-order differential equations in q^i; the equations (2.1.3) are worthy of independent study for a general L, since they are the equations for stationary values of the *action integral*

$$\delta \int_{t_1}^{t_2} L(q^i, \dot{q}^i)\, dt = 0, \tag{2.1.4}$$

as will be discussed in detail later. These *variational principles* play a fundamental role throughout mechanics—both in particle mechanics and field theory.

---

[1]Newton and subsequent workers in mechanics thought of this inertial frame as one "fixed relative to the distant stars." While this raises serious questions about what this could really mean mathematically or physically, it remains a good starting point. Deeper insight is found in Chapter 8 and in courses in general relativity.

It is easily verified that $dE/dt = 0$, where E is the **total energy**:

$$E = \frac{1}{2}m\|\dot{\mathbf{q}}\|^2 + V(\mathbf{q}).$$

Lagrange and Hamilton observed that it is convenient to introduce the momentum $p_i = m\dot{q}^i$ and rewrite E as a function of p_i and q^i by letting

$$H(\mathbf{q}, \mathbf{p}) = \frac{\|\mathbf{p}\|^2}{2m} + V(\mathbf{q}), \tag{2.1.5}$$

for then Newton's second law is equivalent to **Hamilton's canonical equations**

$$\dot{q}^i = \frac{\partial H}{\partial p_i}, \quad \dot{p}_i = -\frac{\partial H}{\partial q^i}, \tag{2.1.6}$$

which is a *first-order* system in $(\mathbf{q}, \mathbf{p})$-space, or **phase space**.

Matrix Notation.   For a deeper understanding of Hamilton's equations, we recall some matrix notation (see Abraham, Marsden, and Ratiu [1988, Section 5.1] for more details). Let E be a real vector space and E^* its dual space. Let $e_1, \dots, e_n$ be a basis of E with the associated dual basis for E^* denoted by $e^1, \dots, e^n$; that is, e^i is defined by

$$\langle e^i, e_j \rangle := e^i(e_j) = \delta^i_j,$$

which equals 1 if $i = j$ and 0 if $i \neq j$. Vectors $v \in E$ are written $v = v^i e_i$ (a sum on i is understood) and covectors $\alpha \in E^*$ as $\alpha = \alpha_i e^i$; v^i and α_i are the **components** of v and α, respectively.

If $A : E \to F$ is a linear transformation, its **matrix** relative to bases $e_1, \dots, e_n$ of E and $f_1, \dots, f_m$ of F is denoted by $A^j_{\ i}$ and is defined by

$$A(e_i) = A^j_{\ i} f_j; \quad \text{i.e.,} \quad [A(v)]^j = A^j_{\ i} v^i. \tag{2.1.7}$$

Thus, the columns of the matrix of A are $A(e_1), \dots, A(e_n)$; the upper index is the row index, and the lower index is the column index. For other linear transformations, we place the indices in their corresponding places. For example, if $A : E^* \to F$ is a linear transformation, its matrix A^{ij} satisfies $A(e^j) = A^{ij} f_i$; that is, $[A(\alpha)]^i = A^{ij}\alpha_j$.

If $B : E \times F \to \mathbb{R}$ is a bilinear form, that is, it is linear separately in each factor, its **matrix** B_{ij} is defined by

$$B_{ij} = B(e_i, f_j); \quad \text{i.e.,} \quad B(v, w) = v^i B_{ij} w^j. \tag{2.1.8}$$

Define the **associated** linear map $B^\flat : E \to F^*$ by

$$B^\flat(v)(w) = B(v, w)$$

and observe that $B^\flat(e_i) = B_{ij}f^j$. Since $B^\flat(e_i)$ is the ith column of the matrix representing the linear map $B^\flat$, it follows that *the matrix of $B^\flat$ in the bases $e_1, \ldots, e_n, f^1, \ldots, f^n$ is the transpose of B_{ij}*; that is,

$$[B^\flat]_{ji} = B_{ij}. \tag{2.1.9}$$

Let Z denote the vector space of (q,p)'s and write $z = (q,p)$. Let the coordinates q^j, p_j be collectively denoted by z^I, $I = 1, \ldots, 2n$. One reason for the notation z is that if one thinks of z as a *complex variable* $z = q + ip$, then Hamilton's equations are equivalent to the following complex form of Hamilton's equations (see Exercise 2.1-1):

$$\dot{z} = -2i\frac{\partial H}{\partial \bar{z}}, \tag{2.1.10}$$

where $\partial/\partial\bar{z} := (\partial/\partial q - i\partial/\partial p)/2$.

Symplectic and Poisson Structures. We can view Hamilton's equations (2.1.6) as follows. Think of the operation

$$\mathbf{d}H(z) = \left(\frac{\partial H}{\partial q^i}, \frac{\partial H}{\partial p_i}\right) \mapsto \left(\frac{\partial H}{\partial p_i}, -\frac{\partial H}{\partial q^i}\right) =: X_H(z), \tag{2.1.11}$$

which forms a vector field X_H, called the **Hamiltonian vector field**, from the differential of H, as the composition of the linear map

$$R : Z^* \to Z$$

with the differential $\mathbf{d}H(z)$ of H. The matrix of R is

$$[R^{AB}] = \begin{bmatrix} 0 & 1 \\ -1 & 0 \end{bmatrix} =: \mathbb{J}, \tag{2.1.12}$$

where we write $\mathbb{J}$ for the specific matrix (2.1.12) sometimes called the *symplectic matrix*. Here, 0 is the $n \times n$ zero matrix and 1 is the $n \times n$ identity matrix. Thus,

$$X_H(z) = R \cdot \mathbf{d}H(z) \tag{2.1.13}$$

or, if the components of X_H are denoted by X^I, $I = 1, \ldots, 2n$,

$$X^I = R^{IJ}\frac{\partial H}{\partial z^J}, \quad \text{i.e.,} \quad X_H = \mathbb{J}\nabla H, \tag{2.1.14}$$

where ∇H is the *naive gradient* of H, that is, the row vector $\mathbf{d}H$ but regarded as a column vector.

Let $B(\alpha,\beta) = \langle \alpha, R(\beta)\rangle$ be the bilinear form associated to R, where $\langle\,,\rangle$ denotes the canonical pairing between Z^* and Z. One calls either the bilinear form B or its associated linear map R the **Poisson structure**. The classical **Poisson bracket** (consistent with what we defined in Chapter 1) is defined by

$$\{F, G\} = B(\mathbf{d}F, \mathbf{d}G) = \mathbf{d}F \cdot \mathbb{J}\nabla G. \tag{2.1.15}$$

The **symplectic structure** Ω is the bilinear form associated to R^{-1} : $Z \to Z^*$, that is, $\Omega(v, w) = \langle R^{-1}(v), w \rangle$, or, equivalently, $\Omega^\flat = R^{-1}$. The matrix of Ω is $\mathbb{J}$ in the sense that

$$\Omega(v, w) = v^T \mathbb{J} w. \tag{2.1.16}$$

To unify notation we shall sometimes write

Ω	for the symplectic form,	$Z \times Z \to \mathbb{R}$	with matrix $\mathbb{J}$,
$\Omega^\flat$	for the associated linear map,	$Z \to Z^*$	with matrix $\mathbb{J}^T$,
$\Omega^\sharp$	for the inverse map $(\Omega^\flat)^{-1} = R$,	$Z^* \to Z$	with matrix $\mathbb{J}$,
B	for the Poisson form,	$Z^* \times Z^* \to \mathbb{R}$	with matrix $\mathbb{J}$.

Hamilton's equations may be written

$$\dot{z} = X_H(z) = \Omega^\sharp \, \mathbf{d}H(z). \tag{2.1.17}$$

Multiplying both sides by $\Omega^\flat$, we get

$$\Omega^\flat X_H(z) = \mathbf{d}H(z). \tag{2.1.18}$$

In terms of the symplectic form, (2.1.18) reads

$$\Omega(X_H(z), v) = \mathbf{d}H(z) \cdot v \tag{2.1.19}$$

for all $z, v \in Z$.

Problems such as rigid-body dynamics, quantum mechanics as a Hamiltonian system, and the motion of a particle in a rotating reference frame motivate the need to generalize these concepts. We shall do this in subsequent chapters and deal with both symplectic and Poisson structures in due course.

Exercises

⋄ **2.1-1.** Writing $z = q + ip$, show that Hamilton's equations are equivalent to

$$\dot{z} = -2i \frac{\partial H}{\partial \bar{z}}.$$

Give a plausible definition of the right-hand side as part of your answer (or consult a book on complex variables theory).

⋄ **2.1-2.** Write the harmonic oscillator $m\ddot{x} + kx = 0$ in the form of Euler-Lagrange equations, as Hamilton's equations, and finally, in the complex form (2.1.10).

⋄ **2.1-3.** Repeat Exercise 2.1-2 for the nonlinear oscillator $m\ddot{x} + kx + \alpha x^3 = 0$.

2.2   Symplectic Forms on Vector Spaces

Let Z be a real Banach space, possibly infinite-dimensional, and let $\Omega :$ $Z \times Z \to \mathbb{R}$ be a continuous bilinear form on Z. The form Ω is said to be **nondegenerate** (**or weakly nondegenerate**) if $\Omega(z_1, z_2) = 0$ for all $z_2 \in Z$ implies $z_1 = 0$. As in §2.1, the induced continuous linear mapping $\Omega^\flat : Z \to Z^*$ is defined by

$$\Omega^\flat(z_1)(z_2) = \Omega(z_1, z_2). \tag{2.2.1}$$

Nondegeneracy of Ω is equivalent to injectivity of $\Omega^\flat$, that is, to the condition "$\Omega^\flat(z) = 0$ implies $z = 0$." The form Ω is said to be **strongly nondegenerate** if $\Omega^\flat$ is an isomorphism, that is, $\Omega^\flat$ is onto as well as being injective. The open mapping theorem guarantees that if Z is a Banach space and $\Omega^\flat$ is one-to-one and onto, then its inverse is continuous. In most of the infinite-dimensional examples discussed in this book Ω will be only (weakly) nondegenerate.

A linear map between finite-dimensional spaces of the same dimension is one-to-one if and only if it is onto. Hence, *when Z is finite-dimensional, weak nondegeneracy and strong nondegeneracy are equivalent*. If Z is finite-dimensional, the matrix elements of Ω relative to a basis $\{e_I\}$ are defined by

$$\Omega_{IJ} = \Omega(e_I, e_J).$$

If $\{e^J\}$ denotes the basis for Z^* that is dual to $\{e_I\}$, that is, $\langle e^J, e_I \rangle = \delta_I^J$, and if we write $z = z^I e_I$ and $w = w^I e_I$, then

$$\Omega(z, w) = z^I \Omega_{IJ} w^J \quad \text{(sum over } I, J\text{)}.$$

Since the matrix of $\Omega^\flat$ relative to the bases $\{e_I\}$ and $\{e^J\}$ equals the transpose of the matrix of Ω relative to $\{e_I\}$, that is $(\Omega^\flat)_{JI} = \Omega_{IJ}$, nondegeneracy is equivalent to $\det[\Omega_{IJ}] \neq 0$. In particular, if Ω is skew and nondegenerate, then Z is even-dimensional, since the determinant of a skew-symmetric matrix with an odd number of rows (and columns) is zero.

Definition 2.2.1.   *A* **symplectic form** Ω *on a vector space Z is a nondegenerate skew-symmetric bilinear form on Z. The pair (Z, Ω) is called a* **symplectic vector space***. If Ω is strongly nondegenerate, (Z, Ω) is called a* **strong symplectic vector space***.*

Examples

We now develop some basic examples of symplectic forms.

(a) Canonical Forms.   Let W be a vector space, and let $Z = W \times W^*$. Define the **canonical symplectic form** Ω on Z by

$$\Omega((w_1, \alpha_1), (w_2, \alpha_2)) = \alpha_2(w_1) - \alpha_1(w_2), \tag{2.2.2}$$

where $w_1, w_2 \in W$ and $\alpha_1, \alpha_2 \in W^*$.

More generally, let W and W' be two vector spaces in duality, that is, there is a weakly nondegenerate pairing $\langle , \rangle : W' \times W \to \mathbb{R}$. Then on $W \times W'$,

$$\Omega((w_1, \alpha_1), (w_2, \alpha_2)) = \langle \alpha_2, w_1 \rangle - \langle \alpha_1, w_2 \rangle \qquad (2.2.3)$$

is a weak symplectic form.   ◆

(b) The Space of Functions. Let $\mathcal{F}(\mathbb{R}^3)$ be the space of smooth functions $\varphi : \mathbb{R}^3 \to \mathbb{R}$, and let $\mathrm{Den}_c(\mathbb{R}^3)$ be the space of smooth densities on $\mathbb{R}^3$ with compact support. We write a density $\pi \in \mathrm{Den}_c(\mathbb{R}^3)$ as a function $\pi' \in \mathcal{F}(\mathbb{R}^3)$ with compact support times the volume element d^3x on $\mathbb{R}^3$ as $\pi = \pi' d^3x$. The spaces $\mathcal{F}$ and Den_c are in weak nondegenerate duality by the pairing $\langle \varphi, \pi \rangle = \int \varphi \pi' d^3x$. Therefore, from (2.2.3) we get the symplectic form Ω on the vector space $Z = \mathcal{F}(\mathbb{R}^3) \times \mathrm{Den}_c(\mathbb{R}^3)$:

$$\Omega((\varphi_1, \pi_1), (\varphi_2, \pi_2)) = \int_{\mathbb{R}^3} \varphi_1 \pi_2 - \int_{\mathbb{R}^3} \varphi_2 \pi_1. \qquad (2.2.4)$$

We choose densities with compact support so that the integrals in this formula will be finite. Other choices of spaces could be used as well.   ◆

(c) Finite-Dimensional Canonical Form. Suppose that W is a real vector space of dimension n. Let $\{e_i\}$ be a basis of W, and let $\{e^i\}$ be the dual basis of W^*. With $Z = W \times W^*$ and defining $\Omega : Z \times Z \to \mathbb{R}$ as in (2.2.2), one computes that the matrix of Ω in the basis

$$\{(e_1, 0), \dots, (e_n, 0), (0, e^1), \dots, (0, e^n)\}$$

is

$$\mathbb{J} = \begin{bmatrix} 0 & 1 \\ -1 & 0 \end{bmatrix}, \qquad (2.2.5)$$

where 1 and 0 are the $n \times n$ identity and zero matrices.   ◆

(d) Symplectic Form Associated to an Inner Product Space. If $(W, \langle , \rangle)$ is a real inner product space, W is in duality with itself, so we obtain a symplectic form on $Z = W \times W$ from (2.2.3):

$$\Omega((w_1, w_2), (z_1, z_2)) = \langle z_2, w_1 \rangle - \langle z_1, w_2 \rangle. \qquad (2.2.6)$$

As a special case of (2.2.6), let $W = \mathbb{R}^3$ with the usual inner product

$$\langle \mathbf{q}, \mathbf{v} \rangle = \mathbf{q} \cdot \mathbf{v} = \sum_{i=1}^{3} q^i v^i.$$

The corresponding symplectic form on $\mathbb{R}^6$ is given by

$$\Omega((\mathbf{q}_1, \mathbf{v}_1), (\mathbf{q}_2, \mathbf{v}_2)) = \mathbf{v}_2 \cdot \mathbf{q}_1 - \mathbf{v}_1 \cdot \mathbf{q}_2, \qquad (2.2.7)$$

where $\mathbf{q}_1, \mathbf{q}_2, \mathbf{v}_1, \mathbf{v}_2 \in \mathbb{R}^3$. This coincides with Ω defined in Example (c) for $W = \mathbb{R}^3$, provided that $\mathbb{R}^3$ is identified with $(\mathbb{R}^3)^*$.   ◆

Bringing Ω to canonical form using elementary linear algebra results in the following statement. *If (Z, Ω) is a p-dimensional symplectic vector space, then p is even. Furthermore, Z is, as a vector space, isomorphic to one of the standard examples, namely $W \times W^*$, and there is a basis of W in which the matrix of Ω is $\mathbb{J}$. Such a basis is called* **canonical**, *as are the corresponding coordinates.* See Exercise 2.2-3.

(e) **Symplectic Form on $\mathbb{C}^n$.** Write elements of complex n-space $\mathbb{C}^n$ as n-tuples $z = (z_1, \ldots, z_n)$ of complex numbers. The **Hermitian inner product** is

$$\langle z, w \rangle = \sum_{j=1}^{n} z_j \overline{w}_j = \sum_{j=1}^{n} (x_j u_j + y_j v_j) + i \sum_{j=1}^{n} (u_j y_j - v_j x_j),$$

where $z_j = x_j + i y_j$ and $w_j = u_j + i v_j$. Thus, $\mathrm{Re}\,\langle z, w \rangle$ is the real inner product and $-\mathrm{Im}\,\langle z, w \rangle$ is the symplectic form if $\mathbb{C}^n$ is identified with $\mathbb{R}^n \times \mathbb{R}^n$.   ◆

(f) **Quantum-Mechanical Symplectic Form.** We now discuss an interesting symplectic vector space that arises in quantum mechanics, as we shall further explain in Chapter 3. Recall that a **Hermitian inner product** $\langle , \rangle : \mathcal{H} \times \mathcal{H} \to \mathbb{C}$ on a complex Hilbert space $\mathcal{H}$ is linear in its first argument and antilinear in its second, and $\langle \psi_1, \psi_2 \rangle$ is the complex conjugate of $\langle \psi_2, \psi_1 \rangle$, where $\psi_1, \psi_2 \in \mathcal{H}$.
Set

$$\Omega(\psi_1, \psi_2) = -2\hbar \, \mathrm{Im}\,\langle \psi_1, \psi_2 \rangle,$$

where $\hbar$ is Planck's constant. One checks that Ω is a strong symplectic form on $\mathcal{H}$.

There is another view of this symplectic form motivated by the preceding Example (d) that is interesting. Let $\mathcal{H}$ be the complexification of a real Hilbert space H, so the complex Hilbert space $\mathcal{H}$ is identified with $H \times H$, and the Hermitian inner product is given by

$$\langle (u_1, u_2), (v_1, v_2) \rangle = \langle u_1, v_1 \rangle + \langle u_2, v_2 \rangle + i(\langle u_2, v_1 \rangle - \langle u_1, v_2 \rangle).$$

The imaginary part of this form coincides with that in (2.2.6).

There is yet another view related to the interpretation of a wave function ψ and its conjugate $\bar{\psi}$ being conjugate variables. Namely, we consider the embedding of $\mathcal{H}$ into $\mathcal{H} \times \mathcal{H}^*$ via $\psi \mapsto (i\psi, \psi)$. Then one checks that the restriction of $\hbar$ times the canonical symplectic form (2.2.6) on $\mathcal{H} \times \mathcal{H}^*$, namely,

$$((\psi_1, \varphi_1), (\psi_2, \varphi_2)) \mapsto \hbar \operatorname{Re}[\langle \varphi_2, \psi_1 \rangle - \langle \varphi_1, \psi_2 \rangle],$$

coincides with Ω.   ◆

Exercises

⋄ **2.2-1.** Verify that the formula for the symplectic form for $\mathbb{R}^{2n}$ as a matrix, namely,

$$\mathbb{J} = \begin{bmatrix} 0 & 1 \\ -1 & 0 \end{bmatrix},$$

coincides with the definition of the symplectic form as the canonical form on $\mathbb{R}^{2n}$ regarded as the product $\mathbb{R}^n \times (\mathbb{R}^n)^*$.

⋄ **2.2-2.** Let (Z, Ω) be a finite-dimensional symplectic vector space and let $V \subset Z$ be a linear subspace. Assume that V is symplectic; that is, Ω restricted to $V \times V$ is nondegenerate. Let

$$V^\Omega = \{ z \in Z \mid \Omega(z, v) = 0 \text{ for all } v \in V \}.$$

Show that V^Ω is symplectic and $Z = V \oplus V^\Omega$.

⋄ **2.2-3.** Find a canonical basis for a symplectic form Ω on Z as follows. Let $e_1 \in Z$, $e_1 \neq 0$. Find $e_2 \in Z$ with $\Omega(e_1, e_2) \neq 0$. By rescaling e_2, assume $\Omega(e_1, e_2) = 1$. Let V be the span of e_1 and e_2. Apply Exercise 2.2-2 and repeat this construction on V^Ω.

⋄ **2.2-4.** Let (Z, Ω) be a finite-dimensional symplectic vector space and $V \subset Z$ a subspace. Define V^Ω as in Exercise 2.2-2. Show that Z/V^Ω and V^* are isomorphic vector spaces.

2.3 Canonical Transformations, or Symplectic Maps

To motivate the definition of symplectic maps (synonymous with canonical transformations), start with Hamilton's equations

$$\dot{q}^i = \frac{\partial H}{\partial p_i}, \quad \dot{p}_i = -\frac{\partial H}{\partial q^i}, \tag{2.3.1}$$

and a transformation $\varphi : Z \to Z$ of phase space to itself. Write

$$(\tilde{q}, \tilde{p}) = \varphi(q, p),$$

that is,

$$\tilde{z} = \varphi(z). \tag{2.3.2}$$

Assume that $z(t) = (q(t), p(t))$ satisfies Hamilton's equations, that is,

$$\dot{z}(t) = X_H(z(t)) = \Omega^\sharp \, \mathbf{d}H(z(t)), \tag{2.3.3}$$

where $\Omega^\sharp : Z^* \to Z$ is the linear map with matrix $\mathbb{J}$ whose entries we denote by B^{JK}. By the chain rule, $\tilde{z} = \varphi(z)$ satisfies

$$\dot{\tilde{z}}^I = \frac{\partial \varphi^I}{\partial z^J} \dot{z}^J =: A^I_J \dot{z}^J \tag{2.3.4}$$

(sum on J). Substituting (2.3.3) into (2.3.4), employing coordinate notation, and using the chain rule, we conclude that

$$\dot{\tilde{z}}^I = A^I_J B^{JK} \frac{\partial H}{\partial z^K} = A^I_J B^{JK} A^L_K \frac{\partial H}{\partial \tilde{z}^L}. \tag{2.3.5}$$

Thus, the equations (2.3.5) are Hamiltonian if and only if

$$A^I_J B^{JK} A^L_K = B^{IL}, \tag{2.3.6}$$

which in matrix notation reads

$$A \mathbb{J} A^T = \mathbb{J}. \tag{2.3.7}$$

In terms of composition of linear maps, (2.3.6) means

$$A \circ \Omega^\sharp \circ A^T = \Omega^\sharp, \tag{2.3.8}$$

since the matrix of $\Omega^\sharp$ in canonical coordinates is $\mathbb{J}$ (see §2.1). A transformation satisfying (2.3.6) is called a **canonical transformation,** a **symplectic transformation**, or a **Poisson transformation.**[2]

Taking determinants of (2.3.7) shows that $\det A = \pm 1$ (we will see in Chapter 9 that $\det A = 1$ is the only possibility) and in particular that A is invertible; taking the inverse of (2.3.8) gives

$$(A^T)^{-1} \circ \Omega^\flat \circ A^{-1} = \Omega^\flat,$$

that is,

$$A^T \circ \Omega^\flat \circ A = \Omega^\flat, \tag{2.3.9}$$

---

[2]In Chapter 10, where Poisson structures can be different from symplectic ones, we will see that (2.3.8) generalizes to the Poisson context.

which has the matrix form

$$A^T \mathbb{J} A = \mathbb{J}, \qquad (2.3.10)$$

since the matrix of Ω^b in canonical coordinates is $-\mathbb{J}$ (see §2.1). Note that (2.3.7) and (2.3.10) are equivalent (the inverse of one gives the other). As bilinear forms, (2.3.9) reads

$$\Omega(\mathbf{D}\varphi(z) \cdot z_1, \mathbf{D}\varphi(z) \cdot z_2) = \Omega(z_1, z_2), \qquad (2.3.11)$$

where $\mathbf{D}\varphi$ is the derivative of φ (the Jacobian matrix in finite dimensions). With (2.3.11) as a guideline, we next write the general condition for a map to be symplectic.

Definition 2.3.1. *If (Z, Ω) and (Y, Ξ) are symplectic vector spaces, a smooth map $f : Z \to Y$ is called **symplectic** or **canonical** if it preserves the symplectic forms, that is, if*

$$\Xi(\mathbf{D}f(z) \cdot z_1, \mathbf{D}f(z) \cdot z_2) = \Omega(z_1, z_2) \qquad (2.3.12)$$

for all $z, z_1, z_2 \in Z$.

We next introduce some notation that will help us write (2.3.12) in a compact and efficient way.

Pull-Back Notation

We introduce a convenient notation for these sorts of transformations.

$\varphi^* f$ ***pull-back of a function:*** $\varphi^* f = f \circ \varphi$.

$\varphi_* g$ ***push-forward of a function:*** $\varphi_* g = g \circ \varphi^{-1}$.

$\varphi_* X$ ***push-forward of a vector field*** X by φ:

$$(\varphi_* X)(\varphi(z)) = \mathbf{D}\varphi(z) \cdot X(z);$$

in components,

$$(\varphi_* X)^I = \frac{\partial \varphi^I}{\partial z^J} X^J.$$

$\varphi^* Y$ ***pull-back of a vector field*** Y by φ: $\varphi^* Y = (\varphi^{-1})_* Y$

$\varphi^* \Omega$ ***pull-back of a bilinear form*** Ω on Z gives a bilinear form $\varphi^* \Omega$ depending on the point $z \in Z$:

$$(\varphi^* \Omega)_z(z_1, z_2) = \Omega(\mathbf{D}\varphi(z) \cdot z_1, \mathbf{D}\varphi(z) \cdot z_2);$$

in components,

$$(\varphi^*\Omega)_{IJ} = \frac{\partial \varphi^K}{\partial z^I} \frac{\partial \varphi^L}{\partial z^J} \Omega_{KL};$$

$\varphi_*\Xi$ **push-forward of a bilinear form** Ξ by φ equals the pull-back by the inverse: $\varphi_*\Xi = (\varphi^{-1})^*\Xi$.

In this pull-back notation, (2.3.12) reads $(f^*\Xi)_z = \Omega_z$, or $f^*\Xi = \Omega$ for short.

The Symplectic Group. It is simple to verify that if (Z, Ω) is a finite-dimensional symplectic vector space, the set of all linear symplectic mappings $T : Z \to Z$ forms a group under composition. It is called the **symplectic group** and is denoted by $\mathrm{Sp}(Z, \Omega)$. As we have seen, in a canonical basis, a matrix A is symplectic if and only if

$$A^T \mathbb{J} A = \mathbb{J}, \tag{2.3.13}$$

where A^T is the transpose of A. For $Z = W \times W^*$ and a canonical basis, if A has the matrix

$$A = \begin{bmatrix} A_{qq} & A_{qp} \\ A_{pq} & A_{pp} \end{bmatrix}, \tag{2.3.14}$$

then one checks (Exercise 2.3-2) that (2.3.13) *is equivalent to either of the following two conditions*:

(1) $A_{qq}A_{qp}^T$ and $A_{pp}A_{pq}^T$ are symmetric and $A_{qq}A_{pp}^T - A_{qp}A_{pq}^T = 1$;

(2) $A_{pq}^T A_{qq}$ and $A_{qp}^T A_{pp}$ are symmetric and $A_{qq}^T A_{pp} - A_{pq}^T A_{pq} = 1$.

In infinite dimensions $\mathrm{Sp}(Z, \Omega)$ is, by definition, the set of elements of $\mathrm{GL}(Z)$ (the group of invertible bounded linear operators of Z to Z) that leave Ω fixed.

Symplectic Orthogonal Complements. If (Z, Ω) is a (weak) symplectic space and E and F are subspaces of Z, we define

$$E^\Omega = \{\, z \in Z \mid \Omega(z, e) = 0 \text{ for all } e \in E \,\},$$

called the **symplectic orthogonal complement** of E. We leave it to the reader to check that

(i) E^Ω is closed;

(ii) $E \subset F$ implies $F^\Omega \subset E^\Omega$;

(iii) $E^\Omega \cap F^\Omega = (E + F)^\Omega$;

(iv) if Z is finite-dimensional, then dim E + dim E^Ω = dim Z (to show this, use the fact that $E^\Omega = \ker(i^* \circ \Omega^b)$, where $i : E \to Z$ is the inclusion and $i^* : Z^* \to E^*$ is its dual, $i^*(\alpha) = \alpha \circ i$, which is surjective; alternatively, use Exercise 2.2-4);

(v) if Z is finite-dimensional, $E^{\Omega\Omega} = E$ (this is also true in infinite dimensions if E is closed); and

(vi) if E and F are closed, then $(E \cap F)^\Omega = E^\Omega + F^\Omega$ (to prove this use (iii) and (v)).

Exercises

⋄ **2.3-1.** Show that a transformation $\varphi : \mathbb{R}^{2n} \to \mathbb{R}^{2n}$ is symplectic in the sense that its derivative matrix $A = \mathbf{D}\varphi(z)$ satisfies the condition $A^T \mathbb{J} A = \mathbb{J}$ if and only if the condition

$$\Omega(Az_1, Az_2) = \Omega(z_1, z_2)$$

holds for all $z_1, z_2 \in \mathbb{R}^{2n}$.

⋄ **2.3-2.** Let $Z = W \times W^*$, let $A : Z \to Z$ be a linear transformation, and, using canonical coordinates, write the matrix of A as

$$A = \begin{bmatrix} A_{qq} & A_{qp} \\ A_{pq} & A_{pp} \end{bmatrix}.$$

Show that A being symplectic is equivalent to either of the two following conditions:

(i) $A_{qq}A_{qp}^T$ and $A_{pp}A_{pq}^T$ are symmetric and $A_{qq}A_{pp}^T - A_{qp}A_{pq}^T = 1$;

(ii) $A_{pq}^T A_{qq}$ and $A_{qp}^T A_{pp}$ are symmetric and $A_{qq}^T A_{pp} - A_{pq}^T A_{qp} = 1$. (Here 1 denotes the $n \times n$ identity.)

⋄ **2.3-3.** Let f be a given function of $\mathbf{q} = (q^1, q^2, \dots, q^n)$. Define the map $\varphi : \mathbb{R}^{2n} \to \mathbb{R}^{2n}$ by $\varphi(\mathbf{q}, \mathbf{p}) = (\mathbf{q}, \mathbf{p} + \mathbf{d}f(\mathbf{q}))$. Show that φ is a canonical (symplectic) transformation.

⋄ **2.3-4.**

(a) Let $A \in GL(n, \mathbb{R})$ be an invertible linear transformation. Show that the map $\varphi : \mathbb{R}^{2n} \to \mathbb{R}^{2n}$ given by $(\mathbf{q}, \mathbf{p}) \mapsto (A\mathbf{q}, (A^{-1})^T \mathbf{p})$ is a canonical transformation.

(b) If $\mathbf{R}$ is a rotation in $\mathbb{R}^3$, show that the map $(\mathbf{q}, \mathbf{p}) \mapsto (\mathbf{Rq}, \mathbf{Rp})$ is a canonical transformation.

◊ **2.3-5.** Let (Z, Ω) be a finite-dimensional symplectic vector space. A subspace $E \subset Z$ is called *isotropic*, *coisotropic*, and **Lagrangian** if $E \subset E^\Omega$, $E^\Omega \subset E$, and $E = E^\Omega$, respectively. Note that E is Lagrangian if and only if it is isotropic and coisotropic at the same time. Show that:

(a) An isotropic (coisotropic) subspace E is Lagrangian if and only if $\dim E = \dim E^\Omega$. In this case necessarily $2 \dim E = \dim Z$.

(b) Every isotropic (coisotropic) subspace is contained in (contains) a Lagrangian subspace.

(c) An isotropic (coisotropic) subspace is Lagrangian if and only if it is a maximal isotropic (minimal coisotropic) subspace.

2.4   The General Hamilton Equations

The concrete form of Hamilton's equations we have already encountered is a special case of a construction on symplectic spaces. Here, we discuss this formulation for systems whose phase space is linear; in subsequent sections we will generalize the setting to phase spaces that are symplectic manifolds and in Chapter 10 to spaces where only a Poisson bracket is given. These generalizations will all be important in our study of specific examples.

Definition 2.4.1. *Let (Z, Ω) be a symplectic vector space. A vector field $X : Z \to Z$ is called* **Hamiltonian** *if*

$$\Omega^\flat(X(z)) = \mathbf{d}H(z), \tag{2.4.1}$$

for all $z \in Z$, for some C^1 function $H : Z \to \mathbb{R}$. Here $\mathbf{d}H(z) = \mathbf{D}H(z)$ is alternative notation for the derivative of H. If such an H exists, we write $X = X_H$ and call H a **Hamiltonian function**, *or* **energy function**, *for the vector field X.*

In a number of important examples, especially infinite-dimensional ones, H need not be defined on all of Z. We shall briefly discuss in §3.3 some of the technicalities involved.

If Z is finite-dimensional, nondegeneracy of Ω implies that $\Omega^\flat : Z \to Z^*$ is an isomorphism, which guarantees that X_H exists for any given function H. However, if Z is infinite-dimensional and Ω is only weakly nondegenerate, we do not know *a priori* that X_H exists for a given H. If it does exist, it is unique, since $\Omega^\flat$ is one-to-one.

The set of Hamiltonian vector fields on Z is denoted by $\mathfrak{X}_{\text{Ham}}(Z)$, or simply $\mathfrak{X}_{\text{Ham}}$. Thus, $X_H \in \mathfrak{X}_{\text{Ham}}$ is the vector field determined by the condition

$$\Omega(X_H(z), \delta z) = \mathbf{d}H(z) \cdot \delta z \quad \text{for all } z, \delta z \in Z. \tag{2.4.2}$$

If X is a vector field, the **interior product** $i_X \Omega$ (also denoted by $X \lrcorner \Omega$) is defined to be the dual vector (also called, a **one-form**) given at a point $z \in Z$ as follows:

$$(i_X \Omega)_z \in Z^*; \quad (i_X \Omega)_z(v) := \Omega(X(z), v),$$

for all $v \in Z$. Then condition (2.4.1) or (2.4.2) may be written as

$$i_X \Omega = dH; \quad \text{i.e.,} \quad X \lrcorner \Omega = dH. \tag{2.4.3}$$

To express H in terms of X_H and Ω, we integrate the identity

$$dH(tz) \cdot z = \Omega(X_H(tz), z)$$

from $t = 0$ to $t = 1$. The fundamental theorem of calculus gives

$$H(z) - H(0) = \int_0^1 \frac{dH(tz)}{dt} dt = \int_0^1 dH(tz) \cdot z \, dt$$

$$= \int_0^1 \Omega(X_H(tz), z) \, dt. \tag{2.4.4}$$

Let us now abstract the calculation we did in arriving at (2.3.7).

Proposition 2.4.2. *Let (Z, Ω) and (Y, Ξ) be symplectic vector spaces and $f : Z \to Y$ a diffeomorphism. Then f is a symplectic transformation if and only if for all Hamiltonian vector fields X_H on Y, we have $f_* X_{H \circ f} = X_H$, that is,*

$$\mathbf{D}f(z) \cdot X_{H \circ f}(z) = X_H(f(z)). \tag{2.4.5}$$

Proof.   Note that for $v \in Z$,

$$\Omega(X_{H \circ f}(z), v) = d(H \circ f)(z) \cdot v = dH(f(z)) \cdot \mathbf{D}f(z) \cdot v$$

$$= \Xi(X_H(f(z)), \mathbf{D}f(z) \cdot v). \tag{2.4.6}$$

If f is symplectic, then

$$\Xi(\mathbf{D}f(z) \cdot X_{H \circ f}(z), \mathbf{D}f(z) \cdot v) = \Omega(X_{H \circ f}(z), v),$$

and thus by nondegeneracy of Ξ and the fact that $\mathbf{D}f(z) \cdot v$ is an arbitrary element of Y (because f is a diffeomorphism and hence $\mathbf{D}f(z)$ is an isomorphism), (2.4.5) holds. Conversely, if (2.4.5) holds, then (2.4.6) implies

$$\Xi(\mathbf{D}f(z) \cdot X_{H \circ f}(z), \mathbf{D}f(z) \cdot v) = \Omega(X_{H \circ f}(z), v)$$

for any $v \in Z$ and any C^1 map $H : Y \to \mathbb{R}$. However, $X_{H \circ f}(z)$ equals an arbitrary element $w \in Z$ for a correct choice of the Hamiltonian function H, namely, $(H \circ f)(z) = \Omega(w, z)$. Thus, f is symplectic. ∎

Definition 2.4.3.   *Hamilton's equations for H is the system of differential equations defined by X_H. Letting $c : \mathbb{R} \to Z$ be a curve, they are the equations*

$$\frac{dc(t)}{dt} = X_H(c(t)). \tag{2.4.7}$$

The Classical Hamilton Equations.   We now relate the abstract form
(2.4.7) to the classical form of Hamilton's equations. In the following, an
n-tuple $(q^1, \ldots, q^n)$ will be denoted simply by (q^i).

Proposition 2.4.4.   *Suppose that* (Z, Ω) *is a $2n$-dimensional symplectic
vector space, and let* $(q^i, p_i) = (q^1, \ldots, q^n, p_1, \ldots, p_n)$ *denote canonical
coordinates, with respect to which* Ω *has matrix* $\mathbb{J}$. *Then in this coordinate
system,* $X_H : Z \to Z$ *is given by*

$$X_H = \left(\frac{\partial H}{\partial p_i}, -\frac{\partial H}{\partial q^i} \right) = \mathbb{J} \cdot \nabla H. \tag{2.4.8}$$

Thus, Hamilton's equations in canonical coordinates are

$$\frac{dq^i}{dt} = \frac{\partial H}{\partial p_i}, \quad \frac{dp_i}{dt} = -\frac{\partial H}{\partial q^i}. \tag{2.4.9}$$

More generally, if $Z = V \times V'$, $\langle \cdot, \cdot \rangle : V \times V' \to \mathbb{R}$ *is a weakly nondegenerate
pairing, and* $\Omega((e_1, \alpha_1), (e_2, \alpha_2)) = \langle \alpha_2, e_1 \rangle - \langle \alpha_1, e_2 \rangle$, *then*

$$X_H(e, \alpha) = \left(\frac{\delta H}{\delta \alpha}, -\frac{\delta H}{\delta e} \right), \tag{2.4.10}$$

where $\delta H / \delta \alpha \in V$ *and* $\delta H / \delta e \in V'$ *are the **partial functional derivatives** defined by*

$$\mathbf{D}_2 H(e, \alpha) \cdot \beta = \left\langle \beta, \frac{\delta H}{\delta \alpha} \right\rangle \tag{2.4.11}$$

for any $\beta \in V'$ *and similarly for* $\delta H / \delta e$; *in* (2.4.10) *it is assumed that the
functional derivatives exist.*

Proof.   If $(f, \beta) \in V \times V'$, then

$$\Omega \left(\left(\frac{\delta H}{\delta \alpha}, -\frac{\delta H}{\delta e} \right), (f, \beta) \right) = \left\langle \beta, \frac{\delta H}{\delta \alpha} \right\rangle + \left\langle \frac{\delta H}{\delta e}, f \right\rangle$$

$$= \mathbf{D}_2 H(e, \alpha) \cdot \beta + \mathbf{D}_1 H(e, \alpha) \cdot f$$

$$= \langle \mathbf{d} H(e, \alpha), (f, \beta) \rangle. \quad \blacksquare$$

Proposition 2.4.5 (Conservation of Energy).   *Let $c(t)$ be an integral curve of X_H. Then $H(c(t))$ is constant in t. If φ_t denotes the flow of X_H,
that is, $\varphi_t(z)$ is the solution of (2.4.7) with initial conditions $z \in Z$, then
$H \circ \varphi_t = H$.*

Proof.   By the chain rule,

$$\frac{d}{dt} H(c(t)) = \mathbf{d} H(c(t)) \cdot \frac{d}{dt} c(t) = \Omega \left(X_H(c(t)), \frac{d}{dt} c(t) \right)$$

$$= \Omega \left(X_H(c(t)), X_H(c(t)) \right) = 0,$$

where the final equality follows from the skew-symmetry of Ω. $\blacksquare$

Exercises

$\diamond$ **2.4-1.** Let the skew-symmetric bilinear form Ω on $\mathbb{R}^{2n}$ have the matrix

$$\begin{bmatrix} \mathbf{B} & \mathbf{1} \\ -\mathbf{1} & \mathbf{0} \end{bmatrix},$$

where $\mathbf{B} = [B_{ij}]$ is a skew-symmetric $n \times n$ matrix, and $\mathbf{1}$ is the identity matrix.

(a) Show that Ω is nondegenerate and hence a symplectic form on $\mathbb{R}^{2n}$.

(b) Show that Hamilton's equations with respect to Ω are, in standard coordinates,

$$\frac{dq^i}{dt} = \frac{\partial H}{\partial p_i}, \quad \frac{dp_i}{dt} = -\frac{\partial H}{\partial q^i} - B_{ij}\frac{\partial H}{\partial p_j}.$$

2.5 When Are Equations Hamiltonian?

Having seen how to derive Hamilton's equations on (Z, Ω) given H, it is natural to consider the converse: When is a given set of equations

$$\frac{dz}{dt} = X(z), \qquad (2.5.1)$$

where $X : Z \to Z$ is a given vector field, Hamilton's equations for some H? If X is linear, the answer is given by the following.

Proposition 2.5.1. Let the vector field $A : Z \to Z$ be linear. Then A is Hamiltonian if and only if A is Ω-skew, that is,

$$\Omega(Az_1, z_2) = -\Omega(z_1, Az_2)$$

for all $z_1, z_2 \in Z$. Furthermore, in this case one can take $H(z) = \frac{1}{2}\Omega(Az, z)$.

Proof. Differentiating the defining relation

$$\Omega(X_H(z), v) = \mathbf{d}H(z) \cdot v \qquad (2.5.2)$$

with respect to z in the direction u and using bilinearity of Ω, one gets

$$\Omega(\mathbf{D}X_H(z) \cdot u, v) = \mathbf{D}^2 H(z)(v, u). \qquad (2.5.3)$$

From this and the symmetry of the second partial derivatives, we get

$$\Omega(\mathbf{D}X_H(z) \cdot u, v) = \mathbf{D}^2 H(z)(u, v) = \Omega(\mathbf{D}X_H(z) \cdot v, u)$$
$$= -\Omega(u, \mathbf{D}X_H(z) \cdot v). \qquad (2.5.4)$$

If $A = X_H$ for some H, then $\mathbf{D}X_H(z) = A$, and (2.5.4) becomes $\Omega(Au, v) = -\Omega(u, Av)$; hence A is Ω-skew.

Conversely, suppose that A is Ω-skew. Defining $H(z) = \frac{1}{2}\Omega(Az, z)$, we claim that $A = X_H$. Indeed,

$$
\begin{aligned}
\mathbf{d}H(z) \cdot u &= \tfrac{1}{2}\Omega(Au, z) + \tfrac{1}{2}\Omega(Az, u) \\
&= -\tfrac{1}{2}\Omega(u, Az) + \tfrac{1}{2}\Omega(Az, u) \\
&= \tfrac{1}{2}\Omega(Az, u) + \tfrac{1}{2}\Omega(Az, u) = \Omega(Az, u). \quad\blacksquare
\end{aligned}
$$

In canonical coordinates, where Ω has matrix $\mathbb{J}$, Ω-skewness of A is equivalent to symmetry of the matrix $\mathbb{J}A$; that is, $\mathbb{J}A + A^T\mathbb{J} = 0$. The vector space of all linear transformations of Z satisfying this condition is denoted by $\mathfrak{sp}(Z, \Omega)$, and its elements are called *infinitesimal symplectic transformations*. In canonical coordinates, if $Z = W \times W^*$ and if A has the matrix

$$
A = \begin{bmatrix} A_{qq} & A_{qp} \\ A_{pq} & A_{pp} \end{bmatrix}, \tag{2.5.5}
$$

then one checks that A is infinitesimally symplectic if and only if A_{qp} and A_{pq} are both symmetric and $A_{qq}^T + A_{pp} = 0$ (see Exercise 2.5-1).

In the complex linear case, we use Example (f) in §2.2 ($2\hbar$ times the negative imaginary part of a Hermitian inner product $\langle\,,\,\rangle$ is the symplectic form) to arrive at the following.

Corollary 2.5.2. *Let $\mathcal{H}$ be a complex Hilbert space with Hermitian inner product $\langle\,,\rangle$ and let $\Omega(\psi_1, \psi_2) = -2\hbar\operatorname{Im}\langle\psi_1, \psi_2\rangle$. Let $A : \mathcal{H} \to \mathcal{H}$ be a complex linear operator. There exists an $H : \mathcal{H} \to \mathbb{R}$ such that $A = X_H$ if and only if iA is symmetric or, equivalently, satisfies*

$$
\langle iA\psi_1, \psi_2 \rangle = \langle \psi_1, iA\psi_2 \rangle. \tag{2.5.6}
$$

In this case, H may be taken to be $H(\psi) = \hbar\langle iA\psi, \psi\rangle$. We let $H_{\mathrm{op}} = i\hbar A$, and thus Hamilton's equation $\dot{\psi} = A\psi$ becomes the Schrödinger equation[3]

$$
i\hbar\frac{\partial\psi}{\partial t} = H_{\mathrm{op}}\psi. \tag{2.5.7}
$$

Proof. The operator A is Ω-skew if and only if the condition

$$
\operatorname{Im}\langle A\psi_1, \psi_2 \rangle = -\operatorname{Im}\langle \psi_1, A\psi_2 \rangle
$$

---

[3]Strictly speaking, equation (2.5.6) is required to hold only on the domain of the operator A, which need not be all of $\mathcal{H}$. We shall ignore these issues for simplicity. This example is continued in §2.6 and in §3.2.

holds for all $\psi_1, \psi_2 \in \mathcal{H}$. Replacing ψ_1 by $i\psi_1$ and using the relation $\operatorname{Im}(iz) = \operatorname{Re} z$, this condition is equivalent to $\operatorname{Re} \langle A\psi_1, \psi_2 \rangle = - \operatorname{Re} \langle \psi_1, A\psi_2 \rangle$. Since

$$\langle iA\psi_1, \psi_2 \rangle = - \operatorname{Im} \langle A\psi_1, \psi_2 \rangle + i \operatorname{Re} \langle A\psi_1, \psi_2 \rangle \tag{2.5.8}$$

and

$$\langle \psi_1, iA\psi_2 \rangle = + \operatorname{Im} \langle \psi_1, A\psi_2 \rangle - i \operatorname{Re} \langle \psi_1, A\psi_2 \rangle \tag{2.5.9}$$

we see that Ω-skewness of A is equivalent to iA being symmetric. Finally,

$$\hbar \langle iA\psi, \psi \rangle = \hbar \operatorname{Re} i \langle A\psi, \psi \rangle = -\hbar \operatorname{Im} \langle A\psi, \psi \rangle = \frac{1}{2} \Omega(A\psi, \psi),$$

and the corollary follows from Proposition 2.5.1. ∎

For nonlinear differential equations, the analogue of Proposition 2.5.1 is the following.

Proposition 2.5.3. *Let $X : Z \to Z$ be a (smooth) vector field on a symplectic vector space (Z, Ω). Then $X = X_H$ for some $H : Z \to \mathbb{R}$ if and only if $\mathbf{D}X(z)$ is Ω-skew for all z.*

Proof. We have seen the "only if" part in the proof of Proposition 2.5.1. Conversely, if $\mathbf{D}X(z)$ is Ω-skew, define[4]

$$H(z) = \int_0^1 \Omega(X(tz), z) \, dt + \text{constant}; \tag{2.5.10}$$

we claim that $X = X_H$. Indeed,

$$\begin{aligned}
\mathbf{d}H(z) \cdot v &= \int_0^1 [\Omega(\mathbf{D}X(tz) \cdot tv, z) + \Omega(X(tz), v)] \, dt \\
&= \int_0^1 [\Omega(t\mathbf{D}X(tz) \cdot z, v) + \Omega(X(tz), v)] \, dt \\
&= \Omega \left(\int_0^1 [t\mathbf{D}X(tz) \cdot z + X(tz)] \, dt, v \right) \\
&= \Omega \left(\int_0^1 \frac{d}{dt} [tX(tz)] \, dt, v \right) = \Omega(X(z), v). \quad \blacksquare
\end{aligned}$$

An interesting characterization of Hamiltonian vector fields involves the Cayley transform. Let (Z, Ω) be a symplectic vector space and $A : Z \to Z$ a

---

[4]Looking ahead to Chapter 4 on differential forms, one can check that (2.5.10) for H is reproduced by the proof of the Poincaré lemma applied to the one-form $\mathbf{i}_X \Omega$. That $\mathbf{D}X(z)$ is Ω-skew is equivalent to $\mathbf{d}(\mathbf{i}_X \Omega) = 0$.

linear transformation such that $I - A$ is invertible. Then A *is Hamiltonian if and only if its **Cayley transform*** $C = (I + A)(I - A)^{-1}$ *is symplectic*. See Exercise 2.5-2. For applications, see Laub and Meyer [1974], Paneitz [1981], Feng [1986], and Austin and Krishnaprasad [1993]. The Cayley transform is useful in some Hamiltonian numerical algorithms, as this last reference and Marsden [1992] show.

Exercises

⋄ **2.5-1.** Let $Z = W \times W^*$ and use a canonical basis to write the matrix of the linear map $A : Z \to Z$ as

$$A = \begin{bmatrix} A_{qq} & A_{qp} \\ A_{pq} & A_{pp} \end{bmatrix}.$$

Show that A is infinitesimally symplectic, that is, $\mathbb{J}A + A^T\mathbb{J} = 0$, if and only if A_{qp} and A_{pq} are both symmetric and $A_{qq}^T + A_{pp} = 0$.

⋄ **2.5-2.** Let (Z, Ω) be a symplectic vector space. Let $A : Z \to Z$ be a linear map and assume that $(I - A)$ is invertible. Show that A is Hamiltonian if and only if its Cayley transform

$$(I + A)(I - A)^{-1}$$

is symplectic. Give an example of a linear Hamiltonian vector field such that $(I - A)$ is not invertible.

⋄ **2.5-3.** Suppose that (Z, Ω) is a finite-dimensional symplectic vector space and let $\varphi : Z \to Z$ be a linear symplectic map with $\det \varphi = 1$ (as mentioned in the text, this assumption is superfluous, as will be shown later). If λ is an eigenvalue of multiplicity k, then so is $1/\lambda$. Prove this using the characteristic polynomial of φ.

⋄ **2.5-4.** Suppose that (Z, Ω) is a finite-dimensional symplectic vector space and let $A : Z \to Z$ be a Hamiltonian vector field.

(a) Show that the **generalized kernel** of A, defined to be the set

$$\{ z \in Z \mid A^k z = 0 \text{ for some integer } k \geq 1 \},$$

is a symplectic subspace.

(b) In general, the literal kernel $\ker A$ is not a symplectic subspace of (Z, Ω). Give a counter example.

2.6 Hamiltonian Flows

This subsection discusses flows of Hamiltonian vector fields a little further. The next subsection gives the abstract definition of the Poisson bracket, relates it to the classical definitions, and then shows how it may be used in describing the dynamics. Later on, Poisson brackets will play an increasingly important role.

Let X_H be a Hamiltonian vector field on a symplectic vector space (Z, Ω) with Hamiltonian $H : Z \to \mathbb{R}$. The **flow** of X_H is the collection of maps $\varphi_t : Z \to Z$ satisfying

$$\frac{d}{dt}\varphi_t(z) = X_H(\varphi_t(z)) \qquad (2.6.1)$$

for each $z \in Z$ and real t and $\varphi_0(z) = z$. Here and in the following, all statements concerning the map $\varphi_t : Z \to Z$ are to be considered only for those z and t such that $\varphi_t(z)$ is defined, as determined by differential equations theory.

Linear Flows.   First consider the case in which A is a (bounded) *linear* vector field. The flow of A may be written as $\varphi_t = e^{tA}$; that is, the solution of $dz/dt = Az$ with initial condition z_0 is given by $z(t) = \varphi_t(z_0) = e^{tA}z_0$.

Proposition 2.6.1.   *The flow φ_t of a linear vector field $A : Z \to Z$ consists of (linear) canonical transformations if and only if A is Hamiltonian.*

Proof.   For all $u, v \in Z$ we have

$$\frac{d}{dt}(\varphi_t^*\Omega)(u, v) = \frac{d}{dt}\Omega(\varphi_t(u), \varphi_t(v))$$

$$= \Omega\left(\frac{d}{dt}\varphi_t(u), \varphi_t(v)\right) + \Omega\left(\varphi_t(u), \frac{d}{dt}\varphi_t(v)\right)$$

$$= \Omega(A\varphi_t(u), \varphi_t(v)) + \Omega(\varphi_t(u), A\varphi_t(v)).$$

Therefore, A is Ω-skew, that is, A is Hamiltonian, if and only if each φ_t is a linear canonical transformation. ∎

Nonlinear Flows.   For nonlinear flows, there is a corresponding result.

Proposition 2.6.2.   *The flow φ_t of a (nonlinear) Hamiltonian vector field X_H consists of canonical transformations. Conversely, if the flow of a vector field X consists of canonical transformations, then it is Hamiltonian.*

Proof.   Let φ_t be the flow of a vector field X. By (2.6.1) and the chain rule,

$$\frac{d}{dt}[\mathbf{D}\varphi_t(z) \cdot v] = \mathbf{D}\left[\frac{d}{dt}\varphi_t(z)\right] \cdot v = \mathbf{D}X(\varphi_t(z)) \cdot (\mathbf{D}\varphi_t(z) \cdot v),$$

which is called the **first variation equation**. Using this, we get

$$\frac{d}{dt}\Omega(\mathbf{D}\varphi_t(z) \cdot u, \mathbf{D}\varphi_t(z) \cdot v) = \Omega(\mathbf{D}X(\varphi_t(z)) \cdot [\mathbf{D}\varphi_t(z) \cdot u], \mathbf{D}\varphi_t(z) \cdot v)$$

$$+ \Omega(\mathbf{D}\varphi_t(z) \cdot u, \mathbf{D}X(\varphi_t(z)) \cdot [\mathbf{D}\varphi_t(z) \cdot v]).$$

If $X = X_H$, then $\mathbf{D}X_H(\varphi_t(z))$ is Ω-skew by Proposition 2.5.3, so

$$\Omega(\mathbf{D}\varphi_t(z) \cdot u, \mathbf{D}\varphi_t(z) \cdot v) = \text{constant.}$$

At $t = 0$ this equals $\Omega(u, v)$, so $\varphi_t^*\Omega = \Omega$. Conversely, if φ_t is canonical, this calculation shows that $\mathbf{D}X(\varphi_t(z))$ is Ω-skew, whence by Proposition 2.5.3, $X = X_H$ for some H. ∎

Later on, we give another proof of Proposition 2.6.2 using differential forms.

Example: The Schrödinger Equation

Recall that if $\mathcal{H}$ is a complex Hilbert space, a complex linear map $U : \mathcal{H} \to \mathcal{H}$ is called **unitary** if it preserves the Hermitian inner product.

Proposition 2.6.3. *Let $A : \mathcal{H} \to \mathcal{H}$ be a complex linear map on a complex Hilbert space $\mathcal{H}$. The flow φ_t of A is canonical, that is, consists of canonical transformations with respect to the symplectic form Ω defined in Example (f) of §2.2, if and only if φ_t is unitary.*

Proof. By definition,

$$\Omega(\psi_1, \psi_2) = -2\hbar \operatorname{Im} \langle \psi_1, \psi_2 \rangle,$$

so

$$\Omega(\varphi_t\psi_1, \varphi_t\psi_2) = -2\hbar \operatorname{Im} \langle \varphi_t\psi_1, \varphi_t\psi_2 \rangle$$

for $\psi_1, \psi_2 \in \mathcal{H}$. Thus, φ_t is canonical if and only if $\operatorname{Im} \langle \varphi_t\psi_1, \varphi_t\psi_2 \rangle = \operatorname{Im} \langle \psi_1, \psi_2 \rangle$, and this in turn is equivalent to unitarity by complex linearity of φ_t, since $\langle \psi_1, \psi_2 \rangle = - \operatorname{Im} \langle i\psi_1, \psi_2 \rangle + i \operatorname{Im} \langle \psi_1, \psi_2 \rangle$. ∎

This shows that the flow of the **Schrödinger equation** $\dot{\psi} = A\psi$ is canonical and unitary and so preserves the probability amplitude of any wave function that is a solution. In other words, we have

$$\langle \varphi_t\psi, \varphi_t\psi \rangle = \langle \psi, \psi \rangle,$$

where φ_t is the flow of A. Later we shall see how this conservation of the norm also results from a symmetry-induced conservation law.

2.7 Poisson Brackets

Definition 2.7.1. *Given a symplectic vector space (Z, Ω) and two functions $F, G : Z \to \mathbb{R}$, the **Poisson bracket** $\{F, G\} : Z \to \mathbb{R}$ of F and G is defined by*

$$\{F, G\}(z) = \Omega(X_F(z), X_G(z)). \tag{2.7.1}$$

Using the definition of a Hamiltonian vector field, we find that equivalent expressions are

$$\{F,G\}(z) = \mathbf{d}F(z) \cdot X_G(z) = -\mathbf{d}G(z) \cdot X_F(z). \qquad (2.7.2)$$

In (2.7.2) we write $\pounds_{X_G} F = \mathbf{d}F \cdot X_G$ for the derivative of F in the direction X_G.

Lie Derivative Notation. The *Lie derivative* of f along X, $\pounds_X f = \mathbf{d}f \cdot X$, is the *directional derivative* of f in the direction X. In coordinates it is given by

$$\pounds_X f = \frac{\partial f}{\partial z^I} X^I \quad \text{(sum on } I\text{)}.$$

Functions F, G such that $\{F, G\} = 0$ are said to be in *involution* or to *Poisson commute*.

Examples

Now we turn to some examples of Poisson brackets.

(a) Canonical Bracket. Suppose that Z is $2n$-dimensional. Then in canonical coordinates $(q^1, \ldots, q^n, p_1, \ldots, p_n)$ we have

$$\{F,G\} = \left[\frac{\partial F}{\partial p_i}, -\frac{\partial F}{\partial q^i} \right] \mathbb{J} \begin{bmatrix} \dfrac{\partial G}{\partial p_i} \\ -\dfrac{\partial G}{\partial q^i} \end{bmatrix}$$

$$= \frac{\partial F}{\partial q^i} \frac{\partial G}{\partial p_i} - \frac{\partial F}{\partial p_i} \frac{\partial G}{\partial q^i} \quad \text{(sum on } i\text{)}. \qquad (2.7.3)$$

From this we get the *fundamental Poisson brackets*

$$\{q^i, q^j\} = 0, \quad \{p_i, p_j\} = 0, \quad \text{and} \quad \{q^i, p_j\} = \delta^i_j. \qquad (2.7.4)$$

In terms of the Poisson structure, that is, the bilinear form B from §2.1, the Poisson bracket takes the form

$$\{F, G\} = B(\mathbf{d}F, \mathbf{d}G). \qquad (2.7.5)$$

◆

(b) The Space of Functions. Let (Z, Ω) be defined as in Example (b) of §2.2 and let $F, G : Z \to \mathbb{R}$. Using equations (2.4.10) and (2.7.1) above, we get

$$\{F,G\} = \Omega(X_F, X_G) = \Omega\left(\left(\frac{\delta F}{\delta \pi}, -\frac{\delta F}{\delta \varphi}\right), \left(\frac{\delta G}{\delta \pi}, -\frac{\delta G}{\delta \varphi}\right)\right)$$
$$= \int_{\mathbb{R}^3} \left(\frac{\delta G}{\delta \pi}\frac{\delta F}{\delta \varphi} - \frac{\delta F}{\delta \pi}\frac{\delta G}{\delta \varphi}\right) d^3x. \tag{2.7.6}$$

This example will be used in the next chapter when we study classical field theory.   ♦

The Jacobi–Lie Bracket.   The *Jacobi–Lie bracket* $[X, Y]$ of two vector fields X and Y on a vector space Z is defined by demanding that

$$\mathbf{d}f \cdot [X, Y] = \mathbf{d}(\mathbf{d}f \cdot Y) \cdot X - \mathbf{d}(\mathbf{d}f \cdot X) \cdot Y$$

for all real-valued functions f. In Lie derivative notation, this reads

$$\pounds_{[X,Y]}f = \pounds_X \pounds_Y f - \pounds_Y \pounds_X f.$$

One checks that this condition becomes, in vector analysis notation,

$$[X, Y] = (X \cdot \nabla)Y - (Y \cdot \nabla)X,$$

and in coordinates,

$$[X, Y]^J = X^I \frac{\partial}{\partial z^I}Y^J - Y^I \frac{\partial}{\partial z^I}X^J.$$

Proposition 2.7.2.   *Let* $[,]$ *denote the Jacobi–Lie bracket of vector fields, and let* $F, G \in \mathcal{F}(Z)$. *Then*

$$X_{\{F,G\}} = -[X_F, X_G]. \tag{2.7.7}$$

Proof.   We calculate as follows:

$$\begin{aligned}
\Omega(X_{\{F,G\}}(z), u) &= \mathbf{d}\{F,G\}(z) \cdot u = \mathbf{d}(\Omega(X_F(z), X_G(z))) \cdot u \\
&= \Omega(\mathbf{D}X_F(z) \cdot u, X_G(z)) + \Omega(X_F(z), \mathbf{D}X_G(z) \cdot u) \\
&= \Omega(\mathbf{D}X_F(z) \cdot X_G(z), u) - \Omega(\mathbf{D}X_G(z) \cdot X_F(z), u) \\
&= \Omega(\mathbf{D}X_F(z) \cdot X_G(z) - \mathbf{D}X_G(z) \cdot X_F(z), u) \\
&= \Omega(-[X_F, X_G](z), u).
\end{aligned}$$

Weak nondegeneracy of Ω implies the result.   ∎

Jacobi's Identity.   We are now ready to prove the Jacobi identity in a fairly general context.

Proposition 2.7.3. *Let (Z, Ω) be a symplectic vector space. Then the Poisson bracket $\{,\} : \mathcal{F}(Z) \times \mathcal{F}(Z) \to \mathcal{F}(Z)$ makes $\mathcal{F}(Z)$ into a **Lie algebra**. That is, this bracket is real bilinear, skew-symmetric, and satisfies Jacobi's identity, that is,*

$$\{F, \{G, H\}\} + \{G, \{H, F\}\} + \{H, \{F, G\}\} = 0.$$

Proof. To verify Jacobi's identity note that for $F, G, H : Z \to \mathbb{R}$, we have

$$\{F, \{G, H\}\} = -\pounds_{X_F}\{G, H\} = \pounds_{X_F}\pounds_{X_G}H,$$
$$\{G, \{H, F\}\} = -\pounds_{X_G}\{H, F\} = -\pounds_{X_G}\pounds_{X_F}H,$$

and

$$\{H, \{F, G\}\} = \pounds_{X_{\{F,G\}}}H,$$

so that

$$\{F, \{G, H\}\} + \{G, \{H, F\}\} + \{H, \{F, G\}\} = \pounds_{X_{\{F,G\}}}H + \pounds_{[X_F, X_G]}H.$$

The result thus follows by (2.7.7). ∎

From Proposition 2.7.2 we see that the Jacobi–Lie bracket of two Hamiltonian vector fields is again Hamiltonian. Thus, we obtain the following corollary.

Corollary 2.7.4. *The set of Hamiltonian vector fields $\mathfrak{X}_{\mathrm{Ham}}(Z)$ forms a Lie subalgebra of $\mathfrak{X}(Z)$.*

Next, we characterize symplectic maps in terms of brackets.

Proposition 2.7.5. *Let $\varphi : Z \to Z$ be a diffeomorphism. Then φ is symplectic if and only if it preserves Poisson brackets, that is,*

$$\{\varphi^*F, \varphi^*G\} = \varphi^*\{F, G\} \qquad (2.7.8)$$

for all $F, G : Z \to \mathbb{R}$.

Proof. We use the identity

$$\varphi^*(\pounds_X f) = \pounds_{\varphi^* X}(\varphi^* f),$$

which follows from the chain rule. Thus,

$$\varphi^*\{F, G\} = \varphi^* \pounds_{X_G} F = \pounds_{\varphi^* X_G}(\varphi^* F)$$

and

$$\{\varphi^*F, \varphi^*G\} = \pounds_{X_{G \circ \varphi}}(\varphi^* F).$$

Thus, φ preserves Poisson brackets if and only if $\varphi^* X_G = X_{G \circ \varphi}$ for every $G : Z \to \mathbb{R}$, that is, if and only if φ is symplectic by Proposition 2.4.2. ∎

Proposition 2.7.6. *Let X_H be a Hamiltonian vector field on Z, with Hamiltonian H and flow φ_t. Then for $F : Z \to \mathbb{R}$,*

$$\frac{d}{dt}(F \circ \varphi_t) = \{F \circ \varphi_t, H\}$$
$$= \{F, H\} \circ \varphi_t. \qquad (2.7.9)$$

Proof. By the chain rule and the definition of X_F,

$$\frac{d}{dt}[(F \circ \varphi_t)(z)] = \mathbf{d}F(\varphi_t(z)) \cdot X_H(\varphi_t(z))$$
$$= \Omega(X_F(\varphi_t(z)), X_H(\varphi_t(z)))$$
$$= \{F, H\}(\varphi_t(z)).$$

By Proposition 2.6.2 and (2.7.8), this equals

$$\{F \circ \varphi_t, H \circ \varphi_t\}(z) = \{F \circ \varphi_t, H\}(z)$$

by conservation of energy.   ∎

Corollary 2.7.7. *Let $F, G : Z \to \mathbb{R}$. Then F is constant along integral curves of X_G if and only if G is constant along integral curves of X_F, and this is true if and only if $\{F, G\} = 0$.*

Proposition 2.7.8. *Let $A, B : Z \to Z$ be linear Hamiltonian vector fields with corresponding energy functions*

$$H_A(z) = \tfrac{1}{2}\Omega(Az, z) \quad and \quad H_B(z) = \tfrac{1}{2}\Omega(Bz, z).$$

Letting

$$[A, B] = A \circ B - B \circ A$$

be the operator commutator, we have

$$\{H_A, H_B\} = H_{[A,B]}. \qquad (2.7.10)$$

Proof. By definition, $X_{H_A} = A$, and so

$$\{H_A, H_B\}(z) = \Omega(Az, Bz).$$

Since A and B are Ω-skew, we get

$$\{H_A, H_B\}(z) = \tfrac{1}{2}\Omega(ABz, z) - \tfrac{1}{2}\Omega(BAz, z)$$
$$= \tfrac{1}{2}\Omega([A, B]z, z) \qquad (2.7.11)$$
$$= H_{[A,B]}(z). \qquad ∎$$

2.8   A Particle in a Rotating Hoop

In this subsection we take a break from the abstract theory to do an example the "old-fashioned" way. This and other examples will also serve as excellent illustrations of the theory we are developing.

Derivation of the Equations.   Consider a particle constrained to move on a circular hoop; for example a bead sliding in a Hula-Hoop. The particle is assumed to have mass m and to be acted on by gravitational and frictional forces, as well as constraint forces that keep it on the hoop. The hoop itself is spun about a vertical axis with constant angular velocity ω, as in Figure 2.8.1.

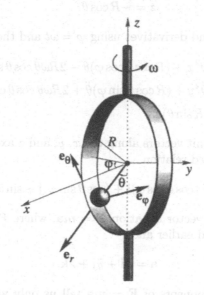

FIGURE 2.8.1. A particle moving in a hoop rotating with angular velocity ω.

The position of the particle in space is specified by the angles θ and φ, as shown in Figure 2.8.1. We can take $\varphi = \omega t$, so the position of the particle becomes determined by θ alone. Let the orthonormal frame along the coordinate directions $\mathbf{e}_\theta$, $\mathbf{e}_\varphi$, and $\mathbf{e}_r$ be as shown.

The forces acting on the particle are:

1. Friction, proportional to the velocity of the particle relative to the hoop: $-\nu R\dot{\theta}\mathbf{e}_\theta$, where $\nu \geq 0$ is a constant.[5]

---

[5]This is a "law of friction" that is more like a viscous fluid friction than a sliding friction in which ν is the ratio of the tangential force to the normal force; in any actual experimental setup (e.g., involving rolling spheres) a realistic modeling of the friction is not a trivial task; see, for example, Lewis and Murray [1995].

2. Gravity: $-mg\mathbf{k}$.

3. Constraint forces in the directions $\mathbf{e}_r$ and $\mathbf{e}_\varphi$ to keep the particle in the hoop.

The equations of motion are derived from Newton's second law $\mathbf{F} = m\mathbf{a}$. To get them, we need to calculate the acceleration $\mathbf{a}$; here $\mathbf{a}$ means the acceleration relative to the *fixed inertial frame xyz* in space; it does not mean $\ddot{\theta}$. Relative to this xyz coordinate system, we have

$$x = R\sin\theta\cos\varphi,$$
$$y = R\sin\theta\sin\varphi, \qquad\qquad (2.8.1)$$
$$z = -R\cos\theta.$$

Calculating the second derivatives using $\varphi = \omega t$ and the chain rule gives

$$\ddot{x} = -\omega^2 x - \dot{\theta}^2 x + (R\cos\theta\cos\varphi)\ddot{\theta} - 2R\omega\dot{\theta}\cos\theta\sin\varphi,$$
$$\ddot{y} = -\omega^2 y - \dot{\theta}^2 y + (R\cos\theta\sin\varphi)\ddot{\theta} + 2R\omega\dot{\theta}\cos\theta\cos\varphi, \qquad (2.8.2)$$
$$\ddot{z} = -z\dot{\theta}^2 + (R\sin\theta)\ddot{\theta}.$$

If $\mathbf{i}$, $\mathbf{j}$, $\mathbf{k}$, denote unit vectors along the x, y, and z axes, respectively, we have the easily verified relation

$$\mathbf{e}_\theta = (\cos\theta\cos\varphi)\mathbf{i} + (\cos\theta\sin\varphi)\mathbf{j} + \sin\theta\mathbf{k}. \qquad (2.8.3)$$

Now consider the vector equation $\mathbf{F} = m\mathbf{a}$, where $\mathbf{F}$ is the sum of the three forces described earlier and

$$\mathbf{a} = \ddot{x}\mathbf{i} + \ddot{y}\mathbf{j} + \ddot{z}\mathbf{k}. \qquad (2.8.4)$$

The $\mathbf{e}_\varphi$ and $\mathbf{e}_r$ components of $\mathbf{F} = m\mathbf{a}$ tell us only what the constraint forces must be; the equation of motion comes from the $\mathbf{e}_\theta$ component:

$$\mathbf{F}\cdot\mathbf{e}_\theta = m\mathbf{a}\cdot\mathbf{e}_\theta. \qquad (2.8.5)$$

Using (2.8.3), the left side of (2.8.5) is

$$\mathbf{F}\cdot\mathbf{e}_\theta = -\nu R\dot{\theta} - mg\sin\theta, \qquad (2.8.6)$$

while from (2.8.2), (2.8.3), and (2.8.4), the right side of (2.8.5) is

$$\begin{aligned}
m\mathbf{a}\cdot\mathbf{e}_\theta &= m\{\ddot{x}\cos\theta\cos\varphi + \ddot{y}\cos\theta\sin\varphi + \ddot{z}\sin\theta\} \\
&= m\{\cos\theta\cos\varphi[-\omega^2 x - \dot{\theta}^2 x + (R\cos\theta\cos\varphi)\ddot{\theta} - 2R\omega\dot{\theta}\cos\theta\sin\varphi] \\
&\quad + \cos\theta\sin\varphi[-\omega^2 y - \dot{\theta}^2 y + (R\cos\theta\sin\varphi)\ddot{\theta} + 2R\omega\dot{\theta}\cos\theta\cos\varphi] \\
&\quad + \sin\theta[-z\dot{\theta}^2 + (R\sin\theta)\ddot{\theta}]\}.
\end{aligned}$$

Using (2.8.1), this simplifies to

$$\mathbf{ma} \cdot \mathbf{e}_\theta = mR\{\ddot{\theta} - \omega^2 \sin\theta \cos\theta\}. \tag{2.8.7}$$

Comparing (2.8.5), (2.8.6), and (2.8.7), we get

$$\ddot{\theta} = \omega^2 \sin\theta \cos\theta - \frac{\nu}{m}\dot{\theta} - \frac{g}{R}\sin\theta \tag{2.8.8}$$

as our final equation of motion. Several remarks concerning it are in order:

(i) If $\omega = 0$ and $\nu = 0$, (2.8.8) reduces to the **pendulum equation**

$$R\ddot{\theta} + g\sin\theta = 0.$$

In fact, our system can be viewed just as well as a **whirling pendulum**.

(ii) For $\nu = 0$, (2.8.8) is Hamiltonian. This is readily verified using the variables $q = \theta$, $p = mR^2\dot{\theta}$, the canonical bracket structure

$$\{F, K\} = \frac{\partial F}{\partial q}\frac{\partial K}{\partial p} - \frac{\partial K}{\partial q}\frac{\partial F}{\partial p}, \tag{2.8.9}$$

and the Hamiltonian

$$H = \frac{p^2}{2mR^2} - mgR\cos\theta - \frac{mR^2\omega^2}{2}\sin^2\theta. \tag{2.8.10}$$

Derivation as Euler–Lagrange Equations. We now use Lagrangian methods to derive (2.8.8). In Figure 2.8.1, the velocity is

$$\mathbf{v} = R\dot{\theta}\mathbf{e}_\theta + (\omega R\sin\theta)\mathbf{e}_\varphi,$$

so the kinetic energy is

$$T = \tfrac{1}{2}m\|\mathbf{v}\|^2 = \tfrac{1}{2}m(R^2\dot{\theta}^2 + [\omega R\sin\theta]^2), \tag{2.8.11}$$

while the potential energy is

$$V = -mgR\cos\theta. \tag{2.8.12}$$

Thus, the Lagrangian is given by

$$L = T - V = \frac{1}{2}mR^2\dot{\theta}^2 + \frac{mR^2\omega^2}{2}\sin^2\theta + mgR\cos\theta, \tag{2.8.13}$$

and the Euler–Lagrange equations, namely,

$$\frac{d}{dt}\frac{\partial L}{\partial\dot{\theta}} = \frac{\partial L}{\partial\theta}$$

(see §1.1 or §2.1), become

$$mR^2\ddot{\theta} = mR^2\omega^2 \sin\theta\cos\theta - mgR\sin\theta,$$

which are the same equations we derived by hand in (2.8.8) for $\nu = 0$. The Legendre transform gives $p = mR^2\dot{\theta}$ and the Hamiltonian (2.8.10). Notice that this Hamiltonian is *not* the kinetic plus potential energy of the particle. In fact, if one postulated this, then Hamilton's equations would give the *incorrect equations*. This has to do with deeper covariance properties of the Lagrangian versus Hamiltonian equations.

Equilibria.   The *equilibrium solutions* are solutions satisfying $\dot{\theta} = 0$, $\ddot{\theta} = 0$; (2.8.8) gives

$$R\omega^2 \sin\theta\cos\theta = g\sin\theta. \qquad (2.8.14)$$

Certainly, $\theta = 0$ and $\theta = \pi$ solve (2.8.14) corresponding to the particle at the bottom or top of the hoop. If $\theta \neq 0$ or π, (2.8.14) becomes

$$R\omega^2 \cos\theta = g, \qquad (2.8.15)$$

which has two solutions when $g/(R\omega^2) < 1$. The value

$$\omega_c = \sqrt{\frac{g}{R}} \qquad (2.8.16)$$

is the *critical rotation rate*. Notice that ω_c is the frequency of linearized oscillations for the simple pendulum, that is, for the equation

$$R\ddot{\theta} + g\theta = 0.$$

For $\omega < \omega_c$ there are only *two* solutions $\theta = 0$, π, while for $\omega > \omega_c$ there are *four solutions*,

$$\theta = 0,\ \pi,\ \pm\cos^{-1}\left(\frac{g}{R\omega^2}\right). \qquad (2.8.17)$$

We say that a *bifurcation* (or a *Hamiltonian pitchfork bifurcation*, to be accurate) has occurred as ω crosses ω_c. We can see this graphically in computer-generated solutions of (2.8.8). Set $x = \theta$, $y = \dot{\theta}$ and rewrite (2.8.8) as

$$\dot{x} = y,$$
$$\dot{y} = \frac{g}{R}(\alpha\cos x - 1)\sin x - \beta y, \qquad (2.8.18)$$

where

$$\alpha = R\omega^2/g \quad \text{and} \quad \beta = \nu/m.$$

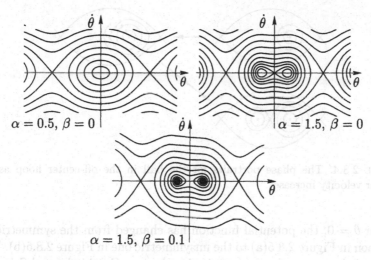

$\alpha = 0.5,\ \beta = 0$ $\alpha = 1.5,\ \beta = 0$

$\alpha = 1.5,\ \beta = 0.1$

FIGURE 2.8.2. Phase portraits of the ball in the rotating hoop.

Taking $g = R$ for illustration, Figure 2.8.2 shows representative orbits in the phase portraits of (2.8.18) for various α, β.

This system with $\nu = 0$, that is, $\beta = 0$, is symmetric in the sense that the $\mathbb{Z}_2$-action given by

$$\theta \mapsto -\theta \quad \text{and} \quad \dot{\theta} \mapsto -\dot{\theta}$$

leaves the phase portrait invariant. If this $\mathbb{Z}_2$ symmetry is broken, by setting the rotation axis a little off center, for example, then one side gets preferred, as in Figure 2.8.3.

FIGURE 2.8.3. A ball in an off-center rotating hoop.

The evolution of the phase portrait for $\nu = 0$ is shown in Figure 2.8.4.

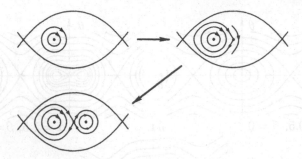

FIGURE 2.8.4. The phase portraits for the ball in the off-center hoop as the angular velocity increases.

Near $\theta = 0$, the potential function has changed from the symmetric bifurcation in Figure 2.8.5(a) to the unsymmetric one in Figure 2.8.5(b). This is what is known as the *cusp catastrophe*; see Golubitsky and Schaeffer [1985] and Arnold [1968, 1984] for more information.

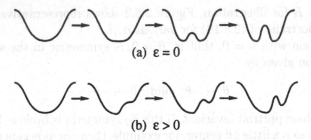

(a) $\varepsilon = 0$

(b) $\varepsilon > 0$

FIGURE 2.8.5. The evolution of the potential for the ball in the (a) centered and (b) off-center hoop as the angular velocity increases.

In (2.8.8), imagine that the hoop is subject to small periodic pulses, say $\omega = \omega_0 + \rho \cos(\eta t)$. Using the Melnikov method described in the introduction and in the following section, it is presumably true (but a messy calculation to prove) that the resulting time-periodic system has horseshoe chaos if ϵ and ν are small (where ϵ measures how off-center the hoop is) but ρ/ν exceeds a critical value. See Exercise 2.8-3 and §2.8.

Exercises

⋄ **2.8-1.**   Derive the equations of motion for a particle in a hoop spinning about a line a distance ϵ off center. What can you say about the equilibria as functions of ϵ and ω?

⋄ **2.8-2.**   Derive the formula of Exercise 1.9-1 for the homoclinic orbit (the orbit tending to the saddle point as $t \to \pm\infty$) of a pendulum $\ddot{\psi} + \sin \psi = 0$.

Do this using conservation of energy, determining the value of the energy on the homoclinic orbit, solving for $\dot\psi$, and then integrating.

◇ **2.8-3.** Using the method of the preceding exercise, derive an integral formula for the homoclinic orbit of the frictionless particle in a rotating hoop.

◇ **2.8-4.** Determine all equilibria of Duffing's equation

$$\ddot x - \beta x + \alpha x^3 = 0,$$

where α and β are positive constants, and study their stability. Derive a formula for the two homoclinic orbits.

◇ **2.8-5.** Determine the equations of motion and bifurcations for a ball in a light rotating hoop, but this time the hoop is not forced to rotate with constant *angular velocity*, but rather is free to rotate so that its *angular momentum* μ is conserved.

◇ **2.8-6.** Consider the pendulum shown in Figure 2.8.6. It is a planar pendulum whose suspension point is being whirled in a circle with angular velocity ω by means of a vertical shaft, as shown. The plane of the pendulum is orthogonal to the radial arm of length R. Ignore frictional effects.

(i) Using the notation in the figure, find the equations of motion of the pendulum.

(ii) Regarding ω as a parameter, show that a supercritical pitchfork bifurcation of equilibria occurs as the angular velocity of the shaft is increased.

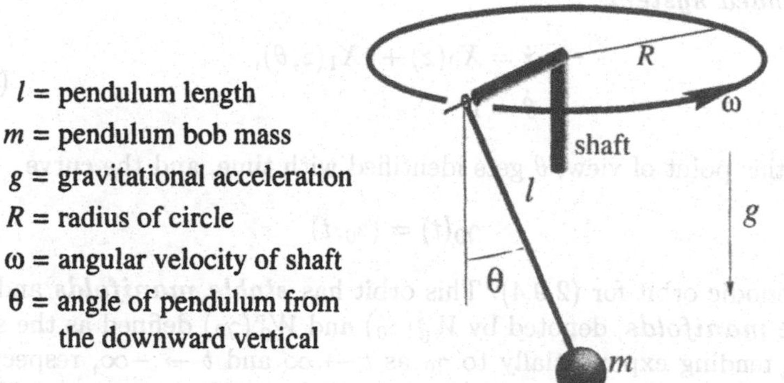

l = pendulum length
m = pendulum bob mass
g = gravitational acceleration
R = radius of circle
ω = angular velocity of shaft
θ = angle of pendulum from the downward vertical

FIGURE 2.8.6. A whirling pendulum.

2.9 The Poincaré–Melnikov Method

Recall from the introduction that in the simplest version of the Poincaré–Melnikov method we are concerned with dynamical equations that perturb a planar Hamiltonian system

$$\dot{z} = X_0(z) \tag{2.9.1}$$

to one of the form

$$\dot{z} = X_0(z) + \epsilon X_1(z, t), \tag{2.9.2}$$

where ϵ is a small parameter, $z \in \mathbb{R}^2$, X_0 is a Hamiltonian vector field with energy H_0, X_1 is periodic with period T and is Hamiltonian with energy a T-periodic function H_1. We assume that X_0 has a homoclinic orbit $\bar{z}(t)$, that is, an orbit such that $\bar{z}(t) \to z_0$, a hyperbolic saddle point, as $t \to \pm\infty$. Define the *Poincaré–Melnikov function* by

$$M(t_0) = \int_{-\infty}^{\infty} \{H_0, H_1\}(\bar{z}(t - t_0), t) \, dt, \tag{2.9.3}$$

where $\{\,,\}$ denotes the Poisson bracket.

There are two convenient ways of visualizing the dynamics of (2.9.2). Introduce the *Poincaré map* $P_\epsilon^s : \mathbb{R}^2 \to \mathbb{R}^2$, which is the time T map for (2.9.2) starting at time s. For $\epsilon = 0$, the point z_0 and the homoclinic orbit are invariant under P_0^s, which is independent of s. The hyperbolic saddle z_0 persists as a nearby family of saddles z_ϵ for $\epsilon > 0$, small, and we are interested in whether or not the stable and unstable manifolds of the point z_ϵ for the map P_ϵ^s intersect transversally (if this holds for one s, it holds for all s). If so, we say that (2.9.2) has *horseshoes for* $\epsilon > 0$.

The second way to study (2.9.2) is to look directly at the suspended system on $\mathbb{R}^2 \times S^1$, where S^1 is the circle; (2.9.2) becomes the autonomous *suspended system*

$$\begin{aligned} \dot{z} &= X_0(z) + \epsilon X_1(z, \theta), \\ \dot{\theta} &= 1. \end{aligned} \tag{2.9.4}$$

From this point of view, θ gets identified with time, and the curve

$$\gamma_0(t) = (z_0, t)$$

is a periodic orbit for (2.9.4). This orbit has **stable manifolds** and **unstable manifolds**, denoted by $W_0^s(\gamma_0)$ and $W_0^u(\gamma_0)$ defined as the sets of points tending exponentially to γ_0 as $t \to \infty$ and $t \to -\infty$, respectively. (See Abraham, Marsden, and Ratiu [1988], Guckenheimer and Holmes [1983], or Wiggins [1988, 1990, 1992] for more details.) In this example, they coincide:

$$W_0^s(\gamma_0) = W_0^u(\gamma_0).$$

For $\epsilon > 0$ the (hyperbolic) closed orbit γ_0 perturbs to a nearby (hyperbolic) closed orbit that has stable and unstable manifolds $W_\epsilon^s(\gamma_\epsilon)$ and $W_\epsilon^u(\gamma_\epsilon)$. If $W_\epsilon^s(\gamma_\epsilon)$ and $W_\epsilon^u(\gamma_\epsilon)$ intersect transversally, we again say that (2.9.2) has **horseshoes**. These two definitions of admitting horseshoes are readily seen to be equivalent.

Theorem 2.9.1 (Poincaré–Melnikov Theorem). *Let the Poincaré–Melnikov function be defined by* (2.9.3). *Assume that $M(t_0)$ has simple zeros as a T-periodic function of t_0. Then for sufficiently small ϵ, equation* (2.9.2) *has horseshoes, that is, homoclinic chaos in the sense of transversal intersecting separatrices.*

Idea of the Proof.   In the suspended picture, we use the energy function H_0 to measure the first-order movement of $W_\epsilon^s(\gamma_\epsilon)$ at $\overline{z}(0)$ at time t_0 as ϵ is varied. Note that points of $\overline{z}(t)$ are regular points for H_0, since H_0 is constant on $\overline{z}(t)$, and $\overline{z}(0)$ is not a fixed point. That is, the differential of H_0 does not vanish at $\overline{z}(0)$. Thus, the values of H_0 give an accurate measure of the distance from the homoclinic orbit. If $(z_\epsilon^s(t, t_0), t)$ is the curve on $W_\epsilon^s(\gamma_\epsilon)$ that is an integral curve of the suspended system and has an condition $z_\epsilon^s(t_0, t_0)$ that is the perturbation of

$$W_0^s(\gamma_0) \cap \{ \text{ the plane } t = t_0 \}$$

in the normal direction to the homoclinic orbit, then $H_0(z_\epsilon^s(t_0, t_0))$ measures the normal distance. But

$$H_0(z_\epsilon^s(\tau_+, t_0)) - H_0(z_\epsilon^s(t_0, t_0)) = \int_{t_0}^{\tau_+} \frac{d}{dt} H_0(z_\epsilon^s(t, t_0))\, dt$$

$$= \int_{t_0}^{\tau_+} \{H_0, H_0 + \epsilon H_1\}(z_\epsilon^s(t, t_0), t)\, dt.$$

$$(2.9.5)$$

From invariant manifold theory one learns that $z_\epsilon^s(t, t_0)$ converges exponentially to $\gamma_\epsilon(t)$, a periodic orbit for the perturbed system as $t \to +\infty$. Notice from the right-hand side of the first equality above that if $z_\epsilon^s(t, t_0)$ were replaced by the periodic orbit $\gamma_\epsilon(t)$, the result would be zero. Since the convergence is exponential, one concludes that the integral is of order ϵ for an interval from some large time to infinity. To handle the finite portion of the integral, we use the fact that $z_\epsilon^s(t, t_0)$ is ϵ-close to $\overline{z}(t - t_0)$ (uniformly as $t \to +\infty$) and that $\{H_0, H_0\} = 0$. Therefore, we see that

$$\{H_0, H_0 + \epsilon H_1\}(z_\epsilon^s(t, t_0), t) = \epsilon\{H_0, H_1\}(\overline{z}(t - t_0), t) + O(\epsilon^2).$$

Using this over a large but finite interval $[t_0, t_1]$ and the exponential closeness over the remaining interval $[t_1, \infty)$, we see that (2.9.5) becomes

$$H_0(z_\epsilon^s(\tau_+, t_0)) - H_0(z_\epsilon^s(t_0, t_0))$$

$$= \epsilon \int_{t_0}^{\tau_+} \{H_0, H_1\}(\overline{z}(t - t_0), t)\, dt + O(\epsilon^2), \quad (2.9.6)$$

where the error is uniformly small as $\tau_+ \to \infty$. Similarly,

$$H_0(z_\epsilon^u(t_0, t_0)) - H_0(z_\epsilon^u(\tau_-, t_0))$$
$$= \epsilon \int_{\tau_-}^{t_0} \{H_0, H_1\}(\bar{z}(t - t_0), t) \, dt + O(\epsilon^2). \quad (2.9.7)$$

Again we use the fact that $z_\epsilon^s(\tau_+, t_0) \to \gamma_\epsilon(\tau_+)$ exponentially fast, a periodic orbit for the perturbed system as $\tau_+ \to +\infty$. Notice that since the orbit is *homoclinic*, the *same* periodic orbit can be used for negative times as well. Using this observation, we can choose τ_+ and τ_- such that

$$H_0(z_\epsilon^s(\tau_+, t_0)) - H_0(z_\epsilon^u(\tau_-, t_0)) \to 0$$

as $\tau_+ \to \infty$ and $\tau_- \to -\infty$. Adding (2.9.6) and (2.9.7), letting $\tau_+ \to \infty$ and $\tau_- \to -\infty$, we get

$$H_0(z_\epsilon^u(t_0, t_0)) - H_0(z_\epsilon^s(t_0, t_0))$$
$$= \epsilon \int_{-\infty}^{\infty} \{H_0, H_1\}(\bar{z}(t - t_0), t) \, dt + O(\epsilon^2). \quad (2.9.8)$$

The integral in this expression is convergent because the curve $\bar{z}(t - t_0)$ tends exponentially to the saddle point as $t \to \pm\infty$ and because the differential of H_0 vanishes at this point. Thus, the integrand tends to zero exponentially fast as t tends to plus and minus infinity.

Since the energy is a "good" measure of the distance between the points $z_\epsilon^u(t_0, t_0))$ and $z_\epsilon^s(t_0, t_0))$, it follows that if $M(t_0)$ has a simple zero at time t_0, then $z_\epsilon^u(t_0, t_0)$ and $z_\epsilon^s(t_0, t_0)$ intersect transversally near the point $\bar{z}(0)$ at time t_0. ∎

If in (2.9.2) only X_0 is Hamiltonian, the same conclusion holds if (2.9.3) is replaced by

$$M(t_0) = \int_{-\infty}^{\infty} (X_0 \times X_1)(\bar{z}(t - t_0), t) \, dt, \quad (2.9.9)$$

where $X_0 \times X_1$ is the (scalar) cross product for planar vector fields. In fact, X_0 need not even be Hamiltonian if an area expansion factor is inserted.

Example A.   Equation (2.9.9) applies to the forced damped Duffing equation

$$\ddot{u} - \beta u + \alpha u^3 = \epsilon(\gamma \cos \omega t - \delta \dot{u}). \quad (2.9.10)$$

Here the homoclinic orbits are given by (see Exercise 2.8-4)

$$u(t) = \pm \sqrt{\frac{2\beta}{\alpha}} \, \text{sech}(\sqrt{\beta} t), \quad (2.9.11)$$

and (2.9.9) becomes, after a residue calculation,

$$M(t_0) = \gamma\pi\omega\sqrt{\frac{2}{\alpha}} \operatorname{sech}\left(\frac{\pi\omega}{2\sqrt{\beta}}\right) \sin(\omega t_0) - \frac{4\delta\beta^{3/2}}{3\alpha}, \qquad (2.9.12)$$

so one has simple zeros and hence chaos of the horseshoe type if

$$\frac{\gamma}{\delta} > \frac{2\sqrt{2}\beta^{3/2}}{3\omega\sqrt{\alpha}} \cosh\left(\frac{\pi\omega}{2\sqrt{\beta}}\right) \qquad (2.9.13)$$

and ϵ is small.   ♦

Example B. Another interesting example, due to Montgomery [1985], concerns the equations for superfluid ^{3}He. These are the Leggett equations, and we shall confine ourselves to what is called the A phase for simplicity (see Montgomery's paper for additional results). The equations are

$$\dot{s} = -\frac{1}{2}\left(\frac{\chi\Omega^2}{\gamma^2}\right)\sin 2\theta$$

and

$$\dot{\theta} = \left(\frac{\gamma^2}{\chi}\right)s - \epsilon\left(\gamma B \sin\omega t + \frac{1}{2}\Gamma\sin 2\theta\right). \qquad (2.9.14)$$

Here s is the spin, θ an angle (describing the "order parameter"), and $\gamma, \chi, \dots$ are physical constants. The homoclinic orbits for $\epsilon = 0$ are given by

$$\bar{\theta}_\pm = 2\tan^{-1}(e^{\pm\Omega t}) - \frac{\pi}{2} \quad \text{and} \quad \bar{s}_\pm = \pm 2\frac{\Omega e^{\pm 2\Omega t}}{1 + e^{\pm 2\Omega t}}. \qquad (2.9.15)$$

One calculates the Poincaré–Melnikov function to be

$$M_\pm(t_0) = \mp\frac{\pi\chi\omega B}{8\gamma} \operatorname{sech}\left(\frac{\omega\pi}{2\Omega}\right) \cos\omega t - \frac{2}{3}\frac{\chi}{\gamma^2}\Omega\Gamma, \qquad (2.9.16)$$

so that (2.9.14) has chaos in the sense of horseshoes if

$$\frac{\gamma B}{\Gamma} > \frac{16}{3\pi}\frac{\Omega}{\omega} \cosh\left(\frac{\pi\omega}{2\Omega}\right) \qquad (2.9.17)$$

and if ϵ is small.   ♦

For references and information on higher-dimensional versions of the method and applications, see Wiggins [1988]. We shall comment on some aspects of this shortly. There is even a version of the Poincaré–Melnikov method applicable to PDEs (due to Holmes and Marsden [1981]). One basically still uses formula (2.9.9) where $X_0 \times X_1$ is replaced by the symplectic pairing between X_0 and X_1. However, there are two new difficulties in addition to standard technical analytic problems that arise with PDEs. The first is that there is a serious problem with resonances. This can be dealt with using the aid of damping. Second, the problem seems to be *not* reducible to two dimensions: The horseshoe involves all the modes. Indeed, the higher modes do seem to be involved in the physical buckling processes for the beam model discussed next.

Example C.   A PDE model for a buckled forced beam is

$$\ddot{w} + w''' + \Gamma w' - \kappa \left(\int_0^1 [w']^2 \, dz \right) w'' = \epsilon (f \cos \omega t - \delta \dot{w}), \qquad (2.9.18)$$

where $w(z,t)$, $0 \le z \le 1$, describes the deflection of the beam,

$$\dot{} = \partial/\partial t, \quad ' = \partial/\partial z,$$

and $\Gamma, \kappa, \ldots$ are physical constants. For this case, one finds that if

(i) $\pi^2 < \Gamma < 4\rho^3$ (first mode is buckled),

(ii) $j^2 \pi^2 (j^2 \pi^2 - \Gamma) \ne \omega^2$, $j = 2, 3, \ldots$ (resonance condition),

(iii) $\dfrac{f}{\delta} > \dfrac{\pi(\Gamma - \pi^2)}{2\omega\sqrt{\kappa}} \cosh\left(\dfrac{\omega}{2\sqrt{\Gamma - \omega^2}} \right)$ (transversal zeros for $M(t_0)$),

(iv) $\delta > 0$,

and ϵ is small, then (2.9.18) has horseshoes. Experiments (see Moon [1987]) showing chaos in a forced buckled beam provided the motivation that led to the study of (2.9.18).   ♦

   This kind of result can also be used for a study of chaos in a van der Waals fluid (Slemrod and Marsden [1985]) and for soliton equations (see Birnir [1986], Ercolani, Forest, and McLaughlin [1990], and Birnir and Grauer [1994]). For example, in the damped, forced sine–Gordon equation one has chaotic transitions between breathers and kink-antikink pairs, and in the Benjamin–Ono equation one can have chaotic transitions between solutions with different numbers of poles.

More Degrees of Freedom.   For Hamiltonian systems with two-degrees-of-freedom, Holmes and Marsden [1982a] show how the Melnikov method may be used to prove the existence of horseshoes on energy surfaces in nearly integrable systems. The class of systems studied have a Hamiltonian of the form

$$H(q, p, \theta, I) = F(q, p) + G(I) + \epsilon H_1(q, p, \theta, I) + O(\epsilon^2), \qquad (2.9.19)$$

where (θ, I) are action-angle coordinates for the oscillator G; we assume that $G(0) = 0$, $G' > 0$. It is also assumed that F has a homoclinic orbit

$$\bar{x}(t) = (\bar{q}(t), \bar{p}(t))$$

and that

$$M(t_0) = \int_{-\infty}^{\infty} \{F, H_1\} \, dt; \qquad (2.9.20)$$

the integral taken along $(\bar{x}(t - t_0), \Omega t, I)$ has simple zeros. Then (2.9.19) has horseshoes on energy surfaces near the surface corresponding to the homoclinic orbit and small I; the horseshoes are taken relative to a Poincaré map strobed to the oscillator G. The paper by Holmes and Marsden [1982a] also studies the effect of positive and negative damping. These results are related to those for forced one-degree-of-freedom systems, since one can often reduce a two-degrees-of-freedom Hamiltonian system to a one-degree-of-freedom forced system.

For some systems in which the variables do not split as in (2.9.19), such as a nearly symmetric heavy top, one needs to exploit a symmetry of the system, and this complicates the situation to some extent. The general theory for this is given in Holmes and Marsden [1983] and was applied to show the existence of horseshoes in the nearly symmetric heavy top; see also some closely related results of Ziglin [1980a].

This theory has been used by Ziglin [1980b] and Koiller [1985] in vortex dynamics, for example, to give a proof of the nonintegrability of the restricted four-vortex problem. Koiller, Soares, and Melo Neto [1985] give applications to the dynamics of general relativity showing the existence of horseshoes in Bianchi IX models. See Oh, Sreenath, Krishnaprasad, and Marsden [1989] for applications to the dynamics of coupled rigid bodies.

Arnold [1964] extended the Poincaré–Melnikov theory to systems with several degrees of freedom. In this case the transverse homoclinic manifolds are based on KAM tori and allow the possibility of chaotic drift from one torus to another. This drift, sometimes known as **Arnold diffusion**, is a much studied topic in Hamiltonian systems, but its theoretical foundations are still the subject of much study.

Instead of a single Melnikov function, in the multidimensional case one has a **Melnikov vector** given schematically by

$$\mathbf{M} = \begin{pmatrix} \int_{-\infty}^{\infty} \{H_0, H_1\} \, dt \\ \int_{-\infty}^{\infty} \{I_1, H_1\} \, dt \\ \cdots \\ \int_{-\infty}^{\infty} \{I_n, H_1\} \, dt \end{pmatrix}, \qquad (2.9.21)$$

where $I_1, \ldots, I_n$ are integrals for the unperturbed (completely integrable) system and where $\mathbf{M}$ depends on t_0 and on angles conjugate to $I_1, \ldots, I_n$.

One requires **M** to have transversal zeros in the vector sense. This result was given by Arnold for forced systems and was extended to the autonomous case by Holmes and Marsden [1982b, 1983]; see also Robinson [1988]. These results apply to systems such as a pendulum coupled to several oscillators and the many-vortex problems. It has also been used in power systems by Salam, Marsden, and Varaiya [1983], building on the horseshoe case treated by Kopell and Washburn [1982]. See also Salam and Sastry [1985]. There have been a number of other directions of research on these techniques. For example, Gruendler [1985] developed a multidimensional version applicable to the spherical pendulum, and Greenspan and Holmes [1983] showed how the Melnikov method can be used to study subharmonic bifurcations. See Wiggins [1988] for more information.

Poincaré and Exponentially Small Terms. In his celebrated memoir on the three-body problem, Poincaré [1890] introduced the mechanism of transversal intersection of separatrices that obstructs the integrability of the system of equations for the three-body problem as well as preventing the convergence of associated series expansions for the solutions. This idea has been developed by Birkhoff and Smale using the horseshoe construction to describe the resulting chaotic dynamics. However, in the region of phase space studied by Poincaré, it has never been proved (except in some generic sense that is not easy to interpret in specific cases) that the equations really are nonintegrable. In fact, Poincaré himself traced the difficulty to the presence of terms in the separatrix splitting that are exponentially small. A crucial component of the measure of the splitting is given by the following formula of Poincaré [1890, p. 223]:

$$J = \frac{-8\pi i}{\exp\left(\frac{\pi}{\sqrt{2\mu}}\right) + \exp\left(-\frac{\pi}{\sqrt{2\mu}}\right)},$$

which is exponentially small (also said to be beyond all orders) in μ. Poincaré was aware of the difficulties that this exponentially small behavior causes; on page 224 of his article, he states, "En d'autres termes, si on regarde μ comme un infiniment petit du premier ordre, la distance BB', sans être nulle, est un infiniment petit d'ordre infini. C'est ainsi que la fonction $e^{-1/\mu}$ est un infiniment petit d'ordre infini sans être nulle ... Dans l'example particulier que nous avons traité plus haut, la distance BB' est du mème ordre de grandeur que l'integral J, c'est à dire que $\exp(-\pi/\sqrt{2\mu})$."

This is a serious difficulty that arises when one uses the Melnikov method near an elliptic fixed point in a Hamiltonian system or in bifurcation problems giving birth to homoclinic orbits. The difficulty is related to those described by Poincaré. Near elliptic points, one sees homoclinic orbits in normal forms, and after a temporal rescaling this leads to a rapidly oscillatory perturbation that is modeled by the following variation of the pendulum equation:

$$\ddot{\phi} + \sin\phi = \epsilon \cos\left(\frac{\omega t}{\epsilon}\right). \tag{2.9.22}$$

If one-formally computes $M(t_0)$, one obtains

$$M(t_0, \epsilon) = \pm 2\pi \operatorname{sech}\left(\frac{\pi\omega}{2\epsilon}\right)\cos\left(\frac{\omega t_0}{\epsilon}\right). \tag{2.9.23}$$

While this has simple zeros, the proof of the Poincaré–Melnikov theorem is no longer valid, since $M(t_0, \epsilon)$ is now of order $\exp(-\pi/(2\epsilon))$, and the error analysis in the proof gives errors only of order ϵ^2. In fact, no expansion in powers of ϵ can detect exponentially small terms like $\exp(-\pi/(2\epsilon))$.

Holmes, Marsden, and Scheurle [1988] and Delshams and Seara [1991] show that (2.9.22) has chaos that is, in a suitable sense, *exponentially small* in ϵ. The idea is to expand expressions for the stable and unstable manifolds in a Perron type series whose terms are of order $\epsilon^k \exp(-\pi/(2\epsilon))$. To do so, the extension of the system to complex time plays a crucial role. One can hope that since such results for (2.9.22) can be proved, it may be possible to return to Poincaré's 1890 work and complete the arguments he left unfinished. In fact, the existence of these exponentially small phenomena is one reason that the problem of Arnold diffusion is both hard and delicate.

To illustrate how exponentially small phenomena enter bifurcation problems, consider the problem of a Hamiltonian saddle node bifurcation

$$\ddot{x} + \mu x + x^2 = 0 \tag{2.9.24}$$

with the addition of higher-order terms and forcing:

$$\ddot{x} + \mu x + x^2 + \text{h.o.t.} = \delta f(t). \tag{2.9.25}$$

The phase portrait of (2.9.24) is shown in Figure 2.9.1.

The system (2.9.24) is Hamiltonian with

$$H(x, \dot{x}) = \frac{1}{2}\dot{x}^2 + \frac{1}{2}\mu x^2 + \frac{1}{3}x^3. \tag{2.9.26}$$

Let us first consider the system without higher-order terms:

$$\ddot{x} + \mu x + x^2 = \delta f(t). \tag{2.9.27}$$

To study it, we rescale to blow up the singularity; let

$$x(t) = \lambda \xi(\tau), \tag{2.9.28}$$

where $\lambda = |\mu|$ and $\tau = t\sqrt{\lambda}$. Letting $' = d/d\tau$, we get

$$\xi'' - \xi + \xi^2 = \frac{\delta}{\mu^2} f\left(\frac{\tau}{\sqrt{-\mu}}\right), \quad \mu < 0,$$

$$\xi'' + \xi + \xi^2 = \frac{\delta}{\mu^2} f\left(\frac{\tau}{\sqrt{\mu}}\right), \quad \mu > 0. \tag{2.9.29}$$

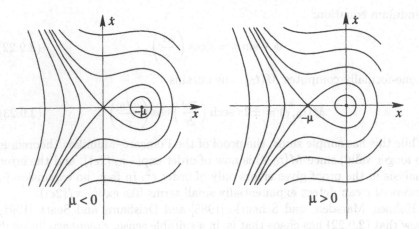

FIGURE 2.9.1. Phase portraits of $\ddot{x} + \mu x + x^2 = 0$.

The exponentially small estimates of Holmes, Marsden, and Scheurle [1988] apply to (2.9.29). One gets exponentially small upper and lower estimates in certain algebraic sectors of the (δ, μ) plane that depend on the nature of f. The estimates for the splitting have the form $C(\delta/\mu^2)\exp(-\pi/\sqrt{|\mu|})$. Now consider

$$\ddot{x} + \mu x + x^2 + x^3 = \delta f(t). \tag{2.9.30}$$

With $\delta = 0$, there are equilibria at the three points with $\dot{x} = 0$ and

$$x = 0,\ -r,\ \text{and}\ -\frac{\mu}{r}, \tag{2.9.31}$$

where

$$r = \frac{1 + \sqrt{1 - 4\mu}}{2}, \tag{2.9.32}$$

which is approximately 1 when $\mu \approx 0$. The phase portrait of (2.9.30) with $\delta = 0$ and $\mu = -1/2$ is shown in Figure 2.9.2. As μ passes through 0, the small lobe in Figure 2.9.2 undergoes the same bifurcation as in Figure 2.9.1, with the large lobe changing only slightly.

Again we rescale, to give

$$\ddot{\xi} - \xi + \xi^2 - \mu\xi^3 = \frac{\delta}{\mu^2} f\left(\frac{\tau}{\sqrt{-\mu}}\right),\quad \mu < 0,$$

$$\ddot{\xi} + \xi + \xi^2 + \mu\xi^3 = \frac{\delta}{\mu^2} f\left(\frac{\tau}{\sqrt{\mu}}\right),\quad \mu > 0. \tag{2.9.33}$$

Notice that for $\delta = 0$, the phase portrait is μ-dependent. The homoclinic orbit surrounding the small lobe for $\mu < 0$ is given explicitly in terms of ξ

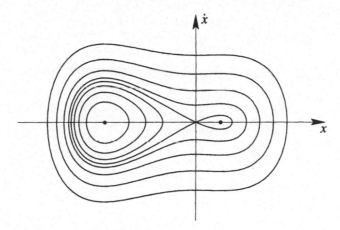

FIGURE 2.9.2. The phase portrait of $\ddot{x} - \frac{1}{2}x + x^2 + x^3 = 0$.

by

$$\xi(\tau) = \frac{4e^\tau}{\left(e^\tau + \frac{2}{3}\right)^2 - 2\mu}, \qquad (2.9.34)$$

which is μ-dependent. An interesting technicality is that without the cubic term, we get μ-independent *double* poles at $t = \pm i\pi + \log 2 - \log 3$ in the complex τ-plane, while (2.9.34) has a pair of simple poles that splits these double poles to the pairs of simple poles at

$$\tau = \pm i\pi + \log\left(\frac{2}{3} \pm i\sqrt{2\lambda}\right), \qquad (2.9.35)$$

where again $\lambda = |\mu|$. (There is no particular significance to the real part, such as $\log 2 - \log 3$ in the case of no cubic term; this can always be gotten rid of by a shift in the base point $\xi(0)$.)

If a quartic term x^4 is added, these pairs of simple poles will split into quartets of branch points, and so on. Thus, while the analysis of higher-order terms has this interesting μ-dependence, it seems that the basic exponential part of the estimates, namely

$$\exp\left(-\frac{\pi}{\sqrt{|\mu|}}\right), \qquad (2.9.36)$$

remains intact.

3
An Introduction to
Infinite-Dimensional Systems

A common choice of configuration space for classical field theory is an infinite-dimensional vector space of functions or tensor fields on space or spacetime, the elements of which are called *fields*. Here we relate our treatment of infinite-dimensional Hamiltonian systems discussed in §2.1 to classical Lagrangian and Hamiltonian field theory and then give examples. Classical field theory is a large subject with many aspects not covered here; we treat only a few topics that are basic to subsequent developments; see Chapters 6 and 7 for additional information and references.

3.1 Lagrange's and Hamilton's Equations for Field Theory

As with finite-dimensional systems, one can begin with a Lagrangian and a variational principle, and then pass to the Hamiltonian via the Legendre transformation. At least formally, all the constructions we did in the finite-dimensional case go over to the infinite-dimensional one.

For instance, suppose we choose our configuration space $Q = \mathcal{F}(\mathbb{R}^3)$ to be the space of fields φ on $\mathbb{R}^3$. Our Lagrangian will be a function $L(\varphi, \dot{\varphi})$ from $Q \times Q$ to $\mathbb{R}$. The variational principle is

$$\delta \int_a^b L(\varphi, \dot{\varphi}) \, dt = 0, \qquad (3.1.1)$$

which is equivalent to the Euler–Lagrange equations

$$\frac{d}{dt}\frac{\delta L}{\delta\dot{\varphi}} = \frac{\delta L}{\delta\varphi} \tag{3.1.2}$$

in the usual way. Here,

$$\pi = \frac{\delta L}{\delta\dot{\varphi}} \tag{3.1.3}$$

is the conjugate momentum, which we regard as a density on $\mathbb{R}^3$ as in Chapter 2. The corresponding Hamiltonian is

$$H(\varphi,\pi) = \int \pi\dot{\varphi} - L(\varphi,\dot{\varphi}), \tag{3.1.4}$$

in accordance with our general theory. We also know that the Hamiltonian should generate the canonical Hamilton equations. We verify this now.

Proposition 3.1.1. *Let $Z = \mathcal{F}(\mathbb{R}^3) \times \mathrm{Den}(\mathbb{R}^3)$, with Ω defined as in Example (b) of §2.2. Then the Hamiltonian vector field $X_H : Z \to Z$ corresponding to a given energy function $H : Z \to \mathbb{R}$ is given by*

$$X_H = \left(\frac{\delta H}{\delta\pi}, -\frac{\delta H}{\delta\varphi}\right). \tag{3.1.5}$$

Hamilton's equations on Z are

$$\frac{\partial\varphi}{\partial t} = \frac{\delta H}{\delta\pi}, \quad \frac{\partial\pi}{\partial t} = -\frac{\delta H}{\delta\varphi}. \tag{3.1.6}$$

Remarks.

1. The symbols $\mathcal{F}$ and Den stand for function spaces included in the space of all functions and densities, chosen to be appropriate to the functional-analytic needs of the particular problem. In practice this often means, among other things, that appropriate conditions at infinity are imposed to permit integration by parts.

2. The equations of motion for a curve $z(t) = (\varphi(t),\pi(t))$ written in the form $\Omega(dz/dt,\delta z) = \mathbf{d}H(z(t)) \cdot \delta z$ for all $\delta z \in Z$ with compact support are called the **weak form of the equations of motion**. They can still be valid when there is not enough smoothness or decay at infinity to justify the literal equality $dz/dt = X_H(z)$; this situation can occur, for example, if one is considering shock waves. ◆

Proof of Proposition 3.1.1. To derive the partial functional derivatives, we use the natural pairing

$$\langle,\rangle : \mathcal{F}(\mathbb{R}^3) \times \mathrm{Den}(\mathbb{R}^3) \to \mathbb{R}, \quad \text{where} \quad \langle\varphi,\pi\rangle = \int \varphi\pi' \, d^3x, \tag{3.1.7}$$

where we write $\pi = \pi' d^3 x \in$ Den. Recalling that $\delta H / \delta \varphi$ is a density, let

$$X = \left(\frac{\delta H}{\delta \pi}, -\frac{\delta H}{\delta \varphi} \right).$$

We need to verify that $\Omega(X(\varphi, \pi), (\delta \varphi, \delta \pi)) = \mathbf{d}H(\varphi, \pi) \cdot (\delta \varphi, \delta \pi)$. Indeed,

$$\begin{aligned}
\Omega(X(\varphi, \pi), (\delta \varphi, \delta \pi)) &= \Omega \left(\left(\frac{\delta H}{\delta \pi}, -\frac{\delta H}{\delta \varphi} \right), (\delta \varphi, \delta \pi) \right) \\
&= \int \frac{\delta H}{\delta \pi} (\delta \pi)' d^3 x + \int \delta \varphi \left(\frac{\delta H}{\delta \varphi} \right)' d^3 x \\
&= \left\langle \frac{\delta H}{\delta \pi}, \delta \pi \right\rangle + \left\langle \delta \varphi, \frac{\delta H}{\delta \varphi} \right\rangle \\
&= \mathbf{D}_\pi H(\varphi, \pi) \cdot \delta \pi + \mathbf{D}_\varphi H(\varphi, \pi) \cdot \delta \varphi \\
&= \mathbf{d}H(\varphi, \pi) \cdot (\delta \varphi, \delta \pi). \qquad \blacksquare
\end{aligned}$$

3.2 Examples: Hamilton's Equations

(a) The Wave Equation. Consider $Z = \mathcal{F}(\mathbb{R}^3) \times \text{Den}(\mathbb{R}^3)$ as above. Let φ denote the configuration variable, that is, the first component in the phase space $\mathcal{F}(\mathbb{R}^3) \times \text{Den}(\mathbb{R}^3)$, and interpret φ as a measure of the displacement from equilibrium of a homogeneous elastic medium. Writing $\pi' = \rho \, d\varphi / dt$, where ρ is the mass density, the **kinetic energy** is

$$T = \frac{1}{2} \int \frac{1}{\rho} [\pi']^2 \, d^3 x.$$

For small displacements φ, one assumes a linear restoring force such as the one given by the **potential energy**

$$\frac{k}{2} \int \|\nabla \varphi\|^2 \, d^3 x,$$

for an (elastic) constant k.

Because we are considering a homogeneous medium, ρ and k are constants, so let us work in units in which they are unity. Nonlinear effects can be modeled in a naive way by introducing a nonlinear term, $U(\varphi)$, into the potential. However, for an elastic medium one really should use constitutive relations based on the principles of continuum mechanics; see Marsden and Hughes [1983]. For the naive model, the Hamiltonian $H : Z \to \mathbb{R}$ is the **total energy**

$$H(\varphi, \pi) = \int \left[\frac{1}{2} (\pi')^2 + \frac{1}{2} \|\nabla \varphi\|^2 + U(\varphi) \right] d^3 x. \qquad (3.2.1)$$

Using the definition of the functional derivative, we find that

$$\frac{\delta H}{\delta \pi} = \pi', \quad \frac{\delta H}{\delta \varphi} = (-\nabla^2 \varphi + U'(\varphi))d^3 x. \tag{3.2.2}$$

Therefore, the equations of motion are

$$\frac{\partial \varphi}{\partial t} = \pi', \quad \frac{\partial \pi'}{\partial t} = \nabla^2 \varphi - U'(\varphi), \tag{3.2.3}$$

or, in second-order form,

$$\frac{\partial^2 \varphi}{\partial t^2} = \nabla^2 \varphi - U'(\varphi). \tag{3.2.4}$$

Various choices of U correspond to various physical applications. When $U' = 0$, we get the linear wave equation, with unit propagation velocity. Another choice, $U(\varphi) = (1/2)m^2\varphi^2 + \lambda\varphi^4$, occurs in the quantum theory of self-interacting mesons; the parameter m is related to the meson mass, and φ^4 governs the nonlinear part of the interaction. When $\lambda = 0$, we get

$$\nabla^2 \varphi - \frac{\partial^2 \varphi}{\partial t^2} = m^2 \varphi, \tag{3.2.5}$$

which is called the **Klein–Gordon equation**.   ♦

Technical Aside. For the wave equation, one appropriate choice of function space is $Z = H^1(\mathbb{R}^3) \times L^2_{\text{Den}}(\mathbb{R}^3)$, where $H^1(\mathbb{R}^3)$ denotes the H^1-functions on $\mathbb{R}^3$, that is, functions that, along with their first derivatives are square integrable, and $L^2_{\text{Den}}(\mathbb{R}^3)$ denotes the space of densities $\pi = \pi' d^3x$, where the function π' on $\mathbb{R}^3$ is square integrable. Note that the Hamiltonian vector field

$$X_H(\varphi, \pi) = (\pi', (\nabla^2 \varphi - U'(\varphi))d^3 x)$$

is defined only on the dense subspace $H^2(\mathbb{R}^3) \times H^1_{\text{Den}}(\mathbb{R}^3)$ of Z. This is a common occurrence in the study of Hamiltonian partial differential equations; we return to this in §3.3.   ♦

In the preceding example, Ω was given by the canonical form with the result that the equations of motion were in the standard form (3.1.5). In addition, the Hamiltonian function was given by the actual energy of the system under consideration. We now give examples in which these statements require reinterpretation but that nevertheless fall into the framework of the general theory developed so far.

(b) The Schrödinger Equation. Let $\mathcal{H}$ be a complex Hilbert space, for example, the space of complex-valued functions ψ on $\mathbb{R}^3$ with the Hermitian inner product

$$\langle \psi_1, \psi_2 \rangle = \int \psi_1(x)\overline{\psi}_2(x)\, d^3x,$$

where the overbar denotes complex conjugation. For a self-adjoint complex-linear operator $H_{\text{op}} : \mathcal{H} \to \mathcal{H}$, the Schrödinger equation is

$$i\hbar \frac{\partial \psi}{\partial t} = H_{\text{op}}\psi, \tag{3.2.6}$$

where $\hbar$ is Planck's constant. Define

$$A = \frac{-i}{\hbar} H_{\text{op}},$$

so that the Schrödinger equation becomes

$$\frac{\partial \psi}{\partial t} = A\psi. \tag{3.2.7}$$

The symplectic form on $\mathcal{H}$ is given by $\Omega(\psi_1, \psi_2) = -2\hbar \operatorname{Im} \langle \psi_1, \psi_2 \rangle$. Self-adjointness of H_{op} is a condition stronger than symmetry and is essential for proving well-posedness of the initial-value problem for (3.2.6); for an exposition, see, for instance, Abraham, Marsden, and Ratiu [1988]. Historically, it was Kato [1950] who established self-adjointness for important problems such as the hydrogen atom.

From §2.5 we know that since H_{op} is symmetric, A is Hamiltonian. The Hamiltonian is

$$H(\psi) = \hbar \langle iA\psi, \psi \rangle = \langle H_{\text{op}}\psi, \psi \rangle, \tag{3.2.8}$$

which is the **expectation value** of H_{op} at ψ, defined by $\langle H_{\text{op}} \rangle (\psi) = \langle H_{\text{op}}\psi, \psi \rangle$. ♦

(c) The Korteweg–de Vries (KdV) Equation. Denote by Z the vector subspace $\mathcal{F}(\mathbb{R})$ consisting of those functions u with $|u(x)|$ decreasing sufficiently fast as $x \to \pm\infty$ that the integrals we will write are defined and integration by parts is justified. As we shall see later, the Poisson brackets for the KdV equation are quite simple, and historically they were found first (see Gardner [1971] and Zakharov [1971, 1974]). To be consistent with our exposition, we begin with the somewhat more complicated symplectic structure. Pair Z with itself using the L^2 inner product. Let the KdV symplectic structure Ω be defined by

$$\Omega(u_1, u_2) = \frac{1}{2}\left(\int_{-\infty}^{\infty} [\hat{u}_1(x)u_2(x) - \hat{u}_2(x)u_1(x)]\, dx \right), \tag{3.2.9}$$

where $\hat{u}$ denotes a primitive of u, that is,

$$\hat{u} = \int_{-\infty}^{x} u(y)\, dy.$$

In §8.5 we shall see a way to *construct* this form. The form Ω is clearly skew-symmetric. Note that if $u_1 = \partial v/\partial x$ for some $v \in Z$, then

$$\int_{-\infty}^{\infty} \hat{u}_2(x)u_1(x)\, dx$$

$$= \int_{-\infty}^{\infty} \hat{u}_2(x)\frac{\partial \hat{u}_1(x)}{\partial x}\, dx$$

$$= \hat{u}_1(x)\hat{u}_2(x)\Big|_{-\infty}^{\infty} - \int_{-\infty}^{\infty} \hat{u}_1(x)u_2(x)\, dx$$

$$= \left(\int_{-\infty}^{\infty} \frac{\partial v(x)}{\partial x}\, dx\right)\left(\int_{-\infty}^{\infty} u_2(x)\, dx\right) - \int_{-\infty}^{\infty} \hat{u}_1(x)u_2(x)\, dx$$

$$= \left(v(x)\Big|_{-\infty}^{\infty}\right)\left(\int_{-\infty}^{\infty} u_2(x)\, dx\right) - \int_{-\infty}^{\infty} \hat{u}_1(x)u_2(x)\, dx$$

$$= -\int_{-\infty}^{\infty} \hat{u}_1(x)u_2(x)\, dx.$$

Thus, if $u_1(x) = \partial v(x)/\partial x$, then Ω can be written as

$$\Omega(u_1, u_2) = \int_{-\infty}^{\infty} \hat{u}_1(x)u_2(x)\, dx = \int_{-\infty}^{\infty} v(x)u_2(x)\, dx. \qquad (3.2.10)$$

To prove weak nondegeneracy of Ω, we check that if $v \neq 0$, there is a w such that $\Omega(w,v) \neq 0$. Indeed, if $v \neq 0$ and we let $w = \partial v/\partial x$, then $w \neq 0$ because $v(x) \to 0$ as $|x| \to \infty$. Hence by (3.2.10),

$$\Omega(w,v) = \Omega\left(\frac{\partial v}{\partial x}, v\right) = \int_{-\infty}^{\infty} (v(x))^2\, dx \neq 0.$$

Suppose that a Hamiltonian $H : Z \to \mathbb{R}$ is given. We claim that the corresponding Hamiltonian vector field X_H is given by

$$X_H(u) = \frac{\partial}{\partial x}\left(\frac{\delta H}{\delta u}\right). \qquad (3.2.11)$$

Indeed, by (3.2.10),

$$\Omega(X_H(v), w) = \int_{-\infty}^{\infty} \frac{\delta H}{\delta v}(x)w(x)\, dx = \mathbf{d}H(v) \cdot w.$$

It follows from (3.2.11) that the corresponding Hamilton equations are

$$u_t = \frac{\partial}{\partial x}\left(\frac{\delta H}{\delta u}\right), \qquad (3.2.12)$$

where, in (3.2.12) and in the following, subscripts denote derivatives with respect to the subscripted variable. As a special case, consider the function

$$H_1(u) = -\frac{1}{6} \int_{-\infty}^{\infty} u^3 \, dx.$$

Then

$$\frac{\partial}{\partial x} \frac{\delta H_1}{\delta u} = -uu_x,$$

and so (3.2.12) becomes the *one-dimensional transport equation*

$$u_t + uu_x = 0. \tag{3.2.13}$$

Next, let

$$H_2(u) = \int_{-\infty}^{\infty} \left(\frac{1}{2} u_x^2 - u^3\right) dx; \tag{3.2.14}$$

then (3.2.12) becomes

$$u_t + 6uu_x + u_{xxx} = 0. \tag{3.2.15}$$

This is the *Korteweg–de Vries (KdV) equation* that describes shallow water waves. For a concise presentation of its famous complete set of integrals, see Abraham and Marsden [1978], §6.5, and for more information, see Newell [1985]. The first few of its integrals are given in Exercise 3.3-1. We will return to this example from time to time in the text, but for now we will find traveling wave solutions of the KdV equation.

Traveling Waves. If we look for traveling wave solutions of (3.2.15), that is, $u(x,t) = \varphi(x - ct)$, for a constant $c > 0$ and a positive function φ, we see that u satisfies the KdV equation if and only if φ satisfies

$$c\varphi' - 6\varphi\varphi' - \varphi''' = 0. \tag{3.2.16}$$

Integrating once gives

$$c\varphi - 3\varphi^2 - \varphi'' = C, \tag{3.2.17}$$

where C is a constant. This equation is Hamiltonian in the canonical variables (φ, φ') with Hamiltonian function

$$h(\varphi, \varphi') = \frac{1}{2}(\varphi')^2 - \frac{c}{2}\varphi^2 + \varphi^3 + C\varphi. \tag{3.2.18}$$

From conservation of energy, $h(\varphi, \varphi') = D$, it follows that

$$\varphi' = \pm\sqrt{c\varphi^2 - 2\varphi^3 - 2C\varphi + 2D}, \tag{3.2.19}$$

or, writing $s = x - ct$, we get

$$s = \pm \int \frac{d\varphi}{\sqrt{c\varphi^2 - 2\varphi^3 - 2C\varphi + 2D}}. \qquad (3.2.20)$$

We seek solutions that together with their derivatives vanish at $\pm\infty$. Then (3.2.17) and (3.2.19) give $C = D = 0$, so

$$s = \pm \int \frac{d\varphi}{\sqrt{c\varphi^2 - 2\varphi^3}} = \pm \frac{1}{\sqrt{c}} \log \left| \frac{\sqrt{c - 2\varphi} - \sqrt{c}}{\sqrt{c - 2\varphi} + \sqrt{c}} \right| + K \qquad (3.2.21)$$

for some constant K that will be determined below.

For $C = D = 0$, the Hamiltonian (3.2.18) becomes

$$h(\varphi, \varphi') = \frac{1}{2}(\varphi')^2 - \frac{c}{2}\varphi^2 + \varphi^3, \qquad (3.2.22)$$

and thus the two equilibria given by $\partial h / \partial \varphi = 0$ and $\partial h / \partial \varphi' = 0$ are $(0,0)$ and $(c/3, 0)$. The matrix of the linearized Hamiltonian system at these equilibria is

$$\begin{bmatrix} 0 & 1 \\ \pm c & 0 \end{bmatrix},$$

which shows that $(0,0)$ is a saddle and $(c/3, 0)$ is spectrally stable. The second variation criterion on the potential energy (see §1.10) $-c\varphi^2/2 + \varphi^3$ at $(c/3, 0)$ shows that this equilibrium is stable. Thus, if $(\varphi(s), \varphi'(s))$ is a homoclinic orbit emanating and ending at $(0,0)$, the value of the Hamiltonian function (3.2.22) on it is $H(0,0) = 0$. From (3.2.22) it follows that $(c/2, 0)$ is a point on this homoclinic orbit, and thus (3.2.20) for $C = D = 0$ is its expression. Taking the initial condition of this orbit at $s = 0$ to be $\varphi(0) = c/2$, $\varphi'(0) = 0$, (3.2.21) forces $K = 0$, and so

$$\left| \frac{\sqrt{c - 2\varphi} - \sqrt{c}}{\sqrt{c - 2\varphi} + \sqrt{c}} \right| = e^{\pm\sqrt{c}s}.$$

Since $\varphi \geq 0$ by hypothesis, the expression in the absolute value is negative, and thus

$$\frac{\sqrt{c - 2\varphi} - \sqrt{c}}{\sqrt{c - 2\varphi} + \sqrt{c}} = -e^{\pm\sqrt{c}s},$$

whose solution is

$$\varphi(s) = \frac{2ce^{\pm\sqrt{c}s}}{(1 + e^{\pm\sqrt{c}s})^2} = \frac{c}{2\cosh^2(\sqrt{c}s/2)}.$$

This produces the *soliton solution*

$$u(x, t) = \frac{c}{2} \operatorname{sech}^2 \left[\frac{\sqrt{c}}{2}(x - ct) \right].$$

♦

(d) Sine–Gordon Equation.   For functions $u(x,t)$, where x and t are real variables, the *sine–Gordon equation* is $u_{tt} = u_{xx} + \sin u$. Equation (3.2.4) shows that it is Hamiltonian with the momentum density $\pi = u_t \, dx$ (and associated function $\pi' = u_t$),

$$H(u) = \int_{-\infty}^{\infty} \left(\frac{1}{2} u_t^2 + \frac{1}{2} u_x^2 + \cos u \right) dx, \qquad (3.2.23)$$

and the canonical bracket structure, as in the wave equation. This equation also has a complete set of integrals; see again Newell [1985]. ◆

(e) Abstract Wave Equation.   Let $\mathcal{H}$ be a real Hilbert space and $B : \mathcal{H} \to \mathcal{H}$ a linear operator. On $\mathcal{H} \times \mathcal{H}$ put the symplectic structure Ω given by (2.2.6). One can check that:

(i) $A = \begin{bmatrix} 0 & I \\ -B & 0 \end{bmatrix}$ is Ω-skew if and only if B is a symmetric operator on $\mathcal{H}$; and

(ii) if B is symmetric, then a Hamiltonian for A is

$$H(x, y) = \frac{1}{2} (\|y\|^2 + \langle Bx, x \rangle). \qquad (3.2.24)$$

The equations of motion (2.4.10) give the *abstract wave equation*

$$\ddot{x} + Bx = 0. \qquad ◆$$

(f) Linear Elastodynamics.   On $\mathbb{R}^3$ consider the equations

$$\rho \mathbf{u}_{tt} = \operatorname{div}(\mathbf{c} \cdot \nabla \mathbf{u}),$$

that is,

$$\rho u_{tt}^i = \frac{\partial}{\partial x^j} \left[c^{ijkl} \frac{\partial u^k}{\partial x^l} \right], \qquad (3.2.25)$$

where ρ is a positive function and $\mathbf{c}$ is a fourth-order tensor field (the *elasticity tensor*) on $\mathbb{R}^3$ with the symmetries $c^{ijkl} = c^{klij} = c^{jikl}$.
On $\mathcal{F}(\mathbb{R}^3; \mathbb{R}^3) \times \mathcal{F}(\mathbb{R}^3; \mathbb{R}^3)$ (or, more precisely, on

$$H^1(\mathbb{R}^3; \mathbb{R}^3) \times L^2(\mathbb{R}^3; \mathbb{R}^3)$$

with suitable decay properties at infinity) define

$$\Omega((\mathbf{u}, \dot{\mathbf{u}}), (\mathbf{v}, \dot{\mathbf{v}})) = \int_{\mathbb{R}^3} \rho(\dot{\mathbf{v}} \cdot \mathbf{u} - \dot{\mathbf{u}} \cdot \mathbf{v}) \, d^3 x. \qquad (3.2.26)$$

The form Ω is the canonical symplectic form (2.2.3) for fields $\mathbf{u}$ and their conjugate momenta $\pi = \rho\dot{\mathbf{u}}$.

On the space of functions $\mathbf{u} : \mathbb{R}^3 \to \mathbb{R}^3$, consider the ρ-weighted L^2-inner product

$$\langle \mathbf{u}, \mathbf{v} \rangle_\rho = \int_{\mathbb{R}^3} \rho \mathbf{u} \cdot \mathbf{v} \, d^3 x. \qquad (3.2.27)$$

Then the operator $B\mathbf{u} = -(1/\rho)\operatorname{div}(\mathbf{c} \cdot \nabla \mathbf{u})$ is symmetric with respect to this inner product, and thus by Example (e) above, the operator $A(\mathbf{u}, \dot{\mathbf{u}}) = (\dot{\mathbf{u}}, (1/\rho)\operatorname{div}(\mathbf{c} \cdot \nabla \mathbf{u}))$ is Ω-skew.

The equations (3.2.25) of linear elastodynamics are checked to be Hamiltonian with respect to Ω given by (3.2.26), and with energy

$$H(\mathbf{u}, \dot{\mathbf{u}}) = \frac{1}{2} \int \rho \|\dot{\mathbf{u}}\|^2 \, d^3 x + \frac{1}{2} \int c^{ijkl} e_{ij} e_{kl} \, d^3 x, \qquad (3.2.28)$$

where

$$e_{ij} = \frac{1}{2} \left(\frac{\partial u^i}{\partial x^j} + \frac{\partial u^j}{\partial x^i} \right). \qquad \blacklozenge$$

Exercises

$\diamond$ **3.2-1.**

(a) Let $\varphi : \mathbb{R}^{n+1} \to \mathbb{R}$. Show directly that the sine–Gordon equation

$$\frac{\partial^2 \varphi}{\partial t^2} - \nabla^2 \varphi + \sin \varphi = 0$$

is the Euler–Lagrange equation of a suitable Lagrangian.

(b) Let $\varphi : \mathbb{R}^{n+1} \to \mathbb{C}$. Write the nonlinear Schrödinger equation

$$i \frac{\partial \varphi}{\partial t} + \nabla^2 \varphi + \beta \varphi |\varphi|^2 = 0$$

as a Hamiltonian system.

$\diamond$ **3.2-2.** Find a "soliton" solution for the sine–Gordon equation

$$\frac{\partial^2 \varphi}{\partial t^2} - \frac{\partial^2 \varphi}{\partial x^2} + \sin \varphi = 0$$

in one spatial dimension.

⋄ **3.2-3.** Consider the complex nonlinear Schrödinger equation in one spatial dimension:

$$i\frac{\partial \varphi}{\partial t} + \frac{\partial^2 \varphi}{\partial x^2} + \beta\varphi|\varphi|^2 = 0, \quad \beta \neq 0.$$

(a) Show that the function $\psi : \mathbb{R} \to \mathbb{C}$ defining the traveling wave solution $\varphi(x,t) = \psi(x - ct)$ for $c > 0$ satisfies a second-order complex differential equation equivalent to a Hamiltonian system in $\mathbb{R}^4$ relative to the noncanonical symplectic form whose matrix is given by

$$\mathbb{J}_c = \begin{bmatrix} 0 & c & 1 & 0 \\ -c & 0 & 0 & 1 \\ -1 & 0 & 0 & 0 \\ 0 & -1 & 0 & 0 \end{bmatrix}.$$

(See Exercise 2.4-1.)

(b) Analyze the equilibria of the resulting Hamiltonian system in $\mathbb{R}^4$ and determine their linear stability properties.

(c) Let $\psi(s) = e^{ics/2}a(s)$ for a real function $a(s)$ and determine a second-order equation for $a(s)$. Show that the resulting equation is Hamiltonian and has heteroclinic orbits for $\beta < 0$. Find them.

(d) Find "soliton" solutions for the complex nonlinear Schrödinger equation.

3.3 Examples: Poisson Brackets and Conserved Quantities

Before proceeding with infinite-dimensional examples, it is first useful to recall some basic facts about angular momentum of particles in $\mathbb{R}^3$. (The reader should supply a corresponding discussion for linear momentum.) Consider a particle moving in $\mathbb{R}^3$ under the influence of a potential V. Let the position coordinate be denoted by $\mathbf{q}$, so that Newton's second law reads

$$m\ddot{\mathbf{q}} = -\nabla V(\mathbf{q}).$$

Let $\mathbf{p} = m\dot{\mathbf{q}}$ be the linear momentum and $\mathbf{J} = \mathbf{q} \times \mathbf{p}$ be the angular momentum. Then

$$\frac{d}{dt}\mathbf{J} = \dot{\mathbf{q}} \times \mathbf{p} + \mathbf{q} \times \dot{\mathbf{p}} = -\mathbf{q} \times \nabla V(\mathbf{q}).$$

If V is radially symmetric, it is a function of $\|\mathbf{q}\|$ alone: assume

$$V(\mathbf{q}) = f(\|\mathbf{q}\|^2),$$

where f is a smooth function (exclude $\mathbf{q} = \mathbf{0}$ if necessary). Then

$$\nabla V(\mathbf{q}) = 2f'(\|\mathbf{q}\|^2)\mathbf{q},$$

so that $\mathbf{q} \times \nabla V(\mathbf{q}) = \mathbf{0}$. Thus, in this case, $d\mathbf{J}/dt = 0$, so $\mathbf{J}$ is conserved.

Alternatively, with

$$H(\mathbf{q}, \mathbf{p}) = \frac{1}{2m}\|\mathbf{p}\|^2 + V(\mathbf{q}),$$

we can check directly that $\{H, J_l\} = 0$ for $l = 1, 2, 3$, where $\mathbf{J} = (J_1, J_2, J_3)$. This also shows that each component J_l is conserved by the Hamiltonian dynamics determined by H.

Additional insight is gained by looking at the components of $\mathbf{J}$ more closely. For example, consider the scalar function

$$F(\mathbf{q}, \mathbf{p}) = \mathbf{J}(\mathbf{q}, \mathbf{p}) \cdot \omega \mathbf{k},$$

where ω is a constant and $\mathbf{k} = (0, 0, 1)$. We find that

$$F(\mathbf{q}, \mathbf{p}) = \omega(q^1 p_2 - p_1 q^2).$$

The Hamiltonian vector field of F is

$$X_F(\mathbf{q}, \mathbf{p}) = \left(\frac{\partial F}{\partial p_1}, \frac{\partial F}{\partial p_2}, \frac{\partial F}{\partial p_3}, -\frac{\partial F}{\partial q^1}, -\frac{\partial F}{\partial q^2}, -\frac{\partial F}{\partial q^3} \right)$$
$$= (-\omega q^2, \omega q^1, 0, -\omega p_2, \omega p_1, 0).$$

Note that X_F is just the vector field corresponding to the flow in the (q^1, q^2) plane and the (p_1, p_2) plane given by rotations about the origin with angular velocity ω. More generally, the Hamiltonian vector field associated with the scalar function defined by $J_\omega := \mathbf{J} \cdot \omega$, where ω is a vector in $\mathbb{R}^3$, has a flow consisting of rotations about the axis ω. As we shall see in Chapters 11 and 12, this is the basis for understanding the link between conservation laws and symmetry more generally.

Another identity is worth noting. Namely, for two vectors ω_1 and ω_2,

$$\{J_{\omega_1}, J_{\omega_2}\} = J_{\omega_1 \times \omega_2},$$

which, as we shall see later, is an important link between the Poisson bracket structure and the structure of the Lie algebra of the rotation group.

(a) **The Schrödinger Bracket.**   In Example (b) of §3.2, we saw that if H_{op} is a self-adjoint complex linear operator on a Hilbert space $\mathcal{H}$, then $A = H_{\mathrm{op}}/(i\hbar)$ is Hamiltonian, and the corresponding energy function H_A is the expectation value $\langle H_{\mathrm{op}} \rangle$ of H_{op}. Letting H_{op} and K_{op} be two such operators, and applying the Poisson bracket–commutator correspondence (2.7.10), or a direct calculation, we get

$$\{\langle H_{\mathrm{op}} \rangle, \langle K_{\mathrm{op}} \rangle\} = \langle [H_{\mathrm{op}}, K_{\mathrm{op}}] \rangle. \tag{3.3.1}$$

In other words, *the expectation value of the commutator is the Poisson bracket of the expectation values.*

Results like this lead one to statements like "Commutators in quantum mechanics are not only *analogous* to Poisson brackets, they *are* Poisson brackets." Even more striking are *true* statements like this: "Don't tell me that quantum mechanics is right and classical mechanics is wrong—after all, quantum mechanics is a *special case* of classical mechanics."

Notice that if we take $K_{op}\psi = \psi$, the identity operator, the corresponding Hamiltonian function is $p(\psi) = \|\psi\|^2$, and from (3.3.1) we see that p is a conserved quantity for any choice of H_{op}, a fact that is central to the probabilistic interpretation of quantum mechanics. Later, we shall see that p is the conserved quantity associated to the **phase symmetry** $\psi \mapsto e^{i\theta}\psi$.

More generally, if F and G are two functions on $\mathcal{H}$ with $\delta F/\delta\psi = \nabla F$, the gradient of F taken relative to the real inner product Re $\langle \, , \rangle$ on H, one finds that

$$X_F = \frac{1}{2i\hbar}\nabla F \qquad (3.3.2)$$

and

$$\{F, G\} = -\frac{1}{2\hbar}\,\text{Im}\,\langle \nabla F, \nabla G\rangle. \qquad (3.3.3)$$

Notice that (3.3.2), (3.3.3), and Im $z = -$ Re(iz) give

$$\begin{aligned}
\mathbf{d}F \cdot X_G &= \text{Re}\,\langle \nabla F, X_G\rangle = \frac{1}{2\hbar}\,\text{Re}\,\langle \nabla F, -i\nabla G\rangle \\
&= \frac{1}{2\hbar}\,\text{Re}\,\langle i\nabla F, \nabla G\rangle \\
&= -\frac{1}{2\hbar}\,\text{Im}\,\langle \nabla F, \nabla G\rangle \\
&= \{F, G\}
\end{aligned}$$

as expected.   ◆

(b) KdV Bracket. Using the definition of the bracket (2.7.1), the symplectic structure, and the Hamiltonian vector field formula from Example (c) of §3.2, one finds that

$$\{F, G\} = \int_{-\infty}^{\infty} \frac{\delta F}{\delta u}\frac{\partial}{\partial x}\left(\frac{\delta G}{\delta u}\right) dx \qquad (3.3.4)$$

for functions F, G of u having functional derivatives that vanish at $\pm\infty$.   ◆

(c) Linear and Angular Momentum for the Wave Equation. The wave equation on $\mathbb{R}^3$ discussed in Example (a) of §3.2 has the Hamiltonian

$$H(\varphi, \pi) = \int_{\mathbb{R}^3}\left[\frac{1}{2}(\pi')^2 + \frac{1}{2}\|\nabla\varphi\|^2 + U(\varphi)\right]d^3x. \qquad (3.3.5)$$

Define the **linear momentum** in the x-direction by

$$P_x(\varphi, \pi) = \int \pi' \frac{\partial \varphi}{\partial x} \, d^3 x. \tag{3.3.6}$$

By (3.3.6), $\delta P_x / \delta \pi = \partial \varphi / \partial x$, and $\delta P_x / \delta \varphi = (-\partial \pi' / \partial x) \, d^3 x$, so we get from (3.2.2)

$$
\begin{aligned}
\{H, P_x\}(\varphi, \pi) &= \int_{\mathbb{R}^3} \left(\frac{\delta P_x}{\delta \pi} \frac{\delta H}{\delta \varphi} - \frac{\delta H}{\delta \pi} \frac{\delta P_x}{\delta \varphi} \right) \\
&= \int_{\mathbb{R}^3} \left[\frac{\partial \varphi}{\partial x} (-\nabla^2 \varphi + U'(\varphi)) + \pi' \frac{\partial \pi'}{\partial x} \right] d^3 x \\
&= \int_{\mathbb{R}^3} \left[-\nabla^2 \varphi \frac{\partial \varphi}{\partial x} + \frac{\partial}{\partial x} \left(U(\varphi) + \frac{1}{2}(\pi')^2 \right) \right] d^3 x \\
&= 0, \tag{3.3.7}
\end{aligned}
$$

assuming that the fields and U vanish appropriately at ∞. (The first term vanishes because it switches sign under integration by parts.) Thus, P_x is conserved. The conservation of P_x is connected with invariance of H under translations in the x-direction. Deeper insights into this connection are explored later. Of course, similar conservation laws hold in the y- and z-directions.

Likewise, the angular momenta $\mathbf{J} = (J_x, J_y, J_z)$, where, for example,

$$J_z(\varphi) = \int_{\mathbb{R}^3} \pi' \left(x \frac{\partial}{\partial y} - y \frac{\partial}{\partial x} \right) \varphi \, d^3 x, \tag{3.3.8}$$

are constants of the motion. This is proved in an analogous way. (For precise function spaces in which these operations can be justified, see Chernoff and Marsden [1974].)   ◆

(d) Linear and Angular Momentum: The Schrödinger Equation.

Linear Momentum. In Example (b) of §3.2, assume that $\mathcal{H}$ is the space of complex-valued L^2-functions on $\mathbb{R}^3$ and that the self-adjoint linear operator $H_{\mathrm{op}}: \mathcal{H} \to \mathcal{H}$ commutes with infinitesimal translations of the argument by a fixed vector $\xi \in \mathbb{R}^3$, that is, $H_{\mathrm{op}}(\mathbf{D}\psi(\cdot) \cdot \xi) = \mathbf{D}(H_{\mathrm{op}}\psi(\cdot)) \cdot \xi$ for any ψ whose derivative is in $\mathcal{H}$. One checks, using (3.3.1), that

$$P_\xi(\psi) = \left\langle \frac{i}{\hbar} \mathbf{D}\psi \cdot \xi, \psi \right\rangle \tag{3.3.9}$$

Poisson commutes with $\langle H_{\mathrm{op}} \rangle$. If ξ is the unit vector along the x-axis, the corresponding conserved quantity is

$$P_x(\psi) = \left\langle \frac{i}{\hbar} \frac{\partial \psi}{\partial x}, \psi \right\rangle.$$

Angular Momentum. Assume that $H_{op} \colon \mathcal{H} \to \mathcal{H}$ commutes with infinitesimal rotations by a fixed skew-symmetric 3×3 matrix $\hat{\omega}$, that is,

$$H_{op}(\mathbf{D}\psi(x) \cdot \hat{\omega}x) = \mathbf{D}((H_{op}\psi)(x)) \cdot \hat{\omega}x \qquad (3.3.10)$$

for every ψ whose derivative is in $\mathcal{H}$, where on the left-hand side, H_{op} is thought of as acting on the function $x \mapsto \mathbf{D}\psi(x) \cdot \hat{\omega}x$. Then the angular momentum function

$$\mathbf{J}(\hat{\omega}) : x \mapsto \langle i\mathbf{D}\psi(x) \cdot \hat{\omega}(x)/\hbar, \psi(x) \rangle \qquad (3.3.11)$$

Poisson commutes with $\mathcal{H}$ so is a conserved quantity. If we choose $\omega = (0, 0, 1)$; that is,

$$\hat{\omega} = \begin{bmatrix} 0 & -1 & 0 \\ 1 & 0 & 0 \\ 0 & 0 & 0 \end{bmatrix},$$

this corresponds to an infinitesimal rotation around the z-axis. Explicitly, the angular momentum around the x^l-axis is given by

$$J_l(\psi) = \left\langle \frac{i}{\hbar} \left(x^j \frac{\partial \psi}{\partial x^k} - x^k \frac{\partial \psi}{\partial x^j} \right), \psi \right\rangle,$$

where (j, k, l) is a cyclic permutation of $(1, 2, 3)$.   ◆

(e) Linear and Angular Momentum for Linear Elastodynamics.
Consider again the equations of linear elastodynamics; see Example (f) of §3.2. Observe that the Hamiltonian is invariant under translations if the elasticity tensor **c** is homogeneous (independent of (x, y, z)); the corresponding conserved linear momentum in the x-direction is

$$P_x = \int_{\mathbb{R}^3} \rho \dot{\mathbf{u}} \cdot \frac{\partial \mathbf{u}}{\partial x} d^3 x. \qquad (3.3.12)$$

Likewise, the Hamiltonian is invariant under rotations if **c** is isotropic, that is, invariant under rotations, which is equivalent to **c** having the form

$$c^{ijkl} = \mu(\delta^{ik}\delta^{jl} + \delta^{il}\delta^{jk}) + \lambda \delta^{ij}\delta^{kl},$$

where μ and λ are constants (see Marsden and Hughes [1983, Section 4.3] for the proof). The conserved angular momentum about the z-axis is

$$J = \int_{\mathbb{R}^3} \rho \dot{\mathbf{u}} \cdot \left(x \frac{\partial \mathbf{u}}{\partial y} - y \frac{\partial \mathbf{u}}{\partial x} \right) d^3 x. \qquad ◆$$

In Chapter 11, we will gain a deeper insight into the significance and construction of these conserved quantities.

Some Technicalities for Infinite-Dimensional Systems. In general, unless the symplectic form on the Banach space Z is strong, the Hamiltonian vector field X_H is *not* defined on the whole of Z but only on a dense subspace. For example, in the case of the wave equation $\partial^2 \varphi / \partial t^2 = \nabla^2 \varphi - U'(\varphi)$, a possible choice of phase space is $H^1(\mathbb{R}^3) \times L^2(\mathbb{R}^3)$, but X_H is defined only on the dense subspace $H^2(\mathbb{R}^3) \times H^1(\mathbb{R}^3)$. It can also happen that the Hamiltonian H is not even defined on the whole of Z. For example, if $H_{\mathrm{op}} = \nabla^2 + V$ for the Schrödinger equation on $L^2(\mathbb{R}^3)$, then H could have domain containing $H^2(\mathbb{R}^3)$, that coincides with the domain of the Hamiltonian vector field iH_{op}. If V is singular, the domain need not be exactly $H^2(\mathbb{R}^3)$. As a quadratic form, H might be extendable to $H^1(\mathbb{R}^3)$. See Reed and Simon [1974, Volume II] or Kato [1984] for details.

The problem of existence and even uniqueness of solutions can be quite delicate. For linear systems one often appeals to Stone's theorem for the Schrödinger and wave equations, and to the Hille–Yosida theorem in the case of more general linear systems. We refer to Marsden and Hughes [1983, Chapter 6], for the theory and examples. In the case of nonlinear Hamiltonian systems, the theorems of Segal [1962], Kato [1975], and Hughes, Kato, and Marsden [1977] are relevant.

For infinite-dimensional nonlinear Hamiltonian systems, technical differentiability conditions on their flows φ_t are needed to ensure that each φ_t is a symplectic map; see Chernoff and Marsden [1974], and especially Marsden and Hughes [1983, Chapter 6]. These technicalities are needed in many interesting examples.   ◆

Exercises

◇ **3.3-1.** Show that $\{F_i, F_j\} = 0$, $i, j = 0, 1, 2, 3$, where the Poisson bracket is the KdV bracket and where

$$F_0(u) = \int_{-\infty}^{\infty} u\, dx,$$

$$F_1(u) = \int_{-\infty}^{\infty} \frac{1}{2} u^2\, dx,$$

$$F_2(u) = \int_{-\infty}^{\infty} \left(-u^3 + \frac{1}{2}(u_x)^2\right) dx \qquad \text{(the KdV Hamiltonian)},$$

$$F_3(u) = \int_{-\infty}^{\infty} \left(\frac{5}{2} u^4 - 5u u_x^2 + \frac{1}{2}(u_{xx})^2\right) dx.$$

4
Manifolds, Vector Fields, and Differential Forms

In preparation for later chapters, it will be necessary for the reader to learn a little bit about manifold theory. We recall a few basic facts here, beginning with the finite-dimensional case. (See Abraham, Marsden, and Ratiu [1988] for a full account.) The reader need not master all of this material now, but it suffices to read through it for general sense and come back to it repeatedly as our development of mechanics proceeds.

4.1   Manifolds

Our first goal is to define the notion of a manifold. Manifolds are, roughly speaking, abstract surfaces that locally look like linear spaces. We shall assume at first that the linear spaces are $\mathbb{R}^n$ for a fixed integer n, which will be the dimension of the manifold.

Coordinate Charts.   Given a set M, a **chart** on M is a subset U of M together with a bijective map $\varphi : U \to \varphi(U) \subset \mathbb{R}^n$. Usually, we denote $\varphi(m)$ by $(x^1, \ldots, x^n)$ and call the x^i the **coordinates** of the point $m \in U \subset M$.

Two charts (U, φ) and (U', φ') such that $U \cap U' \neq \varnothing$ are called **compatible** if $\varphi(U \cap U')$ and $\varphi'(U' \cap U)$ are open subsets of $\mathbb{R}^n$ and the maps

$$\varphi' \circ \varphi^{-1} | \varphi(U \cap U') : \varphi(U \cap U') \longrightarrow \varphi'(U \cap U')$$

and

$$\varphi \circ (\varphi')^{-1} | \varphi'(U \cap U') : \varphi'(U \cap U') \longrightarrow \varphi(U \cap U')$$

are C^∞. Here, $\varphi' \circ \varphi^{-1}|\varphi(U \cap U')$ denotes the restriction of the map $\varphi' \circ \varphi^{-1}$ to the set $\varphi(U \cap U')$. See Figure 4.1.1.

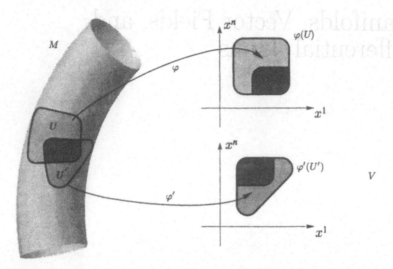

FIGURE 4.1.1. Overlapping charts on a manifold.

We call M a **differentiable n-manifold** if the following hold:

M1. *The set M is covered by a collection of charts, that is, every point is represented in at least one chart.*

M2. *M has an **atlas**; that is, M can be written as a union of compatible charts.*

If a chart is compatible with a given atlas, then it can be included into the atlas itself to produce a new, larger, atlas. One wants to allow such charts, thereby enlarging a given atlas, and so one really wants to define a **differentiable structure** as a **maximal atlas**. We will assume that this is done and resist the temptation to make this process overly formal.

A simple example will make what we have in mind clear. Suppose one considers Euclidean three-space $\mathbb{R}^3$ as a manifold with simply one (identity) chart. Certainly, we want to allow other charts such as those defined by spherical coordinates. Allowing all possible charts whose changes of coordinates with the standard Euclidean coordinates are smooth then gives us a maximal atlas.

A **neighborhood** of a point m in a manifold M is defined to be the inverse image of a Euclidean space neighborhood of the point $\varphi(m)$ under a chart map $\varphi : U \to \mathbb{R}^n$. Neighborhoods define open sets, and one checks that the open sets in M define a topology. *Usually, we assume without explicit mention that the topology is Hausdorff:* Two different points m, m' in M have nonintersecting neighborhoods.

Tangent Vectors.   Two curves $t \mapsto c_1(t)$ and $t \mapsto c_2(t)$ in an n-manifold M are called **equivalent** at the point m if

$$c_1(0) = c_2(0) = m \quad \text{and} \quad (\varphi \circ c_1)'(0) = (\varphi \circ c_2)'(0)$$

in some chart φ. Here the prime denotes the differentiation of curves in Euclidean space. It is easy to check that this definition is chart indepen-dent and that it defines an equivalence relation. A **tangent vector** v to a manifold M at a point $m \in M$ is an equivalence class of curves at m.

It is a theorem that the set of tangent vectors to M at m forms a vector space. It is denoted by $T_m M$ and is called the **tangent space** to M at $m \in M$.

Given a curve $c(t)$, we denote by $c'(s)$ the tangent vector at $c(s)$ defined by the equivalence class of $t \mapsto c(s+t)$ at $t = 0$. We have set things up so that tangent vectors to manifolds are thought of intuitively as tangent vectors to curves in M.

Let $\varphi : U \subset M \to \mathbb{R}^n$ be a chart for the manifold M, so that we get as-sociated coordinates $(x^1, \dots, x^n)$ for points in U. Let v be a tangent vector to M at m; i.e., $v \in T_m M$, and let c be a curve that is a representative of the equivalence class v. The **components** of v are the numbers $v^1, \dots, v^n$ defined by taking the derivatives of the components, in Euclidean space, of the curve $\varphi \circ c$:

$$v^i = \frac{d}{dt}(\varphi \circ c)^i \Big|_{t=0} ,$$

where $i = 1, \dots, n$ and where c is a representative curve for the tangent vector v. From the definition, the components are independent of the repre-sentative curve chosen, but they do, of course, depend on the chart chosen.

Tangent Bundles.   The **tangent bundle** of M, denoted by TM, is the set that is the disjoint union of the tangent spaces to M at the points $m \in M$, that is,

$$TM = \bigcup_{m \in M} T_m M.$$

Thus, a point of TM is a vector v that is tangent to M at some point $m \in M$.

If M is an n-manifold, then TM is a $2n$-manifold. To define the dif-ferentiable structure on TM, we need to specify how to construct local coordinates on TM. To do this, let $x^1, \dots, x^n$ be local coordinates on M and let $v^1, \dots, v^n$ be components of a tangent vector in this coordinate system. Then the $2n$ numbers $x^1, \dots, x^n, v^1, \dots, v^n$ give a local coordi-nate system on TM. This is the basic idea one uses to prove that indeed TM is a $2n$-manifold.

The **natural projection** is the map $\tau_M : TM \to M$ that takes a tangent vector v to the point $m \in M$ at which the vector v is attached (that is,

$v \in T_m M$). The inverse image $\tau_M^{-1}(m)$ of a point $m \in M$ under the natural projection τ_M is the tangent space $T_m M$. This space is called the **fiber** of the tangent bundle over the point $m \in M$.

Differentiable Maps and the Chain Rule.   Let $f : M \to N$ be a map of a manifold M to a manifold N. We call f **differentiable** (resp. C^k) if in local coordinates on M and N, the map f is represented by differentiable (resp. C^k) functions. Here, by "represented" we simply mean that coordinate charts are chosen on both M and N so that in these coordinates f, suitably restricted, becomes a map between Euclidean spaces. One of course has to check that this notion of smoothness is independent of the charts chosen—this follows from the chain rule.

The **derivative** of a differentiable map $f : M \to N$ at a point $m \in M$ is defined to be the linear map

$$T_m f : T_m M \to T_{f(m)} N$$

constructed in the following way. For $v \in T_m M$, choose a curve $c :]-\epsilon, \epsilon[\to M$ with $c(0) = m$, and associated velocity vector $dc/dt|_{t=0} = v$. Then $T_m f \cdot v$ is the velocity vector at $t = 0$ of the curve $f \circ c : \mathbb{R} \to N$, that is,

$$T_m f \cdot v = \frac{d}{dt} f(c(t)) \Big|_{t=0} .$$

The vector $T_m f \cdot v$ does not depend on the curve c but only on the vector v, as is seen using the chain rule. If $f : M \to N$ is of class C^k, then $Tf : TM \to TN$ is a mapping of class C^{k-1}. Note that

$$\frac{dc}{dt} \Big|_{t=0} = T_0 c \cdot 1.$$

If $f : M \to N$ and $g : N \to P$ are differentiable maps (or maps of class C^k), then $g \circ f : M \to P$ is differentiable (or of class C^k), and the **chain rule** holds:

$$T(g \circ f) = Tg \circ Tf.$$

Diffeomorphisms.   A differentiable (or of class C^k) map $f : M \to N$ is called a **diffeomorphism** if it is bijective and its inverse is also differentiable (or of class C^k).

If $T_m f : T_m M \to T_{f(m)} N$ is an isomorphism, the **inverse function theorem** states that f is a **local diffeomorphism** around $m \in M$, that is, there are open neighborhoods U of m in M and V of $f(m)$ in N such that $f|U : U \to V$ is a diffeomorphism. The set of all diffeomorphisms $f : M \to M$ forms a group under composition, and the chain rule shows that $T(f^{-1}) = (Tf)^{-1}$.

Submanifolds and Submersions. A *submanifold* of M is a subset $S \subset M$ with the property that for each $s \in S$ there is a chart (U, φ) in M with the *submanifold property*, namely,

SM. $\varphi : U \to \mathbb{R}^k \times \mathbb{R}^{n-k}$ and $\varphi(U \cap S) = \varphi(U) \cap (\mathbb{R}^k \times \{0\})$.

The number k is called the *dimension* of the submanifold S.

This latter notion is in agreement with the definition of dimension for a general manifold, since S is a manifold in its own right all of whose charts are of the form $(U \cap S, \varphi|(U \cap S))$ for all charts (U, φ) of M having the submanifold property. Note that any open subset of M is a submanifold and that a submanifold is necessarily *locally closed*, that is, every point $s \in S$ admits an open neighborhood U of s in M such that $U \cap S$ is closed in U.

There are convenient ways to construct submanifolds using smooth mappings. If $f : M \to N$ is a smooth map, a point $m \in M$ is a *regular point* if $T_m f$ is surjective; otherwise, m is a *critical point* of f. If $C \subset M$ is the set of critical points of f, then $f(C) \subset N$ is the set of *critical values* of f and $N \backslash f(C)$ is the set of *regular values* of f.[1]

The *submersion theorem* states that if $f : M \to N$ is a smooth map and n is a regular value of f, then $f^{-1}(n)$ is a smooth submanifold of M of dimension $\dim M - \dim N$ and

$$T_m \left(f^{-1}(n) \right) = \ker T_m f.$$

The *local onto theorem* states that $T_m f : T_m M \to T_{f(m)} N$ is surjective if and only if there are charts $\varphi : U \subset M \to U'$ at m in M and $\psi : V \subset N \to V'$ at $f(m)$ in N such that φ maps into the product space $\mathbb{R}^{\dim M - \dim N} \times \mathbb{R}^{\dim N}$; the image of U' correspondingly has the form of a product $U' = U'' \times V'$; the point m gets mapped to the origin $\varphi(m) = (\mathbf{0}, \mathbf{0})$, as does $f(m)$, namely, $\psi(f(m)) = \mathbf{0}$; and the local representative of f is a projection:

$$(\psi \circ f \circ \varphi^{-1})(x, y) = x.$$

In particular, $f|U : U \to V$ is onto. If $T_m f$ is onto for every $m \in M$, then f is called a *submersion*. It follows that submersions are open mappings (the images of open sets are open).

Immersions and Embeddings. A C^k map $f : M \to N$ is called an *immersion* if $T_m f$ is injective for every $m \in M$. The *local 1-to-1 theorem* states that $T_m f$ is injective if and only if there are charts $\varphi : U \subset M \to U'$ at m in M and $\psi : V \subset N \to V'$ at $f(m)$ in N such that V' is a product

---

[1] *Sard's theorem* states that if $f : M \to N$ is a C^k-map, $k \geq 1$, and if M has the property that every open covering has a countable subcovering, then if $k > \max(0, \dim M - \dim N)$, the set of regular values of f is residual and hence dense in N.

$V' = U' \times V'' \subset \mathbb{R}^{\dim M} \times \mathbb{R}^{\dim N - \dim M}$; both m and $f(m)$ get sent to zero, i.e., $\varphi(m) = \mathbf{0}$ and $\psi(f(m)) = (\mathbf{0}, \mathbf{0})$; and the local representative of f is the inclusion

$$(\psi \circ f \circ \varphi^{-1})(x) = (x, \mathbf{0}).$$

In particular, $f|U : U \to V$ is injective. The *immersion theorem* states that $T_m f$ is injective if and only if there is a neighborhood U of m in M such that $f(U)$ is a submanifold of N and $f|U : U \to f(U)$ is a diffeomorphism.

It should be noted that this theorem does *not* say that $f(M)$ is a submanifold of N. For example, f may not be injective and $f(M)$ may thus have self-intersections. Even if f is an injective immersion, the image $f(M)$ may *not* be a submanifold of N. An example is indicated in Figure 4.1.2.

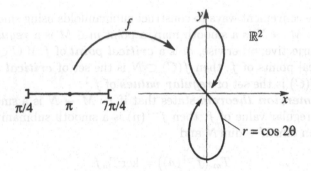

FIGURE 4.1.2. An injective immersion.

The map indicated in the figure (explicitly given by $f :]\pi/4, 7\pi/4[\to \mathbb{R}^2$; $\theta \mapsto (\sin\theta \cos 2\theta, \cos\theta \cos 2\theta)$) is an injective immersion, but the topology induced from $\mathbb{R}^2$ onto its image does not coincide with the usual topology of the open interval: Any neighborhood of the origin in the relative topology consists, in the domain interval, of the union of an open interval about π with two open segments $]\pi/4, \pi/4 + \epsilon[$, $]7\pi/4 - \epsilon, 7\pi/4[$. Thus, the image of f is not a submanifold of $\mathbb{R}^2$, but an *injectively immersed submanifold*.

An immersion $f : M \to N$ that is a homeomorphism onto $f(M)$ with the relative topology induced from N is called an *embedding*. In this case $f(M)$ is a submanifold of N and $f : M \to f(M)$ is a diffeomorphism. For example, if $f : M \to N$ is an injective immersion and if M is compact, then f is an embedding. Thus, the example given in the preceding figure is an example of an injective immersion that is not an embedding (and of course, M is not compact).

Another example of an injective immersion that is not an embedding is the linear flow on the torus $\mathbb{T}^2 = \mathbb{R}^2/\mathbb{Z}^2$ with irrational slope: $f(t) = (t, \alpha t) \pmod{\mathbb{Z}^2}$. However, there is a difference between this injective immersion and the "figure eight" example above: In some sense, the second example is better behaved; it has some "uniformity" about its lack of being an embedding.

An injective immersion $f : M \to N$ is called *regular* if the following property holds: If $g : L \to M$ is any map of the manifold L into M, then g is C^k if and only if $f \circ g : L \to N$ is C^k for any $k \geq 1$. It is easy to see that all embeddings satisfy this property but that the previous example also satisfies it, without being an embedding, and that the "figure eight" example (see Figure 4.1.2) does not satisfy it. Varadarajan [1974] calls such maps *quasi-regular embeddings*; they appear below in the Frobenius theorem and in the study of Lie subgroups.

Vector Fields and Flows. A *vector field* X on a manifold M is a map $X : M \to TM$ that assigns a vector $X(m)$ at the point $m \in M$; that is, $\tau_M \circ X =$ identity. The real vector space of vector fields on M is denoted by $\mathfrak{X}(M)$. An *integral curve* of X with initial condition m_0 at $t = 0$ is a (differentiable) map $c :]a, b[\to M$ such that $]a, b[$ is an open interval containing 0, $c(0) = m_0$, and

$$c'(t) = X(c(t))$$

for all $t \in]a, b[$. In formal presentations we usually suppress the domain of definition, even though this is technically important.

The *flow* of X is the collection of maps $\varphi_t : M \to M$ such that $t \mapsto \varphi_t(m)$ is the integral curve of X with initial condition m. Existence and uniqueness theorems from ordinary differential equations guarantee that φ is smooth in m and t (where defined) if X is. From uniqueness, we get the *flow property*

$$\varphi_{t+s} = \varphi_t \circ \varphi_s$$

along with the initial conditions $\varphi_0 =$ identity. The flow property generalizes the situation where $M = V$ is a *linear* space, $X(m) = Am$ for a (bounded) *linear* operator A, and where

$$\varphi_t(m) = e^{tA}m$$

to the *nonlinear* case.

A *time-dependent vector field* is a map $X : M \times \mathbb{R} \to TM$ such that $X(m, t) \in T_m M$ for each $m \in M$ and $t \in \mathbb{R}$. An *integral curve* of X is a curve $c(t)$ in M such that $c'(t) = X(c(t), t)$. In this case, the flow is the collection of maps

$$\varphi_{t,s} : M \to M$$

such that $t \mapsto \varphi_{t,s}(m)$ is the integral curve $c(t)$ with initial condition $c(s) = m$ at $t = s$. Again, the existence and uniqueness theorem from ODE theory applies, and in particular, uniqueness gives the *time-dependent flow property*

$$\varphi_{t,s} \circ \varphi_{s,r} = \varphi_{t,r}.$$

If X happens to be time independent, the two notions of flows are related by $\varphi_{t,s} = \varphi_{t-s}$.

Differentials and Covectors.   If $f : M \to \mathbb{R}$ is a smooth function, we can differentiate it at any point $m \in M$ to obtain a map $T_m f : T_m M \to T_{f(m)} \mathbb{R}$. Identifying the tangent space of $\mathbb{R}$ at any point with itself (a process we usually do in any vector space), we get a linear map $\mathbf{d}f(m) : T_m M \to \mathbb{R}$. That is, $\mathbf{d}f(m) \in T_m^* M$, the dual of the vector space $T_m M$. We call $\mathbf{d}f$ the *differential* of f. For $v \in T_m M$, we call $\mathbf{d}f(m) \cdot v$ the *directional derivative* of f in the direction v. In a coordinate chart or in linear spaces, this notion coincides with the usual notion of a directional derivative learned in vector calculus.

Explicitly, in coordinates, the directional derivative is given by

$$\mathbf{d}f(m) \cdot v = \sum_{i=1}^{n} \frac{\partial (f \circ \varphi^{-1})}{\partial x^i} v^i,$$

where φ is a chart at m. We will employ the *summation convention* and drop the summation sign when there are repeated indices.

One can show that specifying the directional derivatives completely determines a vector, and so we can identify a basis of $T_m M$ using the operators $\partial / \partial x^i$. We write

$$\{e_1, \dots, e_n\} = \left\{ \frac{\partial}{\partial x^1}, \dots, \frac{\partial}{\partial x^n} \right\}$$

for this basis, so that $v = v^i \partial / \partial x^i$.

If we replace each vector space $T_m M$ with its dual $T_m^* M$, we obtain a new $2n$-manifold called the *cotangent bundle* and denoted by $T^* M$. The dual basis to $\partial / \partial x^i$ is denoted by dx^i. Thus, relative to a choice of local coordinates we get the basic formula

$$\mathbf{d}f(x) = \frac{\partial f}{\partial x^i} dx^i$$

for any smooth function $f : M \to \mathbb{R}$.

Exercises

⋄ **4.1-1.**   Show that the two-sphere $S^2 \subset \mathbb{R}^3$ is a 2-manifold.

⋄ **4.1-2.**   If $\varphi_t : S^2 \to S^2$ rotates points on S^2 about a fixed axis through an angle t, show that φ_t is the flow of a certain vector field on S^2.

⋄ **4.1-3.**   Let $f : S^2 \to \mathbb{R}$ be defined by $f(x, y, z) = z$. Compute $\mathbf{d}f$ relative to spherical coordinates (θ, φ).

4.2 Differential Forms

We next review some of the basic definitions, properties, and operations on differential forms, without proofs (see Abraham, Marsden, and Ratiu [1988] and references therein).

> *The main idea of differential forms is to provide a generalization of the basic operations of vector calculus,* div, grad, *and* curl, *and the integral theorems of Green, Gauss, and Stokes to manifolds of arbitrary dimension.*

Basic Definitions. We have already met one-forms, a term that is used in two ways—they are either members of a particular cotangent space $T_m^* M$ or else, analogous to a vector field, an assignment of a covector in $T_m^* M$ to each $m \in M$. A basic example of a one-form is the differential of a real-valued function.

A 2-*form* Ω on a manifold M is a function $\Omega(m) : T_m M \times T_m M \to \mathbb{R}$ that assigns to each point $m \in M$ a skew-symmetric bilinear form on the tangent space $T_m M$ to M at m. More generally, a k-*form* α (sometimes called a *differential form of degree* k) on a manifold M is a function $\alpha(m) : T_m M \times \cdots \times T_m M$ (there are k factors) $\to \mathbb{R}$ that assigns to each point $m \in M$ a skew-symmetric k-multilinear map on the tangent space $T_m M$ to M at m. Without the skew-symmetry assumption, α would be called a $(0, k)$-*tensor*. A map $\alpha : V \times \cdots \times V$ (there are k factors) $\to \mathbb{R}$ is *multilinear* when it is linear in each of its factors, that is,

$$\alpha(v_1, \ldots, av_j + bv_j', \ldots, v_k)$$
$$= a\alpha(v_1, \ldots, v_j, \ldots, v_k) + b\alpha(v_1, \ldots, v_j', \ldots, v_k)$$

for all j with $1 \le j \le k$. A k-multilinear map $\alpha : V \times \ldots \times V \to \mathbb{R}$ is *skew* (or *alternating*) when it changes sign whenever two of its arguments are interchanged, that is, for all $v_1, \ldots, v_k \in V$,

$$\alpha(v_1, \ldots, v_i, \ldots, v_j, \ldots, v_k) = -\alpha(v_1, \ldots, v_j, \ldots, v_i, \ldots, v_k).$$

Let $x^1, \ldots, x^n$ denote coordinates on M, let

$$\{e_1, \ldots, e_n\} = \{\partial/\partial x^1, \ldots, \partial/\partial x^n\}$$

be the corresponding basis for $T_m M$, and let

$$\{e^1, \ldots, e^n\} = \{dx^1, \ldots, dx^n\}$$

be the dual basis for $T_m^* M$. Then at each $m \in M$, we can write a 2-form as

$$\Omega_m(v, w) = \Omega_{ij}(m)v^i w^j, \quad \text{where} \quad \Omega_{ij}(m) = \Omega_m\left(\frac{\partial}{\partial x^i}, \frac{\partial}{\partial x^j}\right),$$

and more generally, a k-form can be written

$$\alpha_m(v_1,\dots,v_k) = \alpha_{i_1\dots i_k}(m)v_1^{i_1}\cdots v_k^{i_k},$$

where there is a sum on $i_1,\dots,i_k$,

$$\alpha_{i_1\dots i_k}(m) = \alpha_m\left(\frac{\partial}{\partial x^{i_1}},\dots,\frac{\partial}{\partial x^{i_k}}\right),$$

and $v_i = v_i^j\partial/\partial x^j$, with a sum on j understood.

Tensor and Wedge Products.   If α is a $(0,k)$-tensor on a manifold M and β is a $(0,l)$-tensor, their **tensor product** $\alpha\otimes\beta$ is the $(0,k+l)$-tensor on M defined by

$$(\alpha\otimes\beta)_m(v_1,\dots,v_{k+l}) = \alpha_m(v_1,\dots,v_k)\beta_m(v_{k+1},\dots,v_{k+l}) \qquad (4.2.1)$$

at each point $m \in M$.

If t is a $(0,p)$-tensor, define the **alternation operator** $\mathbf{A}$ acting on t by

$$\mathbf{A}(t)(v_1,\dots,v_p) = \frac{1}{p!}\sum_{\pi\in S_p}\operatorname{sgn}(\pi)t(v_{\pi(1)},\dots,v_{\pi(p)}), \qquad (4.2.2)$$

where $\operatorname{sgn}(\pi)$ is the **sign** of the permutation π,

$$\operatorname{sgn}(\pi) = \begin{cases} +1 \text{ if } \pi \text{ is even}, \\ -1 \text{ if } \pi \text{ is odd}, \end{cases} \qquad (4.2.3)$$

and S_p is the group of all permutations of the set $\{1,2,\dots,p\}$. The operator $\mathbf{A}$ therefore skew-symmetrizes p-multilinear maps.

If α is a k-form and β is an l-form on M, their **wedge product** $\alpha\wedge\beta$ is the $(k+l)$-form on M defined by[2]

$$\alpha\wedge\beta = \frac{(k+l)!}{k!\,l!}\mathbf{A}(\alpha\otimes\beta). \qquad (4.2.4)$$

For example, if α and β are one-forms, then

$$(\alpha\wedge\beta)(v_1,v_2) = \alpha(v_1)\beta(v_2) - \alpha(v_2)\beta(v_1),$$

while if α is a 2-form and β is a 1-form,

$$(\alpha\wedge\beta)(v_1,v_2,v_3) = \alpha(v_1,v_2)\beta(v_3) + \alpha(v_3,v_1)\beta(v_2) + \alpha(v_2,v_3)\beta(v_1).$$

We state the following without proof:

---

[2]The numerical factor in (4.2.4) agrees with the convention of Abraham and Marsden [1978], Abraham, Marsden, and Ratiu [1988], and Spivak [1976], but *not* that of Arnold [1989], Guillemin and Pollack [1974], or Kobayashi and Nomizu [1963]; it is the Bourbaki [1971] convention.

Proposition 4.2.1.   *The wedge product has the following properties:*

(i) $\alpha \wedge \beta$ is **associative**: $\alpha \wedge (\beta \wedge \gamma) = (\alpha \wedge \beta) \wedge \gamma$.

(ii) $\alpha \wedge \beta$ is **bilinear** in α, β:

$$(a\alpha_1 + b\alpha_2) \wedge \beta = a(\alpha_1 \wedge \beta) + b(\alpha_2 \wedge \beta),$$
$$\alpha \wedge (c\beta_1 + d\beta_2) = c(\alpha \wedge \beta_1) + d(\alpha \wedge \beta_2).$$

(iii) $\alpha \wedge \beta$ is **anticommutative**: $\alpha \wedge \beta = (-1)^{kl}\beta \wedge \alpha$, where α is a k-form and β is a 1-form.

In terms of the dual basis dx^i, any k-form can be written locally as

$$\alpha = \alpha_{i_1 \dots i_k} dx^{i_1} \wedge \cdots \wedge dx^{i_k},$$

where the sum is over all i_j satisfying $i_1 < \cdots < i_k$.

Pull-Back and Push-Forward.   Let $\varphi : M \to N$ be a C^∞ map from the manifold M to the manifold N and α be a k-form on N. Define the **pull-back** $\varphi^*\alpha$ of α by φ to be the k-form on M given by

$$(\varphi^*\alpha)_m(v_1, \dots, v_k) = \alpha_{\varphi(m)}(T_m\varphi \cdot v_1, \dots, T_m\varphi \cdot v_k). \tag{4.2.5}$$

If φ is a diffeomorphism, the **push-forward** φ_* is defined by $\varphi_* = (\varphi^{-1})^*$.

Here is another basic property.

Proposition 4.2.2.   *The pull-back of a wedge product is the wedge product of the pull-backs:*

$$\varphi^*(\alpha \wedge \beta) = \varphi^*\alpha \wedge \varphi^*\beta. \tag{4.2.6}$$

Interior Products and Exterior Derivatives.   Let α be a k-form on a manifold M and X a vector field. The **interior product** $\mathbf{i}_X\alpha$ (sometimes called the **contraction** of X and α and written, using the "hook" notation, as $X \lrcorner \alpha$) is defined by

$$(\mathbf{i}_X\alpha)_m(v_2, \dots, v_k) = \alpha_m(X(m), v_2, \dots, v_k). \tag{4.2.7}$$

Proposition 4.2.3.   *Let α be a k-form and β a 1-form on a manifold M. Then*

$$\mathbf{i}_X(\alpha \wedge \beta) = (\mathbf{i}_X\alpha) \wedge \beta + (-1)^k\alpha \wedge (\mathbf{i}_X\beta). \tag{4.2.8}$$

In the "hook" notation, this proposition reads

$$X \lrcorner (\alpha \wedge \beta) = (X \lrcorner \alpha) \wedge \beta + (-1)^k\alpha \wedge (X \lrcorner \beta).$$

The **exterior derivative** $\mathbf{d}\alpha$ of a k-form α on a manifold M is the $(k+1)$-form on M determined by the following proposition:

Proposition 4.2.4. *There is a unique mapping* **d** *from* k-*forms on* M *to* $(k+1)$-*forms on* M *such that:*

(i) *If* α *is a* 0-*form* $(k = 0)$, *that is,* $\alpha = f \in \mathcal{F}(M)$, *then* **d**$f$ *is the one-form that is the differential of* f.

(ii) **d**α *is* **linear** *in* α, *that is, for all real numbers* c_1 *and* c_2,

$$\mathbf{d}(c_1\alpha_1 + c_2\alpha_2) = c_1\mathbf{d}\alpha_1 + c_2\mathbf{d}\alpha_2.$$

(iii) **d**α *satisfies the* **product rule**, *that is,*

$$\mathbf{d}(\alpha \wedge \beta) = \mathbf{d}\alpha \wedge \beta + (-1)^k\alpha \wedge \mathbf{d}\beta,$$

where α *is a* k-*form and* β *is a* 1-*form.*

(iv) $\mathbf{d}^2 = 0$, *that is,* $\mathbf{d}(\mathbf{d}\alpha) = 0$ *for any* k-*form* α.

(v) **d** *is a* **local operator**, *that is,* $\mathbf{d}\alpha(m)$ *depends only on* α *restricted to any open neighborhood of* m; *in fact, if* U *is open in* M, *then*

$$\mathbf{d}(\alpha|U) = (\mathbf{d}\alpha)|U.$$

If α is a k-form given in coordinates by

$$\alpha = \alpha_{i_1 \ldots i_k} dx^{i_1} \wedge \cdots \wedge dx^{i_k} \quad (\text{sum on } i_1 < \cdots < i_k),$$

then the coordinate expression for the exterior derivative is

$$\mathbf{d}\alpha = \frac{\partial \alpha_{i_1 \ldots i_k}}{\partial x^j} dx^j \wedge dx^{i_1} \wedge \cdots \wedge dx^{i_k}$$

$$(\text{sum on all } j \text{ and } i_1 < \cdots < i_k). \qquad (4.2.9)$$

Formula (4.2.9) can be taken as the definition of the exterior derivative, provided that one shows that (4.2.9) has the above-described properties and, correspondingly, is independent of the choice of coordinates.

Next is a useful proposition that in essence rests on the chain rule:

Proposition 4.2.5. *Exterior differentiation commutes with pull-back, that is,*

$$\mathbf{d}(\varphi^*\alpha) = \varphi^*(\mathbf{d}\alpha), \qquad (4.2.10)$$

where α *is a* k-*form on a manifold* N *and* $\varphi : M \to N$ *is a smooth map between manifolds.*

A k-form α is called **closed** if $\mathbf{d}\alpha = 0$ and **exact** if there is a $(k-1)$-form β such that $\alpha = \mathbf{d}\beta$. By Proposition 4.2.4(iv) every exact form is closed. Exercise 4.4-2 gives an example of a closed nonexact one-form.

Proposition 4.2.6 (Poincaré Lemma). *A closed form is locally exact; that is, if* $d\alpha = 0$, *there is a neighborhood about each point on which* $\alpha = d\beta$.

See Exercise 4.2-5 for the proof.

The definition and properties of vector-valued forms are direct extensions of those for usual forms on vector spaces and manifolds. One can think of a vector-valued form as an array of usual forms (see Abraham, Marsden, and Ratiu [1988]).

Vector Calculus. The table below entitled "Vector Calculus and Differential Forms" summarizes how forms are related to the usual operations of vector calculus. We now elaborate on a few items in this table. In item 4, note that

$$\mathbf{d}f = \frac{\partial f}{\partial x}dx + \frac{\partial f}{\partial y}dy + \frac{\partial f}{\partial z}dz = (\operatorname{grad} f)^{\flat} = (\nabla f)^{\flat},$$

which is equivalent to $\nabla f = (\mathbf{d}f)^{\sharp}$.

The **Hodge star operator** on $\mathbb{R}^3$ maps k-forms to $(3-k)$-forms and is uniquely determined by linearity and the properties in item 2. (This operator can be defined on general Riemannian manifolds; see Abraham, Marsden, and Ratiu [1988].)

In item 5, if we let $F = F_1\mathbf{e}_1 + F_2\mathbf{e}_2 + F_3\mathbf{e}_3$, so $F^{\flat} = F_1\,dx + F_2\,dy + F_3\,dz$, then

$$
\begin{aligned}
\mathbf{d}(F^{\flat}) &= \mathbf{d}F_1 \wedge dx + F_1\mathbf{d}(dx) + \mathbf{d}F_2 \wedge dy + F_2\mathbf{d}(dy) \\
&\quad + \mathbf{d}F_3 \wedge dz + F_3\mathbf{d}(dz) \\
&= \left(\frac{\partial F_1}{\partial x}dx + \frac{\partial F_1}{\partial y}dy + \frac{\partial F_1}{\partial z}dz\right) \wedge dx \\
&\quad + \left(\frac{\partial F_2}{\partial x}dx + \frac{\partial F_2}{\partial y}dy + \frac{\partial F_2}{\partial z}dz\right) \wedge dy \\
&\quad + \left(\frac{\partial F_3}{\partial x}dx + \frac{\partial F_3}{\partial y}dy + \frac{\partial F_3}{\partial z}dz\right) \wedge dz \\
&= -\frac{\partial F_1}{\partial y}dx \wedge dy + \frac{\partial F_1}{\partial z}dz \wedge dx + \frac{\partial F_2}{\partial x}dx \wedge dy - \frac{\partial F_2}{\partial z}dy \wedge dz \\
&\quad - \frac{\partial F_3}{\partial x}dz \wedge dx + \frac{\partial F_3}{\partial y}dy \wedge dz \\
&= \left(\frac{\partial F_2}{\partial x} - \frac{\partial F_1}{\partial y}\right)dx \wedge dy + \left(\frac{\partial F_1}{\partial z} - \frac{\partial F_3}{\partial x}\right)dz \wedge dx \\
&\quad + \left(\frac{\partial F_3}{\partial y} - \frac{\partial F_2}{\partial z}\right)dy \wedge dz.
\end{aligned}
$$

Hence, using item 2,

$$*(\mathbf{d}(F^{\flat})) = \left(\frac{\partial F_2}{\partial x} - \frac{\partial F_1}{\partial y}\right) dz + \left(\frac{\partial F_1}{\partial z} - \frac{\partial F_3}{\partial x}\right) dy + \left(\frac{\partial F_3}{\partial y} - \frac{\partial F_2}{\partial z}\right) dx,$$

$$(*(\mathbf{d}(F^{\flat})))^{\sharp} = \left(\frac{\partial F_3}{\partial y} - \frac{\partial F_2}{\partial z}\right) \mathbf{e}_1 + \left(\frac{\partial F_1}{\partial z} - \frac{\partial F_3}{\partial x}\right) \mathbf{e}_2 + \left(\frac{\partial F_2}{\partial x} - \frac{\partial F_1}{\partial y}\right) \mathbf{e}_3$$

$$= \operatorname{curl} F = \nabla \times F.$$

With reference to item 6, let $F = F_1 \mathbf{e}_1 + F_2 \mathbf{e}_2 + F_3 \mathbf{e}_3$, so

$$F^{\flat} = F_1 \, dx + F_2 \, dy + F_3 \, dz.$$

Thus $*(F^{\flat}) = F_1 \, dy \wedge dz + F_2(-dx \wedge dz) + F_3 \, dx \wedge dy$, and so

$$\mathbf{d}(*(F^{\flat})) = \mathbf{d}F_1 \wedge dy \wedge dz - \mathbf{d}F_2 \wedge dx \wedge dz + \mathbf{d}F_3 \wedge dx \wedge dy$$

$$= \left(\frac{\partial F_1}{\partial x}dx + \frac{\partial F_1}{\partial y}dy + \frac{\partial F_1}{\partial z}dz\right) \wedge dy \wedge dz$$

$$- \left(\frac{\partial F_2}{\partial x}dx + \frac{\partial F_2}{\partial y}dy + \frac{\partial F_2}{\partial z}dz\right) \wedge dx \wedge dz$$

$$+ \left(\frac{\partial F_3}{\partial x}dx + \frac{\partial F_3}{\partial y}dy + \frac{\partial F_3}{\partial z}dz\right) \wedge dx \wedge dy$$

$$= \frac{\partial F_1}{\partial x}dx \wedge dy \wedge dz + \frac{\partial F_2}{\partial y}dx \wedge dy \wedge dz + \frac{\partial F_3}{\partial z}dx \wedge dy \wedge dz$$

$$= \left(\frac{\partial F_1}{\partial x} + \frac{\partial F_2}{\partial y} + \frac{\partial F_3}{\partial z}\right) dx \wedge dy \wedge dz = (\operatorname{div} F) \, dx \wedge dy \wedge dz.$$

Therefore, $*(\mathbf{d}(*(F^{\flat}))) = \operatorname{div} F = \nabla \cdot F$.

Vector Calculus and Differential Forms

1. Sharp and Flat (Using standard coordinates in $\mathbb{R}^3$)

(a) $v^{\flat} = v^1 \, dx + v^2 \, dy + v^3 \, dz$, the one-form corresponding to the vector $v = v^1 \mathbf{e}_1 + v^2 \mathbf{e}_2 + v^3 \mathbf{e}_3$.

(b) $\alpha^{\sharp} = \alpha_1 \mathbf{e}_1 + \alpha_2 \mathbf{e}_2 + \alpha_3 \mathbf{e}_3$, the vector corresponding to the one-form $\alpha = \alpha_1 \, dx + \alpha_2 \, dy + \alpha_3 \, dz$.

2. Hodge Star Operator

(a) $*1 = dx \wedge dy \wedge dz$.

(b) $*dx = dy \wedge dz$, $*dy = -dx \wedge dz$, $*dz = dx \wedge dy$,
$*(dy \wedge dz) = dx$, $*(dx \wedge dz) = -dy$, $*(dx \wedge dy) = dz$.

(c) $*(dx \wedge dy \wedge dz) = 1$.

3. Cross Product and Dot Product

(a) $v \times w = [*(v^{\flat} \wedge w^{\flat})]^{\sharp}$.

(b) $(v \cdot w)dx \wedge dy \wedge dz = v^{\flat} \wedge *(w^{\flat})$.

4. Gradient $\nabla f = \operatorname{grad} f = (df)^{\sharp}$.

5. Curl $\nabla \times F = \operatorname{curl} F = [*(dF^{\flat})]^{\sharp}$.

6. Divergence $\nabla \cdot F = \operatorname{div} F = *d(*F^{\flat})$.

Exercises

⬦ **4.2-1.**   Let $\varphi : \mathbb{R}^3 \to \mathbb{R}^2$ be given by $\varphi(x, y, z) = (x + z, xy)$. For

$$\alpha = e^{v} \, du + u \, dv \in \Omega^{1}(\mathbb{R}^2) \quad \text{and} \quad \beta = u \, du \wedge dv,$$

compute $\alpha \wedge \beta$, $\varphi^{*}\alpha$, $\varphi^{*}\beta$, and $\varphi^{*}\alpha \wedge \varphi^{*}\beta$.

⬦ **4.2-2.**   Given

$$\alpha = y^2 \, dx \wedge dz + \sin(xy) \, dx \wedge dy + e^{x} \, dy \wedge dz \in \Omega^{2}(\mathbb{R}^3)$$

and

$$X = 3\partial/\partial x + \cos z\partial/\partial y - x^2\partial/\partial z \in \mathfrak{X}(\mathbb{R}^3),$$

compute $d\alpha$ and $\mathbf{i}_X\alpha$.

⬦ **4.2-3.**

(a) Denote by $\bigwedge^{k}(\mathbb{R}^n)$ the vector space of all skew-symmetric k-linear maps on $\mathbb{R}^n$. Prove that this space has dimension $n!/(k! \, (n-k)!)$ by showing that a basis is given by $\{ e^{i_1} \wedge \cdots \wedge e^{i_k} \mid i_1 < \ldots < i_k \}$, where $\{e_1, \ldots, e_n\}$ is a basis of $\mathbb{R}^n$ and $\{e^1, \ldots, e^n\}$ is its dual basis, that is, $e^i(e_j) = \delta^i_j$.

(b) If $\mu \in \bigwedge^{n}(\mathbb{R}^n)$ is nonzero, prove that the map $v \in \mathbb{R}^n \mapsto \mathbf{i}_v\mu \in \bigwedge^{n-1}(\mathbb{R}^n)$ is an isomorphism.

(c) If M is a smooth n-manifold and $\mu \in \Omega^{n}(M)$ is nowhere-vanishing (in which case it is called a volume form), show that the map $X \in \mathfrak{X}(M) \mapsto \mathbf{i}_X\mu \in \Omega^{n-1}(M)$ is an isomorphism.

◇ **4.2-4.** Let $\alpha = \alpha_i \, dx^i$ be a closed one-form in a ball around the origin in $\mathbb{R}^n$. Show that $\alpha = \mathbf{d}f$ for

$$f(x^1, \ldots, x^n) = \int_0^1 \alpha_j(tx^1, \ldots, tx^n)x^j \, dt.$$

◇ **4.2-5.**

(a) Let U be an open ball around the origin in $\mathbb{R}^n$ and $\alpha \in \Omega^k(U)$ a closed form. Verify that $\alpha = \mathbf{d}\beta$, where

$$\beta(x^1, \ldots, x^n)$$
$$= \left(\int_0^1 t^{k-1} \alpha_{j i_1 \ldots i_{k-1}}(tx^1, \ldots, tx^n)x^j \, dt \right) dx^{i_1} \wedge \cdots \wedge dx^{i_{k-1}},$$

and where the sum is over $i_1 < \cdots < i_{k-1}$. Here,

$$\alpha = \alpha_{j_1 \ldots j_k} \, dx^{j_1} \wedge \cdots \wedge dx^{j_k},$$

where $j_1 < \cdots < j_k$ and where α is extended to be skew-symmetric in its lower indices.

(b) Deduce the Poincaré lemma from (a).

◇ **4.2-6** (Construction of a homotopy operator for a retraction). Let M be a smooth manifold and $N \subset M$ a smooth submanifold. A family of smooth maps $r_t : M \to M$, $t \in [0,1]$, is called a **retraction of M onto N** if $r_t|N =$ identity on N for all $t \in [0,1]$, $r_1 =$ identity on M, r_t is a diffeomorphism of M with $r_t(M)$ for every $t \neq 0$, and $r_0(M) = N$. Let X_t be the time-dependent vector field generated by r_t, $t \neq 0$. Show that the operator $\mathbf{H} : \Omega^k(M) \to \Omega^{k-1}(M)$ defined by

$$\mathbf{H} = \int_0^1 (r_t^* \mathbf{i}_{X_t} \alpha) \, dt$$

satisfies
$$\alpha - (r_0^* \alpha) = \mathbf{d}\mathbf{H}\alpha + \mathbf{H}\mathbf{d}\alpha.$$

(a) Deduce the **relative Poincaré lemma** from this formula: If $\alpha \in \Omega^k(M)$ is closed and $\alpha|N = 0$, then there is a neighborhood U of N such that $\alpha|U = \mathbf{d}\beta$ for some $\beta \in \Omega^{k-1}(U)$ and $\beta|N = 0$. (Hint: Use the existence of a tubular neighborhood of N in M.)

(b) Deduce the **global Poincaré lemma** for contractible manifolds: If M is contractible, that is, there is a retraction of M to a point, and if $\alpha \in \Omega^k(M)$ is closed, then α is exact.

4.3   The Lie Derivative

Lie Derivative Theorem.   The *dynamic definition* of the Lie derivative is as follows. Let α be a k-form and let X be a vector field with flow φ_t. The **Lie derivative** of α along X is given by

$$\pounds_X \alpha = \lim_{t \to 0} \frac{1}{t}[(\varphi_t^* \alpha) - \alpha] = \frac{d}{dt}\varphi_t^* \alpha \Big|_{t=0}. \qquad (4.3.1)$$

This definition together with properties of pull-backs yields the following.

Theorem 4.3.1 (Lie Derivative Theorem).

$$\frac{d}{dt}\varphi_t^* \alpha = \varphi_t^* \pounds_X \alpha. \qquad (4.3.2)$$

This formula holds also for *time-dependent* vector fields in the sense that

$$\frac{d}{dt}\varphi_{t,s}^* \alpha = \varphi_{t,s}^* \pounds_X \alpha,$$

and in the expression $\pounds_X \alpha$ the vector field X is evaluated at time t.

If f is a real-valued function on a manifold M and X is a vector field on M, the **Lie derivative of f along X** is the **directional derivative**

$$\pounds_X f = X[f] := \mathbf{d}f \cdot X. \qquad (4.3.3)$$

If M is finite-dimensional, then

$$\pounds_X f = X^i \frac{\partial f}{\partial x^i}. \qquad (4.3.4)$$

For this reason one often writes

$$X = X^i \frac{\partial}{\partial x^i}.$$

If Y is a vector field on a manifold N and $\varphi : M \to N$ is a diffeomorphism, the **pull-back** $\varphi^* Y$ is a vector field on M defined by

$$(\varphi^* Y)(m) = \left(T_m \varphi^{-1} \circ Y \circ \varphi\right)(m). \qquad (4.3.5)$$

Two vector fields X on M and Y on N are said to be φ-**related** if

$$T\varphi \circ X = Y \circ \varphi. \qquad (4.3.6)$$

Clearly, if $\varphi : M \to N$ is a diffeomorphism and Y is a vector field on N, then $\varphi^* Y$ and Y are φ-related. For a diffeomorphism φ, the **push-forward** is defined, as for forms, by $\varphi_* = (\varphi^{-1})^*$.

Jacobi–Lie Brackets. If M is finite-dimensional and C^∞, then the set of vector fields on M coincides with the set of derivations on $\mathcal{F}(M)$. The same result is true for C^k manifolds and vector fields if $k \geq 2$. This property is false for infinite-dimensional manifolds; see Abraham, Marsden, and Ratiu [1988]. If M is C^∞ and smooth, then the derivation $f \mapsto X[Y[f]] - Y[X[f]]$, where $X[f] = \mathbf{d}f \cdot X$, determines a unique vector field denoted by $[X, Y]$ and called the **Jacobi–Lie bracket** of X and Y. Defining $\pounds_X Y = [X, Y]$ gives the **Lie derivative** of Y along X. Then the Lie derivative formula (4.3.2) holds with α replaced by Y, and the pull-back operation given by (4.3.5).

If M is infinite-dimensional, then one defines the Lie derivative of Y along X by

$$\left. \frac{d}{dt} \right|_{t=0} \varphi_t^* Y = \pounds_X Y, \tag{4.3.7}$$

where φ_t is the flow of X. Then formula (4.3.2) with α replaced by Y holds, and the action of the vector field $\pounds_X Y$ on a function f is given by $X[Y[f]] - Y[X[f]]$, which is denoted, as in the finite-dimensional case, by $[X, Y][f]$. As before $[X, Y] = \pounds_X Y$ is also called the Jacobi–Lie bracket of vector fields.

If M is finite-dimensional, then

$$(\pounds_X Y)^j = X^i \frac{\partial Y^j}{\partial x^i} - Y^i \frac{\partial X^j}{\partial x^i} = (X \cdot \nabla)Y^j - (Y \cdot \nabla)X^j, \tag{4.3.8}$$

and in general, where we identify X, Y with their local representatives, we have

$$[X, Y] = \mathbf{D}Y \cdot X - \mathbf{D}X \cdot Y. \tag{4.3.9}$$

The formula for $[X, Y] = \pounds_X Y$ can be remembered by writing

$$\left[X^i \frac{\partial}{\partial x^i}, Y^j \frac{\partial}{\partial x^j} \right] = X^i \frac{\partial Y^j}{\partial x^i} \frac{\partial}{\partial x^j} - Y^j \frac{\partial X^i}{\partial x^j} \frac{\partial}{\partial x^i}.$$

Algebraic Definition of the Lie Derivative. The *algebraic approach* to the Lie derivative on forms or tensors proceeds as follows. Extend the definition of the Lie derivative from functions and vector fields to differential forms, by requiring that the Lie derivative be a derivation; for example, for one-forms α, write

$$\pounds_X \langle \alpha, Y \rangle = \langle \pounds_X \alpha, Y \rangle + \langle \alpha, \pounds_X Y \rangle, \tag{4.3.10}$$

where X, Y are vector fields and $\langle \alpha, Y \rangle = \alpha(Y)$. More generally,

$$\pounds_X (\alpha(Y_1, \dots, Y_k)) = (\pounds_X \alpha)(Y_1, \dots, Y_k) + \sum_{i=1}^{k} \alpha(Y_1, \dots, \pounds_X Y_i, \dots, Y_k),$$
$$\tag{4.3.11}$$

where $X, Y_1, \dots, Y_k$ are vector fields and α is a k-form.

Proposition 4.3.2. *The dynamic and algebraic definitions of the Lie derivative of a differential k-form are equivalent.*

Cartan's Magic Formula. A very important formula for the Lie derivative is given by the following.

Theorem 4.3.3. *For X a vector field and α a k-form on a manifold M, we have*

$$\pounds_X\alpha = \mathbf{d}\,\mathbf{i}_X\alpha + \mathbf{i}_X\,\mathbf{d}\alpha, \tag{4.3.12}$$

or, in the "hook" notation,

$$\pounds_X\alpha = \mathbf{d}(X \lrcorner \alpha) + X \lrcorner \mathbf{d}\alpha.$$

This is proved by a lengthy but straightforward calculation.

Another property of the Lie derivative is the following: If $\varphi : M \to N$ is a diffeomorphism, then

$$\varphi^*\pounds_Y\beta = \pounds_{\varphi^*Y}\varphi^*\beta$$

for $Y \in \mathfrak{X}(N)$ and $\beta \in \Omega^k(M)$. More generally, if $X \in \mathfrak{X}(M)$ and $Y \in \mathfrak{X}(N)$ are ψ related, that is, $T\psi \circ X = Y \circ \psi$ for $\psi : M \to N$ a smooth map, then $\pounds_X\psi^*\beta = \psi^*\pounds_Y\beta$ for all $\beta \in \Omega^k(N)$.

There are a number of valuable identities relating the Lie derivative, the exterior derivative, and the interior product that we record at the end of this chapter. For example, if Θ is a one-form and X and Y are vector fields, identity 6 in the table at the end of §4.4 gives the useful identity

$$\mathbf{d}\Theta(X,Y) = X[\Theta(Y)] - Y[\Theta(X)] - \Theta([X,Y]). \tag{4.3.13}$$

Volume Forms and Divergence. An n-manifold M is said to be *orientable* if there is a nowhere-vanishing n-form μ on it; μ is called a *volume form*, and it is a basis of $\Omega^n(M)$ over $\mathcal{F}(M)$. Two volume forms μ_1 and μ_2 on M are said to define the same *orientation* if there is an $f \in \mathcal{F}(M)$ with $f > 0$ and such that $\mu_2 = f\mu_1$. Connected orientable manifolds admit precisely two orientations. A basis $\{v_1,\dots,v_n\}$ of T_mM is said to be *positively oriented* relative to the volume form μ on M if $\mu(m)(v_1,\dots,v_n) > 0$. Note that the volume forms defining the same orientation form a convex cone in $\Omega^n(M)$, that is, if $a > 0$ and μ is a volume form, then $a\mu$ is again a volume form, and if $t \in [0,1]$ and μ_1, μ_2 are volume forms defining the same orientation, then $t\mu_1 + (1-t)\mu_2$ is again a volume form defining the same orientation as μ_1 or μ_2. The first property is obvious. To prove the second, let $m \in M$ and let $\{v_1,\dots,v_n\}$ be a positively oriented basis of T_mM relative to the orientation defined by μ_1, or equivalently (by hypothesis) by μ_2. Then $\mu_1(m)(v_1,\dots,v_n) > 0$, $\mu_2(m)(v_1,\dots,v_n) > 0$, so that their convex combination is again strictly positive.

If $\mu \in \Omega^n(M)$ is a volume form, since $\pounds_X \mu \in \Omega^n(M)$, there is a function, called the **divergence** of X relative to μ and denoted by $\mathrm{div}_\mu(X)$ or simply $\mathrm{div}(X)$, such that

$$\pounds_X \mu = \mathrm{div}_\mu(X)\mu. \qquad (4.3.14)$$

From the dynamic approach to Lie derivatives it follows that $\mathrm{div}_\mu(X) = 0$ if and only if $F_t^*\mu = \mu$, where F_t is the flow of X. This condition says that F_t is **volume preserving**. If $\varphi : M \to M$, since $\varphi^*\mu \in \Omega^n(M)$ there is a function, called the **Jacobian** of φ and denoted by $J_\mu(\varphi)$ or simply $J(\varphi)$, such that

$$\varphi^*\mu = J_\mu(\varphi)\mu. \qquad (4.3.15)$$

Thus, φ is volume preserving if and only if $J_\mu(\varphi) = 1$. From the inverse function theorem, we see that φ is a local diffeomorphism if and only if $J_\mu(\varphi) \neq 0$ on M.

Frobenius' Theorem.   We also mention a basic result called **Frobenius' theorem**. If $E \subset TM$ is a vector subbundle, it is said to be **involutive** if for any two vector fields X, Y on M with values in E, the Jacobi–Lie bracket $[X, Y]$ is also a vector field with values in E. The subbundle E is said to be **integrable** if for each point $m \in M$ there is a local submanifold of M containing m such that its tangent bundle equals E restricted to this submanifold. If E is integrable, the local integral manifolds can be extended to get, through each $m \in M$, a connected maximal integral manifold, which is unique and is a regularly immersed submanifold of M. The collection of all maximal integral manifolds through all points of M is said to form a **foliation**.

The Frobenius theorem states that the *involutivity of E is equivalent to the integrability of E.*

Exercises

$\diamond$ **4.3-1.**   Let M be an n-manifold, $\mu \in \Omega^n(M)$ a volume form, $X, Y \in \mathfrak{X}(M)$, and $f, g : M \to \mathbb{R}$ smooth functions such that $f(m) \neq 0$ for all m. Prove the following identities:

(a) $\mathrm{div}_{f\mu}(X) = \mathrm{div}_\mu(X) + X[f]/f$;

(b) $\mathrm{div}_\mu(gX) = g\,\mathrm{div}_\mu(X) + X[g]$; and

(c) $\mathrm{div}_\mu([X, Y]) = X[\mathrm{div}_\mu(Y)] - Y[\mathrm{div}_\mu(X)]$.

$\diamond$ **4.3-2.**   Show that the partial differential equation

$$\frac{\partial f}{\partial t} = \sum_{i=1}^{n} X^i(x^1, \ldots, x^n)\frac{\partial f}{\partial x^i}$$

with initial condition $f(x,0) = g(x)$ has the solution $f(x,t) = g(F_t(x))$, where F_t is the flow of the vector field $(X^1, \ldots, X^n)$ in $\mathbb{R}^n$ whose flow is assumed to exist for all time. Show that the solution is *unique*. Generalize this exercise to the equation

$$\frac{\partial f}{\partial t} = X[f]$$

for X a vector field on a manifold M.

◇ **4.3-3.** Show that if M and N are orientable manifolds, so is $M \times N$.

4.4 Stokes' Theorem

The basic idea of the definition of the integral of an n-form μ on an oriented n-manifold M is to pick a covering by coordinate charts and to sum up the ordinary integrals of $f(x^1, \ldots, x^n) \, dx^1 \cdots dx^n$, where

$$\mu = f(x^1, \ldots, x^n) \, dx^1 \wedge \cdots \wedge dx^n$$

is the local representative of μ, being careful not to count overlaps twice. The change of variables formula guarantees that the result, denoted by $\int_M \mu$, is well-defined.

If one has an oriented manifold with boundary, then the boundary, ∂M, inherits a compatible orientation. This proceeds in a way that generalizes the relation between the orientation of a surface and its boundary in the classical Stokes' theorem in $\mathbb{R}^3$.

Theorem 4.4.1 (Stokes' Theorem). *Suppose that M is a compact, oriented k-dimensional manifold with boundary ∂M. Let α be a smooth $(k-1)$-form on M. Then*

$$\int_M d\alpha = \int_{\partial M} \alpha. \tag{4.4.1}$$

Special cases of Stokes' theorem are as follows:

The Integral Theorems of Calculus. Stokes' theorem generalizes and synthesizes the classical theorems of calculus:

(a) Fundamental Theorem of Calculus.

$$\int_a^b f'(x) \, dx = f(b) - f(a). \tag{4.4.2}$$

(b) Green's Theorem. For a region $\Omega \subset \mathbb{R}^2$,

$$\iint_\Omega \left(\frac{\partial Q}{\partial x} - \frac{\partial P}{\partial y} \right) dx \, dy = \int_{\partial \Omega} P \, dx + Q \, dy. \tag{4.4.3}$$

(c) **Divergence Theorem.** For a region $\Omega \subset \mathbb{R}^3$,

$$\iiint_\Omega \operatorname{div} \mathbf{F}\, dV = \iint_{\partial\Omega} \mathbf{F}\cdot\mathbf{n}\, dA. \qquad (4.4.4)$$

(d) **Classical Stokes' Theorem.** For a surface $S \subset \mathbb{R}^3$,

$$\iint_S \left\{ \left(\frac{\partial R}{\partial y} - \frac{\partial Q}{\partial z}\right) dy \wedge dz \right.$$

$$\left. + \left(\frac{\partial P}{\partial z} - \frac{\partial R}{\partial x}\right) dz \wedge dx + \left(\frac{\partial Q}{\partial x} - \frac{\partial P}{\partial y}\right) dx \wedge dy \right\}$$

$$= \iint_S \mathbf{n}\cdot\operatorname{curl}\mathbf{F}\, dA = \int_{\partial S} P\, dx + Q\, dy + R\, dz, \qquad (4.4.5)$$

where $\mathbf{F} = (P, Q, R)$.

Notice that the Poincaré lemma generalizes the vector calculus theorems in $\mathbb{R}^3$, saying that if $\operatorname{curl}\mathbf{F} = 0$, then $\mathbf{F} = \nabla f$, and if $\operatorname{div}\mathbf{F} = 0$, then $\mathbf{F} = \nabla \times \mathbf{G}$. Recall that it states that *if α is closed, then locally α is exact; that is, if $\mathbf{d}\alpha = 0$, then locally $\alpha = \mathbf{d}\beta$ for some β.* On contractible manifolds these statements hold globally.

Cohomology. The failure of closed forms to be globally exact leads to the study of a very important topological invariant of M, the *de Rham cohomology*. The kth de Rham cohomology group, denoted by $H^k(M)$, is defined by

$$H^k(M) := \frac{\ker(\mathbf{d} : \Omega^k(M) \to \Omega^{k+1}(M))}{\operatorname{range}(\mathbf{d} : \Omega^{k-1}(M) \to \Omega^k(M))}.$$

The de Rham theorem states that these Abelian groups are isomorphic to the so-called singular cohomology groups of M defined in algebraic topology in terms of simplices and that depend only on the topological structure of M and not on its differentiable structure. The isomorphism is provided by integration; the fact that the integration map drops to the preceding quotient is guaranteed by Stokes' theorem. A useful particular case of this theorem is the following: If M is an orientable compact boundaryless n-manifold, then $\int_M \mu = 0$ if and only if the n-form μ is exact. This statement is equivalent to $H^n(M) = \mathbb{R}$ for M compact and orientable.

Change of Variables. Another basic result in integration theory is the global change of variables formula.

Theorem 4.4.2 (Change of Variables). *Let M and N be oriented n-manifolds and let $\varphi : M \to N$ be an orientation-preserving diffeomorphism. If α is an n-form on N (with, say, compact support), then*

$$\int_M \varphi^*\alpha = \int_N \alpha.$$

Identities for Vector Fields and Forms

1. Vector fields on M with the bracket $[X, Y]$ form a *Lie algebra*; that is, $[X, Y]$ is real bilinear, skew-symmetric, and *Jacobi's identity* holds:

$$[[X, Y], Z] + [[Z, X], Y] + [[Y, Z], X] = 0.$$

   Locally,

$$[X, Y] = \mathbf{D}Y \cdot X - \mathbf{D}X \cdot Y = (X \cdot \nabla)Y - (Y \cdot \nabla)X,$$

   and on functions,

$$[X, Y][f] = X[Y[f]] - Y[X[f]].$$

2. For diffeomorphisms φ and ψ,

$$\varphi_*[X, Y] = [\varphi_* X, \varphi_* Y] \quad \text{and} \quad (\varphi \circ \psi)_* X = \varphi_* \psi_* X.$$

3. The forms on a manifold comprise a real associative algebra with $\wedge$ as multiplication. Furthermore, $\alpha \wedge \beta = (-1)^{kl} \beta \wedge \alpha$ for k- and l-forms α and β, respectively.

4. For maps φ and ψ,

$$\varphi^*(\alpha \wedge \beta) = \varphi^* \alpha \wedge \varphi^* \beta \quad \text{and} \quad (\varphi \circ \psi)^* \alpha = \psi^* \varphi^* \alpha.$$

5. $\mathbf{d}$ is a real linear map on forms, $\mathbf{dd}\alpha = 0$, and

$$\mathbf{d}(\alpha \wedge \beta) = \mathbf{d}\alpha \wedge \beta + (-1)^k \alpha \wedge \mathbf{d}\beta$$

   for α a k-form.

6. For α a k-form and $X_0, \ldots, X_k$ vector fields,

$$(\mathbf{d}\alpha)(X_0, \ldots, X_k) = \sum_{i=0}^{k} (-1)^i X_i[\alpha(X_0, \ldots, \hat{X}_i, \ldots, X_k)]$$

$$+ \sum_{0 \le i < j \le k} (-1)^{i+j} \alpha([X_i, X_j], X_0, \ldots, \hat{X}_i, \ldots, \hat{X}_j, \ldots, X_k),$$

   where $\hat{X}_i$ means that X_i is omitted. Locally,

$$\mathbf{d}\alpha(x)(v_0, \ldots, v_k) = \sum_{i=0}^{k} (-1)^i \mathbf{D}\alpha(x) \cdot v_i(v_0, \ldots, \hat{v}_i, \ldots, v_k).$$

7. For a map φ,

$$\varphi^* d\alpha = d\varphi^* \alpha.$$

8. **Poincaré Lemma.** If $d\alpha = 0$, then the k-form α is locally exact; that is, there is a neighborhood U about each point on which $\alpha = d\beta$. This statement is global on contractible manifolds or more generally if $H^k(M) = 0$.

9. $i_X \alpha$ is real bilinear in X, α, and for $h : M \to \mathbb{R}$,

$$i_{hX}\alpha = h i_X \alpha = i_X h\alpha.$$

Also, $i_X i_X \alpha = 0$ and

$$i_X(\alpha \wedge \beta) = i_X \alpha \wedge \beta + (-1)^k \alpha \wedge i_X \beta$$

for α a k-form.

10. For a diffeomorphism φ,

$$\varphi^*(i_X \alpha) = i_{\varphi^* X}(\varphi^* \alpha), \quad \text{i.e.,} \quad \varphi^*(X \lrcorner \alpha) = (\varphi^* X) \lrcorner (\varphi^* \alpha).$$

If $f : M \to N$ is a mapping and Y is f-related to X, that is,

$$Tf \circ X = Y \circ f,$$

then

$$i_X f^* \alpha = f^* i_Y \alpha; \quad \text{i.e.,} \quad X \lrcorner (f^* \alpha) = f^*(Y \lrcorner \alpha).$$

11. $\mathcal{L}_X \alpha$ is real bilinear in X, α and

$$\mathcal{L}_X(\alpha \wedge \beta) = \mathcal{L}_X \alpha \wedge \beta + \alpha \wedge \mathcal{L}_X \beta.$$

12. **Cartan's Magic Formula:**

$$\mathcal{L}_X \alpha = d i_X \alpha + i_X d\alpha = d(X \lrcorner \alpha) + X \lrcorner d\alpha.$$

13. For a diffeomorphism φ,

$$\varphi^* \mathcal{L}_X \alpha = \mathcal{L}_{\varphi^* X} \varphi^* \alpha.$$

If $f : M \to N$ is a mapping and Y is f-related to X, then

$$\mathcal{L}_Y f^* \alpha = f^* \mathcal{L}_X \alpha.$$

14. $(\pounds_X \alpha)(X_1, \dots, X_k) = X[\alpha(X_1, \dots, X_k)]$

$$- \sum_{i=0}^{k} \alpha(X_1, \dots, [X, X_i], \dots, X_k).$$

Locally,

$$(\pounds_X \alpha)(x) \cdot (v_1, \dots, v_k) = (\mathbf{D}\alpha_x \cdot X(x))(v_1, \dots, v_k)$$

$$+ \sum_{i=0}^{k} \alpha_x(v_1, \dots, \mathbf{D}X_x \cdot v_i, \dots, v_k).$$

15. The following identities hold:

   (a) $\pounds_{fX} \alpha = f \pounds_X \alpha + df \wedge \mathbf{i}_X \alpha;$

   (b) $\pounds_{[X,Y]} \alpha = \pounds_X \pounds_Y \alpha - \pounds_Y \pounds_X \alpha;$

   (c) $\mathbf{i}_{[X,Y]} \alpha = \pounds_X \mathbf{i}_Y \alpha - \mathbf{i}_Y \pounds_X \alpha;$

   (d) $\pounds_X \mathbf{d}\alpha = \mathbf{d} \pounds_X \alpha;$

   (e) $\pounds_X \mathbf{i}_X \alpha = \mathbf{i}_X \pounds_X \alpha;$

   (f) $\pounds_X (\alpha \wedge \beta) = \pounds_X \alpha \wedge \beta + \alpha \wedge \pounds_X \beta.$

16. If M is a finite-dimensional manifold, $X = X^l \partial/\partial x^l$, and

$$\alpha = \alpha_{i_1 \dots i_k} dx^{i_1} \wedge \dots \wedge dx^{i_k},$$

where $i_1 < \dots < i_k$, then the following formulas hold:

$$\mathbf{d}\alpha = \left(\frac{\partial \alpha_{i_1 \dots i_k}}{\partial x^l} \right) dx^l \wedge dx^{i_1} \wedge \dots \wedge dx^{i_k},$$

$$\mathbf{i}_X \alpha = X^l \alpha_{l i_2 \dots i_k} dx^{i_2} \wedge \dots \wedge dx^{i_k},$$

$$\pounds_X \alpha = X^l \left(\frac{\partial \alpha_{i_1 \dots i_k}}{\partial x^l} \right) dx^{i_1} \wedge \dots \wedge dx^{i_k}$$

$$+ \alpha_{l i_2 \dots i_k} \left(\frac{\partial X^l}{\partial x^{i_1}} \right) dx^{i_1} \wedge dx^{i_2} \wedge \dots \wedge dx^{i_k} + \dots.$$

Exercises

◇ **4.4-1.** Let Ω be a closed bounded region in $\mathbb{R}^2$. Use Green's theorem to show that the area of Ω equals the line integral

$$\frac{1}{2} \int_{\partial \Omega} (x \, dy - y \, dx).$$

◇ **4.4-2.** On $\mathbb{R}^2\backslash\{(0,0)\}$ consider the one-form

$$\alpha = \frac{x\,dy - y\,dx}{x^2 + y^2}.$$

(a) Show that this form is closed.

(b) Using the angle θ as a variable on S^1, compute $i^*\alpha$, where $i : S^1 \to \mathbb{R}^2$ is the standard embedding.

(c) Show that α is not exact.

◇ **4.4-3 (The Magnetic Monopole).** Let $\mathbf{B} = g\mathbf{r}/r^3$ be a vector field on Euclidean three-space minus the origin where $r = \|\mathbf{r}\|$. Show that $\mathbf{B}$ cannot be written as the curl of something.

5

Hamiltonian Systems on Symplectic Manifolds

Now we are ready to geometrize Hamiltonian mechanics to the context of manifolds. First we make phase spaces nonlinear, and then we study Hamiltonian systems in this context.

5.1 Symplectic Manifolds

Definition 5.1.1. *A **symplectic manifold** is a pair* (P, Ω) *where* P *is a manifold and* Ω *is a closed (weakly) nondegenerate two-form on* P. *If* Ω *is strongly nondegenerate, we speak of a **strong symplectic manifold**.*

As in the linear case, strong nondegeneracy of the two-form Ω means that at each $z \in P$, the bilinear form $\Omega_z : T_z P \times T_z P \to \mathbb{R}$ is nondegenerate, that is, Ω_z defines an isomorphism

$$\Omega_z^\flat : T_z P \to T_z^* P.$$

For a (weak) symplectic form, the induced map $\Omega^\flat : \mathfrak{X}(P) \to \mathfrak{X}^*(P)$ between vector fields and one-forms is one-to-one, but in general is not surjective. We will see later that Ω is required to be closed, that is, $\mathbf{d}\Omega = 0$, where $\mathbf{d}$ is the exterior derivative, so that the induced Poisson bracket satisfies the Jacobi identity and so that the flows of Hamiltonian vector fields will consist of canonical transformations. In coordinates z^I on P in the finite-dimensional case, if $\Omega = \Omega_{IJ} \, dz^I \wedge dz^J$ (sum over all $I < J$), then

$d\Omega = 0$ becomes the condition

$$\frac{\partial \Omega_{IJ}}{\partial z^K} + \frac{\partial \Omega_{KI}}{\partial z^J} + \frac{\partial \Omega_{JK}}{\partial z^I} = 0. \qquad (5.1.1)$$

Examples

(a) **Symplectic Vector Spaces.** If (Z, Ω) is a symplectic vector space, then it is also a symplectic manifold. The requirement $d\Omega = 0$ is satisfied automatically, since Ω is a *constant* form (that is, $\Omega(z)$ is independent of $z \in Z$). ♦

(b)   The cylinder $S^1 \times \mathbb{R}$ with coordinates (θ, p) is a symplectic manifold with $\Omega = d\theta \wedge dp$. ♦

(c)   The torus $\mathbb{T}^2$ with periodic coordinates (θ, φ) is a symplectic manifold with $\Omega = d\theta \wedge d\varphi$. ♦

(d)   The two-sphere S^2 of radius r is symplectic with Ω the standard *area element* $\Omega = r^2 \sin\theta \, d\theta \wedge d\varphi$ on the sphere as the symplectic form. ♦

   Given a manifold Q, we will show in Chapter 6 that the cotangent bundle T^*Q has a natural symplectic structure. When Q is the **configuration space** of a mechanical system, T^*Q is called the **momentum phase space**. This important example generalizes the linear examples with phase spaces of the form $W \times W^*$ that we studied in Chapter 2.

Darboux' Theorem.   The next result says that, in principle, every strong symplectic manifold is, in suitable local coordinates, a symplectic vector space. (By contrast, a corresponding result for Riemannian manifolds is not true unless they have zero curvature; that is, are flat.)

Theorem 5.1.2 (Darboux' Theorem).   *Let (P, Ω) be a strong symplectic manifold. Then in a neighborhood of each $z \in P$, there is a local coordinate chart in which Ω is constant.*

Proof.   We can assume $P = E$ and $z = 0 \in E$, where E is a Banach space. Let Ω_1 be the constant form equaling $\Omega(0)$. Let $\Omega' = \Omega_1 - \Omega$ and $\Omega_t = \Omega + t\Omega'$, for $0 \le t \le 1$. For each t, the bilinear form $\Omega_t(0) = \Omega(0)$ is nondegenerate. Hence by openness of the set of linear isomorphisms of E to E^* and compactness of $[0,1]$, there is a neighborhood of 0 on which Ω_t is strongly nondegenerate for all $0 \le t \le 1$. We can assume that this neighborhood is a ball. Thus by the Poincaré lemma, $\Omega' = d\alpha$ for some one-form α. Replacing α by $\alpha - \alpha(0)$, we can suppose $\alpha(0) = 0$. Define a smooth time-dependent vector field X_t by

$$\mathbf{i}_{X_t}\Omega_t = -\alpha,$$

which is possible, since Ω_t is strongly nondegenerate. Since $\alpha(0) = 0$, we get $X_t(0) = 0$, and so from the local existence theory for ordinary differential equations, there is a ball on which the integral curves of X_t are defined for a time at least one; see Abraham, Marsden, and Ratiu [1988, Section 4.1], for the technical theorem. Let F_t be the flow of X_t starting at $F_0 = $ identity. By the Lie derivative formula for *time-dependent* vector fields, we have

$$\frac{d}{dt}(F_t^*\Omega_t) = F_t^*(\pounds_{X_t}\Omega_t) + F_t^*\frac{d}{dt}\Omega_t$$

$$= F_t^*\mathbf{d}\mathbf{i}_{X_t}\Omega_t + F_t^*\Omega' = F_t^*(\mathbf{d}(-\alpha) + \Omega') = 0.$$

Thus, $F_1^*\Omega_1 = F_0^*\Omega_0 = \Omega$, so F_1 provides a chart transforming Ω to the constant form Ω_1.   ∎

This proof is due to Moser [1965]. As was noted by Weinstein [1971], this proof generalizes to the infinite-dimensional *strong* symplectic case. Unfortunately, many interesting infinite-dimensional symplectic manifolds are *not* strong. In fact, the analogue of Darboux's theorem is not valid for weak symplectic forms. For an example, see Exercise 5.1-3, and for conditions under which it is valid, see Marsden [1981], Olver [1988], Bambusi [1998], and references therein. For an equivariant Darboux theorem and references, see Dellnitz and Melbourne [1993] and the discussion in Chapter 9.

Corollary 5.1.3. *If (P, Ω) is a finite-dimensional symplectic manifold, then P is even dimensional, and in a neighborhood of $z \in P$ there are local coordinates $(q^1, \ldots, q^n, p_1, \ldots, p_n)$ (where $\dim P = 2n$) such that*

$$\Omega = \sum_{i=1}^{n} dq^i \wedge dp_i. \tag{5.1.2}$$

This follows from Darboux' theorem and the canonical form for linear symplectic forms. As in the vector space case, coordinates in which Ω takes the above form are called **canonical coordinates.**

Corollary 5.1.4. *If (P, Ω) is a $2n$-dimensional symplectic manifold, then P is oriented by the **Liouville volume** form, defined as*

$$\Lambda = \frac{(-1)^{n(n-1)/2}}{n!}\, \Omega \wedge \cdots \wedge \Omega \quad (n \text{ times}). \tag{5.1.3}$$

In canonical coordinates $(q^1, \ldots, q^n, p_1, \ldots, p_n)$, Λ has the expression

$$\Lambda = dq^1 \wedge \cdots \wedge dq^n \wedge dp_1 \wedge \cdots \wedge dp_n. \tag{5.1.4}$$

Thus, if (P, Ω) is a $2n$-dimensional symplectic manifold, then (P, Λ) is a **volume manifold** (that is, a manifold with a volume element). The measure associated to Λ is called the **Liouville measure**. The factor $(-1)^{n(n-1)/2}/n!$ is chosen so that in canonical coordinates, Λ has the expression (5.1.4).

Exercises

◇ **5.1-1.** Show how to construct (explicitly) canonical coordinates for the symplectic form $\Omega = f\mu$ on S^2, where μ is the standard area element and where $f : S^2 \to \mathbb{R}$ is a positive function.

◇ **5.1-2** (Moser [1965]). Let μ_0 and μ_1 be two volume elements (nowhere-vanishing n-forms) on the compact boundaryless n-manifold M giving M the same orientation. Assume that $\int_M \mu_0 = \int_M \mu_1$. Show that there is a diffeomorphism $\varphi : M \to M$ such that $\varphi^* \mu_1 = \mu_0$.

◇ **5.1-3.** (Requires some functional analysis.) Prove that Darboux' theorem fails for the following weak symplectic form. Let H be a real Hilbert space and $S : H \to H$ a compact, self-adjoint, and positive operator whose range is dense in H but not equal to H. Let $A_x = S + \|x\|^2 I$ and

$$g_x(e, f) = \langle A_x e, f \rangle.$$

Let Ω be the weak symplectic form on $H \times H$ associated to g. Show that there is no coordinate chart about $(0,0) \in H \times H$ on which Ω is constant.

◇ **5.1-4.** Use the method of proof of the Darboux theorem to show the following. Assume that Ω_0 and Ω_1 are two symplectic forms on the compact manifold P such that $[\Omega_0]$, $[\Omega_1]$ are the cohomology classes of Ω_0 and Ω_1, respectively, in $H^2(P; \mathbb{R})$. If for every $t \in [0, 1]$ the form $\Omega_t := (1-t)\Omega_0 + \Omega_1$ is nondegenerate, show that there is a diffeomorphism $\varphi : P \longrightarrow P$ such that $\varphi^* \Omega_1 = \Omega_0$.

◇ **5.1-5.** Prove the following **relative Darboux theorem.** Let S be a submanifold of P and assume that Ω_0 and Ω_1 are two strong symplectic forms on P such that $\Omega_0|S = \Omega_1|S$. Then there is an open neighborhood V of S in P and a diffeomorphism $\varphi : V \longrightarrow \varphi(V) \subset P$ such that $\varphi|S =$ identity on S and $\varphi^* \Omega_1 = \Omega_0$. (Hint: Use Exercise 4.2-6.)

5.2 Symplectic Transformations

Definition 5.2.1. Let (P_1, Ω_1) and (P_2, Ω_2) be symplectic manifolds. A C^∞-mapping $\varphi : P_1 \to P_2$ is called **symplectic** or **canonical** if

$$\varphi^* \Omega_2 = \Omega_1. \tag{5.2.1}$$

Recall that $\Omega_1 = \varphi^* \Omega_2$ means that for each $z \in P_1$, and all $v, w \in T_z P_1$, we have the following identity:

$$\Omega_{1z}(v, w) = \Omega_{2\varphi(z)}(T_z\varphi \cdot v, T_z\varphi \cdot w),$$

where Ω_{1z} means Ω_1 evaluated at the point z and where $T_z\varphi$ is the tangent (derivative) of φ at z.

If $\varphi : (P_1, \Omega_1) \to (P_2, \Omega_2)$ is canonical, the property $\varphi^*(\alpha \wedge \beta) = \varphi^* \alpha \wedge \varphi^* \beta$ implies that $\varphi^* \Lambda = \Lambda$; that is, φ also preserves the Liouville measure. Thus we get the following:

Proposition 5.2.2.   *A smooth canonical transformation between symplectic manifolds of the same dimension is volume preserving and is a local diffeomorphism.*

The last statement comes from the inverse function theorem: If φ is volume preserving, its Jacobian determinant is 1, so φ is locally invertible. It is clear that the set of canonical diffeomorphisms of P form a subgroup of $\mathrm{Diff}(P)$, the group of all diffeomorphisms of P. This group, denoted by $\mathrm{Diff}_{\mathrm{can}}(P)$, plays a key role in the study of plasma dynamics.

If Ω_1 and Ω_2 are exact, say $\Omega_1 = -\mathbf{d}\Theta_1$ and $\Omega_2 = -\mathbf{d}\Theta_2$, then (5.2.1) is equivalent to

$$\mathbf{d}(\varphi^* \Theta_2 - \Theta_1) = 0. \tag{5.2.2}$$

Let $M \subset P_1$ be an oriented two-manifold with boundary ∂M. Then if (5.2.2) holds, we get

$$0 = \int_M \mathbf{d}(\varphi^* \Theta_2 - \Theta_1) = \int_{\partial M} (\varphi^* \Theta_2 - \Theta_1),$$

that is,

$$\int_{\partial M} \varphi^* \Theta_2 = \int_{\partial M} \Theta_1. \tag{5.2.3}$$

Proposition 5.2.3.   *The map $\varphi : P_1 \to P_2$ is canonical if and only if (5.2.3) holds for every oriented two-manifold $M \subset P_1$ with boundary ∂M.*

The converse is proved by choosing M to be a small disk in P_1 and using the fact that if the integral of a two-form over any small disk vanishes, then the form is zero. The latter assertion is proved by contradiction, constructing a two-form on a two-disk whose coefficient is a bump function. Equation (5.2.3) is an example of an **integral invariant**. For more information, see Arnold [1989] and Abraham and Marsden [1978].

Exercises

⋄ **5.2-1.**   Let $\varphi : \mathbb{R}^{2n} \to \mathbb{R}^{2n}$ be a map of the form $\varphi(q, p) = (q, p + \alpha(q))$. Use the canonical one-form to determine when φ is symplectic.

⋄ **5.2-2.**   Let $\mathbb{T}^6$ be the six-torus with symplectic form

$$\Omega = d\theta_1 \wedge d\theta_2 + d\theta_3 \wedge d\theta_4 + d\theta_5 \wedge d\theta_6.$$

Show that if $\varphi : \mathbb{T}^6 \to \mathbb{T}^6$ is symplectic and $M \subset \mathbb{T}^6$ is a compact oriented four-manifold with boundary, then

$$\int_{\partial M} \varphi^*(\Omega \wedge \Theta) = \int_{\partial M} \Omega \wedge \Theta,$$

where $\Theta = \theta_1 \, d\theta_2 + \theta_3 \, d\theta_4 + \theta_5 \, d\theta_6$.

◇ **5.2-3.** Show that any canonical map between finite-dimensional symplectic manifolds is an immersion.

5.3 Complex Structures and Kähler Manifolds

This section develops the relation between complex and symplectic geometry a little further. It may be omitted on a first reading.

Complex Structures. We begin with the case of vector spaces. By a *complex structure* on a real vector space Z, we mean a linear map $\mathbb{J} : Z \to Z$ such that $\mathbb{J}^2 = -\text{Identity}$. Setting $iz = \mathbb{J}(z)$ gives Z the structure of a complex vector space.

Note that if Z is finite-dimensional, the hypothesis on $\mathbb{J}$ implies that $(\det \mathbb{J})^2 = (-1)^{\dim Z}$, so $\dim Z$ must be an even number, since $\det \mathbb{J} \in \mathbb{R}$. The complex dimension of Z is half the real dimension. Conversely, if Z is a complex vector space, it is also a real vector space by restricting scalar multiplication to the real numbers. In this case, $\mathbb{J}z = iz$ is the complex structure on Z. As before, the real dimension of Z is twice the complex dimension, since the vectors z and iz are linearly independent.

We have already seen that the imaginary part of a complex inner product is a symplectic form. Conversely, *if $\mathcal{H}$ is a real Hilbert space and Ω is a skew-symmetric weakly nondegenerate bilinear form on $\mathcal{H}$, then there is a complex structure $\mathbb{J}$ on $\mathcal{H}$ and a real inner product s such that*

$$s(z, w) = -\Omega(\mathbb{J}z, w). \tag{5.3.1}$$

The expression

$$h(z, w) = s(z, w) - i\Omega(z, w) \tag{5.3.2}$$

defines a Hermitian inner product, and h or s is complete on $\mathcal{H}$ if and only if Ω is strongly nondegenerate. (See Abraham and Marsden [1978, p. 173] for the proof.) Moreover, given any two of $(s, \mathbb{J}, \Omega)$, there is at most one third structure such that (5.3.1) holds.

If we identify $\mathbb{C}^n$ with $\mathbb{R}^{2n}$ and write

$$z = (z_1, \ldots, z_n) = (x_1 + iy_1, \ldots, x_n + iy_n) = ((x_1, y_1), \ldots, (x_n, y_n)),$$

then

$$-\operatorname{Im}\langle(z_1,\dots,z_n),(z_1',\dots,z_n')\rangle = -\operatorname{Im}(z_1\overline{z}_1' + \cdots + z_n\overline{z}_n')$$
$$= -(x_1'y_1 - x_1y_1' + \cdots + x_n'y_n - x_ny_n').$$

Thus, the canonical symplectic form on $\mathbb{R}^{2n}$ may be written

$$\Omega(z,z') = -\operatorname{Im}\langle z,z'\rangle = \operatorname{Re}\langle iz,z'\rangle, \qquad (5.3.3)$$

which, by (5.3.1), agrees with the convention that $\mathbb{J} : \mathbb{R}^{2n} \to \mathbb{R}^{2n}$ is multiplication by i.

An *almost complex structure* $\mathbb{J}$ on a manifold M is a smooth tangent bundle isomorphism $\mathbb{J} : TM \to TM$ covering the identity map on M such that for each point $z \in M$, $\mathbb{J}_z = \mathbb{J}(z) : T_zM \to T_zM$ is a complex structure on the vector space T_zM. A manifold with an almost complex structure is called an *almost complex manifold*.

A manifold M is called a *complex manifold* if it admits an atlas $\{(U_\alpha, \varphi_\alpha)\}$ whose charts $\varphi_\alpha : U_\alpha \subset M \to E$ map to a complex Banach space E and the transition functions $\varphi_\beta \circ \varphi_\alpha^{-1} : \varphi_\alpha(U_\alpha \cap U_\beta) \to \varphi_\beta(U_\alpha \cap U_\beta)$ are holomorphic maps. The complex structure on E (multiplication by i) induces via the chart maps φ_α an almost complex structure on each chart domain U_α. Since the transition functions are biholomorphic diffeomorphisms, the almost complex structures on $U_\alpha \cap U_\beta$ induced by φ_α and φ_β coincide. This shows that a complex manifold is also almost complex. The converse is not true.

If M is an almost complex manifold, T_zM is endowed with the structure of a complex vector space. A *Hermitian metric* on M is a smooth assignment of a (possibly weak) complex inner product on T_zM for each $z \in M$. As in the case of vector spaces, the imaginary part of the Hermitian metric defines a nondegenerate (real) two-form on M. The real part of a Hermitian metric is a Riemannian metric on M. If the complex inner product on each tangent space is strongly nondegenerate, the metric is *strong*; in this case both the real and imaginary parts of the Hermitian metric are strongly nondegenerate over $\mathbb{R}$.

Kähler Manifolds. An almost complex manifold M with a Hermitian metric $\langle\,,\,\rangle$ is called a *Kähler manifold* if M is a complex manifold and the two-form $-\operatorname{Im}\langle\,,\,\rangle$ is a closed two-form on M. There is an equivalent definition that is often useful: A Kähler manifold is a smooth manifold with a Riemannian metric g and an almost complex structure $\mathbb{J}$ such that $\mathbb{J}_z$ is g-skew for each $z \in M$ and such that $\mathbb{J}$ is covariantly constant with respect to g. (One requires some Riemannian geometry to understand this definition—it will not be required in what follows.) The important fact used later on is the following:

Any Kähler manifold is also symplectic, with symplectic form given by

$$\Omega_z(v_z, w_z) = \langle \mathbb{J}_z v_z, w_z \rangle . \tag{5.3.4}$$

In this second definition of Kähler manifolds, the condition $d\Omega = 0$ follows from $\mathbb{J}$ being covariantly constant. A *strong Kähler manifold* is a Kähler manifold whose Hermitian inner product is strong.

Projective Spaces. Any complex Hilbert space $\mathcal{H}$ is a strong Kähler manifold. As an example of a more interesting Kähler manifold, we shall consider the projectivization $\mathbb{P}\mathcal{H}$ of a complex Hilbert space $\mathcal{H}$. In particular, *complex projective n-space* $\mathbb{CP}^n$ will result when this construction is applied to $\mathbb{C}^n$. Recall from Example (f) of §2.3 that $\mathcal{H}$ is a symplectic vector space relative to the quantum-mechanical symplectic form

$$\Omega(\psi_1, \psi_2) = -2\hbar \operatorname{Im} \langle \psi_1, \psi_2 \rangle ,$$

where $\langle \, , \, \rangle$ is the Hermitian inner product on $\mathcal{H}$, $\hbar$ is Planck's constant, and $\psi_1, \psi_2 \in \mathcal{H}$. Recall also that $\mathbb{P}\mathcal{H}$ is the space of complex lines through the origin in $\mathcal{H}$. Denote by $\pi : \mathcal{H} \backslash \{0\} \to \mathbb{P}\mathcal{H}$ the canonical projection that sends a vector $\psi \in \mathcal{H} \backslash \{0\}$ to the complex line it spans, denoted by $[\psi]$ when thought of as a point in $\mathbb{P}\mathcal{H}$ and by $\mathbb{C}\psi$ when interpreted as a subspace of $\mathcal{H}$. The space $\mathbb{P}\mathcal{H}$ is a smooth complex manifold, π is a smooth map, and the tangent space $T_{[\psi]}\mathbb{P}\mathcal{H}$ is isomorphic to $\mathcal{H}/\mathbb{C}\psi$. Thus, the map π is a surjective submersion. (Submersions were discussed in Chapter 4, see also Abraham, Marsden, and Ratiu [1988, Chapter 3].) Since the kernel of

$$T_\psi \pi : \mathcal{H} \to T_{[\psi]}\mathbb{P}\mathcal{H}$$

is $\mathbb{C}\psi$, the map $T_\psi \pi | (\mathbb{C}\psi)^\perp$ is a complex linear isomorphism from $(\mathbb{C}\psi)^\perp$ to $T_\psi \mathbb{P}\mathcal{H}$ that depends on the chosen representative ψ in $[\psi]$.

If $U : \mathcal{H} \to \mathcal{H}$ is a unitary operator, that is, U is invertible and

$$\langle U\psi_1, U\psi_2 \rangle = \langle \psi_1, \psi_2 \rangle$$

for all $\psi_1, \psi_2 \in \mathcal{H}$, then the rule $[U][\psi] := [U\psi]$ defines a biholomorphic diffeomorphism on $\mathbb{P}\mathcal{H}$.

Proposition 5.3.1.

(i) *If* $[\psi] \in \mathbb{P}\mathcal{H}$, $\|\psi\| = 1$, *and* $\varphi_1, \varphi_2 \in (\mathbb{C}\psi)^\perp$, *the formula*

$$\langle T_\psi \pi(\varphi_1), T_\psi \pi(\varphi_2) \rangle = 2\hbar \langle \varphi_1, \varphi_2 \rangle \tag{5.3.5}$$

gives a well-defined strong Hermitian inner product on $T_{[\psi]}\mathbb{P}\mathcal{H}$, *that is, the left-hand side does not depend on the choice of* ψ *in* $[\psi]$. *The dependence on* $[\psi]$ *is smooth, and so* (5.3.5) *defines a Hermitian metric on* $\mathbb{P}\mathcal{H}$ *called the* **Fubini–Study metric**. *This metric is invariant under the action of the maps* $[U]$, *for all unitary operators* U *on* $\mathcal{H}$.

(ii) *For* $[\psi] \in \mathbb{P}\mathcal{H}$, $\|\psi\| = 1$, *and* $\varphi_1, \varphi_2 \in (\mathbb{C}\psi)^{\perp}$,

$$g_{[\psi]}(T_{\psi}\pi(\varphi_1), T_{\psi}\pi(\varphi_2)) = 2\hbar \,\mathrm{Re}\, \langle \varphi_1, \varphi_2 \rangle \qquad (5.3.6)$$

defines a strong Riemannian metric on $\mathbb{P}\mathcal{H}$ *invariant under all transformations* $[U]$.

(iii) *For* $[\psi] \in \mathbb{P}\mathcal{H}$, $\|\psi\| = 1$, *and* $\varphi_1, \varphi_2 \in (\mathbb{C}\psi)^{\perp}$,

$$\Omega_{[\psi]}(T_{\psi}\pi(\varphi_1), T_{\psi}\pi(\varphi_2)) = -2\hbar \,\mathrm{Im}\, \langle \varphi_1, \varphi_2 \rangle \qquad (5.3.7)$$

defines a strong symplectic form on $\mathbb{P}\mathcal{H}$ *invariant under all transformations* $[U]$.

Proof. We first prove (i).[1] If $\lambda \in \mathbb{C}\backslash\{0\}$, then $\pi(\lambda(\psi + t\varphi)) = \pi(\psi + t\varphi)$, and since

$$(T_{\lambda\psi}\pi)(\lambda\varphi) = \frac{d}{dt}\pi(\lambda\psi + t\lambda\varphi)\bigg|_{t=0} = \frac{d}{dt}\pi(\psi + t\varphi)\bigg|_{t=0} = (T_{\psi}\pi)(\varphi),$$

we get $(T_{\lambda\psi}\pi)(\lambda\varphi) = (T_{\psi}\pi)(\varphi)$. Thus, if $\|\lambda\psi\| = \|\psi\| = 1$, it follows that $|\lambda| = 1$. We have, by (5.3.5),

$$\langle (T_{\lambda\psi}\pi)(\lambda\varphi_1), (T_{\lambda\psi}\pi)(\lambda\varphi_2) \rangle = 2\hbar \langle \lambda\varphi_1, \lambda\varphi_2 \rangle = 2\hbar|\lambda|^2 \langle \varphi_1, \varphi_2 \rangle$$
$$= 2\hbar \langle \varphi_1, \varphi_2 \rangle = \langle (T_{\psi}\pi)(\varphi_1), (T_{\psi}\pi)(\varphi_2) \rangle.$$

This shows that the definition (5.3.5) of the Hermitian inner product is independent of the normalized representative $\psi \in [\psi]$ chosen in order to define it. This Hermitian inner product is strong, since it coincides with the inner product on the complex Hilbert space $(\mathbb{C}\psi)^{\perp}$.

A straightforward computation (see Exercise 5.3-3) shows that for $\psi \in \mathcal{H}\backslash\{0\}$ and $\varphi_1, \varphi_2 \in \mathcal{H}$ arbitrary, the Hermitian metric is given by

$$\langle T_{\psi}\pi(\varphi_1), T_{\psi}\pi(\varphi_2) \rangle = 2\hbar\|\psi\|^{-2}(\langle \varphi_1, \varphi_2 \rangle - \|\psi\|^{-2} \langle \varphi_1, \psi \rangle \langle \psi, \varphi_2 \rangle). \quad (5.3.8)$$

Since the right-hand side is smooth in $\psi \in \mathcal{H}\backslash\{0\}$ and this formula drops to $\mathbb{P}\mathcal{H}$, it follows that (5.3.5) is smooth in $[\psi]$.

If U is a unitary map on $\mathcal{H}$ and $[U]$ is the induced map on $\mathbb{P}\mathcal{H}$, we have

$$T_{[\psi]}[U] \cdot T_{\psi}\pi(\varphi) = T_{[\psi]}[U] \cdot \frac{d}{dt}[\psi + t\varphi]\bigg|_{t=0} = \frac{d}{dt}[U][\psi + t\varphi]\bigg|_{t=0}$$

$$= \frac{d}{dt}[U(\psi + t\varphi)]\bigg|_{t=0} = T_{U\psi}\pi(U\varphi).$$

---

[1]One can give a conceptually cleaner, but more advanced, approach to this process using general reduction theory. The proof given here is by a direct argument.

Therefore, since $\|U\psi\| = \|\psi\| = 1$ and $\langle U\varphi_j, U\psi \rangle = 0$, we get by (5.3.5),

$$
\begin{aligned}
\langle T_{[\psi]}[U] \cdot T_\psi \pi(\varphi_1), T_{[\psi]}[U] \cdot T_\psi \pi(\varphi_2) \rangle &= \langle T_{U\psi}\pi(U\varphi_1), T_{U\psi}\pi(U\varphi_2) \rangle \\
&= \langle U\varphi_1, U\varphi_2 \rangle = \langle \varphi_1, \varphi_2 \rangle \\
&= \langle T_\psi\pi(\varphi_1), T_\psi\pi(\varphi_2) \rangle,
\end{aligned}
$$

which proves the invariance of the Hermitian metric under the action of the transformation $[U]$.

Part (ii) is obvious as the real part of the Hermitian metric (5.3.5).

Finally, we prove (iii). From the invariance of the metric it follows that the form Ω is also invariant under the action of unitary maps, that is, $[U]^*\Omega = \Omega$. So, also $[U]^*\mathbf{d}\Omega = \mathbf{d}\Omega$. Now consider the unitary map U_0 on $\mathcal{H}$ defined by $U_0\psi = \psi$ and $U_0 = -\text{Identity}$ on $(\mathbb{C}\psi)^\perp$. Then from $[U_0]^*\Omega = \Omega$ we have for $\varphi_1, \varphi_2, \varphi_3 \in (\mathbb{C}\psi)^\perp$,

$$
\begin{aligned}
&\mathbf{d}\Omega([\psi])(T_\psi\pi(\varphi_1), T_\psi\pi(\varphi_2), T_\psi\pi(\varphi_3)) \\
&= \mathbf{d}\Omega([\psi])(T_{[\psi]}[U_0] \cdot T_\psi\pi(\varphi_1), T_{[\psi]}[U_0] \cdot T_\psi\pi(\varphi_2), T_{[\psi]}[U_0] \cdot T_\psi\pi(\varphi_3)).
\end{aligned}
$$

But

$$
T_{[\psi]}[U_0] \cdot T_\psi\pi(\varphi) = T_\psi\pi(-\varphi) = -T_\psi\pi(\varphi),
$$

which implies by trilinearity of $\mathbf{d}\Omega$ that $\mathbf{d}\Omega = 0$.

The symplectic form Ω is strongly nondegenerate, since on $T_{[\psi]}\mathbb{P}\mathcal{H}$ it restricts to the corresponding quantum-mechanical symplectic form on the Hilbert space $(\mathbb{C}\psi)^\perp$. ∎

The results above prove that $\mathbb{P}\mathcal{H}$ is an infinite-dimensional Kähler manifold on which the unitary group $U(\mathcal{H})$ acts by isometries. This can be generalized to Grassmannian manifolds of finite- (or infinite-) dimensional subspaces of $\mathcal{H}$, and even more, to flag manifolds (see Besse [1987] and Pressley and Segal [1986]).

Exercises

⋄ **5.3-1.** On $\mathbb{C}^n$, show that $\Omega = -\mathbf{d}\Theta$, where $\Theta(z) \cdot w = \frac{1}{2}\text{Im}\langle z, w \rangle$.

⋄ **5.3-2.** Let P be a manifold that is both symplectic, with symplectic form Ω, and Riemannian, with metric g.

(a) Show that P has an almost complex structure $\mathbb{J}$ such that $\Omega(u, v) = g(\mathbb{J}u, v)$ if and only if

$$
\Omega(\nabla F, v) = -g(X_F, v)
$$

for all $F \in \mathcal{F}(P)$.

(b) Under the hypothesis of (a), show that a Hamiltonian vector field X_H is locally a gradient if and only if $\mathcal{L}_{\nabla H}\Omega = 0$.

⋄ **5.3-3.** Show that for any vectors $\varphi_1, \varphi_2 \in \mathcal{H}$ and $\psi \neq 0$ the Fubini–Study metric can be written

$$\langle T_\psi \pi(\varphi_1), T_\psi \pi(\varphi_2) \rangle = 2\hbar \|\psi\|^{-2} (\langle \varphi_1, \varphi_2 \rangle - \|\psi\|^{-2} \langle \varphi_1, \psi \rangle \langle \psi, \varphi_2 \rangle).$$

Conclude that the Riemannian metric and symplectic forms are given by

$$g_{[\psi]}(T_\psi \pi(\varphi_1), T_\psi \pi(\varphi_2)) = \frac{2\hbar}{\|\psi\|^4} \operatorname{Re}(\langle \varphi_1, \varphi_2 \rangle \|\psi\|^2 - \langle \varphi_1, \psi \rangle \langle \psi, \varphi_2 \rangle)$$

and

$$\Omega_{[\psi]}(T_\psi \pi(\varphi_1), T_\psi \pi(\varphi_2)) = -\frac{2\hbar}{\|\psi\|^4} \operatorname{Im}(\langle \varphi_1, \varphi_2 \rangle \|\psi\|^2 - \langle \varphi_1, \psi \rangle \langle \psi, \varphi_2 \rangle).$$

⋄ **5.3-4.** Prove that $d\Omega = 0$ on $\mathbb{P}\mathcal{H}$ directly without using the invariance under the maps $[U]$, for U a unitary operator on $\mathcal{H}$.

⋄ **5.3-5.** For $\mathbb{C}^{n+1}$, show that in a projective chart of $\mathbb{C}\mathbb{P}^n$ the symplectic form Ω is determined by

$$\pi^* \Omega = (1 + |z|^2)^{-1}(d\sigma - (1 + |z|^2)^{-1} \sigma \wedge \bar\sigma),$$

where $d|z|^2 = \sigma + \bar\sigma$ (explicitly, $\sigma = \sum_{i=1}^{n+1} z_i d\bar{z}_i$) and $\pi : \mathbb{C}^n \backslash \{0\} \to \mathbb{C}\mathbb{P}^n$ is the projection. Use this to show that $d\Omega = 0$. Note the similarity between this formula and the corresponding one in Exercise 5.3-3.

5.4 Hamiltonian Systems

With the geometry of symplectic manifolds now available, we are ready to study Hamiltonian dynamics in this setting.

Definition 5.4.1. *Let (P, Ω) be a symplectic manifold. A vector field X on P is called **Hamiltonian** if there is a function $H : P \to \mathbb{R}$ such that*

$$i_X \Omega = dH; \tag{5.4.1}$$

that is, for all $v \in T_z P$, we have the identity

$$\Omega_z(X(z), v) = dH(z) \cdot v.$$

*In this case we write X_H for X. The set of all Hamiltonian vector fields on P is denoted by $\mathfrak{X}_{\mathrm{Ham}}(P)$. **Hamilton's equations** are the evolution equations*

$$\dot{z} = X_H(z).$$

In finite dimensions, Hamilton's equations in canonical coordinates are

$$\frac{dq^i}{dt} = \frac{\partial H}{\partial p_i}, \quad \frac{dp^i}{dt} = -\frac{\partial H}{\partial q^i}.$$

Vector Fields and Flows. A vector field X is called *locally Hamiltonian* if $\mathbf{i}_X \Omega$ is closed. This is equivalent to $\pounds_X \Omega = 0$, where $\pounds_X \Omega$ denotes Lie differentiation of Ω along X, because

$$\pounds_X \Omega = \mathbf{i}_X \mathbf{d}\Omega + \mathbf{d}\mathbf{i}_X \Omega = \mathbf{d}\mathbf{i}_X \Omega.$$

If X is locally Hamiltonian, it follows from the Poincaré lemma that there locally exists a function H such that $\mathbf{i}_X \Omega = \mathbf{d}H$, so locally $X = X_H$, and thus the terminology is consistent. In a symplectic vector space, we have seen in Chapter 2 that the condition that $\mathbf{i}_X \Omega$ be closed is equivalent to $DX(z)$ being Ω-skew. Thus, the definition of locally Hamiltonian is an intrinsic generalization of what we did in the vector space case.

The flow φ_t of a locally Hamiltonian vector field X satisfies $\varphi_t^* \Omega = \Omega$, since

$$\frac{d}{dt}\varphi_t^* \Omega = \varphi_t^* \pounds_X \Omega = 0,$$

and thus we have proved the following:

Proposition 5.4.2. *The flow φ_t of a vector field X consists of symplectic transformations (that is, for each t, we have $\varphi_t^* \Omega = \Omega$ where defined) if and only if X is locally Hamiltonian.*

A constant vector field on the torus $\mathbb{T}^2$ gives an example of a locally Hamiltonian vector field that is not Hamiltonian. (See Exercise 5.4-1.)

Using the straightening out theorem (see, for example, Abraham, Marsden, and Ratiu [1988, Section 4.1]) it is easy to see that on an even-dimensional manifold *any* vector field is locally Hamiltonian near points where it is nonzero, relative to *some* symplectic form. However, it is not so simple to get a general criterion of this sort that is global, covering singular points as well.

Energy Conservation. If X_H is Hamiltonian with flow φ_t, then by the chain rule,

$$\frac{d}{dt}(H\varphi_t(z)) = \mathbf{d}H(\varphi_t(z)) \cdot X_H(\varphi_t(z))$$
$$= \Omega\left(X_H(\varphi_t(z)), X_H(\varphi_t(z))\right) = 0, \qquad (5.4.2)$$

since Ω is skew. Thus $H \circ \varphi_t$ is constant in t. We have proved the following:

Proposition 5.4.3 (Conservation of Energy). *If φ_t is the flow of X_H on the symplectic manifold P, then $H \circ \varphi_t = H$ (where defined).*

Transformation of Hamiltonian Systems. As in the vector space case, we have the following results.

Proposition 5.4.4. *A diffeomorphism $\varphi : P_1 \to P_2$ of symplectic manifolds is symplectic if and only if it satisfies*

$$\varphi^* X_H = X_{H \circ \varphi} \qquad (5.4.3)$$

for all functions $H : U \to \mathbb{R}$ (such that X_H is defined) where U is any open subset of P_2.

Proof. The statement (5.4.3) means that for each $z \in P$,

$$T_{\varphi(z)}\varphi^{-1} \cdot X_H(\varphi(z)) = X_{H \circ \varphi}(z),$$

that is,

$$X_H(\varphi(z)) = T_z\varphi \cdot X_{H \circ \varphi}(z).$$

In other words,

$$\Omega(\varphi(z))(X_H(\varphi(z)), T_z\varphi \cdot v) = \Omega(\varphi(z))(T_z\varphi \cdot X_{H \circ \varphi}(z), T_z\varphi \cdot v)$$

for all $v \in T_z P$. If φ is symplectic, this becomes

$$\mathbf{d}H(\varphi(z)) \cdot [T_z\varphi \cdot v] = \mathbf{d}(H \circ \varphi)(z) \cdot v,$$

which is true by the chain rule. Thus, if φ is symplectic, then (5.4.3) holds. The converse is proved in the same way. ∎

The same qualifications on technicalities pertinent to the infinite-dimensional case that were discussed for vector spaces apply to the present context as well. For instance, given H, there is no *a priori* guarantee that X_H exists: We usually assume it abstractly and verify it in examples. Also, we may wish to deal with X_H's that have dense domains rather than everywhere defined smooth vector fields. These technicalities are important, but they do not affect many of the main goals of this book. We shall, for simplicity, deal only with everywhere defined vector fields and refer the reader to Chernoff and Marsden [1974] and Marsden and Hughes [1983] for the general case. We shall also tacitly restrict our attention to functions that *have* Hamiltonian vector fields. Of course, in the finite-dimensional case these technical problems disappear.

Exercises

◇ **5.4-1.** Let X be a constant nonzero vector field on the two-torus. Show that X is locally Hamiltonian but is not globally Hamiltonian.

◇ **5.4-2.** Show that the bracket of two locally Hamiltonian vector fields on a symplectic manifold (P, Ω) is globally Hamiltonian.

◇ **5.4-3.** Consider the equations on $\mathbb{C}^2$ given by

$$\dot{z}_1 = -iw_1 z_1 + ip\bar{z}_2 + iz_1(a|z_1|^2 + b|z_2|^2),$$
$$\dot{z}_2 = -iw_2 z_2 + iq\bar{z}_1 + iz_2(c|z_1|^2 + d|z_2|^2).$$

Show that this system is Hamiltonian if and only if $p = q$ and $b = c$ with

$$H = \frac{1}{2}\left(w_2|z_2|^2 + w_1|z_1|^2\right) - p\operatorname{Re}(z_1 z_2) - \frac{a}{4}|z_1|^4 - \frac{b}{2}|z_1 z_2|^2 - \frac{d}{4}|z_2|^4.$$

◇ **5.4-4.** Let (P, Ω) be a symplectic manifold and $\varphi : S \longrightarrow P$ an immersion. The immersion φ is called a *coisotropic immersion* if $T_s\varphi(T_sS)$ is a coisotropic subspace of $T_{\varphi(s)}P$ for every $s \in S$. This means that

$$[T_s\varphi(T_sS)]^{\Omega(s)} \subset T_s\varphi(T_sS)$$

for every $s \in S$ (see Exercise 2.3-5). If (P, Ω) is a strong symplectic manifold, show that $\varphi : S \longrightarrow P$ is a coisotropic immersion if and only if $X_H(\varphi(s)) \in T_s\varphi(T_sS)$ for all $s \in S$, all open neighborhoods U of $\varphi(s)$ in P, and all smooth functions $H : U \longrightarrow \mathbb{R}$ satisfying $H|\varphi(S) \cap U = $ constant.

5.5 Poisson Brackets on Symplectic Manifolds

Analogous to the vector space treatment, we define the *Poisson bracket* of two functions $F, G : P \to \mathbb{R}$ by

$$\{F, G\}(z) = \Omega(X_F(z), X_G(z)). \qquad (5.5.1)$$

From Proposition 5.4.4 we get (see the proof of Proposition 2.7.5) the following result.

Proposition 5.5.1. *A diffeomorphism* $\varphi : P_1 \to P_2$ *is symplectic if and only if*

$$\{F, G\} \circ \varphi = \{F \circ \varphi, G \circ \varphi\} \qquad (5.5.2)$$

for all functions $F, G \in \mathcal{F}(U)$, *where* U *is an arbitrary open subset of* P_2.

Using this, Proposition 5.4.2 shows that the following statement holds.

Proposition 5.5.2. *If* φ_t *is the flow of a Hamiltonian vector field* X_H (*or a locally Hamiltonian vector field*), *then*

$$\varphi_t^*\{F, G\} = \{\varphi_t^*F, \varphi_t^*G\}$$

for all $F, G \in \mathcal{F}(P)$ (*or restricted to an open set if the flow is not everywhere defined*).

Corollary 5.5.3. *The following derivation identity holds:*

$$X_H[\{F, G\}] = \{X_H[F], G\} + \{F, X_H[G]\}, \qquad (5.5.3)$$

where we use the notation $X_H[F] = \pounds_{X_H}F$ *for the derivative of* F *in the direction* X_H.

Proof. Differentiate the identity

$$\varphi_t^*\{F,G\} = \{\varphi_t^*F, \varphi_t^*G\}$$

in t at $t=0$, where φ_t is the flow of X_H. The left-hand side clearly gives the left side of (5.5.3). To evaluate the right-hand side, first notice that

$$\Omega_z^\flat\left[\frac{d}{dt}\bigg|_{t=0} X_{\varphi_t^*F}(z)\right] = \frac{d}{dt}\bigg|_{t=0}\Omega_z^\flat X_{\varphi_t^*F}(z)$$

$$= \frac{d}{dt}\bigg|_{t=0}\mathbf{d}(\varphi_t^*F)(z)$$

$$= (\mathbf{d}X_H[F])(z) = \Omega_z^\flat(X_{X_H[F]}(z)).$$

Thus,

$$\frac{d}{dt}\bigg|_{t=0} X_{\varphi_t^*F} = X_{X_H[F]}.$$

Therefore,

$$\frac{d}{dt}\bigg|_{t=0}\{\varphi_t^*F, \varphi_t^*G\} = \frac{d}{dt}\bigg|_{t=0}\Omega_z(X_{\varphi_t^*F}(z), X_{\varphi_t^*G}(z))$$

$$= \Omega_z(X_{X_H[F]}, X_G(z)) + \Omega_z(X_F(z), X_{X_H[G]}(z))$$

$$= \{X_H[F], G\}(z) + \{F, X_H[G]\}(z). \qquad \blacksquare$$

Lie Algebras and Jacobi's Identity. The above development leads to important insight into Poisson brackets.

Proposition 5.5.4. *The functions $\mathcal{F}(P)$ form a Lie algebra under the Poisson bracket.*

Proof. Since $\{F,G\}$ is obviously real bilinear and skew-symmetric, the only thing to check is Jacobi's identity. From

$$\{F,G\} = \mathbf{i}_{X_F}\Omega(X_G) = \mathbf{d}F(X_G) = X_G[F],$$

we have

$$\{\{F,G\},H\} = X_H[\{F,G\}],$$

and so by Corollary 5.5.3 we get

$$\{\{F,G\},H\} = \{X_H[F], G\} + \{F, X_H[G]\}$$
$$= \{\{F,H\}, G\} + \{F, \{G,H\}\}, \qquad (5.5.4)$$

which is Jacobi's identity. $\blacksquare$

This derivation gives us additional insight: *Jacobi's identity is just the infinitesimal statement of φ_t being canonical.*

In the same spirit, one can check that *if Ω is a nondegenerate two-form with the Poisson bracket defined by (5.5.1), then the Poisson bracket satisfies the Jacobi identity if and only if Ω is closed* (see Exercise 5.5-1). The **Poisson bracket–Lie derivative identity**

$$\{F, G\} = X_G[F] = -X_F[G] \qquad (5.5.5)$$

we derived in this proof will be useful.

Proposition 5.5.5. *The set of Hamiltonian vector fields $\mathfrak{X}_{\mathrm{Ham}}(P)$ is a Lie subalgebra of $\mathfrak{X}(P)$, and in fact,*

$$[X_F, X_G] = -X_{\{F,G\}}. \qquad (5.5.6)$$

Proof. As derivations,

$$
\begin{aligned}
[X_F, X_G][H] &= X_F X_G[H] - X_G X_F[H] \\
&= X_F[\{H, G\}] - X_G[\{H, F\}] \\
&= \{\{H, G\}, F\} - \{\{H, F\}, G\} \\
&= -\{H, \{F, G\}\} = -X_{\{F,G\}}[H],
\end{aligned}
$$

by Jacobi's identity. ∎

Proposition 5.5.6. *We have*

$$\frac{d}{dt}(F \circ \varphi_t) = \{F \circ \varphi_t, H\} = \{F, H\} \circ \varphi_t, \qquad (5.5.7)$$

where φ_t is the flow of X_H and $F \in \mathcal{F}(P)$.

Proof. By (5.5.5) and the chain rule,

$$\frac{d}{dt}(F \circ \varphi_t)(z) = \mathbf{d}F(\varphi_t(z)) \cdot X_H(\varphi_t(z)) = \{F, H\}(\varphi_t(z)).$$

Since φ_t is symplectic, this becomes

$$\{F \circ \varphi_t, H \circ \varphi_t\}(z),$$

which also equals $\{F \circ \varphi_t, H\}(z)$ by conservation of energy. This proves (5.5.7). ∎

Equations in Poisson Bracket Form. Equation (5.5.7), often written more compactly as

$$\dot{F} = \{F, H\}, \qquad (5.5.8)$$

is called the **equation of motion in Poisson bracket form**. We indicated in Chapter 1 why the formulation (5.5.8) is important.

Corollary 5.5.7.   *$F \in \mathcal{F}(P)$ is a constant of the motion for X_H if and only if $\{F, H\} = 0$.*

Proposition 5.5.8.   *Assume that the functions f, g, and $\{f, g\}$ are integrable relative to the Liouville volume $\Lambda \in \Omega^{2n}(P)$ on a $2n$-dimensional symplectic manifold (P, Ω). Then*

$$\int_P \{f, g\} \Lambda = \int_{\partial P} f \mathbf{i}_{X_g} \Lambda = -\int_{\partial P} g \mathbf{i}_{X_f} \Lambda.$$

Proof.   Since $\pounds_{X_g} \Omega = 0$, it follows that $\pounds_{X_g} \Lambda = 0$, so that $\operatorname{div}(f X_g) = X_g[f] = \{f, g\}$. Therefore, by Stokes' theorem,

$$\int_P \{f, g\} \Lambda = \int_P \operatorname{div}(f X_g) \Lambda = \int_P \pounds_{f X_g} \Lambda = \int_P \mathbf{d} \mathbf{i}_{f X_g} \Lambda = \int_{\partial P} f \mathbf{i}_{X_g} \Lambda,$$

the second equality following by skew-symmetry of the Poisson bracket. ∎

Corollary 5.5.9.   *Assume that $f, g, h \in \mathcal{F}(P)$ have compact support or decay fast enough such that they and their Poisson brackets are L^2 integrable relative to the Liouville volume on a $2n$-dimensional symplectic manifold (P, Ω). Assume also that at least one of f and g vanish on ∂P if $\partial P \neq \varnothing$. Then the L^2-inner product is bi-invariant on the Lie algebra $(\mathcal{F}(P), \{\,,\})$, that is,*

$$\int_P f\{g, h\} \Lambda = \int_P \{f, g\} h \Lambda.$$

Proof.   From $\{hf, g\} = h\{f, g\} + f\{h, g\}$ we get

$$0 = \int_P \{hf, g\} \Lambda = \int_P h\{f, g\} \Lambda + \int_P f\{h, g\} \Lambda.$$

However, from Proposition 5.5.8, the integral of $\{hf, g\}$ over P vanishes, since one of f or g vanishes on ∂P. The corollary then follows. ∎

Exercises

⋄ **5.5-1.**   Let Ω be a nondegenerate two-form on a manifold P. Form Hamiltonian vector fields and the Poisson bracket using the same definitions as in the symplectic case. Show that Jacobi's identity holds if and only if the two-form Ω is closed.

⋄ **5.5-2.**   Let P be a compact boundaryless symplectic manifold. Show that the space of functions $\mathcal{F}_0(P) = \{f \in \mathcal{F}(P) \mid \int_P f \Lambda = 0\}$ is a Lie subalgebra of $(\mathcal{F}(P), \{\,,\})$ isomorphic to the Lie algebra of Hamiltonian vector fields on P.

⋄ **5.5-3.** Using the complex notation $z^j = q^j + ip_j$, show that the symplectic form on $\mathbb{C}^n$ may be written as

$$\Omega = \frac{i}{2} \sum_{k=1}^{n} dz^k \wedge d\bar{z}^k,$$

and the Poisson bracket may be written

$$\{F, G\} = \frac{2}{i} \sum_{k=1}^{n} \left(\frac{\partial F}{\partial z^k} \frac{\partial G}{\partial \bar{z}^k} - \frac{\partial G}{\partial z^k} \frac{\partial F}{\partial \bar{z}^k} \right).$$

⋄ **5.5-4.** Let $J : \mathbb{C}^2 \to \mathbb{R}$ be defined by

$$J = \frac{1}{2}(|z_1|^2 - |z_2|^2).$$

Show that

$$\{H, J\} = 0,$$

where H is given in Exercise 5.4-3.

⋄ **5.5-5.** Let (P, Ω) be a $2n$-dimensional symplectic manifold. Show that the Poisson bracket may be defined by

$$\{F, G\}\Omega^n = \gamma\, dF \wedge dG \wedge \Omega^{n-1}$$

for a suitable constant γ.

⋄ **5.5-6.** Let $\varphi : S \longrightarrow P$ be a coisotropic immersion (see Exercise 5.4-4). Let $F, H : P \longrightarrow \mathbb{R}$ be smooth functions such that $d(\varphi^* F)(s)$, $(\varphi^* H)(s)$ vanish on $(T_s \phi)^{-1}([T_s \varphi(T_s S)]^{\Omega(\varphi(s))})$ for all $s \in S$. Show that $\varphi^*\{F, H\}$ depends only on $\varphi^* F$ and $\varphi^* H$.

6
Cotangent Bundles

In many mechanics problems, the phase space is the cotangent bundle T^*Q of a configuration space Q. There is an "intrinsic" symplectic structure on T^*Q that can be described in various equivalent ways. Assume first that Q is n-dimensional, and pick local coordinates $(q^1, \dots, q^n)$ on Q. Since $(dq^1, \dots, dq^n)$ is a basis of T_q^*Q, we can write any $\alpha \in T_q^*Q$ as $\alpha = p_i \, dq^i$. This procedure defines induced local coordinates $(q^1, \dots, q^n, p_1, \dots, p_n)$ on T^*Q. Define the **canonical symplectic form** on T^*Q by

$$\Omega = dq^i \wedge dp_i.$$

This defines a two-form Ω, which is clearly closed, and in addition, it can be checked to be independent of the choice of coordinates $(q^1, \dots, q^n)$. Furthermore, observe that Ω is locally constant, that is, the coefficient multiplying the basis forms $dq^i \wedge dp_i$, namely the number 1, does not explicitly depend on the coordinates $(q^1, \dots, q^n, p_1, \dots, p_n)$ of phase space points. In this section we show how to construct Ω intrinsically, and then we will study this canonical symplectic structure in some detail.

6.1   The Linear Case

To motivate a coordinate-independent definition of Ω, consider the case in which Q is a vector space W (which could be infinite-dimensional), so that $T^*Q = W \times W^*$. We have already described the canonical two-form on

$W \times W^*$:

$$\Omega_{(w,\alpha)}((u,\beta),(v,\gamma)) = \langle \gamma, u \rangle - \langle \beta, v \rangle, \tag{6.1.1}$$

where $(w, \alpha) \in W \times W^*$ is the base point, $u, v \in W$, and $\beta, \gamma \in W^*$. This canonical two-form will be constructed from the **canonical one-form** Θ, defined as follows:

$$\Theta_{(w,\alpha)}(u,\beta) = \langle \alpha, u \rangle. \tag{6.1.2}$$

The next proposition shows that the canonical two-form (6.1.1) is exact:

$$\Omega = -d\Theta. \tag{6.1.3}$$

We begin with a computation that reconciles these formulas with their coordinate expressions.

Proposition 6.1.1. *In the finite-dimensional case the symplectic form Ω defined by (6.1.1) can be written $\Omega = dq^i \wedge dp_i$ in coordinates $q^1, \ldots, q^n$ on W and corresponding dual coordinates $p_1, \ldots, p_n$ on W^*. The associated canonical one-form is given by $\Theta = p_i \, dq^i$, and (6.1.3) holds.*

Proof. If $(q^1, \ldots, q^n, p_1, \ldots, p_n)$ are coordinates on T^*W, then

$$\left(\frac{\partial}{\partial q^1}, \ldots, \frac{\partial}{\partial q^n}, \frac{\partial}{\partial p_1}, \ldots, \frac{\partial}{\partial p_n} \right)$$

denotes the induced basis for $T_{(w,\alpha)}(T^*W)$, and $(dq^1, \ldots, dq^n, dp_1, \ldots, dp_n)$ denotes the associated dual basis of $T^*_{(w,\alpha)}(T^*W)$. Write

$$(u, \beta) = \left(u^j \frac{\partial}{\partial q^j}, \, \beta_j \frac{\partial}{\partial p_j} \right)$$

and similarly for (v, γ). Hence

$$(dq^i \wedge dp_i)_{(w,\alpha)}((u,\beta),(v,\gamma)) = (dq^i \otimes dp_i - dp_i \otimes dq^i)((u,\beta),(v,\gamma))$$
$$= dq^i(u,\beta)dp_i(v,\gamma) - dp_i(u,\beta)dq^i(v,\gamma)$$
$$= u^i \gamma_i - \beta_i v^i.$$

Also, $\Omega_{(w,\alpha)}((u,\beta),(v,\gamma)) = \gamma(u) - \beta(v) = \gamma_i u^i - \beta_i v^i$. Thus,

$$\Omega = dq^i \wedge dp_i.$$

Similarly,

$$(p_i \, dq^i)_{(w,\alpha)}(u,\beta) = \alpha_i \, dq^i(u,\beta) = \alpha_i u^i,$$

and

$$\Theta_{(w,\alpha)}(u,\beta) = \alpha(u) = \alpha_i u^i.$$

Comparing, we get $\Theta = p_i \, dq^i$. Therefore,

$$-\mathbf{d}\Theta = -\mathbf{d}(p_i \, dq^i) = dq^i \wedge dp_i = \Omega. \qquad \blacksquare$$

To verify (6.1.3) for the infinite-dimensional case, use (6.1.2) and the second identity in item 6 of the table at the end of §4.4 to give

$$\mathbf{d}\Theta_{(w,\alpha)}((u_1,\beta_1),(u_2,\beta_2)) = [\mathbf{D}\Theta_{(w,\alpha)} \cdot (u_1,\beta_1)] \cdot (u_2,\beta_2)$$
$$- [\mathbf{D}\Theta_{(w,\alpha)} \cdot (u_2,\beta_2)] \cdot (u_1,\beta_1)$$
$$= \langle \beta_1, u_2 \rangle - \langle \beta_2, u_1 \rangle,$$

since $\mathbf{D}\Theta_{(w,\alpha)} \cdot (u,\beta) = \langle \beta, \cdot \rangle$. But this equals $-\Omega_{(w,\alpha)}((u_1,\beta_1),(u_2,\beta_2))$.
To give an intrinsic interpretation to Θ, let us prove that

$$\Theta_{(w,\alpha)} \cdot (u,\beta) = \left\langle \alpha, T_{(w,\alpha)} \pi_W (u,\beta) \right\rangle, \qquad (6.1.4)$$

where $\pi_W : W \times W^* \to W$ is the projection. Indeed, (6.1.4) coincides with (6.1.2), since $T_{(w,\alpha)} \pi_W : W \times W^* \to W$ is the projection on the first factor.

Exercises

◇ **6.1-1 (Jacobi–Haretu Coordinates).** Consider the three-particle configuration space $Q = \mathbb{R}^3 \times \mathbb{R}^3 \times \mathbb{R}^3$ with elements denoted by $\mathbf{r}_1, \mathbf{r}_2$, and $\mathbf{r}_3$. Call the conjugate momenta $\mathbf{p}_1, \mathbf{p}_2, \mathbf{p}_3$ and equip the phase space T^*Q with the canonical symplectic structure Ω. Let $\mathbf{j} = \mathbf{p}_1 + \mathbf{p}_2 + \mathbf{p}_3$. Let $\mathbf{r} = \mathbf{r}_2 - \mathbf{r}_1$ and let $\mathbf{s} = \mathbf{r}_3 - \frac{1}{2}(\mathbf{r}_1 + \mathbf{r}_2)$. Show that the form Ω pulled back to the level sets of $\mathbf{j}$ has the form $\Omega = d\mathbf{r} \wedge d\pi + d\mathbf{s} \wedge d\sigma$, where the variables π and σ are defined by $\pi = \frac{1}{2}(\mathbf{p}_2 - \mathbf{p}_1)$ and $\sigma = \mathbf{p}_3$.

6.2 The Nonlinear Case

Definition 6.2.1. *Let Q be a manifold. We define $\Omega = -\mathbf{d}\Theta$, where Θ is the one-form on T^*Q defined analogous to (6.1.4), namely*

$$\Theta_\beta(v) = \langle \beta, T\pi_Q \cdot v \rangle, \qquad (6.2.1)$$

where $\beta \in T^*Q$, $v \in T_\beta(T^*Q)$, $\pi_Q : T^*Q \to Q$ is the projection, and $T\pi_Q : T(T^*Q) \to TQ$ is the tangent map of π_Q.

The computations in Proposition 6.1.1 show that $(T^*Q, \Omega = -\mathbf{d}\Theta)$ is a symplectic manifold; indeed, in local coordinates with $(w,\alpha) \in U \times W^*$, where U is open in W, and where $(u,\beta),(v,\gamma) \in W \times W^*$, the two-form $\Omega = -\mathbf{d}\Theta$ is given by

$$\Omega_{(w,\alpha)}((u,\beta),(v,\gamma)) = \gamma(u) - \beta(v). \qquad (6.2.2)$$

Darboux' theorem and its corollary can be interpreted as asserting that any (strong) symplectic manifold locally looks like $W \times W^*$ in suitable local coordinates.

Hamiltonian Vector Fields. For a function $H : T^*Q \to \mathbb{R}$, the Hamiltonian vector field X_H on the cotangent bundle T^*Q is given in canonical cotangent bundle charts $U \times W^*$, where U is open in W, by

$$X_H(w, \alpha) = \left(\frac{\delta H}{\delta \alpha}, -\frac{\delta H}{\delta w} \right). \tag{6.2.3}$$

Indeed, setting $X_H(w, \alpha) = (w, \alpha, v, \gamma)$, for any $(u, \beta) \in W \times W^*$ we have

$$\begin{aligned}
\mathbf{d}H_{(w,\alpha)} \cdot (u, \beta) &= \mathbf{D}_w H_{(w,\alpha)} \cdot u + \mathbf{D}_\alpha H_{(w,\alpha)} \cdot \beta \\
&= \left\langle \frac{\delta H}{\delta w}, u \right\rangle + \left\langle \beta, \frac{\delta H}{\delta \alpha} \right\rangle,
\end{aligned} \tag{6.2.4}$$

which, by definition and (6.2.2), equals

$$\Omega_{(w,\alpha)}(X_H(w, \alpha), (u, \beta)) = \langle \beta, v \rangle - \langle \gamma, u \rangle. \tag{6.2.5}$$

Comparing (6.2.4) and (6.2.5) gives (6.2.3). In finite dimensions, (6.2.3) is the familiar right-hand side of Hamilton's equations.

Poisson Brackets. Formula (6.2.3) and the definition of the Poisson bracket show that in canonical cotangent bundle charts,

$$\{f, g\}(w, \alpha) = \left\langle \frac{\delta f}{\delta w}, \frac{\delta g}{\delta \alpha} \right\rangle - \left\langle \frac{\delta g}{\delta w}, \frac{\delta f}{\delta \alpha} \right\rangle, \tag{6.2.6}$$

which in finite dimensions becomes

$$\{f, g\}(q^i, p_i) = \sum_{i=1}^{n} \left(\frac{\partial f}{\partial q^i} \frac{\partial g}{\partial p_i} - \frac{\partial f}{\partial p_i} \frac{\partial g}{\partial q^i} \right). \tag{6.2.7}$$

Pull-Back Characterization. Another characterization of the canonical one-form that is sometimes useful is the following:

Proposition 6.2.2. Θ *is the unique one-form on T^*Q such that*

$$\alpha^* \Theta = \alpha \tag{6.2.8}$$

for any local one-form α on Q, where on the left-hand side, α is regarded as a map (of some open subset of) Q to T^*Q.

Proof. In finite dimensions, if $\alpha = \alpha_i(q^j)\, dq^i$ and $\Theta = p_i\, dq^i$, then to calculate $\alpha^* \Theta$ means that we substitute $p_i = \alpha_i(q^j)$ into Θ, a process that clearly gives back α, so $\alpha^* \Theta = \alpha$. The general argument is as follows. If Θ is the canonical one-form on T^*Q, and $v \in T_q Q$, then

$$\begin{aligned}
(\alpha^* \Theta)_q \cdot v &= \Theta_{\alpha(q)} \cdot T_q \alpha(v) = \langle \alpha(q), T_{\alpha(q)} \pi_Q(T_q \alpha(v)) \rangle \\
&= \langle \alpha(q), T_q(\pi_Q \circ \alpha)(v) \rangle = \alpha(q) \cdot v,
\end{aligned}$$

since $\pi_Q \circ \alpha = $ identity on Q.

For the converse, assume that Θ is a one-form on T^*Q satisfying (6.2.8). We will show that it must then be the canonical one-form (6.2.1). In finite dimensions this is straightforward: If $\Theta = A_i \, dq^i + B^i \, dp_i$ for A_i, B^i functions of (q^j, p_j), then

$$\alpha^* \Theta = (A_i \circ \alpha) \, dq^i + (B^i \circ \alpha) \, d\alpha_i = \left(A_j \circ \alpha + (B^i \circ \alpha) \frac{\partial \alpha_i}{\partial q^j} \right) dq^j,$$

which equals $\alpha = \alpha_i \, dq^i$ if and only if

$$A_j \circ \alpha + (B^i \circ \alpha) \frac{\partial \alpha_i}{\partial q^j} = \alpha_j.$$

Since this must hold for all α_j, putting $\alpha_1, \dots, \alpha_n$ constant, it follows that $A_j \circ \alpha = \alpha_j$, that is, $A_j = p_j$. Therefore, the remaining equation is

$$(B^i \circ \alpha) \frac{\partial \alpha_i}{\partial q^j} = 0$$

for any α_i; choosing $\alpha_i(q^1, \dots, q^n) = q_0^i + (q^i - q_0^i)p_i^0$ (no sum) implies $0 = (B^j \circ \alpha)(q_0^1, \dots, q_0^n)p_j^0$ for all (q_0^j, p_j^0); therefore, $B^j = 0$ and thus $\Theta = p_i \, dq^i$.[1]   ∎

Exercises

⋄ **6.2-1.** Let N be a submanifold of M and denote by Θ_N and Θ_M the canonical one-forms on the cotangent bundles $\pi_N : T^*N \to N$ and $\pi_M : T^*M \to M$, respectively. Let $\pi : (T^*M)|N \to T^*N$ be the projection defined by $\pi(\alpha_n) = \alpha_n|T_n N$, where $n \in N$ and $\alpha_n \in T_n^* M$. Show that $\pi^* \Theta_N = i^* \Theta_M$, where $i : (T^*M)|N \to T^*M$ is the inclusion.

⋄ **6.2-2.** Let $f : Q \to \mathbb{R}$ and $X \in \mathfrak{X}(T^*Q)$. Show that

$$\Theta(X) \circ \mathbf{d}f = X[f \circ \pi_Q] \circ \mathbf{d}f.$$

---

[1]In infinite dimensions, the proof is slightly different. We will show that if (6.2.8) holds, then Θ is locally given by (6.1.4), and thus it is the canonical one-form. If $U \subset E$ is the chart domain in the Banach space E modeling Q, then for any $v \in E$ we have

$$(\alpha^* \Theta)_u \cdot (u, v) = \Theta(u, \alpha(u)) \cdot (v, \mathbf{D}\alpha(u) \cdot v),$$

where α is given locally by $u \mapsto (u, \alpha(u))$ for $\alpha : U \to E^*$. Thus (6.2.8) is equivalent to

$$\Theta_{(u, \alpha(u))} \cdot (v, \mathbf{D}\alpha(u) \cdot v) = \langle \alpha(u), v \rangle,$$

which would imply (6.1.4) and hence Θ being the canonical one-form, provided that we can show that for prescribed $\gamma, \delta \in E^*$, $u \in U$, and $v \in E$, there is an $\alpha : U \to E^*$ such that $\alpha(u) = \gamma$, and $\mathbf{D}\alpha(u) \cdot v = \delta$. Such a mapping is constructed in the following way. For $v = 0$ choose $\alpha(u)$ to equal γ for all u. For $v \neq 0$, by the Hahn–Banach theorem one can find a $\varphi \in E^*$ such that $\varphi(v) = 1$. Now set $\alpha(x) = \gamma - \varphi(u)\delta + \varphi(x)\delta$.

⋄ **6.2-3.**   Let Q be a given configuration manifold and let the **extended phase space** be defined by $(T^*Q) \times \mathbb{R}$. Given a time-dependent vector field X on T^*Q, extend it to a vector field $\tilde{X}$ on $(T^*Q) \times \mathbb{R}$ by $\tilde{X} = (X, 1)$. Let H be a (possibly time-dependent) function on $(T^*Q) \times \mathbb{R}$ and set

$$\Omega_H = \Omega + dH \wedge dt,$$

where Ω is the canonical two-form. Show that X is the Hamiltonian vector field for H if and only if

$$\mathbf{i}_{\tilde{X}}\Omega_H = 0.$$

⋄ **6.2-4.**   Give an example of a symplectic manifold (P, Ω), where Ω is exact but P is *not* a cotangent bundle.

6.3   Cotangent Lifts

We now describe an important way to create symplectic transformations on cotangent bundles.

Definition 6.3.1.   *Given two manifolds Q and S and a diffeomorphism $f : Q \to S$, the **cotangent lift** $T^*f : T^*S \to T^*Q$ of f is defined by*

$$\langle T^*f(\alpha_s), v \rangle = \langle \alpha_s, (Tf \cdot v) \rangle, \tag{6.3.1}$$

where

$$\alpha_s \in T_s^*S, \quad v \in T_qQ, \quad and \quad s = f(q).$$

The importance of this construction is that T^*f is guaranteed to be symplectic; it is often called a "point transformation" because it arises from a diffeomorphism on points in configuration space. Notice that while Tf covers f, T^*f covers f^{-1}. Denote by $\pi_Q : T^*Q \to Q$ and $\pi_S : T^*S \to S$ the canonical cotangent bundle projections.

Proposition 6.3.2.   *A diffeomorphism $\varphi : T^*S \to T^*Q$ preserves the canonical one-forms Θ_Q and Θ_S on T^*Q and T^*S, respectively, if and only if φ is the cotangent lift T^*f of some diffeomorphism $f : Q \to S$.*

Proof.   First assume that $f : Q \to S$ is a diffeomorphism. Then for arbitrary $\beta \in T^*S$ and $v \in T_\beta(T^*S)$, we have

$$\begin{aligned}
((T^*f)^*\Theta_Q)_\beta \cdot v &= (\Theta_Q)_{T^*f(\beta)} \cdot TT^*f(v) \\
&= \langle T^*f(\beta), (T\pi_Q \circ TT^*f) \cdot v \rangle \\
&= \langle \beta, T(f \circ \pi_Q \circ T^*f) \cdot v \rangle \\
&= \langle \beta, T\pi_S \cdot v \rangle = \Theta_{S\beta} \cdot v,
\end{aligned}$$

since $f \circ \pi_Q \circ T^* f = \pi_S$.

Conversely, assume that $\varphi^* \Theta_Q = \Theta_S$, that is,

$$\langle \varphi(\beta), T(\pi_Q \circ \varphi)(v) \rangle = \langle \beta, T\pi_S(v) \rangle \tag{6.3.2}$$

for all $\beta \in T^* S$ and $v \in T_\beta(T^* S)$. Since φ is a diffeomorphism, the range of $T_\beta(\pi_Q \circ \varphi)$ is $T_{\pi_Q(\varphi(\beta))}Q$, so that letting $\beta = 0$ in (6.3.2) implies that $\varphi(0) = 0$. Arguing similarly for φ^{-1} instead of φ, we conclude that φ restricted to the zero section S of $T^* S$ is a diffeomorphism onto the zero section Q of $T^* Q$. Define $f : Q \to S$ by $f = \varphi^{-1}|Q$. We will show below that φ is fiber-preserving, or, equivalently, that $f \circ \pi_Q = \pi_S \circ \varphi^{-1}$. For this we use the following:

Lemma 6.3.3. *Define the flow F_t^Q on $T^* Q$ by $F_t^Q(\alpha) = e^t \alpha$ and let V_Q be the vector field it generates. Then*

$$\langle \Theta_Q, V_Q \rangle = 0, \quad \pounds_{V_Q} \Theta_Q = \Theta_Q, \quad \text{and} \quad \mathbf{i}_{V_Q} \Omega_Q = -\Theta_Q. \tag{6.3.3}$$

Proof. Since F_t^Q is fiber-preserving, V_Q will be tangent to the fibers, and hence $T\pi_Q \circ V_Q = 0$. This implies by (6.2.1) that $\langle \Theta_Q, V_Q \rangle = 0$. To prove the second formula, note that $\pi_Q \circ F_t^Q = \pi_Q$. Let $\alpha \in T_q^* Q$, $v \in T_\alpha(T^* Q)$, and Θ_α denote Θ_Q evaluated at α. We have

$$((F_t^Q)^* \Theta)_\alpha \cdot v = \Theta_{F_t^Q(\alpha)} \cdot TF_t^Q(v)$$
$$= \left\langle F_t^Q(\alpha), (T\pi_Q \circ TF_t^Q)(v) \right\rangle$$
$$= \left\langle e^t \alpha, T(\pi_Q \circ F_t^Q)(v) \right\rangle$$
$$= e^t \langle \alpha, T\pi_Q(v) \rangle = e^t \Theta_\alpha \cdot v,$$

that is,

$$(F_t^Q)^* \Theta_Q = e^t \Theta_Q.$$

Taking the derivative relative to t at $t = 0$ yields the second formula. Finally, the first two formulas imply

$$\mathbf{i}_{V_Q} \Omega_Q = -\mathbf{i}_{V_Q} \mathbf{d}\Theta_Q = -\pounds_{V_Q} \Theta_Q + \mathbf{d}\mathbf{i}_{V_Q} \Theta_Q = -\Theta_Q. \qquad \blacktriangledown$$

Continuing the proof of the proposition, note that by (6.3.3) we have

$$\mathbf{i}_{\varphi^* V_Q} \Omega_S = \mathbf{i}_{\varphi^* V_Q} \varphi^* \Omega_Q = \varphi^* (\mathbf{i}_{V_Q} \Omega_Q)$$
$$= -\varphi^* \Theta_Q = -\Theta_S = \mathbf{i}_{V_S} \Omega_S,$$

so that weak nondegeneracy of Ω_S implies $\varphi^* V_Q = V_S$. Thus φ commutes with the flows F_t^Q and F_t^S, that is, for any $\beta \in T^* S$ we have $\varphi(e^t \beta) =$

$e^t \varphi(\beta)$. Letting $t \to -\infty$ in this equality implies $(\varphi \circ \pi_S)(\beta) = (\pi_Q \circ \varphi)(\beta)$, since $e^t \beta \to \pi_S(\beta)$ and $e^t \varphi(\beta) \to (\pi_Q \circ \varphi)(\beta)$ for $t \to -\infty$. Thus

$$\pi_Q \circ \varphi = \varphi \circ \pi_S, \quad \text{or} \quad f \circ \pi_Q = \pi_S \circ \varphi^{-1}.$$

Finally, we show that $T^* f = \varphi$. For $\beta \in T^* S$, $v \in T_\beta(T^* S)$, (6.3.2) gives

$$
\begin{aligned}
\langle T^* f(\beta), T(\pi_Q \circ \varphi)(v) \rangle &= \langle \beta, T(f \circ \pi_Q \circ \varphi)(v) \rangle \\
&= \langle \beta, T \pi_S(v) \rangle = (\Theta_S)_\beta \cdot v \\
&= (\varphi^* \Theta_Q)_\beta \cdot v = (\Theta_Q)_{\varphi(\beta)} \cdot T_\beta \varphi(v) \\
&= \langle \varphi(\beta), T_\beta(\pi_Q \circ \varphi)(v) \rangle,
\end{aligned}
$$

which shows that $T^* f = \varphi$, since the range of $T_\beta(\pi_Q \circ \varphi)$ is the whole tangent space at $(\pi_Q \circ \varphi)(\beta)$ to Q. ■

In finite dimensions, the first part of this proposition can be seen in coordinates as follows. Write $(s^1, \ldots, s^n) = f(q^1, \ldots, q^n)$ and define

$$p_j = \frac{\partial s^i}{\partial q^j} r_i, \tag{6.3.4}$$

where $(q^1, \ldots, q^n, p_1, \ldots, p_n)$ are cotangent bundle coordinates on $T^* Q$ and $(s^1, \ldots, s^n, r_1, \ldots, r_n)$ on $T^* S$. Since f is a diffeomorphism, it determines the q^i in terms of the s^j, say $q^i = q^i(s^1, \ldots, s^n)$, so both q^i and p_j are functions of $(s^1, \ldots, s^n, r_1, \ldots, r_n)$. The map $T^* f$ is given by

$$(s^1, \ldots, s^n, r_1, \ldots, r_n) \mapsto (q^1, \ldots, q^n, p_1, \ldots, p_n). \tag{6.3.5}$$

To see that (6.3.5) preserves the canonical one-form, use the chain rule and (6.3.4):

$$r_i \, ds^i = r_i \frac{\partial s^i}{\partial q^k} \, dq^k = p_k \, dq^k. \tag{6.3.6}$$

Note that if f and g are diffeomorphisms of Q, then

$$T^*(f \circ g) = T^* g \circ T^* f, \tag{6.3.7}$$

that is, the cotangent lift switches the order of composition; in fact, *it is useful to think of $T^* f$ as the **adjoint** of Tf*.

Exercises

◇ **6.3-1.** The **Lorentz group** $\mathcal{L}$ is the group of invertible linear transformations of $\mathbb{R}^4$ to itself that preserve the quadratic form $x^2 + y^2 + z^2 - c^2 t^2$, where c is a constant, the speed of light. Describe all elements of this group. Let Λ_0 denote one of these transformations. Map $\mathcal{L}$ to itself by $\Lambda \mapsto \Lambda_0 \Lambda$. Calculate the cotangent lift of this map.

◇ **6.3-2.** We have shown that a transformation of $T^* Q$ is the cotangent lift of a diffeomorphism of configuration space if and only if it preserves the canonical one-form. Find this result in Whittaker's book.

6.4 Lifts of Actions

A *left action* of a group G on a manifold M associates to each group element $g \in G$ a diffeomorphism Φ_g of M such that $\Phi_{gh} = \Phi_g \circ \Phi_h$. Thus, the collection of Φ_g's is a *group of transformations of M*. If we replace the condition $\Phi_{gh} = \Phi_g \circ \Phi_h$ by $\Psi_{gh} = \Psi_h \circ \Psi_g$, we speak of a *right action*. We often write $\Phi_g(m) = g \cdot m$ and $\Psi_g(m) = m \cdot g$ for $m \in M$.

Definition 6.4.1.   *Let Φ be an action of a group G on a manifold Q. The **right lift** Φ^* of the action Φ to the symplectic manifold T^*Q is the right action defined by the rule*

$$\Phi_g^*(\alpha) = (T_{g^{-1}\cdot q}^* \Phi_g)(\alpha), \tag{6.4.1}$$

where $g \in G$, $\alpha \in T_q^*Q$, and $T^*\Phi_g$ is the cotangent lift of the diffeomorphism $\Phi_g : Q \to Q$.

By (6.3.7), we see that

$$\Phi_{gh}^* = T^*\Phi_{gh} = T^*(\Phi_g \circ \Phi_h) = T^*\Phi_h \circ T^*\Phi_g = \Phi_h^* \circ \Phi_g^*, \tag{6.4.2}$$

so Φ^* is a right action. To get a *left action*, denoted by Φ_* and called the *left lift* of Φ, one sets

$$(\Phi_*)_g = T_{g\cdot q}^*(\Phi_{g^{-1}}). \tag{6.4.3}$$

In either case, these lifted actions are actions by canonical transformations because of Proposition 6.3.2. We shall return to the study of actions of groups after we study Lie groups in Chapter 9.

Examples

(a)   For a system of N particles in $\mathbb{R}^3$, we choose the configuration space $Q = \mathbb{R}^{3N}$. We write $(\mathbf{q}_j)$ for an N-tuple of vectors labeled by $j = 1, \ldots, N$. Similarly, elements of the momentum phase space $P = T^*\mathbb{R}^{3N} \cong \mathbb{R}^{6N} \cong \mathbb{R}^{3N} \times \mathbb{R}^{3N}$ are denoted by $(\mathbf{q}_j, \mathbf{p}^j)$. Let the additive group $G = \mathbb{R}^3$ of translations act on Q according to

$$\Phi_{\mathbf{x}}(\mathbf{q}_j) = \mathbf{q}_j + \mathbf{x}, \quad \text{where } \mathbf{x} \in \mathbb{R}^3. \tag{6.4.4}$$

Each of the N position vectors $\mathbf{q}_j$ is translated by the same vector $\mathbf{x}$.

Lifting the diffeomorphism $\Phi_{\mathbf{x}} : Q \to Q$, we obtain an action Φ^* of G on P. We assert that

$$\Phi_{\mathbf{x}}^*(\mathbf{q}_j, \mathbf{p}^j) = (\mathbf{q}_j - \mathbf{x}, \mathbf{p}^j). \tag{6.4.5}$$

To verify (6.4.5), observe that $T\Phi_{\mathbf{x}} : TQ \to TQ$ is given by

$$(\mathbf{q}_i, \dot{\mathbf{q}}_j) \mapsto (\mathbf{q}_i + \mathbf{x}, \dot{\mathbf{q}}_j), \tag{6.4.6}$$

so its dual is $(\mathbf{q}_i, \mathbf{p}^j) \mapsto (\mathbf{q}_i - \mathbf{x}, \mathbf{p}^j)$.   ♦

(b)   Consider the action of $GL(n, \mathbb{R})$, the group of $n \times n$ invertible matrices, or, more properly, the group of invertible linear transformations of $\mathbb{R}^n$ to itself, on $\mathbb{R}^n$ given by

$$\Phi_A(\mathbf{q}) = A\mathbf{q}. \tag{6.4.7}$$

The group of induced canonical transformations of $T^*\mathbb{R}^n$ to itself is given by

$$\Phi_A^*(\mathbf{q}, \mathbf{p}) = (A^{-1}\mathbf{q}, A^T\mathbf{p}), \tag{6.4.8}$$

which is readily verified. Notice that this reduces to the same transformation of $\mathbf{q}$ and $\mathbf{p}$ when A is orthogonal.   ♦

Exercises

⋄ **6.4-1.**   Let the multiplicative group $\mathbb{R} \setminus \{0\}$ act on $\mathbb{R}^n$ by $\Phi_\lambda(\mathbf{q}) = \lambda\mathbf{q}$. Calculate the cotangent lift of this action.

6.5   Generating Functions

Consider a symplectic diffeomorphism $\varphi : T^*Q_1 \rightarrow T^*Q_2$ described by functions

$$p_i = p_i(q^j, s^j), \quad r_i = r_i(q^j, s^j), \tag{6.5.1}$$

where (q^i, p_i) and (s^j, r_j) are cotangent coordinates on T^*Q_1 and on T^*Q_2, respectively. In other words, assume that we have a map

$$\Gamma : Q_1 \times Q_2 \rightarrow T^*Q_1 \times T^*Q_2 \tag{6.5.2}$$

whose image is the graph of φ. Let Θ_1 be the canonical one-form on T^*Q_1 and Θ_2 be that on T^*Q_2. By definition,

$$\mathbf{d}(\Theta_1 - \varphi^*\Theta_2) = 0. \tag{6.5.3}$$

This implies, in view of (6.5.1), that

$$p_i \, dq^i - r_i \, ds^i \tag{6.5.4}$$

is closed. Restated, $\Gamma^*(\Theta_1 - \Theta_2)$ is closed. This condition holds if $\Gamma^*(\Theta_1 - \Theta_2)$ is exact, namely,

$$\Gamma^*(\Theta_1 - \Theta_2) = \mathbf{d}S \tag{6.5.5}$$

for a function $S(q, s)$. In coordinates, (6.5.5) reads

$$p_i \, dq^i - r_i \, ds^i = \frac{\partial S}{\partial q^i} \, dq^i + \frac{\partial S}{\partial s^i} \, ds^i, \tag{6.5.6}$$

which is equivalent to

$$p_i = \frac{\partial S}{\partial q^i}, \qquad r_i = -\frac{\partial S}{\partial s^i}. \tag{6.5.7}$$

One calls S a *generating function* for the canonical transformation. With generating functions of this sort, one may run into singularities even with the identity map! See Exercise 6.5-1.

Presupposed relations other than (6.5.1) lead to conclusions other than (6.5.7). Point transformations are generated in this sense; if $S(q^i, r_j) = s^j(q)r_j$, then

$$s^i = \frac{\partial S}{\partial r_i} \quad \text{and} \quad p_i = \frac{\partial S}{\partial q^i}. \tag{6.5.8}$$

(Here one writes $p_i \, dq^i + s^i \, dr_i = dS$.)

In general, consider a diffeomorphism $\varphi : P_1 \to P_2$ of one symplectic manifold (P_1, Ω_1) to another (P_2, Ω_2) and denote the graph of φ by

$$\Gamma(\varphi) \subset P_1 \times P_2.$$

Let $i_\varphi : \Gamma(\varphi) \to P_1 \times P_2$ be the inclusion and let $\Omega = \pi_1^* \Omega_1 - \pi_2^* \Omega_2$, where $\pi_i : P_1 \times P_2 \to P_i$ is the projection. One verifies that φ *is symplectic if and only if* $i_\varphi^* \Omega = 0$. Indeed, since $\pi_1 \circ i_\varphi$ is the projection restricted to $\Gamma(\varphi)$ and $\pi_2 \circ i_\varphi = \varphi \circ \pi_1$ on $\Gamma(\varphi)$, it follows that

$$i_\varphi^* \Omega = (\pi_1 | \Gamma(\varphi))^* (\Omega_1 - \varphi^* \Omega_2),$$

and hence $i_\varphi^* \Omega = 0$ if and only if φ is symplectic, because $(\pi_1 | \Gamma(\varphi))^*$ is injective. In this case, one says that $\Gamma(\varphi)$ is an *isotropic* submanifold of $P_1 \times P_2$ (equipped with the symplectic form Ω); in fact, since $\Gamma(\varphi)$ has half the dimension of $P_1 \times P_2$, it is *maximally* isotropic, or a **Lagrangian manifold**.

Now suppose one *chooses* a form Θ such that $\Omega = -d\Theta$. Then $i_\varphi^* \Omega = -d i_\varphi^* \Theta = 0$, so *locally* on $\Gamma(\varphi)$ there is a function $S : \Gamma(\varphi) \to \mathbb{R}$ such that

$$i_\varphi^* \Theta = dS. \tag{6.5.9}$$

This defines the **generating function** of the canonical transformation φ. Since $\Gamma(\varphi)$ is diffeomorphic to P_1 and also to P_2, we can regard S as a function on P_1 or P_2. If $P_1 = T^* Q_1$ and $P_2 = T^* Q_2$, we can equally well regard (at least locally) S as defined on $Q_1 \times Q_2$. In this way, the general construction of generating functions reduces to the case in equations (6.5.7) and (6.5.8) above. By making other choices of Q, the reader can construct other generating functions and reproduce formulas in, for instance, Goldstein [1980] or Whittaker [1927]. The approach here is based on Sniatycki and Tulczyjew [1971].

Generating functions play an important role in Hamilton–Jacobi theory, in the classical–quantum-mechanical relationship (where S plays the role of the quantum-mechanical phase), and in numerical integration schemes for Hamiltonian systems. We shall see a few of these aspects later on.

Exercises

⋄ **6.5-1.** Show that

$$S(q^i, s^j, t) = \frac{1}{2t}\|\mathbf{q} - \mathbf{s}\|^2$$

generates a canonical transformation that is the identity at $t = 0$.

⋄ **6.5-2** (A first-order symplectic integrator).   Given H, let

$$S(q^i, r_j, t) = r_k q^k - tH(q^i, r_j).$$

Show that S generates a canonical transformation that is a first-order approximation to the flow of X_H for small t.

6.6   Fiber Translations and Magnetic Terms

Momentum Shifts.   We saw above that cotangent lifts provide a basic construction of canonical transformations. Fiber translations provide a second.

Proposition 6.6.1 (Momentum Shifting Lemma).   *Let A be a one-form on Q and let $t_A : T^*Q \to T^*Q$ be defined by $\alpha_q \mapsto \alpha_q + A(q)$, where $\alpha_q \in T_q^*Q$. Let Θ be the canonical one-form on T^*Q. Then*

$$t_A^*\Theta = \Theta + \pi_Q^* A, \qquad (6.6.1)$$

where $\pi_Q : T^*Q \to Q$ is the projection. Hence

$$t_A^*\Omega = \Omega - \pi_Q^* dA, \qquad (6.6.2)$$

where $\Omega = -d\Theta$ is the canonical symplectic form. Thus, t_A is a canonical transformation if and only if $dA = 0$.

Proof.   We prove this using a finite-dimensional coordinate computation. The reader is asked to supply the coordinate-free and infinite-dimensional proofs as an exercise. In coordinates, t_A is the map

$$t_A(q^i, p_j) = (q^i, p_j + A_j). \qquad (6.6.3)$$

Thus,

$$t_A^* \Theta = t_A^*(p_i \mathbf{d}q^i) = (p_i + A_i)\mathbf{d}q^i = p_i \mathbf{d}q^i + A_i \mathbf{d}q^i, \tag{6.6.4}$$

which is the coordinate expression for $\Theta + \pi_Q^* A$. The remaining assertions follow directly from this.   ∎

In particular, fiber translation by the differential of a function $A = \mathbf{d}f$ is a canonical transformation; in fact, f induces, in the sense of the preceding section, a generating function (see Exercise 6.6-2). The two basic classes of canonical transformations, lifts, and fiber translations play an important part in mechanics.

Magnetic Terms.   A symplectic form on T^*Q different from the canonical one is obtained in the following way. Let B be a closed two-form on Q. Then $\Omega - \pi_Q^* B$ is a closed two-form on T^*Q, where Ω is the canonical two-form. To see that $\Omega - \pi_Q^* B$ is (weakly) nondegenerate, use the fact that in a local chart this form is given at the point (w, α) by

$$((u, \beta), (v, \gamma)) \mapsto \langle \gamma, u \rangle - \langle \beta, v \rangle - B(w)(u, v). \tag{6.6.5}$$

Proposition 6.6.2.

(i) *Let Ω be the canonical two-form on T^*Q and let $\pi_Q : T^*Q \to Q$ be the projection. If B is a closed two-form on Q, then*

$$\Omega_B = \Omega - \pi_Q^* B \tag{6.6.6}$$

is a (weak) symplectic form on T^*Q.

(ii) *Let B and B' be closed two-forms on Q and assume that $B - B' = \mathbf{d}A$. Then the mapping t_A (fiber translation by A) is a symplectic diffeomorphism of (T^*Q, Ω_B) with $(T^*Q, \Omega_{B'})$.*

Proof.   Part (i) follows by an argument similar to that in the momentum shifting lemma. For (ii), use formula (6.6.2) to get

$$t_A^* \Omega = \Omega - \pi_Q^* \mathbf{d}A = \Omega - \pi_Q^* B + \pi_Q^* B', \tag{6.6.7}$$

so that

$$t_A^*(\Omega - \pi_Q^* B') = \Omega - \pi_Q^* B,$$

since $\pi_Q \circ t_A = \pi_Q$.   ∎

Symplectic forms of the type Ω_B arise in the reduction process.[2] In the following section, we explain why the extra term $\pi_Q^* B$ is called a *magnetic term*.

---

[2]Magnetic terms come up in what is called the **cotangent bundle reduction theorem**; see Smale [1970], Abraham and Marsden [1978], Kummer [1981], Nill [1983], Montgomery, Marsden, and Ratiu [1984], Gozzi and Thacker [1987], and Marsden [1992].

Exercises

◇ **6.6-1.** Provide the intrinsic proof of Proposition 6.6.1.

◇ **6.6-2.** If $A = \mathbf{d}f$, use a coordinate calculation to check that $S(q^i, r_i) = r_i q^i - f(q^i)$ is a generating function for t_A.

6.7 A Particle in a Magnetic Field

Let B be a closed two-form on $\mathbb{R}^3$ and let $\mathbf{B} = B_x \mathbf{i} + B_y \mathbf{j} + B_z \mathbf{k}$ be the associated divergence-free vector field, that is,

$$\mathbf{i}_\mathbf{B}(dx \wedge dy \wedge dz) = B,$$

so that

$$B = B_x \, dy \wedge dz - B_y \, dx \wedge dz + B_z \, dx \wedge dy.$$

Thinking of $\mathbf{B}$ as a magnetic field, the equations of motion for a particle with charge e and mass m are given by the **Lorentz force law**

$$m\frac{d\mathbf{v}}{dt} = \frac{e}{c}\mathbf{v} \times \mathbf{B}, \tag{6.7.1}$$

where $\mathbf{v} = (\dot{x}, \dot{y}, \dot{z})$. On $\mathbb{R}^3 \times \mathbb{R}^3$, that is, $(\mathbf{x}, \mathbf{v})$-space, consider the symplectic form

$$\Omega_B = m(dx \wedge d\dot{x} + dy \wedge d\dot{y} + dz \wedge d\dot{z}) - \frac{e}{c}B, \tag{6.7.2}$$

that is, (6.6.6). As Hamiltonian, take the kinetic energy

$$H = \frac{m}{2}(\dot{x}^2 + \dot{y}^2 + \dot{z}^2). \tag{6.7.3}$$

Writing $X_H(u, v, w) = (u, v, w, \dot{u}, \dot{v}, \dot{w})$, the condition

$$\mathbf{d}H = \mathbf{i}_{X_H}\Omega_B \tag{6.7.4}$$

is the same as

$$m(\dot{x}\,d\dot{x} + \dot{y}\,d\dot{y} + \dot{z}\,d\dot{z})$$
$$= m(u\,d\dot{x} - \dot{u}\,dx + v\,d\dot{y} - \dot{v}\,dy + w\,d\dot{z} - \dot{w}\,dz)$$
$$- \frac{e}{c}[B_x v\,dz - B_x w\,dy - B_y u\,dz + B_y w\,dx + B_z u\,dy - B_z v\,dx],$$

which is equivalent to $u = \dot{x}$, $v = \dot{y}$, and $w = \dot{z}$, together with the equations

$$m\dot{u} = \frac{e}{c}(B_z v - B_y w),$$

$$m\dot{v} = \frac{e}{c}(B_x w - B_z u),$$

$$m\dot{w} = \frac{e}{c}(B_y u - B_x v),$$

that is, to

$$m\ddot{x} = \frac{e}{c}(B_z\dot{y} - B_y\dot{z}),$$

$$m\ddot{y} = \frac{e}{c}(B_x\dot{z} - B_z\dot{x}), \qquad (6.7.5)$$

$$m\ddot{z} = \frac{e}{c}(B_y\dot{x} - B_x\dot{y}),$$

which is the same as (6.7.1). Thus *the equations of motion for a particle in a magnetic field are Hamiltonian, with energy equal to the kinetic energy and with the symplectic form* Ω_B.

If $B = dA$, that is, $\mathbf{B} = \nabla \times \mathbf{A}$, where $\mathbf{A}^\flat = A$, then the map $t_A : (\mathbf{x}, \mathbf{v}) \mapsto (\mathbf{x}, \mathbf{p})$, where $\mathbf{p} = m\mathbf{v} + e\mathbf{A}/c$, pulls back the canonical form to Ω_B by the momentum shifting lemma. Thus, equations (6.7.1) are also Hamiltonian relative to the canonical bracket on $(\mathbf{x}, \mathbf{p})$-space with the Hamiltonian

$$H_A = \frac{1}{2m}\left\|\mathbf{p} - \frac{e}{c}\mathbf{A}\right\|^2. \qquad (6.7.6)$$

Remarks.

1. Not every magnetic field can be written as $\mathbf{B} = \nabla \times \mathbf{A}$ on Euclidean space. For example, the *field of a magnetic monopole of strength* $g \neq 0$, namely

$$\mathbf{B}(\mathbf{r}) = g\frac{\mathbf{r}}{\|\mathbf{r}\|^3}, \qquad (6.7.7)$$

cannot be written this way, since the flux of $\mathbf{B}$ through the unit sphere is $4\pi g$, yet Stokes' theorem applied to the two-sphere would give zero; see Exercise 4.4-3. Thus, one might think that the Hamiltonian formulation involving only $\mathbf{B}$ (that is, using Ω_B and H) is preferable. However, there is a way to recover the magnetic potential $\mathbf{A}$ by regarding it as a connection on a *nontrivial bundle* over $\mathbb{R}^3 \setminus \{0\}$. (This bundle over the sphere S^2 is the **Hopf fibration** $S^3 \to S^2$.) For a readable account of some aspects of this situation, see Yang [1985].

2. When one studies the motion of a particle in a Yang–Mills field, one finds a beautiful generalization of this construction and related ideas using the theory of principal bundles; see Sternberg [1977], Weinstein [1978a], and Montgomery [1984].

3. In Chapter 8 we study centrifugal and Coriolis forces and will see some structures analogous to those here. ♦

Exercises

◇ **6.7-1.** Show that particles in constant magnetic fields move in helixes.

◇ **6.7-2.** Verify "by hand" that $\frac{1}{2}m\|\mathbf{v}\|^2$ is conserved for a particle moving in a magnetic field.

◇ **6.7-3.** Verify "by hand" that Hamilton's equations for H_A are the Lorentz force law equations (6.7.1).

7
Lagrangian Mechanics

Our approach so far has emphasized the Hamiltonian point of view. However, there is an independent point of view, that of Lagrangian mechanics, based on variational principles. This alternative viewpoint, computational convenience, and the fact that the Lagrangian is very useful in covariant relativistic theories can be used as arguments for the importance of the Lagrangian formulation. Ironically, it was Hamilton [1834] who discovered the variational basis of Lagrangian mechanics.

7.1 Hamilton's Principle of Critical Action

Much of mechanics can be based on variational principles. Indeed, it is the variational formulation that is the most covariant, being useful for relativistic systems as well. In the next chapter we shall see the utility of the Lagrangian approach in the study of rotating frames and moving systems, and we will also use it as an important way to approach Hamilton–Jacobi theory.

Consider a **configuration manifold** Q and the velocity phase space TQ. We consider a function $L : TQ \to \mathbb{R}$ called the **Lagrangian**. Speaking informally, Hamilton's **principle of critical action** states that

$$\delta \int L\left(q^i, \frac{dq^i}{dt}\right) dt = 0, \qquad (7.1.1)$$

where we take variations among paths $q^i(t)$ in Q with fixed endpoints. (We will study this process a little more carefully in §8.1.) Taking the variation in (7.1.1), the chain rule gives

$$\int \left[\frac{\partial L}{\partial q^i} \delta q^i + \frac{\partial L}{\partial \dot{q}^i} \frac{d}{dt} \delta q^i \right] dt \qquad (7.1.2)$$

for the left-hand side. Integrating the second term by parts and using the boundary conditions $\delta q^i = 0$ at the endpoints of the time interval in question, we get

$$\int \left[\frac{\partial L}{\partial q^i} - \frac{d}{dt} \left(\frac{\partial L}{\partial \dot{q}^i} \right) \right] \delta q^i \, dt = 0. \qquad (7.1.3)$$

If this is to hold for all such variations $\delta q^i(t)$, then

$$\frac{\partial L}{\partial q^i} - \frac{d}{dt} \frac{\partial L}{\partial \dot{q}^i} = 0, \qquad (7.1.4)$$

which are the *Euler–Lagrange equations*.

We set $p_i = \partial L / \partial \dot{q}^i$, assume that the transformation $(q^i, \dot{q}^j) \mapsto (q^i, p_j)$ is invertible, and define the *Hamiltonian* by

$$H(q^i, p_j) = p_i \dot{q}^i - L(q^i, \dot{q}^i). \qquad (7.1.5)$$

Note that

$$\dot{q}^i = \frac{\partial H}{\partial p_i},$$

since

$$\frac{\partial H}{\partial p_i} = \dot{q}^i + p_j \frac{\partial \dot{q}^j}{\partial p_i} - \frac{\partial L}{\partial \dot{q}^j} \frac{\partial \dot{q}^j}{\partial p_i} = \dot{q}^i$$

from (7.1.5) and the chain rule. Likewise,

$$\dot{p}_i = -\frac{\partial H}{\partial q^i}$$

from (7.1.4) and

$$\frac{\partial H}{\partial q^j} = p_i \frac{\partial \dot{q}^i}{\partial q^j} - \frac{\partial L}{\partial q^j} - \frac{\partial L}{\partial \dot{q}^i} \frac{\partial \dot{q}^i}{\partial q^j} = -\frac{\partial L}{\partial q^j}.$$

In other words, the *Euler–Lagrange equations are equivalent to Hamilton's equations*.

Thus, it is reasonable to explore the geometry of the Euler–Lagrange equations using the canonical form on T^*Q pulled back to TQ using $p_i = \partial L / \partial \dot{q}^i$. We do this in the next sections.

This is one standard way to approach the geometry of the Euler–Lagrange equations. Another *is to use the variational principle itself*. The reader will notice that the canonical one-form $p_i dq^i$ appears as the *boundary terms* when we take the variations. This can, in fact, be used as a basis for the introduction of the canonical one-form in Lagrangian mechanics. We shall develop this approach in Chapter 8. See also Exercise 7.2-2.

Exercises

$\diamond$ **7.1-1.**   Verify that the Euler–Lagrange and Hamilton equations are equivalent, even if L is time-dependent.

$\diamond$ **7.1-2.**   Show that the conservation of energy equation results if in Hamilton's principle, variations corresponding to reparametrizations of the given curve $q(t)$ are chosen.

7.2   The Legendre Transform

Fiber Derivatives.   Given a Lagrangian $L : TQ \to \mathbb{R}$, define a map $\mathbb{F}L : TQ \to T^*Q$, called the *fiber derivative*, by

$$\mathbb{F}L(v) \cdot w = \left.\frac{d}{ds}\right|_{s=0} L(v + sw), \tag{7.2.1}$$

where $v, w \in T_qQ$. Thus, $\mathbb{F}L(v) \cdot w$ is the derivative of L at v along the fiber T_qQ in the direction w. Note that $\mathbb{F}L$ is fiber-preserving; that is, it maps the fiber T_qQ to the fiber T_q^*Q. In a local chart $U \times E$ for TQ, where U is open in the model space E for Q, the fiber derivative is given by

$$\mathbb{F}L(u, e) = (u, \mathbf{D}_2L(u, e)), \tag{7.2.2}$$

where $\mathbf{D}_2L$ denotes the partial derivative of L with respect to its second argument. For finite-dimensional manifolds, with (q^i) denoting coordinates on Q and $(q^i, \dot{q}^i)$ the induced coordinates on TQ, the fiber derivative has the expression

$$\mathbb{F}L(q^i, \dot{q}^i) = \left(q^i, \frac{\partial L}{\partial \dot{q}^i}\right), \tag{7.2.3}$$

that is, $\mathbb{F}L$ is given by

$$p_i = \frac{\partial L}{\partial \dot{q}^i}. \tag{7.2.4}$$

The *associated energy function* is defined by $E(v) = \mathbb{F}L(v) \cdot v - L(v)$.

In many examples it is the relationship (7.2.4) that gives physical meaning to the momentum variables. We call $\mathbb{F}L$ the **Legendre transform**.

Lagrangian Forms.   Let Ω denote the canonical symplectic form on T^*Q. Using $\mathbb{F}L$, we obtain a one-form Θ_L and a closed two-form Ω_L on TQ by setting

$$\Theta_L = (\mathbb{F}L)^*\Theta \quad \text{and} \quad \Omega_L = (\mathbb{F}L)^*\Omega. \tag{7.2.5}$$

We call Θ_L the *Lagrangian one-form* and Ω_L the *Lagrangian two-form*. Since $\mathbf{d}$ commutes with pull-back, we get $\Omega_L = -\mathbf{d}\Theta_L$. Using the local expressions for Θ and Ω, a straightforward pull-back computation yields the following local formula for Θ_L and Ω_L: If E is the model space for Q, U is the range in E of a chart on Q, and $U \times E$ is the corresponding range of the induced chart on TQ, then for $(u, e) \in U \times E$ and tangent vectors $(e_1, e_2), (f_1, f_2)$ in $E \times E$, we have

$$T_{(u,e)}\mathbb{F}L \cdot (e_1, e_2)$$
$$= (u, \mathbf{D}_2 L(u, e), e_1, \mathbf{D}_1(\mathbf{D}_2 L(u, e)) \cdot e_1 + \mathbf{D}_2(\mathbf{D}_2 L(u, e)) \cdot e_2),$$
$$\tag{7.2.6}$$

so that using the local expression for Θ and the definition of pull-back,

$$\Theta_L(u, e) \cdot (e_1, e_2) = \mathbf{D}_2 L(u, e) \cdot e_1. \tag{7.2.7}$$

Similarly, one finds that

$$\Omega_L(u, e) \cdot ((e_1, e_2), (f_1, f_2))$$
$$= \mathbf{D}_1(\mathbf{D}_2 L(u, e) \cdot e_1) \cdot f_1 - \mathbf{D}_1(\mathbf{D}_2 L(u, e) \cdot f_1) \cdot e_1$$
$$+ \mathbf{D}_2 \mathbf{D}_2 L(u, e) \cdot e_1 \cdot f_2 - \mathbf{D}_2 \mathbf{D}_2 L(u, e) \cdot f_1 \cdot e_2, \tag{7.2.8}$$

where $\mathbf{D}_1$ and $\mathbf{D}_2$ denote the first and second partial derivatives. In finite dimensions, formulae (7.2.6) and (7.2.7) or a direct pull-back of $p_i dq^i$ and $dq^i \wedge dp_i$ yields

$$\Theta_L = \frac{\partial L}{\partial \dot{q}^i} dq^i \tag{7.2.9}$$

and

$$\Omega_L = \frac{\partial^2 L}{\partial \dot{q}^i \partial q^j} dq^i \wedge dq^j + \frac{\partial^2 L}{\partial \dot{q}^i \partial \dot{q}^j} dq^i \wedge d\dot{q}^j \tag{7.2.10}$$

(a sum on all i, j is understood). As a $2n \times 2n$ skew-symmetric matrix,

$$\Omega_L = \begin{bmatrix} A & \left[\dfrac{\partial^2 L}{\partial \dot{q}^i \partial \dot{q}^j}\right] \\ \left[-\dfrac{\partial^2 L}{\partial \dot{q}^i \partial \dot{q}^j}\right] & 0 \end{bmatrix}, \tag{7.2.11}$$

where A is the skew-symmetrization of $\partial^2 L/(\partial \dot{q}^i \partial q^j)$. From these expressions, it follows that Ω_L *is (weakly) nondegenerate if and only if the quadratic form* $\mathbf{D}_2 \mathbf{D}_2 L(u, e)$ *is (weakly) nondegenerate.* In this case, we say that L is a *regular* or *nondegenerate* Lagrangian. The implicit function theorem shows that *the fiber derivative is locally invertible if and only if L is regular.*

Exercises

◇ **7.2-1.** Let

$$L(q^1, q^2, q^3, \dot{q}^1, \dot{q}^2, \dot{q}^3) = \frac{m}{2}\left(\left(\dot{q}^1\right)^2 + \left(\dot{q}^2\right)^2 + \left(\dot{q}^3\right)^2 \right) + q^1\dot{q}^1 + q^2\dot{q}^2 + q^3\dot{q}^3.$$

Calculate Θ_L, Ω_L, and the corresponding Hamiltonian.

◇ **7.2-2.** For $v \in T_qQ$, define its **vertical lift** $v^l \in T_v(TQ)$ to be the tangent vector to the curve $v + tv$ at $t = 0$. Show that Θ_L may be defined by

$$w \lrcorner\, \Theta_L = v^l \lrcorner\, \mathbf{d}L,$$

where $w \in T_v(TQ)$ and where $w \lrcorner\, \Theta_L = \mathbf{i}_w \Theta_L$ is the interior product. Also, show that the energy is

$$E(v) = v^l \lrcorner\, \mathbf{d}L - L(v).$$

◇ **7.2-3** (Abstract Legendre Transform). Let V be a vector bundle over a manifold S and let $L : V \to \mathbb{R}$. For $v \in V$, let

$$w = \frac{\partial L}{\partial v} \in v^*$$

denote the fiber derivative. Assume that the map $v \mapsto w$ is a local diffeomorphism and let $H : V^* \to \mathbb{R}$ be defined by

$$H(w) = \langle w, v \rangle - L(v).$$

Show that

$$v = \frac{\partial H}{\partial w}.$$

7.3 Euler–Lagrange Equations

Hyperregular Lagrangians. Given a Lagrangian L, the **action** of L is the map $A : TQ \to \mathbb{R}$ that is defined by $A(v) = \mathbb{F}L(v) \cdot v$, and as we defined above, the **energy** of L is $E = A - L$. In charts,

$$A(u, e) = \mathbf{D}_2 L(u, e) \cdot e, \tag{7.3.1}$$
$$E(u, e) = \mathbf{D}_2 L(u, e) \cdot e - L(u, e), \tag{7.3.2}$$

and in finite dimensions, (7.3.1) and (7.3.2) read

$$A(q^i, \dot{q}^i) = \dot{q}^i \frac{\partial L}{\partial \dot{q}^i} = p_i \dot{q}^i, \tag{7.3.3}$$

$$E(q^i, \dot{q}^i) = \dot{q}^i \frac{\partial L}{\partial \dot{q}^i} - L(q^i, \dot{q}^i) = p_i \dot{q}^i - L(q^i, \dot{q}^i). \tag{7.3.4}$$

If L is a Lagrangian such that $\mathbb{F}L : TQ \to T^*Q$ is a diffeomorphism, we say that L is a **hyperregular** Lagrangian. In this case, set $H = E \circ (\mathbb{F}L)^{-1}$. Then X_H and X_E are $\mathbb{F}L$-related, since $\mathbb{F}L$ is, by construction, symplectic. Thus, hyperregular Lagrangians on TQ induce Hamiltonian systems on T^*Q. Conversely, one can show that hyperregular Hamiltonians on T^*Q come from Lagrangians on TQ (see §7.4 for definitions and details).

Lagrangian Vector Field.   More generally, a vector field Z on TQ is called a **Lagrangian vector field** or a **Lagrangian system** for L if the **Lagrangian condition**

$$\Omega_L(v)(Z(v), w) = \mathbf{d}E(v) \cdot w \tag{7.3.5}$$

holds for all $v \in T_qQ$ and $w \in T_v(TQ)$. If L is regular, so that Ω_L is a (weak) symplectic form, then there would exist at most one such Z, which would be the Hamiltonian vector field of E with respect to the (weak) symplectic form Ω_L. In this case we know that E is conserved on the flow of Z. In fact, the same result holds, even if L is degenerate:

Proposition 7.3.1. *Let Z be a Lagrangian vector field for L and let $v(t) \in TQ$ be an integral curve of Z. Then $E(v(t))$ is constant in t.*

Proof.   By the chain rule,

$$\frac{d}{dt}E(v(t)) = \mathbf{d}E(v(t)) \cdot \dot{v}(t) = \mathbf{d}E(v(t)) \cdot Z(v(t))$$

$$= \Omega_L(v(t))(Z(v(t))), Z(v(t)) = 0 \tag{7.3.6}$$

by skew-symmetry of Ω_L .   ∎

We usually assume that Ω_L is nondegenerate, but the degenerate case comes up in the Dirac theory of constraints (see Dirac [1950, 1964], Kunzle [1969], Hanson, Regge, and Teitelboim [1976], Gotay, Nester, and Hinds [1979], references therein, and §8.5).

Second-Order Equations.   The vector field Z often has a special property, namely, that Z is a second-order equation.

Definition 7.3.2. *A vector field V on TQ is called a **second-order equation** if $T\tau_Q \circ V = $ identity, where $\tau_Q : TQ \to Q$ is the canonical projection. If $c(t)$ is an integral curve of V, then $(\tau_Q \circ c)(t)$ is called the **base integral curve** of $c(t)$.*

It is easy to see that the condition for V being second-order is equivalent to the following: For any chart $U \times E$ on TQ, we can write $V(u,e) = ((u,e),(e,V_2(u,e)))$, for some map $V_2 : U \times E \to E$. Thus, the dynamics are determined by $\dot{u} = e$, and $\dot{e} = V_2(u,e)$; that is, $\ddot{u} = V_2(u,\dot{u})$, a second-order equation in the standard sense. This local computation also shows that the base integral curve uniquely determines an integral curve of V through a given initial condition in TQ.

The Euler–Lagrange Equations. From the point of view of Lagrangian vector fields, the main result concerning the Euler–Lagrange equations is the following.

Theorem 7.3.3. *Let Z be a Lagrangian system for L and suppose Z is a second-order equation. Then in a chart $U \times E$, an integral curve $(u(t), v(t)) \in U \times E$ of Z satisfies the **Euler–Lagrange equations**; that is,*

$$\frac{du(t)}{dt} = v(t),$$

$$\frac{d}{dt} \mathbf{D}_2 L(u(t), v(t)) \cdot w = \mathbf{D}_1 L(u(t), v(t)) \cdot w \qquad (7.3.7)$$

for all $w \in E$. In finite dimensions, the Euler–Lagrange equations take the form

$$\frac{dq^i}{dt} = \dot{q}^i,$$

$$\frac{d}{dt}\left(\frac{\partial L}{\partial \dot{q}^i}\right) = \frac{\partial L}{\partial q^i}, \quad i = 1, \dots, n. \qquad (7.3.8)$$

If L is regular, that is, Ω_L is (weakly) nondegenerate, then Z is automatically second-order, and if it is strongly nondegenerate, then

$$\frac{d^2 u}{dt^2} = \frac{dv}{dt} = [\mathbf{D}_2 \mathbf{D}_2 L(u, v)]^{-1}(\mathbf{D}_1 L(u, v) - \mathbf{D}_1 \mathbf{D}_2 L(u, v) \cdot v), \qquad (7.3.9)$$

or in finite dimensions,

$$\ddot{q}^j = G^{ij}\left(\frac{\partial L}{\partial q^i} - \frac{\partial^2 L}{\partial q^j \partial \dot{q}^i} \dot{q}^j\right), \quad i, j = 1, \dots, n, \qquad (7.3.10)$$

where $[G^{ij}]$ is the inverse of the matrix $(\partial^2 L / \partial q^i \partial \dot{q}^j)$. Thus $u(t)$ and $q^i(t)$ are base integral curves of the Lagrangian vector field Z if and only if they satisfy the Euler–Lagrange equations.

Proof. From the definition of the energy E we have the local expression

$$\mathbf{D}E(u, e) \cdot (e_1, e_2) = \mathbf{D}_1(\mathbf{D}_2 L(u, e) \cdot e) \cdot e_1 + \mathbf{D}_2(\mathbf{D}_2 L(u, e) \cdot e) \cdot e_2$$
$$- \mathbf{D}_1 L(u, e) \cdot e_1 \qquad (7.3.11)$$

(the term $\mathbf{D}_2 L(u, e) \cdot e_2$ has canceled). Locally, we may write

$$Z(u, e) = (u, e, Y_1(u, e), Y_2(u, e)).$$

Using formula (7.2.8) for Ω_L, the condition (7.3.5) on Z may be written

$$\mathbf{D}_1 \mathbf{D}_2 L(u, e) \cdot Y_1(u, e)) \cdot e_1 - \mathbf{D}_1(\mathbf{D}_2 L(u, e) \cdot e_1) \cdot Y_1(u, e)$$
$$+ \mathbf{D}_2 \mathbf{D}_2 L(u, e) \cdot Y_1(u, e) \cdot e_2 - \mathbf{D}_2 \mathbf{D}_2 L(u, e) \cdot e_1 \cdot Y_2(u, e)$$
$$= \mathbf{D}_1(\mathbf{D}_2 L(u, e) \cdot e) \cdot e_1 - \mathbf{D}_1 L(u, e) \cdot e_1 + \mathbf{D}_2 \mathbf{D}_2 L(u, e) \cdot e \cdot e_2.$$
$$(7.3.12)$$

Thus, if Ω_L is a weak symplectic form, then $\mathbf{D}_2\mathbf{D}_2L(u,e)$ is weakly non-degenerate, so setting $e_1 = 0$ we get $Y_1(u,e) = e$; that is, Z is a second-order equation. In any case, if we assume that Z is second-order, condition (7.3.12) becomes

$$\mathbf{D}_1L(u,e) \cdot e_1 = \mathbf{D}_1(\mathbf{D}_2L(u,e) \cdot e_1) \cdot e + \mathbf{D}_2\mathbf{D}_2L(u,e) \cdot e_1 \cdot Y_2(u,e)$$
$$(7.3.13)$$

for all $e_1 \in E$. If $(u(t), v(t))$ is an integral curve of Z, then (using dots to denote time differentiation) $\dot{u} = v$ and $\ddot{u} = Y_2(u,v)$, so (7.3.13) becomes

$$\mathbf{D}_1L(u,\dot{u}) \cdot e_1 = \mathbf{D}_1(\mathbf{D}_2L(u,\dot{u}) \cdot e_1) \cdot \dot{u} + \mathbf{D}_2\mathbf{D}_2L(u,\dot{u}) \cdot e_1 \cdot \ddot{u}$$
$$= \frac{d}{dt}\mathbf{D}_2L(u,\dot{u}) \cdot e_1 \qquad (7.3.14)$$

by the chain rule.

The last statement follows by using the chain rule on the left-hand side of Lagrange's equation and using nondegeneracy of L to solve for $\dot{v}$, that is, $\ddot{q}^j$. ∎

Exercises

⋄ **7.3-1.** Give an explicit example of a degenerate Lagrangian L that has a second-order Lagrangian system Z.

⋄ **7.3-2.** Check directly that the validity of the expression (7.3.8) is coordinate independent. In other words, verify directly that the form of the Euler–Lagrange equations does not depend on the local coordinates chosen to describe them.

7.4   Hyperregular Lagrangians and Hamiltonians

Above, we said that a smooth Lagrangian $L : TQ \to \mathbb{R}$ is *hyperregular* if $\mathbb{F}L : TQ \to T^*Q$ is a diffeomorphism. From (7.2.8) or (7.2.11) it follows that the symmetric bilinear form $\mathbf{D}_2\mathbf{D}_2L(u,e)$ is strongly nondegenerate. As before, let $\pi_Q : T^*Q \to Q$ and $\tau_Q : TQ \to Q$ denote the canonical projections.

Proposition 7.4.1. *Let L be a hyperregular Lagrangian on TQ and let $H = E \circ (\mathbb{F}L)^{-1} \in \mathcal{F}(T^*Q)$, where E is the energy of L. Then the Lagrangian vector field Z on TQ and the Hamiltonian vector field X_H on T^*Q are $\mathbb{F}L$-related, that is,*

$$(\mathbb{F}L)^*X_H = Z.$$

Furthermore, if $c(t)$ is an integral curve of Z and $d(t)$ an integral curve of X_H with $\mathbb{F}L(c(0)) = d(0)$, then

$$\mathbb{F}L(c(t)) = d(t) \quad and \quad (\tau_Q \circ c)(t) = (\pi_Q \circ d)(t).$$

*The curve $(\tau_Q \circ c)(t)$ is called the **base integral curve** of $c(t)$, and similarly, $(\pi_Q \circ d)(t)$ is the **base integral curve** of $d(t)$.*

Proof.   For $v \in TQ$ and $w \in T_v(TQ)$, we have

$$
\begin{aligned}
\Omega(\mathbb{F}L(v))(T_v\mathbb{F}L(Z(v)), T_v\mathbb{F}L(w)) &= ((\mathbb{F}L)^*\Omega)(v)(Z(v), w) \\
&= \Omega_L(v)(Z(v), w) \\
&= \mathbf{d}E(v) \cdot w \\
&= \mathbf{d}(H \circ \mathbb{F}L)(v) \cdot w \\
&= \mathbf{d}H(\mathbb{F}L(v)) \cdot T_v\mathbb{F}L(w) \\
&= \Omega(\mathbb{F}L(v))(X_H(\mathbb{F}L(v)), T_v\mathbb{F}L(w)),
\end{aligned}
$$

so that by weak nondegeneracy of Ω and the fact that $T_v\mathbb{F}L$ is an isomorphism, it follows that

$$T_v\mathbb{F}L(Z(v)) = X_H(\mathbb{F}L(v)).$$

Thus $T\mathbb{F}L \circ Z = X_H \circ \mathbb{F}L$, that is, $Z = (\mathbb{F}L)^*X_H$.

If φ_t denotes the flow of Z and ψ_t the flow of X_H, the relation $Z = (\mathbb{F}L)^*X_H$ is equivalent to $\mathbb{F}L \circ \varphi_t = \psi_t \circ \mathbb{F}L$. Thus, if $c(t) = \varphi_t(v)$, then

$$\mathbb{F}L(c(t)) = \psi_t(\mathbb{F}L(v))$$

is an integral curve of X_H that at $t = 0$ passes through $\mathbb{F}L(v) = \mathbb{F}L(c(0))$, whence $\psi_t(\mathbb{F}L(v)) = d(t)$ by uniqueness of integral curves of smooth vector fields. Finally, since $\tau_Q = \pi_Q \circ \mathbb{F}L$, we get

$$(\tau_Q \circ c)(t) = (\pi_Q \circ \mathbb{F}L \circ c)(t) = (\pi_Q \circ d)(t). \quad \blacksquare$$

The Action.   We claim that the action A of L is related to the Lagrangian vector field Z of L by

$$A(v) = \langle \Theta_L(v), Z(v) \rangle, \quad v \in TQ. \tag{7.4.1}$$

We prove this formula under the assumption that Z is a second-order equation, even if L is not regular. In fact,

$$
\begin{aligned}
\langle \Theta_L(v), Z(v) \rangle &= \langle ((\mathbb{F}L)^*\Theta)(v), Z(v) \rangle \\
&= \langle \Theta(\mathbb{F}L(v)), T_v\mathbb{F}L(Z(v)) \rangle \\
&= \langle \mathbb{F}L(v), T\pi_Q \cdot T_v\mathbb{F}L(Z(v)) \rangle \\
&= \langle \mathbb{F}L(v), T_v(\pi_Q \circ \mathbb{F}L)(Z(v)) \rangle \\
&= \langle \mathbb{F}L(v), T_v\tau_Q(Z(v)) \rangle = \langle \mathbb{F}L(v), v \rangle = A(v),
\end{aligned}
$$

by definition of a second-order equation and the definition of the action. If L is hyperregular and $H = E \circ (\mathbb{F}L)^{-1}$, then

$$A \circ (\mathbb{F}L)^{-1} = \langle \Theta, X_H \rangle . \tag{7.4.2}$$

Indeed, by (7.4.1), the properties of push-forward, and the previous proposition, we have

$$A \circ (\mathbb{F}L)^{-1} = (\mathbb{F}L)_* A = (\mathbb{F}L)_* (\langle \Theta_L, Z \rangle) = \langle (\mathbb{F}L)_* \Theta_L, (\mathbb{F}L)_* Z \rangle = \langle \Theta, X_H \rangle .$$

If $H : T^*Q \to \mathbb{R}$ is a smooth Hamiltonian, the function $G : T^*Q \to \mathbb{R}$ given by $G = \langle \Theta, X_H \rangle$ is called the **action** of H. Thus, (7.4.2) says that the push-forward of the action A of L equals the action G of $H = E \circ (\mathbb{F}L)^{-1}$.

Hyperregular Hamiltonians. A Hamiltonian H is called **hyperregular** if $\mathbb{F}H : T^*Q \to TQ$, defined by

$$\mathbb{F}H(\alpha) \cdot \beta = \frac{d}{ds}\bigg|_{s=0} H(\alpha + s\beta), \tag{7.4.3}$$

where $\alpha, \beta \in T_q^*Q$, is a diffeomorphism; here we must assume that either the model space E of Q is reflexive, so that $T_q^{**}Q = T_qQ$ for all $q \in Q$, or what is more reasonable, that $\mathbb{F}H(\alpha)$ lies in $T_qQ \subset T_q^{**}Q$. As in the case of Lagrangians, hyperregularity of H implies the strong nondegeneracy of $\mathbf{D}_2\mathbf{D}_2H(u, \alpha)$, and the curve $s \mapsto \alpha + s\beta$ appearing in (7.4.3) can be replaced by an arbitrary smooth curve $\alpha(s)$ in T_q^*Q such that

$$\alpha(0) = \alpha \quad \text{and} \quad \alpha'(0) = \beta.$$

Proposition 7.4.2.   (i) *Let $H \in \mathcal{F}(T^*Q)$ be a hyperregular Hamiltonian and define*

$$E = H \circ (\mathbb{F}H)^{-1}, \quad A = G \circ (\mathbb{F}H)^{-1}, \quad \text{and} \quad L = A - E \in \mathcal{F}(TQ).$$

Then L is a hyperregular Lagrangian and $\mathbb{F}L = \mathbb{F}H^{-1}$. Furthermore, A is the action of L, and E the energy of L.

(ii) *Let $L \in \mathcal{F}(TQ)$ be a hyperregular Lagrangian and define*

$$H = E \circ (\mathbb{F}L)^{-1}.$$

Then H is a hyperregular Hamiltonian and $\mathbb{F}H = (\mathbb{F}L)^{-1}$.

Proof.   (i) Locally, $G(u, \alpha) = \langle \alpha, \mathbf{D}_2H(u, \alpha) \rangle$, so that

$$A(u, \mathbf{D}_2H(u, \alpha)) = (A \circ \mathbb{F}H)(u, \alpha) = G(u, \alpha) = \langle \alpha, \mathbf{D}_2H(u, \alpha) \rangle ,$$

whence

$$(L \circ \mathbb{F}H)(u, \alpha) = L(u, \mathbf{D}_2H(u, \alpha)) = \langle \alpha, \mathbf{D}_2H(u, \alpha) \rangle - H(u, \alpha).$$

Let $e = \mathbf{D}_2(\mathbf{D}_2 H(u, \alpha)) \cdot \beta$, and let $e(s) = \mathbf{D}_2 H(u, \alpha + s\beta)$ be a curve that at $s = 0$ passes through $e(0) = \mathbf{D}_2 H(u, \alpha)$ and whose derivative at $s = 0$ equals $e'(0) = \mathbf{D}_2(\mathbf{D}_2 H(u, \alpha)) \cdot \beta = e$. Therefore,

$$
\begin{aligned}
\langle(\mathbb{F}L \circ \mathbb{F}H)(u, \alpha), e\rangle \\
= \langle \mathbb{F}L(u, \mathbf{D}_2 H(u, \alpha)), e\rangle \\
= \frac{d}{dt}\Big|_{s=0} L(u, e(s)) = \frac{d}{dt}\Big|_{s=0} L(u, \mathbf{D}_2 H(u, \alpha + s\beta)) \\
= \frac{d}{dt}\Big|_{s=0} [\langle\alpha + s\beta, \mathbf{D}_2 H(u, \alpha + s\beta)\rangle - H(u, \alpha + s\beta)] \\
= \langle\alpha, \mathbf{D}_2(\mathbf{D}_2 H(u, \alpha)) \cdot \beta\rangle = \langle\alpha, e\rangle.
\end{aligned}
$$

Since $\mathbf{D}_2 \mathbf{D}_2 H(u, \alpha)$ is strongly nondegenerate, this implies that $e \in E$ is arbitrary and hence $\mathbb{F}L \circ \mathbb{F}H = $ identity. Since $\mathbb{F}H$ is a diffeomorphism, this says that $\mathbb{F}L = (\mathbb{F}H)^{-1}$ and hence that L is hyperregular.

To see that A is the action of L, note that since $\mathbb{F}H^{-1} = \mathbb{F}L$, we have by definition of G,

$$
A = G \circ (\mathbb{F}H)^{-1} = \langle\Theta, X_H\rangle \circ \mathbb{F}L,
$$

which by (7.4.2) implies that A is the action of L. Therefore, $E = A - L$ is the energy of L.

(ii) Locally, since we define $H = E \circ (\mathbb{F}L)^{-1}$, we have

$$
\begin{aligned}
(H \circ \mathbb{F}L)(u, e) &= H(u, \mathbf{D}_2 L(u, e)) \\
&= A(u, e) - L(u, e) \\
&= \mathbf{D}_2 L(u, e) \cdot e - L(u, e)
\end{aligned}
$$

and proceed as before. Let

$$
\alpha = \mathbf{D}_2(\mathbf{D}_2 L(u, e)) \cdot f,
$$

where $f \in E$ and $\alpha(s) = \mathbf{D}_2 L(u, e + sf)$; then

$$
\alpha(0) = \mathbf{D}_2 L(u, e) \quad \text{and} \quad \alpha'(0) = \alpha,
$$

so that

$$
\begin{aligned}
\langle\alpha, (\mathbb{F}H \circ \mathbb{F}L)(u, e)\rangle &= \langle\alpha, \mathbb{F}H(u, \mathbf{D}_2 L(u, e))\rangle \\
&= \frac{d}{ds}\Big|_{s=0} H(u, \alpha(s)) \\
&= \frac{d}{ds}\Big|_{s=0} H(u, \mathbf{D}_2 L(u, e + sf)) \\
&= \frac{d}{ds}\Big|_{s=0} [\langle\mathbf{D}_2 L(u, e + sf), e + sf\rangle - L(u, e + sf)] \\
&= \langle\mathbf{D}_2(\mathbf{D}_2 L(u, e)) \cdot f, e\rangle = \langle\alpha, e\rangle,
\end{aligned}
$$

which shows, by strong nondegeneracy of $\mathbf{D}_2\mathbf{D}_2L$, that $\mathbb{F}H \circ \mathbb{F}L =$ identity. Since $\mathbb{F}L$ is a diffeomorphism, it follows that $\mathbb{F}H = (\mathbb{F}L)^{-1}$ and H is hyperregular.   ∎

The main result is summarized in the following.

Theorem 7.4.3. *Hyperregular Lagrangians $L \in \mathcal{F}(TQ)$ and hyperregular Hamiltonians $H \in \mathcal{F}(T^*Q)$ correspond in a bijective manner by the preceding constructions. The following diagram commutes:*

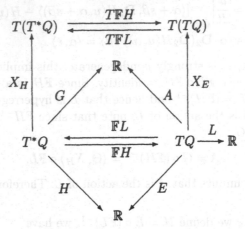

Proof. Let L be a hyperregular Lagrangian and let H be the associated hyperregular Hamiltonian, that is,

$$H = E \circ (\mathbb{F}L)^{-1} = (A - L) \circ (\mathbb{F}L)^{-1} = G - L \circ \mathbb{F}H$$

by Propositions 7.4.1 and 7.4.2. From H we construct a Lagrangian L' by

$$L' = G \circ (\mathbb{F}H)^{-1} - H \circ (\mathbb{F}H)^{-1}$$
$$= G \circ (\mathbb{F}H)^{-1} - (G - L \circ \mathbb{F}H) \circ (\mathbb{F}H)^{-1} = L.$$

Conversely, if H is a given hyperregular Hamiltonian, then the associated Lagrangian L is hyperregular and is given by

$$L = G \circ (\mathbb{F}H)^{-1} - H \circ (\mathbb{F}H)^{-1} = A - H \circ \mathbb{F}L.$$

Thus, the corresponding hyperregular Hamiltonian induced by L is

$$H' = E \circ (\mathbb{F}L)^{-1} = (A - L) \circ (\mathbb{F}L)^{-1}$$
$$= A \circ (\mathbb{F}L)^{-1} - (A - H \circ \mathbb{F}L) \circ (\mathbb{F}L)^{-1} = H.$$

The commutativity of the two diagrams is now a direct consequence of the above and Propositions 7.4.1 and 7.4.2.   ∎

Neighborhood Theorem for Regular Lagrangians. We now prove an important theorem for regular Lagrangians that concerns the structure of solutions near a given one.

Definition 7.4.4. *Let $\bar{q}(t)$ be a given solution of the Euler–Lagrange equations, $\bar{t}_1 \leq t \leq \bar{t}_2$. Let $\bar{q}_1 = \bar{q}\,(\bar{t}_1)$ and $\bar{q}_2 = \bar{q}\,(\bar{t}_2)$. We say that $\bar{q}(t)$ is a **nonconjugate solution** if there is a neighborhood $\mathcal{U}$ of the curve $\bar{q}(t)$ and neighborhoods $\mathcal{U}_1 \subset \mathcal{U}$ of $\bar{q}_1$ and $\mathcal{U}_2 \subset \mathcal{U}$ of $\bar{q}_2$ such that for all $q_1 \in \mathcal{U}_1$ and $q_2 \in \mathcal{U}_2$ and t_1 close to $\bar{t}_1$, t_2 close to $\bar{t}_2$, there exists a unique solution $q(t)$, $t_1 \leq t \leq t_2$, of the Euler–Lagrange equations satisfying the following conditions: $q\,(t_1) = q_1$, $q\,(t_2) = q_2$, and $q(t) \in \mathcal{U}$. See Figure 7.4.1.*

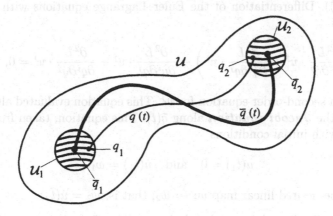

FIGURE 7.4.1. Neighborhood theorem

To determine conditions guaranteeing that a solution is nonconjugate, we shall use the following observation. Let $\bar{v}_1 = \dot{\bar{q}}\,(t_1)$ and $\bar{v}_2 = \dot{\bar{q}}\,(t_2)$. Let F_t be the flow of the Euler–Lagrange equations on TQ. By construction of $F_t(q, v)$, we have $F_{t_2}\,(\bar{q}_1, \bar{v}_1) = (\bar{q}_2, \bar{v}_2)$.

Next, we attempt to apply the implicit function theorem to the flow map. We want to solve

$$(\pi_Q \circ F_{t_2})\,(q_1, v_1) = q_2$$

for v_1, where we regard q_1, t_1, t_2 as parameters. To do this, we form the linearization

$$w_2 := T_{v_1}(\pi_Q \circ F_{\bar{t}_2})\,(\bar{q}_1, \bar{v}_1) \cdot w_1.$$

We require that $w_1 \mapsto w_2$ be invertible. The right-hand side of this equation suggests forming the curve

$$w(t) := T_{v_1} \pi_Q F_t(\bar{q}_1, \bar{v}_1) \cdot w_1, \qquad (7.4.4)$$

which is the solution of the linearized, or first variation, equation of the Euler–Lagrange equations satisfied by $F_t(\bar{q}_1, \bar{v}_1)$. Let us work out the equation satisfied by

$$w(t) := T_{v_1} \pi_Q F_t(\bar{q}_1, \bar{v}_1) \cdot w_1$$

in coordinates. Start with a solution $q(t)$ of the Euler–Lagrange equations

$$\frac{d}{dt} \frac{\partial L}{\partial \dot{q}^i} - \frac{\partial L}{\partial q^i} = 0.$$

Given the curve of initial conditions $\varepsilon \mapsto (q_1, v_1 + \varepsilon w_1)$, we get corresponding solutions $(q_\varepsilon(t), \dot{q}_\varepsilon(t))$, whose derivative with respect to ε we denoted by $(u(t), \dot{u}(t))$. Differentiation of the Euler–Lagrange equations with respect to ε gives

$$\frac{d}{dt} \left(\frac{\partial^2 L}{\partial \dot{q}^i \partial \dot{q}^j} \cdot \dot{u}^j + \frac{\partial^2 L}{\partial \dot{q}^i \partial q^j} \cdot u^j \right) - \frac{\partial^2 L}{\partial q^i \partial q^j} \cdot u^j - \frac{\partial^2 L}{\partial q^i \partial \dot{q}^j} \cdot u^j = 0, \quad (7.4.5)$$

which is a second-order equation for u^j. This equation evaluated along $\bar{q}(t)$ is called the *Jacobi equation* along $\bar{q}(t)$. This equation, taken from $\bar{q}(\bar{t}_1)$ to $\bar{q}(\bar{t}_2)$ with initial conditions

$$u(t_1) = 0 \quad \text{and} \quad \dot{u}(t_1) = w_1,$$

defines the desired linear map $w_1 \mapsto w_2$; that is, $w_2 = \dot{u}(\bar{t}_2)$.

Theorem 7.4.5. *Assume that L is a regular Lagrangian. If the linear map $w_1 \mapsto w_2$ is an isomorphism, then $\bar{q}(t)$ is nonconjugate.*

Proof. This follows directly from the implicit function theorem. Under the hypothesis that $w_1 \mapsto w_2$ is invertible, there are neighborhoods $\mathcal{U}_1$ of $\bar{q}_1$, $\mathcal{U}_2$ of $\bar{q}_2$ and neighborhoods of $\bar{t}_1$ and $\bar{t}_2$ as well as a smooth function $v_1 = v_1(t_1, t_2, q_1, q_2)$ defined on the product of these four neighborhoods such that

$$(\pi_Q \circ F_{t_2})(q_1, v_1(t_1, t_2, q_1, q_2)) = q_2 \quad (7.4.6)$$

is an identity. Then

$$q(t) := (\pi_Q \circ F_t)(q_1, v_1(t_1, t_2, q_1, q_2))$$

is a solution of the Euler–Lagrange equations with initial conditions

$$(q_1, v_1(t_1, t_2, q_1, q_2)) \text{ at } t = t_1.$$

Moreover, $q(t_2) = q_2$ by (7.4.6). The fact that v_1 is close to $\bar{v}_1$ means that the geodesic found lies in a neighborhood of the curve $\bar{q}(t)$; this produces the neighborhood $\mathcal{U}$. ∎

If q_1 and q_2 are close and if t_2 is not much different from t_1, then by continuity, $\dot{u}(t)$ is approximately constant over $[t_1, t_2]$, so that

$$w_2 = \dot{u}(t_2) = (t_2 - t_1)\dot{u}(t_1) + O(t_2 - t_1)^2 = (t_2 - t_1)w_1 + O(t_2 - t_1)^2.$$

Thus, in these circumstances, the map $w_1 \mapsto w_2$ is invertible. Therefore, we get the following corollary.

Corollary 7.4.6. *Let $L : TQ \times \mathbb{R} \to \mathbb{R}$ be a given C^2 regular Lagrangian and let $v_q \in TQ$ and $t_1 \in \mathbb{R}$. Then the solution of the Euler–Lagrange equations with initial condition v_q at $t = t_1$ is nonconjugate for a sufficiently small time interval $[t_1, t_2]$.*

The term "nonconjugate" comes from the study of geodesics, which are considered in the next section.

Exercises

◊ **7.4-1.** Write down the Lagrangian and the equations of motion for a spherical pendulum with S^2 as configuration space. Convert the equations to Hamiltonian form using the Legendre transformation. Find the conservation law corresponding to angular momentum about the axis of gravity by "bare hands" methods.

◊ **7.4-2.** Let $L(q, \dot{q}) = \frac{1}{2}m(q)\dot{q}^2 - V(q)$ on $T\mathbb{R}$, where $m(q) > 0$ and $V(q)$ are smooth. Show that *any* two points $q_1, q_2 \in \mathbb{R}$ can be joined by a solution of the Euler–Lagrange equations. (Hint: Consider the energy equation.)

7.5 Geodesics

Let Q be a weak pseudo-Riemannian manifold whose metric evaluated at $q \in Q$ is denoted interchangeably by $\langle \cdot , \cdot \rangle$ or $g(q)$ or g_q. Consider on TQ the Lagrangian given by the kinetic energy of the metric, that is,

$$L(v) = \tfrac{1}{2} \langle v, v \rangle_q , \tag{7.5.1}$$

or in finite dimensions

$$L(v) = \tfrac{1}{2} g_{ij} v^i v^j . \tag{7.5.2}$$

The fiber derivative of L is given for $v, w \in T_q Q$ by

$$\mathbb{F}L(v) \cdot w = \langle v, w \rangle \tag{7.5.3}$$

or in finite dimensions by

$$\mathbb{F}L(v) \cdot w = g_{ij} v^i w^j, \quad \text{i.e.,} \quad p_i = g_{ij} \dot{q}^j . \tag{7.5.4}$$

From this equation we see that in any chart U for Q,

$$\mathbf{D}_2\mathbf{D}_2L(q,v)\cdot(e_1,e_2) = \langle e_1,e_2\rangle_q,$$

where $\langle\,,\,\rangle_q$ denotes the inner product on E induced by the chart. Thus, L is automatically weakly nondegenerate. Note that the action is given by $A = 2L$, so $E = L$.

The Lagrangian vector field Z in this case is denoted by $S : TQ \to T^2Q$ and is called the **Christoffel map** or **geodesic spray** of the metric $\langle\,,\,\rangle_q$. Thus, S is a second-order equation and hence has a local expression of the form

$$S(q,v) = ((q,v),(v,\gamma(q,v))) \tag{7.5.5}$$

in a chart on Q. To determine the map $\gamma : U \times E \to E$ from Lagrange's equations, note that

$$\mathbf{D}_1L(q,v)\cdot w = \tfrac{1}{2}\mathbf{D}_q\langle v,v\rangle_q\cdot w \quad \text{and} \quad \mathbf{D}_2L(q,v)\cdot w = \langle v,w\rangle_q, \tag{7.5.6}$$

so that the Euler–Lagrange equations (7.3.7) are

$$\dot{q} = v, \tag{7.5.7}$$

$$\frac{d}{dt}(\langle v,w\rangle_q) = \tfrac{1}{2}\mathbf{D}_q\langle v,v\rangle_q\cdot w. \tag{7.5.8}$$

Keeping w fixed and expanding the left-hand side of (7.5.8) yields

$$\mathbf{D}_q\langle v,w\rangle_q\cdot\dot{q} + \langle\dot{v},w\rangle_q. \tag{7.5.9}$$

Taking into account $\dot{q} = v$, we get

$$\langle\ddot{q},w\rangle_q = \tfrac{1}{2}\mathbf{D}_q\langle v,v\rangle_q\cdot w - \mathbf{D}_q\langle v,w\rangle_q\cdot v. \tag{7.5.10}$$

Hence $\gamma : U \times E \to E$ is defined by the equality

$$\langle\gamma(q,v),w\rangle_q = \tfrac{1}{2}\mathbf{D}_q\langle v,v\rangle_q\cdot w - \mathbf{D}_q\langle v,w\rangle_q\cdot v; \tag{7.5.11}$$

note that $\gamma(q,v)$ is a quadratic form in v. If Q is finite-dimensional, we define the **Christoffel symbols** Γ^i_{jk} by putting

$$\gamma^i(q,v) = -\Gamma^i_{jk}(q)v^jv^k \tag{7.5.12}$$

and demanding that $\Gamma^i_{jk} = \Gamma^i_{kj}$. With this notation, the relation (7.5.11) is equivalent to

$$-g_{il}\Gamma^i_{jk}v^jv^kw^l = \frac{1}{2}\frac{\partial g_{jk}}{\partial q^l}v^jv^kw^l - \frac{\partial g_{jl}}{\partial q^k}v^jw^lv^k. \tag{7.5.13}$$

Taking into account the symmetry of Γ^i_{jk}, this gives

$$\Gamma^h_{jk} = \frac{1}{2} g^{hl} \left(\frac{\partial g_{jl}}{\partial q^k} + \frac{\partial g_{kl}}{\partial q^j} - \frac{\partial g_{jk}}{\partial q^l} \right). \qquad (7.5.14)$$

In infinite dimensions, since the metric $\langle\,,\rangle$ is only weakly nondegenerate, (7.5.11) guarantees the uniqueness of γ but not its existence. It exists whenever the Lagrangian vector field S exists.

The integral curves of S projected to Q are called *geodesics* of the metric g. By (7.5.5), their basic governing equation has the local expression

$$\ddot{q} = \gamma(q, \dot{q}), \qquad (7.5.15)$$

which in finite dimensions reads

$$\ddot{q}^i + \Gamma^i_{jk} \dot{q}^j \dot{q}^k = 0, \qquad (7.5.16)$$

where $i, j, k = 1, \dots, n$ and, as usual, there is a sum on j and k. Note that the definition of γ makes sense in both the finite- and infinite-dimensional cases, whereas the Christoffel symbols Γ^i_{jk} are literally defined only for finite-dimensional manifolds. Working intrinsically with g provides a way to deal with geodesics of weak Riemannian (and pseudo-Riemannian) metrics on infinite-dimensional manifolds.

Taking the Lagrangian approach as basic, we see that the Γ^i_{jk} live as geometric objects in $T(TQ)$. This is because they encode the principal part of the Lagrangian vector field Z. If one writes down the transformation properties of Z on $T(TQ)$ in natural charts, the classical transformation rule for the Γ^i_{jk} results:

$$\overline{\Gamma}^k_{ij} = \frac{\partial q^p}{\partial \overline{q}^i} \frac{\partial q^m}{\partial \overline{q}^j} \Gamma^r_{pm} \frac{\partial \overline{q}^k}{\partial q^r} + \frac{\partial \overline{q}^k}{\partial q^l} \frac{\partial^2 q^l}{\partial \overline{q}^i \partial \overline{q}^j}, \qquad (7.5.17)$$

where $(q^1, \dots, q^n), (\overline{q}^1, \dots, \overline{q}^n)$ are two different coordinate systems on an open set of Q. We leave this calculation to the reader.

The Lagrangian approach leads naturally to invariant manifolds for the geodesic flow. For example, for each real $e > 0$, let

$$\Sigma_e = \{\, v \in TQ \mid \|v\| = e \,\}$$

be the *pseudo-sphere bundle* of radius $\sqrt{e}$ in TQ. Then Σ_e is a smooth submanifold of TQ invariant under the geodesic flow. Indeed, if we show that Σ_e is a smooth submanifold, its invariance under the geodesic flow, that is, under the flow of Z, follows by conservation of energy. To show that Σ_e is a smooth submanifold we prove that e is a regular value of L for $e > 0$. This is done locally by (7.5.6):

$$\begin{aligned}
\mathbf{D}L(u, v) \cdot (w_1, w_2) &= \mathbf{D}_1 L(u, v) \cdot w_1 + \mathbf{D}_2 L(u, v) \cdot w_2 \\
&= \tfrac{1}{2} \mathbf{D}_u \langle v, v \rangle_u \cdot w_1 + \langle v, w_2 \rangle_u \\
&= \langle v, w_2 \rangle_u, \qquad (7.5.18)
\end{aligned}$$

since $\langle v, v \rangle = 2e = $ constant. By weak nondegeneracy of the pseudo-metric $\langle , \rangle$, this shows that $\mathbf{D}L(u, v) : E \times E \to \mathbb{R}$ is a surjective linear map, that is, e is a regular value of L.

Convex Neighborhoods and Conjugate Points.   We proved in the last section that short arcs of solutions of the Euler–Lagrange equations are nonconjugate. In the special case of geodesics one can do somewhat better by exploiting the fact, evident from the quadratic nature of (7.5.16), that if $q(t)$ is a solution and $\alpha > 0$, then so is $q(\alpha t)$, so one can "rescale" solutions simply by changing the size of the initial velocity. One finds that locally there are *convex* neighborhoods, that is, neighborhoods U such that for any $q_1, q_2 \in U$ there is a unique geodesic (up to a scaling) joining q_1, q_2 and lying in U. In Riemannian geometry there is another important result, the **Hopf–Rinow theorem**, stating that any two points (in the same connected component) can be joined by some geodesic.

As one follows a geodesic from a given point, there is a first point after which *nearby* geodesics fail to be unique. These are **conjugate points**. They are the zeros of the Jacobi equation discussed earlier. For example, on a great circle on a sphere, pairs of antipodal points are conjugate.

In certain circumstances one can "reduce" the Euler–Lagrange problem to one of geodesics: See the discussion of the Jacobi metric in §7.7.

Covariant derivatives.   We now reconcile the above approach to geodesics via Lagrangian systems to a common approach in differential geometry. Define the *covariant derivative*

$$\nabla : \mathfrak{X}(Q) \times \mathfrak{X}(Q) \to \mathfrak{X}(Q), \quad (X, Y) \mapsto \nabla_X Y$$

locally by

$$(\nabla_X Y)(u) = -\gamma(u)(X(u), Y(u)) + \mathbf{D}Y(u) \cdot X(u), \qquad (7.5.19)$$

where X, Y are the local representatives of X and Y, and $\gamma(u) : E \times E \to E$ denotes the symmetric bilinear form defined by the polarization of $\gamma(u, v)$, which is a quadratic form in v. In local coordinates, the preceding equation becomes

$$\nabla_X Y = X^j Y^k \Gamma^i_{jk} \frac{\partial}{\partial q^i} + X^j \frac{\partial Y^k}{\partial q^j} \frac{\partial}{\partial q^k}. \qquad (7.5.20)$$

It is straightforward to check that this definition is chart independent and that ∇ satisfies the following conditions:

  (i) ∇ is $\mathbb{R}$-bilinear;

  (ii) for $f : Q \to \mathbb{R}$,

$$\nabla_{fX} Y = f \nabla_X Y \quad \text{and} \quad \nabla_X fY = f \nabla_X Y + X[f] Y;$$

and

(iii) for vector fields X and Y,

$$(\nabla_X Y - \nabla_Y X)(u) = \mathbf{D}Y(u) \cdot X(u) - \mathbf{D}X(u) \cdot Y(u)$$
$$= [X, Y](u). \tag{7.5.21}$$

In fact, these three properties characterize covariant derivative operators. The particular covariant derivative determined by (7.5.14) is called the **Levi-Civita covariant derivative**. If $c(t)$ is a curve in Q and $X \in \mathfrak{X}(Q)$, the **covariant derivative of X along** c is defined by

$$\frac{DX}{Dt} = \nabla_u X, \tag{7.5.22}$$

where u is a vector field coinciding with $\dot{c}(t)$ at $c(t)$. This is possible, since by (7.5.19) or (7.5.20), $\nabla_X Y$ depends only on the point values of X. Explicitly, in a local chart, we have

$$\frac{DX}{Dt}(c(t)) = -\gamma_{c(t)}(u(c(t)), X(c(t))) + \frac{d}{dt}X(c(t)), \tag{7.5.23}$$

which shows that DX/Dt depends only on $\dot{c}(t)$ and not on how $\dot{c}(t)$ is extended to a vector field. In finite dimensions,

$$\left(\frac{DX}{Dt}\right)^i = \Gamma^i_{jk}(c(t))\dot{c}^j(t)X^k(c(t)) + \frac{d}{dt}X^i(c(t)). \tag{7.5.24}$$

The vector field X is called **autoparallel** or **parallel transported** along c if $DX/Dt = 0$. Thus $\dot{c}$ is autoparallel along c if and only if

$$\ddot{c}(t) - \gamma(t)(\dot{c}(t), \dot{c}(t)) = 0,$$

that is, $c(t)$ is a geodesic. In finite dimensions, this reads

$$\ddot{c}^i + \Gamma^i_{jk}\dot{c}^j\dot{c}^k = 0.$$

Exercises

⋄ **7.5-1.** Consider the Lagrangian

$$L_\epsilon(x, y, z, \dot{x}, \dot{y}, \dot{z}) = \frac{1}{2}\left(\dot{x}^2 + \dot{y}^2 + \dot{z}^2\right) - \frac{1}{2\epsilon}\left[1 - (x^2 + y^2 + z^2)\right]^2$$

for a particle in $\mathbb{R}^3$. Let $\gamma_\epsilon(t)$ be the curve in $\mathbb{R}^3$ obtained by solving the Euler–Lagrange equations for L_ϵ with the initial conditions $\mathbf{x}_0, \mathbf{v}_0 = \dot{\gamma}_\epsilon(0)$. Show that

$$\lim_{\epsilon \to 0} \gamma_\epsilon(t)$$

is a great circle on the two-sphere S^2, provided that $\mathbf{x}_0$ has length one and that $\mathbf{x}_0 \cdot \mathbf{v}_0 = 0$.

⋄ **7.5-2.** Write out the geodesic equations in terms of q^i and p_i and check directly that Hamilton's equations are satisfied.

7.6   The Kaluza–Klein Approach to Charged Particles

In §6.7 we studied the motion of a charged particle in a magnetic field as a Hamiltonian system. Here we show that this description is the reduction of a larger and, in some sense, simpler system called the **Kaluza–Klein system**.[1]

Physically, we are motivated as follows: Since charge is a basic conserved quantity, we would like to introduce a new cyclic variable whose conjugate momentum is the charge.[2] For a charged particle, the resultant system is in fact geodesic motion!

Recall from §6.7 that if $\mathbf{B} = \nabla \times \mathbf{A}$ is a given magnetic field on $\mathbb{R}^3$, then with respect to canonical variables $(\mathbf{q}, \mathbf{p})$, the Hamiltonian is

$$H(\mathbf{q}, \mathbf{p}) = \frac{1}{2m} \left\| \mathbf{p} - \frac{e}{c} \mathbf{A} \right\|^2 . \qquad (7.6.1)$$

First we claim that we can obtain (7.6.1) via the Legendre transform if we choose

$$L(\mathbf{q}, \dot{\mathbf{q}}) = \frac{1}{2} m \left\| \dot{\mathbf{q}} \right\|^2 + \frac{e}{c} \mathbf{A} \cdot \dot{\mathbf{q}}. \qquad (7.6.2)$$

Indeed, in this case,

$$\mathbf{p} = \frac{\partial L}{\partial \dot{\mathbf{q}}} = m\dot{\mathbf{q}} + \frac{e}{c} \mathbf{A} \qquad (7.6.3)$$

and

$$\begin{aligned} H(\mathbf{q}, \mathbf{p}) &= \mathbf{p} \cdot \dot{\mathbf{q}} - L(\mathbf{q}, \dot{\mathbf{q}}) \\ &= \left(m\dot{\mathbf{q}} + \frac{e}{c} \mathbf{A} \right) \cdot \dot{\mathbf{q}} - \frac{1}{2} m \left\| \dot{\mathbf{q}} \right\|^2 - \frac{e}{c} \mathbf{A} \cdot \dot{\mathbf{q}} \\ &= \frac{1}{2} m \left\| \dot{\mathbf{q}} \right\|^2 = \frac{1}{2m} \left\| \mathbf{p} - \frac{e}{c} \mathbf{A} \right\|^2 . \end{aligned} \qquad (7.6.4)$$

Thus, the Euler–Lagrange equations for (7.6.2) reproduce the equations for a particle in a magnetic field.[3]

Let the configuration space be

$$Q_K = \mathbb{R}^3 \times S^1 \qquad (7.6.5)$$

---

[1]After learning reduction theory (see Abraham and Marsden [1978] or Marsden [1992]), the reader can revisit this construction, but here all the constructions are done directly.

[2]This process is applicable to other situations as well; for example, in fluid dynamics one can profitably introduce a variable conjugate to the conserved mass density or entropy; see Marsden, Ratiu, and Weinstein [1984a, 1984b].

[3]If an electric field $E = -\nabla\varphi$ is also present, one simply subtracts $e\varphi$ from L, treating $e\varphi$ as a potential energy, as in the next section.

with variables $(\mathbf{q}, \theta)$. Define $A = \mathbf{A}^\flat$, a one-form on $\mathbb{R}^3$, and consider the one-form

$$\omega = A + d\theta \tag{7.6.6}$$

on Q_K called the **connection one-form**. Let the **Kaluza–Klein Lagrangian** be defined by

$$
\begin{aligned}
L_K(\mathbf{q}, \dot{\mathbf{q}}, \theta, \dot{\theta}) &= \frac{1}{2} m \|\dot{\mathbf{q}}\|^2 + \frac{1}{2} \left\| \left\langle \omega, (\mathbf{q}, \dot{\mathbf{q}}, \theta, \dot{\theta}) \right\rangle \right\|^2 \\
&= \frac{1}{2} m \|\dot{\mathbf{q}}\|^2 + \frac{1}{2} (\mathbf{A} \cdot \dot{\mathbf{q}} + \dot{\theta})^2.
\end{aligned} \tag{7.6.7}
$$

The corresponding momenta are

$$\mathbf{p} = m\dot{\mathbf{q}} + (\mathbf{A} \cdot \dot{\mathbf{q}} + \dot{\theta})\mathbf{A} \tag{7.6.8}$$

and

$$p = \mathbf{A} \cdot \dot{\mathbf{q}} + \dot{\theta}. \tag{7.6.9}$$

Since L_K is quadratic and positive definite in $\dot{\mathbf{q}}$ and $\dot{\theta}$, the Euler–Lagrange equations are the geodesic equations on $\mathbb{R}^3 \times S^1$ for the metric for which L_K is the kinetic energy. Since p is constant in time, as can be seen from the Euler–Lagrange equation for $(\theta, \dot{\theta})$, we can define the **charge** e by setting

$$p = \frac{e}{c}; \tag{7.6.10}$$

then (7.6.8) coincides with (7.6.3). The corresponding Hamiltonian on T^*Q_K endowed with the canonical symplectic form is

$$H_K(\mathbf{q}, \mathbf{p}, \theta, p) = \frac{1}{2m} \|\mathbf{p} - p\mathbf{A}\|^2 + \frac{1}{2} p^2. \tag{7.6.11}$$

With (7.6.10), (7.6.11) differs from (7.6.1) by the constant $p^2/2$.

These constructions generalize to the case of a particle in a Yang–Mills field, where ω becomes the **connection** of a Yang–Mills field and its **curvature** measures the field strength that, for an electromagnetic field, reproduces the relation $\mathbf{B} = \nabla \times \mathbf{A}$. Also, the possibility of putting the interaction in the Hamiltonian, or via a momentum shift, into the symplectic structure, also generalizes. We refer to Wong [1970], Sternberg [1977], Weinstein [1978a], and Montgomery [1984] for details and further references. Finally, we remark that the relativistic context is the most natural in which to introduce the full electromagnetic field. In that setting the construction we have given for the magnetic field will include both electric and magnetic effects. Consult Misner, Thorne, and Wheeler [1973] for additional information.

Exercises

⋄ **7.6-1.** The bob on a spherical pendulum has a charge e, mass m, and moves under the influence of a constant gravitational field with acceleration g, and a magnetic field **B**. Write down the Lagrangian, the Euler–Lagrange equations, and the variational principle for this system. Transform the system to Hamiltonian form. Find a conserved quantity if the field **B** is symmetric about the axis of gravity.

7.7   Motion in a Potential Field

We now generalize geodesic motion to include potentials $V : Q \to \mathbb{R}$. Recall that the **gradient** of V is the vector field grad $V = \nabla V$ defined by the equality

$$\langle \operatorname{grad} V(q), v \rangle_q = \mathbf{d}V(q) \cdot v, \qquad (7.7.1)$$

for all $v \in T_q Q$. In finite dimensions, this definition becomes

$$(\operatorname{grad} V)^i = g^{ij} \frac{\partial V}{\partial q^j}. \qquad (7.7.2)$$

Define the (weakly nondegenerate) Lagrangian $L(v) = \frac{1}{2} \langle v, v \rangle_q - V(q)$. A computation similar to the one in §7.5 shows that the Euler–Lagrange equations are

$$\ddot{q} = \gamma(q, \dot{q}) - \operatorname{grad} V(q), \qquad (7.7.3)$$

or in finite dimensions,

$$\ddot{q}^i + \Gamma^i_{jk} \dot{q}^j \dot{q}^k + g^{il} \frac{\partial V}{\partial q^l} = 0. \qquad (7.7.4)$$

The action of L is given by

$$A(v) = \langle v, v \rangle_q, \qquad (7.7.5)$$

so that the energy is

$$E(v) = A(v) - L(v) = \frac{1}{2} \langle v, v \rangle_q + V(q). \qquad (7.7.6)$$

The equations (7.7.3) written as

$$\dot{q} = v, \qquad \dot{v} = \gamma(q, v) - \operatorname{grad} V(q) \qquad (7.7.7)$$

are thus Hamiltonian with Hamiltonian function E with respect to the symplectic form Ω_L.

Invariant Form. There are several ways to write equations (7.7.7) in invariant form. Perhaps the simplest is to use the language of covariant derivatives from the last section and to write

$$\frac{D\dot{c}}{Dt} = -\nabla V \qquad (7.7.8)$$

or, what is perhaps better,

$$g^{\flat}\frac{D\dot{c}}{Dt} = -\mathbf{d}V, \qquad (7.7.9)$$

where $g^{\flat} : TQ \to T^*Q$ is the map associated to the Riemannian metric. This last equation is the geometric way of writing $m\mathbf{a} = \mathbf{F}$.

Another method uses the following terminology:

Definition 7.7.1. *Let $v, w \in T_qQ$. The **vertical lift** of w with respect to v is defined by*

$$\mathrm{ver}(w, v) = \left.\frac{d}{dt}\right|_{t=0}(v + tw) \in T_v(TQ).$$

*The **horizontal part** of a vector $U \in T_v(TQ)$ is $T_v\tau_Q(U) \in T_qQ$. A vector field is called **vertical** if its horizontal part is zero.*

In charts, if $v = (u, e)$, $w = (u, f)$, and $U = ((u, e), (e_1, e_2))$, this definition says that

$$\mathrm{ver}(w, v) = ((u, e), (0, f)) \quad \text{and} \quad T_v\tau_Q(U) = (u, e_1).$$

Thus, U is vertical iff $e_1 = 0$. Thus, *any vertical vector $U \in T_v(TQ)$ is the vertical lift of some vector w (which in a natural local chart is (u, e_2)) with respect to v.*

If S denotes the geodesic spray of the metric $\langle , \rangle$ on TQ, equations (7.7.7) say that the Lagrangian vector field Z defined by $L(v) = \frac{1}{2}\langle v, v\rangle_q - V(q)$, where $v \in T_qQ$, is given by

$$Z = S - \mathrm{ver}(\nabla V), \qquad (7.7.10)$$

that is,

$$Z(v) = S(v) - \mathrm{ver}((\nabla V)(q), v). \qquad (7.7.11)$$

Remarks. In general, there is *no* canonical way to take the *vertical part* of a vector $U \in T_v(TQ)$ without extra structure. Having such a structure is what one means by a **connection**. In case Q is pseudo-Riemannian, such a projection can be constructed in the following manner. Suppose, in natural charts, that $U = ((u, e), (e_1, e_2))$. Define

$$U_{\mathrm{ver}} = ((u, e), (0, \gamma(u)(e_1, e_2) + e_2))$$

where $\gamma(u)$ is the bilinear symmetric form associated to the quadratic form $\gamma(u, e)$ in e.   ◆

We conclude with some miscellaneous remarks connecting motion in a potential field with geodesic motion. We confine ourselves to the finite-dimensional case for simplicity.

Definition 7.7.2.   *Let* $g = \langle,\rangle$ *be a pseudo-Riemannian metric on* Q *and let* $V : Q \to \mathbb{R}$ *be bounded above. If* $e > V(q)$ *for all* $q \in Q$, *define the* **Jacobi metric** g_e *by* $g_e = (e - V)g$, *that is,*

$$g_e(v, w) = (e - V(q)) \langle v, w \rangle$$

for all $v, w \in T_qQ$.

Theorem 7.7.3.   *Let* Q *be finite-dimensional. The base integral curves of the Lagrangian* $L(v) = \frac{1}{2} \langle v, v \rangle - V(q)$ *with energy* e *are the same as geodesics of the Jacobi metric with energy 1, up to a reparametrization.*

The proof is based on the following proposition of separate interest.

Proposition 7.7.4.   *Let* (P, Ω) *be a (finite-dimensional) symplectic manifold,* $H, K \in \mathcal{F}(P)$, *and assume that* $\Sigma = H^{-1}(h) = K^{-1}(k)$ *for* $h, k \in \mathbb{R}$ *regular values of* H *and* K, *respectively. Then the integral curves of* X_H *and* X_K *on the invariant submanifold* Σ *of both* X_H *and* X_K *coincide up to a reparametrization.*

Proof.   From $\Omega(X_H(z), v) = \mathbf{d}H(z) \cdot v$, we see that

$$X_H(z) \in (\ker \mathbf{d}H(z))^\Omega = (T_z\Sigma)^\Omega,$$

the symplectic orthogonal complement of $T_z\Sigma$. Since

$$\dim P = \dim T_z\Sigma + \dim(T_z\Sigma)^\Omega$$

(see §2.3) and since $T_z\Sigma$ has codimension one, $(T_z\Sigma)^\Omega$ has dimension one. Thus, the nonzero vectors $X_H(z)$ and $X_K(z)$ are multiples of each other at every point $z \in \Sigma$, that is, there is a smooth nowhere-vanishing function $\lambda : \Sigma \to \mathbb{R}$ such that $X_H(z) = \lambda(z)X_K(z)$ for all $z \in \Sigma$. Let $c(t)$ be the integral curve of X_K with initial condition $c(0) = z_0 \in \Sigma$. The function

$$\varphi \mapsto \int_0^\varphi \frac{dt}{(\lambda \circ c)(t)}$$

is a smooth monotone function and therefore has an inverse $t \mapsto \varphi(t)$. If $d(t) = (c \circ \varphi)(t)$, then $d(0) = z_0$ and

$$d'(t) = \varphi'(t)c'(\varphi(t)) = \frac{1}{t'(\varphi)} X_K(c(\varphi(t))) = (\lambda \circ c)(\varphi)X_K(d(t))$$

$$= \lambda(d(t))X_K(d(t)) = X_H(d(t)),$$

that is, the integral curve of X_H through z_0 is obtained by reparametrizing the integral curve of X_K through z_0. ∎

Proof of Theorem 7.7.3.   Let H be the Hamiltonian for L, namely

$$H(q,p) = \frac{1}{2}\|p\|^2 + V(q),$$

and let H_e be that for the Jacobi metric:

$$H_e(q,p) = \frac{1}{2}(e - V(q))^{-1}\|p\|^2.$$

The factor $(e - V(q))^{-1}$ occurs because the inverse metric is used for the momenta. Clearly, $H = e$ defines the same set as $H_e = 1$, so the result follows from Proposition 7.7.4 if we show that e is a regular value of H and 1 is a regular value of H_e. Note that if $(q,p) \in H^{-1}(e)$, then $p \neq 0$, since $e > V(q)$ for all $q \in Q$. Therefore, $\mathbb{F}H(q,p) \neq 0$ for any $(q,p) \in H^{-1}(e)$, and hence $\mathbf{d}H(q,p) \neq 0$, that is, e is a regular value of H. Since

$$\mathbb{F}H_e(q,\dot{p}) = \frac{1}{2}(e - V(q))^{-1}\mathbb{F}H(q,p),$$

this also shows that

$$\mathbb{F}H_e(q,p) \neq 0 \quad \text{for all} \quad (q,p) \in H^{-1}(e) = H_e^{-1}(1),$$

and thus 1 is a regular value of H_e.   ∎

7.8   The Lagrange–d'Alembert Principle

In this section we study a generalization of Lagrange's equations for mechanical systems with exterior forces. A special class of such forces is dissipative forces, which will be studied at the end of this section.

Force Fields.   Let $L : TQ \to \mathbb{R}$ be a Lagrangian function, let Z be the Lagrangian vector field associated to L, assumed to be a second-order equation, and denote by $\tau_Q : TQ \to Q$ the canonical projection. Recall that a vector field Y on TQ is called *vertical* if $T\tau_Q \circ Y = 0$. Such a vector field Y defines a one-form Δ^Y on TQ by contraction with Ω_L:

$$\Delta^Y = -\mathbf{i}_Y\Omega_L = Y \lrcorner \Omega_L.$$

Proposition 7.8.1.   *If Y is vertical, then Δ^Y is a **horizontal one-form**, that is, $\Delta^Y(U) = 0$ for any vertical vector field U on TQ. Conversely, given a horizontal one-form Δ on TQ, and assuming that L is regular, the vector field Y on TQ, defined by $\Delta = -\mathbf{i}_Y\Omega_L$, is vertical.*

Proof.   This follows from a straightforward calculation in local coordinates. We use the fact that a vector field $Y(u,e) = (Y_1(u,e), Y_2(u,e))$ is vertical if and only if the first component Y_1 is zero, and the local formula for Ω_L derived earlier:

$$\Omega_L(u,e)(Y_1,Y_2),(U_1,U_2))$$
$$= \mathbf{D}_1(\mathbf{D}_2 L(u,e) \cdot Y_1) \cdot U_1 - \mathbf{D}_1(\mathbf{D}_2 L(u,e) \cdot U_1) \cdot Y_1$$
$$+ \mathbf{D}_2 \mathbf{D}_2 L(u,e) \cdot Y_1 \cdot U_2 - \mathbf{D}_2 \mathbf{D}_2 L(u,e) \cdot U_1 \cdot Y_2. \qquad (7.8.1)$$

This shows that $(\mathbf{i}_Y \Omega_L)(U) = 0$ for all vertical U is equivalent to

$$\mathbf{D}_2 \mathbf{D}_2 L(u,e)(U_2, Y_1) = 0.$$

If Y is vertical, this is clearly true. Conversely, if L is regular and the last displayed equation is true, then $Y_1 = 0$, so Y is vertical. ∎

Proposition 7.8.2. *Any fiber-preserving map $F : TQ \to T^*Q$ over the identity induces a horizontal one-form $\tilde{F}$ on TQ by*

$$\tilde{F}(v) \cdot V_v = \langle F(v), T_v \tau_Q(V_v) \rangle, \qquad (7.8.2)$$

*where $v \in TQ$ and $V_v \in T_v(TQ)$. Conversely, formula (7.8.2) defines, for any horizontal one-form $\tilde{F}$, a fiber-preserving map F over the identity. Any such F is called a **force field**, and thus, in the regular case, any vertical vector field Y is induced by a force field.*

Proof. Given F, formula (7.8.2) clearly defines a smooth one-form $\tilde{F}$ on TQ. If V_v is vertical, then the right-hand side of formula (7.8.2) vanishes, and so $\tilde{F}$ is a horizontal one-form. Conversely, given a horizontal one-form $\tilde{F}$ on TQ and given $v, w \in T_q Q$, let $V_v \in T_v(TQ)$ be such that $T_v \tau(V_v) = w$. Then define F by formula (7.8.2); that is, $\langle F(v), w \rangle = \tilde{F}(v) \cdot V_v$. Since $\tilde{F}$ is horizontal, we see that F is well-defined, and its expression in charts shows that it is smooth. ∎

Treating Δ^Y as the exterior force one-form acting on a mechanical system with a Lagrangian L, we now will write the governing equations of motion.

The Lagrange–d'Alembert Principle.   First, we recall the definition from Vershik and Faddeev [1981] and Wang and Krishnaprasad [1992].

Definition 7.8.3. *The **Lagrangian force** associated with a Lagrangian L and a given second-order vector field (the ultimate equations of motion) X is the horizontal one-form on TQ defined by*

$$\Phi_L(X) = \mathbf{i}_X \Omega_L - \mathbf{d}E. \qquad (7.8.3)$$

*Given a horizontal one-form ω (referred to as the **exterior force one-form**), the **local Lagrange–d'Alembert principle** associated with the second-order vector field X on TQ states that*

$$\Phi_L(X) + \omega = 0. \qquad (7.8.4)$$

It is easy to check that $\Phi_L(X)$ is indeed horizontal if X is second-order. Conversely, if L is regular and if $\Phi_L(X)$ is horizontal, then X is second-order.

One can also formulate an equivalent principle in terms of variational principles.

Definition 7.8.4. *Given a Lagrangian L and a force field F, as defined in Proposition 7.8.2, the* **integral Lagrange–d'Alembert principle** *for a curve $q(t)$ in Q is*

$$\delta \int_a^b L(q(t), \dot{q}(t))\, dt + \int_a^b F(q(t), \dot{q}(t)) \cdot \delta q\, dt = 0, \qquad (7.8.5)$$

where the variation is given by the usual expression

$$\delta \int_a^b L(q(t), \dot{q}(t))\, dt = \int_a^b \left(\frac{\partial L}{\partial q^i} \delta q^i + \frac{\partial L}{\partial \dot{q}^i} \frac{d}{dt} \delta q^i \right) dt$$

$$= \int_a^b \left(\frac{\partial L}{\partial q^i} - \frac{d}{dt} \frac{\partial L}{\partial \dot{q}^i} \right) \delta q^i\, dt \qquad (7.8.6)$$

for a given variation δq (vanishing at the endpoints).

The two forms of the Lagrange–d'Alembert principle are in fact equivalent. This will follow from the fact that both give the Euler–Lagrange equations with forcing in local coordinates (provided that Z is second-order). We shall see this in the following development.

Proposition 7.8.5. *Let the exterior force one-form ω be associated to a vertical vector field Y, that is, let $\omega = \Delta^Y = -i_Y \Omega_L$. Then $X = Z + Y$ satisfies the local Lagrange–d'Alembert principle. Conversely, if, in addition, L is regular, the only second-order vector field X satisfying the local Lagrange–d'Alembert principle is $X = Z + Y$.*

Proof. For the first part, the equality $\Phi_L(X) + \omega = 0$ is a simple verification. For the converse, we already know that X is a solution, and uniqueness is guaranteed by regularity. ∎

To develop the differential equations associated to $X = Z + Y$, we take $\omega = -i_Y \Omega_L$ and note that in a coordinate chart, $Y(q, v) = (0, Y_2(q, v))$, since Y is vertical, that is, $Y_1 = 0$. From the local formula for Ω_L, we get

$$\omega(q, v) \cdot (u, w) = D_2 D_2 L(q, v) \cdot Y_2(q, v) \cdot u. \qquad (7.8.7)$$

Letting $X(q, v) = (v, X_2(q, v))$, one finds that

$$\Phi_L(X)(q, v) \cdot (u, w)$$
$$= (-D_1(D_2 L(q, v) \cdot) \cdot v - D_2 D_2 L(q, v) \cdot X_2(q, v) + D_1 L(q, v)) \cdot u. \qquad (7.8.8)$$

Thus, the local Lagrange–d'Alembert principle becomes

$$(-\mathbf{D}_1(\mathbf{D}_2 L(q, v)\cdot) \cdot v - \mathbf{D}_2\mathbf{D}_2 L(q, v) \cdot X_2(q, v) + \mathbf{D}_1 L(q, v)$$
$$+ \mathbf{D}_2\mathbf{D}_2 L(q, v) \cdot Y_2(q, v)) = 0. \quad (7.8.9)$$

Setting $v = dq/dt$ and $X_2(q, v) = dv/dt$, the preceding relation and the chain rule give

$$\frac{d}{dt}\mathbf{D}_2 L(q, v) - \mathbf{D}_1 L(q, v) = \mathbf{D}_2\mathbf{D}_2 L(q, v) \cdot Y_2(q, v), \quad (7.8.10)$$

which in finite dimensions reads

$$\frac{d}{dt}\left(\frac{\partial L}{\partial \dot{q}^i}\right) - \frac{\partial L}{\partial q^i} = \frac{\partial^2 L}{\partial \dot{q}^i \, \partial \dot{q}^j} Y^j(q^k, \dot{q}^k). \quad (7.8.11)$$

The force one-form Δ^Y is therefore given by

$$\Delta^Y(q^k, \dot{q}^k) = \frac{\partial^2 L}{\partial \dot{q}^i \, \partial \dot{q}^j} Y^j(q^k, \dot{q}^k)\, dq^i, \quad (7.8.12)$$

and the corresponding force field is

$$F^Y = \left(q^i, \frac{\partial^2 L}{\partial \dot{q}^i \, \partial \dot{q}^j} Y^j(q^k, \dot{q}^k)\right). \quad (7.8.13)$$

Thus, the condition for an integral curve takes the form of the standard Euler–Lagrange equations with forces:

$$\frac{d}{dt}\left(\frac{\partial L}{\partial \dot{q}^i}\right) - \frac{\partial L}{\partial q^i} = F_i^Y(q^k, \dot{q}^k). \quad (7.8.14)$$

Since the integral Lagrange–d'Alembert principle gives the same equations, it follows that the two principles are equivalent. From now on, we will refer to either one as simply the *Lagrange–d'Alembert principle.*

We summarize the results obtained so far in the following:

Theorem 7.8.6. *Given a regular Lagrangian and a force field $F : TQ \to T^*Q$, for a curve $q(t)$ in Q the following are equivalent:*

(a) *$q(t)$ satisfies the local Lagrange–d'Alembert principle;*

(b) *$q(t)$ satisfies the integral Lagrange–d'Alembert principle; and*

(c) *$q(t)$ is the base integral curve of the second-order equation $Z + Y$, where Y is the vertical vector field on TQ inducing the force field F by (7.8.13), and Z is the Lagrangian vector field on L.*

The Lagrange–d'Alembert principle plays a crucial role in **nonholonomic mechanics**, such as mechanical systems with rolling constraints. See, for example, Bloch, Krishnaprasad, Marsden, and Murray [1996] and references therein.

Dissipative Forces. Let E denote the energy defined by L, that is, $E = A - L$, where $A(v) = \langle \mathbb{F}L(v), v \rangle$ is the action of L.

Definition 7.8.7. *A vertical vector field Y on TQ is called **weakly dissipative** if $\langle dE, Y \rangle \leq 0$ at all points of TQ. If the inequality is strict off the zero section of TQ, then Y is called **dissipative**. A **dissipative Lagrangian system** on TQ is a vector field $Z + Y$, for Z a Lagrangian vector field and Y a dissipative vector field.*

Corollary 7.8.8. *A vertical vector field Y on TQ is dissipative if and only if the force field F^Y that it induces satisfies $\langle F^Y(v), v \rangle < 0$ for all nonzero $v \in TQ$ (≤ 0 for the weakly dissipative case).*

Proof. Let Y be a vertical vector field. By Proposition 7.8.1, Y induces a horizontal one-form $\Delta^Y = -\mathbf{i}_Y \Omega_L$ on TQ, and by Proposition 7.8.2, Δ^Y in turn induces a force field F^Y given by

$$\langle F^Y(v), w \rangle = \Delta^Y(v) \cdot V_v = -\Omega_L(v)(Y(v), V_v), \qquad (7.8.15)$$

where $T\tau_Q(V_v) = w$ and $V_v \in T_v(TQ)$. If Z denotes the Lagrangian system defined by L, we get

$$
\begin{aligned}
(dE \cdot Y)(v) &= (\mathbf{i}_Z \Omega_L)(Y)(v) = \Omega_L(Z, Y)(v) \\
&= -\Omega_L(v)(Y(v), Z(v)) \\
&= \langle F^Y(v), T_v \tau(Z(v)) \rangle \\
&= \langle F^Y(v), v \rangle,
\end{aligned}
$$

since Z is a second-order equation. Thus, $dE \cdot Y < 0$ if and only if $\langle F^Y(v), v \rangle < 0$ for all $v \in TQ$. ∎

Definition 7.8.9. *Given a dissipative vector field Y on TQ, let $F^Y : TQ \to T^*Q$ be the induced force field. If there is a function $R : TQ \to \mathbb{R}$ such that F^Y is the fiber derivative of $-R$, then R is called a **Rayleigh dissipation function**.*

Note that in this case, $\mathbf{D}_2 R(q, v) \cdot v > 0$ for the dissipativity of Y. Thus, if R is linear in the fiber variable, the Rayleigh dissipation function takes on the classical form $\langle \mathcal{R}(q)v, v \rangle$, where $\mathcal{R}(q) : TQ \to T^*Q$ is a bundle map over the identity that defines a symmetric positive definite form on each fiber of TQ.

Finally, if the force field is given by a Rayleigh dissipation function R, then the Euler–Lagrange equations with forcing become

$$\frac{d}{dt}\left(\frac{\partial L}{\partial \dot{q}^i}\right) - \frac{\partial L}{\partial q^i} = -\frac{\partial R}{\partial \dot{q}^i}. \qquad (7.8.16)$$

Combining Corollary 7.8.8 with the fact that the differential of E along Z is zero, we find that under the flow of the Euler–Lagrange equations with forcing of Rayleigh dissipation type, we have

$$\frac{d}{dt}E(q,v) = F(v)\cdot v = -\mathbb{F}R(q,v)\cdot v < 0. \qquad (7.8.17)$$

Exercises

$\diamond$ **7.8-1.** What is the power or rate of work equation (see §2.1) for a system with forces on a Riemannian manifold?

$\diamond$ **7.8-2.** Write the equations for a ball in a rotating hoop, including friction, in the language of this section (see §2.8). Compute the Rayleigh dissipation function.

$\diamond$ **7.8-3.** Consider a Riemannian manifold Q and a potential function $V : Q \to \mathbb{R}$. Let K denote the kinetic energy function and let $\omega = -\mathbf{d}V$. Show that the Lagrange–d'Alembert principle for K with external forces given by the one-form ω produces the same dynamics as the standard kinetic minus potential Lagrangian.

7.9 The Hamilton–Jacobi Equation

In §6.5 we studied generating functions of canonical transformations. Here we link them with the flow of a Hamiltonian system via the Hamilton–Jacobi equation. In this section we approach Hamilton–Jacobi theory from the point of view of extended phase space. In the next chapter we will have another look at Hamilton–Jacobi theory from the variational point of view, as it was originally developed by Jacobi [1866]. In particular, we will show in that section, roughly speaking, that the integral of the Lagrangian along solutions of the Euler–Lagrange equations, but thought of as a function of the endpoints, satisfies the Hamilton–Jacobi equation.

Canonical Transformations and Generating Functions. We consider a symplectic manifold P and form the *extended phase space* $P \times \mathbb{R}$. For our purposes in this section, we will use the following definition. A *time-dependent canonical transformation* is a diffeomorphism

$$\rho : P \times \mathbb{R} \to P \times \mathbb{R}$$

of the form

$$\rho(z,t) = (\rho_t(z), t),$$

where for each $t \in \mathbb{R}$, $\rho_t : P \to P$ is a symplectic diffeomorphism.

In this section we will specialize to the case of cotangent bundles, so assume that $P = T^*Q$ for a configuration manifold Q. For each fixed t, let $S_t : Q \times Q \to \mathbb{R}$ be the generating function for a time-dependent symplectic map, as described in §6.5. Thus, we get a function $S : Q \times Q \times \mathbb{R} \to \mathbb{R}$ defined by $S(q_1, q_2, t) = S_t(q_1, q_2)$. As explained in §6.5, one has to be aware that in general, generating functions are defined only locally, and indeed, the global theory of generating functions and the associated global Hamilton–Jacobi theory is more sophisticated. We will give a brief (optional) introduction to this general theory at the end of this section. See also Abraham and Marsden [1978, Section 5.3] for more information and references. Since our goal in the first part of this section is to give an *introductory presentation of the theory*, we will do many of the calculations in coordinates.

Recall that in local coordinates, the conditions for a generating function are written as follows. If the transformation ψ has the local expression

$$\psi : (q^i, p_i, t) \mapsto (\overline{q}^i, \overline{p}_i, t),$$

with inverse denoted by

$$\phi : (\overline{q}^i, \overline{p}_i, t) \mapsto (q^i, p_i, t),$$

and if $S(q^i, \overline{q}^i, t)$ is a generating function for ψ, we have the relations

$$\overline{p}_i = -\frac{\partial S}{\partial \overline{q}^i} \quad \text{and} \quad p_i = \frac{\partial S}{\partial q^i}. \tag{7.9.1}$$

From (7.9.1) it follows that

$$p_i \, dq^i = \overline{p}_i \, d\overline{q}^i + \frac{\partial S}{\partial q^i} dq^i + \frac{\partial S}{\partial \overline{q}^i} d\overline{q}^i$$

$$= \overline{p}_i \, d\overline{q}^i - \frac{\partial S}{\partial t} dt + \mathbf{d}S, \tag{7.9.2}$$

where $\mathbf{d}S$ is the differential of S as a function on $Q \times Q \times \mathbb{R}$:

$$\mathbf{d}S = \frac{\partial S}{\partial q^i} dq^i + \frac{\partial S}{\partial \overline{q}^i} d\overline{q}^i + \frac{\partial S}{\partial t} dt.$$

Let $K : T^*Q \times \mathbb{R} \to \mathbb{R}$ be an arbitrary function. From (7.9.2) we get the following basic relationship:

$$p_i \, dq^i - K(q^i, p_i, t) \, dt = \overline{p}_i \, d\overline{q}^i - \overline{K}(\overline{q}^i, \overline{p}_i, t) \, dt + \mathbf{d}S(q^i, \overline{q}^i, t), \tag{7.9.3}$$

where $\overline{K}(\overline{q}^i, \overline{p}_i, t) = K(q^i, p_i, t) + \partial S(q^i, \overline{q}^i, t)/\partial t$. If we define

$$\Theta_K = p_i \, dq^i - K \, dt, \tag{7.9.4}$$

then (7.9.3) is equivalent to

$$\Theta_K = \psi^* \Theta_{\overline{K}} + \psi^* \mathbf{d}S, \tag{7.9.5}$$

where $\psi : T^*Q \times \mathbb{R} \to Q \times Q \times \mathbb{R}$ is the map

$$(q^i, p_i, t) \mapsto (q^i, \overline{q}^i(q^j, p_j, t), t).$$

By taking the exterior derivative of (7.9.3) (or (7.9.5)), it follows that

$$dq^i \wedge dp_i + dK \wedge dt = d\overline{q}^i \wedge d\overline{p}_i + d\overline{K} \wedge dt. \qquad (7.9.6)$$

This may be written as

$$\Omega_K = \psi^*\Omega_{\overline{K}}, \qquad (7.9.7)$$

where $\Omega_K = -\mathbf{d}\Theta_K = dq^i \wedge dp_i + dK \wedge dt$.

Recall from Exercise 6.2-3 that given a time-dependent function K and associated time-dependent vector field X_K on T^*Q, the vector field $\tilde{X}_K = (X_K, 1)$ on $T^*Q \times \mathbb{R}$ is uniquely determined (among all vector fields with a 1 in the second component) by the equation $\mathbf{i}_{\tilde{X}_K}\Omega_K = 0$. From this relation and (7.9.7), we get

$$0 = \psi_*(\mathbf{i}_{\tilde{X}_K}\Omega_K) = \mathbf{i}_{\psi_*(\tilde{X}_K)}\psi_*\Omega_K = \mathbf{i}_{\psi_*(\tilde{X}_K)}\Omega_{\overline{K}}.$$

Since ψ is the identity in the second component, that is, it preserves time, the vector field $\psi_*(\tilde{X}_K)$ has a 1 in the second component, and therefore by uniqueness of such vector fields we get the identity

$$\psi_*(\tilde{X}_K) = \tilde{X}_{\overline{K}}. \qquad (7.9.8)$$

The Hamilton–Jacobi Equation.   The data we shall need are a Hamiltonian H and a generating function S, as above.

Definition 7.9.1.   *Given a time-dependent Hamiltonian H and a transformation ψ with generating function S as above, we say that the **Hamilton–Jacobi equation** holds if*

$$H\left(q^1, \ldots, q^n, \frac{\partial S}{\partial q^1}, \ldots, \frac{\partial S}{\partial q^n}, t\right) + \frac{\partial S}{\partial t}(q^i, \overline{q}^i, t) = 0, \qquad (7.9.9)$$

in which $\partial S/\partial q^i$ are evaluated at $(q^i, \overline{q}^i, t)$ and in which the $\overline{q}^i$ are regarded as constants.

The Hamilton–Jacobi equation may be regarded as a nonlinear partial differential equation for the function S relative to the variables $(q^1, \ldots, q^n, t)$ depending parametrically on $(\overline{q}^1, \ldots, \overline{q}^n)$.

Definition 7.9.2.   *We say that the map ψ **transforms a vector field** $\tilde{X}$ **to equilibrium** if*

$$\psi_*\tilde{X} = (0, 1). \qquad (7.9.10)$$

If ψ transforms $\tilde{X}$ to equilibrium, then the integral curves of $\tilde{X}$ with initial conditions (q_0^i, p_i^0, t_0) are given by

$$(q^i(t), p_i(t), t) = \psi^{-1}(\bar{q}^i(q_0^i, p_i^0, t_0), \bar{p}_i(q_0^i, p_i^0, t_0), t + t_0), \qquad (7.9.11)$$

since the integral curves of the constant vector field $(0, 1)$ are just straight lines in the t-direction and since ψ maps integral curves of $\tilde{X}$ to those of $(0, 1)$. In other words, if a map transforms a vector field $\tilde{X}$ to equilibrium, the integral curves of $\tilde{X}$ are represented by straight lines in the image space, and so the vector field has been "integrated."

Notice that if ϕ is the inverse of ψ, then ϕ_t is the flow of the vector field X in the usual sense.

Theorem 7.9.3 (Hamilton–Jacobi).

(i) *Suppose that S satisfies the Hamilton–Jacobi equation for a given time-dependent Hamiltonian H and that S generates a time-dependent canonical transformation ψ. Then ψ transforms $\tilde{X}_H$ to equilibrium. Thus, as explained above, the solution of Hamilton's equations for H are given in terms of ψ by (7.9.11).*

(ii) *Conversely, if ψ is a time-dependent canonical transformation with generating function S that transforms $\tilde{X}_H$ to equilibrium, then there is a function $\hat{S}$, which differs from S only by a function of t that also generates ψ, and satisfies the Hamilton–Jacobi equation for H.*

Proof. To prove (i), assume that S satisfies the Hamilton–Jacobi equation. As we explained above, this means that $\overline{H} = 0$. From (7.9.8) we get

$$\psi_* \tilde{X}_H = \tilde{X}_{\overline{H}} = (0, 1).$$

This proves the first statement.

To prove the converse (ii), assume that

$$\psi_* \tilde{X}_H = (0, 1),$$

and so, again by (7.9.8),

$$\tilde{X}_{\overline{H}} = \tilde{X}_0 = (0, 1),$$

which means that $\overline{H}$ is a constant relative to the variables $(\bar{q}^i, \bar{p}_i)$ (its Hamiltonian vector field at each instant of time is zero) and thus $\overline{H} = f(t)$, a function of time only. We can then modify S to $\hat{S} = S - F$, where $F(t) = \int^t f(s)\,ds$. This function, differing from S by a function of time alone, generates the same map ψ. Since

$$0 = \overline{H} - f(t) = H + \partial S/\partial t - dF/dt = H + \partial \hat{S}/\partial t,$$

and $\partial S/\partial q^i = \partial \hat{S}/\partial q^i$, we see that $\hat{S}$ satisfies the Hamilton–Jacobi equation for H. ∎

Remarks.
1. In general, the function S develops *singularities*, or *caustics*, as time increases, so it must be used with care. This process is, however, fundamental in geometric optics and in quantization. Moreover, one has to be careful with the sense in which S generates the identity at $t = 0$, as it might have singular behavior in t.

2. Here is another link between the Lagrangian and Hamiltonian view of the Hamilton–Jacobi theory. Define S for t close to a fixed time t_0 by the *action integral*

$$S(q^i, \bar{q}^i, t) = \int_{t_0}^{t} L(q^i(s), \dot{q}^i(s), s) \, ds,$$

where $q^i(s)$ is the solution of the Euler–Lagrange equation equaling $\bar{q}^i$ at time t_0 and equaling q^i at time t. We will show in §8.2 that S satisfies the Hamilton–Jacobi equation. See Arnold [1989, Section 4.6] and Abraham and Marsden [1978, Section 5.2] for more information.

3. If H is time-independent and W satisfies the time-independent Hamilton–Jacobi equation

$$H\left(q^i, \frac{\partial W}{\partial q^i}\right) = E,$$

then $S(q^i, \bar{q}^i, t) = W(q^i, \bar{q}^i) - tE$ satisfies the time-dependent Hamilton–Jacobi equation, as is easily checked. When using this remark, it is important to remember that E is not really a "constant," but it equals $H(\bar{q}, \bar{p})$, the energy evaluated at $(\bar{q}, \bar{p})$, which will eventually be the initial conditions. We emphasize that one must generate the time t-map using S rather than W.

4. The Hamilton–Jacobi equation is fundamental in the study of the quantum–classical relationship is described in the Internet supplement for Chapter 7.

5. The action function S is a key tool used in the proof of the *Liouville–Arnold theorem*, which gives the existence of action angle coordinates for systems with integrals in involution; see Arnold [1989] and Abraham and Marsden [1978] for details.

6. The Hamilton–Jacobi equation plays an important role in the development of numerical integrators that preserve the symplectic structure (see de Vogelaére [1956], Channell [1983], Feng [1986], Channell and Scovel [1990], Ge and Marsden [1988], Marsden [1992], and Wendlandt and Marsden [1997]).

7. The method of separation of variables. It is sometimes possible to simplify and even solve the Hamilton–Jacobi equation by what is often called the method of separation of variables. Assume that in the Hamilton–Jacobi equation the coordinate q^1 and the term $\partial S/\partial q^1$ appear jointly in some expression $f(q^1, \partial S/\partial q^1)$ that does not involve $q^2, \ldots, q^n, t$. That is, we can write H in the form

$$H\left(q^1, q^2, \ldots, q^n, p_1, p_2, \ldots, p_n\right) = \tilde{H}(f(q^1, p_1), q^2, \ldots, q^n, p_2, \ldots, p_n)$$

for some smooth functions f and $\tilde{H}$. Then one seeks a solution of the Hamilton–Jacobi equation in the form

$$S(q^i, \overline{q}^i, t) = S_1(q^1, \overline{q}^1) + \tilde{S}(q^2, \ldots, q^n, \overline{q}^2, \ldots, \overline{q}^n).$$

We then note that if S_1 solves

$$f\left(q^1, \frac{\partial S_1}{\partial q^1}\right) = C(\overline{q}^1)$$

for an arbitrary function $C(\overline{q}^1)$ and if $\tilde{S}$ solves

$$\tilde{H}\left(C(\overline{q}^1), q^2, \ldots, q^n, \frac{\partial \tilde{S}}{\partial q^2}, \ldots, \frac{\partial \tilde{S}}{\partial q^n}\right) + \frac{\partial \tilde{S}}{\partial t} = 0,$$

then S solves the original Hamilton–Jacobi equation. In this way, one of the variables is eliminated, and one tries to repeat the procedure.

A closely related situation occurs when H is independent of time and one seeks a solution of the form

$$S(q^i, \overline{q}^i, t) = W(q^i, \overline{q}^i) + S_1(t).$$

The resulting equation for S_1 has the solution $S_1(t) = -Et$, and the remaining equation for W is the time-independent Hamilton–Jacobi equation as in Remark 3.

If q^1 is a cyclic variable, that is, if H does not depend explicitly on q^1, then we can choose $f(q^1, p_1) = p_1$, and correspondingly, we can choose $S_1(q^1) = C(\overline{q}^1)q^1$. In general, if there are k cyclic coordinates $q^1, q^2, \ldots, q^k$, we seek a solution to the Hamilton–Jacobi equation of the form

$$S(q^i, \overline{q}^i, t) = \sum_{j=1}^{k} C_j(\overline{q}^j)q^j + \tilde{S}(q^{k+1}, \ldots, q^n, \overline{q}^{k+1}, \ldots, \overline{q}^n, t),$$

with $p_i = C_i(\overline{q}^i)$, $i = 1, \ldots, k$, being the momenta conjugate to the cyclic variables. ♦

The Geometry of Hamilton–Jacobi Theory (Optional). Now we describe briefly and informally some additional geometry connected with the Hamilton–Jacobi equation (7.9.9). For each $x = (q^i, t) \in \tilde{Q} := Q \times \mathbb{R}$, $dS(x)$ is an element of the cotangent bundle $T^*\tilde{Q}$. We suppress the dependence of S on $\bar{q}^i$ for the moment, since it does not play an immediate role. As x varies in $\tilde{Q}$, the set $\{ dS(x) \mid x \in \tilde{Q} \}$ defines a submanifold of $T^*\tilde{Q}$ that in terms of coordinates is given by $p_j = \partial S/\partial q^j$ and $p = \partial S/\partial t$; here the variables conjugate to q^i are denoted by p_i and that conjugate to t is denoted by p. We will write $\xi_i = p_i$ for $i = 1, 2, \ldots, n$ and $\xi_{n+1} = p$. We call this submanifold the *range*, or *graph*, of dS (either term is appropriate, depending on whether one thinks of dS as a mapping or as a section of a bundle) and denote it by graph $dS \subset T^*\tilde{Q}$. The restriction of the canonical symplectic form on $T^*\tilde{Q}$ to graph dS is zero, since

$$\sum_{j=1}^{n+1} dx^j \wedge d\xi_j = \sum_{j=1}^{n+1} dx^j \wedge d\frac{\partial S}{\partial x_j} = \sum_{j,k=1}^{n+1} dx^j \wedge dx^k \frac{\partial^2 S}{\partial x^j \partial x^k} = 0.$$

Moreover, the dimension of the submanifold graph dS is half of the dimension of the symplectic manifold $T^*\tilde{Q}$. Such a submanifold is called **Lagrangian**, as we already mentioned in connection with generating functions (§6.5). What is important here is that the projection from graph dS to $\tilde{Q}$ is a diffeomorphism, and even more, the converse holds: If $\Lambda \subset T^*\tilde{Q}$ is a Lagrangian submanifold of $T^*\tilde{Q}$ such that the projection on $\tilde{Q}$ is a diffeomorphism in a neighborhood of a point $\lambda \in \Lambda$, then in some neighborhood of λ we can write $\Lambda = $ graph $d\varphi$ for some function φ. To show this, notice that because the projection is a diffeomorphism, Λ is given (around λ) as a submanifold of the form $(x^j, \rho_j(x))$. The condition for Λ to be Lagrangian requires that on Λ,

$$\sum_{j=1}^{n+1} dx^j \wedge d\xi_j = 0,$$

that is,

$$\sum_{j=1}^{n+1} dx^j \wedge d\rho_j(x) = 0, \quad \text{i.e.,} \quad \frac{\partial \rho_j}{\partial x^k} - \frac{\partial \rho_k}{\partial x^j} = 0;$$

thus, there is a φ such that $\rho_j = \partial\varphi/\partial x^j$, which is the same as $\Lambda = $ graph $d\varphi$. The conclusion of these remarks is that Lagrangian submanifolds of $T^*\tilde{Q}$ are natural generalizations of graphs of differentials of functions on $\tilde{Q}$. Note that Lagrangian submanifolds are defined even if the projection to $\tilde{Q}$ is not a diffeomorphism. For more information on Lagrangian manifolds and generating functions, see Abraham and Marsden [1978], Weinstein [1977], and Guillemin and Sternberg [1977].

From the point of view of Lagrangian submanifolds, *the graph of the differential of a solution of the Hamilton–Jacobi equation is a Lagrangian submanifold of $T^*\tilde{Q}$ that is contained in the surface $\tilde{H}_0 \subset T^*\tilde{Q}$ defined by the equation $\tilde{H} := p + H(q^i, p_i, t) = 0$.* Here, as above, $p = \xi_{n+1}$ is the momentum conjugate to t. This point of view allows one to include solutions that are singular in the usual context. This is not the only benefit: We also get more insight in the content of the Hamilton–Jacobi Theorem 7.9.3. The tangent space to $\tilde{H}_0$ has dimension 1 less than the dimension of the symplectic manifold $T^*\tilde{Q}$, and it is given by the set of vectors X such that $(dp + \mathbf{d}H)(X) = 0$. If a vector Y is in the symplectic orthogonal of $T_{(x,\xi)}(\tilde{H}_0)$, that is,

$$\sum_{j=1}^{n+1}(dx^j \wedge d\xi_j)(X, Y) = 0$$

for all $X \in T_{(x,\xi)}(\tilde{H}_0)$, then Y is a multiple of the vector field

$$X_{\tilde{H}} = \frac{\partial}{\partial t} - \frac{\partial H}{\partial t}\frac{\partial}{\partial p} + X_H$$

evaluated at (x, ξ). Moreover, the integral curves of $X_{\tilde{H}}$ projected to (q^i, p_i) are the solutions of Hamilton's equations for H.

The key observation that links Hamilton's equations and the Hamilton–Jacobi equation is that *the vector field $X_{\tilde{H}}$, which is obviously tangent to $\tilde{H}_0$, is, moreover, tangent to any Lagrangian submanifold contained in $\tilde{H}_0$* (the reason for this is a very simple algebraic fact given in Exercise 7.9-3). This is the same as saying that a solution of Hamilton's equations for $\tilde{H}$ is either disjoint from a Lagrangian submanifold contained in $\tilde{H}_0$ or completely contained in it. This gives a way to construct a solution of the Hamilton–Jacobi equation starting from an initial condition at $t = t_0$. Namely, take a Lagrangian submanifold Λ_0 in T^*Q and embed it in $T^*\tilde{Q}$ at $t = t_0$ using

$$(q^i, p_i) \mapsto (q^i, t = t_0, p_i, p = -H(q^i, p_i, t_0)).$$

The result is an isotropic submanifold $\tilde{\Lambda}_0 \subset T^*\tilde{Q}$, that is, a submanifold on which the canonical form vanishes. Now take all integral curves of $X_{\tilde{H}}$ whose initial conditions lie in $\tilde{\Lambda}_0$. The collection of these curves spans a manifold Λ whose dimension is one higher than $\tilde{\Lambda}_0$. It is obtained by flowing $\tilde{\Lambda}_0$ along $X_{\tilde{H}}$; that is, $\Lambda = \cup_t \Lambda_t$, where $\Lambda_t = \Phi_t(\tilde{\Lambda}_0)$ and Φ_t is the flow of $X_{\tilde{H}}$. Since $X_{\tilde{H}}$ is tangent to $\tilde{H}_0$ and $\Lambda_0 \subset \tilde{H}_0$, we get $\Lambda_t \subset \tilde{H}_0$ and hence $\Lambda \subset \tilde{H}_0$. Since the flow Φ_t of $X_{\tilde{H}}$ is a canonical map, it leaves the symplectic form of $T^*\tilde{Q}$ invariant and therefore takes an isotropic submanifold into an isotropic one; in particular, Λ_t is an isotropic submanifold of $T^*\tilde{Q}$. The tangent space of Λ at some $\lambda \in \Lambda_t$ is a direct sum of the tangent space of

Λ_t and the subspace generated by $X_{\tilde{H}}$. Since the first subspace is contained in $T_\lambda \tilde{H}_0$ and the second is symplectically orthogonal to $T_\lambda \tilde{H}_0$, we see that Λ is also an isotropic submanifold of $T^*\tilde{Q}$. But its dimension is half that of $T^*\tilde{Q}$, and therefore Λ is a Lagrangian submanifold contained in $\tilde{H}_0$, that is, it is a solution of the Hamilton–Jacobi equation with initial condition Λ_0 at $t = t_0$.

Using the above point of view it is easy to understand the singularities of a solution of the Hamilton–Jacobi equation. They correspond to those points of the Lagrangian manifold solution where the projection to $\tilde{Q}$ is not a local diffeomorphism. These singularities might be present in the initial condition (that is, Λ_0 might not locally project diffeomorphically to Q), or they might appear at later times by folding the submanifolds Λ_t as t varies. The projection of such a singular point to $\tilde{Q}$ is called a **caustic point** of the solution. Caustic points are of fundamental importance in geometric optics and the semiclassical approximation of quantum mechanics. We refer to Abraham and Marsden [1978, Section 5.3] and Guillemin and Sternberg [1984] for further information.

Exercises

◇ **7.9-1.** Solve the Hamilton–Jacobi equation for the harmonic oscillator. Check directly the validity of the Hamilton–Jacobi theorem (connecting the solution of the Hamilton–Jacobi equation and the flow of the Hamiltonian vector field) for this case.

◇ **7.9-2.** Verify by *direct calculation* the following. Let $W(q,\bar{q})$ and

$$H(q,p) = \frac{p^2}{2m} + V(q)$$

be given, where $q, p \in \mathbb{R}$. Show that for $p \neq 0$,

$$\frac{1}{2m}(W_q)^2 + V = E$$

and $\dot{q} = p/m$ if and only if $(q, W_q(q, \bar{q}))$ satisfies Hamilton's equation with energy E.

◇ **7.9-3.** Let (V, Ω) be a symplectic vector space and $W \subset V$ be a linear subspace. Recall from §2.4 that

$$W^\Omega = \{\, v \in V \mid \Omega(v, w) = 0 \text{ for all } w \in W \,\}$$

denotes the symplectic orthogonal of W. A subspace $L \subset V$ is called **Lagrangian** if $L = L^\Omega$. Show that if $L \subset W$ is a Lagrangian subspace, then $W^\Omega \subset L$.

◇ **7.9-4.** Solve the Hamilton–Jacobi equation for a central force field. Check directly the validity of the Hamilton–Jacobi theorem.

8

Variational Principles, Constraints, and Rotating Systems

This chapter deals with two related topics: constrained Lagrangian (and Hamiltonian) systems and rotating systems. Constrained systems are illustrated by a particle constrained to move on a sphere. Such constraints that involve conditions on the *configuration* variables are called "holonomic."[1] For rotating systems, one needs to distinguish systems that are viewed from rotating coordinate systems (passively rotating systems) and systems that themselves are rotated (actively rotating systems—such as a Foucault pendulum and weather systems rotating with the Earth). We begin with a more detailed look at variational principles, and then we turn to a version of the Lagrange multiplier theorem that will be useful for our analysis of constraints.

8.1 A Return to Variational Principles

In this section we take a closer look at variational principles. Technicalities involving infinite-dimensional manifolds prevent us from presenting the full story from that point of view. For these, we refer to, for example, Smale [1964], Palais [1968], and Klingenberg [1978]. For the classical geometric theory without the infinite-dimensional framework, the reader may consult,

---

[1] In this volume we shall not discuss "nonholonomic" constraints such as rolling constraints. We refer to Bloch, Krishnaprasad, Marsden, and Murray [1996], Koon and Marsden [1997b], and Zenkov, Bloch, and Marsden [1998] for a discussion of nonholonomic systems and further references.

for example, Bolza [1973], Whittaker [1927], Gelfand and Fomin [1963], or Hermann [1968].

Hamilton's Principle. We begin by setting up the space of paths joining two points.

Definition 8.1.1. *Let Q be a manifold and let $L : TQ \to \mathbb{R}$ be a regular Lagrangian. Fix two points q_1 and q_2 in Q and an interval $[a, b]$, and define the **path space** from q_1 to q_2 by*

$$\Omega(q_1, q_2, [a, b])$$
$$= \{ c : [a, b] \to Q \mid c \text{ is a } C^2 \text{ curve}, c(a) = q_1, c(b) = q_2 \} \quad (8.1.1)$$

and the map $\mathfrak{S} : \Omega(q_1, q_2, [a, b]) \to \mathbb{R}$ by

$$\mathfrak{S}(c) = \int_a^b L(c(t), \dot{c}(t)) \, dt.$$

What we shall *not* prove is that $\Omega(q_1, q_2, [a, b])$ is a smooth infinite-dimensional manifold. This is a special case of a general result in the topic of manifolds of mappings, wherein spaces of maps from one manifold to another are shown to be smooth infinite-dimensional manifolds. Accepting this, we can prove the following.

Proposition 8.1.2. *The tangent space $T_c\Omega(q_1, q_2, [a, b])$ to the manifold $\Omega(q_1, q_2, [a, b])$ at a point, that is, a curve $c \in \Omega(q_1, q_2, [a, b])$, is the set of C^2 maps $v : [a, b] \to TQ$ such that $\tau_Q \circ v = c$ and $v(a) = 0$, $v(b) = 0$, where $\tau_Q : TQ \to Q$ denotes the canonical projection.*

Proof. The tangent space to a manifold consists of tangents to smooth curves in the manifold. The tangent vector to a curve $c_\lambda \in \Omega(q_1, q_2, [a, b])$ with $c_0 = c$ is

$$v = \left. \frac{d}{d\lambda} c_\lambda \right|_{\lambda=0}. \quad (8.1.2)$$

However, $c_\lambda(t)$, for each fixed t, is a curve through $c_0(t) = c(t)$. Hence

$$\left. \frac{d}{d\lambda} c_\lambda(t) \right|_{\lambda=0}$$

is a tangent vector to Q based at $c(t)$. Hence $v(t) \in T_{c(t)}Q$; that is, $\tau_Q \circ v = c$. The restrictions $c_\lambda(a) = q_1$ and $c_\lambda(b) = q_2$ lead to $v(a) = 0$ and $v(b) = 0$, but otherwise v is an arbitrary C^2 function. ∎

One refers to v as an **infinitesimal variation** of the curve c subject to fixed endpoints, and we use the notation $v = \delta c$. See Figure 8.1.1.

Now we can state and sketch the proof of a main result in the calculus of variations in a form due to Hamilton [1834].

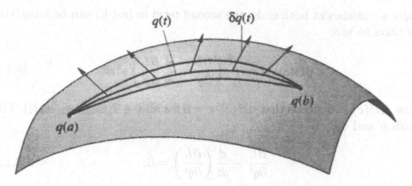

FIGURE 8.1.1. The variation $\delta q(t)$ of a curve $q(t)$ is a field of vectors tangent to the configuration manifold along that curve.

Theorem 8.1.3 (Variational Principle of Hamilton).   *Let L be a Lagrangian on TQ. A curve $c_0 : [a, b] \to Q$ joining $q_1 = c_0(a)$ to $q_2 = c_0(b)$ satisfies the Euler–Lagrange equations*

$$\frac{d}{dt}\left(\frac{\partial L}{\partial \dot{q}^i}\right) = \frac{\partial L}{\partial q^i} \tag{8.1.3}$$

if and only if c_0 is a critical point of the function $\mathfrak{S} : \Omega(q_1, q_2, [a, b]) \to \mathbb{R}$, that is, $\mathbf{d}\mathfrak{S}(c_0) = 0$. If L is regular, either condition is equivalent to c_0 being a base integral curve of X_E.

As in §7.1, the condition $\mathbf{d}\mathfrak{S}(c_0) = 0$ is denoted by

$$\delta \int_a^b L(c_0(t), \dot{c}_0(t)) \, dt = 0; \tag{8.1.4}$$

that is, the integral is stationary when it is differentiated with c regarded as the independent variable.

Proof.   We work out $\mathbf{d}\mathfrak{S}(c) \cdot v$ just as in §7.1. Write v as the tangent to the curve c_λ in $\Omega(q_1, q_2, [a, b])$ as in (8.1.2). By the chain rule,

$$\mathbf{d}\mathfrak{S}(c) \cdot v = \frac{d}{d\lambda}\mathfrak{S}(c_\lambda)\bigg|_{\lambda=0} = \frac{d}{d\lambda}\int_a^b L(c_\lambda(t), \dot{c}_\lambda(t)) \, dt \bigg|_{\lambda=0}. \tag{8.1.5}$$

Differentiating (8.1.5) under the integral sign, and using local coordinates,[2] we get

$$\mathbf{d}\mathfrak{S}(c) \cdot v = \int_a^b \left(\frac{\partial L}{\partial q^i} v^i + \frac{\partial L}{\partial \dot{q}^i} \dot{v}^i\right) dt. \tag{8.1.6}$$

---

[2]If the curve $c_0(t)$ does not lie in a single coordinate chart, divide the curve $c(t)$ into a finite partition each of whose elements lies in a chart and apply the argument below.

Since v vanishes at both ends, the second term in (8.1.6) can be integrated by parts to give

$$\mathbf{d\mathfrak{S}}(c) \cdot v = \int_a^b \left(\frac{\partial L}{\partial q^i} - \frac{d}{dt} \frac{\partial L}{\partial \dot{q}^i} \right) v^i \, dt. \qquad (8.1.7)$$

Now, $\mathbf{d\mathfrak{S}}(c) = 0$ means that $\mathbf{d\mathfrak{S}}(c) \cdot v = 0$ for all $v \in T_c\Omega(q_1, q_2, [a, b])$. This holds if and only if

$$\frac{\partial L}{\partial q^i} - \frac{d}{dt} \left(\frac{\partial L}{\partial \dot{q}^i} \right) = 0, \qquad (8.1.8)$$

since the integrand is continuous and v is arbitrary, except for $v = 0$ at the ends. (This last assertion was proved in Theorem 7.3.3.) ∎

The reader can check that Hamilton's principle proceeds virtually unchanged for time-dependent Lagrangians. We shall use this remark below.

The Principle of Critical Action. Next we discuss variational principles with the constraint of constant energy imposed. To compensate for this constraint, we let the interval $[a, b]$ be variable.

Definition 8.1.4. *Let L be a regular Lagrangian and let Σ_e be a regular energy surface for the energy E of L, that is, e is a regular value of E and $\Sigma_e = E^{-1}(e)$. Let $q_1, q_2 \in Q$ and let $[a, b]$ be a given interval. Define $\Omega(q_1, q_2, [a, b], e)$ to be the set of pairs (τ, c), where $\tau : [a, b] \to \mathbb{R}$ is C^2, satisfies $\dot{\tau} > 0$, and where $c : [\tau(a), \tau(b)] \to Q$ is a C^2 curve with*

$$c(\tau(a)) = q_1, \quad c(\tau(b)) = q_2,$$

and

$$E\left(c(\tau(t)), \dot{c}(\tau(t))\right) = e, \qquad \text{for all } t \in [a, b].$$

Arguing as in Proposition 8.1.2, computation of the derivatives of curves $(\tau_\lambda, c_\lambda)$ in $\Omega(q_1, q_2, [a, b], e)$ shows that the tangent space to $\Omega(q_1, q_2, [a, b], e)$ at (τ, c) consists of the space of pairs of C^2 maps

$$\alpha : [a, b] \to \mathbb{R} \quad \text{and} \quad v : [\tau(a), \tau(b)] \to TQ$$

such that $v(t) \in T_{c(t)}Q$,

$$\begin{aligned}
\dot{c}(\tau(a))\alpha(a) + v(\tau(a)) &= 0, \\
\dot{c}(\tau(b))\alpha(b) + v(\tau(b)) &= 0,
\end{aligned} \qquad (8.1.9)$$

and

$$\mathbf{d}E[c(\tau(t)), \dot{c}(\tau(t))] \cdot [\dot{c}(\tau(t))\alpha(t) + v(\tau(t)), \ddot{c}(\tau(t))\dot{\alpha}(t) + \dot{v}(\tau(t))] = 0. \\ (8.1.10)$$

Theorem 8.1.5 (Principle of Critical Action). *Let $c_0(t)$ be a solution of the Euler–Lagrange equations and let $q_1 = c_0(a)$ and $q_2 = c_0(b)$. Let e be the energy of $c_0(t)$ and assume that it is a regular value of E. Define the map $\mathcal{A} : \Omega(q_1, q_2, [a, b], e) \to \mathbb{R}$ by*

$$\mathcal{A}(\tau, c) = \int_{\tau(a)}^{\tau(b)} A(c(t), \dot{c}(t))\, dt, \qquad (8.1.11)$$

where A is the action of L. Then

$$\mathbf{d}\mathcal{A}(\mathrm{Id}, c_0) = 0, \qquad (8.1.12)$$

where Id is the identity map. Conversely, if (Id, c_0) is a critical point of $\mathcal{A}$ and c_0 has energy e, a regular value of E, then c_0 is a solution of the Euler–Lagrange equations.

In coordinates, (8.1.11) reads

$$\mathcal{A}(\tau, c) = \int_{\tau(a)}^{\tau(b)} \frac{\partial L}{\partial \dot{q}^i} \dot{q}^i\, dt = \int_{\tau(a)}^{\tau(b)} p_i\, dq^i, \qquad (8.1.13)$$

the integral of the canonical one-form along the curve $\gamma = (c, \dot{c})$. Being the line integral of a one-form, $\mathcal{A}(\tau, c)$ is independent of the parametrization τ. Thus, one may think of $\mathcal{A}$ as defined on the space of (unparametrized) curves joining q_1 and q_2.

Proof. If the curve c has energy e, then

$$\mathcal{A}(\tau, c) = \int_{\tau(a)}^{\tau(b)} [L(q^i, \dot{q}^i) + e]\, dt.$$

Differentiating $\mathcal{A}$ with respect to τ and c by the method of Theorem 8.1.3 gives

$$\mathbf{d}\mathcal{A}(\mathrm{Id}, c_0) \cdot (\alpha, v)$$
$$= \alpha(b) [L(c_0(b), \dot{c}_0(b)) + e] - \alpha(a) [L(c_0(a), \dot{c}_0(a)) + e]$$
$$+ \int_a^b \left(\frac{\partial L}{\partial q^i}(c_0(t), \dot{c}_0(t)) v^i(t) + \frac{\partial L}{\partial \dot{q}^i}(c_0(t), \dot{c}_0(t)) \dot{v}^i(t) \right) dt. \quad (8.1.14)$$

Integrating by parts gives

$$\mathbf{d}\mathcal{A}(\mathrm{Id}, c_0) \cdot (\alpha, v)$$
$$= \left[\alpha(t) [L(c_0(t), \dot{c}_0(t)) + e] + \frac{\partial L}{\partial \dot{q}^i}(c_0(t), \dot{c}_0(t)) v^i(t) \right]_a^b$$
$$+ \int_a^b \left(\frac{\partial L}{\partial q^i}(c_0(t), \dot{c}_0(t)) - \frac{d}{dt}\frac{\partial L}{\partial \dot{q}^i}(c_0(t), \dot{c}_0(t)) \right) v^i(t)\, dt. \quad (8.1.15)$$

224   8. Variational Principles, Constraints, & Rotating Systems

Using the boundary conditions $v = -\dot{c}\alpha$, noted in the description of the tangent space $T_{(\mathrm{Id},c_0)}\Omega(q_1, q_2, [a,b], e)$ and the energy constraint $(\partial L/\partial \dot{q}^i)\dot{c}^i - L = e$, the boundary terms cancel, leaving

$$\mathbf{d}\mathcal{A}(\mathrm{Id}, c_0) \cdot (\alpha, v) = \int_a^b \left(\frac{\partial L}{\partial q^i} - \frac{d}{dt}\frac{\partial L}{\partial \dot{q}^i} \right) v^i \, dt. \qquad (8.1.16)$$

However, we can choose v arbitrarily; notice that the presence of α in the linearized energy constraint means that no restrictions are placed on the variations v^i on the open set where $\dot{c} \neq 0$. The result therefore follows. ∎

If $L = K - V$, where K is the kinetic energy of a Riemannian metric, then Theorem 8.1.5 states that a curve c_0 is a solution of the Euler–Lagrange equations if and only if

$$\delta_e \int_a^b 2K(c_0, \dot{c}_0)\, dt = 0, \qquad (8.1.17)$$

where δ_e indicates a variation holding the energy and endpoints but not the parametrization fixed; this is symbolic notation for the precise statement in Theorem 8.1.5. Using the fact that $K \geq 0$, a calculation of the Euler–Lagrange equations (Exercise 8.1-3) shows that (8.1.17) is the same as

$$\delta_e \int_a^b \sqrt{2K(c_0, \dot{c}_0)}\, dt = 0, \qquad (8.1.18)$$

that is, arc length is extremized (subject to constant energy). This is *Jacobi's form of the principle of "least action"* and represents a key to linking mechanics and geometric optics, which was one of Hamilton's original motivations. In particular, geodesics are characterized as extremals of arc length. Using the Jacobi metric (see §7.7) one gets yet another variational principle.[3]

Phase Space Form of the Variational Principle.   The above variational principles for Lagrangian systems carry over to some extent to Hamiltonian systems.

Theorem 8.1.6 (Hamilton's Principle in Phase Space).   *Consider a Hamiltonian H on a given cotangent bundle T^*Q. A curve $(q^i(t), p_i(t))$ in T^*Q satisfies Hamilton's equations iff*

$$\delta \int_a^b [p_i \dot{q}^i - H(q^i, p_i)]\, dt = 0 \qquad (8.1.19)$$

for variations over curves $(q^i(t), p_i(t))$ in phase space, where $\dot{q}^i = dq^i/dt$ and where q^i are fixed at the endpoints.

---

[3]Other interesting variational principles are those of Gauss, Hertz, Gibbs, and Appell. A modern account, along with references, is Lewis [1996].

Proof.   Computing as in (8.1.6), we find that

$$\delta \int_a^b [p_i \dot{q}^i - H(q^i, p_i)]\, dt = \int_a^b \left[(\delta p_i)\dot{q}^i + p_i(\delta \dot{q}^i) - \frac{\partial H}{\partial q^i}\delta q^i - \frac{\partial H}{\partial p_i}\delta p_i \right] dt.$$
(8.1.20)

Since $q^i(t)$ are fixed at the two ends, we have $p_i \delta q^i = 0$ at the two ends, and hence the second term of (8.1.20) can be integrated by parts to give

$$\int_a^b \left[\dot{q}^i(\delta p_i) - \dot{p}_i(\delta q^i) - \frac{\partial H}{\partial q^i}\delta q^i - \frac{\partial H}{\partial p_i}\delta p_i \right] dt,$$
(8.1.21)

which vanishes for all $\delta p_i, \delta q^i$ exactly when Hamilton's equations hold.   ∎

Hamilton's principle in phase space (8.1.19) on an exact symplectic manifold $(P, \Omega = -\mathbf{d}\Theta)$ reads

$$\delta \int_a^b (\Theta - H\,dt) = 0,$$
(8.1.22)

again with suitable boundary conditions. Likewise, if we impose the constraint $H = $ constant, the principle of least action reads

$$\delta \int_{\tau(a)}^{\tau(b)} \Theta = 0.$$
(8.1.23)

In Cendra and Marsden [1987], Cendra, Ibort, and Marsden [1987], Marsden and Scheurle [1993a, 1993b], and Holm, Marsden, and Ratiu [1998a], it is shown how to form variational principles on certain symplectic and Poisson manifolds even when Ω is not exact, but does arise by a reduction process. The variational principle for the Euler–Poincaré equations that was described in the introduction and that we shall encounter again in Chapter 13 is a special instance of this.

The one-form $\Theta_H := \Theta - H\,dt$ in (8.1.22), regarded as a one-form on $P \times \mathbb{R}$, is an example of a *contact form* and plays an important role in time-dependent and relativistic mechanics. Let

$$\Omega_H = -\mathbf{d}\Theta_H = \Omega + dH \wedge dt$$

and observe that the vector field X_H is characterized by the statement that its suspension $\tilde{X}_H = (X_H, 1)$, a vector field on $P \times \mathbb{R}$, lies in the kernel of Ω_H:

$$\mathbf{i}_{\tilde{X}_H} \Omega_H = 0.$$

Exercises

◇ **8.1-1.** In Hamilton's principle, show that the boundary conditions of fixed $q(a)$ and $q(b)$ can be changed to $p(b) \cdot \delta q(b) = p(a) \cdot \delta q(a)$. What is the corresponding statement for Hamilton's principle in phase space?

◇ **8.1-2.** Show that the equations for a particle in a magnetic field B and a potential V can be written as

$$\delta \int (K - V)\, dt = -\frac{e}{c} \int \delta q \cdot (v \times B)\, dt.$$

◇ **8.1-3.** Do the calculation showing that

$$\delta_e \int_a^b 2K(c_0, \dot{c}_0)\, dt = 0$$

and

$$\delta_e \int_a^b \sqrt{2K(c_0, \dot{c}_0)}\, dt = 0$$

are equivalent.

8.2 The Geometry of Variational Principles

In Chapter 7 we derived the "geometry" of Lagrangian systems on TQ by pulling back the geometry from the Hamiltonian side on T^*Q. Now we show how *all of this basic geometry of Lagrangian systems can be derived directly from Hamilton's principle*. The exposition below follows Marsden, Patrick, and Shkoller [1998].

A Brief Review. Recall that given a Lagrangian function $L : TQ \to \mathbb{R}$, we construct the corresponding *action functional* $\mathfrak{S}$ on C^2 curves $q(t)$, $a \le t \le b$, by (using coordinate notation)

$$\mathfrak{S}(q(\cdot)) \equiv \int_a^b L\left(q^i(t), \frac{dq^i}{dt}(t)\right) dt. \tag{8.2.1}$$

Hamilton's principle (Theorem 8.1.3) seeks the curves $q(t)$ for which the functional $\mathfrak{S}$ is stationary under variations of $q^i(t)$ with *fixed endpoints* at *fixed times*. Recall that this calculation gives

$$\mathbf{d}\mathfrak{S}(q(\cdot)) \cdot \delta q(\cdot) = \int_a^b \delta q^i \left(\frac{\partial L}{\partial q^i} - \frac{d}{dt}\frac{\partial L}{\partial \dot{q}^i}\right) dt + \frac{\partial L}{\partial \dot{q}^i}\delta q^i \bigg|_a^b. \tag{8.2.2}$$

The last term in (8.2.2) vanishes, since $\delta q(a) = \delta q(b) = 0$, so that the requirement that $q(t)$ be stationary for $\mathfrak{S}$ yields the Euler–Lagrange equations

$$\frac{\partial L}{\partial q^i} - \frac{d}{dt}\frac{\partial L}{\partial \dot{q}^i} = 0. \tag{8.2.3}$$

Recall that L is called *regular* when the matrix $[\partial^2 L/\partial\dot{q}^i\partial\dot{q}^j]$ is everywhere nonsingular, and in this case the Euler–Lagrange equations are second-order ordinary differential equations for the required curves.

Since the action (8.2.1) is independent of the choice of coordinates, the Euler–Lagrange equations are coordinate-independent as well. Consequently, it is natural that the Euler–Lagrange equations may be intrinsically expressed using the language of differential geometry.

Recall that one defines the *canonical 1-form* Θ on the $2n$-dimensional cotangent bundle T^*Q of Q by

$$\Theta(\alpha_q)\cdot w_{\alpha_q} = \langle\alpha_q, T_{\alpha_q}\pi_Q(w_{\alpha_q})\rangle,$$

where $\alpha_q \in T_q^*Q$, $w_{\alpha_q} \in T_{\alpha_q}T^*Q$, and $\pi_Q : T^*Q \to Q$ is the projection. The Lagrangian L defines a fiber-preserving bundle map $\mathbb{F}L : TQ \to T^*Q$, the Legendre transformation, by fiber differentiation:

$$\mathbb{F}L(v_q)\cdot w_q = \left.\frac{d}{d\epsilon}\right|_{\epsilon=0} L(v_q + \epsilon w_q).$$

One normally defines the *Lagrange 1-form* on TQ by pull-back,

$$\Theta_L = \mathbb{F}L^*\Theta,$$

and the *Lagrange 2-form* by $\Omega_L = -d\Theta_L$. We then seek a vector field X_E (called the *Lagrange vector field*) on TQ such that $X_E \lrcorner\, \Omega_L = dE$, where the *energy* E is defined by

$$E(v_q) = \langle\mathbb{F}L(v_q), v_q\rangle - L(v_q) = \Theta_L(X_E)(v_q) - L(v_q).$$

If $\mathbb{F}L$ is a local diffeomorphism, which is equivalent to L being regular, then X_E exists and is unique, and its integral curves solve the Euler–Lagrange equations. The Euler–Lagrange equations are second-order equations in TQ. In addition, the flow F_t of X_E is symplectic, that is, preserves $\Omega_L: F_t^*\Omega_L = \Omega_L$. These facts were proved using differential forms and Lie derivatives in the last three chapters.

The Variational Approach. Besides being more faithful to history, sometimes there are advantages to staying on the "Lagrangian side." Many examples can be given, but the theory of Lagrangian reduction (the Euler-Poincaré equations being an instance) is one example. Other examples are the direct variational approach to questions in black-hole dynamics given by Wald [1993] and the development of variational asymptotics (see Holm [1996], Holm, Marsden, and Ratiu [1998b], and references therein). In such studies, it is the variational principle that is the center of attention.

The development begins by removing the endpoint condition $\delta q(a) = \delta q(b) = 0$ from (8.2.2) but still keeping the time interval fixed. Equation (8.2.2) becomes

$$\mathbf{d}\Theta\big(q(\cdot)\big)\cdot\delta q(\cdot) = \int_a^b \delta q^i\left(\frac{\partial L}{\partial q^i} - \frac{d}{dt}\frac{\partial L}{\partial\dot{q}^i}\right)dt + \frac{\partial L}{\partial\dot{q}^i}\delta q^i\bigg|_a^b, \tag{8.2.4}$$

but now the left side operates on more general δq, and correspondingly, the last term on the right side need not vanish. That last term of (8.2.4) is a linear pairing of the function $\partial L/\partial \dot{q}^i$, a function of q^i and $\dot{q}^i$, with the tangent vector δq^i. Thus, one may consider it a 1-form on TQ, namely, the Lagrange 1-form $(\partial L/\partial \dot{q}^i) dq^i$.

Theorem 8.2.1. *Given a C^k Lagrangian L, $k \geq 2$, there exists a unique C^{k-2} mapping $D_{EL}L : \ddot{Q} \to T^*Q$, defined on the **second-order submanifold***

$$\ddot{Q} := \left\{ \frac{d^2q}{dt^2}(0) \in T(TQ) \,\middle|\, q \text{ is a } C^2 \text{ curve in } Q \right\}$$

of $T(TQ)$, and a unique C^{k-1} 1-form Θ_L on TQ, such that for all C^2 variations $q_\epsilon(t)$ (on a fixed t-interval) of $q(t)$, where $q_0(t) = q(t)$, we have

$$\mathbf{d}\mathfrak{S}(q(\cdot)) \cdot \delta q(\cdot) = \int_a^b D_{EL}L\left(\frac{d^2q}{dt^2}\right) \cdot \delta q \, dt + \Theta_L\left(\frac{dq}{dt}\right) \cdot \hat{\delta}q \bigg|_a^b, \quad (8.2.5)$$

where

$$\delta q(t) = \frac{d}{d\epsilon}\bigg|_{\epsilon=0} q_\epsilon(t), \qquad \hat{\delta}q(t) = \frac{d}{d\epsilon}\bigg|_{\epsilon=0} \frac{d}{dt} q_\epsilon(t).$$

*The 1-form so defined is a called the **Lagrange 1-form**.*

Indeed, uniqueness and local existence follow from the calculation (8.2.2). The coordinate independence of the action implies the global existence of D_{EL} and the 1-form Θ_L.

Thus, using the variational principle, the Lagrange 1-form Θ_L is the "boundary part" of the functional derivative of the action when the boundary is varied. The analogue of the symplectic form is the negative exterior derivative of Θ_L; that is, $\Omega_L \equiv -\mathbf{d}\Theta_L$.

Lagrangian Flows Are Symplectic. One of Lagrange's basic discoveries was that the solutions of the Euler–Lagrange equations give rise to a symplectic map. It is a curious twist of history that he did this without the machinery of differential forms, the Hamiltonian formalism, or Hamilton's principle itself.

Assuming that L is regular, the variational principle gives coordinate-independent second-order ordinary differential equations. We temporarily denote the vector field on TQ so obtained by X, and its flow by F_t. Now consider the restriction of $\mathfrak{S}$ to the subspace $\mathcal{C}_L$ of solutions of the variational principle. The space $\mathcal{C}_L$ may be identified with the initial conditions for the flow; to $v_q \in TQ$ we associate the integral curve $s \mapsto F_s(v_q)$, $s \in [0, t]$. The value of $\mathfrak{S}$ on the base integral curve $q(s) = \pi_Q(F_s(v_q))$ is denoted by $\mathfrak{S}_t$,

that is,

$$\mathfrak{S}_t = \int_0^t L(F_s(v_q))\, ds, \tag{8.2.6}$$

which is again called the *action*. We regard $\mathfrak{S}_t$ as a real-valued function on TQ. Note that by (8.2.6), $d\mathfrak{S}_t/dt = L(F_t(v_q))$. The fundamental equation (8.2.5) becomes

$$\mathbf{d}\mathfrak{S}_t(v_q) \cdot w_{v_q} = \Theta_L\big(F_t(v_q)\big) \cdot \frac{d}{d\epsilon}\bigg|_{\epsilon=0} F_t(v_q + \epsilon w_{v_q}) - \Theta_L(v_q) \cdot w_{v_q},$$

where $\epsilon \mapsto v_q + \epsilon w_{v_q}$ symbolically represents a curve at v_q in TQ with derivative w_{v_q}. Note that the first term on the right-hand side of (8.2.5) vanishes, since we have restricted $\mathfrak{S}$ to solutions. The second term becomes the one stated, remembering that now $\mathfrak{S}_t$ is regarded as a function on TQ. We have thus derived the equation

$$\mathbf{d}\mathfrak{S}_t = F_t^*\Theta_L - \Theta_L. \tag{8.2.7}$$

Taking the exterior derivative of (8.2.7) yields the fundamental fact that the flow of X is symplectic:

$$0 = \mathbf{dd}\mathfrak{S}_t = \mathbf{d}(F_t^*\Theta_L - \Theta_L) = -F_t^*\Omega_L + \Omega_L,$$

which is equivalent to $F_t^*\Omega_L = \Omega_L$. Thus, *using the variational principle, the analogue that the evolution is symplectic is the equation $\mathbf{d}^2 = 0$, applied to the action restricted to the space of solutions of the variational principle.* Equation (8.2.7) also provides the differential–geometric equations for X. Indeed, taking one time-derivative of (8.2.7) gives $\mathbf{d}L = \pounds_X\Theta_{\mathcal{L}}$, so that

$$X \lrcorner\, \Omega_L = -X \lrcorner\, \mathbf{d}\Theta_L = -\pounds_X\Theta_L + \mathbf{d}(X \lrcorner\, \Theta_L) = \mathbf{d}(X \lrcorner\, \Theta_L - L) = \mathbf{d}E,$$

where we define $E = X \lrcorner\, \Theta_L - L$. Thus, quite naturally, we find that $X = X_E$.

The Hamilton–Jacobi Equation.   Next, we give a derivation of the Hamilton–Jacobi equation from variational principles. *Allowing L to be time-dependent*, Jacobi [1866] showed that the *action integral* defined by

$$S(q^i, \bar{q}^i, t) = \int_{t_0}^t L(q^i(s), \dot{q}^i(s), s)\, ds,$$

where $q^i(s)$ is the solution of the Euler–Lagrange equation subject to the conditions $q^i(t_0) = \bar{q}^i$ and $q^i(t) = q^i$, satisfies the Hamilton–Jacobi equation. There are several implicit assumptions in Jacobi's argument: L is regular and the time $|t - t_0|$ is assumed to be small, so that by the convex neighborhood theorem, S is a well-defined function of the endpoints. We can allow $|t - t_0|$ to be large as long as the solution $q(t)$ is near a nonconjugate solution.

Theorem 8.2.2 (Hamilton–Jacobi).   *With the above assumptions, the function $S(q, \bar{q}, t)$ satisfies the Hamilton–Jacobi equation:*

$$\frac{\partial S}{\partial t} + H\left(q, \frac{\partial S}{\partial q}, t\right) = 0.$$

Proof.   In this equation, $\bar{q}$ is held fixed. Define v, a tangent vector at $\bar{q}$, implicitly by

$$\pi_Q F_t(v) = q, \tag{8.2.8}$$

where $F_t : TQ \to TQ$ is the flow of the Euler–Lagrange equations, as in Theorem 7.4.5. As before, identifying the space of solutions $\mathcal{C}_L$ of the Euler–Lagrange equations with the set of initial conditions, which is TQ, we regard

$$\mathfrak{S}_t(v_q) := S(q, \bar{q}, t) := \int_0^t L(F_s(v_q), s)\, ds \tag{8.2.9}$$

as a real-valued function on TQ. Thus, by the chain rule and our previous calculations for $\mathfrak{S}_t$ (see (8.2.7)), equation (8.2.9) gives

$$\frac{\partial S}{\partial t} = \frac{\partial \mathfrak{S}_t}{\partial t} + \mathbf{d}\mathfrak{S}_t \cdot \frac{\partial v}{\partial t}$$

$$= L(F_t(v), t) + (F_t^* \Theta_L)\left(\frac{\partial v}{\partial t}\right) - \Theta_L\left(\frac{\partial v}{\partial t}\right), \tag{8.2.10}$$

where $\partial v / \partial t$ is computed by keeping $\bar{q}$ and q fixed and only changing t. Notice that in (8.2.10), q and $\bar{q}$ are held fixed on both sides of the equation; $\partial S / \partial t$ is a *partial* and *not* a total time-derivative.

Implicitly differentiating the defining condition (8.2.8) with respect to t gives

$$T\pi_Q \cdot X_E(F_t(v)) + T\pi_Q \cdot TF_t \cdot \frac{\partial v}{\partial t} = 0.$$

Thus, since $T\pi_Q \cdot X_E(u) = u$ by the second-order equation property, we get

$$T\pi_Q \cdot TF_t \cdot \frac{\partial v}{\partial t} = -\dot{q},$$

where $(q, \dot{q}) = F_t(v) \in T_q Q$. Thus,

$$(F_t^* \Theta_L)\left(\frac{\partial v}{\partial t}\right) = \frac{\partial L}{\partial \dot{q}^i} \dot{q}^i.$$

Also, since the base point of v does not change with t, $T\pi_Q \cdot (\partial v / \partial t) = 0$, so $\Theta_L(\partial v / \partial t) = 0$. Thus, (8.2.10) becomes

$$\frac{\partial S}{\partial t} = L(q, \dot{q}, t) - \frac{\partial L}{\partial \dot{q}}\dot{q} = -H(q, p, t),$$

where $p = \partial L / \partial \dot{q}$ as usual.

It remains only to show that $\partial S / \partial q = p$. To do this, we differentiate (8.2.8) implicitly with respect to q to give

$$T\pi_Q \cdot TF_t(v) \cdot (T_q v \cdot u) = u. \tag{8.2.11}$$

Then, from (8.2.9) and (8.2.7),

$$T_q S(q, \overline{q}, t) \cdot u = \mathbf{d}\mathfrak{S}_t(v) \cdot (T_q v \cdot u)$$
$$= (F_t^* \Theta_L)(T_q v \cdot u) - \Theta_L(T_q v \cdot u).$$

As in (8.2.10), the last term vanishes, since the base point $\overline{q}$ of v is fixed. Then, letting $p = \mathbb{F}L(F_t(v))$, we get, from the definition of Θ_L and pullback,

$$(F_t^* \Theta_L)(T_q v \cdot u) = \langle p, T\pi_Q \cdot TF_t(v) \cdot (T_q v \cdot u) \rangle = \langle p, u \rangle$$

in view of (8.2.11).   ∎

The fact that $\partial S / \partial q = p$ also follows from the definition of S and the fundamental formula (8.2.4). Just as we derived $p = \partial S / \partial q$, we can derive $\partial S / \partial \overline{q} = -\overline{p}$; in other words, S *is the generating function for the canonical transformation* $(q, p) \mapsto (\overline{q}, \overline{p})$.

Some History of the Euler–Lagrange Equations. In the following paragraphs we make a few historical remarks concerning the Euler–Lagrange equations.[4] Naturally, much of the story focuses on Lagrange. Section V of Lagrange's *Mécanique Analytique* [1788] contains the equations of motion in Euler–Lagrange form (8.1.3). Lagrange writes $Z = T - V$ for what we would call the Lagrangian today. In the previous section Lagrange came to these equations by asking for a coordinate-invariant expression for mass times acceleration. His conclusion is that it is given (in abbreviated notation) by $(d/dt)(\partial T / \partial v) - \partial T / \partial q$, which transforms under arbitrary substitutions of position variables as a one-form. Lagrange does *not* recognize the equations of motion as being equivalent to the variational principle

$$\delta \int L \, dt = 0.$$

This was observed only a few decades later by Hamilton [1834]. The peculiar fact about this is that Lagrange *did* know the general form of the differential equations for variational problems, and he actually had commented on

---

[4]Many of these interesting historical points were conveyed to us by Hans Duistermaat, to whom we are very grateful. The reader can also profitably consult some of the standard texts such as those of Whittaker [1927], Wintner [1941], and Lanczos [1949] for additional interesting historical information.

Euler's proof of this—his early work on this in 1759 was admired very much by Euler. He immediately applied it to give a proof of the Maupertuis principle of least action, as a consequence of Newton's equations of motion. This principle, apparently having its roots in the early work of Leibnitz, is a less natural principle in the sense that the curves are varied only over those that have a constant energy. It is also Hamilton's principle that applies in the *time-dependent* case, when H is *not* conserved and that also generalizes to allow for certain external forces as well.

This discussion in the *Mécanique Analytique* precedes the equations of motion in general coordinates, and so is written in the case that the kinetic energy is of the form $\sum_i m_i v_i^2$, where the m_i are positive constants. Wintner [1941] is also amazed by the fact that the more complicated Maupertuis principle precedes Hamilton's principle. One possible explanation is that Lagrange did not consider L as an interesting physical quantity—for him it was only a convenient function for writing down the equations of motion in a coordinate-invariant fashion. The time span between his work on variational calculus and the *Mécanique Analytique* (1788, 1808) could also be part of the explanation—he may not have been thinking of the variational calculus when he addressed the question of a coordinate-invariant formulation of the equations of motion.

Section V starts by discussing the evident fact that the position and velocity at time t depend on the initial position and velocity, which can be chosen freely. We might write this as (suppressing the coordinate indices for simplicity) $q = q(t, q_0, v_0)$, $v = v(t, q_0, v_0)$, and in modern terminology we would talk about the flow in $x = (q, v)$-space. One problem in reading Lagrange is that he does not explicitly write the variables on which his quantities depend. In any case, he then makes an infinitesimal variation in the initial condition and looks at the corresponding variations of position and velocity at time t. In our notation, $\delta x = (\partial x / \partial x_0)(t, x_0)\delta x_0$. We would say that he considers the tangent mapping of the flow on the tangent bundle of $X = TQ$. Now comes the first interesting result. He makes two such variations, one denoted by δx and the other by Δx, and he writes down a bilinear form $\omega(\delta x, \Delta x)$, in which we recognize ω as the pull-back of the canonical symplectic form on the cotangent bundle of Q, by means of the fiber derivative $\mathbb{F}L$. What he then shows is that this symplectic product is constant as a function of t. This is nothing other than the *invariance of the symplectic form ω under the flow in TQ*.

It is striking that Lagrange obtains the invariance of the symplectic form in TQ and not in T^*Q just as we do in the text where this is derived from Hamilton's principle. In fact, Lagrange does *not* look at the equations of motion in the cotangent bundle via the transformation $\mathbb{F}L$; again it is Hamilton who observes that these take the canonical Hamiltonian form. This is retrospectively puzzling, since later on in Section V, Lagrange states very explicitly that it is useful to pass to the (q, p)-coordinates by means of the coordinate transformation $\mathbb{F}L$, and one even sees written down a system of ordinary differential equations *in Hamiltonian form*, but with the total energy function H replaced by some other mysterious function $-\Omega$. Lagrange does use the letter H for the constant value of energy, apparently in honor of Huygens. He also knew about the conservation of momentum as a result of translational symmetry.

The part where he does this deals with the case in which he perturbs the system by perturbing the potential from $V(q)$ to $V(q) - \Omega(q)$, leaving the kinetic energy unchanged. To this perturbation problem he applies his famous method of variation of constants, which is presented here in a truly nonlinear framework! In our notation, he keeps $t \mapsto x(t, x_0)$ as the solution of the unperturbed system, and then looks at the differential equations for $x_0(t)$ that make $t \mapsto x(t, x_0(t))$ a solution of the perturbed system. The result is that if V is the vector field of the unperturbed system and $V + W$ is the vector field of the perturbed system, then

$$\frac{dx_0}{dt} = ((e^{tV})^* W)(x_0).$$

In words, $x_0(t)$ is the solution of the time-dependent system, the vector field of which is obtained by pulling back W by means of the flow of V after time t. In the case that Lagrange considers, the dq/dt-component of the perturbation is equal to zero, and the dp/dt-component is equal to $\partial\Omega/\partial q$. Thus, it is obviously in a Hamiltonian form; here one does not use anything about Legendre transformations (which Lagrange does not seem to know). But Lagrange knows already that the flow of the unperturbed system preserves the symplectic form, and he shows that the pull-back of his W under such a transformation is a vector field in Hamiltonian form. Actually, this is a time-dependent vector field, defined by the function

$$G(t, q_0, p_0) = -\Omega(q(t, q_0, p_0)).$$

A potential point of confusion is that Lagrange denotes this by $-\Omega$ and writes down expressions like $d\Omega/dp$, and one might first think that these are zero because Ω was assumed to depend only on q. Lagrange presumably means that

$$\frac{dq_0}{dt} = \frac{\partial G}{\partial p_0}, \qquad \frac{dp_0}{dt} = -\frac{\partial G}{\partial q_0}.$$

Most classical textbooks on mechanics, for example Routh [1877, 1884], correctly point out that Lagrange has the invariance of the symplectic form in (q, v) coordinates (rather than in the canonical (q, p) coordinates). Less attention is usually paid to the variation of constants equation in Hamiltonian form, but it must have been generally known that Lagrange derived these—see, for example, Weinstein [1981]. In fact, we should point out that the whole question of linearizing the Euler–Lagrange and Hamilton equations and retaining the mechanical structure is remarkably subtle (see Marsden, Ratiu, and Raugel [1991], for example).

Lagrange continues by introducing the *Poisson brackets* for arbitrary functions, arguing that these are useful in writing the time-derivative of arbitrary functions of arbitrary variables, along solutions of systems in Hamiltonian form. He also continues by saying that if Ω is small, then $x_0(t)$ in zero-order approximation is a constant, and he obtains the next-order approximation by an integration over t; here Lagrange introduces the first steps of the so-called *method of averaging*. When Lagrange discovered (in 1808) the invariance of the symplectic form, the variations-of-constants equations in Hamiltonian form, and the Poisson brackets, he was already 73 years old. It is quite probable that Lagrange generously gave some of these bracket ideas to Poisson at this time. In any case, it is clear that Lagrange had a surprisingly large part of the symplectic picture of classical mechanics.

Exercises

⬦ **8.2-1.**   Derive the Hamilton–Jacobi equation starting with the phase space version of Hamilton's principle.

8.3 Constrained Systems

We begin this section with the Lagrange multiplier theorem for purposes of studying constrained dynamics.

The Lagrange Multiplier Theorem.   We state the theorem with a sketch of the proof, referring to Abraham, Marsden, and Ratiu [1988] for details. We shall not be absolutely precise about the technicalities (such as how to interpret dual spaces).

First, consider the case of functions defined on linear spaces. Let V and Λ be Banach spaces and let $\varphi : V \to \Lambda$ be a smooth map. Suppose 0 is a regular value of φ, so that $C := \varphi^{-1}(0)$ is a submanifold. Let $h : V \to \mathbb{R}$ be a smooth function and define $\overline{h} : V \times \Lambda^* \to \mathbb{R}$ by

$$\overline{h}(x, \lambda) = h(x) - \langle \lambda, \varphi(x) \rangle. \tag{8.3.1}$$

Theorem 8.3.1 (Lagrange Multiplier Theorem for Linear Spaces).   *The following are equivalent conditions on $x_0 \in C$:*

(i) *x_0 is a critical point of $h|C$; and*

(ii) *there is a $\lambda_0 \in \Lambda^*$ such that (x_0, λ_0) is a critical point of $\overline{h}$.*

Sketch of Proof.   Since

$$\mathbf{D}\overline{h}(x_0, \lambda_0) \cdot (x, \lambda) = \mathbf{D}h(x_0) \cdot x - \langle \lambda_0, \mathbf{D}\varphi(x_0) \cdot x \rangle - \langle \lambda, \varphi(x_0) \rangle$$

and $\varphi(x_0) = 0$, the condition $\mathbf{D}\bar{h}(x_0, \lambda_0) \cdot (x, \lambda) = 0$ is equivalent to

$$\mathbf{D}h(x_0) \cdot x = \langle \lambda_0, \mathbf{D}\varphi(x_0) \cdot x \rangle \qquad (8.3.2)$$

for all $x \in V$ and $\lambda \in \Lambda^*$. The tangent space to C at x_0 is $\ker \mathbf{D}\varphi(x_0)$, so (8.3.2) implies that $h|C$ has a critical point at x_0.

Conversely, if $h|C$ has a critical point at x_0, then $\mathbf{D}h(x_0) \cdot x = 0$ for all x satisfying $\mathbf{D}\varphi(x_0) \cdot x = 0$. By the implicit function theorem, there is a smooth coordinate change that straightens out C; that is, it allows us to assume that $V = W \oplus \Lambda$, $x_0 = 0$, C is (in a neighborhood of 0) equal to W, and φ (in a neighborhood of the origin) is the projection to Λ. With these simplifications, condition (i) means that the first partial derivative of h vanishes. We choose λ_0 to be $\mathbf{D}_2 h(x_0)$ regarded as an element of Λ^*; then (8.3.2) clearly holds. ∎

The Lagrange multiplier theorem is a convenient test for constrained critical points, as we know from calculus. It also leads to a convenient test for constrained maxima and minima. For instance, to test for a minimum, let $\alpha > 0$ be a constant, let (x_0, λ_0) be a critical point of $\bar{h}$, and consider

$$h_\alpha(x, \lambda) = h(x) - \langle \lambda, \varphi(x) \rangle + \alpha \|\lambda - \lambda_0\|^2, \qquad (8.3.3)$$

which also has a critical point at (x_0, λ_0). Clearly, if h_α has a minimum at (x_0, λ_0), then $h|C$ has a minimum at x_0. This observation is convenient, since one can use the unconstrained second derivative test on h_α, which leads to the theory of **bordered Hessians**. (For an elementary discussion, see Marsden and Tromba [1996, p. 220ff].)

A second remark concerns the generalization of the Lagrange multiplier theorem to the case where V is a manifold but h is still real-valued. Such a context is as follows. Let M be a manifold and let $N \subset M$ be a submanifold. Suppose $\pi : E \to M$ is a vector bundle over M and φ is a section of E that is transverse to fibers. Assume $N = \varphi^{-1}(0)$.

Theorem 8.3.2 (Lagrange Multiplier Theorem for Manifolds).   *The following are equivalent for $x_0 \in N$ and $h : M \to \mathbb{R}$ smooth:*

(i) *x_0 is a critical point of $h|N$; and*

(ii) *there is a section λ_0 of the dual bundle E^* such that $\lambda_0(x_0)$ is a critical point of $\bar{h} : E^* \to \mathbb{R}$ defined by*

$$\bar{h}(\lambda_x) = h(x) - \langle \lambda_x, \varphi(x) \rangle. \qquad (8.3.4)$$

In (8.3.4), λ_x denotes an arbitrary element of E_x^*. We leave it to the reader to adapt the proof of the previous theorem to this situation.

Holonomic Constraints.   Many mechanical systems are obtained from higher-dimensional ones by adding constraints. Rigidity in rigid-body mechanics and incompressibility in fluid mechanics are two such examples, while constraining a free particle to move on a sphere is another.

Typically, constraints are of two types. Holonomic constraints are those imposed on the configuration space of a system, such as those mentioned in the preceding paragraph. Others, such as *rolling constraints*, involve the conditions on the velocities and are termed *nonholonomic*.

A *holonomic constraint* can be defined for our purposes as the specification of a submanifold $N \subset Q$ of a given configuration manifold Q. (More generally, a holonomic constraint is an integrable subbundle of TQ.) Since we have the natural inclusion $TN \subset TQ$, a given Lagrangian $L : TQ \to \mathbb{R}$ can be restricted to TN to give a Lagrangian L_N. We now have two Lagrangian systems, namely those associated to L and to L_N, assuming that both are regular. We now relate the associated variational principles and the Hamiltonian vector fields.

Suppose that $N = \varphi^{-1}(0)$ for a section $\varphi : Q \to E^*$, the dual of a vector bundle E over Q. The variational principle for L_N can be phrased as

$$\delta \int L_N(q, \dot{q}) \, dt = 0, \qquad (8.3.5)$$

where the variation is over curves with fixed endpoints and subject to the constraint $\varphi(q(t)) = 0$. By the Lagrange multiplier theorem, (8.3.5) is equivalent to

$$\delta \int [L(q(t), \dot{q}(t)) - \langle \lambda(q(t), t), \varphi(q(t)) \rangle] \, dt = 0 \qquad (8.3.6)$$

for some function $\lambda(q, t)$ taking values in the bundle E and where the variation is over curves q in Q and curves λ in E.[5] In coordinates, (8.3.6) reads

$$\delta \int [L(q^i, \dot{q}^i) - \lambda^a(q^i, t) \varphi_a(q^i)] \, dt = 0. \qquad (8.3.7)$$

The corresponding Euler–Lagrange equations in the variables q^i, λ^a are

$$\frac{d}{dt} \frac{\partial L}{\partial \dot{q}^i} = \frac{\partial L}{\partial q^i} - \lambda^a \frac{\partial \varphi_a}{\partial q^i} \qquad (8.3.8)$$

and

$$\varphi_a = 0. \qquad (8.3.9)$$

---

[5]This conclusion assumes some regularity in t on the Lagrange multiplier λ. One can check (after the fact) that this assumption is justified by relating λ to the forces of constraint, as in the next theorem.

They are viewed as equations in the unknowns $q^i(t)$ and $\lambda^a(q^i, t)$; if E is a trivial bundle, we can take λ to be a function only of t.[6]

We summarize these findings as follows.

Theorem 8.3.3. *The Euler–Lagrange equations for L_N on the manifold $N \subset Q$ are equivalent to the equations* (8.3.8) *together with the constraints* $\varphi = 0$.

We interpret the term $-\lambda^a \partial \varphi_a / \partial q^i$ as the *force of constraint*, since it is the force that is added to the Euler–Lagrange operator (see §7.8) in the *unconstrained space* in order to maintain the constraints. In the next section we will develop the geometric interpretation of these forces of constraint.

Notice that $\mathcal{L} = L - \lambda^a \varphi_a$ as a Lagrangian in q, and λ is degenerate in λ; that is, the time-derivative of λ does not appear, so its conjugate momentum π_a is constrained to be zero. Regarding $\mathcal{L}$ as defined on TE, the corresponding Hamiltonian on T^*E is formally

$$\mathcal{H}(q, p, \lambda, \pi) = H(q, p) + \lambda^a \varphi_a, \tag{8.3.10}$$

where H is the Hamiltonian corresponding to L.

One has to be a little careful in interpreting Hamilton's equations, because $\mathcal{L}$ is degenerate; the general theory appropriate for this situation is the *Dirac theory of constraints*, which we discuss in §8.5. However, in the present context this theory is quite simple and proceeds as follows. One calls $C \subset T^*E$ defined by $\pi_a = 0$ the **primary constraint set**; it is the image of the Legendre transform, provided that the original L was regular. The canonical form Ω is pulled back to C to give a presymplectic form (a closed but possibly degenerate two-form) Ω_C, and one seeks $X_{\mathcal{H}}$ such that

$$\mathbf{i}_{X_{\mathcal{H}}} \Omega_C = \mathbf{d}\mathcal{H}. \tag{8.3.11}$$

In this case, the degeneracy of Ω_C gives no equation for λ; that is, the evolution of λ is indeterminate. The other Hamiltonian equations are equivalent to (8.3.8) and (8.3.9), so in this sense the Lagrangian and Hamiltonian pictures are still equivalent.

Exercises

⋄ **8.3-1.** Write out the second derivative of h_α at (x_0, λ_0) and relate your answer to the bordered Hessian.

⋄ **8.3-2.** Derive the equations for a simple pendulum using the Lagrange multiplier method and compare them with those obtained using generalized coordinates.

---

[6]The combination $\mathcal{L} = L - \lambda^a \varphi_a$ is related to the Routhian construction for a Lagrangian with cyclic variables; see §8.9.

◊ **8.3-3** (Neumann [1859]).

(a) Derive the equations of motion of a particle of unit mass on the sphere S^{n-1} under the influence of a quadratic potential $A\mathbf{q} \cdot \mathbf{q}$, $\mathbf{q} \in \mathbb{R}^n$, where A is a fixed real diagonal matrix.

(b) Form the matrices $X = (q^i q^j)$, $P = (\dot{q}^i q^j - q^j \dot{q}^i)$. Show that the system in (a) is equivalent to $\dot{X} = [P, X]$, $\dot{P} = [X, A]$. (This was observed first by K. Uhlenbeck.) Equivalently, show that

$$(-X + P\lambda + A\lambda^2)^{\cdot} = [-X + P\lambda + A\lambda^2, -P - A\lambda].$$

(c) Verify that

$$E(X, P) = -\frac{1}{4} \operatorname{trace}(P^2) + \frac{1}{2} \operatorname{trace}(AX)$$

is the total energy of this system.

(d) Verify that for $k = 1, \ldots, n-1$,

$$f_k(X, P) = \frac{1}{2(k+1)} \operatorname{trace}\left(-\sum_{i=0}^{k} A^i X A^{k-i} + \sum_{\substack{i+j+l=k-1 \\ i,j,l \geq 0}} A^i P A^j P A^l\right),$$

are conserved on the flow of the C. Neumann problem (Ratiu [1981b]).

8.4   Constrained Motion in a Potential Field

We saw in the preceding section how to write the equations for a constrained system in terms of variables on the containing space. We continue this line of investigation here by specializing to the case of motion in a potential field. In fact, we shall determine by geometric methods the extra terms that need to be added to the Euler–Lagrange equations, that is, the forces of constraint, to ensure that the constraints are maintained.

Let Q be a (weak) Riemannian manifold and let $N \subset Q$ be a submanifold. Let

$$\mathbb{P} : (TQ)|N \to TN \tag{8.4.1}$$

be the orthogonal projection of TQ to TN defined pointwise on N.

Consider a Lagrangian $L : TQ \to \mathbb{R}$ of the form $L = K - V \circ \tau_Q$, that is, kinetic minus potential energy. The Riemannian metric associated to the kinetic energy is denoted by $\langle\!\langle \, , \rangle\!\rangle$. The restriction $L_N = L|TN$ is also of the form kinetic minus potential, using the metric induced on N and the potential $V_N = V|N$. We know from §7.7 that if E_N is the energy of L_N, then

$$X_{E_N} = S_N - \text{ver}(\nabla V_N), \qquad (8.4.2)$$

where S_N is the spray of the metric on N and ver( ) denotes vertical lift. Recall that integral curves of (8.4.2) are solutions of the Euler–Lagrange equations. Let S be the geodesic spray on Q.

First notice that ∇V_N and ∇V are related in a very simple way: For $q \in N$,

$$\nabla V_N(q) = \mathbb{P} \cdot [\nabla V(q)].$$

Thus, the main complication is in the geodesic spray.

Proposition 8.4.1. $S_N = T\mathbb{P} \circ S$ at points of TN.

Proof. For the purpose of this proof we can ignore the potential and let $L = K$. Let $R = TQ|N$, so that $\mathbb{P} : R \to TN$ and therefore

$$T\mathbb{P} : TR \to T(TN), \quad S : R \to T(TQ), \quad \text{and} \quad T\tau_Q \circ S = \text{identity},$$

since S is second-order. But

$$TR = \{\, w \in T(TQ) \mid T\tau_Q(w) \in TN \,\},$$

so $S(TN) \subset TR$, and hence $T\mathbb{P} \circ S$ makes sense at points of TN.

If $v \in TQ$ and $w \in T_v(TQ)$, then $\Theta_L(v) \cdot w = \langle\!\langle v, T_v\tau_Q(w)\rangle\!\rangle$. Letting $i : R \to TQ$ be the inclusion, we claim that

$$\mathbb{P}^*\Theta_{L|TN} = i^*\Theta_L. \qquad (8.4.3)$$

Indeed, for $v \in R$ and $w \in T_vR$, the definition of pull-back gives

$$\mathbb{P}^*\Theta_{L|TN}(v) \cdot w = \langle\!\langle \mathbb{P}v, (T\tau_Q \circ T\mathbb{P})(w)\rangle\!\rangle = \langle\!\langle \mathbb{P}v, T(\tau_Q \circ \mathbb{P})(w)\rangle\!\rangle. \qquad (8.4.4)$$

Since on R, $\tau_Q \circ \mathbb{P} = \tau_Q$, $\mathbb{P}^* = \mathbb{P}$, and $w \in T_vR$, (8.4.4) becomes

$$\mathbb{P}^*\Theta_{L|TN}(v) \cdot w = \langle\!\langle \mathbb{P}v, T\tau_Q(w)\rangle\!\rangle = \langle\!\langle v, \mathbb{P}T\tau_Q(w)\rangle\!\rangle = \langle\!\langle v, T\tau_Q(w)\rangle\!\rangle$$
$$= \Theta_L(v) \cdot w = (i^*\Theta_L)(v) \cdot w.$$

Taking the exterior derivative of (8.4.3) gives

$$\mathbb{P}^*\Omega_{L|TN} = i^*\Omega_L. \qquad (8.4.5)$$

In particular, for $v \in TN$, $w \in T_vR$, and $z \in T_v(TN)$, the definition of pull-back and (8.4.5) give

$$\Omega_L(v)(w, z) = (i^*\Omega_L)(v)(w, z) = (\mathbb{P}^*\Omega_{L|TN})(v)(w, z)$$
$$= \Omega_{L|TN}(\mathbb{P}v)(T\mathbb{P}(w), T\mathbb{P}(z))$$
$$= \Omega_{L|TN}(v)(T\mathbb{P}(w), z). \qquad (8.4.6)$$

But

$$\mathbf{d}E(v) \cdot z = \Omega_L(v)(S(v), z) = \Omega_{L|TN}(v)(S_N(v), z),$$

since S and S_N are Hamiltonian vector fields for E and $E|TN$, respectively. From (8.4.6),

$$\Omega_{L|TN}(v)(T\mathbb{P}(S(v)), z) = \Omega_L(v)(S(v), z) = \Omega_{L|TN}(v)(S_N(v), z),$$

so by weak nondegeneracy of $\Omega_{L|TN}$ we get the desired relation

$$S_N = T\mathbb{P} \circ S.$$ ∎

Corollary 8.4.2.   *For $v \in T_q N$:*

(i) *$(S - S_N)(v)$ is the vertical lift of a vector $Z(v) \in T_q Q$ relative to v;*

(ii) *$Z(v) \perp T_q N$; and*

(iii) *$Z(v) = -\nabla_v v + \mathbb{P}(\nabla_v v)$ is minus the normal component of $\nabla_v v$, where in $\nabla_v v$, v is extended to a vector field on Q tangent to N.*

Proof.   (i) Since $T\tau_Q(S(v)) = v = T\tau_Q(S_N(v))$, we have

$$T\tau_Q(S - S_N)(v) = 0,$$

that is, $(S - S_N)(v)$ is vertical. The statement now follows from the comments following Definition 7.7.1.

(ii) For $u \in T_q Q$, we have $T\mathbb{P} \cdot \mathrm{ver}(u, v) = \mathrm{ver}(\mathbb{P}u, v)$, since

$$\mathrm{ver}(\mathbb{P}u, v) = \left.\frac{d}{dt}(v + t\mathbb{P}u)\right|_{t=0} = \left.\frac{d}{dt}\mathbb{P}(v + tu)\right|_{t=0}$$
$$= T\mathbb{P} \cdot \mathrm{ver}(u, v). \qquad (8.4.7)$$

By Part (i), $S(v) - S_N(v) = \mathrm{ver}(Z(v), v)$ for some $Z(v) \in T_q Q$, so that using the previous theorem, (8.4.7), and $\mathbb{P} \circ \mathbb{P} = \mathbb{P}$, we get

$$\mathrm{ver}(\mathbb{P}Z(v), v) = T\mathbb{P} \cdot \mathrm{ver}(Z(v), v)$$
$$= T\mathbb{P}(S(v) - S_N(v))$$
$$= T\mathbb{P}(S(v) - T\mathbb{P} \circ S(v)) = 0.$$

Therefore, $\mathbb{P}Z(v) = 0$, that is, $Z(v) \perp T_q N$.

(iii) Let $v(t)$ be a curve of tangents to N; $v(t) = \dot{c}(t)$, where $c(t) \in N$. Then in a chart,

$$S(c(t), v(t)) = \big(c(t), v(t), v(t), \gamma_{c(t)}(v(t), v(t))\big)$$

by (7.5.5). Extending $v(t)$ to a vector field v on Q tangent to N we get, in a standard chart,

$$\nabla_v v = -\gamma_c(v, v) + \mathbf{D}v(c) \cdot v = -\gamma_c(v, v) + \frac{dv}{dt}$$

by (7.5.19), so on TN,

$$S(v) = \frac{dv}{dt} - \text{ver}(\nabla_v v, v).$$

Since $dv/dt \in TN$, (8.4.7) and the previous proposition give

$$S_N(v) = T\mathbb{P}\frac{dv}{dt} - \text{ver}(\mathbb{P}(\nabla_v v), v) = \frac{dv}{dt} - \text{ver}(\mathbb{P}(\nabla_v v), v).$$

Thus, by part (i),

$$\text{ver}(Z(v), v) = S(v) - S_N(v) = \text{ver}(-\nabla_v v + \mathbb{P}\nabla_v v, v). \qquad \blacksquare$$

The map $Z : TN \to TQ$ is called the **force of constraint**. We shall prove below that if the codimension of N in Q is one, then

$$Z(v) = -\nabla_v v + \mathbb{P}(\nabla_v v) = -\langle \nabla_v v, n \rangle n,$$

where n is the unit normal vector field to N in Q, equals the negative of the quadratic form associated to the second fundamental form of N in Q, a result due to Gauss. (We shall define the second fundamental form, which measures how "curved" N is within Q, shortly.) It is not obvious at first that the expression $\mathbb{P}(\nabla_v v) - \nabla_v v$ depends only on the pointwise values of v, but this follows from its identification with $Z(v)$.

To prove the above statement, we recall that the Levi-Civita covariant derivative has the property that for vector fields $u, v, w \in \mathfrak{X}(Q)$ the following identity is satisfied:

$$w[\langle u, v \rangle] = \langle \nabla_w u, v \rangle + \langle u, \nabla_w v \rangle, \qquad (8.4.8)$$

as may be easily checked. Assume now that u and v are vector fields tangent to N and n is the unit normal vector field to N in Q. The identity (8.4.8) yields

$$\langle \nabla_v u, n \rangle + \langle u, \nabla_v n \rangle = 0. \qquad (8.4.9)$$

The **second fundamental form** in Riemannian geometry is defined to be the map

$$(u, v) \mapsto -\langle \nabla_u n, v \rangle \qquad (8.4.10)$$

with u, v, n as above. It is a classical result that this bilinear form is symmetric and hence is uniquely determined by polarization from its quadratic form $-\langle \nabla_v n, v \rangle$. In view of equation (8.4.9), this quadratic from has the alternative expression $\langle \nabla_v v, n \rangle$, which, after multiplication by n, equals $-Z(v)$, thereby proving the claim above.

As indicated, this discussion of the second fundamental form is under the assumption that the codimension of N in Q is one—keep in mind that our discussion of forces of constraint requires no such restriction.

As before, interpret $Z(v)$ as the constraining force needed to keep particles in N. Notice that N is totally geodesic (that is, geodesics in N are geodesics in Q) iff $Z = 0$.

Some interesting studies in the problem of showing convergence of solutions in the limit of strong constraining forces are Rubin and Ungar [1957], Ebin [1982], and van Kampen and Lodder [1984].

Exercises

◇ **8.4-1.** Compute the force of constraint Z and the second fundamental form for the sphere of radius R in $\mathbb{R}^3$.

◇ **8.4-2.** Assume that L is a regular Lagrangian on TQ and $N \subset Q$. Let $i : TN \to TQ$ be the embedding obtained from $N \subset Q$ and let Ω_L be the Lagrange two-form on TQ. Show that $i^*\Omega_L$ is the Lagrange two-form $\Omega_{L|TN}$ on TN. Assuming that L is hyperregular, show that the Legendre transform defines a symplectic embedding $T^*N \subset T^*Q$.

◇ **8.4-3.** In $\mathbb{R}^3$, let

$$H(\mathbf{q}, \mathbf{p}) = \frac{1}{2m} \left[\|\mathbf{p}\|^2 - (\mathbf{p} \cdot \mathbf{q})^2 \right] + mgq^3,$$

where $\mathbf{q} = (q^1, q^2, q^3)$. Show that Hamilton's equations in $\mathbb{R}^3$ *automatically* preserve T^*S^2 and give the equations for the spherical pendulum when restricted to this invariant (symplectic) submanifold. (Hint: Use the formulation of Lagrange's equations with constraints in §8.3.)

◇ **8.4-4.** Redo the C. Neumann problem in Exercise 8.3-3 using Corollary 8.4.2 and the interpretation of the constraining force in terms of the second fundamental form.

8.5 Dirac Constraints

If (P, Ω) is a symplectic manifold, a submanifold $S \subset P$ is called a **symplectic submanifold** when $\omega := i^*\Omega$ is a symplectic form on S, $i : S \to P$ being the inclusion. Thus, S inherits a Poisson bracket structure; its relationship to the bracket structure on P is given by a formula of Dirac [1950] that will be derived in this section. Dirac's work was motivated by the study of constrained systems, especially relativistic ones, where one thinks of S as a constraint subspace of phase space (see Gotay, Isenberg, and Marsden [1997] and references therein for more information). Let us work in the finite-dimensional case; the reader is invited to study the intrinsic infinite-dimensional version using Remark 1 below.

Dirac's Formula.   Let $\dim P = 2n$ and $\dim S = 2k$. In a neighborhood of a point z_0 of S, choose coordinates $z^1, \dots, z^{2n}$ on P such that S is given by

$$z^{2k+1} = 0, \dots, z^{2n} = 0,$$

and so $z^1, \dots, z^{2k}$ provide local coordinates for S.

Consider the matrix whose entries are

$$C^{ij}(z) = \{z^i, z^j\}, \quad i, j = 2k+1, \dots, 2n.$$

Assume that the coordinates are chosen such that C^{ij} is an invertible matrix at z_0 and hence in a neighborhood of z_0. (Such coordinates always exist, as is easy to see.) Let the inverse of C^{ij} be denoted by $[C_{ij}(z)]$. Let F be a smooth function on P and $F|S$ its restriction to S. We are interested in relating $X_{F|S}$ and X_F as well as the brackets $\{F, G\}|S$ and $\{F|S, G|S\}$.

Proposition 8.5.1 (Dirac's Bracket Formula).   *In a coordinate neighborhood as described above, and for $z \in S$, we have*

$$X_{F|S}(z) = X_F(z) - \sum_{i,j=2k+1}^{2n} \{F, z^i\} C_{ij}(z) X_{z^j}(z) \tag{8.5.1}$$

and

$$\{F|S, G|S\}(z) = \{F, G\}(z) - \sum_{i,j=2k+1}^{2n} \{F, z^i\} C_{ij}(z) \{z^j, G\}. \tag{8.5.2}$$

Proof.   To verify (8.5.1), we show that the right-hand side satisfies the condition required for $X_{F|S}(z)$, namely that it be a vector field on S and that

$$\omega_z(X_{F|S}(z), v) = \mathbf{d}(F|S)_z \cdot v \tag{8.5.3}$$

for $v \in T_z S$. Since S is symplectic,

$$T_z S \cap (T_z S)^\Omega = \{0\},$$

where $(T_z S)^\Omega$ denotes the Ω-orthogonal complement. Since

$$\dim(T_z S) + \dim(T_z S)^\Omega = 2n,$$

we get

$$T_z P = T_z S \oplus (T_z S)^\Omega. \tag{8.5.4}$$

If $\pi_z : T_z P \to T_z S$ is the associated projection operator, one can verify that

$$X_{F|S}(z) = \pi_z \cdot X_F(z), \tag{8.5.5}$$

so in fact, (8.5.1) is a formula for π_z in coordinates; equivalently,

$$(\mathrm{Id} - \pi_z) X_F(z) = \sum_{i,j=2k+1}^{2n} \{F, z^i\} C_{ij}(z) X_{z^j}(z) \tag{8.5.6}$$

gives the projection to $(T_z S)^\Omega$. To verify (8.5.6), we need to check that the right-hand side

(i) is an element of $(T_z S)^\Omega$;

(ii) equals $X_F(z)$ if $X_F(z) \in (T_z S)^\Omega$; and

(iii) equals 0 if $X_F(z) \in T_z S$.

To prove (i), observe that $X_K(z) \in (T_z S)^\Omega$ means

$$\Omega(X_K(z), v) = 0 \qquad \text{for all } v \in T_z S;$$

that is,

$$\mathbf{d}K(z) \cdot v = 0 \qquad \text{for all } v \in T_z S.$$

But for $K = z^j$, $j = 2k+1, \dots, 2n$, $K \equiv 0$ on S, and hence $\mathbf{d}K(z) \cdot v = 0$. Thus, $X_{z^j}(z) \in (T_z S)^\Omega$, so (i) holds.

For (ii), if $X_F(z) \in (T_z S)^\Omega$, then

$$\mathbf{d}F(z) \cdot v = 0 \qquad \text{for all } v \in T_z S$$

and, in particular, for $v = \partial/\partial z^i$, $i = 1, \dots, 2k$. Therefore, for $z \in S$, we can write

$$\mathbf{d}F(z) = \sum_{j=2k+1}^{2n} a_j \, dz^j \tag{8.5.7}$$

and hence

$$X_F(z) = \sum_{j=2k+1}^{2n} a_j X_{z^j}(z). \tag{8.5.8}$$

The a_j are determined by pairing (8.5.8) with dz^i, $i = 2k+1, \ldots, 2n$, to give

$$-\langle dz^i, X_F(z) \rangle = \{F, z^i\} = \sum_{j=2k+1}^{2n} a_j \{z^j, z^i\} = \sum_{j=2k+1}^{2n} a_j C^{ji},$$

or

$$a_j = \sum_{i=2k+1}^{2n} \{F, z^i\} C_{ij}, \qquad (8.5.9)$$

which proves (ii). Finally, for (iii), $X_F(z) \in T_z S = ((T_z S)^\Omega)^\Omega$ means that $X_F(z)$ is Ω orthogonal to each X_{z^j}, $j = 2k+1, \ldots, 2n$. Thus, $\{F, z^j\} = 0$, so the right-hand side of (8.5.6) vanishes.

Formula (8.5.6) is therefore proved, and so, equivalently, (8.5.1) holds. Formula (8.5.2) follows by writing $\{F|S, G|S\} = \omega(X_{F|S}, X_{G|S})$ and substituting (8.5.1). In doing this, the last two terms cancel.   ∎

In (8.5.2) notice that $\{F|S, G|S\}(z)$ is intrinsic to $F|S$, $G|S$, and S. The bracket does not depend on how $F|S$ and $G|S$ are extended off S to functions F, G on P. This is not true for just $\{F, G\}(z)$, which *does* depend on the extensions, but the extra term in (8.5.2) cancels this dependence.

Remarks.

1.   A coordinate-free way to write (8.5.2) is as follows. Write $S = \psi^{-1}(m_0)$, where $\psi : P \to M$ is a submersion on S. For $z \in S$ and $m = \psi(z)$, let

$$C_m : T_m^* M \times T_m^* M \to \mathbb{R} \qquad (8.5.10)$$

be given by

$$C_m(dF_m, dG_m) = \{F \circ \psi, G \circ \psi\}(z) \qquad (8.5.11)$$

for $F, G \in \mathcal{F}(M)$. Assume that C_m is invertible, with "inverse"

$$C_m^{-1} : T_m M \times T_m M \to \mathbb{R}.$$

Then

$$\{F|S, G|S\}(z) - \{F, G\}(z) - C_m^{-1}(T_z\psi \cdot X_F(z), T_z\psi \cdot X_G(z)). \qquad (8.5.12)$$

2.   There is another way to derive and write Dirac's formula, using complex structures. Suppose $\langle\!\langle\,,\,\rangle\!\rangle_z$ is an inner product on $T_z P$ and

$$\mathbb{J}_z : T_z P \to T_z P$$

is an orthogonal transformation satisfying $\mathbb{J}_z^2 = -\,\text{Identity}$ and, as in §5.3,

$$\Omega_z(u, v) = \langle\!\langle \mathbb{J}_z u, v \rangle\!\rangle \tag{8.5.13}$$

for all $u, v \in T_z P$. With the inclusion $i : S \to P$ as before, we get corresponding structures induced on S; let

$$\omega = i^* \Omega. \tag{8.5.14}$$

If ω is nondegenerate, then (8.5.14) and the induced metric define an associated complex structure $\mathbb{K}$ on S. At a point $z \in S$, suppose one has arranged to choose $\mathbb{J}_z$ to map $T_z S$ to itself, and that $\mathbb{K}_z$ is the restriction of $\mathbb{J}_z$ to $T_z S$. At z, we then get

$$(T_z S)^\perp = (T_z S)^\Omega,$$

and thus symplectic projection coincides with orthogonal projection. From (8.5.5), and using coordinates as described earlier, but for which the $X_{z^j}(z)$ are also orthogonal, we get

$$X_{F|S}(z) = X_F(z) - \sum_{j=2k+1}^{2n} \langle X_F(z), X_{z^j}(z) \rangle X_{z^j}(z)$$

$$= X_F(z) + \sum_{j=2k+1}^{2n} \Omega(X_F(z), \mathbb{J}^{-1} X_{z^j}(z)) X_{z^j}. \tag{8.5.15}$$

This is equivalent to (8.5.1) and so also gives (8.5.2); to see this, one shows that

$$\mathbb{J}^{-1} X_{z^j}(z) = - \sum_{i=2k+1}^{2n} X_{z^i}(z) C_{ij}(z). \tag{8.5.16}$$

Indeed, the symplectic pairing of each side with X_{z^p} gives δ_j^p.

3.   For a relationship between Poisson reduction and Dirac's formula, see Marsden and Ratiu [1986].

Examples

(a) Holonomic Constraints.   To treat *holonomic constraints* by the Dirac formula, proceed as follows. Let $N \subset Q$ be as in §8.4, so that

$TN \subset TQ$; with $i : N \rightarrow Q$ the inclusion, one obtains $(Ti)^* \Theta_L = \Theta_{L_N}$ by considering the following commutative diagram:

$$
\begin{array}{ccc}
TN & \xrightarrow{\;\;Ti\;\;} & TQ|N \\
{\scriptstyle FL_N}\downarrow & & \downarrow{\scriptstyle FL} \\
T^*N & \xleftarrow{\quad\quad} & T^*Q|N \\
& \text{projection} &
\end{array}
$$

This realizes TN as a symplectic submanifold of TQ, and so Dirac's formula can be applied, reproducing (8.4.2). See Exercise 8.4-2. ◆

(b) KdV Equation. Suppose[7] one starts with a Lagrangian of the form

$$ L(v_q) = \langle \alpha(q), v \rangle - h(q), \tag{8.5.17} $$

where α is a one-form on Q, and h is a function on Q. In coordinates, (8.5.17) reads

$$ L(q^i, \dot{q}^i) = \alpha_i(q)\dot{q}^i - h(q^i). \tag{8.5.18} $$

The corresponding momenta are

$$ p_i = \frac{\partial L}{\partial \dot{q}^i} = \alpha_i, \quad \text{i.e.}, \quad p = \alpha(q), \tag{8.5.19} $$

while the Euler–Lagrange equations are

$$ \frac{d}{dt}(\alpha_i(q^j)) = \frac{\partial L}{\partial q^i} = \frac{\partial \alpha_j}{\partial q^i}\dot{q}^j - \frac{\partial h}{\partial q^i}, $$

that is,

$$ \frac{\partial \alpha_i}{\partial q^j}\dot{q}^j - \frac{\partial \alpha_j}{\partial q^i}\dot{q}^j = -\frac{\partial h}{\partial q^i}. \tag{8.5.20} $$

In other words, with $v^i = \dot{q}^i$,

$$ i_v d\alpha = -dh. \tag{8.5.21} $$

If $d\alpha$ is nondegenerate on Q, then (8.5.21) defines Hamilton's equations for a vector field v on Q with Hamiltonian h and symplectic form $\Omega_\alpha = -d\alpha$.

This collapse, or reduction, from TQ to Q is another instance of the Dirac theory and how it deals with degenerate Lagrangians in attempting

---

[7]We thank P. Morrison and M. Gotay for the following comment on how to view the KdV equation using constraints; see Gotay [1988].

to form the corresponding Hamiltonian system. Here the primary constraint manifold is the graph of α. Note that if we form the Hamiltonian on the primaries, then

$$H = p_i \dot{q}^i - L = \alpha_i \dot{q}^i - \alpha_i \dot{q}^i + h(q) = h(q), \qquad (8.5.22)$$

that is, $H = h$, as expected from (8.5.21).

To put the KdV equation $u_t + 6uu_x + u_{xxx} = 0$ in this context, let $u = \psi_x$; that is, ψ is an indefinite integral for u. Observe that the KdV equation is the Euler–Lagrange equation for

$$L(\psi, \psi_t) = \int \left[\tfrac{1}{2} \psi_t \psi_x + \psi_x^3 - \tfrac{1}{2}(\psi_{xx})^2 \right] dx, \qquad (8.5.23)$$

that is, $\delta \int L\,dt = 0$ gives $\psi_{xt} + 6\psi_x \psi_{xx} + \psi_{xxxx} = 0$, which is the KdV equation for u. Here α is given by

$$\langle \alpha(\psi), \varphi \rangle = \tfrac{1}{2} \int \psi_x \varphi \, dx, \qquad (8.5.24)$$

and so by formula 6 in the table in §4.4,

$$-d\alpha(\psi)(\psi_1, \psi_2) = \tfrac{1}{2} \int (\psi_1 \psi_{2x} - \psi_2 \psi_{1x}) \, dx, \qquad (8.5.25)$$

which equals the KdV symplectic structure (3.2.9). Moreover, (8.5.22) gives the Hamiltonian

$$H = \int \left[\tfrac{1}{2}(\psi_{xx})^2 - \psi_x^3 \right] dx = \int \left[\tfrac{1}{2}(u_x)^2 - u^3 \right] dx, \qquad (8.5.26)$$

also coinciding with Example (c) of §3.2.   ♦

Exercises

◇ **8.5-1.** Derive formula (8.4.2) from (8.5.1).

◇ **8.5-2.** Work out Dirac's formula for

(a) $T^* S^1 \subset T^* \mathbb{R}^2$; and

(b) $T^* S^2 \subset T^* \mathbb{R}^3$.

In each case, note that the embedding makes use of the metric. Reconcile your analysis with what you found in Exercise 8.4-2.

8.6   Centrifugal and Coriolis Forces

In this section we discuss, in an elementary way, the basic ideas of centrifugal and Coriolis forces. This section takes the view of rotating *observers*, while the next sections take the view of rotating *systems*.

Rotating Frames. Let V be a three-dimensional oriented inner product space that we regard as "inertial space." Let ψ_t be a curve in $\mathrm{SO}(V)$, the group of orientation-preserving orthogonal linear transformations of V to V, and let X_t be the (possibly time-dependent) vector field generating ψ_t; that is,

$$X_t(\psi_t(\mathbf{v})) = \frac{d}{dt}\psi_t(\mathbf{v}), \tag{8.6.1}$$

or, equivalently,

$$X_t(\mathbf{v}) = (\dot{\psi}_t \circ \psi_t^{-1})(\mathbf{v}). \tag{8.6.2}$$

Differentiation of the orthogonality condition $\psi_t \cdot \psi_t^T = \mathrm{Id}$ shows that X_t is skew-symmetric.

A vector $\boldsymbol{\omega}$ in three-space defines a skew-symmetric 3×3 linear transformation $\hat{\omega}$ using the cross product; specifically, it is defined by the equation

$$\hat{\omega}(\mathbf{v}) = \boldsymbol{\omega} \times \mathbf{v}.$$

Conversely, any skew matrix can be so represented in a unique way. As we shall see later (see §9.2, especially equation (9.2.4)), this is a fundamental link between the Lie algebra of the rotation group and the cross product. This relation also will play a crucial role in the dynamics of a rigid body.

In particular, we can represent the skew matrix X_t this way:

$$X_t(\mathbf{v}) = \boldsymbol{\omega}(t) \times \mathbf{v}, \tag{8.6.3}$$

which defines $\boldsymbol{\omega}(t)$, the ***instantaneous rotation vector***.

Let $\{\mathbf{e}_1, \mathbf{e}_2, \mathbf{e}_3\}$ be a fixed (inertial) orthonormal frame in V and let $\{\,\boldsymbol{\xi}_i = \psi_t(\mathbf{e}_i) \mid i = 1, 2, 3\,\}$ be the corresponding ***rotating frame***. Given a point $\mathbf{v} \in V$, let $\mathbf{q} = (q^1, q^2, q^3)$ denote the vector in $\mathbb{R}^3$ defined by $\mathbf{v} = q^i \mathbf{e}_i$ and let $\mathbf{q}_R \in \mathbb{R}^3$ be the corresponding coordinate vector representing the components of the same vector $\mathbf{v}$ in the rotating frame, so $\mathbf{v} = q_R^i \boldsymbol{\xi}_i$. Let $A_t = A(t)$ be the *matrix* of ψ_t relative to the basis $\mathbf{e}_i$, that is, $\boldsymbol{\xi}_i = A_i^j \mathbf{e}_j$; then

$$\mathbf{q} = A_t \mathbf{q}_R, \quad \text{i.e.,} \quad q^j = A_i^j q_R^i, \tag{8.6.4}$$

and (8.6.2) in matrix notation becomes

$$\hat{\omega} = \dot{A}_t A_t^{-1}. \tag{8.6.5}$$

Newton's Law in a Rotating Frame. Assume that the point $\mathbf{v}(t)$ moves in V according to Newton's second law with a potential energy $U(\mathbf{v})$. Using $U(\mathbf{q})$ for the corresponding function induced on $\mathbb{R}^3$, Newton's law reads

$$m\ddot{\mathbf{q}} = -\nabla U(\mathbf{q}), \tag{8.6.6}$$

which are the Euler–Lagrange equations for

$$L(\mathbf{q}, \dot{\mathbf{q}}) = \frac{m}{2} \langle \dot{\mathbf{q}}, \dot{\mathbf{q}} \rangle - U(\mathbf{q}) \qquad (8.6.7)$$

or Hamilton's equations for

$$H(\mathbf{q}, \mathbf{p}) = \frac{1}{2m} \langle \mathbf{p}, \mathbf{p} \rangle + U(\mathbf{q}). \qquad (8.6.8)$$

To find the equation satisfied by $\mathbf{q}_R$, differentiate (8.6.4) with respect to time,

$$\dot{\mathbf{q}} = \dot{A}_t \mathbf{q}_R + A_t \dot{\mathbf{q}}_R = \dot{A}_t A_t^{-1} \mathbf{q} + A_t \dot{\mathbf{q}}_R, \qquad (8.6.9)$$

that is,

$$\dot{\mathbf{q}} = \boldsymbol{\omega}(t) \times \mathbf{q} + A_t \dot{\mathbf{q}}_R, \qquad (8.6.10)$$

where, by abuse of notation, $\boldsymbol{\omega}$ is also used for the representation of $\boldsymbol{\omega}$ in the inertial frame $\mathbf{e}_i$. Differentiating (8.6.10),

$$\ddot{\mathbf{q}} = \dot{\boldsymbol{\omega}} \times \mathbf{q} + \boldsymbol{\omega} \times \dot{\mathbf{q}} + \dot{A}_t \dot{\mathbf{q}}_R + A_t \ddot{\mathbf{q}}_R$$
$$= \dot{\boldsymbol{\omega}} \times \mathbf{q} + \boldsymbol{\omega} \times (\boldsymbol{\omega} \times \mathbf{q} + A_t \dot{\mathbf{q}}_R) + \dot{A}_t A_t^{-1} A_t \dot{\mathbf{q}}_R + A_t \ddot{\mathbf{q}}_R,$$

that is,

$$\ddot{\mathbf{q}} = \dot{\boldsymbol{\omega}} \times \mathbf{q} + \boldsymbol{\omega} \times (\boldsymbol{\omega} \times \mathbf{q}) + 2(\boldsymbol{\omega} \times A_t \dot{\mathbf{q}}_R) + A_t \ddot{\mathbf{q}}_R. \qquad (8.6.11)$$

The *angular velocity* in the rotating frame is (see (8.6.4))

$$\boldsymbol{\omega}_R = A_t^{-1} \boldsymbol{\omega}, \quad \text{i.e.,} \quad \boldsymbol{\omega} = A_t \boldsymbol{\omega}_R. \qquad (8.6.12)$$

Differentiating (8.6.12) with respect to time gives

$$\dot{\boldsymbol{\omega}} = \dot{A}_t \boldsymbol{\omega}_R + A_t \dot{\boldsymbol{\omega}}_R = \dot{A}_t A_t^{-1} \boldsymbol{\omega} + A_t \dot{\boldsymbol{\omega}}_R = A_t \dot{\boldsymbol{\omega}}_R, \qquad (8.6.13)$$

since $\dot{A}_t A_t^{-1} \boldsymbol{\omega} = \boldsymbol{\omega} \times \boldsymbol{\omega} = 0$. Multiplying (8.6.11) by A_t^{-1} gives

$$A_t^{-1} \ddot{\mathbf{q}} = \dot{\boldsymbol{\omega}}_R \times \mathbf{q}_R + \boldsymbol{\omega}_R \times (\boldsymbol{\omega}_R \times \mathbf{q}_R) + 2(\boldsymbol{\omega}_R \times \dot{\mathbf{q}}_R) + \ddot{\mathbf{q}}_R. \qquad (8.6.14)$$

Since $m\ddot{\mathbf{q}} = -\nabla U(\mathbf{q})$, we have

$$mA_t^{-1} \ddot{\mathbf{q}} = -\nabla U_R(\mathbf{q}_R), \qquad (8.6.15)$$

where the *rotated potential* U_R is the *time-dependent* potential defined by

$$U_R(\mathbf{q}_R, t) = U(A_t \mathbf{q}_R) = U(\mathbf{q}), \qquad (8.6.16)$$

so that $\nabla U(\mathbf{q}) = A_t \nabla U_R(\mathbf{q}_R)$. Therefore, by (8.6.15), Newton's equations (8.6.6) become

$$m\ddot{\mathbf{q}}_R + 2(\boldsymbol{\omega}_R \times m\dot{\mathbf{q}}_R) + m\boldsymbol{\omega}_R \times (\boldsymbol{\omega}_R \times \mathbf{q}_R) + m\dot{\boldsymbol{\omega}}_R \times \mathbf{q}_R$$
$$= -\nabla U_R(\mathbf{q}_R, t),$$

that is,

$$m\ddot{\mathbf{q}}_R = -\nabla U_R(\mathbf{q}_R, t) - m\boldsymbol{\omega}_R \times (\boldsymbol{\omega}_R \times \mathbf{q}_R)$$
$$- 2m(\boldsymbol{\omega}_R \times \dot{\mathbf{q}}_R) - m\dot{\boldsymbol{\omega}}_R \times \mathbf{q}_R, \qquad (8.6.17)$$

which expresses the equations of motion entirely in terms of rotated quantities.

Ficticious Forces.   There are three types of "fictitious forces" that suggest themselves if we try to identify (8.6.17) with $m\mathbf{a} = \mathbf{F}$:

(i) *Centrifugal force* $m\boldsymbol{\omega}_R \times (\mathbf{q}_R \times \boldsymbol{\omega}_R)$;

(ii) *Coriolis force* $2m\dot{\mathbf{q}}_R \times \boldsymbol{\omega}_R$; and

(iii) *Euler force* $m\mathbf{q}_R \times \dot{\boldsymbol{\omega}}_R$.

Note that the Coriolis force $2m\boldsymbol{\omega}_R \times \dot{\mathbf{q}}_R$ is orthogonal to $\boldsymbol{\omega}_R$ and $m\dot{\mathbf{q}}_R$, while the centrifugal force

$$m\boldsymbol{\omega}_R \times (\boldsymbol{\omega}_R \times \mathbf{q}_R) = m[(\boldsymbol{\omega}_R \cdot \mathbf{q}_R)\boldsymbol{\omega}_R - \|\boldsymbol{\omega}_R\|^2 \mathbf{q}_R]$$

is in the plane of $\boldsymbol{\omega}_R$ and $\mathbf{q}_R$. Also note that the Euler force is due to the *nonuniformity* of the rotation rate.

Lagrangian Form.   It is of interest to ask the sense in which (8.6.17) is Lagrangian or Hamiltonian. To answer this, it is useful to begin with the Lagrangian approach, which, we will see, is simpler. Substitute (8.6.10) into (8.6.7) to express the Lagrangian in terms of rotated quantities:

$$L = \frac{m}{2} \langle \boldsymbol{\omega} \times \mathbf{q} + A_t \dot{\mathbf{q}}_R, \boldsymbol{\omega} \times \mathbf{q} + A_t \dot{\mathbf{q}}_R \rangle - U(\mathbf{q})$$
$$= \frac{m}{2} \langle \boldsymbol{\omega}_R \times \mathbf{q}_R + \dot{\mathbf{q}}_R, \boldsymbol{\omega}_R \times \mathbf{q}_R + \dot{\mathbf{q}}_R \rangle - U_R(\mathbf{q}_R, t), \qquad (8.6.18)$$

which defines a new (time-dependent!) Lagrangian $L_R(\mathbf{q}_R, \dot{\mathbf{q}}_R, t)$. Remarkably, (8.6.17) are precisely the Euler–Lagrange equations for L_R; that is, (8.6.17) are equivalent to

$$\frac{d}{dt}\frac{\partial L_R}{\partial \dot{\mathbf{q}}_R^i} = \frac{\partial L_R}{\partial \mathbf{q}_R^i},$$

as is readily verified. If one thinks about performing a time-dependent transformation in the variational principle, then in fact, one sees that this is reasonable.

Hamiltonian Form. To find the sense in which (8.6.17) is Hamiltonian, perform a Legendre transformation on L_R. The conjugate momentum is

$$\mathbf{p}_R = \frac{\partial L_R}{\partial \dot{\mathbf{q}}_R} = m(\boldsymbol{\omega}_R \times \mathbf{q}_R + \dot{\mathbf{q}}_R), \qquad (8.6.19)$$

and so the Hamiltonian has the expression

$$\begin{aligned}
H_R(\mathbf{q}_R, \mathbf{p}_R) &= \langle \mathbf{p}_R, \dot{\mathbf{q}}_R \rangle - L_R \\
&= \frac{1}{m}\langle \mathbf{p}_R, \mathbf{p}_R - m\boldsymbol{\omega}_R \times \mathbf{q}_R \rangle - \frac{1}{2m}\langle \mathbf{p}_R, \mathbf{p}_R \rangle + U_R(\mathbf{q}_R, t) \\
&= \frac{1}{2m}\langle \mathbf{p}_R, \mathbf{p}_R \rangle + U_R(\mathbf{q}_R, t) - \langle \mathbf{p}_R, \boldsymbol{\omega}_R \times \mathbf{q}_R \rangle. \qquad (8.6.20)
\end{aligned}$$

Thus, (8.6.17) are equivalent to Hamilton's canonical equations with Hamiltonian (8.6.20) and with the canonical symplectic form. In general, H_R is time-dependent. Alternatively, if we perform the momentum shift

$$\mathfrak{p}_R = \mathbf{p}_R - m\boldsymbol{\omega}_R \times \mathbf{q}_R = m\dot{\mathbf{q}}_R, \qquad (8.6.21)$$

then we get

$$\begin{aligned}
\tilde{H}_R(\mathbf{q}_R, \mathfrak{p}_R) &:= H_R(\mathbf{q}_R, \mathbf{p}_R) \\
&= \frac{1}{2m}\langle \mathfrak{p}_R, \mathfrak{p}_R \rangle + U_R(\mathbf{q}_R) - \frac{m}{2}\|\boldsymbol{\omega}_R \times \mathbf{q}_R\|^2, \qquad (8.6.22)
\end{aligned}$$

which is in the usual form of kinetic plus potential energy, but now the potential is *amended* by the centrifugal potential $m\|\boldsymbol{\omega}_R \times \mathbf{q}_R\|^2/2$, and the canonical symplectic structure

$$\Omega_{\text{can}} = d\mathbf{q}_R^i \wedge d(\mathbf{p}_R)_i$$

gets transformed, by the momentum shifting lemma, or directly, to

$$d\mathbf{q}_R^i \wedge d(\mathbf{p}_R)_i = d\mathbf{q}_R^i \wedge d(\mathfrak{p}_R)_i + \epsilon_{ijk}\omega_R^i d\mathbf{q}_R^i \wedge d\mathbf{q}_R^j,$$

where ϵ_{ijk} is the alternating tensor. Note that

$$\tilde{\Omega}_R = \tilde{\Omega}_{\text{can}} + *\boldsymbol{\omega}_R, \qquad (8.6.23)$$

where $*\boldsymbol{\omega}_R$ means the two-form associated to the vector $\boldsymbol{\omega}_R$, and that (8.6.23) has the same form as the corresponding expression for a particle in a magnetic field (§6.7).

In general, the momentum shift (8.6.21) is time-dependent, so care is needed in interpreting the sense in which the equations for $\mathfrak{p}_R$ and $\mathbf{q}_R$ are Hamiltonian. In fact, the equations should be computed as follows. Let X_H be a Hamiltonian vector field on P and let $\zeta_t : P \to P$ be a *time-dependent* map with generator Y_t:

$$\frac{d}{dt}\zeta_t(z) = Y_t(\zeta_t(z)). \qquad (8.6.24)$$

Assume that ζ_t is symplectic for each t. If $\dot{z}(t) = X_H(z(t))$ and we let $w(t) = \zeta_t(z(t))$, then w satisfies

$$\dot{w} = T\zeta_t \cdot X_H(z(t)) + Y_t(\zeta_t(z(t))), \tag{8.6.25}$$

that is,

$$\dot{w} = X_K(w) + Y_t(w) \tag{8.6.26}$$

where $K = H \circ \zeta_t^{-1}$. The extra term Y_t in (8.6.26) is, in the example under consideration, the Euler force.

So far we have been considering a fixed system as seen from different rotating observers. Analogously, one can consider systems that themselves are subjected to a superimposed rotation, an example being the Foucault pendulum. It is clear that the physical behavior in the two cases can be different—in fact, the Foucault pendulum and the example in the next section show that one can get a real physical effect from rotating a system—obviously, rotating observers can cause nontrivial changes in the *description* of a system but cannot make any *physical* difference. Nevertheless, the strategy for the analysis of rotating systems is analogous to the above. The easiest approach, as we have seen, is to transform the Lagrangian. The reader may wish to reread §2.10 for an easy and specific instance of this.

Exercises

◇ **8.6-1.**   Generalize the discussion of Newton's law seen in a rotating frame to that of a particle moving in a magnetic field as seen from a rotating observer. Do so first directly and then by Lagrangian methods.

8.7   The Geometric Phase for a Particle in a Hoop

This discussion follows Berry [1985] with some small modifications (due to Marsden, Montgomery, and Ratiu [1990]) necessary for a geometric interpretation of the results. Figure 8.7.1, shows a planar hoop (not necessarily circular) in which a bead slides without friction.

As the bead is sliding, the hoop is rotated in its plane through an angle $\theta(t)$ with angular velocity $\omega(t) = \dot{\theta}(t)\mathbf{k}$. Let s denote the arc length along the hoop, measured from a reference point on the hoop, and let $\mathbf{q}(s)$ be the vector from the origin to the corresponding point on the hoop; thus the shape of the hoop is determined by this function $\mathbf{q}(s)$. The unit tangent vector is $\mathbf{q}'(s)$, and the position of the reference point $\mathbf{q}(s(t))$ relative to an inertial frame in space is $R_{\theta(t)}\mathbf{q}(s(t))$, where R_θ is the rotation in the plane of the hoop through an angle θ. Note that

$$\dot{R}_\theta R_\theta^{-1}\mathbf{q} = \boldsymbol{\omega} \times \mathbf{q} \quad \text{and} \quad R_\theta\boldsymbol{\omega} = \boldsymbol{\omega}.$$

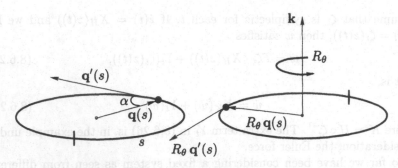

FIGURE 8.7.1. A particle sliding in a rotating hoop.

The Equations of Motion. The configuration space is a fixed closed curve (the hoop) in the plane with length ℓ. The Lagrangian $L(s, \dot{s}, t)$ is simply the kinetic energy of the particle. Since

$$\frac{d}{dt} R_{\theta(t)} \mathbf{q}(s(t)) = R_{\theta(t)} \mathbf{q}'(s(t)) \dot{s}(t) + R_{\theta(t)} [\boldsymbol{\omega}(t) \times \mathbf{q}(s(t))],$$

the Lagrangian is

$$L(s, \dot{s}, t) = \frac{1}{2} m \| \mathbf{q}'(s) \dot{s} + \boldsymbol{\omega} \times \mathbf{q} \|^2. \tag{8.7.1}$$

Note that the momentum conjugate to s is $p = \partial L / \partial \dot{s}$; that is,

$$p = m \mathbf{q}' \cdot [\mathbf{q}' \dot{s} + \boldsymbol{\omega} \times \mathbf{q}] = mv, \tag{8.7.2}$$

where v is the component of the velocity *with respect to the inertial frame* tangent to the curve. The Euler–Lagrange equations

$$\frac{d}{dt} \frac{\partial L}{\partial \dot{s}} = \frac{\partial L}{\partial s}$$

become

$$\frac{d}{dt} [\mathbf{q}' \cdot (\mathbf{q}' \dot{s} + \boldsymbol{\omega} \times \mathbf{q})] = (\mathbf{q}' \dot{s} + \boldsymbol{\omega} \times \mathbf{q}) \cdot (\mathbf{q}'' \dot{s} + \boldsymbol{\omega} \times \mathbf{q}').$$

Using $\| \mathbf{q}' \|^2 = 1$, its consequence $\mathbf{q}' \cdot \mathbf{q}'' = 0$, and simplifying, we get

$$\ddot{s} + \mathbf{q}' \cdot (\dot{\boldsymbol{\omega}} \times \mathbf{q}) - (\boldsymbol{\omega} \times \mathbf{q}) \cdot (\boldsymbol{\omega} \times \mathbf{q}') = 0. \tag{8.7.3}$$

The second and third terms in (8.7.3) are the Euler and centrifugal forces, respectively. Since $\boldsymbol{\omega} = \dot{\theta} \mathbf{k}$, we can rewrite (8.7.3) as

$$\ddot{s} = \dot{\theta}^2 \mathbf{q} \cdot \mathbf{q}' - \ddot{\theta} q \sin \alpha, \tag{8.7.4}$$

where α is as in Figure 8.7.1 and $q = \| \mathbf{q} \|$.

Averaging.   From (8.7.4) and Taylor's formula with remainder, we get

$$s(t) = s_0 + \dot{s}_0 t + \int_0^t (t - \tau)\{\dot{\theta}(\tau)^2 \mathbf{q}(s(\tau)) \cdot \mathbf{q}'(s(\tau))$$
$$- \ddot{\theta}(\tau)q(s(\tau))\sin\alpha(s(\tau))\} \, d\tau. \qquad (8.7.5)$$

The angular velocity $\dot{\theta}$ and acceleration $\ddot{\theta}$ are assumed small with respect to the particle's velocity, so by the averaging theorem (see, for example, Hale [1963]), the s-dependent quantities in (8.7.5) can be replaced by their averages around the hoop:

$$s(t) \approx s_0 + \dot{s}_0 t + \int_0^t (t - \tau) \left\{ \dot{\theta}(\tau)^2 \frac{1}{\ell} \int_0^\ell \mathbf{q} \cdot \mathbf{q}' \, ds \right.$$
$$\left. - \ddot{\theta}(\tau)\frac{1}{\ell} \int_0^\ell q(s)\sin\alpha(s) \, ds \right\} d\tau. \qquad (8.7.6)$$

Technical Aside.   The essence of averaging in this case can be seen as follows. Suppose $g(t)$ is a rapidly varying function whose oscillations are bounded in magnitude by a constant C and $f(t)$ is slowly varying on an interval $[a, b]$. Over one period of g, say $[\alpha, \beta]$, we have

$$\int_\alpha^\beta f(t)g(t) \, dt \approx \bar{g} \int_\alpha^\beta f(t) \, dt, \qquad (8.7.7)$$

where

$$\bar{g} = \frac{1}{\beta - \alpha} \int_\alpha^\beta g(t) \, dt$$

is the average of g. The assumption that the oscillations of g are bounded by C means that

$$|g(t) - \bar{g}| \leq C \quad \text{for all } t \in [\alpha, \beta].$$

The error in (8.7.7) is $\int_\alpha^\beta f(t)(g(t) - \bar{g}) \, dt$, whose absolute value is bounded as follows. Let M be the maximum value of f on $[\alpha, \beta]$ and m be the minimum. Then

$$\left| \int_\alpha^\beta f(t)[g(t) - \bar{g}] \, dt \right| = \left| \int_\alpha^\beta (f(t) - m)[g(t) - \bar{g}] \, dt \right|$$
$$\leq (\beta - \alpha)(M - m)C$$
$$\leq (\beta - \alpha)^2 DC,$$

where D is the maximum of $|f'(t)|$ for $\alpha \leq t \leq \beta$. Now these errors over each period are added up over $[a, b]$. Since the error estimate has the *square* of $\beta - \alpha$ as a factor, one still gets something small as the period of g tends to 0.

In (8.7.5) we change variables from t to s, do the averaging, and then change back.

The Phase Formula. The first inner integral in (8.7.6) over s vanishes (since the integrand is $(d/ds)\|\mathbf{q}(s)\|^2$), and the second is $2A$, where A is the area enclosed by the hoop. Integrating by parts,

$$\int_0^T (T-\tau)\ddot{\theta}(\tau)\,d\tau = -T\dot{\theta}(0) + \int_0^T \dot{\theta}(\tau)\,d\tau = -T\dot{\theta}(0) + 2\pi, \qquad (8.7.8)$$

assuming that the hoop makes one complete revolution in time T. Substituting (8.7.8) in (8.7.6) gives

$$s(T) \approx s_0 + \dot{s}_0 T + \frac{2A}{\ell}\dot{\theta}_0 T - \frac{4\pi A}{\ell}, \qquad (8.7.9)$$

where $\dot{\theta}_0 = \dot{\theta}(0)$. The initial velocity of the bead *relative to the hoop* is $\dot{s}_0$, while its component along the curve *relative to the inertial frame* is (see (8.7.2))

$$v_0 = \mathbf{q}'(0) \cdot [\mathbf{q}'(0)\dot{s}_0 + \boldsymbol{\omega}_0 \times \mathbf{q}(0)] = \dot{s}_0 + \omega_0 q(s_0)\sin\alpha(s_0). \qquad (8.7.10)$$

Now we replace $\dot{s}_0$ in (8.7.9) by its expression in terms of v_0 from (8.7.10) and average over all initial conditions to get

$$\langle s(T) - s_0 - v_0 T\rangle = -\frac{4\pi A}{\ell}, \qquad (8.7.11)$$

which means that *on average*, the shift in position is by $4\pi A/\ell$ between the rotated and nonrotated hoop. Note that if $\dot{\theta}_0 = 0$ (the situation assumed by Berry [1985]), then averaging over initial conditions is not necessary.

This extra length $4\pi A/\ell$ is sometimes called the geometric phase or the **Berry–Hannay phase**. This example is related to a number of interesting effects, both classically and quantum-mechanically, such as the Foucault pendulum and the Aharonov–Bohm effect. The effect is known as *holonomy* and can be viewed as an instance of *reconstruction* in the context of symmetry and reduction. For further information and additional references, see Aharonov and Anandan [1987], Montgomery [1988], Montgomery [1990], and Marsden, Montgomery, and Ratiu [1989, 1990]. For related ideas in soliton dynamics, see Alber and Marsden [1992].

Exercises

⋄ **8.7-1.** Consider the dynamics of a ball in a slowly rotating planar hoop, as in the text. However, this time, consider rotating the hoop about an axis that is not perpendicular to the plane of the hoop, but makes an angle θ with the normal. Compute the geometric phase for this problem.

⋄ **8.7-2.** Study the geometric phase for a particle in a general spatial hoop that is moved through a closed curve in SO(3).

◇ **8.7-3.**   Consider the dynamics of a ball in a slowly rotating planar hoop, as in the text. However, this time, consider a charged particle with charge e and a fixed magnetic field $\mathbf{B} = \nabla \times \mathbf{A}$ in the vicinity of the hoop. Compute the geometric phase for this problem.

8.8   Moving Systems

The particle in the rotating hoop is an example of a rotated or, more generally, a *moving system*. Other examples are a pendulum on a merry-go-round (Exercise 8.8-4) and a fluid on a rotating sphere (like the Earth's ocean and atmosphere). As we have emphasized, systems of this type are not to be confused with rotating observers! Actually rotating a system causes real physical effects, such as the trade winds and hurricanes.

This section develops a general context for such systems. Our purpose is to show how to systematically derive Lagrangians and the resulting equations of motion for moving systems, like the bead in the hoop of the last section. This will also prepare the reader who wants to pursue the question of how moving systems fit in the context of phases (Marsden, Montgomery, and Ratiu [1990]).

The Lagrangian.   Consider a Riemannian manifold S, a submanifold Q, and a space M of embeddings of Q into S. Let $m_t \in M$ be a given curve. If a particle in Q is following a curve $q(t)$, and if Q moves by superposing the motion m_t, then the path of the particle in S is given by $m_t(q(t))$. Thus, its velocity in S is given by

$$T_{q(t)} m_t \cdot \dot{q}(t) + \mathcal{Z}_t(m_t(q(t))), \tag{8.8.1}$$

where $\mathcal{Z}_t(m_t(q)) = (d/dt)m_t(q)$. Consider a Lagrangian on TQ of the usual form of kinetic minus potential energy:

$$L_{m_t}(q, v) = \frac{1}{2}\|T_{q(t)} m_t \cdot v + \mathcal{Z}_t(m_t(q))\|^2 - V(q) - U(m_t(q)), \tag{8.8.2}$$

where V is a given potential on Q, and U is a given potential on S.

The Hamiltonian.   We now compute the Hamiltonian associated to this Lagrangian by taking the associated Legendre transform. If we take the derivative of (8.8.2) with respect to v in the direction of w, we obtain

$$\frac{\partial L_{m_t}}{\partial v} \cdot w = p \cdot w = \left\langle T_{q(t)} m_t \cdot v + \mathcal{Z}_t\left(m_t(q(t))\right)^T, T_{q(t)} m_t \cdot w \right\rangle_{m_t(q(t))}, \tag{8.8.3}$$

where $p \cdot w$ means the natural pairing between the covector $p \in T^*_{q(t)}Q$ and the vector $w \in T_{q(t)}Q$, while $\langle\,,\,\rangle_{m_t(q(t))}$ denotes the metric inner product on S at the point $m_t(q(t))$ and T denotes the orthogonal projection to the tangent space $Tm_t(Q)$ using the metric of S at $m_t(q(t))$. We endow Q with the (possibly time-dependent) metric induced by the mapping m_t. In other words, we choose the metric on Q that makes m_t into an isometry for each t. Using this definition, (8.8.3) gives

$$p \cdot w = \left\langle v + \left(T_{q(t)}m_t\right)^{-1} \cdot \mathcal{Z}_t\left(m_t(q(t))\right)^T, w \right\rangle_{q(t)};$$

that is,

$$p = \left(v + \left(T_{q(t)}m_t\right)^{-1} \cdot \left[\mathcal{Z}_t\left(m_t(q(t))^T\right]\right)^\flat, \qquad (8.8.4)$$

where $\flat$ is the index-lowering operation at $q(t)$ using the metric on Q.

Physically, if S is $\mathbb{R}^3$, then p is the inertial momentum (see the hoop example in the preceding section). This extra term $\mathcal{Z}_t(m_t(q))^T$ is associated with a connection called the **Cartan connection** on the bundle $Q \times M \to M$, with horizontal lift defined to be $\mathcal{Z}(m) \mapsto (Tm^{-1} \cdot \mathcal{Z}(m)^T, \mathcal{Z}(m))$. (See, for example, Marsden and Hughes [1983] for an account of some aspects of Cartan's contributions.)

The corresponding Hamiltonian (given by the standard prescription $H = pv - L$) picks up a cross term and takes the form

$$H_{m_t}(q, p) = \frac{1}{2}\|p\|^2 - \mathcal{P}(\mathcal{Z}_t) - \frac{1}{2}\|\mathcal{Z}_t^\perp\|^2 + V(q) + U(m_t(q)), \qquad (8.8.5)$$

where the time-dependent vector field Z_t on Q is defined by

$$Z_t(q) = \left(T_{q(t)}m_t\right)^{-1} \cdot [\mathcal{Z}_t(m_t(q)]^T$$

and where $\mathcal{P}(Z_t(q))(q, p) = \langle p, Z_t(q) \rangle$ and $\mathcal{Z}_t^\perp$ denotes the component perpendicular to $m_t(Q)$. The Hamiltonian vector field of this cross term, namely $X_{\mathcal{P}(\mathcal{Z}_t)}$, represents the noninertial forces and also has the natural interpretation as a horizontal lift of the vector field $\mathcal{Z}_t$ relative to a certain connection on the bundle $T^*Q \times M \to M$, naturally derived from the Cartan connection.

Remarks on Averaging. Let G be a Lie group that acts on T^*Q in a Hamiltonian fashion and leaves H_0 (defined by setting $\mathcal{Z} = 0$ and $U = 0$ in (8.8.5)) invariant. (Lie groups are discussed in the next chapter, so these remarks can be omitted on a first reading.) In our examples, G is either $\mathbb{R}$ acting on T^*Q by the flow of H_0 (the hoop), or a subgroup of the isometry group of Q that leaves V and U invariant, and acts on T^*Q by cotangent lift (this is appropriate for the Foucault pendulum). In any case, we assume that G has an invariant measure relative to which we can average.

Assuming the "averaging principle" (see Arnold [1989], for example) we replace H_{m_t} by its G-average,

$$\langle H_{m_t}\rangle (q,p) = \frac{1}{2}\|p\|^2 - \langle \mathcal{P}(Z_t)\rangle - \frac{1}{2}\langle \|Z_t^\perp\|^2\rangle + V(q) + \langle U(m_t(q))\rangle .$$

$$(8.8.6)$$

In (8.8.6) we shall assume that the term $\frac{1}{2}\langle \|Z_t^\perp\|^2\rangle$ is small and discard it. Thus, define

$$\mathcal{H}(q,p,t) = \frac{1}{2}\|p\|^2 - \langle \mathcal{P}(Z_t)\rangle + V(q) + \langle U(m_t(q))\rangle$$

$$= \mathcal{H}_0(q,p) - \langle \mathcal{P}(Z_t)\rangle + \langle U(m_t(q))\rangle . \qquad (8.8.7)$$

Consider the dynamics on $T^*Q \times M$ given by the vector field

$$(X_{\mathcal{H}}, Z_t) = (X_{\mathcal{H}_0} - X_{\langle \mathcal{P}(Z_t)\rangle} + X_{\langle U \circ m_t\rangle}, Z_t). \qquad (8.8.8)$$

The vector field, consisting of the extra terms in this representation due to the superposed motion of the system, namely

$$\text{hor}(Z_t) = (-X_{\langle \mathcal{P}(Z_t)\rangle}, Z_t), \qquad (8.8.9)$$

has a natural interpretation as the horizontal lift of Z_t relative to a connection on $T^*Q \times M$, which is obtained by averaging the Cartan connection and is called the **Cartan–Hannay–Berry connection**. The holonomy of this connection is the **Hannay–Berry phase** of a slowly moving constrained system. For details of this approach, see Marsden, Montgomery, and Ratiu [1990].

Exercises

⬦ **8.8-1.**   Consider the particle in a hoop of §8.7. For this problem, identify all the elements of formula (8.8.2) and use that identification to obtain the Lagrangian (8.7.1).

⬦ **8.8-2.**   Consider the particle in a rotating hoop discussed in §2.8.

(a) Use the tools of this section to obtain the Lagrangian given in §2.8.

(b) Suppose that the hoop rotates freely. Can you still use the tools of part (a)? If so, compute the new Lagrangian and point out the differences between the two cases.

(c) Analyze, in the same fashion as in §2.8, the equilibria of the free system. Does this system also bifurcate?

⬦ **8.8-3.**   Set up the equations for the Foucault pendulum using the ideas in this section.

◇ **8.8-4.** Consider again the mechanical system in Exercise 2.8-6, but this time hang a *spherical* pendulum from the rotating arm. Investigate the geometric phase when the arm is swung once around. (Consider doing the experiment!) Is the term $\|Z_i^\perp\|^2$ really small in this example?

8.9   Routh Reduction

An abelian version of Lagrangian reduction was known to Routh by around 1860. A modern account was given in Arnold [1988], and motivated by that, Marsden and Scheurle [1993a] gave a geometrization and a generalization of the Routh procedure to the nonabelian case.

In this section we give an elementary classical description in preparation for more sophisticated reduction procedures, such as Euler–Poincaré reduction in Chapter 13.

We assume that Q is a product of a manifold S and a number, say k, of copies of the circle S^1, namely $Q = S \times (S^1 \times \cdots \times S^1)$. The factor S, called *shape space*, has coordinates denoted by $x^1, \ldots, x^m$, and coordinates on the other factors are written $\theta^1, \ldots, \theta^k$. Some or all of the factors of S^1 can be replaced by $\mathbb{R}$ if desired, with little change. We assume that the variables θ^a, $a = 1, \ldots, k$, are *cyclic*, that is, they do not appear explicitly in the Lagrangian, although their velocities do.

As we shall see after Chapter 9 is studied, invariance of L under the action of the abelian group $G = S^1 \times \cdots \times S^1$ is another way to express that fact that θ^a are cyclic variables. That point of view indeed leads ultimately to deeper insight, but here we focus on some basic calculations done "by hand" in coordinates.

A basic class of examples (for which Exercises 8.9-1 and 8.9-2 provide specific instances) are those for which the Lagrangian L has the form kinetic minus potential energy:

$$L(x, \dot{x}, \dot{\theta}) = \frac{1}{2} g_{\alpha\beta}(x)\dot{x}^\alpha \dot{x}^\beta + g_{a\alpha}(x)\dot{x}^\alpha \dot{\theta}^a + \frac{1}{2} g_{ab}(x)\dot{\theta}^a \dot{\theta}^b - V(x), \quad (8.9.1)$$

where there is a sum over α, β from 1 to m and over a, b from 1 to k. Even in simple examples, such as the double spherical pendulum or the simple pendulum on a cart (Exercise 8.9-2), the matrices $g_{\alpha\beta}$, $g_{a\alpha}$, g_{ab} can depend on x.

Because θ^a are cyclic, the corresponding conjugate momenta

$$p_a = \frac{\partial L}{\partial \dot{\theta}^a} \tag{8.9.2}$$

are conserved quantities. In the case of the Lagrangian (8.9.1), these momenta are given by

$$p_a = g_{a\alpha}\dot{x}^\alpha + g_{ab}\dot{\theta}^b.$$

Definition 8.9.1. *The **classical Routhian** is defined by setting $p_a = \mu_a = $ constant and performing a partial Legendre transformation in the variables θ^a :*

$$R^\mu(x, \dot{x}) = \left[L(x, \dot{x}, \dot{\theta}) - \mu_a \dot{\theta}^a \right]\Big|_{p_a = \mu_a}, \qquad (8.9.3)$$

where it is understood that the variable $\dot{\theta}^a$ is eliminated using the equation $p_a = \mu_a$ and μ_a is regarded as a constant.

Now consider the Euler–Lagrange equations

$$\frac{d}{dt} \frac{\partial L}{\partial \dot{x}^a} - \frac{\partial L}{\partial x^a} = 0; \qquad (8.9.4)$$

we attempt to write these as Euler–Lagrange equations for a function from which $\dot{\theta}^a$ has been eliminated. We claim that the Routhian R^μ does the job. To see this, we compute the Euler–Lagrange expression for R^μ using the chain rule:

$$\frac{d}{dt} \left(\frac{\partial R^\mu}{\partial \dot{x}^\alpha} \right) - \frac{\partial R^\mu}{\partial x^\alpha} = \frac{d}{dt} \left(\frac{\partial L}{\partial \dot{x}^\alpha} + \frac{\partial L}{\partial \dot{\theta}^a} \frac{\partial \dot{\theta}^a}{\partial \dot{x}^\alpha} \right)$$

$$- \left(\frac{\partial L}{\partial x^\alpha} + \frac{\partial L}{\partial \dot{\theta}^a} \frac{\partial \dot{\theta}^a}{\partial x^\alpha} \right) - \frac{d}{dt} \left(\mu_a \frac{\partial \dot{\theta}^a}{\partial \dot{x}^\alpha} \right) + \mu_a \frac{\partial \dot{\theta}^a}{\partial x^\alpha}.$$

The first and third terms vanish by (8.9.4), and the remaining terms vanish using $\mu_a = p_a$. Thus, we have proved the following result.

Proposition 8.9.2. *The Euler–Lagrange equations (8.9.4) for $L(x, \dot{x}, \dot{\theta})$ together with the conservation laws $p_a = \mu_a$ are equivalent to the Euler–Lagrange equations for the Routhian $R^\mu(x, \dot{x})$ together with $p_a = \mu_a$.*

The Euler–Lagrange equations for R^μ are called the **reduced Euler–Lagrange equations**, since the configuration space Q with variables (x^a, θ^a) has been *reduced* to the configuration space S with variables x^α.

In what follows we shall make the following notational conventions: g^{ab} denote the entries of the inverse matrix of the $m \times m$ matrix $[g_{ab}]$, and similarly, $g^{\alpha\beta}$ denote the entries of the inverse of the $k \times k$ matrix $[g_{\alpha\beta}]$. We will not use the entries of the inverse of the whole matrix tensor on Q, so there is no danger of confusion.

Proposition 8.9.3. *For L given by (8.9.1) we have*

$$R^\mu(x, \dot{x}) = g_{a\alpha} g^{ac} \mu_c \dot{x}^\alpha + \frac{1}{2} \left(g_{\alpha\beta} - g_{a\alpha} g^{ac} g_{c\beta} \right) \dot{x}^\alpha \dot{x}^\beta - V_\mu(x), \qquad (8.9.5)$$

where

$$V_\mu(x) = V(x) + \frac{1}{2} g^{ab} \mu_a \mu_b$$

*is the **amended potential.***

Proof. We have $\mu_a = g_{a\alpha}\dot{x}^\alpha + g_{ab}\dot{\theta}^b$, so

$$\dot{\theta}^a = g^{ab}\mu_b - g^{ab}g_{b\alpha}\dot{x}^\alpha. \tag{8.9.6}$$

Substituting this in the definition of R^μ gives

$$
\begin{aligned}
R^\mu(x,\dot{x}) &= \frac{1}{2}g_{\alpha\beta}\dot{x}^\alpha\dot{x}^\beta + (g_{a\alpha}\dot{x}^\alpha)\left(g^{ac}\mu_c - g^{ac}g_{c\beta}\dot{x}^\beta\right) \\
&\quad + \frac{1}{2}g_{ab}\left(g^{ac}\mu_c - g^{ac}g_{c\beta}\dot{x}^\beta\right)\left(g^{bd}\mu_d - g^{bd}g_{d\gamma}\dot{x}^\gamma\right) \\
&\quad - \mu_a\left(g^{ac}\mu_c - g^{ac}g_{c\beta}\dot{x}^\beta\right) - V(x).
\end{aligned}
$$

The terms linear in $\dot{x}$ are

$$g_{a\alpha}g^{ac}\mu_c\dot{x}^\alpha - g_{ab}g^{ac}\mu_c g^{bd}g_{d\gamma}\dot{x}^\gamma + \mu_a g^{ac}g_{c\beta}\dot{x}^\beta = g_{a\alpha}g^{ac}\mu_c\dot{x}^\alpha,$$

while the terms quadratic in $\dot{x}$ are

$$\frac{1}{2}(g_{\alpha\beta} - g_{a\alpha}g^{ac}g_{c\beta})\dot{x}^\alpha\dot{x}^\beta,$$

and the terms dependent only on x are $-V_\mu(x)$, as required.   ∎

Note that R^μ has picked up a term linear in the velocity, and the potential as well as the kinetic energy matrix (the **mass matrix**) have both been modified.

The term linear in the velocities has the form $A_\alpha^a\mu_a\dot{x}^\alpha$, where $A_\alpha^a = g^{ab}g_{b\alpha}$. The Euler–Lagrange expression for this term can be written

$$\frac{d}{dt}A_\alpha^a\mu_a - \frac{\partial}{\partial x^\alpha}A_\beta^a\mu_a\dot{x}^\beta = \left(\frac{\partial A_\alpha^a}{\partial x^\beta} - \frac{\partial A_\beta^a}{\partial x^\alpha}\right)\mu_a\dot{x}^\beta,$$

which is denoted by $B_{\alpha\beta}^a\mu_a\dot{x}^\beta$. If we think of the one-form $A_\alpha^a dx^\alpha$, then $B_{\alpha\beta}^a$ is its exterior derivative. The quantities A_α^a are called **connection coefficients**, and $B_{\alpha\beta}^a$ are called the **curvature coefficients**.

Introducing the modified (simpler) Routhian, obtained by deleting the terms linear in $\dot{x}$,

$$\tilde{R}^\mu = \frac{1}{2}\left(g_{\alpha\beta} - g_{a\alpha}g^{ab}g_{b\beta}\right)\dot{x}^\alpha\dot{x}^\beta - V_\mu(x),$$

the equations take the form

$$\frac{d}{dt}\frac{\partial\tilde{R}^\mu}{\partial\dot{x}^\alpha} - \frac{\partial\tilde{R}^\mu}{\partial x^\alpha} = -B_{\alpha\beta}^a\mu_a\dot{x}^\beta, \tag{8.9.7}$$

which is the form that makes intrinsic sense and generalizes. The extra terms have the structure of magnetic, or Coriolis, terms that we have seen in a variety of earlier contexts.

The above gives a hint of the large amount of geometry hidden behind the apparently simple process of Routh reduction. In particular, *connections* A_α^a and their *curvatures* $B_{\alpha\beta}^a$ play an important role in more general theories, such as those involving nonablelian symmetry groups (like the rotation group).

Another suggestive hint of more general theories is that the kinetic term in (8.9.5) can be written in the following way:

$$\frac{1}{2}(\dot{x}^\alpha, -A_\delta^a\dot{x}^\delta)\begin{pmatrix} g_{\alpha\beta} & g_{\alpha b} \\ g_{a\beta} & g_{ab} \end{pmatrix}\begin{pmatrix} \dot{x}^\beta \\ -A_\gamma^b\dot{x}^\gamma \end{pmatrix},$$

which also exhibits its positive definite nature.

Routh himself (in the mid 1800s) was very interested in rotating mechanical systems, such as those possessing an angular momentum conservation law. In this context, Routh used the term "steady motion" for dynamic motions that were uniform rotations about a fixed axis. *We may identify these with equilibria of the reduced Euler–Lagrange equations.*

Since the Coriolis term does not affect conservation of energy (we have seen this earlier with the dynamics of a particle in a magnetic field), we can apply the Lagrange–Dirichlet test to reach the following conclusion:

Proposition 8.9.4 (Routh's Stability Criterion). *Steady motions correspond to critical points* x_e *of the amended potential* V_μ. *If* $d^2V_\mu(x_e)$ *is positive definite, then the steady motion* x_e *is stable.*

When more general symmetry groups are involved, one speaks of **relative equilibria** rather than steady motions, a change of terminology due to Poincaré around 1890. This is the beginning of a more sophisticated theory of stability, leading up to the **energy–momentum method** outlined in §1.7.

Exercises

◇ **8.9-1.**   Carry out Routh reduction for the spherical pendulum.

◇ **8.9-2.**   Carry out Routh reduction for the planar pendulum on a cart, as in Figure 8.9.1.

◇ **8.9-3** (Two-body problem).   Compute the amended potential for the planar motion of a particle moving in a central potential $V(r)$. Compare the result with the "effective potential" found in, for example, Goldstein [1980].

◇ **8.9-4.**   Let L be a Lagrangian on TQ and let

$$\hat{R}^\mu(q, \dot{q}) = L(q, \dot{q}) + A_{\alpha\, a}^a \mu_a q^a,$$

where A^a is an $\mathbb{R}^k$-valued one-form on TQ and $\mu \in \mathbb{R}^{k*}$.

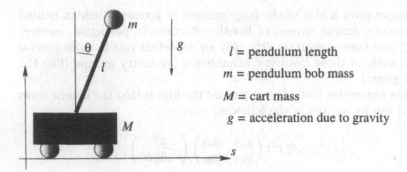

l = pendulum length

m = pendulum bob mass

M = cart mass

g = acceleration due to gravity

FIGURE 8.9.1. A pendulum on a cart.

(a) Write Hamilton's principle for L as a Lagrange–d'Alembert principle for $\hat{R}^\mu$.

(b) Letting $\hat{H}^\mu$ be the Hamiltonian associated with $\hat{R}^\mu$, show that the original Euler–Lagrange equations for L can be written as

$$\dot{q}^\alpha = \frac{\partial \hat{H}^\mu}{\partial p_\alpha},$$

$$\dot{p}_\alpha = \frac{\partial \hat{H}^\mu}{\partial q^\alpha} + \beta^a_{\alpha\beta}\mu_b\frac{\partial \hat{H}^\mu}{\partial p_\beta}.$$

9
An Introduction to Lie Groups

To prepare for the next chapters, we present some basic facts about Lie groups. Alternative expositions and additional details can be obtained from Abraham and Marsden [1978], Olver [1986], and Sattinger and Weaver [1986]. In particular, in this book we shall require only elementary facts about the general theory and a knowledge of a few of the more basic groups, such as the rotation and Euclidean groups.

Here are how some of the basic groups occur in mechanics:

Linear and Angular Momentum. These arise as conserved quantities associated with the groups of translations and rotations in space.

Rigid Body. Consider a free rigid body rotating about its center of mass, taken to be the origin. "Free" means that there are no external forces, and "rigid" means that the distance between any two points of the body is unchanged during the motion. Consider a point X of the body at time $t = 0$, and denote its position at time t by $f(X, t)$. Rigidity of the body and the assumption of a smooth motion imply that $f(X, t) = \mathbf{A}(t)X$, where $\mathbf{A}(t)$ is a proper rotation, that is, $\mathbf{A}(t) \in \mathrm{SO}(3)$, the proper rotation group of $\mathbb{R}^3$, the 3×3 orthogonal matrices with determinant 1. The set $\mathrm{SO}(3)$ will be shown to be a three-dimensional Lie group, and since it describes any possible position of the body, it serves as the *configuration space*. The group $\mathrm{SO}(3)$ also plays a dual role of a *symmetry group*, since the same physical motion is described if we rotate our coordinate axes. Used as a symmetry group, $\mathrm{SO}(3)$ leads to conservation of angular momentum.

Heavy Top.   Consider a rigid body moving with a fixed point but under the influence of gravity. This problem still has a configuration space SO(3), but the symmetry group is only the circle group S^1, consisting of rotations about the direction of gravity. One says that gravity has *broken* the symmetry from SO(3) to S^1. This time, "eliminating" the S^1 symmetry "mysteriously" leads one to the larger Euclidean group SE(3) of rigid motion of $\mathbb{R}^3$. This is a manifestation of the general theory of semidirect products (see the Introduction, where we showed that the heavy top equations are Lie–Poisson for SE(3), and Marsden, Ratiu, and Weinstein [1984a, 1984b]).

Incompressible Fluids.   Let Ω be a region in $\mathbb{R}^3$ that is filled with a moving incompressible fluid and is free of external forces. Denote by $\eta(X, t)$ the trajectory of a fluid particle that at time $t = 0$ is at $X \in \Omega$. For fixed t the map η_t defined by $\eta_t(X) = \eta(X, t)$ is a diffeomorphism of Ω. In fact, since the fluid is incompressible, we have $\eta_t \in \mathrm{Diff}_{\mathrm{vol}}(\Omega)$, the group of volume-preserving diffeomorphisms of Ω. Thus, the configuration space for the problem is the infinite-dimensional Lie group $\mathrm{Diff}_{\mathrm{vol}}(\Omega)$. Using $\mathrm{Diff}_{\mathrm{vol}}(\Omega)$ as a symmetry group leads to Kelvin's circulation theorem as a conservation law. See Marsden and Weinstein [1983].

Compressible Fluids.   In this case the configuration space is the whole diffeomorphism group $\mathrm{Diff}(\Omega)$. The symmetry group consists of density-preserving diffeomorphisms $\mathrm{Diff}_\rho(\Omega)$. The density plays a role similar to that of gravity in the heavy top and again leads to semidirect products, as does the next example.

Magnetohydrodynamics (MHD).   This example is that of a compressible fluid consisting of charged particles with the dominant electromagnetic force being the magnetic field produced by the particles themselves (possibly together with an external field). The configuration space remains $\mathrm{Diff}(\Omega)$, but the fluid motion is coupled with the magnetic field (regarded as a two-form on Ω).

Maxwell–Vlasov Equations.   Let $f(\mathbf{x}, \mathbf{v}, t)$ denote the density function of a collisionless plasma. The function f evolves in time by means of a time-dependent canonical transformation on $\mathbb{R}^6$, that is, $(\mathbf{x}, \mathbf{v})$-space. In other words, the evolution of f can be described by $f_t = \eta_t^* f_0$, where f_0 is the initial value of f, f_t its value at time t, and η_t is a canonical transformation. Thus, $\mathrm{Diff}_{\mathrm{can}}(\mathbb{R}^6)$, the group of canonical transformations, plays an important role.

Maxwell's Equations   Maxwell's equations for electrodynamics are invariant under gauge transformations that transform the magnetic (or 4) potential by $\mathbf{A} \mapsto \mathbf{A} + \nabla\varphi$. This gauge group is an infinite-dimensional Lie group. The conserved quantity associated with the gauge symmetry in this case is the charge.

9.1   Basic Definitions and Properties

Definition 9.1.1. *A **Lie group** is a (Banach) manifold G that has a group structure consistent with its manifold structure in the sense that group multiplication*

$$\mu : G \times G \to G, \quad (g, h) \mapsto gh,$$

is a C^∞ map.

The maps $L_g : G \to G$, $h \mapsto gh$, and $R_h : G \to G$, $g \mapsto gh$, are called the **left and right translation maps**. Note that

$$L_{g_1} \circ L_{g_2} = L_{g_1 g_2} \quad \text{and} \quad R_{h_1} \circ R_{h_2} = R_{h_2 h_1}.$$

If $e \in G$ denotes the identity element, then $L_e = \text{Id} = R_e$, and so

$$(L_g)^{-1} = L_{g^{-1}} \quad \text{and} \quad (R_h)^{-1} = R_{h^{-1}}.$$

Thus, L_g and R_h are diffeomorphisms for each g and h. Notice that

$$L_g \circ R_h = R_h \circ L_g,$$

that is, left and right translation commute. By the chain rule,

$$T_{gh} L_{g^{-1}} \circ T_h L_g = T_h (L_{g^{-1}} \circ L_g) = \text{Id}.$$

Thus, $T_h L_g$ is invertible. Likewise, $T_g R_h$ is an isomorphism.

We now show that the **inversion map** $I : G \to G$; $g \mapsto g^{-1}$ is C^∞. Indeed, consider solving

$$\mu(g, h) = e$$

for h as a function of g. The partial derivative with respect to h is just $T_h L_g$, which is an isomorphism. Thus, the solution g^{-1} is a smooth function of g by the implicit function theorem.

Lie groups can be finite- or infinite-dimensional. For a first reading of this section, the reader may wish to assume that G is finite-dimensional.[1]

Examples

(a)   Any Banach space V is an Abelian Lie group with group operations

$$\mu : V \times V \to V, \quad \mu(x, y) = x + y, \quad \text{and} \quad I : V \to V, \quad I(x) = -x.$$

The identity is just the zero vector. We call such a Lie group a **vector group**.   ◆

---

[1] We caution that some interesting infinite-dimensional groups (such as groups of diffeomorphisms) are *not* Banach–Lie groups in the (naive) sense just given.

(b)   The group of linear isomorphisms of $\mathbb{R}^n$ to $\mathbb{R}^n$ is a Lie group of dimension n^2, called the **general linear group** and denoted by $\mathrm{GL}(n,\mathbb{R})$. It is a smooth manifold, since it is an open subset of the vector space $L(\mathbb{R}^n,\mathbb{R}^n)$ of all linear maps of $\mathbb{R}^n$ to $\mathbb{R}^n$. Indeed, $\mathrm{GL}(n,\mathbb{R})$ is the inverse image of $\mathbb{R}\backslash\{0\}$ under the continuous map $A \mapsto \det A$ of $L(\mathbb{R}^n,\mathbb{R}^n)$ to $\mathbb{R}$. For $A, B \in \mathrm{GL}(n,\mathbb{R})$, the group operation is composition,

$$\mu : \mathrm{GL}(n,\mathbb{R}) \times \mathrm{GL}(n,\mathbb{R}) \to \mathrm{GL}(n,\mathbb{R})$$

given by

$$(A,B) \mapsto A \circ B,$$

and the inversion map is

$$I : \mathrm{GL}(n,\mathbb{R}) \to \mathrm{GL}(n,\mathbb{R})$$

defined by

$$I(A) = A^{-1}.$$

Group multiplication is the restriction of the continuous bilinear map

$$(A,B) \in L(\mathbb{R}^n,\mathbb{R}^n) \times L(\mathbb{R}^n,\mathbb{R}^n) \mapsto A \circ B \in L(\mathbb{R}^n,\mathbb{R}^n).$$

Thus, μ is C^∞, and so $\mathrm{GL}(n,\mathbb{R})$ is a Lie group.

The group identity element e is the identity map on $\mathbb{R}^n$. If we choose a basis in $\mathbb{R}^n$, we can represent each $A \in \mathrm{GL}(n,\mathbb{R})$ by an invertible $n \times n$ matrix. The group operation is then matrix multiplication $\mu(A,B) = AB$, and $I(A) = A^{-1}$ is matrix inversion. The identity element e is the $n \times n$ identity matrix. The group operations are obviously smooth, since the formulas for the product and inverse of matrices are smooth (rational) functions of the matrix components.   ◆

(c)   In the same way, one sees that for a Banach space V, the group $\mathrm{GL}(V,V)$ of invertible elements of $L(V,V)$ is a Banach–Lie group. For the proof that this is open in $L(V,V)$, see Abraham, Marsden, and Ratiu [1988]. Further examples are given in the next section.   ◆

Charts.   Given any local chart on G, one can construct an entire atlas on the Lie group G by use of left (or right) translations. Suppose, for example, that (U,φ) is a chart about $e \in G$, and that $\varphi : U \to V$. Define a chart (U_g,φ_g) about $g \in G$ by letting

$$U_g = L_g(U) = \{\, L_g h \mid h \in U \,\}$$

and defining

$$\varphi_g = \varphi \circ L_{g^{-1}} : U_g \to V, \ h \mapsto \varphi(g^{-1}h).$$

The set of charts $\{(U_g,\varphi_g)\}$ forms an atlas, provided that one can show that the transition maps

$$\varphi_{g_1} \circ \varphi_{g_2}^{-1} = \varphi \circ L_{g_1^{-1}g_2} \circ \varphi^{-1} : \varphi_{g_2}(U_{g_1} \cap U_{g_2}) \to \varphi_{g_1}(U_{g_1} \cap U_{g_2})$$

are diffeomorphisms (between open sets in a Banach space). But this follows from the smoothness of group multiplication and inversion.

Invariant Vector Fields. A vector field X on G is called *left invariant* if for every $g \in G$ we have $L_g^* X = X$, that is, if

$$(T_h L_g) X(h) = X(gh)$$

for every $h \in G$. We have the commutative diagram in Figure 9.1.1 and illustrate the geometry in Figure 9.1.2.

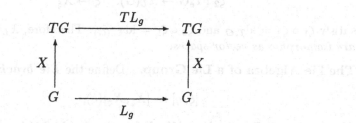

FIGURE 9.1.1. The commutative diagram for a left-invariant vector field.

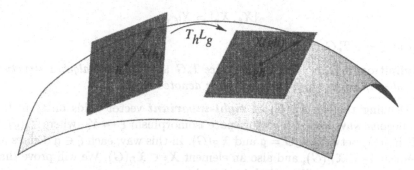

FIGURE 9.1.2. A left-invariant vector field.

Let $\mathfrak{X}_L(G)$ denote the set of left-invariant vector fields on G. If $g \in G$ and $X, Y \in \mathfrak{X}_L(G)$, then

$$L_g^*[X, Y] = [L_g^* X, L_g^* Y] = [X, Y],$$

so $[X, Y] \in \mathfrak{X}_L(G)$. Therefore, $\mathfrak{X}_L(G)$ is a Lie subalgebra of $\mathfrak{X}(G)$, the set of all vector fields on G.

For each $\xi \in T_e G$, we define a vector field X_ξ on G by letting

$$X_\xi(g) = T_e L_g(\xi).$$

Then

$$X_\xi(gh) = T_e L_{gh}(\xi) = T_e(L_g \circ L_h)(\xi)$$
$$= T_h L_g(T_e L_h(\xi)) = T_h L_g(X_\xi(h)),$$

which shows that X_ξ is left invariant. The linear maps

$$\zeta_1 : \mathfrak{X}_L(G) \to T_e G, \quad X \mapsto X(e)$$

and

$$\zeta_2 : T_e G \to \mathfrak{X}_L(G), \quad \xi \mapsto X_\xi$$

satisfy $\zeta_1 \circ \zeta_2 = \mathrm{id}_{T_e G}$ and $\zeta_2 \circ \zeta_1 = \mathrm{id}_{\mathfrak{X}_L(G)}$. Therefore, $\mathfrak{X}_L(G)$ and $T_e G$ are isomorphic as vector spaces.

The Lie Algebra of a Lie Group. Define the *Lie bracket* in $T_e G$ by

$$[\xi, \eta] := [X_\xi, X_\eta](e),$$

where $\xi, \eta \in T_e G$ and where $[X_\xi, X_\eta]$ is the Jacobi–Lie bracket of vector fields. This clearly makes $T_e G$ into a Lie algebra. (Lie algebras were defined in the Introduction.) We say that this defines a bracket in $T_e G$ via *left extension*. Note that by construction,

$$[X_\xi, X_\eta] = X_{[\xi, \eta]}$$

for all $\xi, \eta \in T_e G$.

Definition 9.1.2. *The vector space $T_e G$ with this Lie algebra structure is called the **Lie algebra** of G and is denoted by $\mathfrak{g}$.*

Defining the set $\mathfrak{X}_R(G)$ of *right-invariant* vector fields on G in the analogous way, we get a vector space isomorphism $\xi \mapsto Y_\xi$, where $Y_\xi(g) = (T_e R_g)(\xi)$, between $T_e G = \mathfrak{g}$ and $\mathfrak{X}_R(G)$. In this way, each $\xi \in \mathfrak{g}$ defines an element $Y_\xi \in \mathfrak{X}_R(G)$, and also an element $X_\xi \in \mathfrak{X}_L(G)$. We will prove that a relation between X_ξ and Y_ξ is given by

$$I_* X_\xi = -Y_\xi, \qquad (9.1.1)$$

where $I : G \to G$ is the inversion map: $I(g) = g^{-1}$. Since I is a diffeomorphism, (9.1.1) shows that $I_* : \mathfrak{X}_L(G) \to \mathfrak{X}_R(G)$ is a vector space isomorphism. To prove (9.1.1) notice first that for $u \in T_g G$ and $v \in T_h G$, the derivative of the multiplication map has the expression

$$T_{(g,h)}\mu(u, v) = T_h L_g(v) + T_g R_h(u). \qquad (9.1.2)$$

In addition, differentiating the map $g \mapsto \mu(g, I(g)) = e$ gives

$$T_{(g, g^{-1})}\mu(u, T_g I(u)) = 0$$

for all $u \in T_g G$. This and (9.1.2) yield

$$T_g I(u) = -(T_e R_{g^{-1}} \circ T_g L_{g^{-1}})(u), \qquad (9.1.3)$$

for all $u \in T_g G$. Consequently, if $\xi \in \mathfrak{g}$, and $g \in G$, we have

$$
\begin{aligned}
(I_* X_\xi)(g) &= (TI \circ X_\xi \circ I^{-1})(g) = T_{g^{-1}} I(X_\xi(g^{-1})) \\
&= -(T_e R_g \circ T_{g^{-1}} L_g)(X_\xi(g^{-1})) && \text{(by (9.1.3))} \\
&= -T_e R_g(\xi) = -Y_\xi(g) && \text{(since } X_\xi(g^{-1}) = T_e L_{g^{-1}}(\xi))
\end{aligned}
$$

and (9.1.1) is proved. Hence for $\xi, \eta \in \mathfrak{g}$,

$$
\begin{aligned}
-Y_{[\xi,\eta]} &= I_* X_{[\xi,\eta]} = I_*[X_\xi, X_\eta] = [I_* X_\xi, I_* X_\eta] \\
&= [-Y_\xi, -Y_\eta] = [Y_\xi, Y_\eta],
\end{aligned}
$$

so that

$$-[Y_\xi, Y_\eta](e) = Y_{[\xi,\eta]}(e) = [\xi, \eta] = [X_\xi, X_\eta](e).$$

Therefore, the Lie algebra bracket $[\,,]^R$ in $\mathfrak{g}$ defined by **right extension** of elements in $\mathfrak{g}$,

$$[\xi, \eta]^R := [Y_\xi, Y_\eta](e),$$

is the *negative* of the one defined by left extension, that is,

$$[\xi, \eta]^R := -[\xi, \eta].$$

Examples

(a) For a vector group V, $T_e V \cong V$; it is easy to see that the left-invariant vector field defined by $u \in T_e V$ is the constant vector field $X_u(v) = u$ for all $v \in V$. Therefore, the Lie algebra of a vector group V is V itself, with the trivial bracket $[v, w] = 0$ for all $v, w \in V$. We say that the Lie algebra is **Abelian** in this case. ◆

(b) The Lie algebra of $\mathrm{GL}(n, \mathbb{R})$ is $L(\mathbb{R}^n, \mathbb{R}^n)$, also denoted by $\mathfrak{gl}(n)$, the vector space of all linear transformations of $\mathbb{R}^n$, with the commutator bracket

$$[A, B] = AB - BA.$$

To see this, we recall that $\mathrm{GL}(n, \mathbb{R})$ is open in $L(\mathbb{R}^n, \mathbb{R}^n)$, and so the Lie algebra, as a vector space, is $L(\mathbb{R}^n, \mathbb{R}^n)$. To compute the bracket, note that for any $\xi \in L(\mathbb{R}^n, \mathbb{R}^n)$,

$$X_\xi : \mathrm{GL}(n, \mathbb{R}) \to L(\mathbb{R}^n, \mathbb{R}^n)$$

given by $A \mapsto A\xi$ is a left-invariant vector field on $\mathrm{GL}(n, \mathbb{R})$ because for every $B \in \mathrm{GL}(n, \mathbb{R})$, the map

$$L_B : \mathrm{GL}(n, \mathbb{R}) \to \mathrm{GL}(n, \mathbb{R})$$

defined by $L_B(A) = BA$ is a linear mapping, and hence

$$X_\xi(L_B A) = BA\xi = T_A L_B X_\xi(A).$$

Therefore, by the local formula

$$[X, Y](x) = \mathbf{D}Y(x) \cdot X(x) - \mathbf{D}X(x) \cdot Y(x),$$

we get

$$[\xi, \eta] = [X_\xi, X_\eta](I) = \mathbf{D}X_\eta(I) \cdot X_\xi(I) - \mathbf{D}X_\xi(I) \cdot X_\eta(I).$$

But $X_\eta(A) = A\eta$ is linear in A, so $\mathbf{D}X_\eta(I) \cdot B = B\eta$. Hence

$$\mathbf{D}X_\eta(I) \cdot X_\xi(I) = \xi\eta,$$

and similarly

$$\mathbf{D}X_\xi(I) \cdot X_\eta(I) = \eta\xi.$$

Thus, $L(\mathbb{R}^n, \mathbb{R}^n)$ has the bracket

$$[\xi, \eta] = \xi\eta - \eta\xi. \tag{9.1.4}$$

◆

(c)   We can also establish (9.1.4) by a coordinate calculation. Choosing a basis in $\mathbb{R}^n$, each $A \in \mathrm{GL}(n, \mathbb{R})$ is specified by its components A^i_j such that $(Av)^i = A^i_j v^j$ (sum on j). Thus, a vector field X on $\mathrm{GL}(n, \mathbb{R})$ has the form $X(A) = \sum_{i,j} C^i_j(A)(\partial/\partial A^i_j)$. It is checked to be left invariant, provided that there is a matrix (ξ^i_j) such that for all A,

$$X(A) = \sum_{i,j,k} A^i_k \xi^k_j \frac{\partial}{\partial A^i_j}.$$

If $Y(A) = \sum_{i,j,k} A^i_k \eta^k_j(\partial/\partial A^i_j)$ is another left-invariant vector field, we have

$$(XY)[f] = \sum A^i_k \xi^k_j \frac{\partial}{\partial A^i_j} \left[\sum A^l_m \eta^m_p \frac{\partial f}{\partial A^l_p} \right]$$

$$= \sum A^i_k \xi^k_j \delta^l_i \delta^j_m \eta^m_p \frac{\partial f}{\partial A^l_p} + \text{(second derivatives)}$$

$$= \sum A^i_k \xi^k_j \eta^j_m \frac{\partial f}{\partial A^i_j} + \text{(second derivatives)},$$

where we have used $\partial A^s_m / \partial A^k_j = \delta^k_s \delta^j_m$. Therefore, the bracket is the left-invariant vector field $[X, Y]$ given by

$$[X, Y][f] = (XY - YX)[f] = \sum A^i_k (\xi^k_j \eta^j_m - \eta^k_j \xi^j_m) \frac{\partial f}{\partial A^i_m}.$$

This shows that the vector field bracket is the usual commutator bracket of $n \times n$ matrices, as before.

◆

One-Parameter Subgroups and the Exponential Map. If X_ξ is the left-invariant vector field corresponding to $\xi \in \mathfrak{g}$, there is a unique integral curve $\gamma_\xi : \mathbb{R} \to G$ of X_ξ starting at e, $\gamma_\xi(0) = e$ and $\gamma_\xi'(t) = X_\xi(\gamma_\xi(t))$. We claim that

$$\gamma_\xi(s+t) = \gamma_\xi(s)\gamma_\xi(t),$$

which means that $\gamma_\xi(t)$ is a smooth **one-parameter subgroup**. Indeed, as functions of t, both sides equal $\gamma_\xi(s)$ at $t = 0$ and both satisfy the differential equation $\sigma'(t) = X_\xi(\sigma(t))$ by left invariance of X_ξ, so they are equal. Left invariance or $\gamma_\xi(t+s) = \gamma_\xi(t)\gamma_\xi(s)$ also shows that $\gamma_\xi(t)$ is defined for all $t \in \mathbb{R}$.

Definition 9.1.3. *The **exponential map** $\exp : \mathfrak{g} \to G$ is defined by*

$$\exp(\xi) = \gamma_\xi(1).$$

We claim that

$$\exp(s\xi) = \gamma_\xi(s).$$

Indeed, for fixed $s \in \mathbb{R}$, the curve $t \mapsto \gamma_\xi(ts)$, which at $t = 0$ passes through e, satisfies the differential equation

$$\frac{d}{dt}\gamma_\xi(ts) = sX_\xi(\gamma_\xi(ts)) = X_{s\xi}(\gamma_\xi(ts)).$$

Since $\gamma_{s\xi}(t)$ satisfies the same differential equation and passes through e at $t = 0$, it follows that $\gamma_{s\xi}(t) = \gamma_\xi(ts)$. Putting $t = 1$ yields $\exp(s\xi) = \gamma_\xi(s)$.

Hence the exponential mapping maps the line $s\xi$ in $\mathfrak{g}$ onto the one-parameter subgroup $\gamma_\xi(s)$ of G, which is tangent to ξ at e. It follows from left invariance that the flow F_t^ξ of X_ξ satisfies $F_t^\xi(g) = gF_t^\xi(e) = g\gamma_\xi(t)$, so

$$F_t^\xi(g) = g\exp(t\xi) = R_{\exp t\xi}g.$$

Let $\gamma(t)$ be a smooth one-parameter subgroup of G, so $\gamma(0) = e$ in particular. We claim that $\gamma = \gamma_\xi$, where $\xi = \gamma'(0)$. Indeed, taking the derivative at $s = 0$ in the relation $\gamma(t+s) = \gamma(t)\gamma(s)$ gives

$$\frac{d\gamma(t)}{dt} = \frac{d}{ds}\bigg|_{s=0} L_{\gamma(t)}\gamma(s) = T_eL_{\gamma(t)}\gamma'(0) = X_\xi(\gamma(t)),$$

so that $\gamma = \gamma_\xi$, since both equal e at $t = 0$. In other words, *all smooth one-parameter subgroups of G are of the form $\exp t\xi$ for some $\xi \in \mathfrak{g}$.* Since everything proved above for X_ξ can be repeated for Y_ξ, it follows that *the exponential map is the same for the left and right Lie algebras of a Lie group.*

From smoothness of the group operations and smoothness of the solutions of differential equations with respect to initial conditions, it follows that exp is a C^∞ map. Differentiating the identity $\exp(s\xi) = \gamma_\xi(s)$ with respect to s at $s = 0$ shows that $T_0 \exp = \mathrm{id}_\mathfrak{g}$. Therefore, by the inverse function theorem, exp is a local diffeomorphism from a neighborhood of zero in $\mathfrak{g}$ onto a neighborhood of e in G. In other words, the exponential map defines a local chart for G at e; in finite dimensions, the coordinates associated to this chart are called the **canonical coordinates** of G. By left translation, this chart provides an atlas for G. (For typical infinite-dimensional groups like diffeomorphism groups, exp is *not* locally onto a neighborhood of the identity. It is *also not true* that the exponential map is a local diffeomorphism at any $\xi \neq 0$, even for finite-dimensional Lie groups.)

It turns out that the exponential map characterizes not only the *smooth* one-parameter subgroups of G, but the *continuous* ones as well, as given in the next proposition (see Varadarajan [1974] for the proof).

Proposition 9.1.4. *Let $\gamma : \mathbb{R} \to G$ be a continuous one-parameter subgroup of G. Then γ is automatically smooth, and hence $\gamma(t) = \exp t\xi$, for some $\xi \in \mathfrak{g}$.*

Examples

(a)   Let $G = V$ be a vector group, that is, V is a vector space and the group operation is vector addition. Then $\mathfrak{g} = V$ and $\exp : V \to V$ is the identity mapping.   ◆

(b)   Let $G = \mathrm{GL}(n, \mathbb{R})$; so $\mathfrak{g} = L(\mathbb{R}^n, \mathbb{R}^n)$. For every $A \in L(\mathbb{R}^n, \mathbb{R}^n)$, the mapping $\gamma_A : \mathbb{R} \to \mathrm{GL}(n, \mathbb{R})$ defined by

$$t \mapsto \sum_{i=0}^{\infty} \frac{t^i}{i!} A^i$$

is a one-parameter subgroup, because $\gamma_A(0) = I$ and

$$\gamma_A'(t) = \sum_{i=0}^{\infty} \frac{t^{i-1}}{(i-1)!} A^i = \gamma_A(t)A.$$

Therefore, the exponential mapping is given by

$$\exp : L(\mathbb{R}^n, \mathbb{R}^n) \to \mathrm{GL}(n, \mathbb{R}^n), \quad A \mapsto \gamma_A(1) = \sum_{i=0}^{\infty} \frac{A^i}{i!}.$$

As is customary, we will write

$$e^A = \sum_{i=0}^{\infty} \frac{A^i}{i!}.$$

We sometimes write $\exp_G : \mathfrak{g} \to G$ when there is more than one group involved.   ◆

(c)   Let G_1 and G_2 be Lie groups with Lie algebras $\mathfrak{g}_1$ and $\mathfrak{g}_2$. Then $G_1 \times G_2$ is a Lie group with Lie algebra $\mathfrak{g}_1 \times \mathfrak{g}_2$, and the exponential map is given by

$$\exp : \mathfrak{g}_1 \times \mathfrak{g}_2 \to G_1 \times G_2, \quad (\xi_1, \xi_2) \mapsto (\exp_1(\xi_1), \exp_2(\xi_2)). \qquad \blacklozenge$$

Computing Brackets.   Here is a *computationally useful formula for the bracket*. One follows these three steps:

1. Calculate the *inner automorphisms*

$$I_g : G \to G, \quad \text{where} \quad I_g(h) = ghg^{-1}.$$

2. Differentiate $I_g(h)$ with respect to h at $h = e$ to produce the *adjoint operators*

$$\mathrm{Ad}_g : \mathfrak{g} \to \mathfrak{g}; \quad \mathrm{Ad}_g \, \eta = T_e I_g \cdot \eta.$$

   Note that (see Figure 9.1.3)

$$\mathrm{Ad}_g \, \eta = T_{g^{-1}} L_g \cdot T_e R_{g^{-1}} \cdot \eta.$$

3. Differentiate $\mathrm{Ad}_g \, \eta$ with respect to g at e in the direction ξ to get $[\xi, \eta]$, that is,

$$T_e \varphi^\eta \cdot \xi = [\xi, \eta], \qquad (9.1.5)$$

   where $\varphi^\eta(g) = \mathrm{Ad}_g \, \eta$.

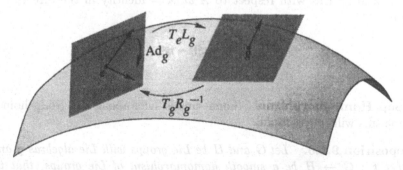

FIGURE 9.1.3. The adjoint mapping is the linearization of conjugation.

Proposition 9.1.5.   *Formula (9.1.5) is valid.*

Proof. Denote by $\varphi_t(g) = g \exp t\xi = R_{\exp t\xi}\, g$ the flow of X_ξ. Then

$$[\xi, \eta] = [X_\xi, X_\eta](e) = \frac{d}{dt} T_{\varphi_t(e)} \varphi_t^{-1} \cdot X_\eta(\varphi_t(e)) \Big|_{t=0}$$

$$= \frac{d}{dt} T_{\exp t\xi}\, R_{\exp(-t\xi)}\, X_\eta(\exp t\xi) \Big|_{t=0}$$

$$= \frac{d}{dt} T_{\exp t\xi}\, R_{\exp(-t\xi)}\, T_e L_{\exp t\xi}\, \eta \Big|_{t=0}$$

$$= \frac{d}{dt} T_e (L_{\exp t\xi} \circ R_{\exp(-t\xi)}) \eta \Big|_{t=0}$$

$$= \frac{d}{dt} \mathrm{Ad}_{\exp t\xi}\, \eta \Big|_{t=0},$$

which is (9.1.5).   ∎

Another way of expressing (9.1.5) is

$$[\xi, \eta] = \frac{d}{dt} \frac{d}{ds} g(t) h(s) g(t)^{-1} \Big|_{s=0, t=0}, \qquad (9.1.6)$$

where $g(t)$ and $h(s)$ are curves in G with $g(0) = e$, $h(0) = e$, and where $g'(0) = \xi$ and $h'(0) = \eta$.

Example. Consider the group $\mathrm{GL}(n, \mathbb{R})$. Formula (9.1.4) also follows from (9.1.5). Here, $I_A B = ABA^{-1}$, and so

$$\mathrm{Ad}_A\, \eta = A \eta A^{-1}.$$

Differentiating this with respect to A at $A = $ Identity in the direction ξ gives

$$[\xi, \eta] = \xi \eta - \eta \xi. \qquad \blacklozenge$$

Group Homomorphisms. Some simple facts about Lie group homomorphisms will prove useful.

Proposition 9.1.6. *Let G and H be Lie groups with Lie algebras $\mathfrak{g}$ and $\mathfrak{h}$. Let $f : G \to H$ be a smooth homomorphism of Lie groups, that is, $f(gh) = f(g)f(h)$, for all $g, h \in G$. Then $T_e f : \mathfrak{g} \to \mathfrak{h}$ is a Lie algebra homomorphism, that is, $(T_e f)[\xi, \eta] = [T_e f(\xi), T_e f(\eta)]$, for all $\xi, \eta \in \mathfrak{g}$. In addition,*

$$f \circ \exp_G = \exp_H \circ T_e f.$$

Proof. Since f is a group homomorphism, $f \circ L_g = L_{f(g)} \circ f$. Thus, $Tf \circ TL_g = TL_{f(g)} \circ Tf$, from which it follows that

$$X_{T_e f(\xi)}(f(g)) = T_g f(X_\xi(g)),$$

that is, X_ξ and $X_{T_e f(\xi)}$ are *f-related*. It follows that the vector fields $[X_\xi, X_\eta]$ and $[X_{T_e f(\xi)}, X_{T_e f(\eta)}]$ are also f-related for all $\xi, \eta \in \mathfrak{g}$ (see Abraham, Marsden, and Ratiu [1988, Section 4.2]). Hence

$$
\begin{aligned}
T_e f([\xi, \eta]) &= (Tf \circ [X_\xi, X_\eta])(e) \quad &&\text{(where } e = e_G) \\
&= [X_{T_e f(\xi)}, X_{T_e f(\eta)}](\bar{e}) \quad &&\text{(where } \bar{e} = e_H = f(e)) \\
&= [T_e f(\xi), T_e f(\eta)].
\end{aligned}
$$

Thus, $T_e f$ is a Lie algebra homomorphism.

Fixing $\xi \in \mathfrak{g}$, note that $\alpha : t \mapsto f(\exp_G(t\xi))$ and $\beta : t \mapsto \exp_H(t T_e f(\xi))$ are one-parameter subgroups of H. Moreover, $\alpha'(0) = T_e f(\xi) = \beta'(0)$, and so $\alpha = \beta$. In particular, $f(\exp_G(\xi)) = \exp_H(T_e f(\xi))$, for all $\xi \in \mathfrak{g}$. ∎

Example. Proposition 9.1.6 applied to the determinant map gives the identity

$$\det(\exp A) = \exp(\operatorname{trace} A)$$

for $A \in \mathrm{GL}(n, \mathbb{R})$. ◆

Corollary 9.1.7. *Assume that $f_1, f_2 : G \to H$ are homomorphisms of Lie groups and that G is connected. If $T_e f_1 = T_e f_2$, then $f_1 = f_2$.*

This follows from Proposition 9.1.6, since a connected Lie group G is generated by a neighborhood of the identity element. This latter fact may be proved following these steps:

1. Show that any open subgroup of a Lie group is closed (since its complement is a union of sets homeomorphic to it).

2. Show that a subgroup of a Lie group is open if and only if it contains a neighborhood of the identity element.

3. Conclude that a Lie group is connected if and only if it is generated by arbitrarily small neighborhoods of the identity element.

From Proposition 9.1.6 and the fact that the inner automorphisms are group homomorphisms, we get the following corollary.

Corollary 9.1.8.

(i) $\exp(\mathrm{Ad}_g \, \xi) = g(\exp \xi)g^{-1}$, *for every $\xi \in \mathfrak{g}$ and $g \in G$; and*

(ii) $\mathrm{Ad}_g[\xi, \eta] = [\mathrm{Ad}_g \, \xi, \mathrm{Ad}_g \, \eta]$.

More Automatic Smoothness Results.   There are some interesting results related in spirit to Proposition 9.1.4 and the preceding discussions. A striking example of this is the following:

Theorem 9.1.9.   *Any continuous homomorphism of finite-dimensional Lie groups is smooth.*

There is a remarkable consequence of this theorem. If G is a topological group (that is, the multiplication and inversion maps are continuous), one could, in principle, have more than one differentiable manifold structure making G into two nonisomorphic Lie groups (i.e., the manifold structures are not diffeomorphic) but both inducing the same topological structure. This phenomenon of "exotic structures" occurs for general manifolds. However, in view of the theorem above, this cannot happen in the case of Lie groups. Indeed, since the identity map is a homeomorphism, it must be a diffeomorphism. Thus, *a topological group that is locally Euclidean (i.e., there is an open neighborhood of the identity homeomorphic to an open ball in $\mathbb{R}^n$) admits at most one smooth manifold structure relative to which it is a Lie group.*

The existence part of this statement is Hilbert's famous fifth problem: Show that a locally Euclidean topological group admits a smooth (actually analytic) structure making it into a Lie group. The solution of this problem was achieved by Gleason and, independently, by Montgomery and Zippin in 1952; see Kaplansky [1971] for an excellent account of this proof.

Abelian Lie Groups.   Since any two elements of an Abelian Lie group G commute, it follows that all adjoint operators Ad_g, $g \in G$, equal the identity. Therefore, by equation (9.1.5), the Lie algebra $\mathfrak{g}$ is Abelian; that is, $[\xi, \eta] = 0$ for all $\xi, \eta \in \mathfrak{g}$.

Examples

(a)   Any finite-dimensional vector space, thought of as an Abelian group under addition, is an Abelian Lie group. The same is true in infinite dimensions for any Banach space. The exponential map is the identity.   ♦

(b)   The unit circle in the complex plane $S^1 = \{ z \in \mathbb{C} \mid |z| = 1 \}$ is an Abelian Lie group under multiplication. The tangent space $T_e S^1$ is the imaginary axis, and we identify $\mathbb{R}$ with $T_e S^1$ by $t \mapsto 2\pi i t$. With this identification, the exponential map $\exp : \mathbb{R} \to S^1$ is given by $\exp(t) = e^{2\pi i t}$. Note that $\exp^{-1}(1) = \mathbb{Z}$.   ♦

(c)   The n-dimensional torus $\mathbb{T}^n = S^1 \times \cdots \times S^1$ (n times) is an Abelian Lie group. The exponential map $\exp : \mathbb{R}^n \to \mathbb{T}^n$ is given by

$$\exp(t_1, \ldots, t_n) = (e^{2\pi i t_1}, \ldots, e^{2\pi i t_n}).$$

Since $S^1 = \mathbb{R}/\mathbb{Z}$, it follows that

$$\mathbb{T}^n = \mathbb{R}/\mathbb{Z}^n,$$

the projection $\mathbb{R}^n \to \mathbb{T}^n$ being given by exp above. ♦

If G is a connected Lie group whose Lie algebra $\mathfrak{g}$ is Abelian, the Lie group homomorphism $g \in G \mapsto \mathrm{Ad}_g \in \mathrm{GL}(\mathfrak{g})$ has induced Lie algebra homomorphism $\xi \in \mathfrak{g} \mapsto \mathrm{ad}_\xi \in \mathfrak{gl}(\mathfrak{g})$ the constant map equal to zero. Therefore, by Corollary 9.1.7, $\mathrm{Ad}_g = $ identity on G, for any $g \in G$. Apply Corollary 9.1.7 again, this time to the conjugation by g on G (whose induced Lie algebra homomorphism is Ad_g), to conclude that it equals the identity map on G. Thus, g commutes with all elements of G; since g was arbitrary, we conclude that G is Abelian. We summarize these observations in the following proposition.

Proposition 9.1.10. *If G is an Abelian Lie group, its Lie algebra $\mathfrak{g}$ is also Abelian. Conversely, if G is connected and $\mathfrak{g}$ is Abelian, then G is Abelian.*

The main structure theorem for Abelian Lie groups is the following, whose proof can be found in Varadarajan [1974] or Knapp [1996].

Theorem 9.1.11. *Every connected Abelian n-dimensional Lie group G is isomorphic to a cylinder, that is, to $\mathbb{T}^k \times \mathbb{R}^{n-k}$ for some $k = 1, \dots, n$.*

Lie Subgroups. It is natural to synthesize the subgroup and submanifold concepts.

Definition 9.1.12. *A **Lie subgroup** H of a Lie group G is a subgroup of G that is also an injectively immersed submanifold of G. If H is a submanifold of G, then H is called a **regular** Lie subgroup.*

For example, the one-parameter subgroups of the torus $\mathbb{T}^2$ that wind densely on the torus are Lie subgroups that are *not* regular.

The Lie algebras $\mathfrak{g}$ and $\mathfrak{h}$ of G and a Lie subgroup H, respectively, are related in the following way:

Proposition 9.1.13. *Let H be a Lie subgroup of G. Then $\mathfrak{h}$ is a Lie subalgebra of $\mathfrak{g}$. Moreover,*

$$\mathfrak{h} = \{\, \xi \in \mathfrak{g} \mid \exp t\xi \in H \text{ for all } t \in \mathbb{R} \,\}.$$

Proof. The first statement is a consequence of Proposition 9.1.6, which also shows that $\exp t\xi \in H$, for all $\xi \in \mathfrak{h}$ and $t \in \mathbb{R}$. Conversely, if $\exp t\xi \in H$, for all $t \in \mathbb{R}$, we have,

$$\left. \frac{d}{dt} \exp t\xi \right|_{t=0} \in \mathfrak{h},$$

since H is a Lie subgroup; but this equals ξ by definition of the exponential map. ∎

The following is a powerful theorem often used to find Lie subgroups.

Theorem 9.1.14. *If H is a closed subgroup of a Lie group G, then H is a regular Lie subgroup. Conversely, if H is a regular Lie subgroup of G, then H is closed.*

The proof of this theorem may be found in Abraham and Marsden [1978], Adams [1969], Varadarajan [1974], or Knapp [1996].

We remind the reader that the Lie algebras appropriate to fluid dynamics and plasma physics are infinite-dimensional. Nevertheless, there is still, with the appropriate technical conditions, a correspondence between Lie groups and Lie algebras analogous to the preceding theorems. The reader should be warned, however, that these theorems do not *naively* generalize to the infinite-dimensional situation, and to prove them for special cases, specialized analytical theorems may be required.

The next result is sometimes called "Lie's third fundamental theorem."

Theorem 9.1.15. *Let G be a Lie group with Lie algebra $\mathfrak{g}$, and let $\mathfrak{h}$ be a Lie subalgebra of $\mathfrak{g}$. Then there exists a unique connected Lie subgroup H of G whose Lie algebra is $\mathfrak{h}$.*

The proof may be found in Knapp [1996] or Varadarajan [1974].

Quotients. If H is a closed subgroup of G, we denote by G/H, the set of left cosets, that is, the collection $\{\, gH \mid g \in G \,\}$. Let $\pi : G \to G/H$ be the projection $g \mapsto gH$.

Theorem 9.1.16. *There is a unique manifold structure on G/H such that the projection $\pi : G \to G/H$ is a smooth surjective submersion. (Recall from Chapter 4 that a smooth map is called a submersion when its derivative is surjective.)*

Again the proof may be found in Abraham and Marsden [1978], Knapp [1996], or Varadarajan [1974].

The Maurer–Cartan Equations. We close this section with a proof of the *Maurer–Cartan structure equations* on a Lie group G. Define $\lambda, \rho \in \Omega^1(G; \mathfrak{g})$, the space of $\mathfrak{g}$-valued one-forms on G, by

$$\lambda(u_g) = T_g L_{g^{-1}}(u_g), \quad \rho(u_g) = T_g R_{g^{-1}}(u_g).$$

Thus, λ and ρ are Lie-algebra-valued one-forms on G that are defined by left and right translation to the identity, respectively. Define the two-form $[\lambda, \lambda]$ by

$$[\lambda, \lambda](u, v) = [\lambda(u), \lambda(v)],$$

and similarly for $[\rho, \rho]$.

Theorem 9.1.17 (Maurer–Cartan Structure Equations).

$$\mathbf{d}\lambda + [\lambda, \lambda] = 0, \quad \mathbf{d}\rho - [\rho, \rho] = 0.$$

Proof.   We use identity 6 from the table in §4.4. Let $X, Y \in \mathfrak{X}(G)$ and let $\xi = T_g L_{g^{-1}}(X(g))$ and $\eta = T_g L_{g^{-1}}(Y(g))$ for fixed $g \in G$. Thus,

$$(\mathbf{d}\lambda)(X_\xi, X_\eta) = X_\xi[\lambda(X_\eta)] - X_\eta[\lambda(X_\xi)] - \lambda([X_\xi, X_\eta]).$$

Since $\lambda(X_\eta)(h) = T_h L_{h^{-1}}(X_\eta(h)) = \eta$ is constant, the first term vanishes. Similarly, the second term vanishes. The third term equals

$$\lambda([X_\xi, X_\eta]) = \lambda(X_{[\xi,\eta]}) = [\xi, \eta],$$

and hence

$$(\mathbf{d}\lambda)(X_\xi, X_\eta) = -[\xi, \eta].$$

Therefore,

$$\begin{aligned}
(\mathbf{d}\lambda + [\lambda, \lambda])(X_\xi, X_\eta) &= -[\xi, \eta] + [\lambda, \lambda](X_\xi, X_\eta) \\
&= -[\xi, \eta] + [\lambda(X_\xi), \lambda(X_\eta)] \\
&= -[\xi, \eta] + [\xi, \eta] = 0.
\end{aligned}$$

This proves that

$$(\mathbf{d}\lambda + [\lambda, \lambda])(X, Y)(g) = 0.$$

Since $g \in G$ was arbitrary as well as X and Y, it follows that $\mathbf{d}\lambda + [\lambda, \lambda] = 0$.

The second relation is proved in the same way but working with the right-invariant vector fields Y_ξ, Y_η. The sign in front of the second term changes, since $[Y_\xi, Y_\eta] = Y_{-[\xi,\eta]}$. ∎

Remark.   If α is a $(0, k)$-tensor with values in a Banach space E_1, and β is a $(0, l)$-tensor with values in a Banach space E_2, and if $B : E_1 \times E_2 \to E_3$ is a bilinear map, then replacing multiplication in (4.2.1) by B, the same formula defines an E_3-valued $(0, k + l)$-tensor on M. Therefore, using Definitions 4.2.2–4.2.4, if

$$\alpha \in \Omega^k(M, E_1) \quad \text{and} \quad \beta \in \Omega^l(M, E_2),$$

then

$$\left[\frac{(k+l)!}{k!l!} \right] \mathbf{A}(\alpha \otimes \beta) \in \Omega^{k+l}(M, E_3).$$

We shall call this expression the **wedge product associated to** B and denote it either by $\alpha \wedge_B \beta$ or $B^\wedge(\alpha, \beta)$.

In particular, if $E_1 = E_2 = E_3 = \mathfrak{g}$ and $B = [\ ,\ ]$ is the Lie algebra bracket, then for $\alpha, \beta \in \Omega^1(M; \mathfrak{g})$, we have

$$[\alpha, \beta]^\wedge(u, v) = [\alpha(u), \beta(v)] - [\alpha(v), \beta(u)] = -[\beta, \alpha]^\wedge(u, v)$$

for any vectors u, v tangent to M. Thus, alternatively, one can write the structure equations as

$$d\lambda + \tfrac{1}{2}[\lambda, \lambda]^\wedge = 0, \quad d\rho - \tfrac{1}{2}[\rho, \rho]^\wedge = 0. \qquad \blacklozenge$$

Haar measure.   One can characterize Lebesgue measure up to a multiplicative constant on $\mathbb{R}^n$ by its invariance under translations. Similarly, on a locally compact group there is a unique (up to a nonzero multiplicative constant) left-invariant measure, called **Haar measure**. For Lie groups the existence of such measures is especially simple.

Proposition 9.1.18.   *Let G be a Lie group. Then there is a volume form μ, unique up to nonzero multiplicative constants, that is left invariant. If G is compact, μ is right invariant as well.*

Proof.   Pick any n-form μ_e on T_eG that is nonzero and define an n-form on T_gG by

$$\mu_g(v_1, \ldots, v_n) = \mu_e \cdot (TL_{g^{-1}}v_1, \ldots, TL_{g^{-1}} \cdot v_n).$$

Then μ_g is left invariant and smooth. For $n = \dim G$, μ_e is unique up to a scalar factor, so μ_g is as well.

Fix $g_0 \in G$ and consider $R_{g_0}^* \mu = c\mu$ for a constant c. If G is compact, this relationship may be integrated, and by the change of variables formula we deduce that $c = 1$. Hence, μ is also right invariant. $\blacksquare$

Exercises

⋄ **9.1-1.**   Verify $\text{Ad}_g[\xi, \eta] = [\text{Ad}_g\ \xi, \text{Ad}_g\ \eta]$ directly for $GL(n)$.

⋄ **9.1-2.**   Let G be a Lie group with group operations $\mu : G \times G \to G$ and $I : G \to G$. Show that the tangent bundle TG is also a Lie group, called the **tangent group** of G with group operations $T\mu : TG \times TG \to TG$, $TI : TG \to TG$.

⋄ **9.1-3** (Defining a Lie group by a chart at the identity).   Let G be a group and suppose that $\varphi : U \to V$ is a one-to-one map from a subset U of G containing the identity element to an open subset V in a Banach space (or Banach manifold). The following conditions are necessary and sufficient for φ to be a chart in a Hausdorff–Banach–Lie group structure on G:

(a) The set $W = \{ (x,y) \in V \times V \mid \varphi^{-1}(y) \in U \}$ is open in $V \times V$, and the map $(x,y) \in W \mapsto \varphi(\varphi^{-1}(x)\varphi^{-1}(y)) \in V$ is smooth.

(b) For every $g \in G$, the set $V_g = \varphi(gUg^{-1} \cap U)$ is open in V and the map $x \in V_g \mapsto \varphi(g\varphi^{-1}(x)g^{-1}) \in V$ is smooth.

◇ **9.1-4 (The Heisenberg group).** Let (Z, Ω) be a symplectic vector space and define on $H := Z \times S^1$ the following operation:

$$(u, \exp i\phi)(v, \exp i\psi) = \big(u + v, \exp i[\phi + \psi + \hbar^{-1}\Omega(u, v)]\big) .$$

(a) Verify that this operation gives H the structure of a noncommutative Lie group.

(b) Show that the Lie algebra of H is given by $\mathfrak{h} = Z \times \mathbb{R}$ with the bracket operation[2]

$$[(u, \phi), (v, \psi)] = (0, 2\hbar^{-1}\Omega(u, v)).$$

(c) Show that $[\mathfrak{h}, [\mathfrak{h}, \mathfrak{h}]] = 0$, that is, $\mathfrak{h}$ is **nilpotent**, and that $\mathbb{R}$ lies in the center of the algebra (i.e., $[\mathfrak{h}, \mathbb{R}] = 0$); one says that $\mathfrak{h}$ is a **central extension** of Z.

9.2 Some Classical Lie Groups

The Real General Linear Group $GL(n, \mathbb{R})$. In the previous section we showed that $GL(n, \mathbb{R})$ is a Lie group, that it is an open subset of the vector space of all linear maps of $\mathbb{R}^n$ into itself, and that its Lie algebra is $\mathfrak{gl}(n, \mathbb{R})$ with the commutator bracket. Since it is open in $L(\mathbb{R}^n, \mathbb{R}^n) = \mathfrak{gl}(n, \mathbb{R})$, the group $GL(n, \mathbb{R})$ is not compact. The determinant function $\det : GL(n, \mathbb{R}) \to \mathbb{R}$ is smooth and maps $GL(n, \mathbb{R})$ onto the two components of $\mathbb{R}\backslash\{0\}$. Thus, $GL(n, \mathbb{R})$ is not connected.
Define

$$GL^+(n, \mathbb{R}) = \{ A \in GL(n, \mathbb{R}) \mid \det(A) > 0 \}$$

and note that it is an open (and hence closed) subgroup of $GL(n, \mathbb{R})$. If

$$GL^-(n, \mathbb{R}) = \{ A \in GL(n, \mathbb{R}) \mid \det(A) < 0 \},$$

the map $A \in GL^+(n, \mathbb{R}) \mapsto I_0 A \in GL^-(n, \mathbb{R})$, where I_0 is the diagonal matrix all of whose entries are 1 except the $(1,1)$-entry, which is -1, is a diffeomorphism. We will show below that $GL^+(n, \mathbb{R})$ is connected, which

---

[2] This formula for the bracket, when applied to the space $Z = \mathbb{R}^{2n}$ of the usual p's and q's , shows that this algebra is the same as that encountered in elementary quantum mechanics via the Heisenberg commutation relations. Hence the name "Heisenberg group."

will prove that $GL^+(n,\mathbb{R})$ is the connected component of the identity in $GL(n,\mathbb{R})$ and that $GL(n,\mathbb{R})$ has exactly two connected components.

To prove this we need a theorem from linear algebra called the polar decomposition theorem. To formulate it, recall that a matrix $R \in GL(n,\mathbb{R})$ is **orthogonal** if $RR^T = R^TR = I$. A matrix $S \in \mathfrak{gl}(n,\mathbb{R})$ is called **symmetric** if $S^T = S$. A symmetric matrix S is called **positive definite**, denoted by $S > 0$, if

$$\langle S\mathbf{v}, \mathbf{v} \rangle > 0$$

for all $\mathbf{v} \in \mathbb{R}^n$, $\mathbf{v} \neq 0$. Note that $S > 0$ implies that S is invertible.

Proposition 9.2.1 (Real Polar Decomposition Theorem). *For any $A \in GL(n,\mathbb{R})$ there exists a unique orthogonal matrix R and positive definite matrices S_1, S_2, such that*

$$A = RS_1 = S_2R. \tag{9.2.1}$$

Proof. Recall first that any positive definite symmetric matrix has a unique square root: If $\lambda_1, \ldots, \lambda_n > 0$ are the eigenvalues of A^TA, diagonalize A^TA by writing

$$A^TA = B\,\mathrm{diag}(\lambda_1, \ldots, \lambda_n)B^{-1},$$

and then define

$$\sqrt{A^TA} = B\,\mathrm{diag}(\sqrt{\lambda_1}, \ldots, \sqrt{\lambda_n})B^{-1}.$$

Let $S_1 = \sqrt{A^TA}$, which is positive definite and symmetric. Define $R = AS_1^{-1}$ and note that

$$R^TR = S_1^{-1}A^TAS_1^{-1} = I,$$

since $S_1^2 = A^TA$ by definition. Since both A and S_1 are invertible, it follows that R is invertible and hence $R^T = R^{-1}$, so R is an orthogonal matrix.

Let us prove uniqueness of the decomposition. If $A = RS_1 = \tilde{R}\tilde{S}_1$, then

$$A^TA = S_1R^T\tilde{R}\tilde{S}_1 = \tilde{S}_1^2.$$

However, the square root of a positive definite matrix is unique, so $S_1 = \tilde{S}_1$, whence also $\tilde{R} = R$.

Now define $S_2 = \sqrt{AA^T}$, and as before, we conclude that $A = S_2R'$ for some orthogonal matrix R'. We prove now that $R' = R$. Indeed, $A = S_2R' = (R'(R')^T)S_2R' = R'((R')^TS_2R')$ and $(R')^TS_2R' > 0$. By uniqueness of the prior polar decomposition, we conclude that $R' = R$ and $(R')^TS_2R' = S_1$. ∎

Now we will use the real polar decomposition theorem to prove that $GL^+(n, \mathbb{R})$ is connected. Let $A \in GL^+(n, \mathbb{R})$ and decompose it as $A = SR$, with S positive definite and R an orthogonal matrix whose determinant is 1. We will prove later that the collection of all orthogonal matrices having determinant equal to 1 is a connected Lie group. Thus there is a continuous path $R(t)$ of orthogonal matrices having determinant 1 such that $R(0) = I$ and $R(1) = R$. Next, define the continuous path of symmetric matrices $S(t) = I + t(S - I)$ and note that $S(0) = I$ and $S(1) = S$. Moreover,

$$\langle S(t)\mathbf{v}, \mathbf{v} \rangle = \langle [I + t(S - I)]\mathbf{v}, \mathbf{v} \rangle$$
$$= \|\mathbf{v}\|^2 + t\langle S\mathbf{v}, \mathbf{v} \rangle - t\|\mathbf{v}\|^2$$
$$= (1 - t)\|\mathbf{v}\|^2 + t\langle S\mathbf{v}, \mathbf{v} \rangle > 0,$$

for all $t \in [0, 1]$, since $\langle S\mathbf{v}, \mathbf{v} \rangle > 0$ by hypothesis. Thus $S(t)$ is a continuous path of positive definite matrices connecting I to S. We conclude that $A(t) := S(t)R(t)$ is a continuous path of matrices whose determinant is strictly positive connecting $A(0) = S(0)R(0) = I$ to $A(1) = S(1)R(1) = SR = A$. Thus, we have proved the following:

Proposition 9.2.2. *The group* $GL(n, \mathbb{R})$ *is a noncompact disconnected* n^2-*dimensional Lie group whose Lie algebra* $\mathfrak{gl}(n, \mathbb{R})$ *consists of all* $n \times n$ *matrices with the bracket*

$$[A, B] = AB - BA.$$

The connected component of the identity is $GL^+(n, \mathbb{R})$, *and* $GL(n, \mathbb{R})$ *has two components.*

The Real Special Linear Group $SL(n, \mathbb{R})$.   Let $\det : L(\mathbb{R}^n, \mathbb{R}^n) \to \mathbb{R}$ be the determinant map and recall that

$$GL(n, \mathbb{R}) = \{ A \in L(\mathbb{R}^n, \mathbb{R}^n) \mid \det A \neq 0 \},$$

so $GL(n, \mathbb{R})$ is open in $L(\mathbb{R}^n, \mathbb{R}^n)$. Notice that $\mathbb{R}\backslash\{0\}$ is a group under multiplication and that

$$\det : GL(n, \mathbb{R}) \to \mathbb{R}\backslash\{0\}$$

is a Lie group homomorphism because

$$\det(AB) = (\det A)(\det B).$$

Lemma 9.2.3. *The map* $\det : GL(n, \mathbb{R}) \to \mathbb{R}\backslash\{0\}$ *is* C^∞, *and its derivative is given by* $\mathbf{D} \det_A \cdot B = (\det A) \operatorname{trace}(A^{-1}B)$.

Proof.   The smoothness of det is clear from its formula in terms of matrix elements. Using the identity

$$\det(A + \lambda B) = (\det A) \det(I + \lambda A^{-1}B),$$

it suffices to prove

$$\frac{d}{d\lambda} \det(I + \lambda C)\Big|_{\lambda=0} = \text{trace}\, C.$$

This follows from the identity for the characteristic polynomial

$$\det(I + \lambda C) = 1 + \lambda\, \text{trace}\, C + \cdots + \lambda^n \det C. \qquad \blacksquare$$

Define the *real special linear group* SL$(n, \mathbb{R})$ by

$$\text{SL}(n, \mathbb{R}) = \{\, A \in \text{GL}(n, \mathbb{R}) \mid \det A = 1 \,\} = \det^{-1}(1). \qquad (9.2.2)$$

From Proposition 9.1.14 it follows that SL$(n, \mathbb{R})$ is a closed Lie subgroup of GL$(n, \mathbb{R})$. However, this method invokes a rather subtle result to prove something that is in reality straightforward. To see this, note that it follows from Lemma 9.2.3 that det : GL$(n, \mathbb{R}) \to \mathbb{R}$ is a submersion, so SL$(n, \mathbb{R}) = \det^{-1}(1)$ is a *smooth* closed submanifold and hence a closed Lie subgroup.

The tangent space to SL$(n, \mathbb{R})$ at $A \in$ SL$(n, \mathbb{R})$ therefore consists of all matrices B such that trace$(A^{-1}B) = 0$. In particular, the tangent space at the identity consists of the matrices with trace zero. We have seen that the Lie algebra of GL$(n, \mathbb{R})$ is $L(\mathbb{R}^n, \mathbb{R}^n) = \mathfrak{gl}(n, \mathbb{R})$ with the Lie bracket given by $[A, B] = AB - BA$. It follows that the *Lie algebra* $\mathfrak{sl}(n, \mathbb{R})$ *of* SL$(n, \mathbb{R})$ *consists of the set of $n \times n$ matrices having trace zero, with the bracket*

$$[A, B] = AB - BA.$$

Since trace$(B) = 0$ imposes one condition on B, it follows that

$$\dim[\mathfrak{sl}(n, \mathbb{R})] = n^2 - 1.$$

In dealing with classical Lie groups it is useful to introduce the following inner product on $\mathfrak{gl}(n, \mathbb{R})$:

$$\langle A, B \rangle = \text{trace}(AB^T). \qquad (9.2.3)$$

Note that

$$\|A\|^2 = \sum_{i,j=1}^{n} a_{ij}^2, \qquad (9.2.4)$$

which shows that this norm on $\mathfrak{gl}(n, \mathbb{R})$ coincides with the Euclidean norm on $\mathbb{R}^{n^2}$.

We shall use this norm to show that $SL(n, \mathbb{R})$ is not compact. Indeed, all matrices of the form

$$
\begin{pmatrix}
1 & 0 & \cdots & t \\
0 & 1 & \cdots & 0 \\
\vdots & \vdots & \ddots & \vdots \\
0 & 0 & \cdots & 1
\end{pmatrix}
$$

are elements of $SL(n, \mathbb{R})$ whose norm equals $\sqrt{n + t^2}$ for any $t \in \mathbb{R}$. Thus, $SL(n, \mathbb{R})$ is not a bounded subset of $\mathfrak{gl}(n, \mathbb{R})$ and hence is not compact.

Finally, let us prove that $SL(n, \mathbb{R})$ is connected. As before, we shall use the real polar decomposition theorem and the fact, to be proved later, that the set of all orthogonal matrices having determinant equal to 1 is a connected Lie group. If $A \in SL(n, \mathbb{R})$, decompose it as $A = SR$, where R is an orthogonal matrix having determinant 1 and S is a positive definite matrix having determinant 1. Since S is symmetric, it can be diagonalized, that is, $S = B \operatorname{diag}(\lambda_1, \ldots, \lambda_n)B^{-1}$ for some orthogonal matrix B and $\lambda_1, \ldots, \lambda_n > 0$. Define the continuous path

$$
S(t) = B \operatorname{diag}\left((1 - t) + t\lambda_1, \ldots, (1 - t) + t\lambda_{n-1}, 1/\prod_{i=1}^{n-1}((1 - t) + t\lambda_i)\right)B^{-1}
$$

for $t \in [0, 1]$ and note that by construction, $\det S(t) = 1$; $S(t)$ is symmetric; $S(t)$ is positive definite, since each entry $(1 - t) + t\lambda_i > 0$ for $t \in [0, 1]$; and $S(0) = I$, $S(1) = S$. Now let $R(t)$ be a continuous path of orthogonal matrices of determinant 1 such that $R(0) = I$ and $R(1) = R$. Therefore, $A(t) = S(t)R(t)$ is a continuous path in $SL(n, \mathbb{R})$ satisfying $A(0) = I$ and $A(1) = SR = A$, thereby showing that $SL(n, \mathbb{R})$ is connected.

Proposition 9.2.4. *The Lie group $SL(n, \mathbb{R})$ is a noncompact connected $(n^2 - 1)$-dimensional Lie group whose Lie algebra $\mathfrak{sl}(n, \mathbb{R})$ consists of the $n \times n$ matrices with trace zero (or linear maps of $\mathbb{R}^n$ to $\mathbb{R}^n$ with trace zero) with the bracket*

$$
[A, B] = AB - BA.
$$

The Orthogonal Group $O(n)$. On $\mathbb{R}^n$ we use the standard inner product

$$
\langle \mathbf{x}, \mathbf{y} \rangle = \sum_{i=1}^{n} x^i y^i,
$$

where $\mathbf{x} = (x^1, \ldots, x^n) \in \mathbb{R}^n$ and $\mathbf{y} = (y^1, \ldots, y^n) \in \mathbb{R}^n$. Recall that a linear map $A \in L(\mathbb{R}^n, \mathbb{R}^n)$ is *orthogonal* if

$$
\langle A\mathbf{x}, A\mathbf{y} \rangle = \langle \mathbf{x}, \mathbf{y} \rangle \tag{9.2.5}
$$

for all $\mathbf{x}, \mathbf{y} \in \mathbb{R}$. In terms of the norm $\|\mathbf{x}\| = \langle \mathbf{x}, \mathbf{x} \rangle^{1/2}$, one sees from the polarization identity that A is orthogonal iff $\|A\mathbf{x}\| = \|\mathbf{x}\|$, for all $\mathbf{x} \in \mathbb{R}^n$, or in terms of the transpose A^T, which is defined by $\langle Ax, y \rangle = \langle x, A^T y \rangle$, we see that A is orthogonal iff $AA^T = I$.

Let $O(n)$ denote the orthogonal elements of $L(\mathbb{R}^n, \mathbb{R}^n)$. For $A \in O(n)$, we see that

$$1 = \det(AA^T) = (\det A)(\det A^T) = (\det A)^2;$$

hence $\det A = \pm 1$, and so $A \in GL(n, \mathbb{R})$. Furthermore, if $A, B \in O(n)$, then

$$\langle ABx, ABy \rangle = \langle Bx, By \rangle = \langle x, y \rangle,$$

and so $AB \in O(n)$. Letting $x' = A^{-1}x$ and $y' = A^{-1}y$, we see that

$$\langle x, y \rangle = \langle Ax', Ay' \rangle = \langle x', y' \rangle,$$

that is,

$$\langle x, y \rangle = \langle A^{-1}x, A^{-1}y \rangle;$$

hence $A^{-1} \in O(n)$.

Let $S(n)$ denote the vector space of symmetric linear maps of $\mathbb{R}^n$ to itself, and let $\psi : GL(n, \mathbb{R}) \to S(n)$ be defined by $\psi(A) = AA^T$. We claim that I is a regular value of ψ. Indeed, if $A \in \psi^{-1}(I) = O(n)$, the derivative of ψ is

$$\mathbf{D}\psi(A) \cdot B = AB^T + BA^T,$$

which is onto (to hit C, take $B = CA/2$). Thus, $\psi^{-1}(I) = O(n)$ is a closed Lie subgroup of $GL(n, \mathbb{R})$, called the **orthogonal group**. The group $O(n)$ is also bounded in $L(\mathbb{R}^n, \mathbb{R}^n)$: The norm of $A \in O(n)$ is

$$\|A\| = \left[\text{trace}(A^T A)\right]^{1/2} = (\text{trace } I)^{1/2} = \sqrt{n}.$$

Therefore, $O(n)$ is compact. We shall see in §9.3 that $O(n)$ is not connected, but has two connected components, one where $\det = +1$ and the other where $\det = -1$.

The Lie algebra $\mathfrak{o}(n)$ of $O(n)$ is $\ker \mathbf{D}\psi(I)$, namely, the skew-symmetric linear maps with the usual commutator bracket $[A, B] = AB - BA$. The space of skew-symmetric $n \times n$ matrices has dimension equal to the number of entries above the diagonal, namely, $n(n-1)/2$. Thus,

$$\dim[O(n)] = \tfrac{1}{2}n(n-1).$$

The **special orthogonal group** is defined as

$$SO(n) = O(n) \cap SL(n, \mathbb{R}),$$

that is,

$$SO(n) = \{ A \in O(n) \mid \det A = +1 \}. \tag{9.2.6}$$

Since $SO(n)$ is the kernel of $\det : O(n) \rightarrow \{-1, 1\}$, that is, $SO(n) = \det^{-1}(1)$, it is an open and closed Lie subgroup of $O(n)$, hence is compact. We shall prove in §9.3 that $SO(n)$ is the connected component of $O(n)$ containing the identity I, and so has the same Lie algebra as $O(n)$. We summarize:

Proposition 9.2.5. *The Lie group $O(n)$ is a compact Lie group of dimension $n(n-1)/2$. Its Lie algebra $\mathfrak{o}(n)$ is the space of skew-symmetric $n \times n$ matrices with bracket $[A, B] = AB - BA$. The connected component of the identity in $O(n)$ is the compact Lie group $SO(n)$, which has the same Lie algebra $\mathfrak{so}(n) = \mathfrak{o}(n)$. The Lie group $O(n)$ has two connected components.*

Rotations in the Plane $SO(2)$.   We parametrize

$$S^1 = \{\mathbf{x} \in \mathbb{R}^2 \mid \|\mathbf{x}\| = 1\}$$

by the polar angle θ, $0 \le \theta < 2\pi$. For each $\theta \in [0, 2\pi]$, let

$$A_\theta = \begin{bmatrix} \cos\theta & -\sin\theta \\ \sin\theta & \cos\theta \end{bmatrix},$$

using the standard basis of $\mathbb{R}^2$. Then $A_\theta \in SO(2)$ represents a counterclockwise rotation through the angle θ. Conversely, if

$$A = \begin{bmatrix} a_1 & a_2 \\ a_3 & a_4 \end{bmatrix}$$

is orthogonal, the relations

$$a_1^2 + a_2^2 = 1, \quad a_3^2 + a_4^2 = 1,$$
$$a_1 a_3 + a_2 a_4 = 0,$$
$$\det A = a_1 a_4 - a_2 a_3 = 1$$

show that $A = A_\theta$ for some θ. Thus, $SO(2)$ can be identified with S^1, that is, with rotations in the plane.

Rotations in Space $SO(3)$.   The Lie algebra $\mathfrak{so}(3)$ of $SO(3)$ may be identified with $\mathbb{R}^3$ as follows. We define the vector space isomorphism $\hat{}\;:\; \mathbb{R}^3 \rightarrow \mathfrak{so}(3)$, called the **hat map**, by

$$\mathbf{v} = (v_1, v_2, v_3) \mapsto \hat{\mathbf{v}} = \begin{bmatrix} 0 & -v_3 & v_2 \\ v_3 & 0 & -v_1 \\ -v_2 & v_1 & 0 \end{bmatrix}. \tag{9.2.7}$$

Note that the identity

$$\hat{\mathbf{v}}\mathbf{w} = \mathbf{v} \times \mathbf{w}$$

characterizes this isomorphism. We get

$$(\hat{\mathbf{u}}\hat{\mathbf{v}} - \hat{\mathbf{v}}\hat{\mathbf{u}})\,\mathbf{w} = \hat{\mathbf{u}}(\mathbf{v} \times \mathbf{w}) - \hat{\mathbf{v}}(\mathbf{u} \times \mathbf{w})$$
$$= \mathbf{u} \times (\mathbf{v} \times \mathbf{w}) - \mathbf{v} \times (\mathbf{u} \times \mathbf{w})$$
$$= (\mathbf{u} \times \mathbf{v}) \times \mathbf{w} = (\mathbf{u} \times \mathbf{v})\hat{\ } \cdot \mathbf{w}.$$

Thus, if we put the cross product on $\mathbb{R}^3$, $\hat{\ }$ becomes a Lie algebra isomorphism, and so *we can identify* so(3) *with* $\mathbb{R}^3$ *carrying the cross product as Lie bracket.*

We also note that the standard dot product may be written

$$\mathbf{v} \cdot \mathbf{w} = \tfrac{1}{2}\,\text{trace}\,(\hat{\mathbf{v}}^T\hat{\mathbf{w}}) = -\tfrac{1}{2}\,\text{trace}\,(\hat{\mathbf{v}}\hat{\mathbf{w}}).$$

Theorem 9.2.6 (Euler's Theorem).   *Every element $A \in$ SO(3), $A \neq I$, is a rotation through an angle θ about an axis $\mathbf{w}$.*

To prove this, we use the following lemma:

Lemma 9.2.7.   *Every $A \in$ SO(3) has an eigenvalue equal to 1.*

Proof.   The eigenvalues of A are given by roots of the third-degree polynomial $\det(A - \lambda I) = 0$. Roots occur in conjugate pairs, so at least one is real. If λ is a real root and x is a nonzero real eigenvector, then $Ax = \lambda x$, so

$$\|Ax\|^2 = \|x\|^2 \quad \text{and} \quad \|Ax\|^2 = |\lambda|^2\,\|x\|^2$$

imply $\lambda = \pm 1$. If all three roots are real, they are $(1,1,1)$ or $(1,-1,-1)$, since $\det A = 1$. If there is one real and two complex conjugate roots, they are $(1, \omega, \bar{\omega})$, since $\det A = 1$. In any case, one real root must be $+1$.   ∎

Proof of Theorem 9.2.6.   By Lemma 9.2.7, the matrix A has an eigenvector $\mathbf{w}$ with eigenvalue 1, say $A\mathbf{w} = \mathbf{w}$. The line spanned by $\mathbf{w}$ is also invariant under A. Let P be the plane perpendicular to $\mathbf{w}$; that is,

$$P = \{\,\mathbf{y} \mid \langle \mathbf{w}, \mathbf{y}\rangle = 0\,\}.$$

Since A is orthogonal, $A(P) = P$. Let $\mathbf{e}_1, \mathbf{e}_2$ be an orthogonal basis in P. Then relative to $(\mathbf{w}, \mathbf{e}_1, \mathbf{e}_2)$, A has the matrix

$$A = \begin{bmatrix} 1 & 0 & 0 \\ 0 & a_1 & a_2 \\ 0 & a_3 & a_4 \end{bmatrix}.$$

Since

$$\begin{bmatrix} a_1 & a_2 \\ a_3 & a_4 \end{bmatrix}$$

lies in SO(2), A is a rotation about the axis $\mathbf{w}$ by some angle.   ∎

Corollary 9.2.8. *Any $A \in SO(3)$ can be written in some orthonormal basis as the matrix*

$$A = \begin{bmatrix} 1 & 0 & 0 \\ 0 & \cos\theta & -\sin\theta \\ 0 & \sin\theta & \cos\theta \end{bmatrix}.$$

The infinitesimal version of Euler's theorem is the following:

Proposition 9.2.9. *Identifying the Lie algebra $\mathfrak{so}(3)$ of $SO(3)$ with the Lie algebra $\mathbb{R}^3$, $\exp(t\hat{\mathbf{w}})$ is a rotation about $\mathbf{w}$ by the angle $t\|\mathbf{w}\|$, where $\mathbf{w} \in \mathbb{R}^3$.*

Proof. To simplify the computation, we pick an orthonormal basis $\{\mathbf{e}_1, \mathbf{e}_2, \mathbf{e}_3\}$ of $\mathbb{R}^3$, with $\mathbf{e}_1 = \mathbf{w}/\|\mathbf{w}\|$. Relative to this basis, $\hat{\mathbf{w}}$ has the matrix

$$\hat{\mathbf{w}} = \|\mathbf{w}\| \begin{bmatrix} 0 & 0 & 0 \\ 0 & 0 & -1 \\ 0 & 1 & 0 \end{bmatrix}.$$

Let

$$c(t) = \begin{bmatrix} 1 & 0 & 0 \\ 0 & \cos t\|\mathbf{w}\| & -\sin t\|\mathbf{w}\| \\ 0 & \sin t\|\mathbf{w}\| & \cos t\|\mathbf{w}\| \end{bmatrix}.$$

Then

$$c'(t) = \begin{bmatrix} 0 & 0 & 0 \\ 0 & -\|\mathbf{w}\|\sin t\|\mathbf{w}\| & -\|\mathbf{w}\|\cos t\|\mathbf{w}\| \\ 0 & \|\mathbf{w}\|\cos t\|\mathbf{w}\| & -\|\mathbf{w}\|\sin t\|\mathbf{w}\| \end{bmatrix}$$

$$= c(t)\hat{\mathbf{w}} = T_I L_{c(t)}(\hat{\mathbf{w}}) = X_{\hat{\mathbf{w}}}(c(t)),$$

where $X_{\hat{\mathbf{w}}}$ is the left-invariant vector field corresponding to $\hat{\mathbf{w}}$. Therefore, $c(t)$ is an integral curve of $X_{\hat{\mathbf{w}}}$; but $\exp(t\hat{\mathbf{w}})$ is also an integral curve of $X_{\hat{\mathbf{w}}}$. Since both agree at $t = 0$, $\exp(t\hat{\mathbf{w}}) = c(t)$, for all $t \in \mathbb{R}$. But the matrix definition of $c(t)$ expresses it as a rotation by an angle $t\|\mathbf{w}\|$ about the axis $\mathbf{w}$. ∎

Despite Euler's theorem, it might be good to recall now that $SO(3)$ *cannot* be written as $S^2 \times S^1$; see Exercise 1.2-4.

Amplifying on Proposition 9.2.9, we give the following explicit formula for $\exp\xi$, where $\xi \in \mathfrak{so}(3)$, which is called *Rodrigues' formula*:

$$\exp[\hat{\mathbf{v}}] = I + \frac{\sin\|\mathbf{v}\|}{\|\mathbf{v}\|}\hat{\mathbf{v}} + \frac{1}{2}\left[\frac{\sin\left(\frac{\|\mathbf{v}\|}{2}\right)}{\frac{\|\mathbf{v}\|}{2}}\right]^2 \hat{\mathbf{v}}^2. \tag{9.2.8}$$

This formula was given by Rodrigues in 1840; see also Exercise 1 in Helgason [1978, p. 249] and see Altmann [1986] for some interesting history of this formula.

Proof of Rodrigues' Formula. By (9.2.7),

$$\hat{v}^2 w = v \times (v \times w) = \langle v, w \rangle v - \|v\|^2 w. \tag{9.2.9}$$

Consequently, we have the recurrence relations

$$\hat{v}^3 = -\|v\|^2 \hat{v}, \quad \hat{v}^4 = -\|v\|^2 \hat{v}^2, \quad \hat{v}^5 = \|v\|^4 \hat{v}, \quad \hat{v}^6 = \|v\|^4 \hat{v}^2, \ldots .$$

Splitting the exponential series in odd and even powers,

$$\exp[\hat{v}] = I + \left[I - \frac{\|v\|^2}{3!} + \frac{\|v\|^4}{5!} - \cdots + (-1)^{n+1} \frac{\|v\|^{2n}}{(2n+1)!} + \cdots \right] \hat{v}$$

$$+ \left[\frac{1}{2!} - \frac{\|v\|^2}{4!} + \frac{\|v\|^4}{6!} + \cdots + (-1)^{n-1} \frac{\|v\|^{n-2}}{(2n)!} + \cdots \right] \hat{v}^2$$

$$= I + \frac{\sin \|v\|}{\|v\|} \hat{v} + \frac{1 - \cos \|v\|}{\|v\|^2} \hat{v}^2, \tag{9.2.10}$$

and so the result follows from the identity $2 \sin^2(\|v\|/2) = 1 - \cos \|v\|$. ∎

The following alternative expression, equivalent to (9.2.8), is often useful. Set $n = v/\|v\|$, so that $\|n\| = 1$. From (9.2.9) and (9.2.10) we obtain

$$\exp[\hat{v}] = I + (\sin \|v\|) \hat{n} + (1 - \cos \|v\|)[n \otimes n - I]. \tag{9.2.11}$$

Here, $n \otimes n$ is the matrix whose entries are $n^i n^j$, or as a bilinear form, $(n \otimes n)(\alpha, \beta) = n(\alpha) n(\beta)$. Therefore, we obtain a rotation about the unit vector $n = v/\|v\|$ of magnitude $\|v\|$.

The results (9.2.8) and (9.2.11) are useful in computational solid mechanics, along with their quaternionic counterparts. We shall return to this point below in connection with SU(2); see Whittaker [1927] and Simo and Fox [1989] for more information.

We next give a topological property of SO(3).

Proposition 9.2.10. *The rotation group* SO(3) *is diffeomorphic to the real projective space* $\mathbb{RP}^3$.

Proof. To see this, map the unit ball D in $\mathbb{R}^3$ to SO(3) by sending (x, y, z) to the rotation about (x, y, z) through the angle $\pi \sqrt{x^2 + y^2 + z^2}$ (and $(0, 0, 0)$ to the identity). This mapping is clearly smooth and surjective. Its restriction to the interior of D is injective. On the boundary of D, this mapping is 2 to 1, so it induces a smooth bijective map from D, with antipodal points on the boundary identified, to SO(3). It is a straightforward exercise to show that the inverse of this map is also smooth. Thus, SO(3) is diffeomorphic with D, with antipodal points on the boundary identified.

However, the mapping

$$(x, y, z) \mapsto (x, y, z, \sqrt{1 - x^2 - y^2 - z^2})$$

is a diffeomorphism between D, with antipodal points on the boundary identified, and the upper unit hemisphere of S^3 with antipodal points on the equator identified. The latter space is clearly diffeomorphic to the unit sphere S^3 with antipodal points identified, which coincides with the space of lines in $\mathbb{R}^4$ through the origin, that is, with $\mathbb{RP}^3$.   ∎

The Real Symplectic Group $\mathrm{Sp}(2n, \mathbb{R})$.   Let

$$\mathbb{J} = \begin{bmatrix} 0 & I \\ -I & 0 \end{bmatrix}.$$

Recall that $A \in L(\mathbb{R}^{2n}, \mathbb{R}^{2n})$ is *symplectic* if $A^T \mathbb{J} A = \mathbb{J}$. Let $\mathrm{Sp}(2n, \mathbb{R})$ be the set of $2n \times 2n$ symplectic matrices. Taking determinants of the condition $A^T \mathbb{J} A = \mathbb{J}$ gives

$$1 = \det \mathbb{J} = (\det A^T) \cdot (\det A\mathbb{J}) \cdot (\det A) = (\det A)^2.$$

Hence,

$$\det A = \pm 1,$$

and so $A \in \mathrm{GL}(2n, \mathbb{R})$. Furthermore, if $A, B \in \mathrm{Sp}(2n, \mathbb{R})$, then

$$(AB)^T \mathbb{J}(AB) = B^T A^T \mathbb{J} A B = \mathbb{J}.$$

Hence, $AB \in \mathrm{Sp}(2n, \mathbb{R})$, and if $A^T \mathbb{J} A = \mathbb{J}$, then

$$\mathbb{J} A = (A^T)^{-1} \mathbb{J} = (A^{-1})^T \mathbb{J},$$

so

$$\mathbb{J} = (A^{-1})^T \mathbb{J} A^{-1}, \quad \text{or} \quad A^{-1} \in \mathrm{Sp}(2n, \mathbb{R}).$$

Thus, $\mathrm{Sp}(2n, \mathbb{R})$ is a group. If

$$A = \begin{bmatrix} a & b \\ c & d \end{bmatrix} \in \mathrm{GL}(2n, \mathbb{R}),$$

then (see Exercise 2.3-2)

$$A \in \mathrm{Sp}(2n, \mathbb{R}) \text{ iff } \begin{cases} a^T c \text{ and } b^T d \text{ are symmetric and} \\ a^T d - c^T b = 1. \end{cases} \tag{9.2.12}$$

Define $\psi : \mathrm{GL}(2n, \mathbb{R}) \to \mathfrak{so}(2n)$ by $\psi(A) = A^T \mathbb{J} A$. Let us show that $\mathbb{J}$ is a regular value of ψ. Indeed, if $A \in \psi^{-1}(\mathbb{J}) = \mathrm{Sp}(2n, \mathbb{R})$, the derivative of ψ is

$$\mathbf{D}\psi(A) \cdot B = B^T \mathbb{J} A + A^T \mathbb{J} B.$$

Now, if $C \in \mathfrak{so}(2n)$, let

$$B = -\tfrac{1}{2} A \mathbb{J} C.$$

We verify, using the identity $A^T \mathbb{J} = \mathbb{J} A^{-1}$, that $\mathbf{D}\psi(A) \cdot B = C$. Indeed,

$$
\begin{aligned}
B^T \mathbb{J} A + A^T \mathbb{J} B &= B^T (A^{-1})^T \mathbb{J} + \mathbb{J} A^{-1} B \\
&= (A^{-1}B)^T \mathbb{J} + \mathbb{J}(A^{-1}B) \\
&= (-\tfrac{1}{2}\mathbb{J}C)^T \mathbb{J} + \mathbb{J}(-\tfrac{1}{2}\mathbb{J}C) \\
&= -\tfrac{1}{2}C^T \mathbb{J}^T \mathbb{J} - \tfrac{1}{2}\mathbb{J}^2 C \\
&= -\tfrac{1}{2}C\mathbb{J}^2 - \tfrac{1}{2}\mathbb{J}^2 C = C,
\end{aligned}
$$

since $\mathbb{J}^T = -\mathbb{J}$ and $\mathbb{J}^2 = -I$. Thus $\mathrm{Sp}(2n, \mathbb{R}) = \psi^{-1}(\mathbb{J})$ is a closed smooth submanifold of $\mathrm{GL}(2n, \mathbb{R})$ whose Lie algebra is

$$
\ker D\psi(\mathbb{J}) = \left\{ B \in L\left(\mathbb{R}^{2n}, \mathbb{R}^{2n}\right) \mid B^T \mathbb{J} + \mathbb{J} B = 0 \right\}.
$$

The Lie group $\mathrm{Sp}(2n, \mathbb{R})$ is called the **symplectic group**, and its Lie algebra

$$
\mathfrak{sp}(2n, \mathbb{R}) = \left\{ A \in L\left(\mathbb{R}^{2n}, \mathbb{R}^{2n}\right) \mid A^T \mathbb{J} + \mathbb{J} A = 0 \right\}
$$

the **symplectic algebra**. Moreover, if

$$
A = \begin{bmatrix} a & b \\ c & d \end{bmatrix} \in \mathfrak{sl}(2n, \mathbb{R}),
$$

then

$$
A \in \mathfrak{sp}(2n, \mathbb{R}) \text{ iff } d = -a^T, \ c = c^T, \text{ and } b = b^T. \tag{9.2.13}
$$

The dimension of $\mathfrak{sp}(2n, \mathbb{R})$ can be readily calculated to be $2n^2 + n$.

Using (9.2.12), it follows that all matrices of the form

$$
\begin{bmatrix} I & 0 \\ tI & I \end{bmatrix}
$$

are symplectic. However, the norm of such a matrix is equal to $\sqrt{2n + t^2 n}$, which is unbounded if $t \in \mathbb{R}$. Therefore, $\mathrm{Sp}(2n, \mathbb{R})$ is not a bounded subset of $\mathfrak{gl}(2n, \mathbb{R})$ and hence is not compact. We next summarize what we have found.

Proposition 9.2.11. *The symplectic group*

$$
\mathrm{Sp}(2n, \mathbb{R}) := \left\{ A \in \mathrm{GL}(2n, \mathbb{R}) \mid A^T \mathbb{J} A = \mathbb{J} \right\}
$$

is a noncompact, connected Lie group of dimension $2n^2 + n$. Its Lie algebra $\mathfrak{sp}(2n, \mathbb{R})$ consists of the $2n \times 2n$ matrices A satisfying $A^T \mathbb{J} + \mathbb{J} A = 0$, where

$$
\mathbb{J} = \begin{bmatrix} 0 & I \\ -I & 0 \end{bmatrix}
$$

with I the $n \times n$ identity matrix.

We shall indicate in §9.3 how one proves that $\mathrm{Sp}(2n, \mathbb{R})$ is connected.

We are ready to prove that symplectic linear maps have determinant 1, a fact that we promised in Chapter 2.

Lemma 9.2.12. *If $A \in \mathrm{Sp}(n, \mathbb{R})$, then $\det A = 1$.*

Proof. Since $A^T \mathbb{J} A = \mathbb{J}$ and $\det \mathbb{J} = 1$, it follows that $(\det A)^2 = 1$. Unfortunately, this still leaves open the possibility that $\det A = -1$. To eliminate it, we proceed in the following way.

Define the symplectic form Ω on $\mathbb{R}^{2n}$ by $\Omega(\mathbf{u}, \mathbf{v}) = \mathbf{u}^T \mathbb{J} \mathbf{v}$, that is, relative to the chosen basis of $\mathbb{R}^{2n}$, the matrix of Ω is $\mathbb{J}$. As we saw in Chapter 5, the standard volume form μ on $\mathbb{R}^{2n}$ is given, up to a factor, by $\mu = \Omega \wedge \Omega \wedge \cdots \wedge \Omega$, or, equivalently,

$$\mu(\mathbf{v}_1, \ldots, \mathbf{v}_{2n}) = \det\left(\Omega(\mathbf{v}_i, \mathbf{v}_j)\right).$$

By the definition of the determinant of a linear map, $(\det A)\mu = A^*\mu$, we get

$$
\begin{aligned}
(\det A)\mu\,(\mathbf{v}_1, \ldots, \mathbf{v}_{2n}) &= (A^*\mu)\,(\mathbf{v}_1, \ldots, \mathbf{v}_{2n}) \\
&= \mu\,(A\mathbf{v}_1, \ldots, A\mathbf{v}_{2n}) = \det\left(\Omega\,(A\mathbf{v}_i, A\mathbf{v}_j)\right) \\
&= \det\left(\Omega\,(\mathbf{v}_i, \mathbf{v}_j)\right) \\
&= \mu\,(\mathbf{v}_1, \ldots, \mathbf{v}_{2n}),
\end{aligned}
$$

since $A \in \mathrm{Sp}(2n, \mathbb{R})$, which is equivalent to $\Omega(A\mathbf{u}, A\mathbf{v}) = \Omega(\mathbf{u}, \mathbf{v})$ for all $\mathbf{u}, \mathbf{v} \in \mathbb{R}^{2n}$. Taking $\mathbf{v}_1, \ldots, \mathbf{v}_{2n}$ to be the standard basis of $\mathbb{R}^{2n}$, we conclude that $\det A = 1$. ∎

Proposition 9.2.13 (Symplectic Eigenvalue Theorem). *If $\lambda_0 \in \mathbb{C}$ is an eigenvalue of $A \in \mathrm{Sp}(2n, \mathbb{R})$ of multiplicity k, then $1/\lambda_0$, $\overline{\lambda}_0$, and $1/\overline{\lambda}_0$ are eigenvalues of A of the same multiplicity k. Moreover, if ± 1 occur as eigenvalues, their multiplicities are even.*

Proof. Since A is a real matrix, if λ_0 is an eigenvalue of A of multiplicity k, so is $\overline{\lambda}_0$ by elementary algebra.

Let us show that $1/\lambda_0$ is also an eigenvalue of A. If $p(\lambda) = \det(A - \lambda I)$ is the characteristic polynomial of A, since

$$\mathbb{J} A \mathbb{J}^{-1} = \left(A^{-1}\right)^T,$$

$\det \mathbb{J} = 1$, $\mathbb{J}^{-1} = -\mathbb{J} = \mathbb{J}^T$, and $\det A = 1$ (by Lemma 9.2.11), we get

$$
\begin{aligned}
p(\lambda) = \det(A - \lambda I) &= \det\left[\mathbb{J}(A - \lambda I)\mathbb{J}^{-1}\right] \\
&= \det(\mathbb{J}A\mathbb{J}^{-1} - \lambda I) = \det\left(\left(A^{-1} - \lambda I\right)^T\right) \\
&= \det(A^{-1} - \lambda I) = \det\left(A^{-1}(I - \lambda A)\right) \\
&= \det(I - \lambda A) = \det\left(\lambda\left(\frac{1}{\lambda}I - A\right)\right) \\
&= \lambda^{2n}\det\left(\frac{1}{\lambda}I - A\right) \\
&= \lambda^{2n}(-1)^{2n}\det\left(A - \frac{1}{\lambda}I\right) \\
&= \lambda^{2n}p\left(\frac{1}{\lambda}\right).
\end{aligned}
\tag{9.2.14}
$$

Since 0 is not an eigenvalue of A, it follows that $p(\lambda) = 0$ iff $p(1/\lambda) = 0$, and hence, λ_0 is an eigenvalue of A iff $1/\lambda_0$ is an eigenvalue of A.

Now assume that λ_0 has multiplicity k, that is,

$$
p(\lambda) = (\lambda - \lambda_0)^k q(\lambda)
$$

for some polynomial $q(\lambda)$ of degree $2n - k$ satisfying $q(\lambda_0) \neq 0$. Since $p(\lambda) = \lambda^{2n}p(1/\lambda)$, we conclude that

$$
p(\lambda) = p\left(\tfrac{1}{\lambda}\right)\lambda^{2n} = (\lambda - \lambda_0)^k q(\lambda) = (\lambda\lambda_0)^k\left(\frac{1}{\lambda_0} - \frac{1}{\lambda}\right)^k q(\lambda).
$$

However,

$$
\frac{\lambda_0^k}{\lambda^{2n-k}}q(\lambda)
$$

is a polynomial in $1/\lambda$, since the degree of $q(\lambda)$ is $2n - k$, $k \leq 2n$. Thus $1/\lambda_0$ is a root of $p(\lambda)$ having multiplicity $l \geq k$. Reversing the roles of λ_0 and $1/\lambda_0$, we similarly conclude that $k \geq l$, and hence it follows that $k = l$.

Finally, note that $\lambda_0 = 1/\lambda_0$ iff $\lambda_0 = \pm 1$. Thus, since all eigenvalues of A occur in pairs whose product is 1 and the size of A is $2n \times 2n$, it follows that the total number of times $+1$ and -1 occur as eigenvalues is even. However, since $\det A = 1$ by Lemma 9.2.12, we conclude that -1 occurs an even number of times as an eigenvalue of A (if it occurs at all). Therefore, the multiplicity of 1 as an eigenvalue of A, if it occurs, is also even. ∎

Figure 9.2.1 illustrates the possible configurations of the eigenvalues of $A \in \mathrm{Sp}(4, \mathbb{R})$.

Next, we study the eigenvalues of matrices in $\mathfrak{sp}(2n, \mathbb{R})$. The following theorem is useful in the stability analysis of relative equilibria. If $A \in$

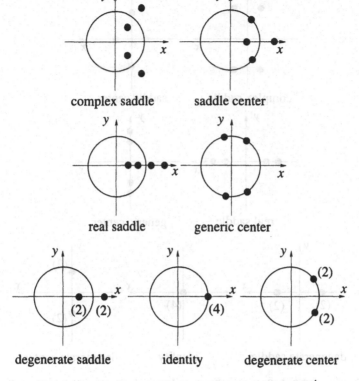

FIGURE 9.2.1. Symplectic eigenvalue theorem on $\mathbb{R}^4$.

$\mathfrak{sp}(2n, \mathbb{R})$, then $A^T \mathbb{J} + \mathbb{J}A = 0$, so that if $p(\lambda) = \det(A - \lambda I)$ is the characteristic polynomial of A, we have

$$
\begin{aligned}
p(\lambda) = \det(A - \lambda I) &= \det(\mathbb{J}(A - \lambda I)\mathbb{J}) \\
&= \det(\mathbb{J}A\mathbb{J} + \lambda I) \\
&= \det(-A^T \mathbb{J}^2 + \lambda I) \\
&= \det(A^T + \lambda I) = \det(A + \lambda I) \\
&= p(-\lambda).
\end{aligned}
$$

In particular, notice that $\operatorname{trace}(A) = 0$. Proceeding as before and using this identity, we conclude the following:

Proposition 9.2.14 (Infinitesimally Symplectic Eigenvalues). *If $\lambda_0 \in \mathbb{C}$ is an eigenvalue of $A \in \mathfrak{sp}(2n, \mathbb{R})$ of multiplicity k, then $-\lambda_0$, $\overline{\lambda}_0$, and $-\overline{\lambda}_0$ are eigenvalues of A of the same multiplicity k. Moreover, if 0 is an eigenvalue, it has even multiplicity.*

Figure 9.2.2 shows the possible infinitesimally symplectic eigenvalue configurations for $A \in \mathfrak{sp}(4, \mathbb{R})$.

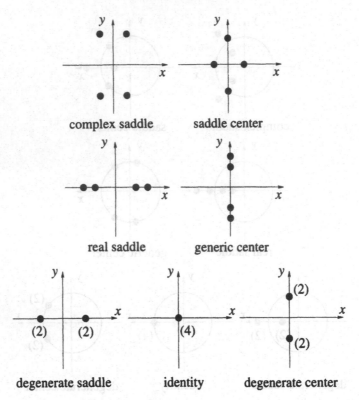

FIGURE 9.2.2. Infinitesimally symplectic eigenvalue theorem on $\mathbb{R}^4$.

The Symplectic Group and Mechanics.   Consider a particle of mass m moving in a potential $V(\mathbf{q})$, where $\mathbf{q} = (q^1, q^2, q^3) \in \mathbb{R}^3$. Newton's second law states that the particle moves along a curve $\mathbf{q}(t)$ in $\mathbb{R}^3$ in such a way that $m\ddot{\mathbf{q}} = -\,\mathrm{grad}\, V(\mathbf{q})$. Introduce the momentum $p_i = m\dot{q}^i$, $i = 1, 2, 3$, and the energy

$$H(\mathbf{q}, \mathbf{p}) = \frac{1}{2m} \sum_{i=1}^{3} p_i^2 + V(\mathbf{q}).$$

Then

$$\frac{\partial H}{\partial q^i} = \frac{\partial V}{\partial q^i} = -m\ddot{q}^i = -\dot{p}_i, \quad \text{and} \quad \frac{\partial H}{\partial p_i} = \frac{1}{m} p_i = \dot{q}^i,$$

and hence *Newton's law* $\mathbf{F} = m\mathbf{a}$ *is equivalent to Hamilton's equations*

$$\dot{q}^i = \frac{\partial H}{\partial p_i}, \quad \dot{p}_i = -\frac{\partial H}{\partial q^i}, \quad i = 1, 2, 3.$$

Writing $z = (\mathbf{q}, \mathbf{p})$,

$$
\mathbb{J} \cdot \operatorname{grad} H(z) = \begin{bmatrix} 0 & I \\ -I & 0 \end{bmatrix} \begin{bmatrix} \dfrac{\partial H}{\partial \mathbf{q}} \\[2mm] \dfrac{\partial H}{\partial \mathbf{p}} \end{bmatrix} = (\dot{\mathbf{q}}, \dot{\mathbf{p}}) = \dot{z},
$$

so Hamilton's equations read $\dot{z} = \mathbb{J} \cdot \operatorname{grad} H(z)$. Now let

$$
f : \mathbb{R}^3 \times \mathbb{R}^3 \to \mathbb{R}^3 \times \mathbb{R}^3
$$

and write $w = f(z)$. If $z(t)$ satisfies Hamilton's equations

$$
\dot{z} = \mathbb{J} \cdot \operatorname{grad} H(z),
$$

then $w(t) = f(z(t))$ satisfies $\dot{w} = A^T \dot{z}$, where $A^T = [\partial w^i / \partial z^j]$ is the Jacobian matrix of f. By the chain rule,

$$
\dot{w} = A^T \mathbb{J} \operatorname{grad}_z H(z) = A^T \mathbb{J} A \operatorname{grad}_w H(z(w)).
$$

Thus, the equations for $w(t)$ have the form of Hamilton's equations with energy $K(w) = H(z(w))$ if and only if $A^T \mathbb{J} A = \mathbb{J}$, that is, iff A is symplectic. A nonlinear transformation f is **canonical** iff its Jacobian matrix is symplectic.

As a special case, consider a linear map $A \in \operatorname{Sp}(2n, \mathbb{R})$ and let $w = Az$. Suppose H is quadratic, that is, of the form $H(z) = \langle z, Bz \rangle / 2$, where B is a symmetric $2n \times 2n$ matrix. Then

$$
\begin{aligned}
\operatorname{grad} H(z) \cdot \delta z &= \tfrac{1}{2} \langle \delta z, Bz \rangle + \langle z, B\delta z \rangle \\
&= \tfrac{1}{2}(\langle \delta z, Bz \rangle + \langle Bz, \delta z \rangle) = \langle \delta z, Bz \rangle,
\end{aligned}
$$

so $\operatorname{grad} H(z) = Bz$ and thus the equations of motion become the linear equations $\dot{z} = \mathbb{J} Bz$. Now

$$
\dot{w} = A\dot{z} = A\mathbb{J}Bz = \mathbb{J}(A^T)^{-1} Bz = \mathbb{J}(A^T)^{-1} B A^{-1} Az = \mathbb{J} B' w,
$$

where $B' = (A^T)^{-1} B A^{-1}$ is symmetric. For the new Hamiltonian we get

$$
\begin{aligned}
H'(w) &= \tfrac{1}{2} \langle w, (A^T)^{-1} B A^{-1} w \rangle = \tfrac{1}{2} \langle A^{-1}w, B A^{-1} w \rangle \\
&= H(A^{-1} w) = H(z).
\end{aligned}
$$

Thus, $\operatorname{Sp}(2n, \mathbb{R})$ *is the linear invariance group of classical mechanics.*

The Complex General Linear Group $\operatorname{GL}(n, \mathbb{C})$. Many important Lie groups involve *complex* matrices. As in the real case,

$$
\operatorname{GL}(n, \mathbb{C}) = \{\, n \times n \text{ invertible complex matrices} \,\}
$$

is an open set in $L(\mathbb{C}^n, \mathbb{C}^n) = \{\, n \times n \text{ complex matrices}\,\}$. Clearly, $\mathrm{GL}(n, \mathbb{C})$ is a group under matrix multiplication. Therefore, $\mathrm{GL}(n, \mathbb{C})$ is a Lie group and has the Lie algebra $\mathfrak{gl}(n, \mathbb{C}) = \{\, n \times n \text{ complex matrices}\,\} = L(\mathbb{C}^n, \mathbb{C}^n)$. Hence $\mathrm{GL}(n, \mathbb{C})$ has complex dimension n^2, that is, real dimension $2n^2$.

We shall prove below that $\mathrm{GL}(n, \mathbb{C})$ is connected (contrast this with the fact that $\mathrm{GL}(n, \mathbb{R})$ has two components). As in the real case, we will need a polar decomposition theorem to do this. A matrix $U \in \mathrm{GL}(n, \mathbb{C})$ is **unitary** if $UU^\dagger = U^\dagger U = I$, where $U^\dagger := \overline{U}^T$. A matrix $P \in \mathfrak{gl}(n, \mathbb{C})$ is called **Hermitian** if $P^\dagger = P$. A Hermitian matrix P is called **positive definite**, denoted by $P > 0$, if $\langle P\mathbf{z}, \mathbf{z} \rangle > 0$ for all $\mathbf{z} \in \mathbb{C}^n$, $\mathbf{z} \neq 0$, where $\langle\,,\,\rangle$ denotes the inner product on $\mathbb{C}^n$. Note that $P > 0$ implies that P is invertible.

Proposition 9.2.15 (Complex Polar Decomposition).   *For any matrix* $A \in \mathrm{GL}(n, \mathbb{C})$, *there exists a unique unitary matrix* U *and positive definite Hermitian matrices* P_1, P_2 *such that*

$$A = UP_1 = P_2 U.$$

The proof is identical to that of Proposition 9.2.1 with the obvious changes. The only additional property needed is the fact that the eigenvalues of a Hermitian matrix are real. As in the proof of the real case, one needs to use the connectedness of the space of unitary matrices (proved in §9.3) to conclude the following:

Proposition 9.2.16.   *The group* $\mathrm{GL}(n, \mathbb{C})$ *is a complex noncompact connected Lie group of complex dimension* n^2 *and real dimension* $2n^2$. *Its Lie algebra* $\mathfrak{gl}(n, \mathbb{C})$ *consists of all* $n \times n$ *complex matrices with the commutator bracket.*

On $\mathfrak{gl}(n, \mathbb{C})$, the inner product is defined by

$$\langle A, B \rangle = \mathrm{trace}(AB^\dagger).$$

The Complex Special Linear Group.   This group is defined by

$$\mathrm{SL}(n, \mathbb{C}) := \{\, A \in \mathrm{GL}(n, \mathbb{C}) \mid \det A = 1 \,\}$$

and is treated as in the real case. In the proof of its connectedness one uses the complex polar decomposition theorem and the fact that any Hermitian matrix can be diagonalized by conjugating it with an appropriate unitary matrix.

Proposition 9.2.17.   *The group* $\mathrm{SL}(n, \mathbb{C})$ *is a complex noncompact Lie group of complex dimension* $n^2 - 1$ *and real dimension* $2(n^2 - 1)$. *Its Lie algebra* $\mathfrak{sl}(n, \mathbb{C})$ *consists of all* $n \times n$ *complex matrices of trace zero with the commutator bracket.*

The Unitary Group U(n). Recall that $\mathbb{C}^n$ has the Hermitian inner product

$$\langle \mathbf{x}, \mathbf{y} \rangle = \sum_{i=0}^{n} x^i \bar{y}^i,$$

where $\mathbf{x} = (x^1, \ldots, x^n) \in \mathbb{C}^n$, $\mathbf{y} = (y^1, \ldots, y^n) \in \mathbb{C}^n$, and $\bar{y}^i$ denotes the complex conjugate. Let

$$U(n) = \{ A \in GL(n, \mathbb{C}) \mid \langle A\mathbf{x}, A\mathbf{y} \rangle = \langle \mathbf{x}, \mathbf{y} \rangle \}.$$

The orthogonality condition $\langle A\mathbf{x}, A\mathbf{y} \rangle = \langle \mathbf{x}, \mathbf{y} \rangle$ is equivalent to $AA^\dagger = A^\dagger A = I$, where $A^\dagger = \bar{A}^T$, that is, $\langle A\mathbf{x}, \mathbf{y} \rangle = \langle \mathbf{x}, A^\dagger \mathbf{y} \rangle$. From $|\det A| = 1$, we see that det maps $U(n)$ into the unit circle $S^1 = \{ z \in \mathbb{C} \mid |z| = 1 \}$. As is to be expected by now, $U(n)$ is a closed Lie subgroup of $GL(n, \mathbb{C})$ with Lie algebra

$$\mathfrak{u}(n) = \{ A \in L(\mathbb{C}^n, \mathbb{C}^n) \mid \langle A\mathbf{x}, \mathbf{y} \rangle = -\langle \mathbf{x}, A\mathbf{y} \rangle \}$$
$$= \{ A \in \mathfrak{gl}(n, \mathbb{C}) \mid A^\dagger = -A \};$$

the proof parallels that for $O(n)$. The elements of $\mathfrak{u}(n)$ are called **skew-Hermitian matrices**. Since the norm of $A \in U(n)$ is

$$\|A\| = \left(\mathrm{trace}(A^\dagger A) \right)^{1/2} = (\mathrm{trace}\, I)^{1/2} = \sqrt{n},$$

it follows that $U(n)$ is closed and bounded, hence compact, in $GL(n, \mathbb{C})$. From the definition of $\mathfrak{u}(n)$ it immediately follows that the real dimension of $U(n)$ is n^2. Thus, even though the entries of the elements of $U(n)$ are complex, $U(n)$ is a *real* Lie group.

In the special case $n = 1$, a complex linear map $\varphi : \mathbb{C} \to \mathbb{C}$ is multiplication by some complex number z, and φ is an isometry if and only if $|z| = 1$. In this way the group $U(1)$ is identified with the unit circle S^1.

The *special unitary group*

$$SU(n) = \{ A \in U(n) \mid \det A = 1 \} = U(n) \cap SL(n, \mathbb{C})$$

is a closed Lie subgroup of $U(n)$ with Lie algebra

$$\mathfrak{su}(n) = \{ A \in L(\mathbb{C}^n, \mathbb{C}^n) \mid \langle A\mathbf{x}, \mathbf{y} \rangle = -\langle \mathbf{x}, A\mathbf{y} \rangle \text{ and trace } A = 0 \}.$$

Hence, $SU(n)$ is compact and has (real) dimension $n^2 - 1$.

We shall prove later that both $U(n)$ and $SU(n)$ are connected.

Proposition 9.2.18. *The group $U(n)$ is a compact real Lie subgroup of $GL(n, \mathbb{C})$ of (real) dimension n^2. Its Lie algebra $\mathfrak{u}(n)$ consists of the space of skew-Hermitian $n \times n$ matrices with the commutator bracket. $SU(n)$ is a closed real Lie subgroup of $U(n)$ of dimension $n^2 - 1$ whose Lie algebra $\mathfrak{su}(n)$ consists of all trace zero skew-Hermitian $n \times n$ matrices.*

In the Internet supplement to this chapter, we shall show that

$$Sp(2n, \mathbb{R}) \cap O(2n, \mathbb{R}) = U(n).$$

We shall also discuss some beautiful generalizations of this fact.

The Group SU(2). This group warrants special attention, since it appears in many physical applications such as the Cayley–Klein parameters for the free rigid body and in the construction of the (nonabelian) gauge group for the Yang–Mills equations in elementary particle physics.

From the general formula for the dimension of SU(n) it follows that $\dim \mathrm{SU}(2) = 3$. The group SU(2) is diffeomorphic to the three-sphere $S^3 = \{\, x \in \mathbb{R}^4 \mid \|\mathbf{x}\| = 1 \,\}$, with the diffeomorphism given by

$$x = (x^0, x^1, x^2, x^3) \in S^3 \subset \mathbb{R}^4 \mapsto \begin{bmatrix} x^0 - ix^3 & -x^2 - ix^1 \\ x^2 - ix^1 & x^0 + ix^3 \end{bmatrix} \in \mathrm{SU}(2).$$

(9.2.15)

Therefore, SU(2) is connected and simply connected.

By Euler's Theorem 9.2.6 every element of SO(3) different from the identity is determined by a vector $\mathbf{v}$, which we can choose to be a unit vector, and an angle of rotation θ about the axis $\mathbf{v}$. The trouble is, the pair $(\mathbf{v}, \theta)$ and $(-\mathbf{v}, -\theta)$ represent the same rotation and there is no consistent way to continuously choose one of these pairs, valid for the entire group SO(3). Such a choice is called, in physics, a choice of *spin*. This immediately suggests the existence of a double cover of SO(3) that, hopefully, should also be a Lie group. We will show below that SU(2) fulfills these requirements. This is based on the following construction.

Let $\sigma_1, \sigma_2, \sigma_3$ be the *Pauli spin matrices*, defined by

$$\sigma_1 = \begin{bmatrix} 0 & 1 \\ 1 & 0 \end{bmatrix}, \quad \sigma_2 = \begin{bmatrix} 0 & -i \\ i & 0 \end{bmatrix}, \quad \text{and} \quad \sigma_3 = \begin{bmatrix} 1 & 0 \\ 0 & -1 \end{bmatrix},$$

and let $\boldsymbol{\sigma} = (\sigma_1, \sigma_2, \sigma_3)$. Then one checks that

$$[\sigma_1, \sigma_2] = 2i\sigma_3 \text{ (plus cyclic permutations)},$$

from which one finds that the map

$$\mathbf{x} \mapsto \tilde{\mathbf{x}} = \frac{1}{2i} \mathbf{x} \cdot \boldsymbol{\sigma} = \frac{1}{2} \begin{bmatrix} -ix^3 & -ix^1 - x^2 \\ -ix^1 + x^2 & ix^3 \end{bmatrix},$$

where $\mathbf{x} \cdot \boldsymbol{\sigma} = x^1 \sigma_1 + x^2 \sigma_2 + x^3 \sigma_3$, is a Lie algebra isomorphism between $\mathbb{R}^3$ and the 2×2 skew-Hermitian traceless matrices (the Lie algebra of SU(2)); that is, $[\tilde{\mathbf{x}}, \tilde{\mathbf{y}}] = (\mathbf{x} \times \mathbf{y})\tilde{}$. Note that

$$-\det(\mathbf{x} \cdot \boldsymbol{\sigma}) = \|\mathbf{x}\|^2, \quad \text{and trace } (\tilde{\mathbf{x}}\tilde{\mathbf{y}}) = -\tfrac{1}{2}\mathbf{x} \cdot \mathbf{y}.$$

Define the Lie group homomorphism $\pi : \mathrm{SU}(2) \to \mathrm{GL}(3, \mathbb{R})$ by

$$(\pi(A)\mathbf{x}) \cdot \boldsymbol{\sigma} = A(\mathbf{x} \cdot \boldsymbol{\sigma})A^\dagger = A(\mathbf{x} \cdot \boldsymbol{\sigma})A^{-1}.$$

(9.2.16)

A straightforward computation, using the expression (9.2.15), shows that $\ker \pi = \{\pm I\}$. Therefore, $\pi(A) = \pi(B)$ if and only if $A = \pm B$. Since

$$\|\pi(A)\mathbf{x}\|^2 = -\det((\pi(A)\mathbf{x}) \cdot \boldsymbol{\sigma})$$
$$= -\det(A(\mathbf{x} \cdot \boldsymbol{\sigma})A^{-1})$$
$$= -\det(\mathbf{x} \cdot \boldsymbol{\sigma}) = \|\mathbf{x}\|^2,$$

it follows that

$$\pi(\mathrm{SU}(2)) \subset \mathrm{O}(3).$$

But $\pi(\mathrm{SU}(2))$ is connected, being the continuous image of a connected space, and so

$$\pi(\mathrm{SU}(2)) \subset \mathrm{SO}(3).$$

Let us show that $\pi : \mathrm{SU}(2) \to \mathrm{SO}(3)$ is a local diffeomorphism. Indeed, if $\tilde{\alpha} \in \mathfrak{su}(2)$, then

$$(T_e\pi(\tilde{\alpha})\mathbf{x}) \cdot \boldsymbol{\sigma} = (\mathbf{x} \cdot \boldsymbol{\sigma})\tilde{\alpha}^\dagger + \tilde{\alpha}(\mathbf{x} \cdot \boldsymbol{\sigma})$$
$$= [\tilde{\alpha}, \mathbf{x} \cdot \boldsymbol{\sigma}] = 2i[\tilde{\alpha}, \tilde{\mathbf{x}}]$$
$$= 2i(\tilde{\alpha} \times \mathbf{x})\tilde{} = (\tilde{\alpha} \times \mathbf{x}) \cdot \boldsymbol{\sigma}$$
$$= (\hat{\alpha}\mathbf{x}) \cdot \boldsymbol{\sigma},$$

that is, $T_e\pi(\tilde{\alpha}) = \hat{\alpha}$. Thus,

$$T_e\pi : \mathfrak{su}(2) \longrightarrow \mathfrak{so}(3)$$

is a Lie algebra isomorphism and hence is a local diffeomorphism in a neighborhood of the identity. Since π is a Lie group homomorphism, it is a local diffeomorphism around every point.

In particular, $\pi(\mathrm{SU}(2))$ is open and hence closed (its complement is a union of open cosets in $\mathrm{SO}(3)$). Since it is nonempty and $\mathrm{SO}(3)$ is connected, we have $\pi(\mathrm{SU}(2)) = \mathrm{SO}(3)$. Therefore,

$$\pi : \mathrm{SU}(2) \to \mathrm{SO}(3)$$

is a 2 to 1 surjective submersion. Summarizing, we have the commutative diagram in Figure 9.2.3.

Proposition 9.2.19. *The Lie group* $\mathrm{SU}(2)$ *is the simply connected 2 to 1 covering group of* $\mathrm{SO}(3)$.

Quaternions. The division ring $\mathbb{H}$ (or, by abuse of language, the non-commutative field) of quaternions is generated over the reals by three elements $\mathbf{i}, \mathbf{j}, \mathbf{k}$ with the relations

$$\mathbf{i}^2 = \mathbf{j}^2 = \mathbf{k}^2 = -1,$$
$$\mathbf{ij} = -\mathbf{ji} = \mathbf{k}, \quad \mathbf{jk} = -\mathbf{kj} = \mathbf{i}, \quad \mathbf{ki} = -\mathbf{ik} = \mathbf{j}.$$

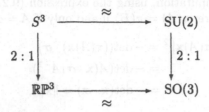

FIGURE 9.2.3. The link between SU(2) and SO(3).

Quaternionic multiplication is performed in the usual manner (like polynomial multiplication) taking the above relations into account. If $a \in \mathbb{H}$, we write

$$a = (a_s, \mathbf{a}_v) = a_s + a_v^1 \mathbf{i} + a_v^2 \mathbf{j} + a_v^3 \mathbf{k}$$

for the *scalar* and *vectorial part of the quaternion*, where a_s, a_v^1, a_v^2, $a_v^3 \in \mathbb{R}$. Quaternions having zero scalar part are also called *pure quaternions*. With this notation, quaternionic multiplication has the expression

$$ab = (a_s b_s - \mathbf{a}_v \cdot \mathbf{b}_v, a_s \mathbf{b}_v + b_s \mathbf{a}_v + \mathbf{a}_v \times \mathbf{b}_v).$$

In addition, every quaternion $a = (a_s, \mathbf{a}_v)$ has a conjugate $\bar{a} := (a_s, -\mathbf{a}_v)$, that is, the real numbers are fixed by the conjugation and $\bar{\mathbf{i}} = -\mathbf{i}$, $\bar{\mathbf{j}} = -\mathbf{j}$, and $\bar{\mathbf{k}} = -\mathbf{k}$. Note that $\overline{ab} = \bar{b}\bar{a}$. Every quaternion $a \neq 0$ has an inverse given by $a^{-1} = \bar{a}/|a|^2$, where

$$|a|^2 := a\bar{a} = \bar{a}a = a_s^2 + \|\mathbf{a}_v\|^2.$$

In particular, the unit quaternions, which, as a set, equal the unit sphere S^3 in $\mathbb{R}^4$, form a group under quaternionic multiplication.

Proposition 9.2.20. *The unit quaternions* $S^3 = \{ a \in \mathbb{H} \mid |a| = 1 \}$ *form a Lie group isomorphic to* SU(2) *via the isomorphism* (9.2.15).

Proof. We already noted that (9.2.15) is a diffeomorphism of S^3 with SU(2), so all that remains to be shown is that it is a group homomorphism, which is a straightforward computation. ∎

Since the Lie algebra of S^3 is the tangent space at 1, it follows that it is isomorphic to the pure quaternions $\mathbb{R}^3$. We begin by determining the adjoint action of S^3 on its Lie algebra.

If $a \in S^3$ and $\mathbf{b}_v$ is a pure quaternion, the derivative of the conjugation is given by

$$\mathrm{Ad}_a \, \mathbf{b}_v = a\mathbf{b}_v a^{-1} = a\mathbf{b}_v \frac{\bar{a}}{|a|^2} = \frac{1}{|a|^2}(-\mathbf{a}_v \cdot \mathbf{b}_v, a_s \mathbf{b}_v + \mathbf{a}_v \times \mathbf{b}_v)(a_s, -\mathbf{a}_v)$$

$$= \frac{1}{|a|^2} \left(0, 2a_s(\mathbf{a}_v \times \mathbf{b}_v) + 2(\mathbf{a}_v \cdot \mathbf{b}_v)\mathbf{a}_v + (a_s^2 - \|\mathbf{a}_v\|^2)\mathbf{b}_v \right).$$

Therefore, if $a(t) = (1, t\mathbf{a}_v)$, we have $a(0) = 1$, $a'(0) = \mathbf{a}_v$, so that the Lie bracket on the pure quaternions $\mathbb{R}^3$ is given by

$$
\begin{aligned}
[\mathbf{a}_v, \mathbf{b}_v] &= \left.\frac{d}{dt}\right|_{t=0} \mathrm{Ad}_{a(t)} \mathbf{b}_v \\
&= \left.\frac{d}{dt}\right|_{t=0} \frac{1}{1 + t^2 \|\mathbf{a}_v\|^2} \left(2t(\mathbf{a}_v \times \mathbf{b}_v) + 2t^2(\mathbf{a}_v \cdot \mathbf{b}_v)\mathbf{a}_v\right. \\
&\qquad\qquad\qquad\qquad \left. + (1 - t^2\|\mathbf{a}_v\|^2)\mathbf{b}_v\right) \\
&= 2\mathbf{a}_v \times \mathbf{b}_v.
\end{aligned}
$$

Thus, *the Lie algebra of S^3 is $\mathbb{R}^3$ relative to the Lie bracket given by twice the cross product of vectors.*

The derivative of the Lie group isomorphism (9.2.15) is given by

$$
\mathbf{x} \in \mathbb{R}^3 \mapsto \begin{bmatrix} -ix^3 & -ix^1 - x^2 \\ -ix^1 + x^2 & ix^3 \end{bmatrix} = 2\tilde{\mathbf{x}} \in \mathfrak{su}(2),
$$

and is thus a Lie algebra isomorphism from $\mathbb{R}^3$ with twice the cross product as bracket to $\mathfrak{su}(2)$, or equivalently to $(\mathbb{R}^3, \times)$.

Let us return to the commutative diagram in Figure 9.2.3 and determine explicitly the 2 to 1 surjective map $S^3 \to SO(3)$ that associates to a quaternion $a \in S^3 \subset \mathbb{H}$ the rotation matrix $A \in SO(3)$. To compute this map, let $a \in S^3$ and associate to it the matrix

$$
U = \begin{bmatrix} a_s - ia_v^3 & -a_v^2 - ia_v^1 \\ a_v^2 - ia_v^1 & a_s + ia_v^3 \end{bmatrix},
$$

where $a = (a_s, \mathbf{a}_v) = (a_s, a_v^1, a_v^2, a_v^3)$. By (9.2.16), the rotation matrix is given by $A = \pi(U)$, namely,

$$
\begin{aligned}
(A\mathbf{x}) \cdot \boldsymbol{\sigma} &= (\pi(U)\mathbf{x}) \cdot \boldsymbol{\sigma} = U(\mathbf{x} \cdot \boldsymbol{\sigma})U^\dagger \\
&= \begin{bmatrix} a_s - ia_v^3 & -a_v^2 - ia_v^1 \\ a_v^2 - ia_v^1 & a_s + ia_v^3 \end{bmatrix} \begin{bmatrix} x^3 & x^1 - ix^2 \\ x^1 + ix^2 & -x^3 \end{bmatrix} \\
&\quad \times \begin{bmatrix} a_s + ia_v^3 & a_v^2 + ia_v^1 \\ -a_v^2 + ia_v^1 & a_s - ia_v^3 \end{bmatrix} \\
&= \left[(a_s^2 + (a_v^1)^2 - (a_v^2)^2 - (a_v^3)^2)\, x^1 + 2(a_v^1 a_v^2 - a_s a_v^3)x^2 \right. \\
&\qquad \left. + 2(a_s a_v^2 + a_v^1 a_v^3)x^3\right]\sigma_1 \\
&\quad + \left[2\left(a_v^1 a_v^2 + a_s a_v^3\right)x^1 + \left(a_s^2 - (a_v^1)^2 + (a_v^2)^2 - (a_v^3)^2\right)x^2 \right. \\
&\qquad \left. + 2\left(a_v^2 a_v^3 - a_s a_v^1\right)x^3\right]\sigma_2 \\
&\quad + \left[2\left(a_v^1 a_v^3 - a_s a_v^2\right)x^1 + 2\left(a_s a_v^1 + a_v^2 a_v^3\right)x^2 \right. \\
&\qquad \left. + \left(a_s^2 - (a_v^1)^2 - (a_v^2)^2 + (a_v^3)^2\right)x^3\right]\sigma_3.
\end{aligned}
$$

Thus, taking into account that $a_s^2 + (a_v^1)^2 + (a_v^2)^2 + (a_v^3)^2 = 1$, we get the expression of the matrix A as

$$
\begin{bmatrix}
2a_s^2 + 2(a_v^1)^2 - 1 & 2(-a_s a_v^3 + a_v^1 a_v^2) & 2(a_s a_v^2 + a_v^1 a_v^3) \\
2(a_s a_v^3 + a_v^1 a_v^2) & 2a_s^2 + 2(a_v^2)^2 - 1 & 2(-a_s a_v^1 + a_v^2 a_v^3) \\
2(-a_s a_v^1 + a_v^2 a_v^3) & 2(a_s a_v^1 + a_v^2 a_v^3) & 2a_s^2 + (a_v^3)^2 - 1
\end{bmatrix}
$$
$$
= (2a_s^2 - 1)I + 2a_s \hat{\mathbf{a}}_v + 2\mathbf{a}_v \otimes \mathbf{a}_v, \quad (9.2.17)
$$

where $\mathbf{a}_v \otimes \mathbf{a}_v$ is the symmetric matrix whose (i,j) entry equals $a_v^i a_v^j$. The map

$$
a \in S^3 \mapsto (2a_s^2 - 1)I + 2a_s \hat{\mathbf{a}}_v + 2\mathbf{a}_v \otimes \mathbf{a}_v
$$

is called the **Euler–Rodrigues parametrization**. It has the advantage, as opposed to the Euler angles parametrization, which has a coordinate singularity, of being global. This is of crucial importance in computational mechanics (see, for example, Marsden and Wendlandt [1997]).

Finally, let us rewrite Rodrigues' formula (9.2.8) in terms of unit quaternions. Let

$$
a = (a_s, \mathbf{a}_v) = \left(\cos \frac{\omega}{2}, \left(\sin \frac{\omega}{2} \right) \mathbf{n} \right),
$$

where $\omega > 0$ is an angle and $\mathbf{n}$ is a unit vector. Since $\hat{\mathbf{n}}^2 = \mathbf{n} \otimes \mathbf{n} - I$, from (9.2.8) we get

$$
\exp(\omega \mathbf{n}) = I + (\sin \omega)\hat{\mathbf{n}} + 2 \left(\sin^2 \frac{\omega}{2} \right) (\mathbf{n} \otimes \mathbf{n} - I)
$$
$$
= \left(1 - 2\sin^2 \frac{\omega}{2} \right) I + 2 \cos \frac{\omega}{2} \sin \frac{\omega}{2} \hat{\mathbf{n}} + 2 \left(\sin^2 \frac{\omega}{2} \right) \mathbf{n} \otimes \mathbf{n}
$$
$$
= \left(2a_s^2 - 1 \right) I + 2a_s \hat{\mathbf{a}}_v + 2\mathbf{a}_v \otimes \mathbf{a}_v.
$$

This expression then produces a rotation associated to each unit quaternion a. In addition, using this parametrization, in 1840 Rodrigues found a beautiful way of expressing the product of two rotations $\exp(\omega_1 \mathbf{n}_1) \cdot \exp(\omega_2 \mathbf{n}_2)$ in terms of the given data. In fact, this was an early exploration of the spin group! We refer to Whittaker [1927, Section 7], Altmann [1986], Enos [1993], Lewis and Simo [1995], and references therein for further information.

SU(2) Conjugacy Classes and the Hopf Fibration. We next determine all conjugacy classes of $S^3 \cong \mathrm{SU}(2)$. If $a \in S^3$, then $a^{-1} = \bar{a}$, and a straightforward computation gives

$$
aba^{-1} = (b_s, 2(\mathbf{a}_v \cdot \mathbf{b}_v)\mathbf{a}_v + 2a_s(\mathbf{a}_v \times \mathbf{b}_v) + (2a_s^2 - 1)\mathbf{b}_v)
$$

for any $b \in S^3$. If $b_s = \pm 1$, that is, $\mathbf{b}_v = 0$, then the above formula shows that $aba^{-1} = b$ for all $a \in S^3$, that is, the classes of I and $-I$, where $I = (1, \mathbf{0})$, each consist of one element, and the center of $\mathrm{SU}(2) \cong S^3$ is $\{\pm I\}$.

In what follows, assume that $b_s \neq \pm 1$, or, equivalently, that $\mathbf{b}_v \neq \mathbf{0}$, and fix this $b \in S^3$ throughout the following discussion. We shall prove that given $\mathbf{x} \in \mathbb{R}^3$ with $\|\mathbf{x}\| = \|\mathbf{b}_v\|$, we can find $a \in S^3$ such that

$$2(\mathbf{a}_v \cdot \mathbf{b}_v)\mathbf{a}_v + 2a_s(\mathbf{a}_v \times \mathbf{b}_v) + (2a_s^2 - 1)\mathbf{b}_v = \mathbf{x}. \qquad (9.2.18)$$

If $\mathbf{x} = c\mathbf{b}_v$ for some $c \neq 0$, then the choice $\mathbf{a}_v = \mathbf{0}$ and $2a_s^2 = 1 + c$ satisfies (9.2.18). Now assume that $\mathbf{x}$ and $\mathbf{b}_v$ are not collinear. Take the dot product of (9.2.18) with $\mathbf{b}_v$ and get

$$2(\mathbf{a}_v \cdot \mathbf{b}_v)^2 + 2a_s^2 \|\mathbf{b}_v\|^2 = \|\mathbf{b}_v\|^2 + \mathbf{x} \cdot \mathbf{b}_v.$$

If $\|\mathbf{b}_v\|^2 + \mathbf{x} \cdot \mathbf{b}_v = 0$, since $\mathbf{b}_v \neq \mathbf{0}$, it follows that $\mathbf{a}_v \cdot \mathbf{b}_v = 0$ and $a_s = 0$. Returning to (9.2.18) it follows that $-\mathbf{b}_v = \mathbf{x}$, which is excluded. Therefore, $\mathbf{x} \cdot \mathbf{b}_v + \|\mathbf{b}_v\|^2 \neq 0$, and searching for $\mathbf{a}_v \in \mathbb{R}^3$ such that $\mathbf{a}_v \cdot \mathbf{b}_v = 0$, it follows that

$$a_s^2 = \frac{\mathbf{x} \cdot \mathbf{b}_v + \|\mathbf{b}_v\|^2}{2\|\mathbf{b}_v\|^2} \neq 0.$$

Now take the cross product of (9.2.18) with $\mathbf{b}_v$ and recall that we assumed $\mathbf{a}_v \cdot \mathbf{b}_v = 0$ to get

$$2a_s\|\mathbf{b}_v\|^2 \mathbf{a}_v = \mathbf{b}_v \times \mathbf{x},$$

whence

$$\mathbf{a}_v = \frac{\mathbf{b}_v \times \mathbf{x}}{2a_s\|\mathbf{b}_v\|^2},$$

which is allowed, since $\mathbf{b}_v \neq \mathbf{0}$ and $a_s \neq 0$. Note that $a = (a_s, \mathbf{a}_v)$ just determined satisfies $\mathbf{a}_v \cdot \mathbf{b}_v = 0$ and

$$|a|^2 = a_s^2 + \|\mathbf{a}_v\|^2 = 1,$$

since $\|\mathbf{x}\| = \|\mathbf{b}_v\|$.

Proposition 9.2.21. *The conjugacy classes of $S^3 \cong \mathrm{SU}(2)$ are the two-spheres*

$$\{\, \mathbf{b}_v \in \mathbb{R}^3 \mid \|\mathbf{b}_v\|^2 = 1 - b_s^2 \,\}$$

for each $b_s \in [-1, 1]$, which degenerate to the north and south poles $(\pm 1, 0, 0, 0)$ comprising the center of $\mathrm{SU}(2)$.

The above proof shows that any unit quaternion is conjugate in S^3 to a quaternion of the form $a_s + a_v^3 \mathbf{k}$, a_s, $a_v^3 \in \mathbb{R}$, which in terms of matrices and the isomorphism (9.2.15) says that *any $\mathrm{SU}(2)$ matrix is conjugate to a diagonal matrix.*

The conjugacy class of $\mathbf{k}$ is the unit sphere S^2, and the orbit map

$$\pi : S^3 \to S^2, \qquad \pi(a) = a\mathbf{k}\bar{a},$$

is the **Hopf fibration**.

The subgroup

$$H = \left\{ a_s + a_v^3 \mathbf{k} \in S^3 \mid a_s, a_v^3 \in \mathbb{R} \right\} \subset S^3$$

is a closed, one-dimensional Abelian Lie subgroup of S^3 isomorphic via (9.2.15) to the set of diagonal matrices in SU(2) and is hence the circle S^1. Note that the isotropy of $\mathbf{k}$ in S^3 consists of H, as an easy computation using (9.2.18) shows. Therefore, since the orbit of $\mathbf{k}$ is diffeomorphic to S^3/H, it follows that *the fibers of the Hopf fibration equal the left cosets aH for $a \in S^3$.*

Finally, we shall give an expression of the Hopf fibration in terms of complex variables. In the representation (9.2.15), set

$$w_1 = x^2 + ix^1, \quad w_2 = x^0 + ix^3,$$

and note that if

$$a = (x^0, x^1, x^2, x^3) \in S^3 \subset \mathbb{H},$$

then $ak\bar{a}$ corresponds to

$$\begin{bmatrix} x^0 - ix^3 & -x^2 - ix^1 \\ x^2 - ix^1 & x^0 + ix^3 \end{bmatrix} \begin{bmatrix} -i & 0 \\ 0 & i \end{bmatrix} \begin{bmatrix} x^0 + ix^3 & x^2 + ix^1 \\ -x^2 + ix^1 & x^0 - ix^3 \end{bmatrix}$$

$$= \begin{bmatrix} -i \left(|x^0 + ix^3|^2 - |x^2 + ix^1|^2 \right) & -2i \left(x^2 + ix^1 \right) \left(x^0 - ix^3 \right) \\ -2i(x^2 - ix^1)(x^0 + ix^3) & i \left(|x^0 + ix^3|^2 - |x^2 + ix^1|^2 \right) \end{bmatrix}.$$

Thus, if we consider the diffeomorphisms

$$(x^0, x^1, x^2, x^3) \in S^3 \subset \mathbb{H} \mapsto \begin{bmatrix} x^0 - ix^3 & -x^2 - ix^1 \\ x^2 - ix^1 & x^0 + ix^3 \end{bmatrix} \in \mathrm{SU}(2)$$

$$\mapsto \left(-i(x^2 + ix^1), -i(x^0 + ix^3) \right) \in S^3 \subset \mathbb{C}^2,$$

the above orbit map, that is, the Hopf fibration, becomes

$$(w_1, w_2) \in S^3 \mapsto \left(2w_1 \bar{w}_2, |w_2|^2 - |w_1|^2 \right) \in S^2.$$

Exercises

⋄ **9.2-1.** Describe the set of matrices in SO(3) that are also *symmetric*.

⋄ **9.2-2.** If $A \in \mathrm{Sp}(2n, \mathbb{R})$, show that $A^T \in \mathrm{Sp}(2n, \mathbb{R})$ as well.

⋄ **9.2-3.** Show that $\mathfrak{sp}(2n, \mathbb{R})$ is isomorphic, as a Lie algebra, to the space of homogeneous quadratic functions on $\mathbb{R}^{2n}$ under the Poisson bracket.

⋄ **9.2-4.** A map $f : \mathbb{R}^n \to \mathbb{R}^n$ preserving the distance between any two points, that is, $\|f(\mathbf{x}) - f(\mathbf{y})\| = \|\mathbf{x} - \mathbf{y}\|$ for all $\mathbf{x}, \mathbf{y} \in \mathbb{R}^n$, is called an *isometry*. Show that f is an isometry preserving the origin if and only if $f \in \mathrm{O}(n)$.

9.3   Actions of Lie Groups

In this section we develop some basic facts about actions of Lie groups on manifolds. One of our main applications later will be the description of Hamiltonian systems with symmetry groups.

Basic Definitions.   We begin with the definition of the action of a Lie group G on a manifold M.

Definition 9.3.1.   *Let M be a manifold and let G be a Lie group. A (left)* **action** *of a Lie group G on M is a smooth mapping $\Phi : G \times M \to M$ such that:*

(i) $\Phi(e, x) = x$ *for all $x \in M$; and*

(ii) $\Phi(g, \Phi(h, x)) = \Phi(gh, x)$ *for all $g, h \in G$ and $x \in M$.*

A **right action** is a map $\Psi : M \times G \to M$ that satisfies $\Psi(x, e) = x$ and $\Psi(\Psi(x, g), h) = \Psi(x, gh)$. We sometimes use the notation $g \cdot x = \Phi(g, x)$ for left actions, and $x \cdot g = \Psi(x, g)$ for right actions. In the infinite-dimensional case there are important situations where care with the smoothness is needed. For the formal development we assume that we are in the Banach–Lie group context.

For every $g \in G$ let $\Phi_g : M \to M$ be given by $x \mapsto \Phi(g, x)$. Then (i) becomes $\Phi_e = \mathrm{id}_M$, while (ii) becomes $\Phi_{gh} = \Phi_g \circ \Phi_h$. Definition 9.3.1 can now be rephrased by saying that the map $g \mapsto \Phi_g$ is a homomorphism of G into $\mathrm{Diff}(M)$, the group of diffeomorphisms of M. In the special but important case where M is a Banach space V and each $\Phi_g : V \to V$ is a continuous linear transformation, the action Φ of G on V is called a **representation** of G on V.

Examples

(a)   SO(3) acts on $\mathbb{R}^3$ by $(A, x) \mapsto Ax$. This action leaves the two-sphere S^2 invariant, so the same formula defines an action of SO(3) on S^2.   ◆

(b)   GL$(n, \mathbb{R})$ acts on $\mathbb{R}^n$ by $(A, x) \mapsto Ax$.   ◆

(c)   Let X be a complete vector field on M, that is, one for which the flow F_t of X is defined for all $t \in \mathbb{R}$. Then $F_t : M \to M$ defines an action of $\mathbb{R}$ on M.   ◆

Orbits and Isotropy.   If Φ is an action of G on M and $x \in M$, the **orbit** of x is defined by

$$\mathrm{Orb}(x) = \{\, \Phi_g(x) \mid g \in G \,\} \subset M.$$

In finite dimensions one can show that $\mathrm{Orb}(x)$ is an immersed submanifold of M (Abraham and Marsden [1978, p. 265]). For $x \in M$, the *isotropy* (or *stabilizer* or *symmetry*) group of Φ at x is given by

$$G_x := \{\, g \in G \mid \Phi_g(x) = x \,\} \subset G.$$

Since the map $\Phi^x : G \to M$ defined by $\Phi^x(g) = \Phi(g, x)$ is continuous, $G_x = (\Phi^x)^{-1}(x)$ is a closed subgroup and hence a Lie subgroup of G. The manifold structure of $\mathrm{Orb}(x)$ is defined by requiring the bijective map $[g] \in G/G_x \mapsto g \cdot x \in \mathrm{Orb}(x)$ to be a diffeomorphism. That G/G_x is a smooth manifold follows from Proposition 9.3.2, which is discussed below.

An action is said to be:

1. *transitive* if there is only one orbit or, equivalently, if for every $x, y \in M$ there is a $g \in G$ such that $g \cdot x = y$;

2. *effective* (or *faithful*) if $\Phi_g = \mathrm{id}_M$ implies $g = e$; that is, $g \mapsto \Phi_g$ is one-to-one; and

3. *free* if it has no fixed points, that is, $\Phi_g(x) = x$ implies $g = e$ or, equivalently, if for each $x \in M$, $g \mapsto \Phi_g(x)$ is one-to-one. Note that an action is free iff $G_x = \{e\}$, for all $x \in M$ and that every free action is faithful.

Examples

(a) Left translation. $L_g : G \to G, h \mapsto gh$, defines a transitive and free action of G on itself. Note that right multiplication $R_g : G \to G, h \mapsto hg$, does not define a left action because $R_{gh} = R_h \circ R_g$, so that $g \mapsto R_g$ is an antihomomorphism. However, $g \mapsto R_g$ does define a right action, while $g \mapsto R_{g^{-1}}$ defines a left action of G on itself. ◆

(b) $g \mapsto I_g = R_{g^{-1}} \circ L_g$. The map $I_g : G \to G$ given by $h \mapsto ghg^{-1}$ is the *inner automorphism* associated with g. Orbits of this action are called *conjugacy classes* or, in the case of matrix groups, *similarity classes*. ◆

(c) Adjoint Action. Differentiating conjugation at e, we get the *adjoint representation* of G on $\mathfrak{g}$:

$$\mathrm{Ad}_g := T_e I_g : T_e G = \mathfrak{g} \to T_e G = \mathfrak{g}.$$

Explicitly, the adjoint action of G on $\mathfrak{g}$ is given by

$$\mathrm{Ad} : G \times \mathfrak{g} \to \mathfrak{g}, \quad \mathrm{Ad}_g(\xi) = T_e(R_{g^{-1}} \circ L_g)\xi.$$

For example, for SO(3) we have $I_A(B) = ABA^{-1}$, so differentiating with respect to B at $B = $ identity gives $\mathrm{Ad}_A \, \hat{\mathbf{v}} = A\hat{\mathbf{v}}A^{-1}$. However,

$$(\mathrm{Ad}_A \, \hat{\mathbf{v}})(\mathbf{w}) = A\hat{\mathbf{v}}(A^{-1}\mathbf{w}) = A(\mathbf{v} \times A^{-1}\mathbf{w}) = A\mathbf{v} \times \mathbf{w},$$

so

$$(\text{Ad}_A \, \hat{\mathbf{v}}) = (A\mathbf{v})\hat{\ }.$$

Identifying $\mathfrak{so}(3) \cong \mathbb{R}^3$, we get $\text{Ad}_A \, \mathbf{v} = A\mathbf{v}$.   ♦

(d) Coadjoint Action.   The *coadjoint action* of G on $\mathfrak{g}^*$, the dual of the Lie algebra $\mathfrak{g}$ of G, is defined as follows. Let $\text{Ad}_g^* : \mathfrak{g}^* \to \mathfrak{g}^*$ be the dual of Ad_g, defined by

$$\langle \text{Ad}_g^* \, \alpha, \xi \rangle = \langle \alpha, \text{Ad}_g \, \xi \rangle$$

for $\alpha \in \mathfrak{g}^*$ and $\xi \in \mathfrak{g}$. Then the map

$$\Phi^* : G \times \mathfrak{g}^* \to \mathfrak{g}^* \quad \text{given by} \quad (g, \alpha) \mapsto \text{Ad}_{g^{-1}}^* \, \alpha$$

is the coadjoint action of G on $\mathfrak{g}^*$. The corresponding *coadjoint representation* of G on $\mathfrak{g}^*$ is denoted by

$$\text{Ad}^* : G \to \text{GL}(\mathfrak{g}^*, \mathfrak{g}^*), \quad \text{Ad}_{g^{-1}}^* = \left(T_e(R_g \circ L_{g^{-1}}) \right)^*.$$

We will avoid the introduction of yet another $*$ by writing $(\text{Ad}_{g^{-1}})^*$ or simply $\text{Ad}_{g^{-1}}^*$, where $*$ denotes the usual linear-algebraic dual, rather than $\text{Ad}^*(g)$, in which $*$ is simply part of the name of the function Ad^*. Any representation of G on a vector space V similarly induces a *contragredient representation* of G on V^*.   ♦

Quotient (Orbit) Spaces.   An action of Φ of G on a manifold M defines an equivalence relation on M by the relation of belonging to the same orbit; explicitly, for $x, y \in M$, we write $x \sim y$ if there exists a $g \in G$ such that $g \cdot x = y$, that is, if $y \in \text{Orb}(x)$ (and hence $x \in \text{Orb}(y)$). We let M/G be the set of these equivalence classes, that is, the set of orbits, sometimes called the *orbit space*. Let

$$\pi : M \to M/G, \quad x \mapsto \text{Orb}(x),$$

and give M/G the quotient topology by defining $U \subset M/G$ to be open if and only if $\pi^{-1}(U)$ is open in M. To guarantee that the orbit space M/G has a smooth manifold structure, further conditions on the action are required.

An action $\Phi : G \times M \to M$ is called *proper* if the mapping

$$\tilde{\Phi} : G \times M \to M \times M,$$

defined by

$$\tilde{\Phi}(g, x) = (x, \Phi(g, x)),$$

is proper. In finite dimensions this means that if $K \subset M \times M$ is compact, then $\tilde{\Phi}^{-1}(K)$ is compact. In general, this means that if $\{x_n\}$ is a convergent sequence in M and $\Phi_{g_n}(x_n)$ converges in M, then $\{g_n\}$ has a convergent subsequence in G. For instance, if G is compact, this condition is automatically satisfied. Orbits of proper Lie group actions are closed and hence embedded submanifolds. The next proposition gives a useful sufficient condition for M/G to be a smooth manifold.

Proposition 9.3.2. *If $\Phi : G \times M \to M$ is a proper and free action, then M/G is a smooth manifold and $\pi : M \to M/G$ is a smooth submersion.*

For the proof, see Proposition 4.2.23 in Abraham and Marsden [1978]. (In infinite dimensions one uses these ideas, but additional technicalities often arise; see Ebin [1970] and Isenberg and Marsden [1982].) The idea of the chart construction for M/G is based on the following observation. If $x \in M$, then there is an isomorphism φ_x of $T_{\pi(x)}(M/G)$ with the quotient space $T_x M/T_x \operatorname{Orb}(x)$. Moreover, if $y = \Phi_g(x)$, then $T_x \Phi_g$ induces an isomorphism

$$\psi_{x,y} : T_x M/T_x \operatorname{Orb}(x) \to T_y M/T_y \operatorname{Orb}(y)$$

satisfying $\varphi_y \circ \psi_{x,y} = \varphi_x$.

Examples

(a) $G = \mathbb{R}$ acts on $M = \mathbb{R}$ by translations; explicitly,

$$\Phi : G \times M \to M, \quad \Phi(s, x) = x + s.$$

Then for $x \in \mathbb{R}$, $\operatorname{Orb}(x) = \mathbb{R}$. Hence M/G is a single point, and the action is transitive, proper, and free.   ♦

(b) $G = \mathrm{SO}(3)$, $M = \mathbb{R}^3$ ($\cong \mathfrak{so}(3)^*$). Consider the action for $\mathbf{x} \in \mathbb{R}^3$ and $A \in \mathrm{SO}(3)$ given by $\Phi_A \mathbf{x} = A\mathbf{x}$. Then

$$\operatorname{Orb}(x) = \{\, \mathbf{y} \in \mathbb{R}^3 \mid \|\mathbf{y}\| = \|\mathbf{x}\| \,\} = \text{a sphere of radius } \|\mathbf{x}\|.$$

Hence $M/G \cong \mathbb{R}^+$. The set

$$\mathbb{R}^+ = \{\, r \in \mathbb{R} \mid r \geq 0 \,\}$$

is not a manifold because it includes the endpoint $r = 0$. Indeed, the action is not free, since it has the fixed point $\mathbf{0} \in \mathbb{R}^3$.   ♦

(c)   Let G be Abelian. Then $\operatorname{Ad}_g = \operatorname{id}_{\mathfrak{g}}$, $\operatorname{Ad}^*_{g^{-1}} = \operatorname{id}_{\mathfrak{g}^*}$, and the adjoint and coadjoint orbits of $\xi \in \mathfrak{g}$ and $\alpha \in \mathfrak{g}^*$, respectively, are the one-point sets $\{\xi\}$ and $\{\alpha\}$.   ♦

We will see later that coadjoint orbits can be natural phase spaces for some mechanical systems like the rigid body; in particular, they are always even-dimensional.

Infinitesimal Generators.   Next we turn to the infinitesimal description of an action, which will be a crucial concept for mechanics.

Definition 9.3.3.   *Suppose* $\Phi : G \times M \to M$ *is an action. For* $\xi \in \mathfrak{g}$, *the map* $\Phi^{\xi} : \mathbb{R} \times M \to M$, *defined by*

$$\Phi^{\xi}(t, x) = \Phi(\exp t\xi, x),$$

is an $\mathbb{R}$-*action on* M. *In other words,* $\Phi_{\exp t\xi} : M \to M$ *is a flow on* M. *The corresponding vector field on* M, *given by*

$$\xi_M(x) := \left. \frac{d}{dt} \right|_{t=0} \Phi_{\exp t\xi}(x),$$

is called the **infinitesimal generator** *of the action corresponding to* ξ.

Proposition 9.3.4.   *The tangent space at* x *to an orbit* $\mathrm{Orb}(x_0)$ *is*

$$T_x \, \mathrm{Orb}(x_0) = \{ \, \xi_M(x) \mid \xi \in \mathfrak{g} \, \},$$

where $\mathrm{Orb}(x_0)$ *is endowed with the manifold structure making* $G/G_{x_0} \to \mathrm{Orb}(x_0)$ *into a diffeomorphism.*

The idea is as follows: Let $\sigma_\xi(t)$ be a curve in G with $\sigma_\xi(0) = e$ that is tangent to ξ at $t = 0$. Then the map $\Phi^{x,\xi}(t) = \Phi_{\sigma_\xi(t)}(x)$ is a smooth curve in $\mathrm{Orb}(x_0)$ with $\Phi^{x,\xi}(0) = x$. Hence by the chain rule (see also Lemma 9.3.7 below),

$$\left. \frac{d}{dt} \right|_{t=0} \Phi^{x,\xi}(t) = \left. \frac{d}{dt} \right|_{t=0} \Phi_{\sigma_\xi(t)}(x) = \xi_M(x)$$

is a tangent vector at x to $\mathrm{Orb}(x_0)$. Furthermore, each tangent vector is obtained in this way, since tangent vectors are equivalence classes of such curves.

The Lie algebra of the isotropy group G_x, $x \in M$, called the **isotropy** (or **stabilizer**, or **symmetry**) **algebra** at x, equals, by Proposition 9.1.13, $\mathfrak{g}_x = \{ \xi \in \mathfrak{g} \mid \xi_M(x) = 0 \}$.

Examples

(a)   The infinitesimal generators for the adjoint action are computed as follows. Let

$$\mathrm{Ad} : G \times \mathfrak{g} \to \mathfrak{g}, \quad \mathrm{Ad}_g(\eta) = T_e(R_{g^{-1}} \circ L_g)(\eta).$$

For $\xi \in \mathfrak{g}$, we compute the corresponding infinitesimal generator $\xi_{\mathfrak{g}}$. By definition,

$$\xi_{\mathfrak{g}}(\eta) = \left(\frac{d}{dt} \right) \Bigg|_{t=0} \mathrm{Ad}_{\exp t\xi}(\eta).$$

By (9.1.5), this equals $[\xi, \eta]$. Thus, for the adjoint action,

$$\xi_\mathfrak{g} = \mathrm{ad}_\xi, \quad \text{i.e.,} \quad \xi_\mathfrak{g}(\eta) = [\xi, \eta]. \tag{9.3.1}$$

This operation deserves a special name. We define the **ad operator** $\mathrm{ad}_\xi : \mathfrak{g} \to \mathfrak{g}$ by $\eta \mapsto [\xi, \eta]$. Thus,

$$\xi_\mathfrak{g} = \mathrm{ad}_\xi. \qquad \blacklozenge$$

(b) We illustrate (a) for the group SO(3) as follows. Let $A(t) = \exp(tC)$, where $C \in \mathfrak{so}(3)$; then $A(0) = I$ and $A'(0) = C$. Thus, with $B \in \mathfrak{so}(3)$,

$$
\begin{aligned}
\left.\frac{d}{dt}\right|_{t=0} (\mathrm{Ad}_{\exp tC} B) &= \left.\frac{d}{dt}\right|_{t=0} (\exp(tC)B(\exp(tC))^{-1}) \\
&= \left.\frac{d}{dt}\right|_{t=0} (A(t)BA(t)^{-1}) \\
&= A'(0)BA^{-1}(0) + A(0)BA^{-1\prime}(0).
\end{aligned}
$$

Differentiating $A(t)A^{-1}(t) = I$, we obtain

$$\frac{d}{dt}(A^{-1}(t)) = -A^{-1}(t)A'(t)A^{-1}(t),$$

so that

$$A^{-1\prime}(0) = -A'(0) = -C.$$

Then the preceding equation becomes

$$\left.\frac{d}{dt}\right|_{t=0} (\mathrm{Ad}_{\exp tC} B) = CB - BC = [C, B],$$

as expected. $\qquad \blacklozenge$

(c) Let $\mathrm{Ad}^* : G \times \mathfrak{g}^* \to \mathfrak{g}^*$ be the coadjoint action $(g, \alpha) \mapsto \mathrm{Ad}^*_{g^{-1}} \alpha$. If $\xi \in \mathfrak{g}$, we compute for $\alpha \in \mathfrak{g}^*$ and $\eta \in \mathfrak{g}$

$$
\begin{aligned}
\langle \xi_{\mathfrak{g}^*}(\alpha), \eta \rangle &= \left\langle \left.\frac{d}{dt}\right|_{t=0} \mathrm{Ad}^*_{\exp(-t\xi)}(\alpha), \eta \right\rangle \\
&= \left.\frac{d}{dt}\right|_{t=0} \left\langle \mathrm{Ad}^*_{\exp(-t\xi)}(\alpha), \eta \right\rangle = \left.\frac{d}{dt}\right|_{t=0} \left\langle \alpha, \mathrm{Ad}_{\exp(-t\xi)} \eta \right\rangle \\
&= \left\langle \alpha, \left.\frac{d}{dt}\right|_{t=0} \mathrm{Ad}_{\exp(-t\xi)} \eta \right\rangle \\
&= \langle \alpha, -[\xi, \eta] \rangle = - \langle \alpha, \mathrm{ad}_\xi(\eta) \rangle = - \langle \mathrm{ad}^*_\xi(\alpha), \eta \rangle.
\end{aligned}
$$

Hence

$$\xi_{\mathfrak{g}^*} = -\mathrm{ad}^*_\xi, \quad \text{or} \quad \xi_{\mathfrak{g}^*}(\alpha) = -\langle \alpha, [\xi, \cdot] \rangle. \tag{9.3.2}$$

$\blacklozenge$

(d)   Identifying $\mathfrak{so}(3) \cong (\mathbb{R}^3, \times)$ and $\mathfrak{so}(3)^* \cong \mathbb{R}^{3^*}$, using the pairing given by the standard Euclidean inner product, (9.3.2) reads

$$\xi_{\mathfrak{so}(3)^*}(l) = -l \cdot (\xi \times \cdot),$$

for $l \in \mathfrak{so}(3)^*$ and $\xi \in \mathfrak{so}(3)$. For $\eta \in \mathfrak{so}(3)$, we have

$$\langle \xi_{\mathfrak{so}(3)^*}(l), \eta \rangle = -l \cdot (\xi \times \eta) = -(l \times \xi) \cdot \eta = -\langle l \times \xi, \eta \rangle,$$

so that

$$\xi_{\mathbb{R}^3}(l) = -l \times \xi = \xi \times l.$$

As expected, $\xi_{\mathbb{R}^3}(l) \in T_l \, \mathrm{Orb}(l)$ is tangent to $\mathrm{Orb}(l)$ (see Figure 9.3.1). Allowing ξ to vary in $\mathfrak{so}(3) \cong \mathbb{R}^3$, one obtains all of $T_l \, \mathrm{Orb}(l)$, consistent with Proposition 9.3.4.   ◆

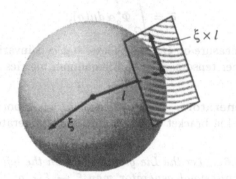

FIGURE 9.3.1. $\xi_{\mathbb{R}^3}(l)$ is tangent to $\mathrm{Orb}(l)$.

Equivariance.   A map between two spaces is equivariant when it respects group actions on these spaces. We state this more precisely:

Definition 9.3.5.   *Let M and N be manifolds and let G be a Lie group that acts on M by $\Phi_g : M \to M$, and on N by $\Psi_g : N \to N$. A smooth map $f : M \to N$ is called **equivariant** with respect to these actions if for all $g \in G$,*

$$f \circ \Phi_g = \Psi_g \circ f, \qquad (9.3.3)$$

that is, if the diagram in Figure 9.3.2 commutes.

Setting $g = \exp(t\xi)$ and differentiating (9.3.3) with respect to t at $t = 0$ gives $Tf \circ \xi_M = \xi_N \circ f$. In other words, ξ_M and ξ_N are f-related. In particular, *if f is an equivariant diffeomorphism, then $f^*\xi_N = \xi_M$.*

Also note that if M/G and N/G are both smooth manifolds with the canonical projections smooth submersions, an equivariant map $f : M \to N$ induces a smooth map $f_G : M/G \to N/G$.

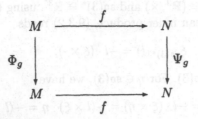

FIGURE 9.3.2. Commutative diagram for equivariance.

Averaging. A useful device for constructing invariant objects is by *averaging*. For example, let G be a compact group acting on a manifold M and let α be a differential form on M. Then we form

$$\overline{\alpha} = \int_G \Phi_g^* \alpha \, d\mu(g),$$

where μ is Haar measure on G. One checks that $\overline{\alpha}$ is invariant. One can do the same with other tensors, such as Riemannian metrics on M, to obtain invariant ones.

Brackets of Generators. Now we come to an important formula relating the Jacobi–Lie bracket of two infinitesimal generators with the Lie algebra bracket.

Proposition 9.3.6. *Let the Lie group G act on the left on the manifold M. Then the infinitesimal generator map $\xi \mapsto \xi_M$ of the Lie algebra $\mathfrak{g}$ of G into the Lie algebra $\mathfrak{X}(M)$ of vector fields of M is a Lie algebra antihomomorphism; that is,*

$$(a\xi + b\eta)_M = a\xi_M + b\eta_M$$

and

$$[\xi_M, \eta_M] = -[\xi, \eta]_M$$

for all $\xi, \eta \in \mathfrak{g}$ and $a, b \in \mathbb{R}$.

To prove this, we use the following lemma:

Lemma 9.3.7.   (i) *Let $c(t)$ be a curve in G, $c(0) = e$, $c'(0) = \xi \in \mathfrak{g}$. Then*

$$\xi_M(x) = \left.\frac{d}{dt}\right|_{t=0} \Phi_{c(t)}(x).$$

(ii) *For every $g \in G$,*

$$(\mathrm{Ad}_g \, \xi)_M = \Phi_{g^{-1}}^* \xi_M.$$

Proof.   (i) Let $\Phi^x : G \to M$ be the map $\Phi^x(g) = \Phi(g, x)$. Since Φ^x is smooth, the definition of the infinitesimal generator says that $T_e\Phi^x(\xi) = \xi_M(x)$. Thus, (i) follows by the chain rule.

(ii) We have

$$
\begin{aligned}
(\mathrm{Ad}_g \, \xi)_M(x) &= \frac{d}{dt}\bigg|_{t=0} \Phi(\exp(t \, \mathrm{Ad}_g \, \xi), x) \\
&= \frac{d}{dt}\bigg|_{t=0} \Phi(g(\exp t\xi)g^{-1}, x) \text{ (by Corollary 9.1.8)} \\
&= \frac{d}{dt}\bigg|_{t=0} (\Phi_g \circ \Phi_{\exp t\xi} \circ \Phi_{g^{-1}}(x)) \\
&= T_{\Phi_g^{-1}(x)} \Phi_g \left(\xi_M \left(\Phi_{g^{-1}}(x) \right) \right) \\
&= \left(\Phi_{g^{-1}}^* \xi_M \right)(x).
\end{aligned}
$$
∎

Proof of Proposition 9.3.6. Linearity follows, since $\xi_M(x) = T_e \Phi_x(\xi)$. To prove the second relation, put $g = \exp t\eta$ in (ii) of the lemma to get

$$
(\mathrm{Ad}_{\exp t\eta} \, \xi)_M = \Phi_{\exp(-t\eta)}^* \xi_M.
$$

But $\Phi_{\exp(-t\eta)}$ is the flow of $-\eta_M$, so differentiating at $t = 0$ the right-hand side gives $[\xi_M, \eta_M]$. The derivative of the left-hand side at $t = 0$ equals $[\eta, \xi]_M$ by the preceding Example (a). ∎

In view of this proposition one defines a left *Lie algebra action* of a manifold M as a Lie algebra antihomomorphism $\xi \in \mathfrak{g} \mapsto \xi_M \in \mathfrak{X}(M)$, such that the mapping $(\xi, x) \in \mathfrak{g} \times M \mapsto \xi_M(x) \in TM$ is smooth.

Let $\Phi : G \times G \to G$ denote the action of G on itself by left translation: $\Phi(g, h) = L_g h$. For $\xi \in \mathfrak{g}$, let Y_ξ be the corresponding *right*-invariant vector field on G. Then

$$
\xi_G(g) = Y_\xi(g) = T_e R_g(\xi),
$$

and similarly, the *infinitesimal generator of right translation is the left-invariant vector field* $g \mapsto T_e L_g(\xi)$.

Derivatives of Curves. It is convenient to have formulas for the derivatives of curves associated with the adjoint and coadjoint actions. For example, let $g(t)$ be a (smooth) curve in G and $\eta(t)$ a (smooth) curve in $\mathfrak{g}$. Let the action be denoted by concatenation:

$$
g(t)\eta(t) = \mathrm{Ad}_{g(t)} \, \eta(t).
$$

Proposition 9.3.8. *The following holds:*

$$
\frac{d}{dt} g(t)\eta(t) = g(t) \left\{ [\xi(t), \eta(t)] + \frac{d\eta}{dt} \right\}, \tag{9.3.4}
$$

where

$$
\xi(t) = g(t)^{-1}\dot{g}(t) := T_{g(t)} L_{g(t)}^{-1} \frac{dg}{dt} \in \mathfrak{g}.
$$

Proof.   We have

$$\left.\frac{d}{dt}\right|_{t=t_0} \mathrm{Ad}_{g(t)}\,\eta(t) = \left.\frac{d}{dt}\right|_{t=t_0} \{g(t_0)[g(t_0)^{-1}g(t)]\eta(t)\}$$

$$= g(t_0)\left.\frac{d}{dt}\right|_{t=t_0} \{[g(t_0)^{-1}g(t)]\eta(t)\},$$

where the first $g(t_0)$ denotes the Ad-action, which is *linear*. Now, $g(t_0)^{-1}g(t)$ is a curve through the identity at $t = t_0$ with tangent vector $\xi(t_0)$, so the above becomes

$$g(t_0)\left\{[\xi(t_0),\eta(t_0)] + \frac{d\eta(t_0)}{dt}\right\}.$$

∎

Similarly, for the coadjoint action we write

$$g(t)\mu(t) = \mathrm{Ad}^*_{g(t)^{-1}}\,\mu(t),$$

and then, as above, one proves that

$$\frac{d}{dt}[g(t)\mu(t)] = g(t)\left\{-\mathrm{ad}^*_{\xi(t)}\,\mu(t) + \frac{d\mu}{dt}\right\},$$

which we could write, extending our concatenation notation to Lie algebra actions as well,

$$\frac{d}{dt}[g(t)\mu(t)] = g(t)\left\{\xi(t)\mu(t) + \frac{d\mu}{dt}\right\}, \tag{9.3.5}$$

where $\xi(t) = g(t)^{-1}\dot{g}(t)$. For right actions, these become

$$\frac{d}{dt}[\eta(t)g(t)] = \left\{\eta(t)\zeta(t) + \frac{d\eta}{dt}\right\}g(t) \tag{9.3.6}$$

and

$$\frac{d}{dt}[\mu(t)g(t)] = \left\{\mu(t)\zeta(t) + \frac{d\mu}{dt}\right\}g(t), \tag{9.3.7}$$

where $\zeta(t) = \dot{g}(t)g(t)^{-1}$,

$$\eta(t)g(t) = \mathrm{Ad}_{g(t)^{-1}}\,\eta(t), \quad \text{and} \quad \eta(t)\zeta(t) = -[\zeta(t),\eta(t)],$$

and where

$$\mu(t)g(t) = \mathrm{Ad}^*_{g(t)}\,\mu(t) \quad \text{and} \quad \mu(t)\zeta(t) = \mathrm{ad}^*_{\zeta(t)}\,\mu(t).$$

Connectivity of Some Classical Groups. First we state two facts about homogeneous spaces:

1. If H is a closed normal subgroup of the Lie group G (that is, if $h \in H$ and $g \in G$, then $ghg^{-1} \in H$), then the quotient G/H is a Lie group and the natural projection $\pi : G \to G/H$ is a smooth group homomorphism. (This follows from Proposition 9.3.2; see also Theorem 2.9.6 in Varadarajan [1974, p. 80].) Moreover, if H and G/H are connected, then G is connected. Similarly, if H and G/H are simply connected, then G is simply connected.

2. Let G, M be finite-dimensional and second countable and let $\Phi : G \times M \to M$ be a transitive action of G on M, and for $x \in M$, let G_x be the isotropy subgroup of x. Then the map $gG_x \mapsto \Phi_g(x)$ is a diffeomorphism of G/G_x onto M. (This follows from Proposition 9.3.2; see also Theorem 2.9.4 in Varadarajan [1974, p. 77].)

The action

$$\Phi : \mathrm{GL}(n, \mathbb{R}) \times \mathbb{R}^n \to \mathbb{R}^n, \quad \Phi(A, x) = Ax,$$

restricted to $\mathrm{O}(n) \times S^{n-1}$ induces a transitive action. The isotropy subgroup of $\mathrm{O}(n)$ at $e_n \in S^{n-1}$ is $\mathrm{O}(n-1)$. Clearly, $\mathrm{O}(n-1)$ is a closed subgroup of $\mathrm{O}(n)$ by embedding any $A \in \mathrm{O}(n-1)$ as

$$\tilde{A} = \begin{bmatrix} A & 0 \\ 0 & 1 \end{bmatrix} \in \mathrm{O}(n),$$

and the elements of $\mathrm{O}(n-1)$ leave e_n fixed. On the other hand, if $A \in \mathrm{O}(n)$ and $Ae_n = e_n$, then $A \in \mathrm{O}(n-1)$. It follows from fact 2 above that the map

$$\mathrm{O}(n)/\mathrm{O}(n-1) \to S^{n-1}, \quad A \cdot \mathrm{O}(n-1) \mapsto Ae_n,$$

is a diffeomorphism. By a similar argument, there is a diffeomorphism

$$S^{n-1} \cong \mathrm{SO}(n)/\mathrm{SO}(n-1).$$

The natural action of $\mathrm{GL}(n, \mathbb{C})$ on $\mathbb{C}^n$ similarly induces a diffeomorphism of $S^{2n-1} \subset \mathbb{R}^{2n}$ with the homogeneous space $\mathrm{U}(n)/\mathrm{U}(n-1)$. Moreover, we get $S^{2n-1} \cong \mathrm{SU}(n)/\mathrm{SU}(n-1)$. In particular, since $\mathrm{SU}(1)$ consists only of the 1×1 identity matrix, S^3 is diffeomorphic with $\mathrm{SU}(2)$, a fact already proved at the end of §9.2.

Proposition 9.3.9. *Each of the Lie groups* $\mathrm{SO}(n)$, $\mathrm{SU}(n)$, *and* $\mathrm{U}(n)$ *is connected for* $n \geq 1$, *and* $\mathrm{O}(n)$ *has two components. The group* $\mathrm{SU}(n)$ *is simply connected.*

Proof. The groups SO(1) and SU(1) are connected, since both consist only of the 1×1 identity matrix, and U(1) is connected, since

$$U(1) = \{\, z \in C \mid |z| = 1 \,\} = S^1.$$

That SO(n), SU(n), and U(n) are connected for all n now follows from fact 1 above, using induction on n and the representation of the spheres as homogeneous spaces. Since every matrix A in O(n) has determinant ± 1, the orthogonal group can be written as the union of two nonempty disjoint connected open subsets as follows:

$$O(n) = SO(n) \cup A \cdot SO(n),$$

where $A = \mathrm{diag}(-1, 1, 1, \ldots, 1)$. Thus, O($n$) has two components. ■

Here is a general strategy for proving the connectivity of the classical groups; see, for example Knapp [1996, p 72]. This works, in particular, for Sp($2n, \mathbb{R}$) (and the groups Sp($2n, \mathbb{C}$), SP$^*(2n)$ discussed in the Internet supplement). Let G be a subgroup of GL($n, \mathbb{R}$) (resp. GL($n, \mathbb{C}$)) defined as the zero set of a collection of real-valued polynomials in the (real and imaginary parts) of the matrix entries. Assume also that G is closed under taking adjoints (see Exercise 9.2-2 for the case of Sp($2n, \mathbb{R}$)). Let $K = G \cap O(n)$ (resp. U(n)) and let $\mathfrak{p}$ be the set of Hermitian matrices in $\mathfrak{g}$. The polar decomposition says that

$$(k, \xi) \in K \times \mathfrak{p} \mapsto k \exp(\xi) \in G$$

is a homeomorphism. It follows that since ξ lies in a connected space, G is connected iff K is connected. For Sp($2m, \mathbb{R}$) our results above show that U(m) is connected, so Sp($2m, \mathbb{R}$) is connected.

Examples

(a) Isometry groups. Let E be a finite-dimensional vector space with a bilinear form $\langle\,,\rangle$. Let G be the group of *isometries* of E, that is, F is an isomorphism of E onto E and $\langle Fe, Fe' \rangle = \langle e, e' \rangle$, for all e and $e' \in E$. Then G is a subgroup and a closed submanifold of GL(E). The Lie algebra of G is

$$\{\, K \in L(E) \mid \langle Ke, e' \rangle + \langle e, Ke' \rangle = 0 \text{ for all } e, e' \in E \,\}. \qquad \blacklozenge$$

(b) Lorentz group. If $\langle\,,\rangle$ denotes the Minkowski metric on $\mathbb{R}^4$, that is,

$$\langle x, y \rangle = \sum_{i=1}^{3} x^i y^i - x^4 y^4,$$

then the group of linear isometries is called the **Lorentz group** L. The dimension of L is six, and L has four connected components. If

$$S = \begin{bmatrix} I_3 & 0 \\ 0 & -1 \end{bmatrix} \in \mathrm{GL}(4, \mathbb{R}),$$

then

$$L = \{ A \in \mathrm{GL}(4, \mathbb{R}) \mid A^T S A = S \},$$

and so the Lie algebra of L is

$$\mathfrak{l} = \{ A \in L(\mathbb{R}^4, \mathbb{R}^4) \mid SA + A^T S = 0 \}.$$

The identity component of L is

$$\{ A \in L \mid \det A > 0 \text{ and } A_{44} > 0 \} = L_\uparrow^+;$$

L and $L_\uparrow^+$ are not compact. ◆

(c) Galilean group. Consider the (closed) subgroup G of $\mathrm{GL}(5, \mathbb{R})$ that consists of matrices with the following block structure:

$$\{ \mathbf{R}, \mathbf{v}, \mathbf{a}, \tau \} := \begin{bmatrix} \mathbf{R} & \mathbf{v} & \mathbf{a} \\ 0 & 1 & \tau \\ 0 & 0 & 1 \end{bmatrix},$$

where $\mathbf{R} \in \mathrm{SO}(3)$, $\mathbf{v}, \mathbf{a} \in \mathbb{R}^3$, and $\tau \in \mathbb{R}$. This group is called the **Galilean group**. Its Lie algebra is a subalgebra of $L(\mathbb{R}^5, \mathbb{R}^5)$ given by the set of matrices of the form

$$\{ \omega, \mathbf{u}, \alpha, \theta \} := \begin{bmatrix} \hat{\omega} & \mathbf{u} & \alpha \\ 0 & 0 & \theta \\ 0 & 0 & 0 \end{bmatrix},$$

where $\omega, \mathbf{u}, \alpha \in \mathbb{R}^3$ and $\theta \in \mathbb{R}$. Obviously the Galilean group acts naturally on $\mathbb{R}^5$; moreover, it acts naturally on $\mathbb{R}^4$, embedded as the following G-invariant subset of $\mathbb{R}^5$:

$$\begin{bmatrix} \mathbf{x} \\ t \end{bmatrix} \mapsto \begin{bmatrix} \mathbf{x} \\ t \\ 1 \end{bmatrix},$$

where $\mathbf{x} \in \mathbb{R}^3$ and $t \in \mathbb{R}$. Concretely, the action of $\{ \mathbf{R}, \mathbf{v}, \mathbf{a}, \tau \}$ on $(\mathbf{x}, t)$ is given by

$$(\mathbf{x}, t) \mapsto (\mathbf{R}\mathbf{x} + t\mathbf{v} + \mathbf{a}, t + \tau).$$

Thus, the Galilean group gives a change of frame of reference (not affecting the "absolute time" variable) by rotations ($\mathbf{R}$), space translations ($\mathbf{a}$), time translations (τ), and going to a moving frame, or boosts ($\mathbf{v}$). ◆

(d) Unitary Group of Hilbert Space.   Another basic example of an infinite-dimensional group is the unitary group $U(\mathcal{H})$ of a complex Hilbert space $\mathcal{H}$. If G is a Lie group and $\rho : G \to U(\mathcal{H})$ is a group homomorphism, we call ρ a ***unitary representation***. In other words, ρ is an action of G on $\mathcal{H}$ by unitary maps.

As with the diffeomorphism group, questions of smoothness regarding $U(\mathcal{H})$ need to be dealt with carefully, and in this book we shall give only a brief indication of what is involved. The reason for care is, for one thing, that one ultimately is dealing with PDEs rather than ODEs and the hypotheses made must be such that PDEs are not excluded. For example, for a unitary representation one assumes that for each $\psi, \varphi \in \mathcal{H}$, the map $g \mapsto \langle \psi, \rho(g)\varphi \rangle$ of G to $\mathbb{C}$ is continuous. In particular, for $G = \mathbb{R}$ one has the notion of a continuous one-parameter group $U(t)$ so that $U(0) =$ identity and

$$U(t + s) = U(t) \circ U(s).$$

Stone's theorem says that in an appropriate sense we can write $U(t) = e^{tA}$, where A is an (unbounded) skew-adjoint operator defined on a dense domain $D(A) \subset \mathcal{H}$. See, for example, Abraham, Marsden, and Ratiu [1988, Section 7.4B] for the proof. Conversely each skew-adjoint operator defines a one-parameter subgroup. Thus, Stone's theorem gives precise meaning to the statement that the Lie algebra $\mathfrak{u}(\mathcal{H})$ of $U(\mathcal{H})$ consists of the skew-adjoint operators. The Lie bracket is the commutator, as long as one is careful with domains.

If ρ is a unitary representation of a finite-dimensional Lie group G on $\mathcal{H}$, then $\rho(\exp(t\xi))$ is a one-parameter subgroup of $U(\mathcal{H})$, so Stone's theorem guarantees that there is a map $\xi \mapsto A(\xi)$ associating a skew-adjoint operator $A(\xi)$ to each $\xi \in \mathfrak{g}$. Formally, we have

$$[A(\xi), A(\eta)] = [\xi, \eta].$$

Results like this are aided by a theorem of Nelson [1959] guaranteeing a dense subspace $D_G \subset \mathcal{H}$ such that

  (i) $A(\xi)$ is well-defined on D_G,

  (ii) $A(\xi)$ maps D_G to D_G, and

  (iii) for $\psi \in D_G$, $[\exp tA(\xi)]\psi$ is C^∞ in t with derivative at $t = 0$ given by $A(\xi)\psi$.

This space is called an ***essential G-smooth part of*** $\mathcal{H}$, and on D_G the above commutator relation and the linearity

$$A(\alpha\xi + \beta\eta) = \alpha A(\xi) + \beta A(\eta)$$

become *literally* true. Moreover, we lose little by using D_G, since $A(\xi)$ is uniquely determined by what it is on D_G.

We identify U(1) with the unit circle in $\mathbb{C}$, and each such complex number determines an element of U($\mathcal{H}$) by multiplication. Thus, we regard U(1) $\subset$ U($\mathcal{H}$). As such, it is a normal subgroup (in fact, elements of U(1) commute with elements of U($\mathcal{H}$)), so the quotient is a group, called the *projective unitary group of* $\mathcal{H}$. We write it as U($\mathbb{P}\mathcal{H}$) = U($\mathcal{H}$)/ U(1). We write elements of U($\mathbb{P}\mathcal{H}$) as $[U]$ regarded as an equivalence class of $U \in$ U($\mathcal{H}$). The group U($\mathbb{P}\mathcal{H}$) acts on projective Hilbert space $\mathbb{P}\mathcal{H} = \mathcal{H}/\mathbb{C}$, as in §5.3, by $[U][\varphi] = [U\varphi]$.

One-parameter subgroups of U($\mathbb{P}\mathcal{H}$) are of the form $[U(t)]$ for a one-parameter subgroup $U(t)$ of U($\mathcal{H}$). This is a particularly simple case of the general problem considered by Bargmann and Wigner of lifting projective representations, a topic we return to later. In any case, this means that we can identify the Lie algebra as $\mathfrak{u}(\mathbb{P}\mathcal{H}) = \mathfrak{u}(\mathcal{H})/i\mathbb{R}$, where we identify the two skew-adjoint operators A and $A + \lambda i$, for λ real.

A *projective representation* of a group G is a homomorphism τ : $G \to$ U($\mathbb{P}\mathcal{H}$); we require continuity of $|\langle \psi, \tau(g)\varphi \rangle|$, which is well-defined for $[\psi], [\varphi] \in \mathbb{P}\mathcal{H}$. There is an analogue of Nelson's theorem that guarantees an *essential G-smooth part* $\mathbb{P}D_G$ of $\mathbb{P}\mathcal{H}$ with properties like those of D_G. ◆

Miscellany. We conclude this section with a variety of remarks.

1. Coadjoint Isotropy. The first remark concerns coadjoint orbit isotropy groups. The main result here is the following theorem, due to Duflo and Vergne [1969]. We give a proof following Rais [1972] in the Internet supplement.

Theorem 9.3.10 (Duflo and Vergne). *Let $\mathfrak{g}$ be a finite-dimensional Lie algebra with dual $\mathfrak{g}^*$ and let $r = \min \{ \dim \mathfrak{g}_\mu \mid \mu \in \mathfrak{g}^* \}$. The set $\{ \mu \in \mathfrak{g}^* \mid \dim \mathfrak{g}_\mu = r \}$ is open and dense in $\mathfrak{g}^*$. If $\dim \mathfrak{g}_\mu = r$, then $\mathfrak{g}_\mu$ is Abelian.*

A simple example is the rotation group SO(3) in which the coadjoint isotropy at each nonzero point is the Abelian group S^1, whereas at the origin it is the nonabelian group SO(3).

2. More on Infinite-Dimensional Groups. We can use a slight reinterpretation of the formulae in this section to calculate the Lie algebra structure of some infinite-dimensional groups. Here we will treat this topic only formally, that is, we assume that the spaces involved are manifolds and do not specify the function-space topologies. For the formal calculations, these structures are not needed, but the reader should be aware that there is a mathematical gap here. (See Ebin and Marsden [1970] and Adams, Ratiu, and Schmid [1986a, 1986b] for more information.)

Given a manifold M, let Diff(M) denote the group of all diffeomorphisms of M. The group operation is composition. The Lie algebra of Diff(M), as a vector space, consists of vector fields on M; indeed, the flow of a vector field is a curve in $\mathrm{Diff}(M)$, and its tangent vector at $t = 0$ is the given vector field.

To determine the Lie algebra bracket, we consider the action of an arbitrary Lie group G on M. Such an action of G on M may be regarded as a homomorphism $\Phi : G \to \mathrm{Diff}(M)$. By Proposition 9.1.5, its derivative at the identity $T_e\Phi$ should be a Lie algebra homomorphism. From the definition of infinitesimal generator, we see that $T_e\Phi \cdot \xi = \xi_M$. Thus, Proposition 9.1.5 suggests that

$$[\xi_M, \eta_M]_{\text{Lie bracket}} = [\xi, \eta]_M.$$

However, by Proposition 9.3.6, $[\xi, \eta]_M = -[\xi_M, \eta_M]$. Thus,

$$[\xi_M, \eta_M]_{\text{Lie bracket}} = -[\xi_M, \eta_M].$$

This suggests that *the Lie algebra bracket on $\mathfrak{X}(M)$ is minus the Jacobi–Lie bracket*.

Another way to arrive at the same conclusion is to use the method of computing brackets in the table in §9.1. To do this, we first compute, according to step 1, the inner automorphism to be

$$I_\eta(\varphi) = \eta \circ \varphi \circ \eta^{-1}.$$

By step 2, we differentiate with respect to φ to compute the Ad map. Letting X be the time derivative at $t = 0$ of a curve φ_t in $\mathrm{Diff}(M)$ with $\varphi_0 = \mathrm{Identity}$, we have

$$\mathrm{Ad}_\eta(X) = (T_e I_\eta)(X) = T_e I_\eta\left[\left.\frac{d}{dt}\right|_{t=0}\varphi_t\right] = \left.\frac{d}{dt}\right|_{t=0} I_\eta(\varphi_t)$$

$$= \left.\frac{d}{dt}\right|_{t=0}(\eta \circ \varphi_t \circ \eta^{-1}) = T\eta \circ X \circ \eta^{-1} = \eta_*X.$$

Hence $\mathrm{Ad}_\eta(X) = \eta_*X$. Thus, *the adjoint action of $\mathrm{Diff}(M)$ on its Lie algebra is just the push-forward operation on vector fields*. Finally, as in step 3, we compute the bracket by differentiating $\mathrm{Ad}_\eta(X)$ with respect to η. But by the Lie derivative characterization of brackets and the fact that push-forward is the inverse of pull-back, we arrive at the same conclusion. In summary, either method suggests that

The Lie algebra bracket on $\mathrm{Diff}(M)$ is minus the Jacobi–Lie bracket of vector fields.

One can also say that the Jacobi–Lie bracket gives the *right* (as opposed to *left*) Lie algebra structure on $\mathrm{Diff}(M)$.

If one restricts to the group of volume-preserving (or symplectic) diffeomorphisms, then the Lie bracket is again minus the Jacobi–Lie bracket on the space of divergence-free (or locally Hamiltonian) vector fields.

Here are three examples of actions of $\text{Diff}(M)$. Firstly, $\text{Diff}(M)$ acts on M by evaluation: The action $\Phi : \text{Diff}(M) \times M \to M$ is given by $\Phi(\varphi, x) = \varphi(x)$. Secondly, the calculations we did for Ad_η show that the adjoint action of $\text{Diff}(M)$ on its Lie algebra is given by push-forward. Thirdly, if we identify the dual space $\mathfrak{X}(M)^*$ with one-form densities by means of integration, then the change-of-variables formula shows that the *coadjoint action is given by push-forward of one-form densities*.

3. Equivariant Darboux Theorem.   In Chapter 5 we studied the Darboux theorem. It is natural to ask the sense in which this theorem holds in the presence of a group action. That is, suppose that one has a Lie group G (say compact) acting symplectically on a symplectic manifold (P, Ω) and that, for example, the group action leaves a point $x_0 \in P$ fixed (one can consider the more general case of an invariant manifold). We ask to what extent one can put the symplectic form into a canonical form in an equivariant way?

This question is best broken up into two parts. The first is whether or not one can find a local equivariant representation in which the symplectic form is constant. This is true and can be proved by establishing an equivariant diffeomorphism between the manifold and its tangent space at x_0 carrying the constant symplectic form, which is just Ω evaluated at $T_{x_0}P$. This is done by checking that Moser's proof given in Chapter 5 can be made equivariant at each stage (see Exercise 9.3-5).

A more subtle question is that of putting the symplectic form into a canonical form equivariantly. For this, one needs first to understand the equivariant classification of normal forms for symplectic structures. This was done in Dellnitz and Melbourne [1993]. For the related question of classifying equivariant normal forms for linear Hamiltonian systems, see Williamson [1936], Melbourne and Dellnitz [1993], and Hörmander [1995].

Exercises

⋄ **9.3-1.**   Let a Lie group G act linearly on a vector space V. Define a group structure on $G \times V$ by

$$(g_1, v_1) \cdot (g_2, v_2) = (g_1 g_2, g_1 v_2 + v_1).$$

Show that this makes $G \times V$ into a Lie group—it is called the **semidirect product** and is denoted by $G \circledS V$. Determine its Lie algebra $\mathfrak{g} \circledS V$.

⋄ **9.3-2.**

(a) Show that the Euclidean group E(3) can be written as $O(3) \circledS \mathbb{R}^3$ in the sense of the preceding exercise.

(b) Show that E(3) is isomorphic to the group of 4×4 matrices of the form

$$\begin{bmatrix} A & \mathbf{b} \\ 0 & 1 \end{bmatrix},$$

where $A \in O(3)$ and $\mathbf{b} \in \mathbb{R}^3$.

⋄ **9.3-3.**   Show that the Galilean group may be written as a semidirect product $G = (SO(3) \circledS \mathbb{R}^3) \circledS \mathbb{R}^4$. Compute explicitly the inverse of a group element, and the adjoint and the coadjoint actions.

⋄ **9.3-4.**   If G is a Lie group, show that TG is isomorphic (as a Lie group) with $G \circledS \mathfrak{g}$ (see Exercise 9.1-2).

⋄ **9.3-5.**   In the relative Darboux theorem of Exercise 5.1-5, assume that a compact Lie group G acts on P, that S is a G-invariant submanifold, and that both Ω_0 and Ω_1 are G-invariant. Conclude that the diffeomorphism $\varphi : U \longrightarrow \varphi(U)$ can be chosen to commute with the G-action and that V, $\varphi(U)$ can be chosen to be a G-invariant.

⋄ **9.3-6.**   Verify, using standard vector notation, the four "derivative of curves" formulas for SO(3).

⋄ **9.3-7.**   Use the complex polar decomposition theorem (Theorem 9.2.15) and simple connectedness of $SU(n)$ to show that $SL(n, \mathbb{C})$ is also simply connected.

⋄ **9.3-8.**   Show that $SL(2, \mathbb{C})$ is the simply connected covering group of the identity component $L_\uparrow^\dagger$ of the Lorentz group.

10
Poisson Manifolds

The dual $\mathfrak{g}^*$ of a Lie algebra $\mathfrak{g}$ carries a Poisson bracket given by

$$\{F, G\}(\mu) = \left\langle \mu, \left[\frac{\delta F}{\delta \mu}, \frac{\delta G}{\delta \mu}\right]\right\rangle$$

for $\mu \in \mathfrak{g}^*$, a formula found by Lie, [1890, Section 75]. As we saw in the Introduction, this *Lie–Poisson bracket* description of many physical systems. This bracket is not the bracket associated with any symplectic structure on $\mathfrak{g}^*$, but is an example of the more general concept of a *Poisson manifold*. On the other hand, we do want to understand how this bracket *is* associated with a symplectic structure on coadjoint orbits and with the canonical symplectic structure on T^*G. These facts are developed in Chapters 13 and 14. Chapter 15 shows how this works in detail for the rigid body.

10.1   The Definition of Poisson Manifolds

This section generalizes the notion of a symplectic manifold by keeping just enough of the properties of Poisson brackets to describe Hamiltonian systems. The history of Poisson manifolds is complicated by the fact that the notion was rediscovered many times under different names; they occur in the works of Lie [1890], Dirac [1930,1964], Pauli [1953], Martin [1959], Jost [1964], Arens [1970], Hermann [1973], Sudarshan and Mukunda [1974], Vinogradov and Krasilshchik [1975], and Lichnerowicz [1975b]. The name *Poisson manifold* was coined by Lichnerowicz. Further historical comments are given in §10.3.

Definition 10.1.1. *A **Poisson bracket** (or a **Poisson structure**) on a manifold P is a bilinear operation $\{\,,\}$ on $\mathcal{F}(P) = C^\infty(P)$ such that:*

(i) $(\mathcal{F}(P), \{\,,\})$ *is a Lie algebra; and*

(ii) $\{\,,\}$ *is a derivation in each factor, that is,*

$$\{FG, H\} = \{F, H\}\, G + F\, \{G, H\}$$

for all F, G, and $H \in \mathcal{F}(P)$.

*A manifold P endowed with a Poisson bracket on $\mathcal{F}(P)$ is called a **Poisson manifold**.*

A Poisson manifold is denoted by $(P, \{\,,\})$, or simply by P if there is no danger of confusion. Note that any manifold has the **trivial Poisson structure**, which is defined by setting $\{F, G\} = 0$, for all $F, G \in \mathcal{F}(P)$. Occasionally, we consider two different Poisson brackets $\{\,,\}_1$ and $\{\,,\}_2$ on the same manifold; the two distinct Poisson manifolds are then denoted by $(P, \{\,,\}_1)$ and $(P, \{\,,\}_2)$. The notation $\{\,,\}_P$ for the bracket on P is also used when confusion might arise.

Examples

(a) Symplectic Bracket. *Any symplectic manifold is a Poisson manifold.* The Poisson bracket is defined by the symplectic form, as was shown in §5.5. Condition (ii) of the definition is satisfied as a consequence of the derivation property of vector fields:

$$\{FG, H\} = X_H[FG] = F X_H[G] + G X_H[F] = F\{G, H\} + G\{F, H\}. \quad \blacklozenge$$

(b) Lie–Poisson Bracket. If $\mathfrak{g}$ is a Lie algebra, then its dual $\mathfrak{g}^*$ is a Poisson manifold with respect to each of the **Lie–Poisson brackets** $\{\,,\}_+$ and $\{\,,\}_-$ defined by

$$\{F, G\}_\pm(\mu) = \pm \left\langle \mu, \left[\frac{\delta F}{\delta \mu}, \frac{\delta G}{\delta \mu} \right] \right\rangle \tag{10.1.1}$$

for $\mu \in \mathfrak{g}^*$ and $F, G \in \mathcal{F}(\mathfrak{g}^*)$. The properties of a Poisson bracket can be easily verified. Bilinearity and skew-symmetry are obvious. The derivation property of the bracket follows from the Leibnitz rule for functional derivatives

$$\frac{\delta(FG)}{\delta \mu} = F(\mu) \frac{\delta G}{\delta \mu} + \frac{\delta F}{\delta \mu} G(\mu).$$

The Jacobi identity for the Lie–Poisson bracket follows from the Jacobi identity for the Lie algebra bracket and the formula

$$\pm \frac{\delta}{\delta \mu} \{F, G\}_{\pm} = \left[\frac{\delta F}{\delta \mu}, \frac{\delta G}{\delta \mu} \right] - \mathbf{D}^2 F(\mu) \left(\mathrm{ad}^*_{\delta G / \delta \mu} \, \mu, \cdot \right)$$

$$+ \mathbf{D}^2 G(\mu) \left(\mathrm{ad}^*_{\delta F / \delta \mu} \, \mu, \cdot \right), \qquad (10.1.2)$$

where we recall from the preceding chapter that for each $\xi \in \mathfrak{g}$, $\mathrm{ad}_\xi : \mathfrak{g} \to \mathfrak{g}$ denotes the map $\mathrm{ad}_\xi(\eta) = [\xi, \eta]$ and $\mathrm{ad}^*_\xi : \mathfrak{g}^* \to \mathfrak{g}^*$ is its dual. We give a different proof that (10.1.1) is a Poisson bracket in Chapter 13. ◆

(c) Rigid-Body Bracket. Specializing Example (b) to the Lie algebra of the rotation group $\mathfrak{so}(3) \cong \mathbb{R}^3$ and identifying $\mathbb{R}^3$ and $(\mathbb{R}^3)^*$ via the standard inner product, we get the following Poisson structure on $\mathbb{R}^3$:

$$\{F, G\}_-(\mathbf{\Pi}) = -\mathbf{\Pi} \cdot (\nabla F \times \nabla G), \qquad (10.1.3)$$

where $\mathbf{\Pi} \in \mathbb{R}^3$ and ∇F, the gradient of F, is evaluated at $\mathbf{\Pi}$. The Poisson bracket properties can be verified by direct computation in this case; see Exercise 1.2-1. We call (10.1.3) the *rigid-body bracket*. ◆

(d) Ideal Fluid Bracket. Specialize the Lie–Poisson bracket to the Lie algebra $\mathfrak{X}_{\mathrm{div}}(\Omega)$ of divergence-free vector fields defined in a region Ω of $\mathbb{R}^3$ and tangent to $\partial\Omega$, with the Lie bracket being the *negative* of the Jacobi–Lie bracket. Identify $\mathfrak{X}^*_{\mathrm{div}}(\Omega)$ with $\mathfrak{X}_{\mathrm{div}}(\Omega)$ using the L^2 pairing

$$\langle \mathbf{v}, \mathbf{w} \rangle = \int_\Omega \mathbf{v} \cdot \mathbf{w} \, d^3x, \qquad (10.1.4)$$

where $\mathbf{v} \cdot \mathbf{w}$ is the ordinary dot product in $\mathbb{R}^3$. Thus, the *plus* Lie–Poisson bracket is

$$\{F, G\}(\mathbf{v}) = -\int_\Omega \mathbf{v} \cdot \left[\frac{\delta F}{\delta \mathbf{v}}, \frac{\delta G}{\delta \mathbf{v}} \right] d^3x, \qquad (10.1.5)$$

where the functional derivative $\delta F / \delta \mathbf{v}$ is the element of $\mathfrak{X}_{\mathrm{div}}(\Omega)$ defined by

$$\lim_{\varepsilon \to 0} \frac{1}{\varepsilon} [F(\mathbf{v} + \varepsilon \delta \mathbf{v}) - F(\mathbf{v})] = \int_\Omega \frac{\delta F}{\delta \mathbf{v}} \cdot \delta \mathbf{v} \, d^3x. \qquad ◆$$

(e) Poisson–Vlasov Bracket. Let $(P, \{\,,\}_P)$ be a Poisson manifold and let $\mathcal{F}(P)$ be the Lie algebra of functions under the Poisson bracket. Identify $\mathcal{F}(P)^*$ with densities f on P. Then the Lie–Poisson bracket has the expression

$$\{F, G\}(f) = \int_P f \left\{ \frac{\delta F}{\delta f}, \frac{\delta G}{\delta f} \right\}_P. \qquad (10.1.6)$$

◆

(f) Frozen Lie–Poisson Bracket. Fix (or "freeze") $\nu \in \mathfrak{g}^*$ and define for any $F, G \in \mathcal{F}(\mathfrak{g}^*)$ the bracket

$$\{F, G\}_{\pm}^{\nu}(\mu) = \pm \left\langle \nu, \left[\frac{\delta F}{\delta \mu}, \frac{\delta G}{\delta \mu} \right] \right\rangle. \tag{10.1.7}$$

The properties of a Poisson bracket are verified as in the case of the Lie–Poisson bracket, the only difference being that (10.1.2) is replaced by

$$\pm \frac{\delta}{\delta \mu} \{F, G\}_{\pm}^{\nu} = -\mathbf{D}^2 F(\nu) \left(\mathrm{ad}_{\delta G/\delta \mu}^* \mu, \cdot \right) + \mathbf{D}^2 G(\nu) \left(\mathrm{ad}_{\delta F/\delta \mu}^* \mu, \cdot \right). \tag{10.1.8}$$

This bracket is useful in the description of the Lie–Poisson equations linearized at an equilibrium point.[1]   ◆

(g) KdV Bracket. Let $S = [S^{ij}]$ be a symmetric matrix. On $\mathcal{F}(\mathbb{R}^n, \mathbb{R}^n)$, set

$$\{F, G\}(u) = \int_{-\infty}^{\infty} \sum_{i,j=1}^{n} S^{ij} \left[\frac{\delta F}{\delta u^i} \frac{d}{dx} \left(\frac{\delta G}{\delta u^j} \right) - \frac{d}{dx} \left(\frac{\delta G}{\delta u^j} \right) \frac{\delta F}{\delta u^i} \right] dx \tag{10.1.9}$$

for functions F, G satisfying $\delta F/\delta u$ and $\delta G/\delta u \to 0$ as $x \to \pm\infty$. This is a Poisson structure that is useful for the KdV equation and for gas dynamics (see Benjamin [1984]).[2] If S is invertible and $S^{-1} = [S_{ij}]$, then (10.1.9) is the Poisson bracket associated with the weak symplectic form

$$\Omega(u, v) = \frac{1}{2} \int_{-\infty}^{\infty} \sum_{i,j=l}^{n} S_{ij} \left[\left(\int_{-\infty}^{y} u^i(x)\, dx \right) v^j(y) \right.$$
$$\left. - \left(\int_{-\infty}^{y} v^j(x)\, dx \right) u^i(y) \right] dy. \tag{10.1.10}$$

This is easily seen by noting that $X_H(u)$ is given by

$$X_H^i(u) = S^{ij} \frac{d}{dx} \frac{\delta H}{\delta u^j}. \qquad\qquad ◆$$

(h) Toda Lattice Bracket. Let

$$P = \left\{ (\mathbf{a}, \mathbf{b}) \in \mathbb{R}^{2n} \mid a^i > 0, \ i = 1, \dots, n \right\}$$

---

[1]See, for example, Abarbanel, Holm, Marsden, and Ratiu [1986].

[2]This is a particular case of Example (f), the Lie algebra being the pseudo-differential operators on the line of order ≤ -1 and $\nu = dS/dx$.

and consider the bracket

$$\{F,G\}(\mathbf{a},\mathbf{b}) = \left[\left(\frac{\partial F}{\partial \mathbf{a}}\right)^T , \left(\frac{\partial F}{\partial \mathbf{b}}\right)^T \right] \mathbf{W} \begin{bmatrix} \dfrac{\partial G}{\partial \mathbf{a}} \\ \dfrac{\partial G}{\partial \mathbf{b}} \end{bmatrix}, \tag{10.1.11}$$

where $(\partial F/\partial \mathbf{a})^T$ is the row vector $(\partial F/\partial a^1, \dots, \partial F/\partial a^n)$, etc., and

$$\mathbf{W} = \begin{bmatrix} 0 & \mathbf{A} \\ -\mathbf{A} & 0 \end{bmatrix}, \quad \text{where } \mathbf{A} = \begin{bmatrix} a^1 & & 0 \\ & \ddots & \\ 0 & & a^n \end{bmatrix}. \tag{10.1.12}$$

In terms of the coordinate functions a_i, b_j, the bracket (10.1.11) is given by

$$\begin{aligned} \{a^i, a^j\} &= 0, \\ \{b^i, b^j\} &= 0, &&\tag{10.1.13} \\ \{a^i, b^j\} &= 0 && \text{if } i \neq j, \\ \{a^i, b^j\} &= a^i && \text{if } i = j. \end{aligned}$$

This Poisson bracket is determined by the symplectic form

$$\Omega = -\sum_{i=1}^{n} \frac{1}{a^i} da^i \wedge db^i \tag{10.1.14}$$

as an easy verification shows. The mapping $(\mathbf{a}, \mathbf{b}) \mapsto (\log \mathbf{a}^{-1}, \mathbf{b})$ is a symplectic diffeomorphism of P with $\mathbb{R}^{2n}$ endowed with the canonical symplectic structure. This symplectic structure is known as the *first Poisson structure of the non-periodic Toda lattice*. We shall not study this example in any detail in this book, but we point out that its bracket is the restriction of a Lie–Poisson bracket to a certain coadjoint orbit of the group of lower triangular matrices; we refer the interested reader to §14.5 of Kostant [1979] and Symes [1980, 1982a, 1982b] for further information. ◆

Exercises

◇ **10.1-1.** If P_1 and P_2 are Poisson manifolds, show how to make $P_1 \times P_2$ into a Poisson manifold.

◇ **10.1-2.** Verify directly that the Lie–Poisson bracket satisfies Jacobi's identity.

◇ **10.1-3 (A Quadratic Bracket).** Let $A = [A^{ij}]$ be a skew-symmetric matrix. On $\mathbb{R}^n$, define $B^{ij} = A^{ij} x^i x^j$ (no sum). Show that the following defines a Poisson structure:

$$\{F, G\} = \sum_{i,j=1}^{n} B^{ij} \frac{\partial F}{\partial x^i} \frac{\partial G}{\partial x^j}.$$

⋄ **10.1-4** (A Cubic Bracket).   For $\mathbf{x} = (x^1, x^2, x^3) \in \mathbb{R}^3$, put

$$\{x^1, x^2\} = \|\mathbf{x}\|^2 x^3,$$
$$\{x^2, x^3\} = \|\mathbf{x}\|^2 x^1,$$
$$\{x^3, x^1\} = \|\mathbf{x}\|^2 x^2.$$

Let $B^{ij} = \{x^i, x^j\}$, for $i < j$ and $i, j = 1, 2, 3$. Set $B^{ji} = -B^{ij}$ and define

$$\{F, G\} = \sum_{i,j=1}^{n} B^{ij} \frac{\partial F}{\partial x^i} \frac{\partial G}{\partial x^j}.$$

Check that this makes $\mathbb{R}^3$ into a Poisson manifold.

⋄ **10.1-5.**   Let $\Phi : \mathfrak{g}^* \to \mathfrak{g}^*$ be a smooth map and define for $F, H : \mathfrak{g}^* \to \mathbb{R}$,

$$\{F, H\}_\Phi (\mu) = \left\langle \Phi(\mu), \left[\frac{\delta F}{\delta \mu}, \frac{\delta H}{\delta \mu} \right] \right\rangle.$$

(a) Show that this rule defines a Poisson bracket on $\mathfrak{g}^*$ if and only if Φ satisfies the following identity:

$$\left\langle \mathbf{D}\Phi(\mu) \cdot \mathrm{ad}_\zeta^*(\mu), [\eta, \xi] \right\rangle + \left\langle \mathbf{D}\Phi(\mu) \cdot \mathrm{ad}_\eta^* \Phi(\mu), [\xi, \zeta] \right\rangle$$
$$+ \left\langle \mathbf{D}\Phi(\mu) \cdot \mathrm{ad}_\xi^* \Phi(\mu), [\zeta, \eta] \right\rangle = 0,$$

for all $\xi, \eta, \zeta \in \mathfrak{g}$ and all $\mu \in \mathfrak{g}^*$.

(b) Show that this relation holds if $\Phi(\mu) = \mu$ and $\Phi(\mu) = \nu$, a fixed element of $\mathfrak{g}^*$, thereby obtaining the Lie–Poisson structure (10.1.1) and the linearized Lie–Poisson structure (10.1.7) on $\mathfrak{g}^*$. Show that it also holds if $\Phi(\mu) = a\mu + \nu$ for fixed $a \in \mathbb{R}$ and $\nu \in \mathfrak{g}^*$.

(c) Assume that $\mathfrak{g}$ has a weakly nondegenerate invariant bilinear form $\kappa : \mathfrak{g} \times \mathfrak{g} \to \mathbb{R}$ and identify $\mathfrak{g}^*$ with $\mathfrak{g}$ by κ. If $\Psi : \mathfrak{g} \to \mathfrak{g}$ is smooth, show that

$$\{F, H\}_\Psi (\xi) = \kappa(\Psi(\xi), [\nabla F(\xi), \nabla H(\xi)])$$

is a Poisson bracket if and only if

$$\kappa(\mathbf{D}\Psi(\lambda) \cdot [\Psi(\lambda), \zeta], [\eta, \xi]) + \kappa(\mathbf{D}\Psi(\lambda) \cdot [\Psi(\lambda), \eta], [\xi, \zeta])$$
$$+ \kappa(\mathbf{D}\Psi(\lambda) \cdot [\Psi(\lambda), \xi], [\zeta, \eta]) = 0,$$

for all $\lambda, \xi, \eta, \zeta \in \mathfrak{g}$. Here, $\nabla F(\xi), \nabla H(\xi) \in \mathfrak{g}$ are the gradients of F and H at $\xi \in \mathfrak{g}$ relative to κ.

Conclude as in (b) that this relation holds if $\Psi(\lambda) = a\lambda + \chi$ for $a \in \mathbb{R}$ and $\chi \in \mathfrak{g}$.

(d) In the hypothesis of (c), let $\Psi(\lambda) = \nabla\psi(\lambda)$ for some smooth $\psi : \mathfrak{g} \to \mathbb{R}$. Show that $\{\ ,\ \}_\Psi$ is a Poisson bracket if and only if

$$\mathbf{D}^2\psi(\lambda)([\nabla\psi(\lambda), \zeta], [\eta, \xi]) - \mathbf{D}^2\psi(\lambda)(\nabla\psi(\lambda), [\zeta, [\eta, \xi]])$$
$$+ \mathbf{D}^2\psi(\lambda)([\nabla\psi(\lambda), \eta], [\xi, \zeta]) - \mathbf{D}^2\psi(\lambda)(\nabla\psi(\lambda), [\eta, [\xi, \zeta]])$$
$$+ \mathbf{D}^2\psi(\lambda)([\nabla\psi(\lambda), \xi], [\zeta, \eta]) - \mathbf{D}^2\psi(\lambda)(\nabla\psi(\lambda), [\xi, [\zeta, \eta]]) = 0,$$

for all $\lambda, \xi, \eta, \zeta \in \mathfrak{g}$. In particular, if $\mathbf{D}^2\psi(\lambda)$ is an invariant bilinear form for all λ, this condition holds. However, if $\mathfrak{g} = \mathfrak{so}(3)$ and ψ is arbitrary, then this condition also holds (see Exercise 1.3-2).

10.2 Hamiltonian Vector Fields and Casimir Functions

Hamiltonian Vector Fields. We begin by extending the notion of a Hamiltonian vector field from the symplectic to the Poisson context.

Proposition 10.2.1. *Let P be a Poisson manifold. If $H \in \mathcal{F}(P)$, then there is a unique vector field X_H on P such that*

$$X_H[G] = \{G, H\} \tag{10.2.1}$$

*for all $G \in \mathcal{F}(P)$. We call X_H the **Hamiltonian vector field** of H.*

Proof. This is a consequence of the fact that any derivation on $\mathcal{F}(P)$ is represented by a vector field. Fixing H, the map $G \mapsto \{G, H\}$ is a derivation, and so it uniquely determines X_H satisfying (10.3.1). (In infinite dimensions some technical conditions are needed for this proof, which are deliberately ignored here; see Abraham, Marsden, and Ratiu [1988, Section 4.2].) ∎

Notice that (10.2.1) agrees with our definition of Poisson brackets in the symplectic case, so if the Poisson manifold P is symplectic, X_H defined here agrees with the definition in §5.5.

Proposition 10.2.2. *The map $H \mapsto X_H$ of $\mathcal{F}(P)$ to $\mathfrak{X}(P)$ is a Lie algebra antihomomorphism; that is,*

$$[X_H, X_K] = -X_{\{H,K\}}.$$

Proof. Using Jacobi's identity, we find that

$$[X_H, X_K][F] = X_H[X_K[F]] - X_K[X_H[F]]$$
$$= \{\{F, K\}, H\} - \{\{F, H\}, K\}$$
$$= -\{F, \{H, K\}\}$$
$$= -X_{\{H,K\}}[F]. \qquad \blacksquare$$

Equations of Motion in Poisson Bracket Form. Next, we establish
the equation $\dot{F} = \{F, H\}$ in the Poisson context.

Proposition 10.2.3. *Let φ_t be a flow on a Poisson manifold P and let
$H : P \to \mathbb{R}$ be a smooth function on P. Then*

(i) *for any $F \in \mathcal{F}(U)$, U open in P,*

$$\frac{d}{dt}(F \circ \varphi_t) = \{F, H\} \circ \varphi_t = \{F \circ \varphi_t, H\},$$

or, for short,

$$\dot{F} = \{F, H\}, \quad \text{for any } F \in \mathcal{F}(U), \ U \text{ open in } P,$$

if and only if φ_t is the flow of X_H.

(ii) *If φ_t is the flow of X_H, then $H \circ \varphi_t = H$.*

Proof. (i) Let $z \in P$. Then

$$\frac{d}{dt}F(\varphi_t(z)) = \mathbf{d}F(\varphi_t(z)) \cdot \frac{d}{dt}\varphi_t(z)$$

and

$$\{F, H\}(\varphi_t(z)) = \mathbf{d}F(\varphi_t(z)) \cdot X_H(\varphi_t(z)).$$

The two expressions are equal for any $F \in \mathcal{F}(U)$, U open in P, if and only
if

$$\frac{d}{dt}\varphi_t(z) = X_H(\varphi_t(z)),$$

by the Hahn–Banach theorem. This is equivalent to $t \mapsto \varphi_t(z)$ being the
integral curve of X_H with initial condition z, that is, φ_t is the flow of X_H.

On the other hand, if φ_t is the flow of X_H, then we have

$$X_H(\varphi_t(z)) = T_z\varphi_t(X_H(z)),$$

so that by the chain rule,

$$\begin{aligned}
\frac{d}{dt}F(\varphi_t(z)) &= \mathbf{d}F(\varphi_t(z)) \cdot X_H(\varphi_t(z)) \\
&= \mathbf{d}F(\varphi_t(z)) \cdot T_z\varphi_t(X_H(z)) \\
&= \mathbf{d}(F \circ \varphi_t)(z) \cdot X_H(z) \\
&= \{F \circ \varphi_t, H\}(z).
\end{aligned}$$

(ii) For the proof of (ii), let $H = F$ in (i). ∎

Corollary 10.2.4. *Let $G, H \in \mathcal{F}(P)$. Then G is constant along the integral curves of X_H if and only if $\{G, H\} = 0$. Either statement is equivalent
to H being constant along the integral curves of X_G.*

Among the elements of $\mathcal{F}(P)$ are functions C such that $\{C, F\} = 0$ for all $F \in \mathcal{F}(P)$, that is, C is constant along the flow of all Hamiltonian vector fields, or, equivalently, $X_C = 0$, that is, C generates trivial dynamics. Such functions are called *Casimir functions* of the Poisson structure. They form the center of the Poisson algebra.[3] This terminology is used in, for example, Sudarshan and Mukunda [1974]. H. B. G. Casimir is a prominent physicist who wrote his thesis (Casimir [1931]) on the quantum mechanics of the rigid body, under the direction of Paul Ehrenfest. Recall that it was Ehrenfest who, in *his* thesis, worked on the variational structure of ideal flow in Lagrangian or material representation.

Examples

(a) Symplectic Case. On a symplectic manifold P, any Casimir function is constant on connected components of P. This holds, since in the symplectic case, $X_C = 0$ implies $dC = 0$, and hence C is locally constant. ◆

(b) Rigid-Body Casimirs. In the context of Example (c) of §10.1, let $C(\mathbf{\Pi}) = \|\mathbf{\Pi}\|^2/2$. Then $\nabla C(\mathbf{\Pi}) = \mathbf{\Pi}$, and by the properties of the triple product, we have for any $F \in \mathcal{F}(\mathbb{R}^3)$,

$$\{C, F\}(\mathbf{\Pi}) = -\mathbf{\Pi} \cdot (\nabla C \times \nabla F) = -\mathbf{\Pi} \cdot (\mathbf{\Pi} \times \nabla F)$$
$$= -\nabla F \cdot (\mathbf{\Pi} \times \mathbf{\Pi}) = 0.$$

This shows that $C(\mathbf{\Pi}) = \|\mathbf{\Pi}\|^2/2$ is a Casimir function. A similar argument shows that

$$C_\Phi(\mathbf{\Pi}) = \Phi\left(\tfrac{1}{2}\|\mathbf{\Pi}\|^2\right) \tag{10.2.2}$$

is a Casimir function, where Φ is an arbitrary (differentiable) function of one variable; this is proved by noting that

$$\nabla C_\Phi(\mathbf{\Pi}) = \Phi'\left(\tfrac{1}{2}\|\mathbf{\Pi}\|^2\right) \mathbf{\Pi}.$$
◆

(c) Helicity. In Example (d) of §10.1, the *helicity*

$$C(\mathbf{v}) = \int_\Omega \mathbf{v} \cdot (\nabla \times \mathbf{v}) \, d^3x \tag{10.2.3}$$

can be checked to be a Casimir function if $\partial\Omega = \varnothing$. ◆

---

[3]The *center* of a group (or algebra) is the set of elements that commute with all elements of the group (or algebra).

(d) Poisson–Vlasov Casimirs. In Example (e) of §10.1, given a differentiable function $\Phi : \mathbb{R} \to \mathbb{R}$, the map $C : \mathcal{F}(P) \to \mathbb{R}$ defined by

$$C(f) = \int \Phi(f(q, p)) \, dq \, dp \qquad (10.2.4)$$

is a Casimir function. Here we choose P to be symplectic, have written $dq \, dp = dz$ for the Liouville measure, and have used it to identify functions and densities. ◆

Some History of Poisson Structures.[4] Following from the work of Lagrange and Poisson discussed at the end of §8.1, the general concept of a Poisson manifold should be credited to Sophus Lie in his treatise on transformation groups written around 1880 in the chapter on "function groups." Lie uses the word "group" for both "group" and "algebra." For example, a "function group" should really be translated as "function algebra."

On page 237, Lie defines what today is called a Poisson structure. The title of Chapter 19 is *The Coadjoint Group*, which is explicitly identified on page 334. Chapter 17, pages 294–298, defines a linear Poisson structure on the dual of a Lie algebra, today called the Lie–Poisson structure, and "Lie's third theorem" is proved for the set of regular elements. On page 349, together with a remark on page 367, it is shown that the Lie–Poisson structure naturally induces a symplectic structure on each coadjoint orbit. As we shall point out in §11.2, Lie also had many of the ideas of momentum maps. For many years this work appears to have been forgotten.

Because of the above history, Marsden and Weinstein [1983] coined the phrase "Lie–Poisson bracket" for this object, and this terminology is now in common use. However, it is not clear that Lie understood the fact that the Lie–Poisson bracket is obtained by a simple reduction process, namely, that it is induced from the canonical cotangent Poisson bracket on T^*G by passing to $\mathfrak{g}^*$ regarded as the quotient T^*G/G, as will be explained in Chapter 13. The link between the closedness of the symplectic form and the Jacobi identity is a little harder to trace explicitly; some comments in this direction are given in Souriau [1970], who gives credit to Maxwell.

Lie's work starts by taking functions $F_1, \ldots, F_r$ on a symplectic manifold M, with the property that there exist functions G_{ij} of r variables such that

$$\{F_i, F_j\} = G_{ij}(F_1, \ldots, F_r).$$

In Lie's time, all functions in sight are implicitly assumed to be analytic. The collection of all functions ϕ of $F_1, \ldots, F_r$ is the "function group"; it is

---

[4]We thank Hans Duistermaat and Alan Weinstein for their help with the comments in this section; the paper of Weinstein [1983a] should also be consulted by the interested reader.

provided with the bracket

$$[\phi, \psi] = \sum_{ij} G_{ij} \phi_i \psi_j, \qquad (10.2.5)$$

where

$$\phi_i = \frac{\partial \phi}{\partial F_i} \quad \text{and} \quad \psi_j = \frac{\partial \psi}{\partial F_j}.$$

Considering $F = (F_1, \ldots, F_r)$ as a map from M to an r-dimensional space P, and ϕ and ψ as functions on P, one may formulate this as saying that $[\phi, \psi]$ is a Poisson structure on P, with the property that

$$F^*[\phi, \psi] = \{F^*\phi, F^*\psi\}.$$

Lie writes down the equations for the G_{ij} that follow from the antisymmetry and the Jacobi identity for the bracket $\{,\}$ on M. He continues with the question, If a given system of functions G_{ij} in r variables satisfies these equations, is it induced, as above, from a function group of functions of $2n$ variables? He shows that under suitable rank conditions the answer is yes. As we shall see below, this result is the precursor to many of the fundamental results about the geometry of Poisson manifolds.

It is obvious that if G_{ij} is a system that satisfies the equations that Lie writes down, then (10.2.5) is a Poisson structure in r-dimensional space. Conversely, for any Poisson structure $[\phi, \psi]$, the functions

$$G_{ij} = [F_i, F_j]$$

satisfy Lie's equations.

Lie continues with more remarks, that are not always stated as explicitly as one would like, on local normal forms of function groups (i.e., of Poisson structures) under suitable rank conditions. These amount to the following: A Poisson structure of constant rank is the same as a foliation with symplectic leaves. It is this characterization that Lie uses to get the symplectic form on the coadjoint orbits. On the other hand, Lie does not apply the symplectic form on the coadjoint orbits to representation theory.

Representation theory of Lie groups started only later with Schur on $GL(n)$, and was continued by Elie Cartan with representations of semisimple Lie algebras, and in the 1930s by Weyl with the representation of compact Lie groups. The coadjoint orbit symplectic structure was connected with representation theory in the work of Kirillov and Kostant. On the other hand, Lie *did* apply the Poisson structure on the dual of the Lie algebra to prove that every abstract Lie algebra can be realized as a Lie algebra of Hamiltonian vector fields, or as a Lie subalgebra of the Poisson algebra of functions on some symplectic manifold. This is "Lie's third fundamental theorem" in the form given by Lie.

In geometry, people like Engel, Study, and, in particular, Elie Cartan studied Lie's work intensely and propagated it very actively. However, through the tainted glasses of retrospection, Lie's work on Poisson structures did not appear to receive as much attention in mechanics as it deserved; for example, even though Cartan himself did very important work in mechanics (such as Cartan [1923, 1928a, 1928b]), he did not seem to realize that the Lie–Poisson bracket was central to the Hamiltonian description of some of the rotating fluid systems he was studying. However, others, such as Hamel [1904, 1949], did study Lie intensively and used his work to make substantial contributions and extensions (such as to the study of nonholonomic systems, including rolling constraints), but many other active schools seem to have missed it. Even more surprising in this context is the contribution of Poincaré [1901b, 1910] to the Lagrangian side of the story, a tale to which we shall come in Chapter 13.

Exercises

◇ **10.2-1.** Verify the relation $[X_H, X_K] = -X_{\{H,K\}}$ directly for the rigid-body bracket.

◇ **10.2-2.** Verify that

$$C(f) = \int \Phi(f(q,p)) \, dq \, dp,$$

defines a Casimir function for the Poisson–Vlasov bracket.

◇ **10.2-3.** Let P be a Poisson manifold and let $M \subset P$ be a connected submanifold with the property that for each $v \in T_x M$ there is a Hamiltonian vector field X_H on P such that $v = X_H(x)$; that is, $T_x M$ is spanned by Hamiltonian vector fields. Prove that any Casimir function is constant on M.

10.3   Properties of Hamiltonian Flows

Hamiltonian Flows Are Poisson.   Now we establish the Poisson analogue of the symplectic nature of the flows of Hamiltonian vector fields.

Proposition 10.3.1.   *If φ_t is the flow of X_H, then*

$$\varphi_t^* \{F, G\} = \{\varphi_t^* F, \varphi_t^* G\};$$

in other words,

$$\{F, G\} \circ \varphi_t = \{F \circ \varphi_t, G \circ \varphi_t\}.$$

Thus, the flows of Hamiltonian vector fields preserve the Poisson structure.

Proof. This is actually true even for time-dependent Hamiltonian systems (as we will see later), but here we will prove it only in the time-independent case. Let $F, K \in \mathcal{F}(P)$ and let φ_t be the flow of X_H. Let

$$u = \{F \circ \varphi_t, K \circ \varphi_t\} - \{F, K\} \circ \varphi_t.$$

Because of the bilinearity of the Poisson bracket,

$$\frac{du}{dt} = \left\{\frac{d}{dt}F \circ \varphi_t, K \circ \varphi_t\right\} + \left\{F \circ \varphi_t, \frac{d}{dt}K \circ \varphi_t\right\} - \frac{d}{dt}\{F, K\} \circ \varphi_t.$$

Using Proposition 10.2.3, this becomes

$$\frac{du}{dt} = \{\{F \circ \varphi_t, H\}, K \circ \varphi_t\} + \{F \circ \varphi_t, \{K \circ \varphi_t, H\}\} - \{\{F, K\} \circ \varphi_t, H\},$$

which, by Jacobi's identity, gives

$$\frac{du}{dt} = \{u, H\} = X_H[u].$$

The unique solution of this equation is $u_t = u_0 \circ \varphi_t$. Since $u_0 = 0$, we get $u = 0$, which is the result. ∎

As in the symplectic case, with which this is, of course, consistent, this argument shows how Jacobi's identity plays a crucial role.

Poisson Maps. A smooth mapping $f : P_1 \to P_2$ between the two Poisson manifolds $(P_1, \{,\}_1)$ and $(P_2, \{,\}_2)$ is called **canonical** or **Poisson** if

$$f^* \{F, G\}_2 = \{f^*F, f^*G\}_1,$$

for all $F, G \in \mathcal{F}(P_2)$. Proposition 10.3.1 shows that flows of Hamiltonian vector fields are canonical maps. We saw already in Chapter 5 that if P_1 and P_2 are symplectic manifolds, a map $f : P_1 \to P_2$ is canonical if and only if it is symplectic.

Properties of Poisson Maps. The next proposition shows that Poisson maps push Hamiltonian flows to Hamiltonian flows.

Proposition 10.3.2. *Let $f : P_1 \to P_2$ be a Poisson map and let $H \in \mathcal{F}(P_2)$. If φ_t is the flow of X_H and ψ_t is the flow of $X_{H \circ f}$, then*

$$\varphi_t \circ f = f \circ \psi_t \quad \text{and} \quad Tf \circ X_{H \circ f} = X_H \circ f.$$

Conversely, if f is a map from P_1 to P_2 and for all $H \in \mathcal{F}(P_2)$ the Hamiltonian vector fields $X_{H \circ f} \in \mathfrak{X}(P_1)$ and $X_H \in \mathfrak{X}(P_2)$ are f-related, that is,

$$Tf \circ X_{H \circ f} = X_H \circ f,$$

then f is canonical.

Proof. For any $G \in \mathcal{F}(P_2)$ and $z \in P_1$, Proposition 10.2.3(i) and the definition of Poisson maps yield

$$\frac{d}{dt} G((f \circ \psi_t)(z)) = \frac{d}{dt}(G \circ f)(\psi_t(z))$$
$$= \{G \circ f, H \circ f\}(\psi_t(z)) = \{G, H\}(f \circ \psi_t)(z),$$

that is, $(f \circ \psi_t)(z)$ is an integral curve of X_H on P_2 through the point $f(z)$. Since $(\varphi_t \circ f)(z)$ is another such curve, uniqueness of integral curves implies that

$$(f \circ \psi_t)(z) = (\varphi_t \circ f)(z).$$

The relation $Tf \circ X_{H \circ f} = X_H \circ f$ follows from $f \circ \psi_t = \varphi_t \circ f$ by taking the time-derivative.

Conversely, assume that for any $H \in \mathcal{F}(P_2)$ we have $Tf \circ X_{H \circ f} = X_H \circ f$. Therefore, by the chain rule,

$$X_{H \circ f}[F \circ f](z) = \mathbf{d}F(f(z)) \cdot T_z f(X_{H \circ f}(z))$$
$$= \mathbf{d}F(f(z)) \cdot X_H(f(z)) = X_H[F](f(z)),$$

that is, $X_{H \circ f}[f^*F] = f^*(X_H[F])$. Thus, for $G \in \mathcal{F}(P_2)$,

$$\{G, H\} \circ f = f^*(X_H[G]) = X_{H \circ f}[f^*G] = \{G \circ f, H \circ f\},$$

and so f is canonical. ∎

Exercises

◇ **10.3-1.** Verify directly that a rotation $R : \mathbb{R}^3 \to \mathbb{R}^3$ is a Poisson map for the rigid-body bracket.

◇ **10.3-2.** If P_1 and P_2 are Poisson manifolds, show that the projection $\pi_1 : P_1 \times P_2 \to P_1$ is a Poisson map. Is the corresponding statement true for symplectic maps?

10.4 The Poisson Tensor

Definition of the Poisson Tensor. By the derivation property of the Poisson bracket, the value of the bracket $\{F, G\}$ at $z \in P$ (and thus $X_F(z)$ as well) depends on F only through $\mathbf{d}F(z)$ (see Theorem 4.2.16 in Abraham, Marsden, and Ratiu [1988] for this type of argument). Thus, there is a contravariant antisymmetric two-tensor

$$B : T^*P \times T^*P \to \mathbb{R}$$

such that

$$B(z)(\alpha_z, \beta_z) = \{F, G\}(z),$$

where $\mathbf{d}F(z) = \alpha_z$ and $\mathbf{d}G(z) = \beta_z \in T_z^*P$. This tensor B is called a *cosymplectic* or **Poisson structure**. In local coordinates $(z^1, \ldots, z^n)$, B is determined by its matrix elements $\{z^I, z^J\} = B^{IJ}(z)$, and the bracket becomes

$$\{F, G\} = B^{IJ}(z)\frac{\partial F}{\partial z^I}\frac{\partial G}{\partial z^J}. \tag{10.4.1}$$

Let $B^\sharp : T^*P \to TP$ be the vector bundle map associated to B, that is,

$$B(z)(\alpha_z, \beta_z) = \langle \alpha_z, B^\sharp(z)(\beta_z) \rangle.$$

Consistent with our conventions, $\dot{F} = \{F, H\}$, the Hamiltonian vector field, is given by $X_H(z) = B_z^\sharp \cdot \mathbf{d}H(z)$. Indeed, $\dot{F}(z) = \mathbf{d}F(z) \cdot X_H(z)$ and

$$\{F, H\}(z) = B(z)(\mathbf{d}F(z), \mathbf{d}H(z)) = \langle \mathbf{d}F(z), B^\sharp(z)(\mathbf{d}H(z)) \rangle.$$

Comparing these expressions gives the stated result.

Coordinate Representation. A convenient way to specify a bracket in finite dimensions is by giving the coordinate relations $\{z^I, z^J\} = B^{IJ}(z)$. The Jacobi identity is then implied by the special cases

$$\{\{z^I, z^J\}, z^K\} + \{\{z^K, z^I\}, z^J\} + \{\{z^J, z^K\}, z^I\} = 0,$$

which are equivalent to the differential equations

$$B^{LI}\frac{\partial B^{JK}}{\partial z^L} + B^{LJ}\frac{\partial B^{KI}}{\partial z^L} + B^{LK}\frac{\partial B^{IJ}}{\partial z^L} = 0 \tag{10.4.2}$$

(the terms are cyclic in I, J, K). Writing $X_H[F] = \{F, H\}$ in coordinates gives

$$X_H^I\frac{\partial F}{\partial z^I} = B^{JK}\frac{\partial F}{\partial z^J}\frac{\partial H}{\partial z^K},$$

and so

$$X_H^I = B^{IJ}\frac{\partial H}{\partial z^J}. \tag{10.4.3}$$

This expression tells us that B^{IJ} should be thought of as the negative inverse of the symplectic matrix, which is literally correct in the nondegenerate case. Indeed, if we write out

$$\Omega(X_H, v) = \mathbf{d}H \cdot v$$

in coordinates, we get

$$\Omega_{IJ}X_H^I v^J = \frac{\partial H}{\partial z^J}v^J, \quad \text{i.e.,} \quad \Omega_{IJ}X_H^I = \frac{\partial H}{\partial z^J}.$$

If $[\Omega^{IJ}]$ denotes the inverse of $[\Omega_{IJ}]$, we get

$$X_H^I = \Omega^{JI} \frac{\partial H}{\partial z^J}, \qquad (10.4.4)$$

so comparing (10.4.3) and (10.4.4), we see that

$$B^{IJ} = -\Omega^{IJ}.$$

Recalling that the matrix of $\Omega^\sharp$ is the inverse of that of $\Omega^\flat$ and that the matrix of $\Omega^\flat$ is the *negative* of that of Ω, we see that $B^\sharp = \Omega^\sharp$.

Let us prove this abstractly. The basic link between the Poisson tensor B and the symplectic form Ω is that they give the same Poisson bracket:

$$\{F, H\} = B(\mathbf{d}F, \mathbf{d}H) = \Omega(X_F, X_H),$$

that is,

$$\langle \mathbf{d}F, B^\sharp \mathbf{d}H \rangle = \langle \mathbf{d}F, X_H \rangle .$$

But

$$\Omega(X_H, v) = \mathbf{d}H \cdot v,$$

and so

$$\langle \Omega^\flat X_H, v \rangle = \langle \mathbf{d}H, v \rangle ,$$

whence

$$X_H = \Omega^\sharp \mathbf{d}H,$$

since $\Omega^\sharp = (\Omega^\flat)^{-1}$. Thus, $B^\sharp \mathbf{d}H = \Omega^\sharp \mathbf{d}H$, for all H, and thus

$$B^\sharp = \Omega^\sharp.$$

Coordinate Representation of Poisson Maps. We have seen that the matrix $[B^{IJ}]$ of the Poisson tensor B converts the differential

$$\mathbf{d}H = \frac{\partial H}{\partial z^I} dz^I$$

of a function to the corresponding Hamiltonian vector field; this is consistent with our treatment in the Introduction and Overview. Another basic concept, that of a Poisson map, is also worthwhile to work out in coordinates.

Let $f : P_1 \rightarrow P_2$ be a Poisson map, so $\{F \circ f, G \circ f\}_1 = \{F, G\}_2 \circ f$. In coordinates z^I on P_1 and w^K on P_2, and writing $w^K = w^K(z^I)$ for the map f, this reads

$$\frac{\partial}{\partial z^I}(F \circ f)\frac{\partial}{\partial z^J}(G \circ f)B_1^{IJ}(z) = \frac{\partial F}{\partial w^K} \frac{\partial G}{\partial w^L} B_2^{KL}(w).$$

By the chain rule, this is equivalent to

$$\frac{\partial F}{\partial w^K}\frac{\partial w^K}{\partial z^I}\frac{\partial G}{\partial w^L}\frac{\partial w^L}{\partial z^J}B_1^{IJ}(z) = \frac{\partial F}{\partial w^K}\frac{\partial G}{\partial w^L}B_2^{KL}(w).$$

Since F and G are arbitrary, f is Poisson iff

$$B_1^{IJ}(z)\frac{\partial w^K}{\partial z^I}\frac{\partial w^L}{\partial z^J} = B_2^{KL}(w).$$

Intrinsically, regarding $B_1(z)$ as a map $B_1(z) : T_z^*P_1 \times T_z^*P_1 \to \mathbb{R}$, this reads

$$B_1(z)(T_z^*f \cdot \alpha_w, T_z^*f \cdot \beta_w) = B_2(w)(\alpha_w, \beta_w), \tag{10.4.5}$$

where $\alpha_w, \beta_w \in T_w^*P_2$ and $f(z) = w$. In analogy with the case of vector fields, we shall say that if equation (10.4.5) holds, then B_1 and B_2 are *f-related* and denote it by $B_1 \sim_f B_2$. In other words, *f is Poisson iff*

$$B_1 \sim_f B_2. \tag{10.4.6}$$

Lie Derivative of the Poisson Tensor.   The next proposition is equivalent to the fact that the flows of Hamiltonian vector fields are Poisson maps.

Proposition 10.4.1.   *For any function $H \in \mathcal{F}(P)$, we have $\pounds_{X_H}B = 0$.*

Proof.   By definition, we have

$$B(dF, dG) = \{F, G\} = X_G[F]$$

for any locally defined functions F and G on P. Therefore,

$$\pounds_{X_H}(B(dF, dG)) = \pounds_{X_H}\{F, G\} = \{\{F, G\}, H\}.$$

However, since the Lie derivative is a derivation,

$$\begin{aligned}
\pounds_{X_H}&(B(dF, dG)) \\
&= (\pounds_{X_H}B)(dF, dG) + B(\pounds_{X_H}dF, dG) + B(dF, \pounds_{X_H}dG) \\
&= (\pounds_{X_H}B)(dF, dG) + B(d\{F, H\}, dG) + B(dF, d\{G, H\}) \\
&= (\pounds_{X_H}B)(dF, dG) + \{\{F, H\}, G\} + \{F, \{G, H\}\} \\
&= (\pounds_{X_H}B)(dF, dG) + \{\{F, G\}, H\},
\end{aligned}$$

by the Jacobi identity. It follows that $(\pounds_{X_H}B)(dF, dG) = 0$ for any locally defined functions $F, G \in \mathcal{F}(U)$. Since any element of T_z^*P can be written as $dF(z)$ for some $F \in \mathcal{F}(U)$, U open in P, it follows that $\pounds_{X_H}B = 0$. ∎

Pauli–Jost Theorem. Suppose that the Poisson tensor B is strongly nondegenerate, that is, it defines an isomorphism $B^\sharp : \mathbf{d}F(z) \mapsto X_F(z)$ of $T_z^* P$ with $T_z P$, for all $z \in P$. Then P is symplectic, and the symplectic form Ω is defined by the formula $\Omega(X_F, X_G) = \{F, G\}$ for any locally defined Hamiltonian vector fields X_F and X_G. One gets $\mathbf{d}\Omega = 0$ from Jacobi's identity—see Exercise 5.5-1. This is the *Pauli–Jost theorem*, due to Pauli [1953] and Jost [1964].

One may be tempted to formulate the above nondegeneracy assumption in a slightly weaker form involving only the Poisson bracket: *Suppose that for every open subset V of P, if $F \in \mathcal{F}(V)$ and $\{F, G\} = 0$ for all $G \in \mathcal{F}(U)$ and all open subsets U of V, then $\mathbf{d}F = 0$ on V, that is, F is constant on the connected components of V.* This condition does *not* imply that P is symplectic, as the following counterexample shows. Let $P = \mathbb{R}^2$ with Poisson bracket

$$\{F, G\}(x, y) = y \left(\frac{\partial F}{\partial x} \frac{\partial G}{\partial y} - \frac{\partial F}{\partial y} \frac{\partial G}{\partial x} \right).$$

If $\{F, G\} = 0$ for all G, then F must be constant on both the upper and lower half-planes, and hence by continuity it must be constant on $\mathbb{R}^2$. However, $\mathbb{R}^2$ with this Poisson structure is clearly not symplectic.

Characteristic Distribution. The subset $B^\sharp(T^* P)$ of TP is called the *characteristic field* or *distribution* of the Poisson structure; it need not be a subbundle of TP in general. Note that skew-symmetry of the tensor B is equivalent to $(B^\sharp)^* = -B^\sharp$, where $(B^\sharp)^* : T^* P \to TP$ is the dual of $B^\sharp$. If P is finite-dimensional, the *rank* of the Poisson structure at a point $z \in P$ is defined to be the rank of $B^\sharp(z) : T_z^* P \to T_z P$; in local coordinates, it is the rank of the matrix $\left[B^{IJ}(z) \right]$. Since the flows of Hamiltonian vector fields preserve the Poisson structure, the rank is constant along such a flow. A Poisson structure for which the rank is everywhere equal to the dimension of the manifold is nondegenerate and hence symplectic.

Poisson Immersions and Submanifolds. An injectively immersed submanifold $i : S \to P$ is called a *Poisson immersion* if any Hamiltonian vector field defined on an open subset of P containing $i(S)$ is in the range of $T_z i$ at all points $i(z)$ for $z \in S$. This is equivalent to the following assertion:

Proposition 10.4.2. *An immersion $i : S \to P$ is Poisson iff it satisfies the following condition. If $F, G : V \subset S \to \mathbb{R}$, where V is open in S, and if $\overline{F}, \overline{G} : U \to \mathbb{R}$ are extensions of $F \circ i^{-1}, G \circ i^{-1} : i(V) \to \mathbb{R}$ to an open neighborhood U of $i(V)$ in P, then $\{\overline{F}, \overline{G}\}|i(V)$ is well-defined and independent of the extensions. The immersed submanifold S is thus endowed with an induced Poisson structure, and $i : S \to P$ becomes a Poisson map.*

Proof.   If $i : S \rightarrow P$ is an injectively immersed Poisson manifold, then

$$\{\overline{F}, \overline{G}\}(i(z)) = \mathbf{d}\overline{F}(i(z)) \cdot X_{\overline{G}}(i(z)) = \mathbf{d}\overline{F}(i(z)) \cdot T_z i(v)$$
$$= \mathbf{d}(\overline{F} \circ i)(z) \cdot v = \mathbf{d}F(z) \cdot v,$$

where $v \in T_z S$ is the unique vector satisfying $X_{\overline{G}}(i(z)) = T_z i(v)$. Thus, $\{\overline{F}, \overline{G}\}(i(z))$ is independent of the extension $\overline{F}$ of $F \circ i^{-1}$. By skew-symmetry of the bracket, it is also independent of the extension $\overline{G}$ of $G \circ i^{-1}$. Then one can define a Poisson structure on S by setting

$$\{F, G\} = \{\overline{F}, \overline{G}\}|i(V)$$

for any open subset V of S. In this way $i : S \rightarrow P$ becomes a Poisson map, since by the computation above we have $X_{\overline{G}}(i(z)) = T_z i(X_G)$.

Conversely, assume that the condition on the bracket stated above holds and let $H : U \rightarrow P$ be a Hamiltonian defined on an open subset U of P intersecting $i(S)$. Then by what was already shown, S is a Poisson manifold, and $i : S \rightarrow P$ is a Poisson map. Because i is Poisson, if $z \in S$ is such that $i(z) \in U$, we have

$$X_H(i(z)) = T_z i(X_{H \circ i}(z)),$$

and thus $X_H(i(z)) \in \text{range } T_z i$, thereby showing that $i : S \rightarrow P$ is a Poisson immersion. ∎

If $S \subset P$ is a submanifold of P and the inclusion i is Poisson, we say that S is a **Poisson submanifold** of P. Note that the only immersed Poisson submanifolds of a symplectic manifold are those whose range in P is open, since for any (weak) symplectic manifold P, we have

$$T_z P = \{ X_H(z) \mid H \in \mathcal{F}(U), \ U \text{ open in } P \}.$$

Note that any Hamiltonian vector field must be tangent to a Poisson submanifold. Also note that the only Poisson submanifolds of a symplectic manifold P are its open sets.

Symplectic Stratifications.   Now we come to an important result that states that every Poisson manifold is a union of symplectic manifolds, each of which is a Poisson submanifold.

Definition 10.4.3.   *Let P be a Poisson manifold. We say that $z_1, z_2 \in P$ are **on the same symplectic leaf** of P if there is a piecewise smooth curve in P joining z_1 and z_2, each segment of which is a trajectory of a locally defined Hamiltonian vector field. This is clearly an equivalence relation, and an equivalence class is called a **symplectic leaf**. The symplectic leaf containing the point z is denoted by Σ_z.*

Theorem 10.4.4 (Symplectic Stratification Theorem).   *Let P be a finite-dimensional Poisson manifold. Then P is the disjoint union of its symplectic leaves. Each symplectic leaf in P is an injectively immersed Poisson* *submanifold, and the induced Poisson structure on the leaf is symplectic. The dimension of the leaf through a point z equals the rank of the Poisson structure at that point, and the tangent space to the leaf at z equals*

$$B^{\#}(z)(T_z^* P) = \{ X_H(z) \mid H \in \mathcal{F}(U),\ U \text{ open in } P \}.$$

The picture one should have in mind is shown in Figure 10.4.1. Note in particular that the dimension of the symplectic leaf through a point can change dimension as the point varies.

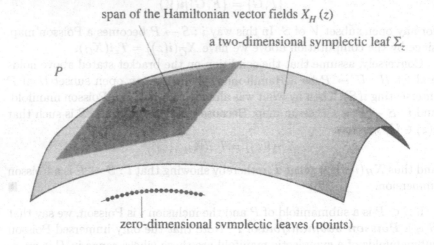

span of the Hamiltonian vector fields $X_H(z)$

a two-dimensional symplectic leaf Σ_z

P

zero-dimensional symplectic leaves (points)

FIGURE 10.4.1. The symplectic leaves of a Poisson manifold.

The Poisson bracket on P can be alternatively described as follows.

To evaluate the Poisson bracket of F and G at $z \in P$, restrict F and G to the symplectic leaf Σ through z, take their bracket on Σ (in the sense of brackets on a symplectic manifold), and evaluate at z.

Also note that since the Casimir functions have differentials that annihilate the characteristic field, they are constant on symplectic leaves.

To get a feeling for the geometric content of the symplectic stratification theorem, let us first prove it under the assumption that the characteristic field is a smooth vector subbundle of TP, which is the case considered originally by Lie [1890]. In finite dimensions, this is guaranteed if the rank of the Poisson structure is constant. Jacobi's identity shows that the characteristic field is involutive, and thus by the Frobenius theorem, it is integrable. Therefore, P is foliated by injectively immersed submanifolds whose tangent space at any point coincides with the subspace of all Hamiltonian vector fields evaluated at z. Thus, each such leaf Σ is an immersed Poisson submanifold of P. Define the two-form Ω on Σ by

$$\Omega(z)(X_F(z), X_G(z)) = \{F, G\}(z)$$

for any functions F, G defined on a neighborhood of z in P. Note that Ω is closed by the Jacobi identity (Exercise 5.5-1). Also, if

$$0 = \Omega(z)(X_F(z), X_G(z)) = \mathbf{d}F(z) \cdot X_G(z)$$

for all locally defined G, then

$$\mathbf{d}F(z)|T_z\Sigma = \mathbf{d}(F \circ i)(z) = 0$$

by the Hahn–Banach theorem. Therefore,

$$0 = X_{F \circ i}(z) = T_z i(X_F(z)) = X_F(z),$$

since Σ is a Poisson submanifold of P and the inclusion $i : \Sigma \to P$ is a Poisson map, thus showing that Ω is weakly nondegenerate and thereby proving the theorem for the constant-rank case.

The general case, proved by Kirillov [1976a], is more subtle, since for differentiable distributions that are not subbundles, integrability and involutivity are not equivalent. We shall prove this case in the Internet supplement.

Proposition 10.4.5. *If P is a Poisson manifold, $\Sigma \subset P$ is a symplectic leaf, and C is a Casimir function, then C is constant on Σ.*

Proof. If C were not locally constant on Σ, then there would exist a point $z \in \Sigma$ such that $\mathbf{d}C(z) \cdot v \neq 0$ for some $v \in T_z\Sigma$. But $T_z\Sigma$ is spanned by $X_k(z)$ for $k \in \mathcal{F}(P)$, and hence $\mathbf{d}C(z) \cdot X_k(z) = \{C, K\}(z) = 0$, which implies that $\mathbf{d}C(z) \cdot v = 0$, which is a contradiction. Thus C is locally constant on Σ and hence constant by connectedness of the leaf Σ. ∎

Examples

(a) Let $P = \mathbb{R}^3$ with the rigid-body bracket. Then the symplectic leaves are spheres centered at the origin. The single point at the origin is the singular leaf in the sense that the Poisson structure has rank zero there. As we shall see later, it is true more generally that the symplectic leaves in $\mathfrak{g}^*$ with the Lie–Poisson bracket are the coadjoint orbits.   ◆

(b) Symplectic leaves need not be submanifolds, and *one cannot conclude that if all the Casimir functions are constants then the Poisson structure is nondegenerate.* For example, consider the three torus $\mathbb{T}^3$ with a codimension 1 foliation with dense leaves, such as obtained by taking the leaves to be the product of $\mathbb{T}^1$ with a leaf of the irrational flow on $\mathbb{T}^2$. Put the usual area element on these leaves and define a Poisson structure on $\mathbb{T}^3$ by declaring these to be the symplectic leaves. Any Casimir function is constant, yet the Poisson structure is degenerate.   ◆

Poisson–Darboux Theorem. Related to the stratification theorem is an analogue of Darboux' theorem. To state it, first recall from Exercise 10.3-2 that we define the product Poisson structure on $P_1 \times P_2$ where P_1, P_2 are Poisson manifolds by the requirements that the projections $\pi_1 : P_1 \times P_2 \to P$ and $\pi_2 : P_1 \times P_2 \to P_2$ be Poisson mappings, and $\pi_1^*(\mathcal{F}(P_1))$ and $\pi_2^*(\mathcal{F}(P_2))$ be commuting subalgebras of $\mathcal{F}(P_1 \times P_2)$. In terms of coordinates, if bracket relations $\{z^I, z^J\} = B^{IJ}(z)$ and $\{w^I, w^J\} = C^{IJ}(w)$ are given on P_1 and P_2, respectively, then these define a bracket on functions of z^I and w^J when augmented by the relations $\{z^I, w^J\} = 0$.

Theorem 10.4.6 (Lie–Weinstein). *Let z_0 be a point in a Poisson manifold P. There is a neighborhood U of z_0 in P and an isomorphism $\varphi = \varphi_S \times \varphi_N : U \to S \times N$, where S is symplectic, N is Poisson, and the rank of N at $\varphi_N(z_0)$ is zero. The factors S and N are unique up to local isomorphism. Moreover, if the rank of the Poisson manifold is constant near z_0, there are coordinates $(q^1, \dots, q^k, p_1, \dots, p_k, y^1, \dots, y^l)$ near x_0 satisfying the canonical bracket relations*

$$\{q^i, q^j\} = \{p_i, p_j\} = \{q^i, y^j\} = \{p_i, y^j\} = 0, \ \{q^i, p_j\} = \delta^i_j.$$

When one is proving this theorem, the manifold S can be taken to be the symplectic leaf of P through z_0, and N is, locally, any submanifold of P, transverse to S, and such that $S \cap N = \{z_0\}$. In many cases the transverse structure on N is of Lie–Poisson type. For the proof of this theorem and related results, see Weinstein [1983b]; the second part of the theorem is due to Lie [1890]. For the main examples in this book we shall not require a detailed local analysis of their Poisson structure, so we shall forgo a more detailed study of the local structure of Poisson manifolds.

Exercises

⋄ **10.4-1.** If $H \in \mathcal{F}(P)$, where P is a Poisson manifold, show that the flow φ_t of X_H preserves the symplectic leaves of P.

⋄ **10.4-2.** Let $(P, \{\,,\,\})$ be a Poisson manifold with Poisson tensor $B \in \Omega_2(P)$. Let

$$B^\sharp : T^*P \to TP, \quad B^\sharp(\mathbf{d}H) = X_H,$$

be the induced bundle map. We shall denote by the same symbol $B^\sharp : \Omega^1(P) \to \mathfrak{X}(P)$ the induced map on the sections. The definitions give

$$B(\mathbf{d}F, \mathbf{d}H) = \langle \mathbf{d}F, B^\sharp(\mathbf{d}H)\rangle = \{F, H\}.$$

Define $\alpha^\sharp := B^\sharp(\alpha)$. Define for any $\alpha, \beta \in \Omega^1(P)$,

$$\{\alpha, \beta\} = -\mathcal{L}_{\alpha^\sharp}\beta + \mathcal{L}_{\beta^\sharp}\alpha - \mathbf{d}(B(\alpha, \beta)).$$

(a) Show that if the Poisson bracket on P is induced by a symplectic form Ω, that is, if $B^\sharp = \Omega^\sharp$, then

$$B(\alpha, \beta) = \Omega(\alpha^\sharp, \beta^\sharp).$$

(b) Show that for any $F, G \in \mathcal{F}(P)$, we have

$$\{F\alpha, G\beta\} = FG\{\alpha, \beta\} - F\alpha^\sharp[G]\beta + G\beta^\sharp[F]\alpha.$$

(c) Show that for any $F, G \in \mathcal{F}(P)$, we have

$$\mathbf{d}\{F, G\} = \{\mathbf{d}F, \mathbf{d}G\}.$$

(d) Show that if $\alpha, \beta \in \Omega^1(P)$ are closed, then $\{\alpha, \beta\} = \mathbf{d}(B(\alpha, \beta))$.

(e) Use $\pounds_{X_H} B = 0$ to show that $\{\alpha, \beta\}^\sharp = -[\alpha^\sharp, \beta^\sharp]$.

(f) Show that $(\Omega^1(P), \{\ ,\ \})$ is a Lie algebra; that is, prove Jacobi's identity.

◇ **10.4-3** (Weinstein [1983b]). Let P be a manifold and X, Y be two linearly independent commuting vector fields. Show that

$$\{F, K\} = X[F]Y[K] - Y[F]X[K]$$

defines a Poisson bracket on P. Show that

$$X_H = Y[H]X - X[H]Y.$$

Show that the symplectic leaves are two-dimensional and that their tangent spaces are spanned by X and Y. Show how to get Example (b) preceding Theorem 10.4.6 from this construction.

10.5 Quotients of Poisson Manifolds

Here we shall give the simplest version of a general construction of Poisson manifolds based on symmetry. This construction represents the first steps in a general procedure called **reduction**.

Poisson Reduction Theorem. Suppose that G is a Lie group that acts on a Poisson manifold and that each map $\Phi_g : P \to P$ is a Poisson map. Let us also suppose that the action is free and proper, so that the quotient space P/G is a smooth manifold and the projection $\pi : P \to P/G$ is a submersion (see the discussion of this point in §9.3).

Theorem 10.5.1. *Under these hypotheses, there is a unique Poisson structure on P/G such that π is a Poisson map. (See Figure 10.5.1.)*

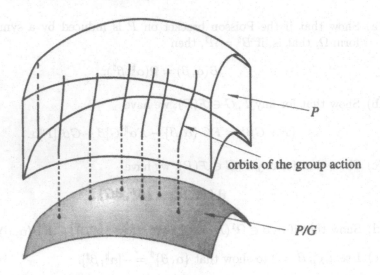

P

orbits of the group action

P/G

FIGURE 10.5.1. The quotient of a Poisson manifold by a group action is a Poisson manifold in a natural way.

Proof. Let us first assume that P/G is Poisson and show uniqueness. The condition that π be Poisson is that for two functions $f, k : P/G \to \mathbb{R}$,

$$\{f, k\} \circ \pi = \{f \circ \pi, k \circ \pi\}, \qquad (10.5.1)$$

where the brackets are on P/G and P, respectively. The function $\overline{f} = f \circ \pi$ is the unique G-invariant function that projects to f. In other words, if $[z] \in P/G$ is an equivalence class, whereby $g_1 \cdot z$ and $g_2 \cdot z$ are equivalent, we let $\overline{f}(g \cdot z) = f([z])$ for all $g \in G$. Obviously, this defines $\overline{f}$ unambiguously, so that $\overline{f} = f \circ \pi$. We can also characterize this as saying that $\overline{f}$ assigns the value $f([z])$ to the whole orbit $G \cdot z$. We can write (10.5.1) as

$$\{f, k\} \circ \pi = \{\overline{f}, \overline{k}\}.$$

Since π is onto, this determines $\{f, k\}$ uniquely.

We can also use (10.5.1) to *define* $\{f, k\}$. First, note that

$$
\begin{aligned}
\{\overline{f}, \overline{k}\}(g \cdot z) &= \left(\{\overline{f}, \overline{k}\} \circ \Phi_g\right)(z) \\
&= \{\overline{f} \circ \Phi_g, \overline{k} \circ \Phi_g\}(z) \\
&= \{\overline{f}, \overline{k}\}(z),
\end{aligned}
$$

since Φ_g is Poisson and since $\overline{f}$ and $\overline{k}$ are constant on orbits. Thus, $\{\overline{f}, \overline{k}\}$ is constant on orbits, too, and so it defines $\{f, k\}$ uniquely.

It remains to show that $\{f, k\}$ so defined satisfies the properties of a Poisson structure. However, these all follow from their counterparts on P. For example, if we write Jacobi's identity on P, namely

$$0 = \{\{\overline{f}, \overline{k}\}, \overline{l}\} + \{\{\overline{l}, \overline{f}\}, \overline{k}\} + \{\{\overline{k}, \overline{l}\}, \overline{f}\},$$

it gives, by construction,

$$0 = \{\{f,k\} \circ \pi, l \circ \pi\} + \{\{l,f\} \circ \pi, k \circ \pi\} + \{\{k,l\} \circ \pi, f \circ \pi\}$$
$$= \{\{f,k\},l\} \circ \pi + \{\{l,f\},k\} \circ \pi + \{\{k,l\},f\} \circ \pi,$$

and thus by surjectivity of π, Jacobi's identity holds on P/G. ∎

This construction is just one of many that produce new Poisson and symplectic manifolds from old ones. We refer to Marsden and Ratiu [1986] and Vaisman [1996] for generalizations of the construction here.

Reduction of Dynamics. If H is a G-invariant Hamiltonian on P, it defines a corresponding function h on P/G such that $H = h \circ \pi$. Since π is a Poisson map, it transforms X_H on P to X_h on P/G; that is, $T\pi \circ X_H = X_h \circ \pi$, or X_H and X_h are π-related. We say that the Hamiltonian system X_H on P **reduces** to that on P/G.

As we shall see in the next chapter, G-invariance of H may be associated with a conserved quantity $J : P \rightarrow \mathbb{R}$. If it is also G-invariant, the corresponding function j on P/G is conserved for X_h, since

$$\{h,j\} \circ \pi = \{H,J\} = 0$$

and so $\{h,j\} = 0$.

Example. Consider the differential equations on $\mathbb{C}^2$ given by

$$\dot{z}_1 = -i\omega_1 z_1 + i\epsilon p \bar{z}_2 + i z_1(s_{11}|z_1|^2 + s_{12}|z_2|^2),$$
$$\dot{z}_2 = -i\omega_2 z_2 + i\epsilon q \bar{z}_1 - i z_2(s_{21}|z_1|^2 + s_{22}|z_2|^2). \tag{10.5.2}$$

Use the standard Hamiltonian structure obtained by taking the real and imaginary parts of z_i as conjugate variables. For example, we write $z_1 = q_1 + ip_1$ and require $\dot{q}_1 = \partial H/\partial p_1$ and $\dot{p}_1 = -\partial H/\partial q_1$. Recall from Chapter 5 that a useful trick in this regard that enables one to work in complex notation is to write Hamilton's equations as $\dot{z}_k = -2i\partial H/\partial \bar{z}_k$. Using this, one readily finds that (see Exercise 5.4-3) *the system* (10.5.2) *is Hamiltonian if and only if* $s_{12} = -s_{21}$ *and* $p = q$. In this case we can choose

$$H(z_1, z_2) = \frac{1}{2}(\omega_2|z_2|^2 + \omega_1|z_1|^2) - \epsilon p \, \mathrm{Re}(z_1 z_2) - \frac{s_{11}}{4}|z_1|^4$$
$$- \frac{s_{12}}{2}|z_1 z_2|^2 + \frac{s_{22}}{4}|z_2|^4. \tag{10.5.3}$$

Note that for equation (10.5.2) with $\epsilon = 0$ there are two copies of S^1 acting on z_1 and z_2 independently; corresponding conserved quantities are $|z_1|^2$ and $|z_2|^2$. However, for $\epsilon \neq 0$, the symmetry action is

$$(z_1, z_2) \mapsto (e^{i\theta} z_1, e^{-i\theta} z_2) \tag{10.5.4}$$

with the conserved quantity (Exercise 5.5-4)

$$J(z_1, z_2) = \tfrac{1}{2}(|z_1|^2 - |z_2|^2). \tag{10.5.5}$$

Let $\phi = (\pi/2) - \theta_1 - \theta_2$, where $z_1 = r_1 \exp(i\theta_1)$, $z_2 = r_2 \exp(i\theta_2)$. We know that the Hamiltonian structure for (10.5.2) on $\mathbb{C}^2$ described above induces one on $\mathbb{C}^2/S^1$ (exclude points where r_1 or r_2 vanishes), and that the two integrals (energy and the conserved quantity) descend to the quotient space, as does the Poisson bracket. The quotient space $\mathbb{C}^2/S^1$ is parametrized by (r_1, r_2, ϕ), and H and J can be dropped to the quotient. Concretely, the process of dropping to the quotient is very simple: If $F(z_1, z_2) = F(r_1, \theta_1, r_2, \theta_2)$ is S^1 invariant, then it can be written (uniquely) as a function f of (r_1, r_2, ϕ).

By Theorem 10.5.1, one can also drop the Poisson bracket to the quotient. Consequently, the equations in (r_1, r_2, ϕ) can be cast in Hamiltonian form $\dot{f} = \{f, h\}$ for the induced Poisson bracket. This bracket is obtained by using the chain rule to relate the complex variables and the polar coordinates. One finds that

$$
\begin{aligned}
&\{f, k\}(r_1, r_2, \phi) \\
&= -\frac{1}{r_1} \left(\frac{\partial f}{\partial r_1} \frac{\partial k}{\partial \phi} - \frac{\partial f}{\partial \phi} \frac{\partial k}{\partial r_1} \right) - \frac{1}{r_2} \left(\frac{\partial f}{\partial r_2} \frac{\partial k}{\partial \phi} - \frac{\partial f}{\partial \phi} \frac{\partial k}{\partial r_2} \right).
\end{aligned} \tag{10.5.6}
$$

The (noncanonical) Poisson bracket (10.5.6) is, of course, the reduction of the original *canonical* Poisson bracket on the space of q and p variables, written in the new polar coordinate variables. Theorem 10.5.1 shows that Jacobi's identity is automatic for this reduced bracket. (See Knobloch, Mahalov, and Marsden [1994] for further examples of this type.)   ◆

As we shall see in Chapter 13, a key example of the Poisson reduction given in 10.5.1 is that in which $P = T^*G$ and G acts on itself by left translations. Then $P/G \cong \mathfrak{g}^*$, and the reduced Poisson bracket is none other than the Lie–Poisson bracket!

Exercises

⋄ **10.5-1.** Let $\mathbb{R}^3$ be equipped with the rigid-body bracket and let $G = S^1$ act on $P = \mathbb{R}^3 \setminus (z\text{-axis})$ by rotation about the z-axis. Compute the induced bracket on P/G.

⋄ **10.5-2.** Compute explicitly the reduced Hamiltonian h in the example in the text and verify directly that the equations for $\dot{r}_1, \dot{r}_2, \phi$ are Hamiltonian on $\mathbb{C}^2$ with Hamiltonian h. Also check that the function j induced by J is a constant of the motion.

10.6   The Schouten Bracket

The goal of this section is to express the Jacobi identity for a Poisson structure in geometric terms analogous to $d\Omega$ for symplectic structures. This will be done in terms of a bracket defined on contravariant antisymmetric tensors generalizing the Lie bracket of vector fields (see, for example, Schouten [1940], Nijenhuis [1953], Lichnerowicz [1978], Olver [1984, 1986], Koszul [1985], Libermann and Marle [1987], Bhaskara and Viswanath [1988], Kosmann-Schwarzbach and Magri [1990], Vaisman [1994], and references therein).

Multivectors. A *contravariant antisymmetric q-tensor* on a finite-dimensional vector space V is a q-linear map

$$A : V^* \times V^* \times \cdots \times V^* \,(q \text{ times}) \to \mathbb{R}$$

that is antisymmetric in each pair of arguments. The space of these tensors will be denoted by $\bigwedge_q(V)$. Thus, each element $\bigwedge_q(V)$ is a finite linear combination of terms of the form $v_1 \wedge \cdots \wedge v_q$, called a *q-vector*, for $v_1, \ldots, v_q \in V$. If V is an infinite-dimensional Banach space, we define $\bigwedge_q(V)$ to be the span of all elements of the form $v_1 \wedge \cdots \wedge v_q$ with $v_1, \ldots, v_q \in V$, where the exterior product is defined in the usual manner relative to a weakly nondegenerate pairing $\langle , \rangle : V^* \times V \to \mathbb{R}$. Thus, $\bigwedge_0(V) = \mathbb{R}$ and $\bigwedge_1(V) = V$. If P is a smooth manifold, let

$$\bigwedge(P) = \bigcup_q \bigwedge_{z \in P} {}_q(T_z P),$$

a smooth vector bundle with fiber over $z \in P$ equal to $\bigwedge_q(T_z P)$. Let $\Omega_q(P)$ denote the smooth sections of $\bigwedge_q(P)$, that is, the elements of $\Omega_q(P)$ are smooth contravariant antisymmetric q-tensor fields on P. Let $\Omega_*(P)$ be the direct sum of the spaces $\Omega_q(P)$, where $\Omega_0(P) = \mathcal{F}(P)$. Note that

$$\Omega_q(P) = 0 \qquad \text{for } q > \dim(P),$$

and that

$$\Omega_1(P) = \mathfrak{X}(P).$$

If $X_1, \ldots, X_q \in \mathfrak{X}(P)$, then $X_1 \wedge \cdots \wedge X_q$ is called a *q-vector field*, or a *multivector field*.

On the manifold P, consider a $(q + p)$-form α and a contravariant antisymmetric q-tensor A. The *interior product* $\mathbf{i}_A \alpha$ of A with α is defined as follows. If $q = 0$, so $A \in \mathbb{R}$, let $\mathbf{i}_A \alpha = A\alpha$. If $q \geq 1$ and if $A = v_1 \wedge \cdots \wedge v_q$, where $v_i \in T_z P$, $i = 1, \ldots, q$, define $\mathbf{i}_A \alpha \in \Omega^p(P)$ by

$$(\mathbf{i}_A \alpha)(v_{q+1}, \ldots, v_{q+p}) = \alpha(v_1, \ldots, v_{q+p}) \tag{10.6.1}$$

for arbitrary $v_{q+1}, \ldots, v_{q+p} \in T_z P$. One checks that the definition does not depend on the representation of A as a q-vector, so $\mathbf{i}_A \alpha$ is well-defined on $\bigwedge_q(P)$ by linear extension. In local coordinates, for finite-dimensional P,

$$(\mathbf{i}_A \alpha)_{i_{q+1} \ldots i_{q+p}} = A^{i_1 \ldots i_q} \alpha_{i_1 \ldots i_{q+p}}, \qquad (10.6.2)$$

where all components are nonstrict. If P is finite-dimensional and $p = 0$, then (10.6.1) defines an isomorphism of $\Omega_q(P)$ with $\Omega^q(P)$. If P is a Banach manifold, then (10.6.1) defines a weakly nondegenerate pairing of $\Omega_q(P)$ with $\Omega^q(P)$. If $A \in \Omega_q(P)$, then q is called the **degree** of A and is denoted by $\deg A$. One checks that

$$\mathbf{i}_{A \wedge B} \alpha = \mathbf{i}_B \mathbf{i}_A \alpha. \qquad (10.6.3)$$

The Lie derivative $\pounds_X$ is a derivation relative to $\wedge$, that is,

$$\pounds_X(A \wedge B) = (\pounds_X A) \wedge B + A \wedge (\pounds_X B)$$

for any $A, B \in \Omega_*(P)$.

The Schouten Bracket.   The next theorem produces an interesting bracket on multivectors.

Theorem 10.6.1 (Schouten Bracket Theorem).   *There is a unique bilinear operation* $[\,,] : \Omega_*(P) \times \Omega_*(P) \to \Omega_*(P)$ *natural with respect to restriction to open sets, called the* **Schouten bracket**, *that satisfies the following properties:*

   (i) *It is a* **biderivation of degree** -1, *that is, it is bilinear,*

$$\deg[A, B] = \deg A + \deg B - 1, \qquad (10.6.4)$$

   and for $A, B, C \in \Omega_*(P)$,

$$[A, B \wedge C] = [A, B] \wedge C + (-1)^{(\deg A + 1) \deg B} B \wedge [A, C]. \qquad (10.6.5)$$

  (ii) *It is determined on* $\mathcal{F}(P)$ *and* $\mathfrak{X}(P)$ *by*

      (a) $[F, G] = 0$, *for all* $F, G \in \mathcal{F}(P)$;
      (b) $[X, F] = X[F]$, *for all* $F \in \mathcal{F}(P)$, $X \in \mathfrak{X}(P)$;
      (c) $[X, Y]$ *for all* $X, Y \in \mathfrak{X}(P)$ *is the usual Jacobi–Lie bracket of vector fields.*

 (iii) $[A, B] = (-1)^{\deg A \deg B}[B, A]$.

In addition, the Schouten bracket satisfies the **graded Jacobi identity**

$$(-1)^{\deg A \deg C}[[A, B], C] + (-1)^{\deg B \deg A}[[B, C], A]$$
$$+ (-1)^{\deg C \deg B}[[C, A], B] = 0. \qquad (10.6.6)$$

Proof. The proof proceeds in standard fashion and is similar to that characterizing the exterior or Lie derivative by its properties (see Abraham, Marsden, and Ratiu [1988]): On functions and vector fields it is given by (ii); then (i) and linear extension determine it on any skew-symmetric contravariant tensor in the second variable and a function and vector field in the first; (iii) tells how to switch such variables, and finally (i) again defines it on any pair of skew-symmetric contravariant tensors. The operation so defined satisfies (i), (ii), and (iii) by construction. Uniqueness is a consequence of the fact that the skew-symmetric contravariant tensors are generated as an exterior algebra locally by functions and vector fields, and (ii) gives these. The graded Jacobi identity is verified on an arbitrary triple of q-, p-, and r-vectors using (i), (ii), and (iii) and then invoking trilinearity of the identity. ∎

Properties. The following formulas are useful in computing with the Schouten bracket. If $X \in \mathfrak{X}(P)$ and $A \in \Omega_p(P)$, induction on the degree of A and the use of property (i) show that

$$[X, A] = \pounds_X A. \tag{10.6.7}$$

An immediate consequence of this formula and the graded Jacobi identity is the *derivation property of the Lie derivative relative to the Schouten bracket*, that is,

$$\pounds_X[A, B] = [\pounds_X A, B] + [A, \pounds_X B], \tag{10.6.8}$$

for $A \in \Omega_p(P), B \in \Omega_q(P)$, and $X \in \mathfrak{X}(P)$. Using induction on the number of vector fields, (10.6.7), and the properties in Theorem 10.6.1, one can prove that

$$[X_1 \wedge \cdots \wedge X_r, A] = \sum_{i=1}^{r} (-1)^{i+1} X_1 \wedge \cdots \wedge \check{X}_i \wedge \cdots \wedge X_r \wedge (\pounds_{X_i} A),$$

$$\tag{10.6.9}$$

where $X_1, \ldots, X_r \in \mathfrak{X}(P)$ and $\check{X}_i$ means that X_i has been omitted. The last formula plus linear extension can be taken as the definition of the Schouten bracket, and one can deduce Theorem 10.6.1 from it; see Vaisman [1994] for this approach. If $A = Y_1 \wedge \cdots \wedge Y_s$ for $Y_1, \ldots, Y_s \in \mathfrak{X}(P)$, the formula above plus the derivation property of the Lie derivative give

$$[X_1 \wedge \cdots \wedge X_r, Y_1 \wedge \cdots \wedge Y_s]$$

$$= (-1)^{r+1} \sum_{i=1}^{r} \sum_{j=1}^{s} (-1)^{i+j} [X_i, Y_j] \wedge X_1 \wedge \cdots \wedge \check{X}_i \wedge \cdots$$

$$\wedge X_r \wedge Y_1 \wedge \cdots \wedge \check{Y}_j \wedge \cdots \wedge Y_s. \tag{10.6.10}$$

Finally, if $A \in \Omega_p(P)$, $B \in \Omega_q(P)$, and $\alpha \in \Omega^{p+q-1}(P)$, the formula

$$i_{[A,B]}\alpha = (-1)^{q(p+1)} i_A d\, i_B \alpha + (-1)^p i_B d\, i_A \alpha - i_B i_A d\alpha \qquad (10.6.11)$$

(which is a direct consequence of (10.6.10) and Cartan's formula for $d\alpha$) can be taken as the definition of $[A, B] \in \Omega_{p+q-1}(P)$; this is the approach taken originally in Nijenhuis [1955].

Coordinate Formulas. In local coordinates, setting $\partial/\partial z^i = \partial_i$, the formulas (10.6.9) and (10.6.10) imply that

1. for any function f,

$$[f, \partial_{i_1} \wedge \cdots \wedge \partial_{i_p}] = \sum_{k=1}^{p} (-1)^{k-1} (\partial_{i_k} f)\, \partial_{i_1} \wedge \cdots \wedge \check{\partial}_{i_k} \wedge \cdots \wedge \partial_{i_p},$$

where ˘ over a symbol means that it is omitted, and

2. $[\partial_{i_1} \wedge \cdots \wedge \partial_{i_p}, \partial_{j_1} \wedge \cdots \wedge \partial_{j_q}] = 0$.

Therefore, if

$$A = A^{i_1 \cdots i_p} \partial_{i_1} \wedge \cdots \wedge \partial_{i_p} \quad \text{and} \quad B = B^{j_1 \cdots j_q} \partial_{j_1} \wedge \cdots \wedge \partial_{j_q},$$

we get

$$[A, B] = A^{\ell i_1 \cdots i_{\ell-1} i_{\ell+1} \cdots i_p} \partial_\ell B^{j_1 \cdots j_q} \partial_{i_1} \wedge \cdots \wedge \partial_{i_{\ell-1}} \wedge \partial_{i_{\ell+1}}$$
$$\wedge \partial_{j_1} \wedge \cdots \wedge \partial_{j_q}$$
$$+ (-1)^p B^{\ell j_1 \cdots j_{\ell-1} j_{\ell+1} \cdots j_q} \partial_\ell A^{i_1 \cdots i_p} \partial_{i_1} \wedge \cdots \wedge \partial_{i_p}$$
$$\wedge \partial_{j_1} \wedge \cdots \wedge \partial_{j_{\ell-1}} \wedge \partial_{j_{\ell+1}} \wedge \cdots \wedge \partial_{j_q} \qquad (10.6.12)$$

or, more succinctly,

$$[A, B]^{k_2 \cdots k_{p+q}} = \varepsilon^{k_2 \cdots k_{p+q}}_{i_2 \cdots i_p j_1 \cdots j_q} A^{\ell i_2 \cdots i_p} \frac{\partial}{\partial x^\ell} B^{j_1 \cdots j_q}$$
$$+ (-1)^p \varepsilon^{k_2 \cdots k_{p+q}}_{i_1 \cdots i_p j_2 \cdots j_q} B^{\ell j_2 \cdots j_p} \frac{\partial}{\partial x^\ell} A^{i_1 \cdots i_q}, \qquad (10.6.13)$$

where all components are nonstrict. Here

$$\varepsilon^{i_1 \cdots i_{p+q}}_{j_1 \cdots j_{p+q}}$$

is the **Kronecker symbol**: It is zero if $(i_1, \ldots, i_{p+q}) \neq (j_1, \ldots, j_{p+q})$, and is 1 (resp., -1) if $j_1, \ldots, j_{p+q}$ is an even (resp., odd) permutation of $i_1, \ldots, i_{p+q}$.

From §10.6 the Poisson tensor $B \in \Omega_2(P)$ defined by a Poisson bracket $\{,\}$ on P satisfies $B(dF, dG) = \{F, G\}$ for any $F, G \in \mathcal{F}(P)$. By (10.6.2), this can be written

$$\{F, G\} = i_B(dF \wedge dG), \qquad (10.6.14)$$

or in local coordinates,

$$\{F, G\} = B^{IJ} \frac{\partial F}{\partial z^I} \frac{\partial G}{\partial z^J}.$$

Writing B locally as a sum of terms of the form $X \wedge Y$ for some $X, Y \in \mathfrak{X}(P)$ and taking $Z \in \mathfrak{X}(P)$ arbitrarily, by (10.6.1) we have for $F, G, H \in \mathcal{F}(P)$,

$$\mathbf{i}_B(\mathbf{d}F \wedge \mathbf{d}G \wedge \mathbf{d}H)(Z)$$

$$= (\mathbf{d}F \wedge \mathbf{d}G \wedge \mathbf{d}H)(X, Y, Z)$$

$$= \det \begin{bmatrix} \mathbf{d}F(X) & \mathbf{d}F(Y) & \mathbf{d}F(Z) \\ \mathbf{d}G(X) & \mathbf{d}G(Y) & \mathbf{d}G(Z) \\ \mathbf{d}H(X) & \mathbf{d}H(Y) & \mathbf{d}H(Z) \end{bmatrix}$$

$$= \det \begin{bmatrix} \mathbf{d}F(X) & \mathbf{d}F(Y) \\ \mathbf{d}G(X) & \mathbf{d}G(Y) \end{bmatrix} \mathbf{d}H(Z) + \det \begin{bmatrix} \mathbf{d}H(X) & \mathbf{d}H(Y) \\ \mathbf{d}F(X) & \mathbf{d}F(Y) \end{bmatrix} \mathbf{d}G(Z)$$

$$+ \det \begin{bmatrix} \mathbf{d}G(X) & \mathbf{d}G(Y) \\ \mathbf{d}H(X) & \mathbf{d}H(Y) \end{bmatrix} \mathbf{d}F(Z)$$

$$= \mathbf{i}_B(\mathbf{d}F \wedge \mathbf{d}G)\mathbf{d}H(Z) + \mathbf{i}_B(\mathbf{d}H \wedge \mathbf{d}F)\mathbf{d}G(Z) + \mathbf{i}_B(\mathbf{d}G \wedge \mathbf{d}H)\mathbf{d}F(Z),$$

that is,

$$\mathbf{i}_B(\mathbf{d}F \wedge \mathbf{d}G \wedge \mathbf{d}H)$$
$$= \mathbf{i}_B(\mathbf{d}F \wedge \mathbf{d}G)\mathbf{d}H + \mathbf{i}_B(\mathbf{d}H \wedge \mathbf{d}F)\mathbf{d}G + \mathbf{i}_B(\mathbf{d}G \wedge \mathbf{d}H)\mathbf{d}F. \quad (10.6.15)$$

The Jacobi–Schouten Identity.   Equations (10.6.14) and (10.6.15) imply

$$\{\{F, G\}, H\} + \{\{H, F\}, G\} + \{\{G, H\}, F\}$$
$$= \mathbf{i}_B(\mathbf{d}\{F, G\} \wedge \mathbf{d}H) + \mathbf{i}_B(\mathbf{d}\{H, F\} \wedge \mathbf{d}G) + \mathbf{i}_B(\mathbf{d}\{G, H\} \wedge \mathbf{d}F)$$
$$= \mathbf{i}_B \mathbf{d}(\mathbf{i}_B(\mathbf{d}F \wedge \mathbf{d}G)\mathbf{d}H + \mathbf{i}_B(\mathbf{d}H \wedge \mathbf{d}F)\mathbf{d}G + \mathbf{i}_B(\mathbf{d}G \wedge \mathbf{d}H)\mathbf{d}F)$$
$$= \mathbf{i}_B \mathbf{d} \mathbf{i}_B(\mathbf{d}F \wedge \mathbf{d}G \wedge \mathbf{d}H)$$
$$= \tfrac{1}{2} \mathbf{i}_{[B,B]}(\mathbf{d}F \wedge \mathbf{d}G \wedge \mathbf{d}H),$$

the last equality being a consequence of (10.6.11). We summarize what we have proved.

Theorem 10.6.2.   *The following identity holds:*

$$\{\{F, G\}, H\} + \{\{H, F\}, G\} + \{\{G, H\}, F\}$$
$$= \frac{1}{2} \mathbf{i}_{[B,B]}(\mathbf{d}F \wedge \mathbf{d}G \wedge \mathbf{d}H). \quad (10.6.16)$$

This result shows that Jacobi's identity for $\{\,,\,\}$ is equivalent to $[B, B] = 0$. Thus, *a Poisson structure is uniquely defined by a contravariant antisymmetric two-tensor whose Schouten bracket with itself vanishes.* The local formula (10.6.13) becomes

$$[B,B]^{IJK} = \sum_{L=1}^{n} \left(B^{LK}\frac{\partial B^{IJ}}{\partial z^L} + B^{LI}\frac{\partial B^{JK}}{\partial z^L} + B^{LJ}\frac{\partial B^{KI}}{\partial z^L} \right),$$

which coincides with our earlier expression (10.4.2).

The Lie–Schouten Identity.   There is another interesting identity that gives the Lie derivative of the Poisson tensor along a Hamiltonian vector field.

Theorem 10.6.3.   *The following identity holds:*

$$\pounds_{X_H} B = \mathbf{i}_{[B,B]}\mathbf{d}H. \tag{10.6.17}$$

Proof.   In coordinates,

$$(\pounds_X B)^{IJ} = X^K\frac{\partial B^{IJ}}{\partial z^K} - B^{IK}\frac{\partial X^J}{\partial z^K} - B^{KJ}\frac{\partial X^I}{\partial z^K},$$

so if $X^I = B^{IJ}(\partial H/\partial z^J)$, this becomes

$$(\pounds_{X_H}B)^{IJ} = B^{KL}\frac{\partial B^{IJ}}{\partial z^K}\frac{\partial H}{\partial z^L} - B^{IK}\frac{\partial}{\partial z^K}\left(B^{JL}\frac{\partial H}{\partial z^L}\right)$$

$$+ B^{JK}\frac{\partial}{\partial z^K}\left(B^{IL}\frac{\partial H}{\partial z^L}\right)$$

$$= \left(B^{KL}\frac{\partial B^{IJ}}{\partial z^K} - B^{IK}\frac{\partial B^{JL}}{\partial z^K} - B^{KJ}\frac{\partial B^{IL}}{\partial z^K}\right)\frac{\partial H}{\partial z^L}$$

$$= [B,B]^{LIJ}\frac{\partial H}{\partial z^L} = \left(\mathbf{i}_{[B,B]}\mathbf{d}H\right)^{IJ},$$

so (10.6.17) follows.   ∎

This identity shows how Jacobi's identity $[B,B] = 0$ *is directly used to show that the flow φ_t of a Hamiltonian vector field is Poisson.* The above derivation shows that *the flow of a time-dependent Hamiltonian vector field consists of Poisson maps*; indeed, even in this case,

$$\frac{d}{dt}\left(\varphi_t^* B\right) = \varphi_t^*\left(\pounds_{X_H}B\right) = \varphi_t^*\left(\mathbf{i}_{[B,B]}\mathbf{d}H\right) = 0$$

is valid.

Exercises

⋄ **10.6-1.**   Prove the following formulas by the method indicated in the text.

(a) If $A \in \Omega_q(P)$ and $X \in \mathfrak{X}(P)$, then $[X, A] = \pounds_X A$.

(b) If $A \in \Omega_q(P)$ and $X_1, \ldots, X_r \in \mathfrak{X}(P)$, then

$$[X_1 \wedge \cdots \wedge X_r, A] = \sum_{i=1}^{r} (-1)^{i+1} X_1 \wedge \cdots \wedge \check{X}_i \wedge \cdots \wedge X_r \wedge (\pounds_{X_i} A).$$

(c) If $X_1, \ldots, X_r, Y_1, \ldots, Y_s \in \mathfrak{X}(P)$, then

$$[X_1 \wedge \cdots \wedge X_r, Y_1 \wedge \cdots \wedge Y_s]$$
$$= (-1)^{r+1} \sum_{i=1}^{r} \sum_{j=1}^{s} (-1)^{i+j} [X_i, Y_i] \wedge X_1 \wedge \cdots \wedge \check{X}_i$$
$$\wedge \cdots \wedge X_r \wedge Y_1 \wedge \cdots \wedge \check{Y}_j \wedge \cdots \wedge Y_s.$$

(d) If $A \in \Omega_p(P)$, $B \in \Omega_q(P)$, and $\alpha \in \Omega^{p+q-1}(P)$, then

$$\mathbf{i}_{[A,B]}\alpha = (-1)^{q(p+1)} \mathbf{i}_A \mathbf{d} \mathbf{i}_B \alpha + (-1)^p \mathbf{i}_B \mathbf{d} \mathbf{i}_A \alpha - \mathbf{i}_B \mathbf{i}_A \mathbf{d}\alpha.$$

$\diamond$ **10.6-2.**   Let M be a finite-dimensional manifold. A *k-vector field* is a skew-symmetric contravariant tensor field $A(x) : T_x^* M \times \cdots \times T_x^* M \to \mathbb{R}$ (k copies of $T_x^* M$). Let $x_0 \in M$ be such that $A(x_0) = 0$.

(a) If $X \in \mathfrak{X}(M)$, show that $(\pounds_X A)(x_0)$ depends only on $X(x_0)$, thereby defining a map $\mathbf{d}_{x_0} A : T_{x_0} M \to T_{x_0} M \wedge \cdots \wedge T_{x_0} M$ (k times), called the *intrinsic derivative* of A at x_0.

(b) If $\alpha_1, \ldots, \alpha_k \in T_x^* M$, $v_1, \ldots, v_k \in T_x M$, show that

$$\langle \alpha_1 \wedge \cdots \wedge \alpha_k, v_1 \wedge \cdots \wedge v_k \rangle := \det [\langle \alpha_i, v_j \rangle]$$

defines a nondegenerate pairing between $T_x^* M \wedge \cdots \wedge T_x^* M$ and $T_x M \wedge \cdots \wedge T_x M$. Conclude that these two spaces are dual to each other, that the space $\Omega^k(M)$ of k-forms is dual to the space of k-contravariant skew-symmetric tensor fields $\Omega_k(M)$, and that the bases

$$\left\{ \, \mathbf{d}x^{i_1} \wedge \cdots \wedge \mathbf{d}x^{i_k} \mid i_1 < \cdots < i_k \, \right\}$$

and

$$\left\{ \frac{\partial}{\partial x^{i_1}} \wedge \cdots \wedge \frac{\partial}{\partial x^{i_k}} \,\middle|\, i_1 < \cdots < i_k \right\}$$

are dual to each other.

(c) Show that the dual map

$$(\mathbf{d}_{x_0} A)^* : T_{x_0}^* M \wedge \cdots \wedge T_{x_0}^* M \to T_{x_0}^* M$$

is given by

$$(\mathbf{d}_{x_0} A)^* (\alpha_1 \wedge \cdots \wedge \alpha_k) = \mathbf{d}(A(\tilde{\alpha}_1, \ldots, \tilde{\alpha}_k))(x_0),$$

where $\tilde{\alpha}_1, \ldots, \tilde{\alpha}_k \in \Omega^1(M)$ are arbitrary one-forms whose values at x_0 are $\alpha_1, \ldots, \alpha_k$.

⋄ **10.6-3** (Weinstein [1983b]).   Let $(P, \{\ ,\ \})$ be a finite-dimensional Poisson manifold with Poisson tensor $B \in \Omega_2(P)$. Let $z_0 \in P$ be such that $B(z_0) = 0$. For $\alpha, \beta \in T_{z_0}^* P$, define

$$[\alpha, \beta]_B = (\mathbf{d}_{z_0} B)^*(\alpha \wedge \beta) = \mathbf{d}(B(\tilde\alpha, \tilde\beta))(z_0)$$

where $\mathbf{d}_{z_0} B$ is the intrinsic derivative of B and $\tilde\alpha, \tilde\beta \in \Omega^1(P)$ are such that $\tilde\alpha(z_0) = \alpha$, $\tilde\beta(z_0) = \beta$. (See Exercise 10.6-2.) Show that $(\alpha, \beta) \mapsto [\alpha, \beta]_B$ defines a bilinear skew-symmetric map $T_{z_0}^* P \times T_{z_0}^* P \to T_{z_0}^* P$. Show that the Jacobi identity for the Poisson bracket implies that $[\ ,\ ]_B$ is a Lie bracket on $T_{z_0}^* P$. Since $(T_{z_0}^* P, [\ ,\ ]_B)$ is a Lie algebra, its dual $T_{z_0} P$ naturally carries the induced Lie–Poisson structure, called the *linearization* of the given Poisson bracket at z_0. Show that the linearization in local coordinates has the expression

$$\{F, G\}(v) = \frac{\partial B^{ij}(z_0)}{\partial z^k} \frac{\partial F}{\partial v^i} \frac{\partial G}{\partial v^j} v^k,$$

for $F, G : T_{z_0} P \to \mathbb{R}$ and $v \in T_{z_0} P$.

⋄ **10.6-4** (Magri–Weinstein).   On the finite-dimensional manifold P, assume that one has a symplectic form Ω and a Poisson structure B. Define $K = B^\sharp \circ \Omega^\flat : TP \to TP$. Show that $(\Omega^\flat)^{-1} + B^\sharp : T^* P \to TP$ defines a new Poisson structure on P if and only if $\Omega^\flat \circ K^n$ induces a closed two-form (called a **presymplectic form**) on P for all $n \in \mathbb{N}$.

10.7   Generalities on Lie–Poisson Structures

The Lie–Poisson Equations.   We begin by working out Hamilton's equations for the Lie–Poisson bracket.

Proposition 10.7.1.   *Let G be a Lie group. The equations of motion for the Hamiltonian H with respect to the $\pm$ Lie–Poisson brackets on $\mathfrak{g}^*$ are*

$$\frac{d\mu}{dt} = \mp \operatorname{ad}^*_{\delta H/\delta\mu} \mu. \qquad (10.7.1)$$

Proof.   Let $F \in \mathcal{F}(\mathfrak{g}^*)$ be an arbitrary function. By the chain rule,

$$\frac{dF}{dt} = \mathbf{D}F(\mu) \cdot \dot\mu = \left\langle \dot\mu, \frac{\delta F}{\delta\mu} \right\rangle, \qquad (10.7.2)$$

while

$$\{F, H\}_\pm(\mu) = \pm \left\langle \mu, \left[\frac{\delta F}{\delta\mu}, \frac{\delta H}{\delta\mu}\right] \right\rangle = \pm \left\langle \mu, -\operatorname{ad}_{\delta H/\delta\mu} \frac{\delta F}{\delta\mu} \right\rangle$$

$$= \mp \left\langle \operatorname{ad}^*_{\delta H/\delta\mu} \mu, \frac{\delta F}{\delta\mu} \right\rangle. \qquad (10.7.3)$$

Nondegeneracy of the pairing and arbitrariness of F imply the result.   ∎

Caution.   In infinite dimensions, $\mathfrak{g}^*$ does not necessarily mean the literal functional-analytic dual of $\mathfrak{g}$, but rather a space in (nondegenerate) duality with $\mathfrak{g}$. In this case, care must be taken with the definition of $\delta F/\delta \mu$.   ◆

Formula (10.7.1) says that on $\mathfrak{g}^*_\pm$, *the Hamiltonian vector field of H* : $\mathfrak{g}^* \to \mathbb{R}$ *is given by*

$$X_H(\mu) = \mp \mathrm{ad}^*_{\delta H/\delta\mu}\, \mu. \tag{10.7.4}$$

For example, for $G = \mathrm{SO}(3)$, formula (10.1.3) for the Lie–Poisson bracket gives

$$X_H(\mathbf{\Pi}) = \mathbf{\Pi} \times \nabla H. \tag{10.7.5}$$

Historical Note.   Lagrange devoted a good deal of attention in Volume 2 of *Mécanique Analytique* to the study of rotational motion of mechanical systems. In fact, in equation A on page 212 he gives the reduced Lie–Poisson equations for $\mathrm{SO}(3)$ for a rather general Lagrangian. This equation is essentially the same as (10.7.5). His derivation was just how we would do it today—by reduction from material to spatial representation. Formula (10.7.5) actually hides a subtle point in that it identifies $\mathfrak{g}$ and $\mathfrak{g}^*$. Indeed, the way Lagrange wrote the equations, they are much more like their counterpart on $\mathfrak{g}$, which are called the *Euler–Poincaré equations*. We will come to these in Chapter 13, where additional historical information may be found.

Coordinate Formulas.   In finite dimensions, if ξ_a, $a = 1, 2, \ldots, l$, is a basis for $\mathfrak{g}$, the structure constants C^d_{ab} are defined by

$$[\xi_a, \xi_b] = C^d_{ab}\xi_d \tag{10.7.6}$$

(a sum on "d" is understood). Thus, the Lie–Poisson bracket becomes

$$\{F, K\}_\pm(\mu) = \pm\mu_d \frac{\partial F}{\partial\mu_a} \frac{\partial K}{\partial\mu_b} C^d_{ab}, \tag{10.7.7}$$

where $\mu = \mu_a\xi^a$, $\{\xi^a\}$ is the basis of $\mathfrak{g}^*$ dual to $\{\xi_a\}$, and summation on repeated indices is understood. Taking F and K to be components of μ, (10.7.7) becomes

$$\{\mu_a, \mu_b\}_\pm = \pm C^d_{ab}\mu_d. \tag{10.7.8}$$

The equations of motion for a Hamiltonian H likewise become

$$\dot{\mu}_a = \mp\mu_d C^d_{ab} \frac{\partial H}{\partial\mu_b}. \tag{10.7.9}$$

Poisson Maps.   In the Lie–Poisson reduction theorem in Chapter 13 we
will show that the maps from T^*G to $\mathfrak{g}^*_-$ (resp., $\mathfrak{g}^*_+$) given by $\alpha_g \mapsto T^*_e L_g \cdot \alpha_g$
(resp., $\alpha_g \mapsto T^*_e R_g \cdot \alpha_g$) are Poisson maps. We will show in the next chapter
that this is a general property of momentum maps. Here is another class
of Poisson maps that will also turn out to be momentum maps.

Proposition 10.7.2.   *Let G and H be Lie groups and let $\mathfrak{g}$ and $\mathfrak{h}$ be their
Lie algebras. Let $\alpha : \mathfrak{g} \to \mathfrak{h}$ be a linear map. The map α is a homomorphism
of Lie algebras if and only if its dual $\alpha^* : \mathfrak{h}^*_\pm \to \mathfrak{g}^*_\pm$ is a (linear) Poisson
map.*

Proof.   Let $F, K \in \mathcal{F}(\mathfrak{g}^*)$. To compute $\delta(F \circ \alpha^*)/\delta\mu$, we let $\nu = \alpha^*(\mu)$
and use the definition of the functional derivative and the chain rule to get

$$\left\langle \frac{\delta}{\delta\mu}(F \circ \alpha^*), \delta\mu \right\rangle = \mathbf{D}(F \circ \alpha^*)(\mu) \cdot \delta\mu = \mathbf{D}F(\alpha^*(\mu)) \cdot \alpha^*(\delta\mu)$$

$$= \left\langle \alpha^*(\delta\mu), \frac{\delta F}{\delta\nu} \right\rangle = \left\langle \delta\mu, \alpha \cdot \frac{\delta F}{\delta\nu} \right\rangle. \qquad (10.7.10)$$

Thus,

$$\frac{\delta}{\delta\mu}(F \circ \alpha^*) = \alpha \cdot \frac{\delta F}{\delta\nu}. \qquad (10.7.11)$$

Hence,

$$\{F \circ \alpha^*, K \circ \alpha^*\}_+ (\mu) = \left\langle \mu, \left[\frac{\delta}{\delta\mu}(F \circ \alpha^*), \frac{\delta}{\delta\mu}(K \circ \alpha^*) \right] \right\rangle$$

$$= \left\langle \mu, \left[\alpha \cdot \frac{\delta F}{\delta\nu}, \alpha \cdot \frac{\delta K}{\delta\nu} \right] \right\rangle. \qquad (10.7.12)$$

The expression (10.7.12) equals

$$\left\langle \mu, \alpha \cdot \left[\frac{\delta F}{\delta\nu}, \frac{\delta G}{\delta\nu} \right] \right\rangle \qquad (10.7.13)$$

for all F and K if and only if α is a Lie algebra homomorphism.   ■

This theorem applies to the case $\alpha = T_e\sigma$ for $\sigma : G \to H$ a Lie group
homomorphism, as one may see by studying the reduction diagram in Figure 10.7.1 (and being cautious that σ need not be a diffeomorphism).

Examples

(a) Plasma to Fluid Poisson Map for the Momentum Variables.
Let G be the group of diffeomorphisms of a manifold Q and let H be the

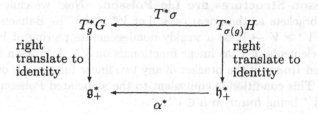

FIGURE 10.7.1. Lie group homomorphisms induce Poisson maps.

group of canonical transformations of $P = T^*Q$. We assume that the topology of Q is such that all locally Hamiltonian vector fields on T^*Q are globally Hamiltonian.[5] Thus, the Lie algebra $\mathfrak{h}$ consists of functions on T^*Q modulo constants. Its dual is identified with itself via the L^2-inner product relative to the Liouville measure $dq\,dp$ on T^*Q. Let $\sigma : G \to H$ be the map $\eta \mapsto T^*\eta^{-1}$, which is a group homomorphism, and let $\alpha = T_e\sigma : \mathfrak{g} \to \mathfrak{h}$. We claim that $\alpha^* : \mathcal{F}(T^*Q)/\mathbb{R} \to \mathfrak{g}^*$ is given by

$$\alpha^*(F) = \int pf(q,p)\,dp, \qquad (10.7.14)$$

where we regard $\mathfrak{g}^*$ as the space of one-form densities on Q, and the integral denotes fiber integration for each fixed $q \in Q$. Indeed, α is the map taking vector fields X on Q to their lifts $X_{\mathcal{P}(X)}$ on T^*Q. Thus, as a map of $\mathfrak{X}(Q)$ to $\mathcal{F}(T^*Q)/\mathbb{R}$, α is given by $X \mapsto \mathcal{P}(X)$. Its dual is given by

$$\langle \alpha^*(f), X \rangle = \langle f, \alpha(X) \rangle = \int_P f\mathcal{P}(X)\,dq\,dp$$

$$= \int_P f(q,p)p \cdot X(q)\,dq\,dp, \qquad (10.7.15)$$

so $\alpha^*(F)$ is given by (10.7.14), as claimed.   ◆

(b) **Plasma to Fluid Map for the Density Variable.** Let $G = \mathcal{F}(Q)$ regarded as an abelian group and let the map $\sigma : G \to \mathrm{Diff}_{\mathrm{can}}(T^*Q)$ be given by $\sigma(\varphi) =$ fiber translation by $d\varphi$. A computation similar to that above gives the Poisson map

$$\alpha^*(f)(q) = \int f(q,p)\,dp \qquad (10.7.16)$$

from $\mathcal{F}(T^*Q)$ to $\mathrm{Den}(Q) = \mathcal{F}(Q)^*$. The integral in (10.7.16) denotes the fiber integration of $f(q,p)$ for fixed $q \in Q$.   ◆

---

[5]For example, this holds if the first cohomology group $H^1(Q)$ is trivial.

Linear Poisson Structures are Lie–Poisson. Next we characterize Lie–Poisson brackets as the linear ones. Let V^* and V be Banach spaces and let $\langle , \rangle : V^* \times V \to \mathbb{R}$ be a weakly nondegenerate pairing of V^* with V. Think of elements of V as linear functionals on V^*. A Poisson bracket on V^* is called *linear* if the bracket of any two linear functionals on V^* is again linear. This condition is equivalent to the associated Poisson tensor $B(\mu) : V \to V^*$ being *linear* in $\mu \in V^*$.

Proposition 10.7.3. *Let $\langle , \rangle : V^* \times V \to \mathbb{R}$ be a (weakly) nondegenerate pairing of the Banach spaces V^* and V, and let V^* have a linear Poisson bracket. Assume that the bracket of any two linear functionals on V^* is in the range of $\langle \mu, \cdot \rangle$ for all $\mu \in V^*$ (this condition is automatically satisfied if V is finite-dimensional). Then V is a Lie algebra, and the Poisson bracket on V^* is the corresponding Lie–Poisson bracket.*

Proof. If $x \in V$, we denote by x' the functional $x'(\mu) = \langle \mu, x \rangle$ on V^*. By hypothesis, the Poisson bracket $\{x', y'\}$ is a linear functional on V^*. By assumption this bracket is represented by an element that we denote by $[x, y]'$ in V, that is, we can write $\{x', y'\} = [x, y]'$. (The element $[x, y]$ is unique, since $\langle , \rangle$ is weakly nondegenerate.) It is straightforward to check that the operation $[,]$ on V so defined is a Lie algebra bracket. Thus, V is a Lie algebra, and one then checks that the given Poisson bracket is the Lie–Poisson bracket for this algebra. ∎

Exercises

⋄ **10.7-1.** Let $\sigma : SO(3) \to GL(3)$ be the inclusion map. Identify $\mathfrak{so}(3)^* = \mathbb{R}^3$ with the rigid-body bracket and identify $\mathfrak{gl}(3)^*$ with $\mathfrak{gl}(3)$ using $\langle A, B \rangle = \text{trace}(AB^T)$. Compute the induced map $\alpha^* : \mathfrak{gl}(3) \to \mathbb{R}^3$ and verify directly that it is Poisson.

11
Momentum Maps

In this chapter we show how to obtain conserved quantities for Lagrangian and Hamiltonian systems with symmetries. This is done using the concept of a momentum mapping, which is a geometric generalization of the classical linear and angular momentum. This concept is more than a mathematical reformulation of a concept that simply describes the well-known Noether theorem. Rather, it is a rich concept that is ubiquitous in the modern developments of geometric mechanics. It has led to surprising insights into many areas of mechanics and geometry.

11.1 Canonical Actions and Their Infinitesimal Generators

Canonical Actions. Let P be a Poisson manifold, let G be a Lie group, and let $\Phi : G \times P \to P$ be a smooth left action of G on P by canonical transformations. If we denote the action by $g \cdot z = \Phi_g(z)$, so that $\Phi_g : P \to P$, then the action being **canonical** means that

$$\Phi_g^* \{F_1, F_2\} = \{\Phi_g^* F_1, \Phi_g^* F_2\} \qquad (11.1.1)$$

for any $F_1, F_2 \in \mathcal{F}(P)$ and any $g \in G$. If P is a symplectic manifold with symplectic form Ω, then the action is canonical if and only if it is symplectic, that is, $\Phi_g^* \Omega = \Omega$ for all $g \in G$.

Infinitesimal Generators. Recall from Chapter 9 on Lie groups that the **infinitesimal generator** of the action corresponding to a Lie algebra element $\xi \in \mathfrak{g}$ is the vector field ξ_P on P obtained by differentiating the action with respect to g at the identity in the direction ξ. By the chain rule,

$$\xi_P(z) = \frac{d}{dt}\left[\exp(t\xi) \cdot z\right]\bigg|_{t=0}. \tag{11.1.2}$$

We will need two general identities, both of which were proved in Chapter 9. First, the flow of the vector field ξ_P is

$$\varphi_t = \Phi_{\exp t\xi}. \tag{11.1.3}$$

Second, we have

$$\Phi_{g^{-1}}^* \xi_P = (\mathrm{Ad}_g\, \xi)_P \tag{11.1.4}$$

and its differentiated companion

$$[\xi_P, \eta_P] = -[\xi, \eta]_P. \tag{11.1.5}$$

The Rotation Group. To illustrate these identities, consider the action of SO(3) on $\mathbb{R}^3$. As was explained in Chapter 9, the Lie algebra $\mathfrak{so}(3)$ of SO(3) is identified with $\mathbb{R}^3$, and the Lie bracket is identified with the cross product. For the action of SO(3) on $\mathbb{R}^3$ given by rotations, the infinitesimal generator of $\omega \in \mathbb{R}^3$ is

$$\omega_{\mathbb{R}^3}(\mathbf{x}) = \omega \times \mathbf{x} = \hat{\omega}(\mathbf{x}). \tag{11.1.6}$$

Then (11.1.4) becomes the identity

$$(\mathbf{A}\omega \times \mathbf{x}) = \mathbf{A}(\omega \times \mathbf{A}^{-1}\mathbf{x}) \tag{11.1.7}$$

for $A \in$ SO(3), while (11.1.5) becomes the Jacobi identity for the vector product.

Poisson Automorphisms. Returning to the general case, differentiate (11.1.1) with respect to g in the direction ξ, to give

$$\xi_P[\{F_1, F_2\}] = \{\xi_P[F_1], F_2\} + \{F_1, \xi_P[F_2]\}. \tag{11.1.8}$$

In the symplectic case, differentiating $\Phi_g^* \Omega = \Omega$ gives

$$\pounds_{\xi_P}\Omega = 0, \tag{11.1.9}$$

that is, ξ_P is *locally Hamiltonian*. For Poisson manifolds, a vector field satisfying (11.1.8) is called an *infinitesimal Poisson automorphism*. Such a vector field need not be locally Hamiltonian (that is, locally of the form X_H). For example, consider the Poisson structure

$$\{F, H\} = x\left(\frac{\partial F}{\partial x}\frac{\partial H}{\partial y} - \frac{\partial H}{\partial x}\frac{\partial F}{\partial y}\right) \tag{11.1.10}$$

on $\mathbb{R}^2$ and $X = \partial/\partial y$ in a neighborhood of a point of the y-axis.

We are interested in the case in which ξ_P is globally Hamiltonian, a condition stronger than (11.1.8). Thus, *assume that there is a global Hamiltonian* $J(\xi) \in \mathcal{F}(P)$ for ξ_P, that is,

$$X_{J(\xi)} = \xi_P. \qquad (11.1.11)$$

Does this equation determine $J(\xi)$? Obviously not, for if $J_1(\xi)$ and $J_2(\xi)$ both satisfy (11.1.11), then

$$X_{J_1(\xi)-J_2(\xi)} = 0; \quad \text{i.e.,} \quad J_1(\xi) - J_2(\xi) \in \mathcal{C}(P),$$

the space of Casimir functions on P. If P is symplectic and connected, then $J(\xi)$ is determined by (11.1.11) up to a constant.

Exercises

◇ **11.1-1.** Verify (11.1.4), namely, $\Phi_{g^{-1}}^* \xi_P = (\mathrm{Ad}_g\, \xi)_P$ and its differentiated companion (11.1.5) $[\xi_P, \eta_P] = -[\xi, \eta]_P$, for the action of $\mathrm{GL}(n)$ on itself by conjugation.

◇ **11.1-2.** Let S^1 act on S^2 by rotations about the z-axis. Compute $J(\xi)$.

11.2 Momentum Maps

Since the right-hand side of (11.1.11) is linear in ξ, by using a basis in the finite-dimensional case *we can modify any given $J(\xi)$ so it too is linear in ξ, and still retain condition* (11.1.11). Indeed, if $e_1, \ldots, e_r$ is a basis of $\mathfrak{g}$, let the new momentum map $\tilde{J}$ be defined by $\tilde{J}(\xi) = \xi^a J(e_a)$.

In the definition of the momentum map, we can replace the left *Lie group* action by a canonical left *Lie algebra* action $\xi \mapsto \xi_P$. In the Poisson manifold context, canonical means that (11.1.8) is satisfied and, in the symplectic manifold context, that (11.1.9) is satisfied. (Recall that for a left Lie algebra action, the map $\xi \in \mathfrak{g} \mapsto \xi_P \in \mathfrak{X}(P)$ is a Lie algebra antihomomorphism.) Thus, we make the following definition:

Definition 11.2.1. *Let a Lie algebra $\mathfrak{g}$ act canonically (on the left) on the Poisson manifold P. Suppose there is a linear map $J : \mathfrak{g} \to \mathcal{F}(P)$ such that*

$$X_{J(\xi)} = \xi_P \qquad (11.2.1)$$

for all $\xi \in \mathfrak{g}$. The map $\mathbf{J} : P \to \mathfrak{g}^*$ defined by

$$\langle \mathbf{J}(z), \xi \rangle = J(\xi)(z) \qquad (11.2.2)$$

*for all $\xi \in \mathfrak{g}$ and $z \in P$ is called a **momentum mapping** of the action.*

Angular Momentum. Consider the angular momentum function for a particle in Euclidean three-space, $\mathbf{J}(z) = \mathbf{q} \times \mathbf{p}$, where $z = (\mathbf{q}, \mathbf{p})$. Let $\xi \in \mathbb{R}^3$ and consider the component of $\mathbf{J}$ around the axis ξ, namely, $\langle \mathbf{J}(z), \xi \rangle = \xi \cdot (\mathbf{q} \times \mathbf{p})$. One checks that Hamilton's equations determined by this function of $\mathbf{q}$ and $\mathbf{p}$ describe infinitesimal rotations about the axis ξ. The defining condition (11.2.1) is a generalization of this elementary statement about angular momentum.

Momentum Maps and Poisson Brackets. Recalling that $X_H[F] = \{F, H\}$, we see that (11.2.1) can be phrased in terms of the Poisson bracket as follows: *For any function F on P and any $\xi \in \mathfrak{g}$,*

$$\{F, J(\xi)\} = \xi_P[F]. \tag{11.2.3}$$

Equation (11.2.2) defines an isomorphism between the space of smooth maps $\mathbf{J}$ from P to $\mathfrak{g}^*$ and the space of linear maps J from $\mathfrak{g}$ to $\mathcal{F}(P)$. We think of the collection of functions $J(\xi)$ as ξ varies in $\mathfrak{g}$ as the components of $\mathbf{J}$. Denote by

$$\mathcal{H}(P) = \{ X_F \in \mathfrak{X}(P) \mid F \in \mathcal{F}(P) \} \tag{11.2.4}$$

the Lie algebra of Hamiltonian vector fields on P and by

$$\mathcal{P}(P) = \{ X \in \mathfrak{X}(P) \mid X[\{F_1, F_2\}] = \{X[F_1], F_2\} + \{F_1, X[F_2]\} \} \tag{11.2.5}$$

the Lie algebra of infinitesimal Poisson automorphisms of P. By (11.1.8), for any $\xi \in \mathfrak{g}$ we have $\xi_P \in \mathcal{P}(P)$. Therefore, giving a momentum map $\mathbf{J}$ is equivalent to specifying a linear map $J : \mathfrak{g} \to \mathcal{F}(P)$ making the diagram in Figure 11.2.1 commute.

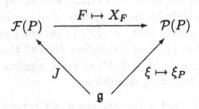

FIGURE 11.2.1. The commutative diagram defining a momentum map.

Since both $\xi \mapsto \xi_P$ and $F \mapsto X_F$ are Lie algebra antihomomorphisms, for $\xi, \eta \in \mathfrak{g}$ we get

$$X_{J([\xi,\eta])} = [\xi, \eta]_P = -[\xi_P, \eta_P] = -\big[X_{J(\xi)}, X_{J(\eta)}\big] = X_{\{J(\xi), J(\eta)\}}, \tag{11.2.6}$$

and so we have the basic identity

$$X_{J([\xi,\eta])} = X_{\{J(\xi),J(\eta)\}}. \tag{11.2.7}$$

The preceding development *defines* momentum maps but does not tell us how to *compute* them in examples. We shall concentrate on that aspect in Chapter 12.

Building on the above commutative diagram, §11.3 discusses an alternative approach to the definition of the momentum map, but it will not be used subsequently in the main text. Rather, we shall give the formulas that will be most important for later applications; the interested reader is referred to Souriau [1970], Weinstein [1977], Abraham and Marsden [1978], Guillemin and Sternberg [1984], and Libermann and Marle [1987] for more information.

Some History of the Momentum Map   The momentum map can be found in the second volume of Lie [1890], where it appears in the context of homogeneous canonical transformations, in which case its expression is given as the contraction of the canonical one-form with the infinitesimal generator of the action. On page 300 it is shown that the momentum map is canonical and on page 329 that it is equivariant with respect to some linear action whose generators are identified on page 331. On page 338 it is proved that if the momentum map has constant rank (a hypothesis that seems to be implicit in all of Lie's work in this area), its image is Ad*-invariant, and on page 343, actions are classified by Ad*-invariant submanifolds.

We now present the modern history of the momentum map based on information and references provided to us by B. Kostant and J.-M. Souriau. We would like to thank them for all their help.

In Kostant's 1965 Phillips lectures at Haverford (the notes of which were written by Dale Husemoller), and in the 1965 U.S.–Japan Seminar (see Kostant [1966]), Kostant introduced the momentum map to generalize a theorem of Wang and thereby classified all homogeneous symplectic manifolds; this is called today "Kostant's coadjoint orbit covering theorem." These lectures also contained the key points of geometric quantization. Souriau introduced the momentum map in his 1965 Marseille lecture notes and put it in print in Souriau [1966]. The momentum map finally got its formal definition and its name, based on its physical interpretation, in Souriau [1967]. Souriau also studied its properties of equivariance, and formulated the coadjoint orbit theorem. The momentum map appeared as a key tool in Kostant's quantization lectures (see, e.g., Theorem 5.4.1 in Kostant [1970]), and Souriau [1970] discussed it at length in his book. Kostant and Souriau realized its importance for linear representations, a fact apparently not foreseen by Lie (Weinstein [1983a]). Independently, work on the momentum map and the coadjoint orbit covering theorem was done by A. Kirillov. This is described in Kirillov [1976b]. This book was first published in 1972 and states that his work on the classification theorem was done about five years earlier (page 301). The modern formulation of the momentum map was developed in the context of classical mechanics in the work of Smale [1970], who applied it extensively in his topological program for the planar n-body problem. Marsden and Weinstein [1974] and other authors quickly seized on the treasures of these ideas.

Exercises

⋄ **11.2-1.** Verify that Hamilton's equations determined by the function $\langle \mathbf{J}(z), \xi \rangle = \xi \cdot (\mathbf{q} \times \mathbf{p})$ give the infinitesimal generator of rotations about the ξ-axis.

⋄ **11.2-2.** Verify that $J([\xi, \eta]) = \{J(\xi), J(\eta)\}$ for angular momentum.

⋄ **11.2-3.**

(a) Let P be a symplectic manifold and G a Lie group acting canonically on P, with an associated momentum map $\mathbf{J} : P \longrightarrow \mathfrak{g}^*$. Let S be a symplectic submanifold of P that is invariant under the G-action. Show that the G-action on S admits a momentum map given by $\mathbf{J}|_S$.

(b) Generalize (a) to the case in which P is a Poisson manifold and S is an immersed G-invariant Poisson submanifold.

11.3   An Algebraic Definition of the Momentum Map

This section gives an optional approach to momentum maps and may be skipped on a first reading.[1] The point of departure is the commutative diagram in Figure 11.2.1 plus the observation that the following sequence is *exact* (that is, the range of each map equals the kernel of the following one):

$$0 \xrightarrow{} \mathcal{C}(P) \xrightarrow{\ i\ } \mathcal{F}(P) \xrightarrow{\ \mathcal{H}\ } \mathcal{P}(P) \xrightarrow{\ \pi\ } \mathcal{P}(P)/(\mathcal{H}(P)) \xrightarrow{} 0.$$

Here, i is the inclusion, π the projection, $\mathcal{H}(F) = X_F$, and $\mathcal{H}(P)$ denotes the Lie algebra of globally Hamiltonian vector fields on P. Let us investigate conditions under which a left Lie algebra action, that is, an antihomomorphism $\rho : \mathfrak{g} \to \mathcal{P}(P)$, lifts through $\mathcal{H}$ to a linear map $J : \mathfrak{g} \to \mathcal{F}(P)$. As we have already seen, this is equivalent to $\mathbf{J}$ being a momentum map. (The requirement that J be a Lie algebra homomorphism will be discussed later.)

---

[1]This section assumes that the reader knows some topology and a little more Lie theory than we have actually covered; this material is *not* needed later on.

If $\mathcal{H} \circ J = \rho$, then $\pi \circ \rho = \pi \circ \mathcal{H} \circ J = 0$. Conversely, if $\pi \circ \rho = 0$, then $\rho(\mathfrak{g}) \subset \mathcal{H}(P)$, so there is a linear map $J : \mathfrak{g} \to \mathcal{F}(P)$ such that $\mathcal{H} \circ J = \rho$. Thus, the obstruction to the existence of J is $\pi \circ \rho = 0$. If P is symplectic, then $\mathcal{P}(P)$ coincides with the Lie algebra of locally Hamiltonian vector fields and thus $\mathcal{P}(P)/\mathcal{H}(P)$ is isomorphic to the first cohomology space $H^1(P)$ regarded as an abelian group. *Thus, in the symplectic case, $\pi \circ \rho = 0$ if and only if the induced mapping $\rho' : \mathfrak{g}/[\mathfrak{g}, \mathfrak{g}] \to H^1(P)$ vanishes.* Here is a list of cases that guarantee that $\pi \circ \rho = 0$:

1. *P is symplectic and $\mathfrak{g}/[\mathfrak{g}, \mathfrak{g}] = 0$.* By the first Whitehead lemma, this is the case whenever $\mathfrak{g}$ is semisimple (see Jacobson [1962] and Guillemin and Sternberg [1984]).

2. $\mathcal{P}(P)/\mathcal{H}(P) = 0$. If P is symplectic, this is equivalent to the vanishing of the first cohomology group $H^1(P)$.

3. If P is exact symplectic, that is, $\Omega = -\mathbf{d}\Theta$, and Θ is invariant under the $\mathfrak{g}$ action, that is,

$$\pounds_{\xi_P} \Theta = 0. \tag{11.3.1}$$

Case 3 occurs, for example, when $P = T^*Q$ and the action is a lift. In Case 3, there is an explicit formula for the momentum map. Since

$$0 = \pounds_{\xi_P} \Theta = \mathbf{d}\mathbf{i}_{\xi_P}\Theta + \mathbf{i}_{\xi_P}\mathbf{d}\Theta, \tag{11.3.2}$$

it follows that

$$\mathbf{d}(\mathbf{i}_{\xi_P}\Theta) = \mathbf{i}_{\xi_P}\Omega, \tag{11.3.3}$$

that is, the interior product of ξ_P with Θ satisfies (11.2.1), and hence the momentum map $\mathbf{J} : P \to \mathfrak{g}^*$ is given by

$$\langle \mathbf{J}(z), \xi \rangle = (\mathbf{i}_{\xi_P}\Theta)(z). \tag{11.3.4}$$

In coordinates, write $\Theta = p_i \, dq^i$ and define $A^j{}_a$ and B_{aj} by

$$\xi_P = \xi^a A^j{}_a \frac{\partial}{\partial q^j} + \xi^a B_{aj} \frac{\partial}{\partial p_j}. \tag{11.3.5}$$

Then (11.3.4) reads

$$J_a(q, p) = p_i A^i{}_a(q, p). \tag{11.3.6}$$

The following example shows that ρ' does not always vanish. Consider the phase space $P = S^1 \times S^1$, with the symplectic form $\Omega = d\theta_1 \wedge d\theta_2$, the Lie algebra $\mathfrak{g} = \mathbb{R}^2$, and the action

$$\rho(x_1, x_2) = x_1 \frac{\partial}{\partial \theta_1} + x_2 \frac{\partial}{\partial \theta_2}. \tag{11.3.7}$$

In this case $[\mathfrak{g}, \mathfrak{g}] = 0$ and $\rho' : \mathbb{R}^2 \to H^1(S^1 \times S^1)$ is an isomorphism, as can be easily checked.

11.4   Conservation of Momentum Maps

One reason that momentum maps are important in mechanics is that they are conserved quantities.

Theorem 11.4.1 (Hamiltonian Version of Noether's Theorem).   *If the Lie algebra $\mathfrak{g}$ acts canonically on the Poisson manifold P and admits a momentum mapping $\mathbf{J} : P \to \mathfrak{g}^*$, and if $H \in \mathcal{F}(P)$ is $\mathfrak{g}$-invariant, that is, $\xi_P[H] = 0$ for all $\xi \in \mathfrak{g}$, then $\mathbf{J}$ is a constant of the motion for H, that is,*

$$\mathbf{J} \circ \varphi_t = \mathbf{J},$$

where φ_t is the flow of X_H. If the Lie algebra action comes from a canonical left Lie group action Φ, then the invariance hypothesis on H is implied by the invariance condition $H \circ \Phi_g = H$ for all $g \in G$.

Proof.   The condition $\xi_P[H] = 0$ implies that the Poisson bracket of $J(\xi)$, the Hamiltonian function for ξ_P, and H vanishes: $\{J(\xi), H\} = 0$. This implies that for each Lie algebra element ξ, $J(\xi)$ is a conserved quantity along the flow of X_H. This means that the values of the corresponding $\mathfrak{g}^*$-valued momentum map $\mathbf{J}$ are conserved. The last assertion of the theorem follows by differentiating the condition $H \circ \Phi_g = H$ with respect to g at the identity e in the direction ξ to obtain $\xi_P[H] = 0$.   ∎

We dedicate the rest of this section to a list of concrete examples of momentum maps.

Examples

(a) The Hamiltonian.   On a Poisson manifold P, consider the $\mathbb{R}$-action given by the flow of a complete Hamiltonian vector field X_H. A corresponding momentum map $\mathbf{J} : P \to \mathbb{R}$ (where we identify $\mathbb{R}^*$ with $\mathbb{R}$ via the usual dot product) equals H.   ◆

(b) Linear Momentum.   In §6.4 we discussed the N-particle system and constructed the cotangent lift of the $\mathbb{R}^3$-action on $\mathbb{R}^{3N}$ (translation on every factor) to be the action on $T^*\mathbb{R}^{3N} \cong \mathbb{R}^{6N}$ given by

$$\mathbf{x} \cdot (\mathbf{q}_i, \mathbf{p}^j) = (\mathbf{q}_j + \mathbf{x}, \mathbf{p}^j), \quad j = 1, \ldots, N. \tag{11.4.1}$$

We show that this action has a momentum map and compute it from the definition. In the next chapter, we shall recompute it more easily utilizing further developments of the theory. Let $\xi \in \mathfrak{g} = \mathbb{R}^3$; the infinitesimal generator ξ_P at a point $(\mathbf{q}_j, \mathbf{p}^j) \in \mathbb{R}^{6N} = P$ is given by differentiating (11.4.1) with respect to $\mathbf{x}$ in the direction ξ:

$$\xi_P(\mathbf{q}_j, \mathbf{p}^j) = (\xi, \xi, \ldots, \xi, 0, 0, \ldots, 0). \tag{11.4.2}$$

On the other hand, by definition of the canonical symplectic structure Ω on P, any candidate $J(\xi)$ has a Hamiltonian vector field given by

$$X_{J(\xi)}(\mathbf{q}_j, \mathbf{p}^j) = \left(\frac{\partial J(\xi)}{\partial \mathbf{p}^j}, -\frac{\partial J(\xi)}{\partial \mathbf{q}_j} \right). \tag{11.4.3}$$

Then, $X_{J(\xi)} = \xi_P$ implies that

$$\frac{\partial J(\xi)}{\partial \mathbf{p}^j} = \xi \quad \text{and} \quad \frac{\partial J(\xi)}{\partial \mathbf{q}_j} = 0, \quad 1 \le j \le N. \tag{11.4.4}$$

Solving these equations and choosing constants such that J is linear, we get

$$J(\xi)(\mathbf{q}_j, \mathbf{p}^j) = \left(\sum_{j=1}^{N} \mathbf{p}^j \right) \cdot \xi, \quad \text{i.e.,} \quad \mathbf{J}(\mathbf{q}_j, \mathbf{p}^j) = \sum_{j=1}^{N} \mathbf{p}^j. \tag{11.4.5}$$

This expression is called the **total linear momentum** of the N-particle system. In this example, Noether's theorem can be deduced directly as follows. Denote by $J_\alpha, q_j^\alpha, p_\alpha^j$, the αth components of $\mathbf{J}$, $\mathbf{q}_j$, and $\mathbf{p}^j$, $\alpha = 1, 2, 3$. Given a Hamiltonian H, determining the evolution of the N-particle system by Hamilton's equations, we get

$$\frac{dJ_\alpha}{dt} = \sum_{j=1}^{N} \frac{dp_\alpha^j}{dt} = -\sum_{j=1}^{N} \frac{\partial H}{\partial q_\alpha^j} = -\left[\sum_{j=1}^{N} \frac{\partial}{\partial q_\alpha^j} \right] H. \tag{11.4.6}$$

The bracket on the right is an operator that evaluates the variation of the scalar function H under a spatial translation, that is, under the action of the translation group $\mathbb{R}^3$ on each of the N coordinate directions. Obviously, J_α is conserved if H is translation-invariant, which is exactly the statement of Noether's theorem. ◆

(c) Angular Momentum. Let SO(3) act on the configuration space $Q = \mathbb{R}^3$ by $\Phi(\mathbf{A}, \mathbf{q}) = \mathbf{Aq}$. We show that the lifted action to $P = T^*\mathbb{R}^3$ has a momentum map and compute it. First note that if $(\mathbf{q}, \mathbf{v}) \in T_\mathbf{q}\mathbb{R}^3$, then $T_\mathbf{q}\Phi_\mathbf{A}(\mathbf{q}, \mathbf{v}) = (\mathbf{Aq}, \mathbf{Av})$. Let $\mathbf{A} \cdot (\mathbf{q}, \mathbf{p}) = T_{\mathbf{Aq}}^* \Phi_{\mathbf{A}^{-1}}(\mathbf{q}, \mathbf{p})$ denote the lift of the SO(3) action to P, and identify covectors with vectors using the Euclidean inner product. If $(\mathbf{q}, \mathbf{p}) \in T_\mathbf{q}^*\mathbb{R}^3$, then $(\mathbf{Aq}, \mathbf{v}) \in T_{\mathbf{Aq}}\mathbb{R}^3$, so

$$\langle \mathbf{A} \cdot (\mathbf{q}, \mathbf{p}), (\mathbf{Aq}, \mathbf{v}) \rangle = \langle (\mathbf{q}, \mathbf{p}), \mathbf{A}^{-1} \cdot (\mathbf{Aq}, \mathbf{v}) \rangle$$
$$= \langle \mathbf{p}, \mathbf{A}^{-1}\mathbf{v} \rangle$$
$$= \langle \mathbf{Ap}, \mathbf{v} \rangle = \langle (\mathbf{Aq}, \mathbf{Ap}), (\mathbf{Aq}, \mathbf{v}) \rangle,$$

that is,

$$\mathbf{A} \cdot (\mathbf{q}, \mathbf{p}) = (\mathbf{Aq}, \mathbf{Ap}). \tag{11.4.7}$$

Differentiating with respect to $\mathbf{A}$, we find that the infinitesimal generator corresponding to $\xi = \hat{\omega} \in \mathfrak{so}(3)$ is

$$\hat{\omega}_P(\mathbf{q}, \mathbf{p}) = (\xi \mathbf{q}, \xi \mathbf{p}) = (\omega \times \mathbf{q}, \omega \times \mathbf{p}). \tag{11.4.8}$$

As in the previous example, to find the momentum map, we solve

$$\frac{\partial J(\xi)}{\partial \mathbf{p}} = \xi \mathbf{q} \quad \text{and} \quad -\frac{\partial J(\xi)}{\partial \mathbf{q}} = \xi \mathbf{p}, \tag{11.4.9}$$

such that $J(\xi)$ is linear in ξ. A solution is given by

$$J(\xi)(\mathbf{q}, \mathbf{p}) = (\xi \mathbf{q}) \cdot \mathbf{p} = (\omega \times \mathbf{q}) \cdot \mathbf{p} = (\mathbf{q} \times \mathbf{p}) \cdot \omega,$$

so that

$$\mathbf{J}(\mathbf{q}, \mathbf{p}) = \mathbf{q} \times \mathbf{p}. \tag{11.4.10}$$

Of course, (11.4.10) is the standard formula for the *angular momentum* of a particle.

In this case, Noether's theorem states that a Hamiltonian that is rotationally invariant has the three components of $\mathbf{J}$ as constants of the motion. This example can be generalized as follows. ♦

(d) Momentum for Matrix Groups. Let $G \subset GL(n, \mathbb{R})$ be a subgroup of the general linear group of $\mathbb{R}^n$. We let G act on $\mathbb{R}^n$ by matrix multiplication on the left, that is, $\Phi_{\mathbf{A}}(\mathbf{q}) = \mathbf{Aq}$. As in the previous example, the induced action on $P = T^*\mathbb{R}^n$ is given by

$$\mathbf{A} \cdot (\mathbf{q}, \mathbf{p}) = (\mathbf{Aq}, (\mathbf{A}^T)^{-1}\mathbf{p}) \tag{11.4.11}$$

and the infinitesimal generator corresponding to $\xi \in \mathfrak{g}$ by

$$\xi_P(\mathbf{q}, \mathbf{p}) = (\xi \mathbf{q}, -\xi^T \mathbf{p}). \tag{11.4.12}$$

To find the momentum map, we solve

$$\frac{\partial J(\xi)}{\partial \mathbf{p}} = \xi \mathbf{q} \quad \text{and} \quad \frac{\partial J(\xi)}{\partial \mathbf{q}} = \xi^T \mathbf{p}, \tag{11.4.13}$$

which we can do by choosing $J(\xi)(\mathbf{q}, \mathbf{p}) = (\xi \mathbf{q}) \cdot \mathbf{p}$, that is,

$$\langle \mathbf{J}(\mathbf{q}, \mathbf{p}), \xi \rangle = (\xi \mathbf{q}) \cdot \mathbf{p}. \tag{11.4.14}$$

If $n = 3$ and $G = SO(3)$, (11.4.14) is equivalent to (11.4.10). In coordinates, $(\xi \mathbf{q}) \cdot \mathbf{p} = \xi^i{}_j q^j p_i$, so

$$[\mathbf{J}(\mathbf{q}, \mathbf{p})]^i_j = q^i p_j.$$

If we identify $\mathfrak{g}$ and $\mathfrak{g}^*$ using $\langle A, B \rangle = \text{trace}(AB^T)$, then $\mathbf{J}(\mathbf{q}, \mathbf{p})$ is the projection of the matrix $q^j p_i$ onto the subspace $\mathfrak{g}$. ♦

(e) Canonical Momentum on $\mathfrak{g}^*$. Let the Lie group G with Lie algebra $\mathfrak{g}$ act by the coadjoint action on $\mathfrak{g}^*$ endowed with the $\pm$ Lie–Poisson structure. Since $\mathrm{Ad}_{g^{-1}} : \mathfrak{g} \to \mathfrak{g}$ is a Lie algebra isomorphism, its dual $\mathrm{Ad}^*_{g^{-1}} : \mathfrak{g}^* \to \mathfrak{g}^*$ is a canonical map by Proposition 10.7.2. Let us prove this fact directly. A computation shows that

$$\frac{\delta F}{\delta(\mathrm{Ad}^*_{g^{-1}}\,\mu)} = \mathrm{Ad}_g\,\frac{\delta\left(F \circ \mathrm{Ad}^*_{g^{-1}}\right)}{\delta\mu}, \tag{11.4.15}$$

whence

$$\{F, H\}_{\pm}\left(\mathrm{Ad}^*_{g^{-1}}\,\mu\right)$$

$$= \pm\left\langle \mathrm{Ad}^*_{g^{-1}}\,\mu, \left[\frac{\delta F}{\delta\left(\mathrm{Ad}^*_{g^{-1}}\,\mu\right)}, \frac{\delta H}{\delta\left(\mathrm{Ad}^*_{g^{-1}}\,\mu\right)}\right]\right\rangle$$

$$= \pm\left\langle \mathrm{Ad}^*_{g^{-1}}\,\mu, \left[\mathrm{Ad}_g\,\frac{\delta\left(F \circ \mathrm{Ad}^*_{g^{-1}}\right)}{\delta\mu}, \mathrm{Ad}_g\,\frac{\delta\left(H \circ \mathrm{Ad}^*_{g^{-1}}\right)}{\delta\mu}\right]\right\rangle$$

$$= \pm\left\langle \mu, \left[\frac{\delta\left(F \circ \mathrm{Ad}^*_{g^{-1}}\right)}{\delta\mu}, \frac{\delta\left(H \circ \mathrm{Ad}^*_{g^{-1}}\right)}{\delta\mu}\right]\right\rangle$$

$$= \left\{F \circ \mathrm{Ad}^*_{g^{-1}}, H \circ \mathrm{Ad}^*_{g^{-1}}\right\}_{\pm}(\mu),$$

that is, the coadjoint action of G on $\mathfrak{g}^*$ is canonical. From Proposition 10.7.1, the Hamiltonian vector field for $H \in \mathcal{F}(\mathfrak{g}^*)$ is given by

$$X_H(\mu) = \mp\,\mathrm{ad}^*_{(\delta H/\delta\mu)}\,\mu. \tag{11.4.16}$$

Since the infinitesimal generator of the coadjoint action corresponding to $\xi \in \mathfrak{g}$ is given by $\xi_{\mathfrak{g}^*} = -\mathrm{ad}^*_\xi$, it follows that the momentum map of the coadjoint action, if it exists, must satisfy

$$\mp\,\mathrm{ad}^*_{(\delta J(\xi)/\delta\mu)}\,\mu = -\mathrm{ad}^*_\xi\,\mu \tag{11.4.17}$$

for every $\mu \in \mathfrak{g}^*$, that is, $J(\xi)(\mu) = \pm\langle\mu, \xi\rangle$, which means that

$$\mathbf{J} = \pm\,\text{identity on } \mathfrak{g}^*. \qquad\blacklozenge$$

(f) Dual of a Lie Algebra Homomorphism. The plasma to fluid map and averaging over a symmetry group in fluid flows are duals of Lie algebra homomorphisms and provide examples of interesting Poisson maps (see §1.7). Let us now show that all such maps are momentum maps.

Let H and G be Lie groups, let $A : H \to G$ be a Lie group homomorphism, and suppose that $\alpha : \mathfrak{h} \to \mathfrak{g}$ is the induced Lie algebra homomorphism, so its dual $\alpha^* : \mathfrak{g}^* \to \mathfrak{h}^*$ is a Poisson map. We assert that α^* *is also a momentum map.* Let H act on $\mathfrak{g}^*_+$ by

$$h \cdot \mu = \mathrm{Ad}^*_{A(h)^{-1}}\,\mu,$$

that is,

$$\langle h \cdot \mu, \xi \rangle = \langle \mu, \mathrm{Ad}_{A(h)^{-1}} \xi \rangle. \tag{11.4.18}$$

Differentiating (11.4.18) with respect to h at e in the direction $\eta \in \mathfrak{h}$ gives the infinitesimal generator

$$\langle \eta_{\mathfrak{g}^*}(\mu), \xi \rangle = - \langle \mu, \mathrm{ad}_{\alpha(\eta)} \xi \rangle = - \langle \mathrm{ad}^*_{\alpha(\eta)} \mu, \xi \rangle. \tag{11.4.19}$$

Setting $\mathbf{J}(\mu) = \alpha^*(\mu)$, that is,

$$J(\eta)(\mu) = \langle \mathbf{J}(\mu), \eta \rangle = \langle \alpha^*(\mu), \eta \rangle = \langle \mu, \alpha(\eta) \rangle, \tag{11.4.20}$$

we get

$$\frac{\delta J(\eta)}{\delta \mu} = \alpha(\eta),$$

and so on $\mathfrak{g}^*_+$,

$$X_{J(\eta)}(\mu) = - \mathrm{ad}^*_{\delta J(\eta)/\delta\mu} \, \mu = - \mathrm{ad}^*_{\alpha(\eta)} \, \mu = \eta_{\mathfrak{g}^*}(\mu), \tag{11.4.21}$$

so we have proved the assertion.   ♦

(g) Momentum Maps for Subalgebras. Assume that $\mathbf{J}_{\mathfrak{g}} : P \to \mathfrak{g}^*$ is a momentum map of a canonical left Lie algebra action of $\mathfrak{g}$ on the Poisson manifold P and let $\mathfrak{h} \subset \mathfrak{g}$ be a subalgebra. Then $\mathfrak{h}$ also acts canonically on P, and this action admits a momentum map $\mathbf{J}_{\mathfrak{h}} : P \to \mathfrak{h}^*$ given by

$$\mathbf{J}_{\mathfrak{h}}(z) = \mathbf{J}_{\mathfrak{g}}(z)|\mathfrak{h}. \tag{11.4.22}$$

Indeed, if $\eta \in \mathfrak{h}$, we have $\eta_P = X_{J_{\mathfrak{g}}(\eta)}$, since the $\mathfrak{g}$-action admits the momentum map $\mathbf{J}_{\mathfrak{g}}$ and $\eta \in \mathfrak{g}$. Therefore, $J_{\mathfrak{h}}(\eta) = J_{\mathfrak{g}}(\eta)$ for all $\eta \in \mathfrak{h}$ defines the induced $\mathfrak{h}$-momentum map on P. This is equivalent to

$$\langle \mathbf{J}_{\mathfrak{h}}(z), \eta \rangle = \langle \mathbf{J}_{\mathfrak{g}}(z), \eta \rangle$$

for all $z \in P$ and $\eta \in \mathfrak{g}$, which proves formula (11.4.22) .   ♦

(h) Momentum Maps for Projective Representations. This example deals with the momentum map for an action of a finite-dimensional Lie group G on projective space that is induced by a unitary representation on the underlying Hilbert space. Recall from §5.3 that the unitary group $U(\mathcal{H})$ acts on $\mathbb{P}\mathcal{H}$ by symplectomorphisms. Due to the difficulties in defining the Lie algebra of $U(\mathcal{H})$ (see Example (d) at the end of §9.3), we cannot define the momentum map for the whole unitary group.

Let $\rho : G \to U(\mathcal{H})$ be a unitary representation of G. We can define the infinitesimal action of its Lie algebra $\mathfrak{g}$ on $\mathbb{P}\mathcal{D}_G$, the essential G-smooth part of $\mathbb{P}\mathcal{H}$, by

$$\xi_{\mathbb{P}\mathcal{H}}([\psi]) = \frac{d}{dt}[(\exp(tA(\xi)))\psi]\Big|_{t=0} = T_\psi \pi(A(\xi)\psi), \tag{11.4.23}$$

where the infinitesimal generator $A(\xi)$ was defined in §9.3, where $[\psi] \in \mathbb{PD}_G$, and where the projection is denoted by $\pi : \mathcal{H} \setminus \{0\} \to \mathbb{PH}$. Let $\varphi \in (\mathbb{C}\psi)^\perp$ and $\|\psi\| = 1$. Since $A(\xi)\psi - \langle A(\xi)\psi, \psi\rangle\psi \in (\mathbb{C}\psi)^\perp$, we have

$$(\mathbf{i}_{\xi_{\mathbb{PH}}}\Omega)(T_\psi\pi(\varphi)) = -2\hbar\,\mathrm{Im}\langle A(\xi)\psi - \langle A(\xi)\psi, \psi\rangle\psi, \varphi\rangle$$
$$= -2\hbar\,\mathrm{Im}\langle A(\xi)\psi, \varphi\rangle.$$

On the other hand, if $\mathbf{J} : \mathbb{PD}_G \to \mathfrak{g}^*$ is defined by

$$\langle \mathbf{J}([\psi]), \xi\rangle = J(\xi)([\psi]) = -i\hbar\frac{\langle\psi, A(\xi)\psi\rangle}{\|\psi\|^2}, \qquad (11.4.24)$$

then for $\varphi \in (\mathbb{C}\psi)^\perp$ and $\|\psi\| = 1$, a short computation gives

$$\mathbf{d}(J(\xi))([\psi])(T_\psi\pi(\varphi)) = \frac{d}{dt}J(\xi)([\psi + t\varphi])\Big|_{t=0}$$
$$= -2\hbar\,\mathrm{Im}\langle A(\xi)\psi, \varphi\rangle.$$

This shows that the map $\mathbf{J}$ defined in (11.4.24) is the momentum map of the G-action on $\mathbb{PH}$. We caution that this momentum map is defined only on a dense subset of the symplectic manifold. Recall that a similar thing happened when we discussed the angular momentum for quantum mechanics in §3.3. ♦

Exercises

⋄ **11.4-1.**   For the action of S^1 on $\mathbb{C}^2$ given by

$$e^{i\theta}(z_1, z_2) = (e^{i\theta}z_1, e^{-i\theta}z_2),$$

show that the momentum map is $J = (|z_1|^2 - |z_2|^2)/2$. Show that the Hamiltonian given in equation (10.5.3) is invariant under S^1, so that Theorem 11.4.1 applies.

⋄ **11.4-2 (Momentum Maps Induced by Subgroups).**   Consider a Poisson action of a Lie group G on the Poisson manifold P with a momentum map $\mathbf{J}$ and let H be a Lie subgroup of G. Denote by $i : \mathfrak{h} \to \mathfrak{g}$ the inclusion between the corresponding Lie algebras and $i^* : \mathfrak{g}^* \to \mathfrak{h}^*$ the dual map. Check that the induced H-action on P has a momentum map given by $\mathbf{K} = i^* \circ \mathbf{J}$, that is, $\mathbf{K} = \mathbf{J}|\mathfrak{h}$.

⋄ **11.4-3 (Euclidean Group in the Plane).**   The special Euclidean group SE(2) consists of all transformations of $\mathbb{R}^2$ of the form $\mathbf{A}\mathbf{z} + \mathbf{a}$, where $\mathbf{z}, \mathbf{a} \in \mathbb{R}^2$, and

$$\mathbf{A} \in \mathrm{SO}(2) = \left\{\text{matrices of the form } \begin{bmatrix} \cos\theta & -\sin\theta \\ \sin\theta & \cos\theta \end{bmatrix}\right\}. \qquad (11.4.25)$$

This group is three-dimensional, with the composition law

$$(\mathbf{A}, \mathbf{a}) \cdot (\mathbf{B}, \mathbf{b}) = (\mathbf{AB}, \mathbf{Ab} + \mathbf{a}), \qquad (11.4.26)$$

identity element $(\mathbf{I}, \mathbf{0})$, and inverse $(\mathbf{A}, \mathbf{a})^{-1} = (\mathbf{A}^{-1}, -\mathbf{A}^{-1}\mathbf{a})$. We let SE(2) act on $\mathbb{R}^2$ by $(\mathbf{A}, \mathbf{a}) \cdot \mathbf{z} = \mathbf{Az} + \mathbf{a}$. Let $\mathbf{z} = (q, p)$ denote coordinates on $\mathbb{R}^2$. Since $\det \mathbf{A} = 1$, we get $\Phi^*_{(\mathbf{A}, \mathbf{a})}(dq \wedge dp) = dq \wedge dp$, that is, SE(2) acts canonically on the symplectic manifold $\mathbb{R}^2$. Show that this action has a momentum map given by $\mathbf{J}(q, p) = \left(-\frac{1}{2}(q^2 + p^2), p, -q\right)$.

11.5 Equivariance of Momentum Maps

Infinitesimal Equivariance. Return to the commutative diagram in §11.2 and the relations (11.1.8). Since two of the maps in the diagram are Lie algebra antihomomorphisms, it is natural to ask whether J is a Lie algebra homomorphism. Equivalently, since $X_{J[\xi,\eta]} = X_{\{J(\xi), J(\eta)\}}$, it follows that

$$J([\xi, \eta]) - \{J(\xi), J(\eta)\} =: \Sigma(\xi, \eta)$$

is a Casimir function on P and hence is constant on every symplectic leaf of P. As a function on $\mathfrak{g} \times \mathfrak{g}$ with values in the vector space $\mathcal{C}(P)$ of Casimir functions on P, Σ is bilinear, antisymmetric, and satisfies

$$\Sigma(\xi, [\eta, \zeta]) + \Sigma(\eta, [\zeta, \xi]) + \Sigma(\zeta, [\xi, \eta]) = 0 \qquad (11.5.1)$$

for all $\xi, \eta, \zeta \in \mathfrak{g}$. One says that Σ is a $\mathcal{C}(P)$-*valued* **2-*cocycle*** of $\mathfrak{g}$; see Souriau [1970] and Guillemin and Sternberg [1984, p. 170], for more information.

It is natural to ask when $\Sigma(\xi, \eta) = 0$ for all $\xi, \eta \in \mathfrak{g}$. In general, this does not happen, and one is led to the study of this invariant. We shall derive an equivalent condition for $J : \mathfrak{g} \to \mathcal{F}(P)$ to be a Lie algebra homomorphism, that is, for $\Sigma = 0$, or, in other words, for the following ***commutation relations*** to hold:

$$J([\xi, \eta]) = \{J(\xi), J(\eta)\}. \qquad (11.5.2)$$

Differentiating relation (11.2.2) with respect to z in the direction $v_z \in T_z P$, we get

$$\mathbf{d}(J(\xi))(z) \cdot v_z = \langle T_z \mathbf{J} \cdot v_z, \xi \rangle \qquad (11.5.3)$$

for all $z \in P$, $v_z \in T_z P$, and $\xi \in \mathfrak{g}$. Thus, for $\xi, \eta \in \mathfrak{g}$,

$$\{J(\xi), J(\eta)\}(z) = X_{J(\eta)}[J(\xi)](z) = \mathbf{d}(J(\xi))(z) \cdot X_{J(\eta)}(z)$$
$$= \langle T_z \mathbf{J} \cdot X_{J(\eta)}(z), \xi \rangle = \langle T_z \mathbf{J} \cdot \eta_P(z), \xi \rangle. \qquad (11.5.4)$$

Note that

$$J([\xi,\eta])(z) = \langle \mathbf{J}(z), [\xi,\eta] \rangle = -\langle \mathbf{J}(z), \mathrm{ad}_\eta\, \xi \rangle = -\langle \mathrm{ad}_\eta^*\, \mathbf{J}(z), \xi \rangle. \quad (11.5.5)$$

Consequently, J is a Lie algebra homomorphism if and only if

$$T_z \mathbf{J} \cdot \eta_P(z) = -\,\mathrm{ad}_\eta^*\, \mathbf{J}(z) \quad (11.5.6)$$

for all $\eta \in \mathfrak{g}$, that is, (11.5.2) and (11.5.6) are equivalent. Momentum maps satisfying (11.5.2) (or (11.5.6)) are called **infinitesimally equivariant momentum maps,** and canonical (left) Lie algebra actions admitting infinitesimally equivariant momentum maps are called **Hamiltonian actions.** With this terminology, we have proved the following proposition:

Theorem 11.5.1. *A canonical left Lie algebra action is Hamiltonian if and only if there is a Lie algebra homomorphism $\psi : \mathfrak{g} \to \mathcal{F}(P)$ such that $X_{\psi(\xi)} = \xi_P$ for all $\xi \in \mathfrak{g}$. If ψ exists, an infinitesimally equivariant momentum map $\mathbf{J}$ is determined by $J = \psi$. Conversely, if $\mathbf{J}$ is infinitesimally equivariant, we can take $\psi = J$.*

Equivariance. Let us justify the terminology "infinitesimally equivariant momentum map." Suppose the canonical left Lie algebra action of $\mathfrak{g}$ on P arises from a canonical left Lie group action of G on P, where $\mathfrak{g}$ is the Lie algebra of G. We say that $\mathbf{J}$ is **equivariant** if

$$\mathrm{Ad}_{g^{-1}}^* \circ \mathbf{J} = \mathbf{J} \circ \Phi_g \quad (11.5.7)$$

for all $g \in G$, that is, the diagram in Figure 11.5.1 commutes.

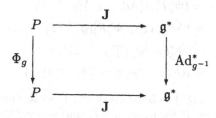

FIGURE 11.5.1. Equivariance of momentum maps.

Equivariance can be reformulated as the identity

$$J(\mathrm{Ad}_g\, \xi)(g \cdot z) = J(\xi)(z) \quad (11.5.8)$$

for all $g \in G$, $\xi \in \mathfrak{g}$, and $z \in P$. A (left) canonical Lie group action is called **globally Hamiltonian** if it has an equivariant momentum map. Differentiating (11.5.7) with respect to g at $g = e$ in the direction $\eta \in \mathfrak{g}$ shows that *equivariance implies infinitesimal equivariance.* We shall see shortly that all the preceding examples (except the one in Exercise 11.4-3) have equivariant momentum maps. Another case of interest occurs in Yang–Mills theory, where the 2-cocycle Σ is related to the **anomaly** (see Bao and Nair [1985] and references therein). The converse question, "When does infinitesimal equivariance imply equivariance?" is treated in §12.4.

Momentum Maps for Compact Groups.   In the next chapter we shall see that many momentum maps that occur in examples are equivariant. The next result shows that for *compact* groups one can *always* choose them to be equivariant.[2]

Theorem 11.5.2.   *Let G be a compact Lie group acting in a canonical fashion on the Poisson manifold P and having a momentum map $\mathbf{J} : P \to \mathfrak{g}^*$. Then $\mathbf{J}$ can be changed by addition of an element of $L(\mathfrak{g}, C(P))$ such that the resulting map is an equivariant momentum map. In particular, if P is symplectic, then $\mathbf{J}$ can be changed by the addition of an element of $\mathfrak{g}^*$ on each connected component so that the resulting map is an equivariant momentum map.*

Proof.   For each $g \in G$ define $\mathbf{J}^g(z) = \mathrm{Ad}^*_{g^{-1}} \mathbf{J}(g^{-1} \cdot z)$ or, equivalently, $J^g(\xi) = J(\mathrm{Ad}_{g^{-1}} \xi) \circ \Phi_{g^{-1}}$. Then $\mathbf{J}^g$ is also a momentum map for the G-action on P. Indeed, if $z \in P$, $\xi \in \mathfrak{g}$, and $F : P \to \mathbb{R}$, we have

$$
\begin{aligned}
\{F, J^g(\xi)\}(z) &= -\mathbf{d}J^g(\xi)(z) \cdot X_F(z) \\
&= -\mathbf{d}J(\mathrm{Ad}_{g^{-1}} \xi)(g^{-1} \cdot z) \cdot T_z \Phi_{g^{-1}} \cdot X_F(z) \\
&= -\mathbf{d}J(\mathrm{Ad}_{g^{-1}} \xi)(g^{-1} \cdot z) \cdot (\Phi^*_g X_F)(g^{-1} \cdot z) \\
&= -\mathbf{d}J(\mathrm{Ad}_{g^{-1}} \xi)(g^{-1} \cdot z) \cdot X_{\Phi^*_g F}(g^{-1} \cdot z) \\
&= \{\Phi^*_g F, J(\mathrm{Ad}_{g^{-1}} \xi)\}(g^{-1} \cdot z) \\
&= (\mathrm{Ad}_{g^{-1}} \xi)_P[\Phi^*_g F](g^{-1} \cdot z) \\
&= (\Phi^*_g \xi_P)[\Phi^*_g F](g^{-1} \cdot z) \\
&= \mathbf{d}F(z) \cdot \xi_P(z) = \{F, J(\xi)\}(z).
\end{aligned}
$$

Therefore, $\{F, J^g(\xi) - J(\xi)\} = 0$ for every $F : P \to \mathbb{R}$, that is, $J^g(\xi) - J(\xi)$ is a Casimir function on P for every $g \in G$ and every $\xi \in \mathfrak{g}$. Now define

$$
\langle \mathbf{J} \rangle = \int_G \mathbf{J}^g \, dg,
$$

where dg denotes the Haar measure on G normalized such that the total volume of G is 1. Equivalently, this definition states that

$$
\langle J \rangle(\xi) = \int_G J^g(\xi) \, dg
$$

---

[2]A fairly general context in which nonequivariant momentum maps are unavoidable is given in Marsden, Misiolek, Perlmutter, and Ratiu [1998].

for every $\xi \in \mathfrak{g}$. By linearity of the Poisson bracket in each factor, it follows that

$$\{F, \langle J \rangle(\xi)\} = \int_G \{F, J^g(\xi)\} \, dg = \int_G \{F, J(\xi)\} \, dg = \{F, J(\xi)\}.$$

Thus $\langle J \rangle$ is also a momentum map for the G-action on P, and $\langle J \rangle(\xi) - J(\xi)$ is a Casimir function on P for every $\xi \in \mathfrak{g}$, that is, $\langle J \rangle - J \in L(\mathfrak{g}, \mathcal{C}(P))$.

The momentum map $\langle J \rangle$ is equivariant. Indeed, noting that

$$J^g(h \cdot z) = \mathrm{Ad}^*_{h^{-1}} J^{h^{-1}g}(z)$$

and using invariance of the Haar measure on G under translations and inversion, for any $h \in G$ we have, after changing variables $g = hk$ in the third equality below,

$$\langle J \rangle(h \cdot z) = \int_G \mathrm{Ad}^*_{h^{-1}} J^{h^{-1}g}(z) \, dg = \mathrm{Ad}^*_{h^{-1}} \int_G J^{h^{-1}g}(z) \, dg$$

$$= \mathrm{Ad}^*_{h^{-1}} \int_G J^k(z) \, dk = \mathrm{Ad}^*_{h^{-1}} \langle J \rangle(z). \qquad \blacksquare$$

Exercises

⋄ **11.5-1.** Show that the map $J : S^2 \to \mathbb{R}$ given by $(x, y, z) \mapsto z$ is a momentum map.

⋄ **11.5-2.** Check directly that angular momentum is an equivariant momentum map, whereas the momentum map in Exercise 11.4-3 is *not* equivariant.

⋄ **11.5-3.** Prove that the momentum map determined by (11.3.4), namely,

$$\langle J(z), \xi \rangle = (\mathbf{i}_{\xi_P} \Theta)(z),$$

is equivariant.

⋄ **11.5-4.** Let $V(n, k)$ denote the vector space of complex $n \times k$ matrices (n rows, k columns). If $A \in V(n, k)$, we denote by $A^\dagger$ its adjoint (transpose conjugate).

(i) Show that

$$\langle A, B \rangle = \mathrm{trace}(AB^\dagger)$$

is a Hermitian inner product on $V(n, k)$.

(ii) Conclude that $V(n, k)$ is a symplectic vector space and determine the symplectic form.

(iii) Show that the action

$$(U, V) \cdot A = UAV^{-1}$$

of $U(n) \times U(k)$ on $V(n, k)$ is a canonical action.

(iv) Compute the infinitesimal generators of this action.

(v) Show that $\mathbf{J} : V(n, k) \to \mathfrak{u}(n)^* \times \mathfrak{u}(k)^*$ given by

$$\langle \mathbf{J}(A), (\xi, \eta) \rangle = \frac{1}{2}\operatorname{trace}(AA^\dagger \xi) - \frac{1}{2}\operatorname{trace}(A^\dagger A\eta)$$

is the momentum map of this action. Identify $\mathfrak{u}(n)^*$ with $\mathfrak{u}(n)$ by the pairing

$$\langle \xi_1, \xi_2 \rangle = -\operatorname{Re}[\operatorname{trace}(\xi_1 \xi_2)] = -\operatorname{trace}(\xi_1 \xi_2),$$

and similarly, for $\mathfrak{u}(k)^* \cong \mathfrak{u}(k)$; conclude that

$$\mathbf{J}(A) = \frac{1}{2}(-iAA^\dagger, A^\dagger A) \in \mathfrak{u}(n) \times \mathfrak{u}(k).$$

(vi) Show that $\mathbf{J}$ is equivariant.

12

Computation and Properties of Momentum Maps

The previous chapter gave the general theory of momentum maps. In this chapter we develop techniques for computing them. One of the most important cases is that in which we have a group action on a cotangent bundle that is obtained from lifting an action on the base via the operation of cotangent lift. These transformations are called *extended point transformations*. As we shall see, in this case there is an explicit formula for the momentum map, and it is always equivariant. Many of the momentum maps one meets in practical examples are of this sort.

12.1 Momentum Maps on Cotangent Bundles

Momentum Functions. We begin by defining functions on cotangent bundles associated to vector fields on the base.

Given a manifold Q, Define the map $\mathcal{P} : \mathfrak{X}(Q) \to \mathcal{F}(T^*Q)$ by

$$\mathcal{P}(X)(\alpha_q) = \langle \alpha_q, X(q) \rangle,$$

for $q \in Q$ and $\alpha_q \in T_q^*Q$, where $\langle\,,\rangle$ denotes the pairing between covectors $\alpha \in T_q^*Q$ and vectors. We call $\mathcal{P}(X)$ the **momentum function of** X.

Notice that in coordinates we have

$$\mathcal{P}(X)(q^i, p_i) = X^j(q^i)p_j. \tag{12.1.1}$$

Definition 12.1.1. *Given a manifold Q, let $\mathcal{L}(T^*Q)$ denote the space of smooth functions $F : T^*Q \to \mathbb{R}$ that are linear on fibers of T^*Q.*

Using coordinates and working in finite dimensions, we can write $F, H \in \mathcal{L}(T^*Q)$ as

$$F(q,p) = \sum_{i=1}^{n} X^i(q)p_i \quad \text{and} \quad H(q,p) = \sum_{i=1}^{n} Y^i(q)p_i$$

for functions X^i and Y^i. We claim that the standard Poisson bracket $\{F, H\}$ is again linear on the fibers. Indeed, using summations on repeated indices,

$$\{F,H\}(q,p) = \frac{\partial F}{\partial q^j}\frac{\partial H}{\partial p_j} - \frac{\partial H}{\partial q^j}\frac{\partial F}{\partial p_j} = \frac{\partial X^i}{\partial q^j}p_i Y^k \delta_k^j - \frac{\partial Y^i}{\partial q^j}p_i X^k \delta_k^j,$$

so

$$\{F,H\} = \left(\frac{\partial X^i}{\partial q^j}Y^j - \frac{\partial Y^i}{\partial q^j}X^j\right)p_i. \qquad (12.1.2)$$

Hence $\mathcal{L}(T^*Q)$ is a Lie subalgebra of $\mathcal{F}(T^*Q)$. If Q is infinite-dimensional, a similar proof, using canonical cotangent bundle charts, works.

Lemma 12.1.2 (Momentum Commutator Lemma). *The Lie algebras*

(i) *$(\mathfrak{X}(Q), [\,,])$ of vector fields on Q; and*

(ii) *Hamiltonian vector fields X_F on T^*Q with $F \in \mathcal{L}(T^*Q)$*

are isomorphic. Moreover, each of these algebras is anti-isomorphic to $(\mathcal{L}(T^*Q), \{\,,\})$. In particular, we have

$$\{\mathcal{P}(X), \mathcal{P}(Y)\} = -\mathcal{P}([X,Y]). \qquad (12.1.3)$$

Proof. Since $\mathcal{P}(X) : T^*Q \to \mathbb{R}$ is linear on fibers, it follows that $\mathcal{P}$ maps $\mathfrak{X}(Q)$ to the space $\mathcal{L}(T^*Q)$. The map $\mathcal{P}$ is linear and satisfies (12.1.3), since

$$[X,Y]^i = \frac{\partial Y^i}{\partial q^j}X^j - \frac{\partial X^i}{\partial q^j}Y^j$$

implies that

$$-\mathcal{P}([X,Y]) = \left(\frac{\partial X^i}{\partial q^j}Y^j - \frac{\partial Y^i}{\partial q^j}X^j\right)p_i,$$

which coincides with $\{\mathcal{P}(X), \mathcal{P}(Y)\}$ by (12.1.2). (We leave it to the reader to write out the infinite-dimensional proof.) Furthermore, $\mathcal{P}(X) = 0$ implies that $X = 0$ by the Hahn–Banach theorem. Finally (assuming that our model space is reflexive), for each $F \in \mathcal{L}(T^*Q)$, we can define $X(F) \in \mathfrak{X}(Q)$ by

$$\langle \alpha_q, X(F)(q)\rangle = F(\alpha_q),$$

where $\alpha_q \in T_q^* Q$. Then $\mathcal{P}(X(F)) = F$, so $\mathcal{P}$ is also surjective, thus proving that $(\mathfrak{X}(Q), [\,,])$ and $(\mathcal{L}(T^*Q), \{\,,\})$ are anti-isomorphic Lie algebras.

The map $F \mapsto X_F$ is a Lie algebra antihomomorphism from the algebra $(\mathcal{L}(T^*Q), \{\,,\})$ to $(\{\, X_F \mid F \in \mathcal{L}(T^*Q)\,\}, [\,,])$ by (5.5.6). This map is surjective by definition. Moreover, if $X_F = 0$, then F is constant on T^*Q, hence equal to zero since it is linear on the fibers. ∎

In quantum mechanics, the **Dirac rule** associates the differential operator

$$X = \frac{\hbar}{i} X^j \frac{\partial}{\partial q^j} \tag{12.1.4}$$

with the momentum function $\mathcal{P}(X)$ (Dirac [1930, Sections 21 and 22]). Thus, if we define $P_X = \mathcal{P}(X)$, then (12.1.3) gives

$$i\hbar\{P_X, P_Y\} = i\hbar\{\mathcal{P}(X), \mathcal{P}(Y)\} = -i\hbar\mathcal{P}([X,Y]) = P_{[X,Y]}. \tag{12.1.5}$$

One can augment (12.1.5) by including lifts of functions on Q. Given $f \in \mathcal{F}(Q)$, let $f^* = f \circ \pi_Q$ where $\pi_Q : T^*Q \to Q$ is the projection, so f^* is constant on fibers. One finds that

$$\{f^*, g^*\} = 0 \tag{12.1.6}$$

and

$$\{f^*, \mathcal{P}(X)\} = X[f]. \tag{12.1.7}$$

The Hamiltonian flow φ_t of X_{f^*} is fiber translation by $-t\,\mathbf{d}f$, that is, $(q, p) \mapsto (q, p - t\,\mathbf{d}f(q))$.

Hamiltonian Flows of Momentum Functions. The flow of $X_{\mathcal{P}(X)}$ is given by the following:

Proposition 12.1.3. *If* $X \in \mathfrak{X}(Q)$ *has flow* φ_t, *then the flow of* $X_{\mathcal{P}(X)}$ *on* T^*Q *is* $T^*\varphi_{-t}$.

Proof. If $\pi_Q : T^*Q \to Q$ denotes the canonical projection, differentiating the relation

$$\pi_Q \circ T^*\varphi_{-t} = \varphi_t \circ \pi_Q \tag{12.1.8}$$

at $t = 0$ gives

$$T\pi_Q \circ Y = X \circ \pi_Q, \tag{12.1.9}$$

where

$$Y(\alpha_q) = \frac{d}{dt} T^*\varphi_{-t}(\alpha_q)\Big|_{t=0}, \tag{12.1.10}$$

so $T^*\varphi_{-t}$ is the flow of Y. Since $T^*\varphi_{-t}$ preserves the canonical one-form Θ on T^*Q, it follows that $\mathcal{L}_Y\Theta = 0$. Hence

$$\mathbf{i}_Y\Omega = -\mathbf{i}_Y\mathbf{d}\Theta = \mathbf{d}\mathbf{i}_Y\Theta. \qquad (12.1.11)$$

By definition of the canonical one-form,

$$\begin{aligned}
\mathbf{i}_Y\Theta(\alpha_q) &= \langle \Theta(\alpha_q), Y(\alpha_q)\rangle = \langle \alpha_q, T\pi_Q(Y(\alpha_q))\rangle \\
&= \langle \alpha_q, X(q)\rangle = \mathcal{P}(X)(\alpha_q),
\end{aligned} \qquad (12.1.12)$$

that is, $\mathbf{i}_Y\Omega = \mathbf{d}\mathcal{P}(X)$, so that $Y = X_{\mathcal{P}(X)}$. ∎

Because of this proposition, the Hamiltonian vector field $X_{\mathcal{P}(X)}$ on T^*Q is called the *cotangent lift* of $X \in \mathfrak{X}(Q)$ to T^*Q. We also use the notation $X' := X_{\mathcal{P}(X)}$ for the cotangent lift of X. From $X_{\{F,H\}} = -[X_F, X_H]$ and (12.1.3), we get

$$\begin{aligned}
[X', Y'] &= [X_{\mathcal{P}(X)}, X_{\mathcal{P}(Y)}] = -X_{\{\mathcal{P}(X),\,\mathcal{P}(Y)\}} \\
&= -X_{-\mathcal{P}[X,Y]} = [X,Y]'.
\end{aligned} \qquad (12.1.13)$$

For finite-dimensional Q, in local coordinates we have

$$\begin{aligned}
X' := X_{\mathcal{P}(X)} &= \sum_{i=1}^{n}\left(\frac{\partial\mathcal{P}(X)}{\partial p_i}\frac{\partial}{\partial q^i} - \frac{\partial\mathcal{P}(X)}{\partial q^i}\frac{\partial}{\partial p_i}\right) \\
&= X^i\frac{\partial}{\partial q^i} - \frac{\partial X^i}{\partial q^j}p_i\frac{\partial}{\partial p_j}.
\end{aligned} \qquad (12.1.14)$$

Cotangent Momentum Maps.   Perhaps the most important result for the computation of momentum maps is the following.

Theorem 12.1.4 (Momentum Maps for Lifted Actions).   *Suppose that the Lie algebra $\mathfrak{g}$ acts on the left on the manifold Q, so that $\mathfrak{g}$ acts on $P = T^*Q$ on the left by the canonical action $\xi_P = \xi'_Q$, where ξ'_Q is the cotangent lift of ξ_Q to P and $\xi \in \mathfrak{g}$. This $\mathfrak{g}$-action on P is Hamiltonian with infinitesimally equivariant momentum map $\mathbf{J} : P \to \mathfrak{g}^*$ given by*

$$\langle \mathbf{J}(\alpha_q), \xi\rangle = \langle \alpha_q, \xi_Q(q)\rangle = \mathcal{P}(\xi_Q)(\alpha_q). \qquad (12.1.15)$$

If $\mathfrak{g}$ is the Lie algebra of a Lie group G that acts on Q and hence on T^*Q by cotangent lift, then $\mathbf{J}$ is equivariant.

In coordinates q^i, p_j on T^*Q and ξ^a on $\mathfrak{g}$, (12.1.15) reads

$$J_a\xi^a = p_i\xi_Q^i = p_i A^i_a\xi^a,$$

where $\xi_Q^i = \xi^a A^i_a$ denote the components of ξ_Q; thus,

$$J_a(q,p) = p_i A^i_a(q). \qquad (12.1.16)$$

Proof. For the case of Lie group actions, the theorem follows directly from Proposition 12.1.3 that the infinitesimal generator is given by $\xi_P = X_{\mathcal{P}(\xi_Q)}$ and thus a momentum map is given by $J(\xi) = \mathcal{P}(\xi_Q)$.

For the case of Lie algebra actions, we need first to verify that the cotangent lift indeed gives a canonical action. For $\xi, \eta \in \mathfrak{g}$, (12.1.13) gives

$$[\xi, \eta]_P = [\xi, \eta]_Q' = -[\xi_Q, \eta_Q]' = -[\xi_Q', \eta_Q'] = -[\xi_P, \eta_P],$$

and hence $\xi \mapsto \xi_P$ is a left algebra action. This action is also canonical, for if $F, H \in \mathcal{F}(P)$, then

$$\begin{aligned}
\xi_P[\{F, H\}] &= X_{\mathcal{P}(\xi_Q)}[\{F, H\}] \\
&= \{X_{\mathcal{P}(\xi_Q)}[F], H\} + \{F, X_{\mathcal{P}(\xi_Q)}[H]\} \\
&= \{\xi_P[F], H\} + \{F, \xi_P[H]\}
\end{aligned}$$

by the Jacobi identity for the Poisson bracket. If φ_t is the flow of ξ_Q, the flow of $\xi_Q' = X_{\mathcal{P}(\xi_Q)}$ is $T^*\varphi_{-t}$. Consequently, $\xi_P = X_{\mathcal{P}(\xi_Q)}$, and thus the $\mathfrak{g}$-action on P admits a momentum map given by $J(\xi) = \mathcal{P}(\xi_Q)$.

Next we turn to the question of equivariance. Since $\xi \in \mathfrak{g} \mapsto \mathcal{P}(\xi_Q) = J(\xi) \in \mathcal{F}(P)$ is a Lie algebra homomorphism by (11.1.5) and (12.1.13), it follows that $\mathbf{J}$ is an infinitesimally equivariant momentum map (Theorem 11.5.1).

Equivariance under G is proved directly in the following way. For any $g \in G$, we have

$$\begin{aligned}
\langle \mathbf{J}(g \cdot \alpha_q), \xi \rangle &= \langle g \cdot \alpha_q, \xi_Q(g \cdot q) \rangle \\
&= \langle \alpha_q, (T_{g \cdot q} \Phi_g^{-1} \circ \xi_Q \circ \Phi_g)(q) \rangle \\
&= \langle \alpha_q, (\Phi_g^* \xi_Q)(q) \rangle \\
&= \langle \alpha_q, (\mathrm{Ad}_{g^{-1}} \xi)_Q(q) \rangle \quad \text{(by Lemma 9.3.7(ii))} \\
&= \langle \mathbf{J}(\alpha_q), \mathrm{Ad}_{g^{-1}} \xi \rangle = \langle \mathrm{Ad}_{g^{-1}}^*(\mathbf{J}(\alpha_q)), \xi \rangle.
\end{aligned}$$ ∎

Remarks.

1. Let $G = \mathrm{Diff}(Q)$ act on T^*Q by cotangent lift. Then the infinitesimal generator of $X \in \mathfrak{X}(Q) = \mathfrak{g}$ is $X_{\mathcal{P}(X)}$ by Proposition 12.1.3, so that the associated momentum map is $\mathbf{J}: T^*Q \to \mathfrak{X}(Q)^*$, which is defined through $J(X) = \mathcal{P}(X)$ by the above calculations.

2. **Momentum Fiber Translations.** Let $G = \mathcal{F}(Q)$ act on T^*Q by fiber translations by $\mathbf{d}f$, that is,

$$f \cdot \alpha_q = \alpha_q + \mathbf{d}f(q). \tag{12.1.17}$$

Since the infinitesimal generator of $\xi \in \mathcal{F}(Q) = \mathfrak{g}$ is the vertical lift of $\mathbf{d}\xi(q)$ at α_q and this in turn equals the Hamiltonian vector field $-X_{\xi \circ \pi_Q}$, we see that the momentum map $\mathbf{J} : T^*Q \to \mathcal{F}(Q)^*$ is given by

$$J(\xi) = -\xi \circ \pi_Q. \tag{12.1.18}$$

This momentum map is equivariant, since π_Q is constant on fiber translations.

3. The commutation relations

$$\begin{aligned}
\{\mathcal{P}(X), \mathcal{P}(Y)\} &= -\mathcal{P}([X,Y]), \\
\{\mathcal{P}(X), \xi \circ \pi_Q\} &= -X[\xi] \circ \pi_Q, \\
\{\xi \circ \pi_Q, \eta \circ \pi_Q\} &= 0,
\end{aligned} \tag{12.1.19}$$

can be rephrased as saying that the pair $(\mathbf{J}(X), \mathbf{J}(f))$ fit together to form a momentum map for the semidirect product group

$$\text{Diff}(Q) \,\circledS\, \mathcal{F}(Q).$$

This plays an important role in the general theory of semidirect products, for which we refer the reader to Marsden, Weinstein, Ratiu, Schmid, and Spencer [1983] and Marsden, Ratiu, and Weinstein [1984a, 1984b].   ♦

The terminology **extended point transformations** arises for the following reasons. Let $\Phi : G \times Q \to Q$ be a smooth action and consider its lift $\Phi : G \times T^*Q \to T^*Q$ to the cotangent bundle. The action Φ moves points in the configuration space Q, and Φ is its natural extension to phase space T^*Q; in coordinates, the action on configuration points $q^i \mapsto \bar{q}^i$ induces the following action on momenta:

$$p_i \mapsto \bar{p}_i = \frac{\partial \bar{q}^j}{\partial q^i} p_j. \tag{12.1.20}$$

Exercises

⋄ **12.1-1.**   What is the analogue of (12.1.19), namely

$$\begin{aligned}
\{\mathcal{P}(X), \mathcal{P}(Y)\} &= -\mathcal{P}([X,Y]), \\
\{\mathcal{P}(X), \xi \circ \pi_Q\} &= -X[\xi] \circ \pi_Q, \\
\{\xi \circ \pi_Q, \eta \circ \pi_Q\} &= 0,
\end{aligned}$$

for rotations and translations on $\mathbb{R}^3$?

⋄ **12.1-2.**   Prove $\{\mathcal{P}(X), \mathcal{P}(Y)\} = -\mathcal{P}([X,Y])$ in infinite dimensions.

⋄ **12.1-3.**   Prove Theorem 12.1.4 as a consequence of $\langle \mathbf{J}(z), \xi \rangle = (\mathbf{i}_{\xi_P} \Theta)(z)$ and Exercise 11.6-3.

12.2 Examples of Momentum Maps

We begin this section with the study of momentum maps on tangent bundles.

Proposition 12.2.1. *Let the Lie algebra* $\mathfrak{g}$ *act on the left on the manifold* Q *and assume that* $L : TQ \to \mathbb{R}$ *is a regular Lagrangian that is invariant under the action of* $\mathfrak{g}$. *Endow* TQ *with the symplectic form* $\Omega_L = (\mathbb{F}L)^*\Omega$, *where* $\Omega = -d\Theta$ *is the canonical symplectic form on* T^*Q. *Then* $\mathfrak{g}$ *acts canonically on* $P = TQ$ *by*

$$\xi_P(v_q) = \frac{d}{dt}\bigg|_{t=0} T_q\varphi_t(v_q),$$

where φ_t *is the flow of* ξ_Q *and has the infinitesimally equivariant momentum map* $\mathbf{J} : TQ \to \mathfrak{g}^*$ *given by*

$$\langle \mathbf{J}(v_q), \xi \rangle = \langle \mathbb{F}L(v_q), \xi_Q(q) \rangle. \tag{12.2.1}$$

If $\mathfrak{g}$ *is the Lie algebra of a Lie group* G *and* G *acts on* Q *and hence on* TQ *by tangent lift, then* $\mathbf{J}$ *is equivariant.*

Proof. Use (11.3.4), a direct calculation, or, if L is hyperregular, the following argument. Since $\mathbb{F}L$ is a symplectic diffeomorphism, $\xi \mapsto \xi_P = (\mathbb{F}L)^*\xi_{T^*Q}$ is a canonical left Lie algebra action. Therefore, the composition of $\mathbb{F}L$ with the momentum map (12.1.15) is the momentum map of the $\mathfrak{g}$-action on TQ. ∎

In coordinates $(q^i, \dot{q}^i)$ on TQ and (ξ^a) on $\mathfrak{g}$, (12.2.1) reads

$$J_a(q^i, \dot{q}^i) = \frac{\partial L}{\partial \dot{q}^i} A^i{}_a(q), \tag{12.2.2}$$

where $\xi_Q^i(q) = \xi^a A^i{}_a(q)$ are the components of ξ_Q.

Next, we shall give a series of examples of momentum maps.

Examples

(a) The Hamiltonian. A Hamiltonian $H : P \to \mathbb{R}$ on a Poisson manifold P having a complete vector field X_H is an equivariant momentum map for the $\mathbb{R}$-action given by the flow of X_H. ◆

(b) Linear Momentum. In the notation of Example (b) of §11.4 we recompute the linear momentum of the N-particle system. Since $\mathbb{R}^3$ acts on points $(\mathbf{q}_1, \dots, \mathbf{q}_N)$ in $\mathbb{R}^{3N}$ by $\mathbf{x} \cdot (\mathbf{q}_j) = (\mathbf{q}_j + \mathbf{x})$, the infinitesimal generator is

$$\xi_{\mathbb{R}^{3N}}(\mathbf{q}_j) = (\mathbf{q}_1, \dots, \mathbf{q}_N, \xi, \dots, \xi) \tag{12.2.3}$$

(this has the base point $(\mathbf{q}_1, \dots, \mathbf{q}_N)$ and vector part $(\xi, \dots, \xi)$ (N times)). Consequently, by (12.1.15), an equivariant momentum map $\mathbf{J} : T^*\mathbb{R}^{3N} \to \mathbb{R}^3$ is given by

$$J(\xi)(\mathbf{q}_j, \mathbf{p}^j) = \sum_{j=1}^N \mathbf{p}^j \cdot \xi, \quad \text{i.e.,} \quad \mathbf{J}(\mathbf{q}_j, \mathbf{p}^j) = \sum_{j=1}^N \mathbf{p}^j. \qquad \blacklozenge$$

(c) **Angular Momentum.**   In the notation of Example (c) of §11.4, let $SO(3)$ act on $\mathbb{R}^3$ by matrix multiplication $\mathbf{A} \cdot \mathbf{q} = \mathbf{Aq}$. The infinitesimal generator is given by $\hat{\omega}_{\mathbb{R}^3}(\mathbf{q}) = \hat{\omega}\mathbf{q} = \omega \times \mathbf{q}$ where $\omega \in \mathbb{R}^3$. Consequently, by (12.1.15), an equivariant momentum map $\mathbf{J} : T^*\mathbb{R}^3 \to \mathfrak{so}(3)^* \cong \mathbb{R}^3$ is given by

$$\langle \mathbf{J}(\mathbf{q}, \mathbf{p}), \omega \rangle = \mathbf{p} \cdot \hat{\omega}\mathbf{q} = \omega \cdot (\mathbf{q} \times \mathbf{p}),$$

that is,

$$\mathbf{J}(\mathbf{q}, \mathbf{p}) = \mathbf{q} \times \mathbf{p}. \qquad (12.2.4)$$

Equivariance in this case reduces to the relation $\mathbf{Aq} \times \mathbf{Ap} = \mathbf{A}(\mathbf{q} \times \mathbf{p})$ for any $A \in SO(3)$. If $A \in O(3) \backslash SO(3)$, such as a reflection, this relation is no longer satisfied; a minus sign appears on the right-hand side, a fact sometimes phrased by stating that *angular momentum is a pseudo-vector*. On the other hand, letting $O(3)$ act on $\mathbb{R}^3$ by matrix multiplication, $\mathbf{J}$ is given by the same formula and so is the momentum map of a lifted action, and these are *always* equivariant. We have an apparent contradiction. What is wrong? The answer is that the adjoint action and the isomorphism $\hat{} : \mathbb{R}^3 \to \mathfrak{so}(3)$ are related, on the component of $-(\text{Identity})$ in $O(3)$, by $A\hat{x}A^{-1} = -(Ax)\hat{}$. Thus, $\mathbf{J}(\mathbf{q}, \mathbf{p})$ is indeed equivariant as it stands. (One does not need a separate terminology like "pseudo-vector" to see what is going on.) $\qquad \blacklozenge$

(d) **Momentum for Matrix Groups.**   In the notation of Example (d) of §11.4, let the Lie group $G \subset GL(n, \mathbb{R})$ act on $\mathbb{R}^n$ by $\mathbf{A} \cdot \mathbf{q} = \mathbf{Aq}$. The infinitesimal generator of this action is given by

$$\xi_{\mathbb{R}^N}(\mathbf{q}) = \xi\mathbf{q}$$

for $\xi \in \mathfrak{g}$, the Lie algebra of G, regarded as a subalgebra $\mathfrak{g} \subset \mathfrak{gl}(n, \mathbb{R})$. By (12.1.15), the lift of the G-action on $\mathbb{R}^3$ to $T^*\mathbb{R}^n$ has an equivariant momentum map $\mathbf{J} : T^*\mathbb{R}^n \to \mathfrak{g}^*$ given by

$$\langle \mathbf{J}(\mathbf{q}, \mathbf{p}), \xi \rangle = \mathbf{p} \cdot (\xi\mathbf{q}), \qquad (12.2.5)$$

which coincides with (11.4.14). $\qquad \blacklozenge$

(e) The Dual of a Lie Algebra Homomorphism. From §11.4 Example (f) it follows that the dual of a Lie algebra homomorphism $\alpha : \mathfrak{h} \to \mathfrak{g}$ is an equivariant momentum map that does not arise from an action that is an extended point transformation. Recall that a *linear map* $\alpha : \mathfrak{h} \to \mathfrak{g}$ *is a Lie algebra homomorphism if and only if the dual map* $\alpha^* : \mathfrak{g}^* \to \mathfrak{h}^*$ *is Poisson.* ◆

(f) Momentum Maps Induced by Subgroups. If a Lie group action of G on P admits an equivariant momentum map $\mathbf{J}$, and if H is a Lie subgroup of G, then in the notation of Exercise 11.4-2, $i^* \circ \mathbf{J} : P \to \mathfrak{h}^*$ is an equivariant momentum map of the induced H-action on P. ◆

(g) Products. Let P_1 and P_2 be Poisson manifolds and let $P_1 \times P_2$ be the product manifold endowed with the product Poisson structure; that is, if $F, G : P_1 \times P_2 \to \mathbb{R}$, then

$$\{F, G\}(z_1, z_2) = \{F_{z_2}, G_{z_2}\}_1(z_1) + \{F_{z_1}, G_{z_1}\}_2(z_2),$$

where $\{\,,\}_i$ is the Poisson bracket on P_i, and $F_{z_1} : P_2 \to \mathbb{R}$ is the function obtained by freezing $z_1 \in P_1$, and similarly for $F_{z_2} : P_1 \to \mathbb{R}$. Let the Lie algebra $\mathfrak{g}$ act canonically on P_1 and P_2 with (equivariant) momentum mappings $\mathbf{J}_1 : P_1 \to \mathfrak{g}^*$ and $\mathbf{J}_2 : P_2 \to \mathfrak{g}^*$. Then

$$\mathbf{J} = \mathbf{J}_1 + \mathbf{J}_2 : P_1 \times P_2 \to \mathfrak{g}^*, \quad \mathbf{J}(z_1, z_2) = \mathbf{J}(z_1) + \mathbf{J}(z_2),$$

is an (equivariant) momentum mapping of the canonical $\mathfrak{g}$-action on the product $P_1 \times P_2$. There is an obvious generalization to the product of N Poisson manifolds. Note that Example (b) is a special case of this for $G = \mathbb{R}^3$ for all factors in the product manifold equal to $T^*\mathbb{R}^3$. ◆

(h) Cotangent Lift on T^*G. The momentum map for the cotangent lift of the *left* translation action of G on G is, by (12.1.15), equal to

$$\langle \mathbf{J}_L(\alpha_g), \xi \rangle = \langle \alpha_g, \xi_G(g) \rangle = \langle \alpha_g, T_e R_g(\xi) \rangle = \langle T_e^* R_g(\alpha_g), \xi \rangle,$$

that is,

$$\mathbf{J}_L(\alpha_g) = T_e^* R_g(\alpha_g). \tag{12.2.6}$$

Similarly, the momentum map for the lift to T^*G of *right* translation of G on G equals

$$\mathbf{J}_R(\alpha_g) = T_e^* L_g(\alpha_g). \tag{12.2.7}$$

Notice that $\mathbf{J}_L$ is *right* invariant, whereas $\mathbf{J}_R$ is *left* invariant. Both are equivariant momentum maps ($\mathbf{J}_R$ with respect to Ad_g^*, which is a *right* action), so they are Poisson maps. The diagram in Figure 12.2.1 summarizes the situation.

This diagram is an example of what is called a ***dual pair***; these illuminate the relation between the body and spatial descriptions of both rigid bodies and fluids; see Chapter 15 for more information. ◆

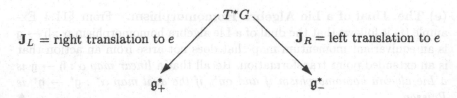

FIGURE 12.2.1. Momentum maps for left and right translations.

(i) Momentum Translation on Functions. Let $P = \mathcal{F}(T^*Q)^*$ with the Lie–Poisson bracket given in Example (e) of §10.1. Using the Liouville measure on T^*Q and assuming that elements of $\mathcal{F}(T^*Q)$ fall off rapidly enough at infinity, we identify $\mathcal{F}(T^*Q)^*$ with $\mathcal{F}(T^*Q)$ using the L^2-pairing. Let $G = \mathcal{F}(Q)$ (with the group operation being addition) act on P by

$$(\varphi \cdot f)(\alpha_q) = f(\alpha_q + \mathbf{d}\varphi(q)), \qquad (12.2.8)$$

that is, in coordinates,

$$f(q^i, p_j) \mapsto f\left(q^i, p_j + \frac{\partial\varphi}{\partial q^i}\right).$$

The infinitesimal generator is

$$\xi_P(f)(\alpha_q) = \mathbb{F}f(\alpha_q) \cdot \mathbf{d}\xi(q), \qquad (12.2.9)$$

where $\mathbb{F}f$ is the fiber derivative of f. In coordinates, (12.2.9) reads

$$\xi_P(f)(q^i, p_j) = \frac{\partial f}{\partial p_j} \cdot \frac{\partial\xi}{\partial q^j}.$$

Since G is a vector space group, its Lie algebra is also $\mathcal{F}(Q)$, and we identify $\mathcal{F}(Q)^*$ with one-form densities on Q. If $f, g, h \in \mathcal{F}(T^*Q)$, we have by Corollary 5.5.9

$$\int_{T^*Q} f\{g, h\} \, dq \, dp = \int_{T^*Q} g\{h, f\} \, dq \, dp. \qquad (12.2.10)$$

Next, note that if $F, H : P = \mathcal{F}(T^*Q) \to \mathbb{R}$, then we get by (12.2.10)

$$X_H[F](f) = \{F, H\}(f) = \int_{T^*Q} f\left\{\frac{\delta F}{\delta f}, \frac{\delta H}{\delta f}\right\} dq \, dp$$

$$= \int_{T^*Q} \frac{\delta F}{\delta f} \left\{\frac{\delta H}{\delta f}, f\right\} dq \, dp. \qquad (12.2.11)$$

On the other hand, by (12.2.9), we have

$$\xi_P[F](f) = \int_{T^*Q} \frac{\delta F}{\delta f} (\mathbb{F}f \cdot (\mathbf{d}\xi \circ \pi_Q)) \, dq \, dp, \qquad (12.2.12)$$

which suggests that the definition of $\mathbf{J}$ should be

$$\langle \mathbf{J}(f), \xi \rangle = \int_{T^*Q} f(\alpha_q)\xi(q)\, dq\, dp. \tag{12.2.13}$$

Indeed, by (12.2.13), we have $\delta J(\xi)/\delta f = \xi \circ \pi_Q$, so that

$$\left\{ \frac{\delta J(\xi)}{\delta f}, f \right\} = \{\xi \circ \pi_Q, f\} = \mathbb{F}f \cdot (\mathbf{d}\xi \circ \pi_Q),$$

and hence by (12.2.11),

$$X_{J(\xi)}[F](f) = \int_{T^*Q} \frac{\delta F}{\delta f} \left\{ \frac{\delta J(\xi)}{\delta f}, f \right\} dq\, dp$$

$$= \int_{T^*Q} \frac{\delta F}{\delta f} (\mathbb{F}f \cdot (\mathbf{d}\xi \circ \pi_Q))\, dq\, dp,$$

which coincides with (12.2.12), thereby proving that $\mathbf{J}$ given by (12.2.13) is the momentum map. In other words, the fiber integral

$$\mathbf{J}(f) = \int_{T^*Q} f(q, p)\, dp, \tag{12.2.14}$$

thought of as a one-form density on Q via (12.2.13), is the momentum map in this case. This momentum map is infinitesimally equivariant. Indeed, if $\xi, \eta \in \mathcal{F}(Q)$, we have for $f \in P$,

$$\{J(\xi), J(\eta)\}(f) = \int_{T^*Q} f \left\{ \frac{\delta J(\xi)}{\delta f}, \frac{\delta J(\eta)}{\delta f} \right\} dq\, dp$$

$$= \int_{T^*Q} f \{\xi \circ \pi_Q, \eta \circ \pi_Q\}\, dq\, dp = 0 = J([\xi, \eta])(f). \quad \blacklozenge$$

(j) More Momentum Translations. Let $\mathrm{Diff}_{\mathrm{can}}(T^*Q)$ be the group of symplectic diffeomorphisms of T^*Q, and as above, let $G = \mathcal{F}(Q)$ act on T^*Q by translation with $\mathbf{d}f$ along the fiber, that is, $f \cdot \alpha_q = \alpha_q + \mathbf{d}f(q)$. Since the action of the additive group $\mathcal{F}(Q)$ is Hamiltonian, $\mathcal{F}(Q)$ acts on $\mathrm{Diff}_{\mathrm{can}}(T^*Q)$ by composition on the right with translations, that is, the action is $(f, \varphi) \in \mathcal{F}(Q) \times \mathrm{Diff}_{\mathrm{can}}(T^*Q) \mapsto \varphi \circ \rho_f \in \mathrm{Diff}_{\mathrm{can}}(T^*Q)$, where $\rho_f(\alpha_q) = \alpha_q + \mathbf{d}f(q)$. The infinitesimal generator of this action is given by (see the comment preceding (12.1.18))

$$\xi_{\mathrm{Diff}_{\mathrm{can}}(T^*Q)}(\varphi) = -T\varphi \circ X_{\xi \circ \pi_Q} \tag{12.2.15}$$

for $\xi \in \mathcal{F}(Q) = \mathfrak{g}$, so that the equivariant momentum map of the lifted action $\mathbf{J} : T^*(\mathrm{Diff}_{\mathrm{can}}(T^*Q)) \to \mathcal{F}(Q)^*$ given by (12.1.15) is in this case

$$J(\xi)(\alpha_\varphi) = - \langle \alpha_\varphi, T\varphi \circ X_{\xi \circ \pi_Q} \rangle, \tag{12.2.16}$$

where the pairing on the right is between vector fields and one-form densities α_φ. $\quad \blacklozenge$

(k) The Divergence of the Electric Field. Let $\mathcal{A}$ be the space of vector potentials $\mathbf{A}$ on $\mathbb{R}^3$ and $P = T^*\mathcal{A}$, whose elements are denoted by $(\mathbf{A}, -\mathbf{E})$ with $\mathbf{A}$ and $\mathbf{E}$ vector fields. Let $G = \mathcal{F}(\mathbb{R}^3)$ act on $\mathcal{A}$ by $\varphi \cdot \mathbf{A} = \mathbf{A} + \nabla\varphi$. Thus, the infinitesimal generator is

$$\xi_{\mathcal{A}}(\mathbf{A}) = \nabla\xi.$$

Hence the momentum map is

$$\langle \mathbf{J}(\mathbf{A}, -\mathbf{E}), \xi \rangle = \int -\mathbf{E} \cdot \nabla\xi \, d^3x = \int (\text{div } \mathbf{E})\xi \, d^3x \qquad (12.2.17)$$

(assuming fast enough falloff to justify integration by parts). Thus,

$$\mathbf{J}(\mathbf{A}, -\mathbf{E}) = \text{div } \mathbf{E} \qquad (12.2.18)$$

is the equivariant momentum map.   ♦

(l) Virtual Work. We usually think of covectors as momenta conjugate to configuration variables. However, covectors can also be thought of as forces. Indeed, if $\alpha_q \in T_q^*Q$ and $w_q \in T_qQ$, we think of

$$\langle \alpha_q, w_q \rangle = \text{force} \times \text{infinitesimal displacement}$$

as the **virtual work**. We now give an example of a momentum map in this context.

Consider a region $\mathcal{B} \subset \mathbb{R}^3$ with boundary $\partial\mathcal{B}$. Let $\mathcal{C}$ be the space of maps $\varphi : \mathcal{B} \rightarrow \mathbb{R}^3$. Regard $T_{\varphi}^*\mathcal{C}$ as the space of **loads**; that is, pairs of maps $\mathbf{b} : \mathcal{B} \rightarrow \mathbb{R}^3$, $\tau : \partial\mathcal{B} \rightarrow \mathbb{R}^3$ paired with a tangent vector $\mathbf{V} \in T_{\varphi}\mathcal{C}$ by

$$\langle (\mathbf{b}, \tau), \mathbf{V} \rangle = \iiint_{\mathcal{B}} \mathbf{b} \cdot \mathbf{V} \, d^3x + \iint_{\partial\mathcal{B}} \tau \cdot \mathbf{V} \, dA.$$

Let $\mathbf{A} \in GL(3, \mathbb{R})$ act on $\mathcal{C}$ by $\varphi \mapsto \mathbf{A} \circ \varphi$. The infinitesimal generator of this action is $\xi_{\mathcal{C}}(\varphi)(X) = \xi\varphi(X)$ for $\xi \in \mathfrak{gl}(3)$ and $X \in \mathcal{B}$. Pair $\mathfrak{gl}(3, \mathbb{R})$ with itself via $\langle \mathbf{A}, \mathbf{B} \rangle = \frac{1}{2}\text{trace}\,(\mathbf{AB})$. The induced momentum map $\mathbf{J} : T^*\mathcal{C} \rightarrow \mathfrak{gl}(3, \mathbb{R})$ is given by

$$\mathbf{J}(\varphi, (\mathbf{b}, \tau)) = \iiint_{\mathcal{B}} \varphi \otimes \mathbf{b} \, d^3x + \iint_{\partial\mathcal{B}} \varphi \otimes \tau \, dA. \qquad (12.2.19)$$

(This is the "astatic load," a concept from elasticity; see, for example, Marsden and Hughes [1983].) If we take $SO(3)$ rather than $GL(3, \mathbb{R})$, we get the angular momentum.   ♦

(m) Momentum Maps for Unitary Representations on Projective Space.

Here we show that the momentum map discussed in Example (h) of §11.4 is equivariant. Recall from the discussion at the end of §9.3 that associated to a unitary representation ρ of a Lie group G on a complex Hilbert space $\mathcal{H}$ there are skew-adjoint operators $A(\xi)$ for each $\xi \in \mathfrak{g}$ depending linearly on ξ and such that $\rho(\exp(t\xi)) = \exp(tA(\xi))$. Thus, taking the t-derivative in the formula $\rho(g)\rho(\exp(t\xi))\rho(g^{-1}) = \exp(t\rho(g)A(\xi)\rho(g)^{-1})$, we get

$$A(\mathrm{Ad}_g \, \xi) = \rho(g)A(\xi)\rho(g)^{-1}. \qquad (12.2.20)$$

Using formula (11.4.24), namely

$$\langle \mathbf{J}([\psi]), \xi \rangle = J(\xi)([\psi]) = -i\hbar \frac{\langle \psi, A(\xi)\psi \rangle}{\|\psi\|^2}, \qquad (12.2.21)$$

we get

$$
\begin{aligned}
J(\mathrm{Ad}_g \, \xi)([\psi]) &= -i\hbar \frac{\langle \psi, \rho(g)A(\xi)\rho(g)^{-1}\psi \rangle}{\|\psi\|^2} \\
&= J(\xi)([\rho(g)^{-1}\psi]) = J(\xi)(g^{-1} \cdot [\psi]),
\end{aligned}
$$

which shows that $\mathbf{J} : \mathbb{P}\mathcal{H} \to \mathfrak{g}^*$ is equivariant. ◆

Exercises

⋄ **12.2-1.** Derive the conservation of $\mathbf{J}$ given by

$$\langle \mathbf{J}(v_q), \xi \rangle = \langle \mathbb{F}L(v_q), \xi_Q(q) \rangle$$

directly from Hamilton's variational principle. (This is the way Noether originally derived conserved quantities.)

⋄ **12.2-2.** If L is independent of one of the coordinates q^i, then it is clear that $p_i = \partial L/\partial \dot{q}^i$ is a constant of the motion from the Euler–Lagrange equations. Derive this from Proposition 12.2.1.

⋄ **12.2-3.** Compute $\mathbf{J}_L$ and $\mathbf{J}_R$ for $G = SO(3)$.

⋄ **12.2-4.** Compute the momentum maps determined by spatial translations and rotations for Maxwell's equations.

⋄ **12.2-5.** Repeat Exercise 12.2-4 for elasticity (the context of Example (l)).

⋄ **12.2-6.** Let P be a symplectic manifold and $\mathbf{J} : P \to \mathfrak{g}^*$ be an (equivariant) momentum map for the symplectic action of a group G on P. Let $\mathcal{F}$ be the space of (smooth) functions on P identified with its dual via integration and equipped with the Lie–Poisson bracket. Let $\mathcal{J} : \mathcal{F} \to \mathfrak{g}^*$ be defined by

$$\langle \mathcal{J}(f), \xi) \rangle = \int f \, \langle \mathbf{J}, \xi \rangle \, d\mu,$$

where μ is Liouville measure. Show that $\mathcal{J}$ is an (equivariant) momentum map.

⋄ **12.2-7.**

(i) Let G act on itself by conjugation. Compute the momentum map of its cotangent lift.

(ii) Let $N \subset G$ be a normal subgroup, so that G acts on N by conjugation. Again, compute the momentum map of the cotangent lift of this conjugation action.

12.3   Equivariance and Infinitesimal Equivariance

This optional section explores the equivariance of momentum maps a little more deeply. We have just seen that equivariance implies infinitesimal equivariance. In this section, we prove, among other things, the converse if G is connected.

A Family of Casimir Functions.   Introduce the map $\Gamma_\eta : G \times P \to \mathbb{R}$ defined by

$$\Gamma_\eta(g, z) = \langle \mathbf{J}(\Phi_g(z)), \eta \rangle - \langle \mathrm{Ad}_{g^{-1}}^* \mathbf{J}(z), \eta \rangle \quad \text{for } \eta \in \mathfrak{g}. \tag{12.3.1}$$

Since

$$\Gamma_{\eta,g}(z) := \Gamma_\eta(g, z) = \left(\Phi_g^* J(\eta) \right)(z) - J \left(\mathrm{Ad}_{g^{-1}} \eta \right)(z), \tag{12.3.2}$$

we get

$$X_{\Gamma_{\eta,g}} = X_{\Phi_g^* J(\eta)} - X_{J(\mathrm{Ad}_{g^{-1}} \eta)} = \Phi_g^* X_{J(\eta)} - \left(\mathrm{Ad}_{g^{-1}} \eta \right)_P$$
$$= \Phi_g^* \eta_P - \left(\mathrm{Ad}_{g^{-1}} \eta \right)_P = 0 \tag{12.3.3}$$

by (11.1.4). Therefore, $\Gamma_{\eta,g}$ is a Casimir function on P and so is constant on every symplectic leaf of P. Since $\eta \mapsto \Gamma_\eta(g, z)$ is linear for every $g \in G$ and $z \in P$, we can define the map $\sigma : G \to L(\mathfrak{g}, \mathcal{C}(P))$, from G to the vector space of all linear maps of $\mathfrak{g}$ into the space of Casimir functions $\mathcal{C}(P)$ on P, by $\sigma(g) \cdot \eta = \Gamma_{\eta,g}$. The behavior of σ under group multiplication is the following. For $\xi \in \mathfrak{g}$, $z \in P$, and $g, h \in G$, we have

$$(\sigma(gh) \cdot \xi)(z) = \Gamma_\xi(gh, z) = \langle \mathbf{J}(\Phi_{gh}(z)), \xi \rangle - \langle \mathrm{Ad}_{(gh)^{-1}}^* \mathbf{J}(z), \xi \rangle$$
$$= \langle \mathbf{J}(\Phi_g(\Phi_h(z))), \xi \rangle - \langle \mathrm{Ad}_{g^{-1}}^* \mathbf{J}((\Phi_h(z)), \xi \rangle$$
$$+ \langle \mathbf{J}(\Phi_h(z)), \mathrm{Ad}_{g^{-1}} \xi \rangle - \langle \mathrm{Ad}_{h^{-1}}^* \mathbf{J}(z), \mathrm{Ad}_{g^{-1}} \xi \rangle$$
$$= \Gamma_\xi(g, \Phi_h(z)) + \Gamma_{\mathrm{Ad}_{g^{-1}} \xi}(h, z)$$
$$= (\sigma(g) \cdot \xi)(\Phi_h(z)) + (\sigma(h) \cdot \mathrm{Ad}_{g^{-1}} \xi)(z). \tag{12.3.4}$$

Connected Lie group actions admitting momentum maps preserve symplectic leaves. This is because G is generated by a neighborhood of the identity in which each element has the form $\exp t\xi$; since $(t, z) \mapsto (\exp t\xi) \cdot z$ is a Hamiltonian flow, it follows that z and $\Phi_h(z)$ are on the same leaf. Thus,

$$(\sigma(g) \cdot \xi)(z) = (\sigma(g) \cdot \xi)(\Phi_h(z))$$

because Casimir functions are constant on leaves. Therefore,

$$\sigma(gh) = \sigma(g) + \text{Ad}^\dagger_{g^{-1}} \sigma(h), \qquad (12.3.5)$$

where $\text{Ad}^\dagger_g$ denotes the action of G on $L(\mathfrak{g}, \mathcal{C}(P))$ induced via the adjoint action by

$$(\text{Ad}^\dagger_g \lambda)(\xi) = \lambda(\text{Ad}_{g^{-1}} \xi) \qquad (12.3.6)$$

for $g \in G$, $\xi \in \mathfrak{g}$, and $\lambda \in L(\mathfrak{g}, \mathcal{C}(P))$.

Cocycles.   Mappings $\sigma : G \to L(\mathfrak{g}, \mathcal{C}(P))$ behaving under group multiplication as in (12.3.5) are called $L(\mathfrak{g}, \mathcal{C}(P))$-valued **one-cocycles** of the group G. A one-cocycle σ is called a **one-coboundary** if there is a $\lambda \in L(\mathfrak{g}, \mathcal{C}(P))$ such that

$$\sigma(g) = \lambda - \text{Ad}^\dagger_{g^{-1}} \lambda \quad \text{for all } g \in G. \qquad (12.3.7)$$

The quotient space of one-cocycles modulo one-coboundaries is called the *first* $L(\mathfrak{g}, \mathcal{C}(P))$-*valued group cohomology of* G and is denoted by $H^1(G, L(\mathfrak{g}, \mathcal{C}(P)))$; its elements are denoted by $[\sigma]$, for σ a one-cocycle.

At the Lie algebra level, bilinear skew-symmetric maps $\Sigma : \mathfrak{g} \times \mathfrak{g} \to \mathcal{C}(P)$ satisfying the Jacobi-type identity (11.6.1) are called $\mathcal{C}(P)$-*valued two-cocycles of* $\mathfrak{g}$. A cocycle Σ is called a *coboundary* if there is a $\lambda \in L(\mathfrak{g}, \mathcal{C}(P))$ such that

$$\Sigma(\xi, \eta) = \lambda([\xi, \eta]) \quad \text{for all } \xi, \eta \in \mathfrak{g}. \qquad (12.3.8)$$

The quotient space of two-cocycles by two-coboundaries is called the *second cohomology of* $\mathfrak{g}$ *with values in* $\mathcal{C}(P)$. It is denoted by $H^2(\mathfrak{g}, \mathcal{C}(P))$ and its elements by $[\Sigma]$. With this notation we have proved the first two parts of the following proposition:

Proposition 12.3.1.   *Let the connected Lie group G act canonically on the Poisson manifold P and have a momentum map $\mathbf{J}$. For $g \in G$ and $\xi \in \mathfrak{g}$, define*

$$\Gamma_{\xi,g} : P \to \mathbb{R}, \quad \Gamma_{\xi,g}(z) = \langle \mathbf{J}(\Phi_g(z)), \xi \rangle - \langle \text{Ad}^*_{g^{-1}} \mathbf{J}(z), \xi \rangle. \qquad (12.3.9)$$

Then

(i) $\Gamma_{\xi,g}$ *is a Casimir on P for every $\xi \in \mathfrak{g}$ and $g \in G$.*

(ii) *Defining* $\sigma : G \to L(\mathfrak{g}, C(P))$ *by* $\sigma(g) \cdot \xi = \Gamma_{\xi, g}$, *we have the identity*

$$\sigma(gh) = \sigma(g) + \operatorname{Ad}_{g^{-1}}^{\dagger} \sigma(h). \tag{12.3.10}$$

(iii) *Defining* $\sigma_\eta : G \to C(P)$ *by* $\sigma_\eta(g) := \sigma(g) \cdot \eta$ *for* $\eta \in \mathfrak{g}$, *we have*

$$T_e \sigma_\eta(\xi) = \Sigma(\xi, \eta) := J([\xi, \eta]) - \{J(\xi), J(\eta)\}. \tag{12.3.11}$$

If $[\sigma] = 0$, *then* $[\Sigma] = 0$.

(iv) *If* $\mathbf{J}_1$ *and* $\mathbf{J}_2$ *are two momentum mappings of the same action with cocycles* σ_1 *and* σ_2, *then* $[\sigma_1] = [\sigma_2]$.

Proof. Since $\sigma_\eta(g)(z) = J(\eta)(g \cdot z) - J(\operatorname{Ad}_{g^{-1}} \eta)(z)$, taking the derivative at $g = e$, we get

$$\begin{aligned}
T_e \sigma_\eta(\xi)(z) &= \mathbf{d}J(\eta)(\xi_P(z)) + J([\xi, \eta])(z) \\
&= X_{J(\xi)}[J(\eta)](z) + J([\xi, \eta])(z) \\
&= -\{J(\xi), J(\eta)\}(z) + J([\xi, \eta])(z).
\end{aligned} \tag{12.3.12}$$

This proves (12.3.11). The second statement in (iii) is a consequence of the definition. To prove (iv) we note that

$$\sigma_1(g)(z) - \sigma_2(g)(z) = \mathbf{J}_1(g \cdot z) - \mathbf{J}_2(g \cdot z) - \operatorname{Ad}_{g^{-1}}^*(\mathbf{J}_1(z) - \mathbf{J}_2(z)). \tag{12.3.13}$$

However, $\mathbf{J}_1$ and $\mathbf{J}_2$ are momentum mappings of the same action, and therefore $J_1(\xi)$ and $J_2(\xi)$ generate the same Hamiltonian vector field for all $\xi \in \mathfrak{g}$, so $J_1 - J_2$ is constant as an element of $L(\mathfrak{g}, C(P))$. Calling this element λ, we have

$$\sigma_1(g) - \sigma_2(g) = \lambda - \operatorname{Ad}_{g^{-1}}^{\dagger} \lambda, \tag{12.3.14}$$

so $\sigma_1 - \sigma_2$ is a coboundary. ∎

Remarks.

1.   Part (iv) of this proposition also holds for Lie algebra actions admitting momentum maps with all σ's replaced by Σ's; indeed,

$$\{J_1(\xi), J_1(\eta)\} = \{J_2(\xi), J_2(\eta)\}$$

because $J_1(\xi) - J_2(\xi)$ and $J_1(\eta) - J_2(\eta)$ are Casimir functions.

2.   If $[\Sigma] = 0$, then *the momentum map* $\mathbf{J} : P \to \mathfrak{g}^*$ *of the canonical Lie algebra action of* $\mathfrak{g}$ *on* P *can be always chosen to be infinitesimally equivariant*, a result due to Souriau [1970] for the symplectic case. To see this, note first that momentum maps are determined only up to elements of $L(\mathfrak{g}, C(P))$. Therefore, if $\lambda \in L(\mathfrak{g}, C(P))$ denotes the element determined by the condition $[\Sigma] = 0$, then $\mathbf{J} + \lambda$ is an infinitesimally equivariant momentum map.

3. *The cohomology class* $[\Sigma]$ *depends only on the Lie algebra action* ρ : $\mathfrak{g} \to \mathfrak{X}(P)$ *and not on the momentum map.* Indeed, because J is determined only up to the addition of a linear map $\lambda : \mathfrak{g} \to C(P)$, defining

$$\Sigma_\lambda(\xi, \eta) := (J + \lambda)([\xi, \eta]) - \{(J + \lambda)(\xi), (J + \lambda)(\eta)\}, \qquad (12.3.15)$$

we obtain

$$\begin{aligned}\Sigma_\lambda(\xi, \eta) &= J([\xi, \eta]) + \lambda([\xi, \eta]) - \{J(\xi), J(\eta)\} \\ &= \Sigma(\xi, \eta) + \lambda([\xi, \eta]),\end{aligned} \qquad (12.3.16)$$

that is, $[\Sigma_\lambda] = [\Sigma]$. Letting $\rho' \in H^2(\mathfrak{g}, C(P))$ denote this cohomology class, J is infinitesimally equivariant if and only if ρ' vanishes. There are some cases in which one can predict that ρ' is zero:

(a) *Assume that P is symplectic and connected (so $C(P) = \mathbb{R}$) and suppose that $H^2(\mathfrak{g}, \mathbb{R}) = 0$. By the second Whitehead lemma (see Jacobson [1962] or Guillemin and Sternberg [1984]), this is the case whenever $\mathfrak{g}$ is semisimple; thus semisimple symplectic Lie algebra actions on symplectic manifolds are Hamiltonian.*

(b) *Suppose P is exact symplectic, $-d\Theta = \Omega$, and*

$$\pounds_{\xi_P}\Theta = 0. \qquad (12.3.17)$$

The proof of equivariance in this case is the following. Assume first that the Lie algebra $\mathfrak{g}$ has an underlying Lie group G that leaves θ invariant. Since $\left(\mathrm{Ad}_{g^{-1}} \xi\right)_P = \Phi_g^* \xi_P$, we get from (11.3.4)

$$\begin{aligned}J(\xi)(g \cdot z) &= (\mathbf{i}_{\xi_P}\Theta)(g \cdot z) = \left(\mathbf{i}_{\left(\mathrm{Ad}_{g^{-1}} \xi\right)_P}\Theta\right)(z) \\ &= J\left(\mathrm{Ad}_{g^{-1}} \xi\right)(z).\end{aligned} \qquad (12.3.18)$$

The proof without the assumption of the existence of the group G is obtained by differentiating the above string of equalities with respect to g at $g = e$.

A simple example in which $\rho' \neq 0$ is provided by phase-space translations on $\mathbb{R}^2$ defined by $\mathfrak{g} = \mathbb{R}^2 = \{(a, b)\}$, $P = \mathbb{R}^2 = \{(q, p)\}$, and

$$(a, b)_P = a\frac{\partial}{\partial q} + b\frac{\partial}{\partial p}. \qquad (12.3.19)$$

This action has a momentum map given by $\langle \mathbf{J}(q, p), (a, b) \rangle = ap - bq$ and

$$\begin{aligned}\Sigma\left((a_1, b_1), (a_2, b_2)\right) &= J\left([(a_1, b_1), (a_2, b_2)]\right) - \{J(a_1, b_1), J(a_2, b_2)\} \\ &= -\{a_1 p - b_1 q, a_2 p - b_2 q\} \\ &= b_1 a_2 - a_1 b_2.\end{aligned} \qquad (12.3.20)$$

Since $[\mathfrak{g}, \mathfrak{g}] = \{0\}$, the only coboundary is zero, so $\rho' \neq 0$. This example is amplified in Example (b) of §12.4.

4. *If P is symplectic and connected and σ is a one-cocycle of the G-action on P, then*

(a) $g \cdot \mu = \mathrm{Ad}^*_{g^{-1}} \mu + \sigma(g)$ *is an action of G on $\mathfrak{g}^*$; and*

(b) **J** *is equivariant with respect to this action.*

Indeed, since P is symplectic and connected, $\mathcal{C}(P) = \mathbb{R}$, and thus $\sigma :$ $G \rightarrow \mathfrak{g}^*$. By Proposition 12.3.1,

$$
\begin{aligned}
(gh) \cdot \mu &= \mathrm{Ad}^*_{(gh)^{-1}} \mu + \sigma(gh) \\
&= \mathrm{Ad}^*_{g^{-1}} \mathrm{Ad}^*_{h^{-1}} \mu + \sigma(g) + \mathrm{Ad}^*_{g^{-1}} \sigma(h) \\
&= \mathrm{Ad}^*_{g^{-1}}(h \cdot \mu) + \sigma(g) = g \cdot (h \cdot \mu), \qquad (12.3.21)
\end{aligned}
$$

which proves (a); (b) is a consequence of the definition.

5. *If P is symplectic and connected, $\mathbf{J} : P \rightarrow \mathfrak{g}^*$ is a momentum map, and Σ is the associated real-valued Lie algebra two-cocycle, then the momentum map $\mathbf{J}$ can be explicitly adjusted to be infinitesimally equivariant by enlarging $\mathfrak{g}$ to the central extension defined by Σ.*

Indeed, the **central extension defined by** Σ is the Lie algebra $\mathfrak{g}' :=$ $\mathfrak{g} \oplus \mathbb{R}$ with the bracket given by

$$
[(\xi, a), (\eta, b)] = ([\xi, \eta], \Sigma(\xi, \eta)). \qquad (12.3.22)
$$

Let $\mathfrak{g}'$ act on P by $\rho(\xi, a)(z) = \xi_P(z)$ and let $\mathbf{J}' : P \rightarrow (\mathfrak{g}')^* = \mathfrak{g}^* \oplus \mathbb{R}$ be the induced momentum map, that is, it satisfies

$$
X_{J'(\xi, a)} = (\xi, a)_P = X_{J(\xi)}, \qquad (12.3.23)
$$

so that

$$
J'(\xi, a) - J(\xi) = \ell(\xi, a), \qquad (12.3.24)
$$

where $\ell(\xi, a)$ is a constant on P and is linear in (ξ, a). Therefore,

$$
\begin{aligned}
J'([(\xi, a), (\eta, b)]) &- \{J'(\xi, a), J'(\eta, a)\} \\
&= J'([\xi, \eta], \Sigma(\xi, \eta)) - \{J(\xi) + \ell(\xi, a), J(\eta) + \ell(\eta, b)\} \\
&= J([\xi, \eta]) + \ell([\xi, \eta], \Sigma(\xi, \eta)) - \{J(\xi), J(\eta)\} \\
&= \Sigma(\xi, \eta) + \ell([(\xi, a), (\eta, b)]) \\
&= (\lambda + \ell)([(\xi, a), (\eta, b)]), \qquad (12.3.25)
\end{aligned}
$$

where $\lambda(\xi, a) = a$. Thus, the real-valued two-cocycle of the $\mathfrak{g}'$ action is a coboundary, and hence J' can be adjusted to become infinitesimally equivariant. Thus,

$$
J'(\xi, a) = J(\xi) - a \qquad (12.3.26)
$$

is the desired infinitesimally equivariant momentum map of $\mathfrak{g}'$ on P.

For example, the action of $\mathbb{R}^2$ on itself by translations has the nonequivariant momentum map $\langle \mathbf{J}(q,p), (\xi,\eta) \rangle = \xi p - \eta q$ with group one-cocycle $\sigma(x,y) \cdot (\xi,\eta) = \xi y - \eta x$; here we think of $\mathbb{R}^2$ endowed with the symplectic form $dq \wedge dp$. The corresponding infinitesimally equivariant momentum map of the central extension is given by (12.3.26), that is, by the expression

$$\langle \mathbf{J}'(q,p), (\xi,\eta,a) \rangle = \xi p - \eta q - a.$$

For more examples, see §12.4.

Consider the situation for the corresponding action of the central extension G' of G on P if $G = E$, a topological vector space regarded as an abelian Lie group. Then $\mathfrak{g} = E$, and $T\sigma_\eta = \sigma_\eta$ by linearity of σ_η, so that $\Sigma(\xi,\eta) = \sigma(\xi) \cdot \eta$, with ξ on the right-hand side thought of as an element of the Lie group G. One defines the central extension G' of G by the circle group S^1 as the Lie group having an underlying manifold $E \times S^1$ and whose multiplication is given by (see Souriau [1970])

$$\left(q_1, e^{i\theta_1}\right) \cdot \left(q_2, e^{i\theta_2}\right) = \left(q_1 + q_2, \exp\left\{i\left[\theta_1 + \theta_2 + \tfrac{1}{2}\Sigma(q_1,q_2)\right]\right\}\right), \quad (12.3.27)$$

identity element by $(0,1)$, and inverse by

$$\left(q, e^{i\theta}\right)^{-1} = \left(-q, e^{-i\theta}\right).$$

Then the Lie algebra of G' is $\mathfrak{g}' = E \oplus \mathbb{R}$ with the bracket given by (12.3.22), and thus the G'-action on P given by $(q, e^{i\theta}) \cdot z = q \cdot z$ has an equivariant momentum map $\mathbf{J}$ given by (12.3.26). If $E = \mathbb{R}^2$, the group G' is the **Heisenberg group** (see Exercise 9.1-4).   ◆

Global Equivariance.   Assume that J is a Lie algebra homomorphism. Since $\Gamma_{\eta,g}$ is a Casimir function on P for every $g \in G$ and $\eta \in \mathfrak{g}$, it follows that $\Gamma_\eta | G \times S$ is independent of $z \in S$, where S is a symplectic leaf. Denote this function that depends only on the leaf S by $\Gamma_\eta^S : G \to \mathbb{R}$. Fixing $z \in S$ and taking the derivative of the map $g \mapsto \Gamma_\eta^S(g,z)$ at $g = e$ in the direction $\xi \in \mathfrak{g}$ gives

$$\langle -(\mathrm{ad}\, \xi)^* \mathbf{J}(z), \eta \rangle - \langle T_z \mathbf{J} \cdot \xi_P(z), \eta \rangle = 0, \quad (12.3.28)$$

that is, $T_e \Gamma_\eta^S = 0$ for all $\eta \in \mathfrak{g}$. By Proposition 12.4.1(ii), we have

$$\Gamma_\eta(gh) = \Gamma_\eta(g) + \Gamma_{\mathrm{Ad}_{g^{-1}}\eta}(h). \quad (12.3.29)$$

Taking the derivative of (12.3.29) with respect to g in the direction ξ at $h = e$ on the leaf S and using $T_e \Gamma_\eta^S = 0$, we get

$$T_g \Gamma_\eta^S(T_e L_g(\xi)) = T_e \Gamma_{\mathrm{Ad}_{g^{-1}}\eta}^S(\xi) = 0. \quad (12.3.30)$$

Thus, Γ_η is constant on $G \times S$ (recall that both G and the symplectic leaves are, by definition, connected). Since $\Gamma_\eta(e, z) = 0$, it follows that $\Gamma_\eta | G \times S = 0$ for any leaf S and hence $\Gamma_\eta = 0$ on $G \times P$. But $\Gamma_\eta = 0$ for every $\eta \in \mathfrak{g}$ is equivalent to equivariance. Together with Theorem 11.5.1 this proves the following:

Theorem 12.3.2. *Let the connected Lie group G act canonically on the left on the Poisson manifold P. The action of G is globally Hamiltonian if and only if there is a Lie algebra homomorphism $\psi : \mathfrak{g} \to \mathcal{F}(P)$ such that $X_{\psi(\xi)} = \xi_P$ for all $\xi \in \mathfrak{g}$, where ξ_P is the infinitesimal generator of the G-action. If $\mathbf{J}$ is the equivariant momentum map of the action, then we can take $\psi = \mathbf{J}$.*

The converse question of the construction of a group action whose momentum map equals a given set of conserved quantities closed under bracketing is addressed in Fong and Meyer [1975]. See also Vinogradov and Krasilshchik [1975] and Conn [1984] for the related question of when (the germs of) Poisson vector fields are Hamiltonian.

Exercises

◇ **12.3-1.** Let G be a Lie group, $\mathfrak{g}$ its Lie algebra, and $\mathfrak{g}^*$ its dual. Let $\wedge^k(\mathfrak{g}^*)$ be the space of maps

$$\alpha : \mathfrak{g}^* \times \cdots \times \mathfrak{g}^* \ (k \text{ times}) \longrightarrow \mathbb{R}$$

such that α is k-linear and skew-symmetric. Define, for each $k \geq 1$, the map

$$\mathbf{d} : \wedge^k(\mathfrak{g}^*) \longrightarrow \wedge^{k+1}(\mathfrak{g}^*)$$

by

$$\mathbf{d}\alpha(\xi_0, \xi_1, \dots, \xi_k) = \sum_{0 \leq i < j \leq k} (-1)^{i+j} \alpha([\xi_i, \xi_j], \xi_0, \dots, \hat{\xi}_i, \dots, \hat{\xi}_j, \dots, \xi_k),$$

where $\hat{\xi}_i$ means that ξ_i is omitted.

(a) Work out $\mathbf{d}\alpha$ explicitly if $\alpha \in \wedge^1(\mathfrak{g}^*)$ and $\alpha \in \wedge^2(\mathfrak{g}^*)$.

(b) Show that if we identify $\alpha \in \wedge^k(\mathfrak{g}^*)$ with its left-invariant extension $\alpha_L \in \Omega^k(G)$ given by

$$\alpha_L(g)(v_1, \dots, v_k) = \alpha(T_e L_{g^{-1}} v_1, \dots, T_e L_{g^{-1}} v_k),$$

where $v_1, \dots, v_k \in T_g G$, then $\mathbf{d}\alpha_L$ is the left-invariant extension of $\mathbf{d}\alpha$, that is, $\mathbf{d}\alpha_L = (\mathbf{d}\alpha)_L$.

(c) Conclude that indeed $\mathbf{d}\alpha \in \wedge^{k+1}(\mathfrak{g}^*)$ if $\alpha \in \wedge^k(\mathfrak{g}^*)$ and that $\mathbf{d} \circ \mathbf{d} = 0$.

(d) Letting

$$Z^k(\mathfrak{g}) = \ker \left(\mathbf{d} : \wedge^k(\mathfrak{g}^*) \longrightarrow \wedge^{k+1}(\mathfrak{g}^*) \right)$$

be the subspace of k-cocycles and

$$B^k(\mathfrak{g}) = \text{range} \left(\mathbf{d} : \wedge^{k-1}(\mathfrak{g}^*) \longrightarrow \wedge^k(\mathfrak{g}^*) \right)$$

be the space of k-coboundaries, show that $B^k(\mathfrak{g}) \subset Z^k(\mathfrak{g})$. The quotient $H^k(\mathfrak{g})/B^k(\mathfrak{g})$ is the kth **Lie algebra cohomology** group of $\mathfrak{g}$ with real coefficients.

⋄ **12.3-2.** Compute the group and Lie algebra cocycles for the momentum map of SE(2) on $\mathbb{R}^2$ given in Exercise 11.4-3.

12.4 Equivariant Momentum Maps Are Poisson

We next show that equivariant momentum maps are Poisson maps. This provides a fundamental method for finding canonical maps between Poisson manifolds. This result is partly contained in Lie [1890], is implicit in Guillemin and Sternberg [1980], and is explicit in Holmes and Marsden [1983] and Guillemin and Sternberg [1984].

Theorem 12.4.1 (Canonical Momentum Maps). *If* $\mathbf{J} : P \to \mathfrak{g}^*$ *is an infinitesimally equivariant momentum map for a left Hamiltonian action of* $\mathfrak{g}$ *on a Poisson manifold* P, *then* $\mathbf{J}$ *is a Poisson map:*

$$\mathbf{J}^* \{F_1, F_2\}_+ = \{\mathbf{J}^* F_1, \mathbf{J}^* F_2\}, \qquad (12.4.1)$$

that is,

$$\{F_1, F_2\}_+ \circ \mathbf{J} = \{F_1 \circ \mathbf{J}, F_2 \circ \mathbf{J}\}$$

for all $F_1, F_2 \in \mathcal{F}(\mathfrak{g}^*)$, *where* $\{\,,\,\}_+$ *denotes the "+" Lie–Poisson bracket.*

Proof. Infinitesimal equivariance means that $J([\xi, \eta]) = \{J(\xi), J(\eta)\}$. For $F_1, F_2 \in \mathcal{F}(\mathfrak{g}^*)$, let $z \in P$, $\xi = \delta F_1/\delta\mu$, and $\eta = \delta F_2/\delta\mu$ evaluated at the particular point $\mu = \mathbf{J}(z) \in \mathfrak{g}^*$. Then

$$\mathbf{J}^* \{F_1, F_2\}_+ (z) = \left\langle \mu, \left[\frac{\delta F_1}{\delta\mu}, \frac{\delta F_2}{\delta\mu} \right] \right\rangle = \langle \mu, [\xi, \eta] \rangle$$

$$= J([\xi, \eta])(z) = \{J(\xi), J(\eta)\}(z).$$

But for any $z \in P$ and $v_z \in T_z P$,

$$\mathbf{d}(F_1 \circ \mathbf{J})(z) \cdot v_z = \mathbf{d}F_1(\mu) \cdot T_z \mathbf{J}(v_z) = \left\langle T_z \mathbf{J}(v_z), \frac{\delta F_1}{\delta\mu} \right\rangle$$

$$= \mathbf{d}J(\xi)(z) \cdot v_z,$$

that is, $(F_1 \circ \mathbf{J})(z)$ and $J(\xi)(z)$ have equal z-derivatives. Since the Poisson bracket on P depends only on the point values of the first derivatives, we conclude that

$$\{F_1 \circ \mathbf{J}, F_2 \circ \mathbf{J}\}(z) = \{J(\xi), J(\eta)\}(z). \qquad \blacksquare$$

Theorem 12.4.2 (Collective Hamiltonian Theorem). *Let* $\mathbf{J} : P \to \mathfrak{g}^*$ *be a momentum map. Let* $z \in P$ *and* $\mu = \mathbf{J}(z) \in \mathfrak{g}^*$. *Then for any* $F \in \mathcal{F}(\mathfrak{g}_+^*)$,

$$X_{F \circ \mathbf{J}}(z) = X_{J(\delta F/\delta\mu)}(z) = \left(\frac{\delta F}{\delta\mu}\right)_P(z). \qquad (12.4.2)$$

Proof.   For any $H \in \mathcal{F}(P)$,

$$
\begin{aligned}
X_{F \circ \mathbf{J}}[H](z) &= -X_H[F \circ \mathbf{J}](z) = -\mathbf{d}(F \circ \mathbf{J})(z) \cdot X_H(z) \\
&= -\mathbf{d}F(\mu)(T_z\mathbf{J} \cdot X_H(z)) = -\left\langle T_z\mathbf{J}(X_H(z)), \frac{\delta F}{\delta\mu}\right\rangle \\
&= -\mathbf{d}J\left(\frac{\delta F}{\delta\mu}\right)(z) \cdot X_H(z) = -X_H\left[J\left(\frac{\delta F}{\delta\mu}\right)\right](z) \\
&= X_{J(\delta F/\delta\mu)}[H](z).
\end{aligned}
$$

This proves the first equality in (12.4.2), and the second results from the definition of the momentum map. $\blacksquare$

Functions on P of the form $F \circ \mathbf{J}$ are called **collective**. Note that if F is the linear function determined by $\xi \in \mathfrak{g}$, then (12.4.2) reduces to $X_{J(\xi)}(z) = \xi_P(z)$, the definition of the momentum map. To demonstrate the relation between these results, let us derive Theorem 12.4.1 from Theorem 12.4.2. Let $\mu = \mathbf{J}(z)$ and $F, H \in \mathcal{F}(\mathfrak{g}_+^*)$. Then

$$
\begin{aligned}
\mathbf{J}^*\{F, H\}_+(z) &= \{F, H\}_+(\mathbf{J}(z)) = \left\langle \mathbf{J}(z), \left[\frac{\delta F}{\delta\mu}, \frac{\delta H}{\delta\mu}\right]\right\rangle \\
&= J\left(\left[\frac{\delta F}{\delta\mu}, \frac{\delta H}{\delta\mu}\right]\right)(z) = \left\{J\left(\frac{\delta F}{\delta\mu}\right), J\left(\frac{\delta H}{\delta\mu}\right)\right\}(z)
\end{aligned}
$$

$$\text{(by infinitesimal equivariance)}$$

$$= X_{J(\delta H/\delta\mu)}\left[J\left(\frac{\delta F}{\delta\mu}\right)\right](z) = X_{H \circ \mathbf{J}}\left[J\left(\frac{\delta F}{\delta\mu}\right)\right](z)$$

$$\text{(by the collective Hamiltonian theorem)}$$

$$= -X_{J(\delta F/\delta\mu)}[H \circ \mathbf{J}](z) = -X_{F \circ \mathbf{J}}[H \circ \mathbf{J}](z)$$

$$\text{(again by the collective Hamiltonian theorem)}$$

$$= \{F \circ \mathbf{J}, H \circ \mathbf{J}\}(z). \qquad \blacksquare$$

Remarks.

1. Let $i : \mathfrak{g} \to \mathcal{F}(\mathfrak{g}^*)$ denote the natural embedding of $\mathfrak{g}$ in its bidual; that is, $i(\xi) \cdot \mu = \langle \mu, \xi \rangle$. Since $\delta i(\xi)/\delta \mu = \xi$, it follows that i is a Lie algebra homomorphism, that is,

$$i([\xi, \eta]) = \{i(\xi), i(\eta)\}_+ . \tag{12.4.3}$$

We claim that *a canonical left Lie algebra action of* $\mathfrak{g}$ *on a Poisson manifold* P *is Hamiltonian if and only if there is a Poisson algebra homomorphism* $\chi : \mathcal{F}(\mathfrak{g}^*_+) \to \mathcal{F}(P)$ *such that* $X_{(\chi \circ i)(\xi)} = \xi_P$ *for all* $\xi \in \mathfrak{g}$. Indeed, if the action is Hamiltonian, let $\chi = \mathbf{J}^*$ (pull-back on functions), and the assertion follows from the definition of momentum maps. The converse relies on the following fact. Let M, N be finite-dimensional manifolds and $\chi : \mathcal{F}(N) \to \mathcal{F}(M)$ be a ring homomorphism. Then there exists a unique smooth map $\varphi : M \to N$ such that $\chi = \varphi^*$. (A similar statement holds for infinite-dimensional manifolds in the presence of some additional technical conditions. See Abraham, Marsden, and Ratiu [1988, Supplement 4.2C].) Therefore, if a ring and Lie algebra homomorphism $\chi : \mathcal{F}(\mathfrak{g}^*_+) \to \mathcal{F}(P)$ is given, there is a unique map $\mathbf{J} : P \to \mathfrak{g}^*$ such that $\chi = \mathbf{J}^*$. But for $\xi, \mu \in \mathfrak{g}^*$ we have

$$\begin{aligned}[(\chi \circ i)(\xi)](z) = \mathbf{J}^*(i(\xi))(z) &= i(\xi)(\mathbf{J}(z)) \\ &= \langle \mathbf{J}(z), \xi \rangle = J(\xi)(z), \tag{12.4.4}\end{aligned}$$

that is, $\chi \circ i = J$, which is a Lie algebra homomorphism because χ is by hypothesis. Since $X_{J(\xi)} = \xi_P$, again by hypothesis, it follows that $\mathbf{J}$ is an infinitesimally equivariant momentum map.

2. Here we have worked with left actions. If in all statements one changes left actions to right actions and "$+$" to "$-$" in the Lie–Poisson structures on $\mathfrak{g}^*$, the resulting statements are true.   ◆

Examples

(a) Phase Space Rotations.   Let (P, Ω) be a linear symplectic space and let G be a subgroup of the linear symplectic group acting on P by matrix multiplication. The infinitesimal generator of $\xi \in \mathfrak{g}$ at $z \in P$ is

$$\xi_P(z) = \xi z, \tag{12.4.5}$$

where ξz is matrix multiplication. This vector field is Hamiltonian with Hamiltonian $\Omega(\xi z, z)/2$ by Proposition 2.7.1. Thus, a momentum map is

$$\langle \mathbf{J}(z), \xi \rangle = \tfrac{1}{2}\Omega(\xi z, z). \tag{12.4.6}$$

For $S \in G$, the adjoint action is

$$\mathrm{Ad}_S \, \xi = S\xi S^{-1}, \qquad (12.4.7)$$

and hence

$$\langle \mathbf{J}(Sz), S\xi S^{-1} \rangle = \frac{1}{2} \, \Omega(S\xi S^{-1} Sz, Sz)$$

$$= \frac{1}{2} \, \Omega(S\xi z, Sz) = \frac{1}{2} \, \Omega(\xi z, z), \qquad (12.4.8)$$

so $\mathbf{J}$ is equivariant. Infinitesimal equivariance is a reformulation of (2.7.10). Notice that this momentum map is not of the cotangent lift type.   ◆

(b) **Phase Space Translations.**   Let (P, Ω) be a linear symplectic space and let G be a subgroup of the translation group of P, with $\mathfrak{g}$ identified with a linear subspace of P. Clearly,

$$\xi_P(z) = \xi$$

in this case. The vector field is Hamiltonian with Hamiltonian given by the linear function

$$J(\xi)(z) = \Omega(\xi, z), \qquad (12.4.9)$$

as is easily checked. This is therefore a momentum map for the action. This momentum map is not equivariant, however. The action of $\mathbb{R}^2$ on $\mathbb{R}^2$ by translation is a specific example; see the end of Remark 3 of §12.3.   ◆

(c) **Lifted Actions and Magnetic Terms.**   Another way nonequivariance of momentum maps comes up is with lifted cotangent actions, but with symplectic forms that are the canonical ones modified by the addition of a magnetic term. For example, endow $P = T^*\mathbb{R}^2$ with the symplectic form

$$\Omega_B = dq^1 \wedge dp_1 + dq^2 \wedge dp_2 + B \, dq^1 \wedge dq^2$$

where B is a function of q^1 and q^2. Consider the action of $\mathbb{R}^2$ on $\mathbb{R}^2$ by translations and lift this to an action of $\mathbb{R}^2$ on P. Note that this action preserves Ω_B if and only if B is constant, which will be assumed from now on. By (12.4.9) the momentum map is

$$\langle \mathbf{J}(\mathbf{q}, \mathbf{p}), \xi \rangle = \mathbf{p} \cdot \xi + B(\xi^1 q^2 - \xi^2 q^1). \qquad (12.4.10)$$

This momentum map is not equivariant; in fact, since $\mathbb{R}^2$ is Abelian, its Lie algebra two-cocycle is given by

$$\Sigma(\xi, \eta) = -\{J(\xi), J(\eta)\} = -2B(\xi^1 \eta^2 - \xi^2 \eta^1).$$

Let us assume from now on that B is nonzero. Viewed in different coordinates, the form Ω_B can be made canonical, and the action by $\mathbb{R}^2$ is still translation by a canonical transformation. To do this, one switches to *guiding center coordinates* $(\mathbf{R}, \mathbf{P})$ defined by $\mathbf{P} = \mathbf{p}$ and $\mathbf{R} = (q^1 - p_2/B, q^2 + p_1/B)$. The physical interpretation of these coordinates is the following: $\mathbf{P}$ is the momentum of the particle, while $\mathbf{R}$ is the center of the nearly circular orbit pursued by the particle with coordinates $(\mathbf{q}, \mathbf{p})$ when the magnetic field is strong (Littlejohn [1983, 1984]). In these coordinates, Ω_B takes the form

$$\Omega_B = B dR^1 \wedge dR^2 - \frac{1}{B} dP_1 \wedge dP_2,$$

and the $\mathbb{R}^2$-action on $T^*\mathbb{R}^2$ becomes translation in the $\mathbf{R}$-variable. The momentum map (12.4.10) becomes

$$\langle \mathbf{J}(\mathbf{R}, \mathbf{P}), \xi \rangle = B(\xi^1 R^2 - \xi^2 R^1), \qquad (12.4.11)$$

which is again a special case of (12.2.5).

The cohomology class $[\Sigma]$ is not equal to 0, as the following argument shows. If Σ were exact, there would exist a linear functional $\lambda : \mathbb{R}^2 \to \mathbb{R}$ such that $\Sigma(\xi, \eta) = \lambda([\xi, \eta]) = 0$ for all ξ, η; this is clearly false. Thus, $\mathbf{J}$ *cannot* be adjusted to obtain an equivariant momentum map.

Following Remark 5 of §12.3, the nonequivariance of the momentum map can be removed by passing to a central extension of $\mathbb{R}^2$. Namely, let $G' = \mathbb{R}^2 \times S^1$ with multiplication given by

$$(\mathbf{a}, e^{i\theta})(\mathbf{b}, e^{i\varphi}) = \left(\mathbf{a} + \mathbf{b}, e^{i(\theta + \varphi + B(a^1 b^2 - a^2 b^1))} \right) \qquad (12.4.12)$$

and letting G' act on $T^*\mathbb{R}^2$ as before by

$$(\mathbf{a}, e^{i\theta}) \cdot (\mathbf{q}, \mathbf{p}) = (\mathbf{q} + \mathbf{a}, \mathbf{p}).$$

Then the momentum map $\mathbf{J} : T^*\mathbb{R}^2 \to \mathfrak{g}'^* = \mathbb{R}^3$ is given by

$$\langle \mathbf{J}(\mathbf{q}, \mathbf{p}), (\xi, a) \rangle = \mathbf{p} \cdot \xi + B(\xi^1 q^2 - \xi^2 q^1) - a. \qquad (12.4.13)$$

$\blacklozenge$

(d) Clairaut's Theorem. Let M be a surface of revolution in $\mathbb{R}^3$ obtained by revolving a graph $r = f(z)$ about the z-axis, where f is a smooth positive function. Pull back the usual metric of $\mathbb{R}^3$ to M and note that it is invariant under rotations about the z-axis. Consider the geodesic flow on M. The momentum map associated with the S^1 symmetry is $\mathbf{J} : TM \to \mathbb{R}$ given by $\langle \mathbf{J}(\mathbf{q}, \mathbf{v}), \xi \rangle = \langle (\mathbf{q}, \mathbf{v}), \xi_M(\mathbf{q}) \rangle$, as usual. Here, ξ_M is the vector field on $\mathbb{R}^3$ associated with a rotation with angular velocity ξ about the z-axis, so $\xi_M(\mathbf{q}) = \xi \mathbf{k} \times \mathbf{q}$. Thus,

$$\langle \mathbf{J}(\mathbf{q}, \mathbf{v}), \xi \rangle = \xi r \|\mathbf{v}\| \cos \theta,$$

where r is the distance to the z-axis and θ is the angle between $\mathbf{v}$ and the horizontal plane. Thus, as $\|\mathbf{v}\|$ is conserved, by conservation of energy, $r\cos\theta$ is *conserved along any geodesic on a surface of revolution*, a statement known as **Clairaut's theorem**. ◆

(e) Mass of a nonrelativistic free quantum particle. Here we show by means of an example the relation between (genuine) projective unitary representations and nonequivariance of the momentum map for the action on the projective space. This complements the discussion in Example (m) of §12.2, where we have shown that for unitary representations the momentum map is equivariant.

Let G be the Galilean group introduced in Example (c) following Proposition 9.3.10, that is, the subgroup of $\mathrm{GL}(5, \mathbb{R})$ consisting of matrices

$$g = \begin{bmatrix} \mathbf{R} & \mathbf{v} & \mathbf{a} \\ 0 & 1 & \tau \\ 0 & 0 & 1 \end{bmatrix},$$

where $\mathbf{R} \in \mathrm{SO}(3)$, $\mathbf{v}, \mathbf{a} \in \mathbb{R}^3$, and $\tau \in \mathbb{R}$. Let $\mathcal{H} = L^2(\mathbb{R}^3; \mathbb{C})$ be the Hilbert space of square (Lebesgue) integrable complex functions on $\mathbb{R}^3$.

Fix a real number $m \neq 0$; for each $g = \{R, \mathbf{v}, \mathbf{a}, \tau\} \in G$, define the following unitary operator in $\mathcal{H}$:

$$(U_m(g)f)(\mathbf{p}) = \exp\left(i\left(\frac{\tau}{2m}|\mathbf{p}|^2 + (\mathbf{p} + m\mathbf{v}) \cdot \mathbf{a}\right)\right) f(R^{-1}(\mathbf{p} + m\mathbf{v})).$$

$$(12.4.14)$$

We can check by direct computation that

$$U_m(g_1)U_m(g_2) = \exp(-im\sigma(g_1, g_2))U_m(g_1g_2), \qquad (12.4.15)$$

where (with $g_j = \{R_j, \mathbf{v}_j, \mathbf{a}_j, \tau_j\}$)

$$\sigma(g_1, g_2) = \tfrac{1}{2}|\mathbf{v}_1|^2\tau_2 + (R_1\mathbf{v}_2) \cdot (\mathbf{v}_1\tau_2 + \mathbf{a}_1). \qquad (12.4.16)$$

Note that $\sigma(e, g) = \sigma(g, e) = 0$, $\sigma(g, g^{-1}) = \sigma(g^{-1}, g)$, and $U_m(g^{-1}) = \exp(-im\sigma(g, g^{-1}))U_m(g)^{-1}$.

From (12.4.15) we see that the map $g \mapsto U_m(g)$ is not a group homomorphism, because of an overall factor in S^1. Each $e^{i\phi} \in S^1$ defines the unitary operator $f \mapsto e^{i\phi}f$ on $\mathcal{H} = L^2(\mathbb{R}^3; \mathbb{C})$. This map assigns to each element of S^1 a unitary operator on $\mathcal{H}$ is clearly an injective group homomorphism. We shall regard the circle group S^1 as embedded in $\mathrm{U}(\mathcal{H})$ in this fashion and note that it is a normal subgroup of $\mathrm{U}(\mathcal{H})$ (since every element of S^1 commutes with every element of $\mathrm{U}(\mathcal{H})$). Define, as in §9.3, the **projective unitary group** of $\mathcal{H}$ by $\mathrm{U}(\mathbb{P}\mathcal{H}) = \mathrm{U}(\mathcal{H})/S^1$. Then (12.4.15) induces a group homomorphism $g \in G \mapsto [U_m(g)] \in \mathrm{U}(\mathbb{P}\mathcal{H})$, that is, we have a *projective unitary representation* of the Galilean group on $\mathcal{H} = L^2(\mathbb{R}^3; \mathbb{C})$. It is easy to see that this action of the Galilean group G on $\mathbb{P}\mathcal{H}$ is symplectic (use the formula in Proposition 5.3.1).

Next, we compute the infinitesimal generators of this action. Note that for any smooth $f \in \mathcal{H} = L^2(\mathbb{R}^3; \mathbb{C})$, the map $g \mapsto U_m(g)f$ is also smooth, so $\mathcal{D} := C^\infty(\mathbb{R}^3; \mathbb{C})$ is invariant under the group action. Thus, it makes sense to define for any $f \in \mathcal{D}$,

$$(a(\xi))f = T_e(U_m(\cdot)f) \cdot \xi, \qquad (12.4.17)$$

where e is the identity matrix in G and $\xi \in \mathfrak{g}$ is arbitrary. This formula shows that $a(\xi)$ is linear in ξ, thereby defining a linear operator $a : \mathcal{D} = C^\infty(\mathbb{R}^3; \mathbb{C}) \to \mathcal{H} = L^2(\mathbb{R}^3; \mathbb{C})$. Because $U_m(g)$ is unitary and $U_m(e) =$ identity operator on $\mathcal{H}$, it follows that $a(\xi)$ is formally skew adjoint on $\mathcal{D}$ for any $\xi \in \mathfrak{g}$. Explicitly, if

$$\xi = \begin{bmatrix} \hat{\omega} & \mathbf{u} & \boldsymbol{\alpha} \\ 0 & 0 & \theta \\ 0 & 0 & 0 \end{bmatrix}$$

(see Example (c) following Proposition 9.3.10), we get

$$(a(\omega)f)(\mathbf{p}) = i\left(\frac{\theta}{2m}|\mathbf{p}|^2 + \mathbf{p} \cdot \boldsymbol{\alpha}\right)f(\mathbf{p}) + (m\mathbf{u} - \omega \times \mathbf{p}) \cdot \frac{\partial f}{\partial \mathbf{p}}, \qquad (12.4.18)$$

or, expressed as a collection of four operators corresponding to $\omega, \mathbf{u}, \boldsymbol{\alpha}$, and θ,

$$(a(\omega)f)(\mathbf{p}) = -\omega \cdot \left(\mathbf{p} \times \frac{\partial f}{\partial \mathbf{p}}\right), \quad (a(\mathbf{u})f)(\mathbf{p}) = m\mathbf{u} \cdot \frac{\partial f}{\partial \mathbf{p}},$$

$$(a(\boldsymbol{\alpha})f)(\mathbf{p}) = i(\boldsymbol{\alpha} \cdot \mathbf{p})f(\mathbf{p}), \qquad (a(\theta)f)(\mathbf{p}) = i\theta\frac{|\mathbf{p}|^2}{2m}f(\mathbf{p}).$$

From these formulas we see that $a(\xi)f$ is well-defined for $f \in \mathcal{D}$ and that $\mathcal{D}$ is invariant under all $a(\xi)$ for $\xi \in \mathcal{D}$. Thus, $a(\xi)$ is uniquely determined as an unbounded skew-adjoint operator on $\mathcal{H}$. Stone's theorem (see Abraham, Marsden, and Ratiu [1988]) guarantees that

$$[\exp ta(\xi)]f = U_m(\exp t\xi)f \qquad (12.4.19)$$

is C^∞ in t with derivative at $t = 0$ equal to $a(\xi)f$, for all $f \in \mathcal{D}$. Clearly, the obvious formulas (taking equivalence classes) define $a(\xi)$ on $\mathbb{P}\mathcal{D}$, and hence conditions (i), (ii), and (iii) of Example (f) at the end of §9.3 hold; therefore, $\mathbb{P}\mathcal{D}$ is an essential G-smooth part of $\mathbb{P}\mathcal{H}$. The momentum map of the projective unitary representation of the Galilean group G on $\mathbb{P}\mathcal{H}$ can thus be defined on $\mathbb{P}\mathcal{D}$. By Example (h) of §11.5, this momentum map is induced from that of the G-action on $\mathcal{H}$ and has thus the expression

$$J(\xi)([f]) = -\frac{i}{2}\frac{\langle f, a(\xi)f\rangle}{\|f\|^2} \quad \text{for } f \neq 0. \qquad (12.4.20)$$

In spite of the fact that (12.4.20) and (11.4.24) look practically the same, the corresponding momentum maps have different properties because the infinitesimal generators $a(\xi)$ behave differently from $A(\xi)$: In (11.4.24), $A(\xi)$ is uniquely determined by ξ, but here $a(\xi)$ is given by the projective representation only up to a linear functional on $\mathfrak{g}$. More crucially, the equivariance relation (12.2.20), which holds for the unitary representation, fails for projective representation. Indeed, let us show that

$$a(\mathrm{Ad}_g \xi) = U_m(g)a(\xi)U_m(g)^{-1} + 2i\Gamma_\xi(g^{-1})1_{\mathcal{H}}, \qquad (12.4.21)$$

where $1_{\mathcal{H}}$ is the identity operator on $\mathcal{H}$ and $\Gamma_\xi(g^{-1}) \in \mathbb{R}$ is a number explicitly computed below. To show this, note that from (12.4.19) and (12.4.15) we get

$$
\begin{aligned}
e^{ta(\mathrm{Ad}_g \xi)} &= U_m(\exp t \, \mathrm{Ad}_g \xi) = U_m(g(\exp t\xi)g^{-1}) \\
&= U_m(g)U_m(\exp t\xi)U_m(g)^{-1}\exp(im\gamma(g,t\xi)), \qquad (12.4.22)
\end{aligned}
$$

where

$$\gamma(g,t\xi) = \sigma(g,(\exp t\xi)g^{-1}) + \sigma(\exp t\xi, g^{-1}) - \sigma(g,g^{-1}). \qquad (12.4.23)$$

Note that $\gamma(g,0) = 0$. Taking the derivative of (12.4.22) with respect to t at $t = 0$ and using Stone's theorem, we get (12.4.21) with

$$\Gamma_\xi(g^{-1}) = \frac{m}{2}\frac{d}{dt}\gamma(g,t\xi)\Big|_{t=0}. \qquad (12.4.24)$$

Using the notation in §9.3, (12.4.16), and (12.4.23), we have for

$$\xi = \{\omega, \mathbf{u}, \alpha, \theta\} \quad \text{and} \quad g = \{\mathbf{R}, \mathbf{v}, \mathbf{a}, \tau\},$$

$$\Gamma_\xi(g^{-1}) = \frac{m}{2}\left(-\frac{1}{2}|\mathbf{v}|^2\theta + (\mathbf{R}\omega)\cdot(\mathbf{a}\times\mathbf{v}) + \mathbf{a}\cdot\mathbf{R}\mathbf{u} - \mathbf{v}\cdot\mathbf{R}\alpha\right), \qquad (12.4.25)$$

which implies, using

$$
g^{-1} = \begin{bmatrix} \mathbf{R}^{-1} & -\mathbf{R}^{-1}\mathbf{v} & \mathbf{R}^{-1}(\tau\mathbf{v}-\mathbf{a}) \\ 0 & 1 & -\tau \\ 0 & 0 & 1 \end{bmatrix},
$$

that

$$\Gamma_\xi(g) = \frac{m}{2}\left(-\frac{1}{2}|\mathbf{v}|^2\theta + \omega\cdot(\mathbf{a}\times\mathbf{v}) + (\tau\mathbf{v}-\mathbf{a})\cdot\mathbf{u} + \mathbf{v}\cdot\alpha\right). \qquad (12.4.26)$$

The $\mathfrak{g}^*$-valued group one-cocycle defined by the momentum map (12.4.20) is thus given by

$$J(\xi)(g\cdot[f]) - J(\mathrm{Ad}_{g^{-1}}\xi)([f]) = \Gamma_\xi(g),$$

in agreement with the notation of Proposition 12.3.1. The real-valued Lie algebra two-cocycle is thus given by (see 12.3.11)

$$\Sigma(\xi,\eta) = T_e\Gamma_\eta(\xi) = \left.\frac{d}{dt}\right|_{t=0} \Gamma_\eta(c(t))$$

$$= \frac{m}{2}(\mathbf{u}\cdot\boldsymbol{\alpha}' - \mathbf{u}'\cdot\boldsymbol{\alpha}), \qquad (12.4.27)$$

where $\xi = \{\omega, \mathbf{u}, \boldsymbol{\alpha}, \theta\}$, $\eta = \{\omega', \mathbf{u}', \boldsymbol{\alpha}', \theta'\}$, and $c(t) = \{e^{t\hat{\omega}}, t\mathbf{u}, t\boldsymbol{\alpha}, t\theta\}$. This cocycle on the Lie algebra is nontrivial, that is, its cohomology class is nonzero (see Exercise 12.4-6). Therefore, the mass of the particle measures the obstruction to equivariance for the momentum map (or for the projective representation to be a unitary representation) in $H^2(\mathfrak{g},\mathbb{R})$. ◆

Exercises

◇ **12.4-1.** Verify directly that angular momentum is a Poisson map.

◇ **12.4-2.** What does the collective Hamiltonian theorem state for angular momentum? Is the result obvious?

◇ **12.4-3.** If $z(t)$ is an integral curve of $X_{F\circ\mathbf{J}}$, show that $\mu(t) = \mathbf{J}(z(t))$ satisfies $\dot{\mu} = \mathrm{ad}^*_{\delta F/\delta\mu}\,\mu$.

◇ **12.4-4.** Consider an ellipsoid of revolution in $\mathbb{R}^3$ and a geodesic starting at the "equator" making an angle of α with the equator. Use Clairaut's theorem to derive a bound on how high the geodesic climbs up the ellipse.

◇ **12.4-5.** Consider the action of SE(2) on $\mathbb{R}^2$ as described in Exercise 11.4-3. Since this action was not defined as a lift, Theorem 12.1.4 is not applicable. In fact, in Exercise 11.6-2 it was shown that this momentum map is not equivariant. Compute the group and Lie algebra cocycles defined by this momentum map. Find the Lie algebra central extension making the momentum map equivariant.

◇ **12.4-6.** Using Exercise 12.4-1, show that for the Galilean algebra, any 2-coboundary has the form

$$\lambda(\xi,\xi') = \mathbf{x}\cdot(\omega\times\omega') + \mathbf{y}\cdot(\omega\times\mathbf{u}' - \omega'\times\mathbf{u}) + \mathbf{z}\cdot(\omega\times\boldsymbol{\alpha}' - \omega'\times\boldsymbol{\alpha} + u\theta' - u'\theta),$$

for some $\mathbf{x}, \mathbf{y}, \mathbf{z} \in \mathbb{R}^3$, where

$$\xi = \{\omega, \mathbf{u}, \boldsymbol{\alpha}, \theta\} \quad \text{and} \quad \xi' = \{\omega', \mathbf{u}', \boldsymbol{\alpha}', \theta'\}.$$

Conclude that the cocycle Σ in Example (e) (see 12.4.27) is not a coboundary. (It can be proven that $H^2(\mathfrak{g},\mathbb{R}) \cong \mathbb{R}$, that is, it is 1-dimensional, but this requires more algebraic work (Guillemin and Sternberg [1977, 1984]).)

◇ **12.4-7.** Deduce the formula for the momentum map in Exercise 11.5-4(v) from (12.4.6) given in Example (a).

12.5   Poisson Automorphisms

Here are some miscellaneous facts about Poisson automorphisms, symplectic leaves, and momentum maps. For a Poisson manifold P, define the following Lie subalgebras of $\mathfrak{X}(P)$:

- **Infinitesimal Poisson Automorphisms.** Let $\mathcal{P}(P)$ be the set of $X \in \mathfrak{X}(P)$ such that
$$X[\{F_1, F_2\}] = \{X[F_1], F_2\} + \{F_1, X[F_2]\}.$$

- **Infinitesimal Poisson Automorphisms Preserving Leaves.** Let $\mathcal{PL}(P)$ be the set of $X \in \mathcal{P}(P)$ such that $X(z) \in T_z S$, where S is the symplectic leaf containing $z \in P$.

- **Locally Hamiltonian Vector Fields.** Let $\mathcal{LH}(P)$ be the set of $X \in \mathfrak{X}(P)$ such that for each $z \in P$, there is an open neighborhood U of z and an $F \in \mathcal{F}(U)$ such that $X|U = X_F|U$.

- **Hamiltonian Vector Fields.** Let $\mathcal{H}(P)$ be the set of Hamiltonian vector fields X_F for $F \in \mathcal{F}(P)$.

Then one has the following facts (references are given if the verification is not straightforward):

1. $\mathcal{H}(P) \subset \mathcal{LH}(P) \subset \mathcal{PL}(P) \subset \mathcal{P}(P)$.

2. If P is symplectic, then $\mathcal{LH}(P) = \mathcal{PL}(P) = \mathcal{P}(P)$, and if $H^1(P) = 0$, then $\mathcal{LH}(P) = \mathcal{H}(P)$.

3. Let P be the trivial Poisson manifold, that is, $\{F, G\} = 0$ for all $F, G \in \mathcal{F}(P)$. Then $\mathcal{P}(P) \neq \mathcal{PL}(P)$.

4. Let $P = \mathbb{R}^2$ with the bracket
$$\{F, G\}(x, y) = x \left(\frac{\partial F}{\partial x} \frac{\partial G}{\partial y} - \frac{\partial G}{\partial x} \frac{\partial F}{\partial y} \right).$$

   This is, in fact, a Lie–Poisson bracket. The vector field
$$X(x, y) = xy \frac{\partial}{\partial y}$$

   is an example of an element of $\mathcal{PL}(P)$ that is not in $\mathcal{LH}(P)$.

5. $\mathcal{H}(P)$ is an ideal in any of the three Lie algebras including it. Indeed, if $Y \in \mathcal{P}(P)$ and $H \in \mathcal{F}(R)$, then $[Y, X_H] = X_{Y[H]}$.

6. If P is symplectic, then $[\mathcal{LH}(P), \mathcal{LH}(P)] \subset \mathcal{H}(P)$. (The Hamiltonian for $[X, Y]$ is $-\Omega(X, Y)$.) This is false for Poisson manifolds in general. If P is symplectic, Calabi [1970] and Lichnerowicz [1973] showed that $[\mathcal{LH}(P), \mathcal{LH}(P)] = \mathcal{H}(P)$.

7. If the Lie algebra $\mathfrak{g}$ admits a momentum map on P, then $\mathfrak{g}_P \subset \mathcal{H}(P)$.

8. Let G be a connected Lie group. If the action admits a momentum map, it preserves the leaves of P. The proof was given in §12.4.

12.6 Momentum Maps and Casimir Functions

In this section we return to Casimir functions studied in Chapter 10 and link them with momentum maps. We will do this in the context of the Poisson manifolds P/G studied in §10.7.

We start with a Poisson manifold P and a free and proper Poisson action of a Lie group G on P admitting an equivariant momentum mapping $\mathbf{J} : P \to \mathfrak{g}^*$. We want to link $\mathbf{J}$ with a Casimir function $C : P/G \to \mathbb{R}$.

Proposition 12.6.1.   *Let $\Phi : \mathfrak{g}^* \to \mathbb{R}$ be a function that is invariant under the coadjoint action. Then:*

 (i) *Φ is a Casimir function for the Lie–Poisson bracket;*

 (ii) *$\Phi \circ \mathbf{J}$ is G-invariant on P and so defines a function $C : P/G \to \mathbb{R}$ such that $\Phi \circ \mathbf{J} = C \circ \pi$, as in Figure 12.8.1; and*

 (iii) *the function C is a Casimir function on P/G.*

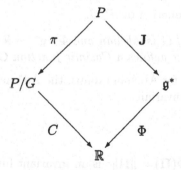

FIGURE 12.6.1. Casimir functions and momentum maps.

Proof.   To prove the first part, we write down the condition of Ad*-invariance as

$$\Phi(\mathrm{Ad}^*_{g^{-1}} \mu) = \Phi(\mu). \qquad (12.6.1)$$

Differentiate this relation with respect to g at $g = e$ in the direction η to get (see equation (9.3.2)),

$$0 = \left.\frac{d}{dt}\right|_{t=0} \Phi\left(\mathrm{Ad}^*_{\exp(-t\eta)}\,\mu\right) = -\mathbf{D}\Phi(\mu) \cdot \mathrm{ad}^*_\eta\,\mu, \qquad (12.6.2)$$

for all $\eta \in \mathfrak{g}$. Thus, by definition of $\delta\Phi/\delta\mu$,

$$0 = \left\langle \mathrm{ad}^*_\eta\,\mu, \frac{\delta\Phi}{\delta\mu} \right\rangle = \left\langle \mu, \mathrm{ad}_\eta\,\frac{\delta\Phi}{\delta\mu} \right\rangle = -\langle \mathrm{ad}^*_{\delta\Phi/\delta\mu}\,\mu, \eta \rangle$$

for all $\eta \in \mathfrak{g}$. In other words,

$$\mathrm{ad}^*_{\delta\Phi/\delta\mu}\,\mu = 0,$$

so by Proposition 10.7.1, $X_\Phi = 0$ and thus Φ is a Casimir function.

To prove the second part, note that by *equivariance* of $\mathbf{J}$ and *invariance* of Φ,

$$\Phi(\mathbf{J}(g \cdot z)) = \Phi(\mathrm{Ad}^*_{g^{-1}}\,\mathbf{J}(z)) = \Phi(\mathbf{J}(z)),$$

so $\Phi \circ \mathbf{J}$ is G-invariant.

Finally, for the third part, we use the collective Hamiltonian theorem 12.4.2 to get for $\mu = \mathbf{J}(z)$,

$$X_{\Phi \circ \mathbf{J}}(z) = \left(\frac{\delta\Phi}{\delta\mu}\right)_P (z),$$

and so $T_z\pi \cdot X_{\Phi \circ \mathbf{J}}(z) = 0$, since infinitesimal generators are tangent to orbits, so project to zero under π. But π is Poisson, so

$$0 = T_z\pi \cdot X_{\Phi \circ \mathbf{J}}(z) = T_z\pi \cdot X_{C \circ \pi}(z) = X_C(\pi(z)).$$

Thus, C is a Casimir function on P/G.   ∎

Corollary 12.6.2. *If G is Abelian and $\Phi : \mathfrak{g}^* \to \mathbb{R}$ is any smooth function, then $\Phi \circ \mathbf{J} = C \circ \pi$ defines a Casimir function C on P/G.*

This follows because for Abelian groups, the Ad^*-action is trivial, so any function on $\mathfrak{g}^*$ is Ad^*-invariant.

Exercises

⋄ **12.6-1.**   Verify that $\Phi(\mathbf{\Pi}) = \|\mathbf{\Pi}\|^2$ is an invariant function on $\mathfrak{so}(3)^*$.

⋄ **12.6-2.**   Use Corollary 12.6.2 to find the Casimir functions for the bracket (10.5.6).

⋄ **12.6-3.**   Show that a left-invariant Hamiltonian $H : T^*G \to \mathbb{R}$ collectivizes relative to the momentum map for the *right* action but need not collectivize for the momentum map of the *left* action.

13
Lie–Poisson and Euler–Poincaré Reduction

Besides the Poisson structure on a symplectic manifold, the Lie–Poisson bracket on $\mathfrak{g}^*$, the dual of a Lie algebra, is perhaps the most fundamental example of a Poisson structure. We shall obtain it in the following manner. Given two smooth functions $F, H \in \mathcal{F}(\mathfrak{g}^*)$, we extend them to functions F_L, H_L (respectively, F_R, H_R) on all T^*G by left (respectively, right) translations. The bracket $\{F_L, H_L\}$ (respectively, $\{F_R, H_R\}$) is taken in the canonical symplectic structure Ω on T^*G. The result is then restricted to $\mathfrak{g}^*$ regarded as the cotangent space at the identity; this defines $\{F, H\}$. We shall prove that one gets the Lie–Poisson bracket this way. This process is called *Lie–Poisson reduction*. In §14.6 we show that the symplectic leaves of this bracket are the coadjoint orbits in $\mathfrak{g}^*$.

There is another side to the story, where the basic objects that are reduced are not Poisson brackets but rather are variational principles. This aspect, which takes place on $\mathfrak{g}$ rather than on $\mathfrak{g}^*$, will be told as well. The passage of a variational principle from TG to $\mathfrak{g}$ is called *Euler–Poincaré reduction*.

13.1   The Lie–Poisson Reduction Theorem

We begin by studying the way the canonical Poisson bracket on T^*G is related to the Lie–Poisson bracket on $\mathfrak{g}^*$.

Theorem 13.1.1 (The Lie–Poisson Reduction Theorem). *Identifying the set of functions on $\mathfrak{g}^*$ with the set of left (respectively, right) invariant* *functions on T^*G endows $\mathfrak{g}^*$ with Poisson structures given by*

$$\{F, H\}_{\pm}(\mu) = \pm \left\langle \mu, \left[\frac{\delta F}{\delta \mu}, \frac{\delta H}{\delta \mu} \right] \right\rangle. \tag{13.1.1}$$

The space $\mathfrak{g}^*$ with this Poisson structure is denoted by $\mathfrak{g}^*_-$ (respectively, $\mathfrak{g}^*_+$). In contexts where the choice of left or right is clear, we shall drop the " $-$ " or " $+$ " from $\{F, H\}_-$ and $\{F, H\}_+$.

Following the terminology introduced by Marsden and Weinstein [1983], this bracket on $\mathfrak{g}^*$ is called the **Lie–Poisson bracket**; the bracket is given explicitly in Lie [1890, p. 204]. See Weinstein [1983a] and §13.7 below for more historical information. In fact, there are already some hints of this structure in Jacobi [1866, p. 7]. It was rediscovered several times since Lie's work. For example, it appears explicitly in Berezin [1967]. It is closely related to results of Arnold, Kirillov, Kostant, and Souriau in the 1960s.

Some Terminology.   Before proving the theorem, we explain the terminology used in its statement. First, recall from Chapter 9 how the Lie algebra of a Lie group G is constructed. We define $\mathfrak{g} = T_eG$, the tangent space at the identity. For $\xi \in \mathfrak{g}$ we define a left-invariant vector field $\xi_L = X_\xi$ on G by setting

$$\xi_L(g) = T_eL_g \cdot \xi, \tag{13.1.2}$$

where $L_g : G \to G$ denotes left translation by $g \in G$ and is defined by $L_g h = gh$. Given $\xi, \eta \in \mathfrak{g}$, define

$$[\xi, \eta] = [\xi_L, \eta_L](e), \tag{13.1.3}$$

where the bracket on the right-hand side is the Jacobi–Lie bracket on vector fields. The bracket (13.1.3) makes $\mathfrak{g}$ into a Lie algebra, that is, $[\,,]$ is bilinear and antisymmetric, and it satisfies Jacobi's identity. For example, if G is a subgroup of $\mathrm{GL}(n)$, the group of invertible $n \times n$ matrices, we identify $\mathfrak{g} = T_eG$ with a vector space of matrices, and then, as we calculated in Chapter 9,

$$[\xi, \eta] = \xi\eta - \eta\xi \tag{13.1.4}$$

is the usual commutator of matrices.

A function $F_L : T^*G \to \mathbb{R}$ is called **left invariant** if for all $g \in G$,

$$F_L \circ T^*L_g = F_L, \tag{13.1.5}$$

where T^*L_g denotes the cotangent lift of L_g, so T^*L_g is the pointwise adjoint of TL_g. Let $\mathcal{F}_L(T^*G)$ denote the space of all smooth left-invariant functions on T^*G. One similarly defines **right-invariant** functions on T^*G and the space $\mathcal{F}_R(T^*G)$. Given $F : \mathfrak{g}^* \to \mathbb{R}$ and $\alpha_g \in T^*G$, set

$$F_L(\alpha_g) = F(T_e^*L_g \cdot \alpha_g) = (F \circ \mathbf{J}_R)(\alpha_g), \tag{13.1.6}$$

where $\mathbf{J}_R : T^*G \to \mathfrak{g}^*$, $\mathbf{J}_R(\alpha_g) = T_e^* L_g \cdot \alpha_g$ is the momentum map of the lift of right translation on G (see (12.2.8). The function $F_L = F \circ \mathbf{J}_R$ is called the **left-invariant extension** of F from $\mathfrak{g}^*$ to T^*G. One similarly defines the **right-invariant extension** by

$$F_R(\alpha_g) = F(T_e^* R_g \cdot \alpha_g) = (F \circ \mathbf{J}_L)(\alpha_g), \tag{13.1.7}$$

where $\mathbf{J}_L : T^*G \to \mathfrak{g}^*$, $\mathbf{J}_L(\alpha_g) = T_e^* R_g \cdot \alpha_g$, is the momentum map of the lift of left translation on G (see (12.2.7)).

Right composition with $\mathbf{J}_R$ (respectively, $\mathbf{J}_L$) thus defines an isomorphism $\mathcal{F}(\mathfrak{g}^*) \to \mathcal{F}_L(T^*G)$ (respectively, $\mathcal{F}(\mathfrak{g}^*) \to \mathcal{F}_R(T^*G)$) whose inverse is restriction to the fiber $T_e^* G = \mathfrak{g}^*$.

Since $T^* L_g$ and $T^* R_g$ are symplectic maps on T^*G, it follows that $\mathcal{F}_L(T^*G)$ and $\mathcal{F}_R(T^*G)$ are closed under the canonical Poisson bracket on T^*G. Thus, one way of rephrasing the Lie–Poisson reduction theorem (we will see another way, using quotients, in §13.3) is to say that the above isomorphisms of $\mathcal{F}(\mathfrak{g}^*)$ with $\mathcal{F}_L(T^*G)$ and $\mathcal{F}_R(T^*G)$, respectively, are also isomorphisms of Lie algebras; that is, the following formulas are valid:

$$\{F, H\}_- = \{F_L, H_L\}|\mathfrak{g}^* \tag{13.1.8}$$

and

$$\{F, H\}_+ = \{F_R, H_R\}|\mathfrak{g}^*, \tag{13.1.9}$$

where $\{\,,\}_\pm$ is the Lie–Poisson bracket on $\mathfrak{g}^*$ and $\{\,,\}$ is the canonical bracket on T^*G.

Proof of the Lie–Poisson Reduction Theorem. The map

$$\mathbf{J}_R : T^*G \to \mathfrak{g}_-^*$$

is a Poisson map by Theorem 12.4.1. Therefore,

$$\{F, H\}_- \circ \mathbf{J}_R = \{F \circ \mathbf{J}_R, H \circ \mathbf{J}_R\} = \{F_L, H_L\}.$$

Restriction of this relation to $\mathfrak{g}^*$ gives (13.1.8). One similarly proves (13.1.9) using the Poisson property of the map $\mathbf{J}_L : T^*G \to \mathfrak{g}_+^*$. ∎

The proof above was *a posteriori*, that is, one had to "already know" the formula for the Lie–Poisson bracket. In §13.3 we will prove this theorem again using momentum functions and quotienting by G (see §10.7). This will represent an *a priori* proof, in the sense that the formula for the Lie–Poisson bracket will be deduced as part of the proof. To gain further insight into this, the next two sections will give constructive proofs of this theorem, in special cases.

Exercises

$\diamond$ **13.1-1.** Let $\mathbf{u}, \mathbf{v} \in \mathbb{R}^3$ and define $F^{\mathbf{u}} : so(3)^* \cong \mathbb{R}^3 \to \mathbb{R}$ by $F_{\mathbf{u}}(\mathbf{x}) = \langle \mathbf{x}, \mathbf{u} \rangle$ and similarly for $F^{\mathbf{v}}$. Let $F_L^{\mathbf{u}} : T^* SO(3) \to \mathbb{R}$ be the left-invariant extension of $F^{\mathbf{u}}$, and similarly for $F_L^{\mathbf{v}}$. Compute the Poisson bracket $\{F_L^{\mathbf{u}}, F_L^{\mathbf{v}}\}$.

13.2 Proof of the Lie–Poisson Reduction Theorem for $GL(n)$

We now prove the Lie–Poisson reduction theorem for the special case of the Lie group $G = GL(n)$ of real invertible $n \times n$ matrices. This is a purely pedagogical exercise, since we know by the general theory that it is true. Nevertheless, by looking at how the proof proceeds in special cases can be instructive. In the Internet supplement we also give a direct proof for the case of the group of volume-preserving diffeomorphisms as well as the group of symplectic diffeomorphisms.

Left translation by $\mathbf{U} \in G$ is given by matrix multiplication: $L_{\mathbf{U}}\mathbf{A} = \mathbf{U}\mathbf{A}$. Identify the tangent space to G at $\mathbf{A}$ with the vector space of all $n \times n$ matrices, so for $\mathbf{B} \in T_{\mathbf{A}}G$,

$$T_{\mathbf{A}} L_{\mathbf{U}} \cdot \mathbf{B} = \mathbf{U}\mathbf{B}$$

as well, since $L_{\mathbf{U}}\mathbf{A}$ is linear in $\mathbf{A}$. The cotangent space is identified with the tangent space via the pairing

$$\langle \pi, \mathbf{B} \rangle = \operatorname{trace}(\pi^T \mathbf{B}), \tag{13.2.1}$$

where π^T is the transpose of π. The cotangent lift of $L_{\mathbf{U}}$ is thus given by

$$\langle T^* L_{\mathbf{U}} \pi, \mathbf{B} \rangle = \langle \pi, T L_{\mathbf{U}} \cdot \mathbf{B} \rangle = \operatorname{trace}(\pi^T \mathbf{U}\mathbf{B});$$

that is,

$$T^* L_{\mathbf{U}} \pi = \mathbf{U}^T \pi. \tag{13.2.2}$$

Given functions $F, G : \mathfrak{g}^* \to \mathbb{R}$, let

$$F_L(\mathbf{A}, \pi) = F(\mathbf{A}^T \pi) \quad \text{and} \quad G_L(\mathbf{A}, \pi) = G(\mathbf{A}^T \pi) \tag{13.2.3}$$

be their left-invariant extensions. By the chain rule, letting $\mu = \mathbf{A}^T \pi$, we get

$$\mathbf{D}_{\mathbf{A}} F_L(\mathbf{A}, \pi) \cdot \delta \mathbf{A} = \mathbf{D}F(\mathbf{A}^T \pi) \cdot (\delta \mathbf{A})^T \pi = \left\langle (\delta \mathbf{A})^T \pi, \frac{\delta F}{\delta \mu} \right\rangle$$

$$= \operatorname{trace}\left(\pi^T \delta \mathbf{A} \frac{\delta F}{\delta \mu} \right). \tag{13.2.4}$$

The canonical bracket is therefore

$$\{F_L, G_L\} = \left\langle \frac{\delta F_L}{\delta \mathbf{A}}, \frac{\delta G_L}{\delta \pi} \right\rangle - \left\langle \frac{\delta G_L}{\delta \mathbf{A}}, \frac{\delta F_L}{\delta \pi} \right\rangle$$

$$= \mathbf{D_A} F_L(\mathbf{A}, \pi) \cdot \frac{\delta G_L}{\delta \pi} - \mathbf{D_A} G_L(\mathbf{A}, \pi) \cdot \frac{\delta F_L}{\delta \pi}. \qquad (13.2.5)$$

Since $\delta F_L/\delta \pi = \delta F/\delta \mu$ at the identity $\mathbf{A} = \mathrm{Id}$, where $\pi = \mu$, using (13.2.4) the Poisson bracket (13.2.5) becomes

$$\{F_L, G_L\}(\mu) = \mathrm{trace}\left(\mu^T \frac{\delta G}{\delta \mu} \frac{\delta F}{\delta \mu} - \mu^T \frac{\delta F}{\delta \mu} \frac{\delta G}{\delta \mu} \right)$$

$$= -\left\langle \mu, \frac{\delta F}{\delta \mu} \frac{\delta G}{\delta \mu} - \frac{\delta G}{\delta \mu} \frac{\delta F}{\delta \mu} \right\rangle = -\left\langle \mu, \left[\frac{\delta F}{\delta \mu}, \frac{\delta G}{\delta \mu} \right] \right\rangle, \qquad (13.2.6)$$

which is the $(-)$ Lie–Poisson bracket. This derivation can be adapted for other matrix groups, including the rotation group $SO(3)$, as special cases. However, in the latter case one has to be extremely careful to treat the orthogonality constraint properly.

Exercises

$\diamond$ **13.2-1.** Let F_L and G_L have the form (13.2.3), so that it makes sense to restrict F_L and G_L to $T^* SO(3)$. Is the bracket of their restrictions given by the restriction of (13.2.5)?

13.3 Lie–Poisson Reduction Using Momentum Functions

Identifying the Quotient as $\mathfrak{g}^*$. Now we turn to a *constructive proof* of the Lie–Poisson reduction theorem using momentum functions. We begin by observing that T^*G/G is diffeomorphic to $\mathfrak{g}^*$. To see this, note that the trivialization of T^*G by left translations given by

$$\lambda : \alpha_g \in T_g^* G \mapsto (g, T_e^* L_g(\alpha_g)) = (g, \mathbf{J}_R(\alpha_g)) \in G \times \mathfrak{g}^*$$

transforms the usual cotangent lift of left translation on G into the G-action on $G \times \mathfrak{g}^*$ given by

$$g \cdot (h, \mu) = (gh, \mu), \qquad (13.3.1)$$

for $g, h \in G$ and $\mu \in \mathfrak{g}^*$. Therefore, T^*G/G is diffeomorphic to $(G \times \mathfrak{g}^*)/G$, which in turn equals $\mathfrak{g}^*$, since G does not act on $\mathfrak{g}^*$ (see (13.3.1)). Thus, we can regard $\mathbf{J}_R : T^*G \to \mathfrak{g}^*$ as the canonical projection $T^*G \to T^*G/G$, and as a consequence of the Poisson reduction theorem (Chapter 10), $\mathfrak{g}^*$ inherits a Poisson bracket, which we will call $\{\,,\,\}_-$ for the time being, uniquely characterized by the relation

$$\{F, H\}_- \circ \mathbf{J}_R = \{F \circ \mathbf{J}_R, H \circ \mathbf{J}_R\} \tag{13.3.2}$$

for any functions $F, H \in \mathcal{F}(\mathfrak{g}^*)$. The goal of this section is to explicitly compute this bracket $\{\,,\,\}_-$ and to discover at the end that it equals the $(-)$ Lie–Poisson bracket.

Before beginning the proof, it is useful to recall that the Poisson bracket $\{F, H\}_-$ for $F, H \in \mathcal{F}(\mathfrak{g}^*)$ depends only on the differentials of F and H at each point. Thus, in determining the bracket $\{\,,\,\}_-$ on $\mathfrak{g}^*$, it is enough to assume that F and H are linear functions on $\mathfrak{g}^*$.

Proof of the Lie–Poisson Reduction Theorem.   The space $\mathcal{F}_L(T^*G)$ of left-invariant functions on T^*G is isomorphic (as a vector space) to $\mathcal{F}(\mathfrak{g}^*)$, the space of all functions on the dual $\mathfrak{g}^*$ of the Lie algebra $\mathfrak{g}$ of G. This isomorphism is given by $F \in \mathcal{F}(\mathfrak{g}^*) \leftrightarrow F_L \in \mathcal{F}_L(T^*G)$, where

$$F_L(\alpha_g) = F(T_e^* L_g \cdot \alpha_g). \tag{13.3.3}$$

Since $\mathcal{F}_L(T^*G)$ is closed under bracketing (which follows because $T^* L_g$ is a symplectic map), $\mathcal{F}(\mathfrak{g}^*)$ gets endowed with a unique Poisson structure. As we remarked just above, it is enough to consider the case in which F is replaced by its linearization at a particular point. This means that it is enough to prove the Lie–Poisson reduction theorem for linear functions on $\mathfrak{g}^*$. If F is linear, we can write $F(\mu) = \langle \mu, \delta F/\delta\mu \rangle$, where $\delta F/\delta\mu$ is a constant in $\mathfrak{g}$, so that letting $\mu = T_e^* L_g \cdot \alpha_g$, we get

$$F_L(\alpha_g) = F(T_e^* L_g \cdot \alpha_g) = \left\langle T_e^* L_g \cdot \alpha_g, \frac{\delta F}{\delta\mu} \right\rangle$$

$$= \left\langle \alpha_g, T_e L_g \cdot \frac{\delta F}{\delta\mu} \right\rangle = \mathcal{P}\left(\left(\frac{\delta F}{\delta\mu} \right)_L \right)(\alpha_g), \tag{13.3.4}$$

where $\xi_L(g) = T_e L_g(\xi)$ is the left-invariant vector field on G whose value at e is $\xi \in \mathfrak{g}$. Thus, by (12.1.3), (13.3.4), and the definition of the Lie algebra bracket, we have

$$\{F_L, H_L\}(\mu) = \left\{ \mathcal{P}\left(\left(\frac{\delta F}{\delta\mu} \right)_L \right), \mathcal{P}\left(\left(\frac{\delta H}{\delta\mu} \right)_L \right) \right\}(\mu)$$

$$= -\mathcal{P}\left(\left[\left(\frac{\delta F}{\delta\mu} \right)_L, \left(\frac{\delta H}{\delta\mu} \right)_L \right] \right)(\mu)$$

$$= -\mathcal{P}\left(\left[\frac{\delta F}{\delta\mu}, \frac{\delta H}{\delta\mu} \right]_L \right)(\mu) = -\left\langle \mu, \left[\frac{\delta F}{\delta\mu}, \frac{\delta H}{\delta\mu} \right] \right\rangle, \tag{13.3.5}$$

as required. Since

$$F \circ \mathbf{J}_R = F_L \quad \text{and} \quad H \circ \mathbf{J}_R = H_L,$$

formulas (13.4.2) and (13.4.6) give

$$\{F, H\}_-(\mu) = \{F_L, H_L\}(\mu) = - \left\langle \mu, \left[\frac{\partial F}{\partial \mu}, \frac{\partial H}{\partial \mu} \right] \right\rangle,$$

that is, the bracket $\{ , \}_-$ introduced by identifying T^*G/G with $\mathfrak{g}^*$ equals the $(-)$ Lie–Poisson bracket.

The formula with "$+$" follows in a similar way by making use of right-invariant extensions of linear functions, since the Lie bracket of two right-invariant vector fields equals minus the Lie algebra bracket of their generators. ∎

13.4   Reduction and Reconstruction of Dynamics

Reduction of Dynamics.   In the last sections we have focused on reducing the *Poisson structure* from T^*G to $\mathfrak{g}^*$. However, it is also very important to reduce the *dynamics* of a given Hamiltonian. The next theorem treats this, which is very useful in examples.

Theorem 13.4.1 (Lie–Poisson Reduction of Dynamics).   *Let G be a Lie group and $H : T^*G \to \mathbb{R}$. Assume that H is left (respectively, right) invariant. Then the function $H^- := H|\mathfrak{g}^*$ (respectively, $H^+ := H|\mathfrak{g}^*$) on $\mathfrak{g}^*$ satisfies $H = H^- \circ \mathbf{J}_R$, that is,*

$$H(\alpha_g) = H^-(\mathbf{J}_R(\alpha_g)) \quad \text{for all } \alpha_g \in T_g^*G, \tag{13.4.1}$$

where $\mathbf{J}_R : T^*G \to \mathfrak{g}_-^*$ is given by $\mathbf{J}_R(\alpha_g) = T^*L_g \cdot \alpha_g$ (respectively, $H = H^+ \circ \mathbf{J}_L$, that is,

$$H(\alpha_g) = H^+(\mathbf{J}_L(\alpha_g)) \quad \text{for all } \alpha_g \in T_g^*G, \tag{13.4.2}$$

where $\mathbf{J}_L : T^*G \to \mathfrak{g}_+^*$ is given by $\mathbf{J}_L(\alpha_g) = T^*R_g \cdot \alpha_g$).

The flow F_t of X_H on T^*G and the flow F_t^- (respectively, F_t^+) of X_{H^-} (respectively, X_{H^+}) on $\mathfrak{g}_-^*$ (respectively, $\mathfrak{g}_+^*$) are related by

$$\mathbf{J}_R(F_t(\alpha_g)) = F_t^-(\mathbf{J}_R(\alpha_g)), \tag{13.4.3}$$

$$\mathbf{J}_L(F_t(\alpha_g)) = F_t^+(\mathbf{J}_L(\alpha_g)). \tag{13.4.4}$$

In other words, a left-invariant Hamiltonian on T^*G induces Lie–Poisson dynamics on $\mathfrak{g}_-^*$, while a right-invariant one induces Lie–Poisson dynamics on $\mathfrak{g}_+^*$. The result is a direct consequence of the Lie–Poisson reduction theorem and the fact that a Poisson map relates Hamiltonian systems and their integral curves to Hamiltonian systems.

Left and Right Reductions.   Above we saw that *left reduction is implemented by the right momentum map*. That is, H and H^- as well as X_H and X_{H^-} are $\mathbf{J}_R$-related if H is left invariant. We can get additional information using the fact that $\mathbf{J}_L$ is conserved.

Proposition 13.4.2.   *Let $H : T_e^*G \to \mathbb{R}$ be left invariant and H^- be its restriction to $\mathfrak{g}^*$ as above. Let $\alpha(t) \in T_{g(t)}^*G$ be an integral curve of X_H and let $\mu(t) = \mathbf{J}_R(\alpha(t))$ and $\nu(t) = \mathbf{J}_L(\alpha(t))$, so that ν is constant in time. Then*

$$\nu = g(t) \cdot \mu(t) := \mathrm{Ad}_{g(t)^{-1}}^* \mu(t). \tag{13.4.5}$$

Proof.   This follows from $\nu = T_e^* R_{g(t)} \alpha(t)$, $\mu(t) = T_e^* L_{g(t)} \alpha(t)$, the definition of the coadjoint action, and the fact that $\mathbf{J}_L$ is conserved. ■

Equation (13.4.5) already determines $g(t)$ in terms of ν and $\mu(t)$ to some extent; for example, for SO(3) it says that $g(t)$ rotates the vector $\mu(t)$ to the fixed vector ν.

The Reconstruction Equation.   Differentiating (13.4.5) in t and using the formulas for differentiating curves from §9.3, we get

$$0 = g(t) \cdot \left\{ \xi(t) \cdot \mu(t) + \frac{d\mu}{dt} \right\},$$

where $\xi(t) = g(t)^{-1} \dot{g}(t)$ and $\xi \cdot \mu = -\mathrm{ad}_\xi^* \mu$.

On the other hand, $\mu(t)$ satisfies the Lie–Poisson equations

$$\frac{d\mu}{dt} = \mathrm{ad}_{\delta H^-/\delta \mu}^* \mu,$$

and so

$$\xi(t) \cdot \mu(t) + \mathrm{ad}_{\delta H^-/\delta \mu}^* \mu(t) = 0;$$

that is,

$$\mathrm{ad}_{(-\xi(t)+\delta H^-/\delta \mu)}^* \mu(t) = 0.$$

A sufficient condition for this is that $\xi(t) = \delta H^-/\delta \mu$; that is,

$$g(t)^{-1} \dot{g}(t) = \frac{\delta H^-}{\delta \mu}, \tag{13.4.6}$$

which is called the **reconstruction equation**. Thus, it is plausible that we can reconstruct $\alpha(t)$ from $\mu(t)$ by first solving (13.4.6) with appropriate initial conditions and then letting

$$\alpha(t) = T_{g(t)}^* L_{g(t)^{-1}} \mu(t). \tag{13.4.7}$$

This gives us a way to go back and forth between T^*G and $\mathfrak{g}^*$, as in Figure 13.4.1.

We now look at the reconstruction procedure a little more closely and from a slightly different point of view.

Lie-Poisson reduction

$$T^*G \underset{\text{Lie-Poisson reconstruction}}{\overset{\longrightarrow}{\underset{\longleftarrow}{}}} \mathfrak{g}^*$$

FIGURE 13.4.1. Lie–Poisson reduction and reconstruction.

Left Trivialization of Dynamics. The next proposition describes the vector field X_H in the left trivialization of T^*G as $G \times \mathfrak{g}^*$. Let $\lambda : T^*G \longrightarrow G \times \mathfrak{g}^*$ be the diffeomorphism defined by

$$\lambda(\alpha_g) = (g, T_e^* L_g(\alpha_g)) = (g, \mathbf{J}_R(\alpha_g)). \tag{13.4.8}$$

It is easily verified that λ is equivariant relative to the cotangent lift of left translations on G and the G-action on $G \times \mathfrak{g}^*$ given by

$$g \cdot (h, \mu) = \Lambda_g(h, \mu) = (gh, \mu), \tag{13.4.9}$$

where $g, h \in G$ and $\mu \in \mathfrak{g}^*$. Let $p_1 : G \times \mathfrak{g}^* \to G$ denote the projection to the first factor. Note that $p_1 \circ \lambda = \pi$, where $\pi : T^*G \to G$ is the canonical cotangent bundle projection.

Proposition 13.4.3. *For $g \in G$, $\mu \in \mathfrak{g}^*$, the push-forward of X_H by λ to $G \times \mathfrak{g}^*$ is the vector field given by*

$$(\lambda_* X_H)(g, \mu) = \left(T_e L_g \frac{\delta H^-}{\delta \mu}, \mu, \operatorname{ad}^*_{\delta H^-/\delta \mu} \mu \right) \in T_g G \times T_\mu \mathfrak{g}^*, \tag{13.4.10}$$

where $H^- = H|\mathfrak{g}^*$.

Proof. As we have already shown, the map $\mathbf{J}_R : T^*G \longrightarrow \mathfrak{g}^*$ can be regarded as the standard projection to the quotient $T^*G \longrightarrow T^*G/G$ for the left action, so that the second component of $\lambda_* X_H$ is the Lie–Poisson reduction of X_H and hence equals the Hamiltonian vector field X_{H^-} on $\mathfrak{g}^*_-$. By Proposition 10.7.1 we can conclude that

$$(\lambda_* X_H)(g, \mu) = (X^\mu(g), \mu, \operatorname{ad}^*_{\delta H^-/\delta \mu} \mu), \tag{13.4.11}$$

where $X^\mu \in \mathfrak{X}(G)$ is a vector field on G depending smoothly on the parameter $\mu \in \mathfrak{g}^*$.

Since H is left-invariant, so is X_H, and by equivariance of the diffeomorphism λ, we also have $\Lambda_g^* \lambda_* X_H = \lambda_* X_H$ for any $g \in G$. This, in turn, is equivalent to

$$T_{gh} L_{g^{-1}} X^\mu(gh) = X^\mu(h)$$

for all $g, h \in G$ and $\mu \in \mathfrak{g}^*$; that is,

$$X^\mu(g) = T_e L_g X^\mu(e). \tag{13.4.12}$$

In view of (13.4.11) and (13.4.12), the proposition is proved if we show that

$$X^\mu(e) = \frac{\delta H^-}{\delta \mu}. \tag{13.4.13}$$

To prove this, we begin by noting that

$$\begin{aligned} X^\mu(e) &= T_{(e,\mu)}p_1(\lambda_* X_H(\mu)) = (T_{(e,\mu)}p_1 \circ T_\mu \lambda) X_H(\mu) \\ &= T_\mu(p_1 \circ \lambda) X_H(\mu) = T_\mu \pi(X_H(\mu)). \end{aligned} \tag{13.4.14}$$

For a fixed $\nu \in \mathfrak{g}^*$, introduce the flow

$$F_t^\nu(\alpha_g) = \alpha_g + t T_g^* L_{g^{-1}}(\nu), \tag{13.4.15}$$

which leaves the fibers of T^*G invariant and therefore defines a vertical vector field V_ν on T^*G (that is, $T\pi \circ V_\nu = 0$) given by

$$V_\nu(\alpha_g) = \left.\frac{d}{dt}\right|_{t=0} (\alpha_g + t T_g^* L_{g^{-1}}(\nu)). \tag{13.4.16}$$

The defining relation $\mathbf{i}_{X_H}\Omega = \mathbf{d}H$ of X_H evaluated at μ in the direction $V_\nu(\mu)$ gives

$$\begin{aligned} \Omega(\mu)(X_H(\mu), V_\nu(\mu)) &= \mathbf{d}H(\mu) \cdot V_\nu(\mu) \\ &= \left.\frac{d}{dt}\right|_{t=o} H(\mu + t\nu) = \left\langle \nu, \frac{\delta H^-}{\delta \mu} \right\rangle, \end{aligned}$$

so that using $\Omega = -\mathbf{d}\Theta$, we get

$$-X_H[\Theta(V_\nu)](\mu) + V_\nu[\Theta(X_H)](\mu) + \Theta([X_H, V_\nu])(\mu) = \left\langle \nu, \frac{dH^-}{\delta \mu} \right\rangle. \tag{13.4.17}$$

We will compute each term on the left-hand side of (13.4.17). Since V_ν is vertical, $T\pi \circ V_\nu = 0$, and so by the defining formula for the canonical one-form, $\Theta(V_\nu) = 0$. The first term thus vanishes. To compute the second term, we use the definition of Θ and (13.4.14) to get

$$\begin{aligned} V_\nu[\Theta(X_H)](\mu) &= \left.\frac{d}{dt}\right|_{t=0} \Theta(X_H)(\mu + t\nu) \\ &= \left.\frac{d}{dt}\right|_{t=0} \langle \mu + t\nu, T_{\mu+t\nu}\pi\,(X_H(\mu+t\nu)) \rangle \\ &= \left.\frac{d}{dt}\right|_{t=0} \langle \mu + t\nu, X^{\mu+t\nu}(e) \rangle \\ &= \langle \nu, X^\mu(e) \rangle + \left\langle \mu, \left.\frac{d}{dt}\right|_{t=0} X^{\mu+t\nu}(e) \right\rangle. \tag{13.4.18} \end{aligned}$$

Finally, to compute the third term, we again use the definition of Θ, the linearity of $T_\mu\pi$ to interchange the order of $T_\mu\pi$ and d/dt, the relation $\pi \circ F_t^\nu = \pi$, and (13.4.14) to get

$$
\begin{aligned}
\Theta([X_H, V_\nu])(\mu) &= \langle \mu, T_\mu\pi \cdot [X_H, V_\nu](\mu)\rangle \\
&= -\left\langle \mu, T_\mu\pi \cdot \frac{d}{dt}\bigg|_{t=0} ((F_t^\nu)^* X_H)(\mu)\right\rangle \\
&= -\left\langle \mu, \frac{d}{dt}\bigg|_{t=0} T_\mu\pi \cdot T_{\mu+t\nu} F_{-t}^\nu (X_H(\mu+t\nu))\right\rangle \\
&= -\left\langle \mu, \frac{d}{dt}\bigg|_{t=0} T_{\mu+t\nu}(\pi \circ F_{-t}^\nu)(X_H(\mu+t\nu))\right\rangle \\
&= -\left\langle \mu, \frac{d}{dt}\bigg|_{t=0} T_{\mu+t\nu}\pi \cdot X_H(\mu+t\nu)\right\rangle \\
&= -\left\langle \mu, \frac{d}{dt}\bigg|_{t=0} X^{\mu+t\nu}(e)\right\rangle. \tag{13.4.19}
\end{aligned}
$$

Adding (13.4.18) and (13.4.19) and using (13.4.17) gives

$$
\langle \nu, X^\mu(e)\rangle = \left\langle \nu, \frac{\delta H^-}{\delta\mu}\right\rangle,
$$

and thus (13.4.13) follows, thereby proving the proposition. ∎

The Reconstruction Theorem. This result now follows from what we have done.

Theorem 13.4.4 (Lie–Poisson Reconstruction of Dynamics). *Let G be a Lie group and suppose $H : T^*G \to \mathbb{R}$ is a left-invariant Hamiltonian. Let $H^- = H|\mathfrak{g}^*$ and let $\mu(t)$ be the integral curve of the Lie–Poisson equations*

$$
\frac{d\mu}{dt} = \operatorname{ad}^*_{\delta H^-/\delta\mu}\mu \tag{13.4.20}
$$

with initial condition $\mu(0) = T_e^* L_{g_0}(\alpha_{g_0})$. Then the integral curve $\alpha(t) \in T_{g(t)}^* G$ of X_H with initial condition $\alpha(0) = \alpha_{g_0}$ is given by

$$
\alpha(t) = T_{g(t)}^* L_{g(t)^{-1}}\mu(t), \tag{13.4.21}
$$

where $g(t)$ is the solution of the equation $g^{-1}\dot{g} = \delta H^-/\delta\mu$, that is,

$$
\frac{dg(t)}{dt} = T_e L_{g(t)}\frac{\delta H^-}{\delta\mu(t)}, \tag{13.4.22}
$$

with initial condition $g(0) = g_0$.

Proof. The curve $\alpha(t)$ is the unique integral curve of X_H with initial condition $\alpha(0) = \alpha_{g_0}$ if and only if

$$\lambda(\alpha(t)) = (g(t), T_e^* L_{g(t)} \alpha(t)) = (g(t), \mathbf{J}_R(\alpha(t)))$$
$$=: (g(t), \mu(t))$$

is the integral curve of $\lambda_* X_H$ with initial condition

$$\lambda(\alpha(0)) = (g_0, T_e^* L_{g_0}(\alpha_{g_0})),$$

which is equivalent to the statement in the theorem in view of (13.4.10). ∎

For right-invariant Hamiltonians $H : T^*G \to \mathbb{R}$, we let $H^+ = H|\mathfrak{g}^*$; the Lie–Poisson equations are

$$\frac{d\mu}{dt} = -\operatorname{ad}_{\delta H^+/\delta\mu}^* \mu, \tag{13.4.23}$$

the reconstruction formula is

$$\alpha(t) = T_{g(t)}^* R_{g(t)^{-1}} \mu(t), \tag{13.4.24}$$

and the equation that $g(t)$ satisfies is $\dot{g}g^{-1} = \delta H^+/\delta\mu$, that is,

$$\frac{dg(t)}{dt} = T_e R_{g(t)} \frac{\delta H^+}{\delta\mu(t)}; \tag{13.4.25}$$

the initial conditions remain unchanged.

Lie–Poisson Reconstruction and Lagrangians. It is useful to keep in mind that the Hamiltonian H on T^*G often arises from a Lagrangian $L :$ $TG \to \mathbb{R}$ via a Legendre transform $\mathbb{F}L$. In fact, many of the constructions and verifications are *simpler* using the Lagrangian formalism. Assume that L is left invariant (respectively, right invariant); that is,

$$L(TL_g \cdot v) = L(v), \tag{13.4.26}$$

respectively,

$$L(TR_g \cdot v) = L(v) \tag{13.4.27}$$

for all $g \in G$ and $v \in T_h G$. Differentiating (13.4.26) and (13.4.27), we obtain

$$\mathbb{F}L(TL_g \cdot v) \cdot (TL_g \cdot w) = \mathbb{F}L(v) \cdot w, \tag{13.4.28}$$

respectively,

$$\mathbb{F}L(TR_g \cdot v) \cdot (TR_g \cdot w) = \mathbb{F}L(v) \cdot w \tag{13.4.29}$$

for all $v, w \in T_h G$ and $g \in G$. In other words,

$$T^* L_g \circ \mathbb{F}L \circ T L_g = \mathbb{F}L, \qquad (13.4.30)$$

respectively,

$$T^* R_g \circ \mathbb{F}L \circ T R_g = \mathbb{F}L. \qquad (13.4.31)$$

Note that the action of L is also left (respectively, right) invariant

$$A(TL_g \cdot v) = A(v), \qquad (13.4.32)$$

respectively,

$$A(TR_g \cdot v) = A(v), \qquad (13.4.33)$$

since

$$A(TL_g \cdot v) = \mathbb{F}L(TL_g \cdot v) \cdot (TL_g \cdot v) = \mathbb{F}L(v) \cdot v = A(v)$$

by (13.4.28). Thus, the energy $E = A - L$ is left (respectively, right) invariant on TG. If L is hyperregular, so $\mathbb{F}L : TG \to T^* G$ is a diffeomorphism, then $H = E \circ (\mathbb{F}L)^{-1}$ is left (respectively, right) invariant on $T^* G$.

Theorem 13.4.5 (Alternative Lie–Poisson Reconstruction). *Let $L : TG \to \mathbb{R}$ be a hyperregular Lagrangian that is left (respectively, right) invariant on TG. Let $H : T^* G \to \mathbb{R}$ be the associated Hamiltonian and $H^- : \mathfrak{g}^*_- \to \mathbb{R}$ (respectively, $H^+ : \mathfrak{g}^*_+ \to \mathbb{R}$) be the induced Hamiltonian on $\mathfrak{g}^*$. Let $\mu(t) \in \mathfrak{g}^*$ be an integral curve for H^- (respectively, H^+) with initial condition $\mu(0) = T^*_e L_{g_0} \cdot \alpha_{g_0}$ (respectively, $\mu(0) = T^*_e R_{g_0} \cdot \alpha_{g_0}$) and let $\xi(t) = \mathbb{F}L^{-1} \mu(t) \in \mathfrak{g}$. Let*

$$v_0 = T_e L_{g_0} \cdot \xi(0) \in T_{g_0} G.$$

The integral curve for the Lagrangian vector field associated with L with initial condition (g_0, v_0) is given by

$$V_L(t) = T_e L_{g(t)} \cdot \xi(t), \qquad (13.4.34)$$

respectively,

$$V_R(t) = T_e R_g(t) \cdot \xi(t), \qquad (13.4.35)$$

where $g(t)$ solves the equation $g^{-1} \dot{g} = \xi$; that is,

$$\frac{dg}{dt} = T_e L_{g(t)} \cdot \xi(t), \quad g(0) = g_0, \qquad (13.4.36)$$

respectively, $\dot{g}^{-1} g = \xi$, that is,

$$\frac{dg}{dt} = T_e R_{g(t)} \cdot \xi(t), \quad g(0) = g_0. \qquad (13.4.37)$$

The corresponding integral curve of X_H on T^*G with initial condition α_{g_0} and covering $\mu(t)$ is

$$\alpha(t) = \mathbb{F}L(V_L(t)) = T^*_{g(t)}L_{(g(t))^{-1}}\mu(t), \qquad (13.4.38)$$

respectively,

$$\alpha(t) = \mathbb{F}L(V_R(t)) = T^*_{g(t)}R_{(g(t))^{-1}}\mu(t). \qquad (13.4.39)$$

Proof.   This follows from Theorem 13.4.4 by applying $\mathbb{F}L^{-1}$ to (13.5.21) and (13.5.24), respectively. As for the equation satisfied by $g(t)$, since the Lagrangian vector field X_E is a second-order equation, we necessarily have

$$\frac{dg}{dt} = V_L(t) = T_e L_{g(t)}\xi(t)$$

and

$$\frac{dg}{dt} = V_R(t) = T_e R_{g(t)}\xi(t),$$

respectively.   ∎

Thus, given $\xi(t)$, one solves (13.4.36) for $g(t)$ and then constructs $V_L(t)$ or $\alpha(t)$ from (13.4.34) and (13.4.38). As we shall see in the examples, this procedure has a natural physical interpretation. The previous theorem generalizes to arbitrary Lagrangian systems in the following way. In fact, Theorem 13.4.5 is a corollary of the next theorem.

Theorem 13.4.6 (Lagrangian Lie–Poisson Reconstruction).   *Let*

$$L : TG \to \mathbb{R}$$

be a left-invariant Lagrangian such that its Lagrangian vector field $Z \in \mathfrak{X}(TG)$ is a second-order equation and is left invariant. Let $Z_G \in \mathfrak{X}(\mathfrak{g})$ be the induced vector field on $(TG)/G \cong \mathfrak{g}$ and let $\xi(t)$ be an integral curve of Z_G. If $g(t) \in G$ is the solution of the nonautonomous ordinary differential equation

$$\dot{g}(t) = T_e L_{g(t)}\xi(t), \quad g(0) = e, \quad g \in G,$$

then

$$V(t) = T_e L_{gg(t)}\xi(t)$$

is the integral curve of Z satisfying $V(0) = T_e L_g\xi(0)$, and $V(t)$ projects to $\xi(t)$, that is,

$$TL_{\tau(V(t))^{-1}}V(t) = \xi(t),$$

where $\tau : TG \to G$ is the tangent bundle projection.

Proof.   Let $V(t)$ be the integral curve of Z satisfying $V(0) = T_e L_g \xi(0)$ for a given element $\xi(0) \in \mathfrak{g}$. Since $\xi(t)$ is the integral curve of Z_G whose flow is conjugated to the flow of Z by left translation, we have

$$TL_{\tau(V(t))^{-1}} V(t) = \xi(t).$$

If $h(t) = \tau(V(t))$, since Z is a second-order equation, we have

$$V(t) = \dot{h}(t) = T_e L_{h(t)} \xi(t), \quad h(0) = \tau(V(0)) = g,$$

so that letting $g(t) = g^{-1} h(t)$ we get $g(0) = e$ and

$$\dot{g}(t) = TL_{g^{-1}} \dot{h}(t) = TL_{g^{-1}} TL_{h(t)} \xi(t) = TL_{g(t)} \xi(t).$$

This determines $g(t)$ uniquely from $\xi(t)$, and so

$$V(t) = T_e L_{h(t)} \xi(t) = T_e L_{gg(t)} \xi(t). \qquad \blacksquare$$

These calculations suggest rather strongly that one should examine the Lagrangian (rather than the Hamiltonian) side of the story on an independent footing. We will do exactly that shortly.

The Lie–Poisson–Hamilton–Jacobi Equation.   Since Poisson brackets and Hamilton's equations naturally drop from T^*G to $\mathfrak{g}^*$, it is natural to ask whether other structures do too, such as those of Hamilton–Jacobi theory. We investigate this question now, leaving the proofs and related remarks to the Internet supplement.

Let H be a G-invariant function on T^*G and let H^- be the corresponding left-reduced Hamiltonian on $\mathfrak{g}^*$. (To be specific, we deal with left actions; of course, there are similar statements for right-reduced Hamiltonians.) If S is invariant, there is a unique function S^- such that $S(g, g_0) = S^-(g^{-1} g_0)$. (One gets a slightly different representation for S by writing $g_0^{-1} g$ in place of $g^{-1} g_0$.)

Proposition 13.4.7 (Ge and Marsden [1988]).   *The left-reduced Hamilton–Jacobi equation, for a function $S^- : G \to \mathbb{R}$, is the following:*

$$\frac{\partial S^-}{\partial t} + H^-(-TR_g^* \cdot \mathbf{d}S^-(g)) = 0, \qquad (13.4.40)$$

*which is called the **Lie–Poisson–Hamilton–Jacobi equation**. The Lie–Poisson flow of the Hamiltonian H^- is generated by the solution S^- of (13.4.40) in the sense that the flow is given by the Poisson transformation $\Pi_0 \mapsto \Pi$ of $\mathfrak{g}^*$ defined as follows. Define $g \in G$ by solving the equation*

$$\Pi_0 = -TL_g^* \cdot \mathbf{d}_g S^- \qquad (13.4.41)$$

for $g \in G$ and then set

$$\Pi = g \cdot \Pi_0 = \mathrm{Ad}_{g^{-1}}^* \Pi_0. \qquad (13.4.42)$$

The action in (13.4.42) is the coadjoint action. Note that (13.4.42) and (13.4.41) give $\Pi = -TR_g^* \cdot \mathbf{d}S^-(g)$.

Exercises

◇ **13.4-1.** Write out the reconstruction equations for the group $G = SO(3)$.

◇ **13.4-2.** Write out the reconstruction equations for $G = \mathrm{Diff}_{\mathrm{vol}}(\Omega)$.

◇ **13.4-3.** Write out the Lie–Poisson–Hamilton–Jacobi equation for $SO(3)$.

13.5   The Euler–Poincaré Equations

Some History of Lie–Poisson and Euler–Poincaré Equations.   We continue with some comments on the history of Poisson structures that we began in §10.3. Recall that we pointed out how Lie, in his work up to 1890 on function groups, had many of the essential ideas of general Poisson manifolds and, in particular, had explicitly studied the Lie–Poisson bracket on duals of Lie algebras.

The theory developed so far in this chapter describes the adaptation of the concepts of Hamiltonian mechanics to the context of the duals of Lie algebras. This theory could easily have been given shortly after Lie's work, but evidently it was not observed for the rigid body or ideal fluids until the work of Pauli [1953], Martin [1959], Arnold [1966a], Ebin and Marsden [1970], Nambu [1973], and Sudarshan and Mukunda [1974], all of whom were apparently unaware of Lie's work on the Lie–Poisson bracket. It seems that even Elie Cartan was unaware of this aspect of Lie's work, which does seem surprising. Perhaps it is less surprising when one thinks for a moment about how many other things Cartan was involved in at the time. Nevertheless, one is struck by the amount of rediscovery and confusion in this subject. Evidently, this situation is not unique to mechanics.

Meanwhile, as Arnold [1988] and Chetaev [1989] pointed out, one can also write the equations directly on the Lie algebra, bypassing the Lie–Poisson equations on the dual. The resulting equations were first written down on a general Lie algebra by Poincaré [1901b]; we refer to these as the Euler–Poincaré equations. We shall develop them from a modern point of view in the next section. Poincaré [1910] goes on to study the effects of the deformation of the earth on its precession—he apparently recognizes the equations as Euler equations on a semidirect product Lie algebra. In general, the command that Poincaré had of the subject is most impressive, and is hard to match in his near contemporaries, except perhaps Riemann [1860, 1861] and Routh [1877, 1884]. It is noteworthy that in Poincaré [1901b] there are no references, so it is rather hard to trace his train of thought or his sources; compare this style with that of Hamel [1904]! In particular, Poincaré gives no hint that he understood the work of Lie on the Lie–Poisson structure, but of course, Poincaré understood the Lie group and the Lie algebra machine very well indeed.

Our derivation of the Euler–Poincaré equations in the next section is based on a reduction of variational principles, not on a reduction of the symplectic or Poisson structure, which is natural for the dual. We also show that the Lie–Poisson equations are related to the Euler–Poincaré equations by the "fiber derivative," in the same way as one gets from the ordinary Euler–Lagrange equations to the Hamilton equations. Even though this is relatively trivial, it does not appear to have been written down before. In the dynamics of ideal fluids, the resulting variational principle is related to what has been known as "Lin constraints" (see also Newcomb [1962] and Bretherton [1970]). This itself has an interesting history, going back to Ehrenfest, Boltzman, and Clebsch; but again, there was little if any contact with the heritage of Lie and Poincaré on the subject. One person who was well aware of the work of both Lie and Poincaré was Hamel.

How does Lagrange fit into this story? In *Mécanique Analytique*, Volume 2, equations A on page 212, are the Euler–Poincaré equations for the rotation group written out explicitly for a reasonably general Lagrangian. Lagrange eventually specializes them to the rigid-body equations, of course. We should remember that Lagrange also developed the key concept of the Lagrangian representation of fluid motion, but it is not clear that he understood that both systems are special instances of one theory. Lagrange spends a large number of pages on his derivation of the Euler–Poincaré equations for SO(3), in fact, a good chunk of Volume 2. His derivation is not as clean as we would give today, but it seems to have the right spirit of a reduction method. That is, he tries to get the equations from the Euler–Lagrange equations on T SO(3) by passing to the Lie algebra.

In view of the historical situation described above, one might argue that the term "Euler–Lagrange-Poincaré" equations is right for these equations. Since Poincaré noted the generalization to arbitrary Lie algebras and applied it to interesting fluid problems, it is clear that his name belongs, but in light of other uses of the term "Euler–Lagrange," it seems that "Euler–Poincaré" is a reasonable choice.

Marsden and Scheurle [1993a, 1993b] and Weinstein [1996] have studied a more general version of Lagrangian reduction whereby one drops the Euler–Lagrange equations from TQ to TQ/G. This is a nonabelian generalization of the classical Routh method and leads to a very interesting coupling of the Euler–Lagrange and Euler–Poincaré equations that we shall briefly sketch in the next section. This problem was also studied by Hamel [1904] in connection with his work on nonholonomic systems (see Koiller [1992] and Bloch, Krishnaprasad, Marsden, and Murray [1996] for more information).

The current vitality of mechanics, including the investigation of fundamental questions, is quite remarkable, given its long history and development. This vitality comes about through rich interactions with pure mathematics (from topology and geometry to group representation theory), and through new and exciting applications to areas like control theory. It is perhaps even more remarkable that absolutely fundamental points, such as a clear and unambiguous linking of Lie's work on the Lie–Poisson bracket on the dual of a Lie algebra and Poincaré's work on the Euler–Poincaré equations on the Lie algebra itself, with the most basic of examples in mechanics, such as the rigid body and the motion of ideal fluids, took nearly a century to complete. The attendant lessons to be learned about communication between pure mathematics and the other mathematical sciences are, hopefully, obvious.

Rigid-Body Dynamics. To understand this section, it will be helpful to develop some more of the basics about rigid-body dynamics from the Introduction (further details are given in Chapter 15). We regard an element $R \in SO(3)$ giving the configuration of the body as a map of a reference configuration $\mathcal{B} \subset \mathbb{R}^3$ to the current configuration $R(\mathcal{B})$; the map R takes a reference or label point $X \in \mathcal{B}$ to a current point $x = RX \in R(\mathcal{B})$. See Figure 13.5.1.

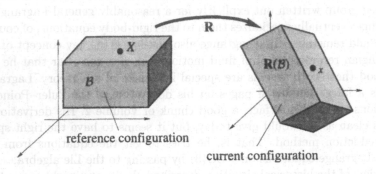

FIGURE 13.5.1. The rotation R takes the reference configuration to the current configuration.

When the rigid body is in motion, the matrix R is time-dependent, and the velocity of a point of the body is $\dot{x} = \dot{R}X = \dot{R}R^{-1}x$. Since R is an orthogonal matrix, $R^{-1}\dot{R}$ and $\dot{R}R^{-1}$ are skew matrices, and so we can write

$$\dot{x} = \dot{R}R^{-1}x = \omega \times x, \tag{13.5.1}$$

which defines the *spatial angular velocity vector* ω. Thus, ω is given by *right* translation of $\dot{R}$ to the identity.

The corresponding *body angular velocity* is defined by

$$\Omega = R^{-1}\omega, \tag{13.5.2}$$

so that Ω is the angular velocity relative to a body fixed frame. Notice that

$$R^{-1}\dot{R}X = R^{-1}\dot{R}R^{-1}x = R^{-1}(\omega \times x)$$
$$= R^{-1}\omega \times R^{-1}x = \Omega \times X, \tag{13.5.3}$$

so that Ω is given by *left* translations of $\dot{\mathbf{R}}$ to the identity. The kinetic energy is obtained by summing up $m\|\dot{x}\|^2/2$ over the body:

$$K = \frac{1}{2} \int_B \rho(X)\|\dot{\mathbf{R}}X\|^2 \, d^3X, \qquad (13.5.4)$$

where ρ is a given mass density in the reference configuration. Since

$$\|\dot{\mathbf{R}}X\| = \|\omega \times x\| = \|\mathbf{R}^{-1}(\omega \times x)\| = \|\Omega \times X\|,$$

K is a quadratic function of Ω. Writing

$$K = \tfrac{1}{2}\Omega^T \mathbb{I}\Omega \qquad (13.5.5)$$

defines the **moment of inertia tensor** $\mathbb{I}$, which, if the body does not degenerate to a line, is a positive definite (3×3) matrix, or better, a quadratic form. This quadratic form can be diagonalized, and this defines the principal axes and moments of inertia. In this basis, we write $\mathbb{I} = \mathrm{diag}(I_1, I_2, I_3)$. The function K is taken to be the Lagrangian of the system on $T\,\mathrm{SO}(3)$ (and by means of the Legendre transformation we get the corresponding Hamiltonian description on $T^*\,\mathrm{SO}(3)$). Notice directly from (13.5.4) that K is *left* (not right) invariant on $T\,\mathrm{SO}(3)$. It follows that the corresponding Hamiltonian is also *left* invariant.

Dynamics in the Group vs. the Algebra.   From the Lagrangian point of view, the relation between the motion in $\mathbf{R}$ space and that in body angular velocity (or Ω) space is as follows:

Theorem 13.5.1.   *The curve $\mathbf{R}(t) \in \mathrm{SO}(3)$ satisfies the Euler–Lagrange equations for the Lagrangian*

$$L(\mathbf{R}, \dot{\mathbf{R}}) = \frac{1}{2} \int_B \rho(X)\|\dot{\mathbf{R}}X\|^2 \, d^3X \qquad (13.5.6)$$

if and only if $\Omega(t)$ defined by $\mathbf{R}^{-1}\dot{\mathbf{R}}\mathbf{v} = \Omega \times \mathbf{v}$ for all $\mathbf{v} \in \mathbb{R}^3$ satisfies Euler's equations

$$\mathbb{I}\dot{\Omega} = \mathbb{I}\Omega \times \Omega. \qquad (13.5.7)$$

One instructive way to prove this *indirectly* is to pass to the Hamiltonian formulation and to use Lie–Poisson reduction. One way to do it *directly* is to use variational principles. By Hamilton's principle, $\mathbf{R}(t)$ satisfies the Euler–Lagrange equations if and only if

$$\delta \int L\, dt = 0.$$

Let $l(\Omega) = \tfrac{1}{2}(\mathbb{I}\Omega) \cdot \Omega$, so that $l(\Omega) = L(\mathbf{R}, \dot{\mathbf{R}})$ if $\mathbf{R}$ and Ω are related as above. To see how we should transform Hamilton's principle, we differentiate the relation $\mathbf{R}^{-1}\dot{\mathbf{R}} = \hat{\Omega}$ with respect to $\mathbf{R}$ to get

$$-\mathbf{R}^{-1}(\delta\mathbf{R})\mathbf{R}^{-1}\dot{\mathbf{R}} + \mathbf{R}^{-1}(\delta\dot{\mathbf{R}}) = \widehat{\delta\Omega}. \qquad (13.5.8)$$

Let the skew matrix $\hat{\Sigma}$ be defined by

$$\hat{\Sigma} = \mathbf{R}^{-1}\delta\mathbf{R} \qquad (13.5.9)$$

and define the corresponding vector Σ, as usual, by

$$\hat{\Sigma}\mathbf{v} = \Sigma \times \mathbf{v}. \qquad (13.5.10)$$

Note that $\dot{\hat{\Sigma}} = -\mathbf{R}^{-1}\dot{\mathbf{R}}\mathbf{R}^{-1}\delta\mathbf{R} + \mathbf{R}^{-1}\delta\dot{\mathbf{R}}$, so

$$\mathbf{R}^{-1}\delta\dot{\mathbf{R}} = \dot{\hat{\Sigma}} + \mathbf{R}^{-1}\dot{\mathbf{R}}\hat{\Sigma}. \qquad (13.5.11)$$

Substituting (13.5.11) and (13.5.9) into (13.5.8) gives $-\hat{\Sigma}\hat{\Omega}+\dot{\hat{\Sigma}}+\hat{\Omega}\hat{\Sigma} = \widehat{\delta\Omega}$, that is,

$$\widehat{\delta\Omega} = \dot{\hat{\Sigma}} + [\hat{\Omega}, \hat{\Sigma}]. \qquad (13.5.12)$$

The identity $[\hat{\Omega}, \hat{\Sigma}] = (\Omega \times \Sigma)\hat{}$ holds by Jacobi's identity for the cross product, and so

$$\delta\Omega = \dot{\Sigma} + \Omega \times \Sigma. \qquad (13.5.13)$$

These calculations prove the following:

Theorem 13.5.2. *Hamilton's variational principle*

$$\delta \int_a^b L\, dt = 0 \qquad (13.5.14)$$

on $T\,\mathrm{SO}(3)$ is equivalent to the **reduced variational principle**

$$\delta \int_a^b l\, dt = 0 \qquad (13.5.15)$$

on $\mathbb{R}^3$ where the variations $\delta\Omega$ are of the form (13.5.13) with $\Sigma(a) = \Sigma(b) = 0$.

Proof of Theorem 13.5.1. It suffices to work out the equations equivalent to the reduced variational principle (13.5.15). Since $l(\Omega) = \langle\mathbb{I}\Omega, \Omega\rangle/2$ and $\mathbb{I}$ is symmetric, we get

$$\delta \int_a^b l\, dt = \int_a^b \langle\mathbb{I}\Omega, \delta\Omega\rangle\, dt = \int_a^b \langle\mathbb{I}\Omega, \dot{\Sigma} + \Omega \times \Sigma\rangle\, dt$$

$$= \int_a^b \left[\left\langle-\frac{d}{dt}\mathbb{I}\Omega, \Sigma\right\rangle + \langle\mathbb{I}\Omega, \Omega \times \Sigma\rangle\right] dt$$

$$= \int_a^b \left\langle-\frac{d}{dt}\mathbb{I}\Omega + \mathbb{I}\Omega \times \Omega, \Sigma\right\rangle dt,$$

where we have integrated by parts and used the boundary conditions $\Sigma(b) = \Sigma(a) = 0$. Since Σ is otherwise arbitrary, (13.5.15) is equivalent to

$$-\frac{d}{dt}(\mathbb{I}\Omega) + \mathbb{I}\Omega \times \Omega = 0,$$

which are Euler's equations. ∎

Euler–Poincaré Reduction. We now generalize this procedure to an arbitrary Lie group, and later will make the direct link with the Lie–Poisson equations.

Theorem 13.5.3. *Let G be a Lie group and let $L : TG \to \mathbb{R}$ be a left-invariant Lagrangian. Let $l : \mathfrak{g} \to \mathbb{R}$ be its restriction to the identity. For a curve $g(t) \in G$, let $\xi(t) = g(t)^{-1} \cdot \dot{g}(t)$; that is, $\xi(t) = T_{g(t)}L_{g(t)^{-1}}\dot{g}(t)$. Then the following are equivalent:*

(i) *$g(t)$ satisfies the Euler–Lagrange equations for L on G;*

(ii) *the variational principle*

$$\delta \int L(g(t), \dot{g}(t))\, dt = 0 \tag{13.5.16}$$

holds, for variations with fixed endpoints;

(iii) *the **Euler–Poincaré equations** hold:*

$$\frac{d}{dt}\frac{\delta l}{\delta \xi} = \mathrm{ad}_\xi^* \frac{\delta l}{\delta \xi}; \tag{13.5.17}$$

(iv) *the variational principle*

$$\delta \int l(\xi(t))\, dt = 0 \tag{13.5.18}$$

holds on $\mathfrak{g}$, using variations of the form

$$\delta \xi = \dot{\eta} + [\xi, \eta], \tag{13.5.19}$$

where η vanishes at the endpoints.

Proof. First of all, the equivalence of (i) and (ii) holds on the tangent bundle of any configuration manifold Q, as we know from Chapter 8. To see that (ii) and (iv) are equivalent, one needs to compute the variations $\delta \xi$ induced on $\xi = g^{-1}\dot{g} = TL_{g^{-1}}\dot{g}$ by a variation of g. We will do this for matrix groups; see Bloch, Krishnaprasad, Marsden, and Ratiu [1996] for the general case. To calculate this, we need to differentiate $g^{-1}\dot{g}$ in the direction of a variation δg. If $\delta g = dg/d\epsilon$ at $\epsilon = 0$, where g is extended to a curve g_ϵ, then

$$\delta \xi = \frac{d}{d\epsilon}\left(g^{-1}\frac{d}{dt}g\right)\bigg|_{\epsilon=0} = -\left(g^{-1}\delta g g^{-1}\right)\dot{g} + g^{-1}\frac{d^2 g}{dt\, d\epsilon}\bigg|_{\epsilon=0},$$

while if $\eta = g^{-1}\delta g$, then

$$\dot{\eta} = \frac{d}{dt}\left(g^{-1}\frac{d}{d\epsilon}g\right)\bigg|_{\epsilon=0} = -\left(g^{-1}\dot{g}g^{-1}\right)\delta g + g^{-1}\frac{d^2 g}{dt\, d\epsilon}\bigg|_{\epsilon=0}.$$

The difference $\delta\xi - \dot{\eta}$ is thus the commutator $[\xi, \eta]$.

To complete the proof, we show the equivalence of (iii) and (iv). Indeed, using the definitions and integrating by parts, we obtain

$$\delta \int l(\xi)dt = \int \left\langle \frac{\delta l}{\delta \xi}, \delta\xi \right\rangle dt = \int \left\langle \frac{\delta l}{\delta \xi}, (\dot{\eta} + \mathrm{ad}_\xi \eta) \right\rangle dt$$

$$= \int \left\langle \left[-\frac{d}{dt}\left(\frac{\delta l}{\delta \xi}\right) + \mathrm{ad}_\xi^* \frac{\delta l}{\delta \xi}\right], \eta \right\rangle dt,$$

so the result follows.   ∎

There is, of course, a right-invariant version of this theorem in which $\xi = \dot{g}g^{-1}$ and (13.5.17), (13.5.19) acquire minus signs, that is,

$$\frac{d}{dt}\frac{\delta l}{\delta \xi} = -\mathrm{ad}_\xi^* \frac{\delta l}{\delta \xi} \quad \text{and} \quad \delta\xi = \dot{\eta} - [\xi, \eta].$$

In coordinates, (13.5.17), reads as follows:

$$\frac{d}{dt}\frac{\partial l}{\partial \xi^a} = C_{da}^b \xi^d \frac{\partial l}{\partial \xi^b}. \tag{13.5.20}$$

Euler–Poincaré Reconstruction.   On the Lagrangian side, reconstruction is very simple and centers on the *reconstruction equation*, which for left-invariant systems reads

$$g(t)^{-1}\dot{g}(t) = \xi(t). \tag{13.5.21}$$

For the rigid body, this is just the definition of the body angular velocity $\mathbf{\Omega}(t)$:

$$\mathbf{R}(t)^{-1}\dot{\mathbf{R}}(t) = \hat{\mathbf{\Omega}}(t). \tag{13.5.22}$$

Reconstruction is read off Theorem 13.5.3 as follows.

Proposition 13.5.4.   *Let $v_0 \in T_{g_0}G$, $\xi_0 = g_0^{-1}v_0 \in \mathfrak{g}$ and let $\xi(t)$ be the solution of the Euler–Poincaré equations with initial condition ξ_0. Solve the reconstruction equation (13.5.21) for $g(t)$ with $g(0) = g_0$. Then the solution of the Euler–Lagrange equations with initial condition v_0 is $v(t) \in T_{g(t)}G$, given by*

$$v(t) = \dot{g}(t) = g(t)\xi(t). \tag{13.5.23}$$

As mentioned earlier, to carry this out in examples it is useful to make use of the conservation law to help solve the reconstruction equation. We shall see this in the case of the rigid body in Chapter 15.

The Legendre Transformation. Since in the hyperregular case the Euler–Lagrange and Hamilton equations on TQ and T^*Q are equivalent, it follows that the Lie–Poisson and Euler–Poincaré equations are also equivalent. To see this *directly*, we make the following Legendre transformation from $\mathfrak{g}$ to $\mathfrak{g}^*$:

$$\mu = \frac{\delta l}{\delta \xi}, \quad h(\mu) = \langle \mu, \xi \rangle - l(\xi).$$

Assuming that the map $\xi \mapsto \mu$ is a diffeomorphism of $\mathfrak{g}$ to $\mathfrak{g}^*$, note that

$$\frac{\delta h}{\delta \mu} = \xi + \left\langle \mu, \frac{\delta \xi}{\delta \mu} \right\rangle - \left\langle \frac{\delta l}{\delta \xi}, \frac{\delta \xi}{\delta \mu} \right\rangle = \xi,$$

so it is now clear that the Lie–Poisson and Euler–Poincaré equations are equivalent.

The Virasoro Algebra. We close this section by showing that the periodic KdV equation (see Example (c) in §3.2)

$$u_t + 6uu_x + u_{xxx} = 0$$

is an Euler–Poincaré equation on a certain Lie algebra called the **Virasoro algebra** $\mathfrak{v}$. These results were obtained in the Lie–Poisson context by Gelfand and Dorfman [1979], Kirillov [1981], Ovsienko and Khesin [1987], and Segal [1991]. See also Pressley and Segal [1986] and references therein.

We begin with the construction of the Virasoro algebra $\mathfrak{v}$. If one identifies elements of $\mathfrak{X}(S^1)$ with periodic functions of period 1 endowed with the Jacobi–Lie bracket $[u, v] = uv' - u'v$, the **Gelfand–Fuchs cocycle** is defined by the expression

$$\Sigma(u, v) = \gamma \int_0^1 u'(x)v''(x)dx,$$

where $\gamma \in \mathbb{R}$ is a constant (to be determined later). The Lie algebra $\mathfrak{X}(S^1)$ of vector fields on the circle has a unique central extension by $\mathbb{R}$ determined by the Gelfand–Fuchs cocycle. Therefore (see (12.3.22) in Remark 5 of §12.3), the Lie algebra bracket on $\mathfrak{v} := \{ (u, a) \mid u \in \mathfrak{X}(S^1), \ a \in \mathbb{R} \}$ is given by

$$[(u, a), (v, b)] = \left(-uv' + u'v, \gamma \int_0^1 u'(x)v''(x) \, dx \right),$$

since the *left* Lie bracket on $\mathfrak{X}(S^1)$ is given by the negative of the Jacobi–Lie bracket for vector fields. Identify the dual of $\mathfrak{v}$ with $\mathfrak{v}$ by the L^2-inner product

$$\langle (u, a), (v, b) \rangle = ab + \int_0^1 u(x)v(x)\, dx.$$

We claim that the coadjoint action $\mathrm{ad}^*_{(u,a)}$ is given by

$$\mathrm{ad}^*_{(u,a)}(v, b) = (b\gamma u''' + 2u'v + uv', 0).$$

Indeed, if $(u, a), (v, b), (w, c) \in \mathfrak{v}$, we have

$$\left\langle \mathrm{ad}^*_{(u,a)}(v, b), (w, c) \right\rangle = \langle (v, b), [(u, a), (w, c)] \rangle$$

$$= \left\langle (v, b), \left(-uw' + u'w, \gamma \int_0^1 u'(x)w''(x)\, dx \right) \right\rangle$$

$$= b\gamma \int_0^1 u'(x)w''(x)\, dx - \int_0^1 v(x)u(x)w'(x)\, dx + \int_0^1 v(x)u'(x)w(x)\, dx.$$

By integrating the first term twice and the second term once by parts and remembering that the boundary terms vanish by periodicity, this expression becomes

$$b\gamma \int_0^1 u'''(x)w(x)\, dx + \int_0^1 (v(x)u(x))'w(x)\, dx + \int_0^1 v(x)u'(x)w(x)\, dx$$

$$= \int_0^1 (b\gamma u'''(x) + 2u'(x)v(x) + u(x)v'(x))w(x)\, dx$$

$$= \langle (b\gamma u''' + 2u'v + uv', 0), (w, c) \rangle.$$

The Euler–Poincaré Form of the KdV Equation. If $F : \mathfrak{v} \to \mathbb{R}$, its functional derivative relative to the L^2-pairing is given by

$$\frac{\delta F}{\delta(u, a)} = \left(\frac{\delta F}{\delta u}, \frac{\partial F}{\partial a} \right),$$

where $\delta F/\delta u$ is the usual L^2-functional derivative of F keeping $a \in \mathbb{R}$ fixed, and $\partial F/\partial a$ is the standard partial derivative of F keeping u fixed. The Euler–Poincaré equations for *right*-invariant systems with Lagrangian $l : \mathfrak{v} \to \mathbb{R}$ become

$$\frac{d}{dt} \frac{\delta l}{\delta(u, a)} = -\mathrm{ad}^*_{(u,a)} \frac{\delta l}{\delta(u, a)}.$$

However,

$$\mathrm{ad}^*_{(u,a)} \frac{\delta l}{\delta(u, a)} = \mathrm{ad}^*_{(u,a)} \left(\frac{\delta l}{\delta u}, \frac{\partial l}{\partial a} \right)$$

$$= \left(\gamma \frac{\partial l}{\partial a} u''' + 2u' \frac{\delta l}{\delta u} + u \left(\frac{\delta l}{\delta u} \right)', 0 \right),$$

so that the Euler–Poincaré equations become the system

$$\frac{d}{dt}\frac{\partial l}{\partial a} = 0$$

$$\frac{d}{dt}\frac{\delta l}{\delta u} = -\gamma \frac{\partial l}{\partial a}u''' - 2u'\frac{\delta l}{\delta u} - u\left(\frac{\delta l}{\delta u}\right)'.$$

If

$$l(u,a) = \frac{1}{2}\left(a^2 + \int_0^1 u^2(x)\,dx\right),$$

then $\partial l/\partial a = a$, $\delta l/\delta u = u$, and the above equations become $da/dt = 0$ together with

$$\frac{du}{dt} = -\gamma a u''' - 3u'u. \tag{13.5.24}$$

Since a is constant, we get

$$u_t + 3u_x u + \gamma a u''' = 0. \tag{13.5.25}$$

This equation is equivalent to the KdV equation upon rescaling time and choosing the constant a appropriately. Indeed, let $u(t,x) = v(\tau(t),x)$ for $\tau(t) = t/2$. Then $u_x = v_x$ and $u_t = v_\tau/2$, so that (13.5.25) can be written as $v_\tau + 6vv_x + 2\gamma a v_{xxx} = 0$, which becomes the KdV equation (see §3.2) if we choose $a = 1/(2\gamma)$.

The Lie–Poisson form of the KdV equation. The $(+)$ Lie–Poisson bracket is given by

$$\{f,h\}(u,a) = \left\langle (u,a), \left[\frac{\delta}{\delta(u,a)}, \frac{\delta h}{\delta(u,a)}\right]\right\rangle$$

$$= \int \left[u\left(\left(\frac{\delta f}{\delta u}\right)'\frac{\delta h}{\delta u} - \frac{\delta f}{\delta u}\left(\frac{\delta h}{\delta u}\right)'\right) + a\gamma\left(\frac{\delta f}{\delta u}\right)'\left(\frac{\delta h}{\delta u}\right)''\right]dx,$$

so that the Lie–Poisson equations $\dot{f} = \{f,h\}$ become $da/dt = 0$ together with

$$\frac{du}{dt} = -u'\left(\frac{\delta h}{\delta u}\right) - 2u\left(\frac{\delta h}{\delta u}\right)' - a\gamma\left(\frac{\delta h}{\delta u}\right)'''. \tag{13.5.26}$$

Taking

$$h(u,a) = \frac{1}{2}a^2 + \frac{1}{2}\int_0^1 u^2(x)\,dx,$$

we get $\partial h/\partial a = a$, $\delta h/\delta u = u$, and so (13.5.26) becomes (13.5.25), as was to be expected and could have been directly obtained by a Legendre transform.

The conclusion is that the KdV equation is the expression in space coordinates of the geodesic equations on the Virasoro group V endowed with the right-invariant metric whose value at the identity is the L^2 inner product. We shall not describe here the Virasoro group, which is a central extension of the diffeomorphism group on S^1; we refer the reader to Pressley and Segal [1986].

Exercises

◇ **13.5-1.** Verify the coordinate form of the Euler–Poincaré equations.

◇ **13.5-2.** Show that the Euler equations for a perfect fluid are Euler–Poincaré equations. Find the variational principle (3) in Newcomb [1962] and Bretherton [1970].

◇ **13.5-3.** Derive the rigid-body Euler equations $\dot{\Pi} = \Pi \times \Omega$ directly from the momentum conservation law $\dot{\pi} = 0$ and the relation $\pi = R\Pi$.

13.6   The Lagrange–Poincaré Equations

As we have mentioned, the Lie–Poisson and Euler–Poincaré equations occur for many systems besides the rigid-body equations. They include the equations of fluid and plasma dynamics, for example. For many other systems, such as a rotating molecule or a spacecraft with movable internal parts, one has a combination of equations of Euler–Poincaré type and Euler–Lagrange type. Indeed, on the Hamiltonian side, this process has undergone development for quite some time. On the Lagrangian side, this process is also very interesting, and it has been recently developed by, among others, Marsden and Scheurle [1993a, 1993b], Holm, Marsden, and Ratiu [1998a], and Cendra, Marsden, and Ratiu [1999]. In this section we give a few *indications* of how this more general theory proceeds.

The general problem is to drop Euler–Lagrange equations and variational principles from a general velocity phase-space TQ to the quotient TQ/G by a Lie group action of G on Q. If L is a G-invariant Lagrangian on TQ, it induces a reduced Lagrangian l on TQ/G. We give a brief *preview* of the general theory in this section. In fact, the material below can also act as motivation for the general theory of connections.

An important ingredient in this work is to introduce a connection A on the principal bundle $Q \to S = Q/G$, assuming that this quotient is nonsingular. For example, the mechanical connection (see Kummer [1981], Marsden [1992] and references therein) may be chosen for A. This connection allows one to split the variables into a horizontal and vertical part. Let x^α, also called "internal variables," be coordinates for shape-space Q/G, let η^a be coordinates for the Lie algebra $\mathfrak{g}$ relative to a chosen basis, let l

be the Lagrangian regarded as a function of the variables $x^\alpha, \dot{x}^\alpha, \eta^a$, and let C^a_{db} be the structure constants of the Lie algebra $\mathfrak{g}$ of G.

If one writes the Euler–Lagrange equations on TQ in a local principal bundle trivialization, with coordinates x^α on the base and η^a in the fiber, then one gets the following system of **Hamel equations**:

$$\frac{d}{dt}\frac{\partial l}{\partial \dot{x}^\alpha} - \frac{\partial l}{\partial x^\alpha} = 0, \tag{13.6.1}$$

$$\frac{d}{dt}\frac{\partial l}{\partial \eta^b} - \frac{\partial l}{\partial \eta^a}C^a_{db}\eta^d = 0. \tag{13.6.2}$$

However, this representation of the equations does not make global intrinsic sense (unless $Q \to S$ admits a global flat connection). The introduction of a connection overcomes this, and one can intrinsically and globally split the original variational principle relative to horizontal and vertical variations. One gets from one form to the other by means of the velocity shift given by replacing η by the vertical part relative to the connection $\xi^a = A^a_\alpha \dot{x}^\alpha + \eta^a$. Here A^d_α are the local coordinates of the connection A. This change of coordinates is motivated from the mechanical point of view, since the variables ξ have the interpretation of the locked angular velocity. The resulting **Lagrange–Poincaré equations** have the following form:

$$\frac{d}{dt}\frac{\partial l}{\partial \dot{x}^\alpha} - \frac{\partial l}{\partial x^\alpha} = \frac{\partial l}{\partial \xi^a}\left(B^a_{\alpha\beta}\dot{x}^\beta + B^a_{\alpha d}\xi^d\right), \tag{13.6.3}$$

$$\frac{d}{dt}\frac{\partial l}{\partial \xi^b} = \frac{\partial l}{\partial \xi^a}(B^a_{b\alpha}\dot{x}^\alpha + C^a_{db}\xi^d). \tag{13.6.4}$$

In these equations, $B^a_{\alpha\beta}$ are the coordinates of the curvature B of A,

$$B^a_{d\alpha} = C^a_{bd}A^b_\alpha \quad \text{and} \quad B^a_{b\alpha} = -B^a_{\alpha b}.$$

The variables ξ^a may be regarded as the rigid part of the variables on the original configuration space, while x^α are the internal variables. As in Simo, Lewis, and Marsden [1991], the division of variables into internal and rigid parts has deep implications for both stability theory and bifurcation theory, again, continuing along lines developed originally by Riemann, Poincaré, and others. The main way this new insight is achieved is through a careful split of the variables, using the (mechanical) connection as one of the main ingredients. This split puts the second variation of the augmented Hamiltonian at a relative equilibrium as well as the symplectic form into "normal form." It is somewhat remarkable that they are *simultaneously* put into a simple form. This link helps considerably with an eigenvalue analysis of the linearized equations and in Hamiltonian bifurcation theory; see, for example, Bloch, Krishnaprasad, Marsden, and Ratiu [1996].

One of the key results in Hamiltonian reduction theory says that the reduction of a cotangent bundle T^*Q by a symmetry group G is a bundle over

T^*S, where $S = Q/G$ is shape-space and where the fiber is either $\mathfrak{g}^*$, the dual of the Lie algebra of G, or is a coadjoint orbit, depending on whether one is doing Poisson or symplectic reduction. We refer to Montgomery, Marsden, and Ratiu [1984], Marsden [1992], and Cendra, Marsden, and Ratiu [1999] for details and references. The Lagrange–Poincaré equations give the analogue of this structure on the tangent bundle.

Remarkably, equations (13.6.3) are very close in form to the equations for a mechanical system with classical nonholonomic velocity constraints (see Naimark and Fufaev [1972] and Koiller [1992]). The connection chosen in that case is the one-form that determines the constraints. This link is made precise in Bloch, Krishnaprasad, Marsden, and Murray [1996]. In addition, this structure appears in several control problems, especially the problem of stabilizing controls considered by Bloch, Krishnaprasad, Marsden, and Sánchez de Alvarez [1992].

For systems with a momentum map $\mathbf{J}$ constrained to a specific value μ, the key to the construction of a reduced Lagrangian system is the modification of the Lagrangian L to the Routhian R^μ, which is obtained from the Lagrangian by subtracting off the mechanical connection paired with the constraining value μ of the momentum map. On the other hand, a basic ingredient needed for the Lagrange–Poincaré equations is a velocity shift in the Lagrangian, the shift being determined by the connection, so this velocity-shifted Lagrangian plays the role that the Routhian does in the constrained theory.

14

Coadjoint Orbits

In this chapter we prove, among other things, that *the coadjoint orbits of a Lie group are symplectic manifolds*. These symplectic manifolds are, in fact, the symplectic leaves for the Lie–Poisson bracket. This result was developed and used by Kirillov, Arnold, Kostant, and Souriau in the early to mid 1960s, although it had important roots going back to the work of Lie, Borel, and Weil. (See Kirillov [1962, 1976b], Arnold [1966a], Kostant [1970], and Souriau [1970].) Here we give a direct proof. Alternatively, one can give a proof using general reduction theory, as in Marsden and Weinstein [1974] and Abraham and Marsden [1978].

Recall from Chapter 9 that the *adjoint representation* of a Lie group G is defined by

$$\mathrm{Ad}_g = T_e I_g : \mathfrak{g} \to \mathfrak{g},$$

where $I_g : G \to G$ is the inner automorphism $I_g(h) = ghg^{-1}$. The *coadjoint action* is given by

$$\mathrm{Ad}^*_{g^{-1}} : \mathfrak{g}^* \to \mathfrak{g}^*,$$

where $\mathrm{Ad}^*_{g^{-1}}$ is the dual of the linear map $\mathrm{Ad}_{g^{-1}}$, that is, it is defined by

$$\langle \mathrm{Ad}^*_{g^{-1}}(\mu), \xi \rangle = \langle \mu, \mathrm{Ad}_{g^{-1}}(\xi) \rangle,$$

where $\mu \in \mathfrak{g}^*$, $\xi \in \mathfrak{g}$, and $\langle , \rangle$ denotes the pairing between $\mathfrak{g}^*$ and $\mathfrak{g}$. The *coadjoint orbit*, $\mathrm{Orb}(\mu)$, through $\mu \in \mathfrak{g}^*$ is the subset of $\mathfrak{g}^*$ defined by

$$\mathrm{Orb}(\mu) := \{ \mathrm{Ad}^*_{g^{-1}}(\mu) \mid g \in G \} := G \cdot \mu.$$

Like the orbit of any group action, $\mathrm{Orb}(\mu)$ is an *immersed submanifold* of $\mathfrak{g}^*$, and *if G is compact, $\mathrm{Orb}(\mu)$ is a closed embedded submanifold.*[1]

14.1   Examples of Coadjoint Orbits

(a) Rotation Group.   As we saw in §9.3, the adjoint action for SO(3) is $\mathrm{Ad}_{\mathbf{A}}(\mathbf{v}) = \mathbf{A}\mathbf{v}$, where $\mathbf{A} \in \mathrm{SO}(3)$ and $\mathbf{v} \in \mathbb{R}^3 \cong \mathfrak{so}(3)$. Identify $\mathfrak{so}(3)^*$ with $\mathbb{R}^3$ by the usual dot product, that is, if $\mathbf{\Pi}, \mathbf{v} \in \mathbb{R}^3$, we have $\langle \mathbf{\Pi}, \hat{\mathbf{v}} \rangle = \mathbf{\Pi} \cdot \mathbf{v}$. Thus, for $\mathbf{\Pi} \in \mathfrak{so}(3)^*$ and $\mathbf{A} \in \mathrm{SO}(3)$,

$$\langle \mathrm{Ad}^*_{\mathbf{A}^{-1}}(\mathbf{\Pi}), \hat{\mathbf{v}} \rangle = \langle \mathbf{\Pi}, \mathrm{Ad}_{\mathbf{A}^{-1}}(\mathbf{v}) \rangle = \langle \mathbf{\Pi}, (\mathbf{A}^{-1}\mathbf{v}) \rangle = \mathbf{\Pi} \cdot \mathbf{A}^{-1}\mathbf{v}$$
$$= (\mathbf{A}^{-1})^T \mathbf{\Pi} \cdot \mathbf{v} = \mathbf{A}\mathbf{\Pi} \cdot \mathbf{v}, \tag{14.1.1}$$

since $\mathbf{A}$ is orthogonal. Hence, with $\mathfrak{so}(3)^*$ identified with $\mathbb{R}^3$, $\mathrm{Ad}^*_{\mathbf{A}^{-1}} = \mathbf{A}$, and so

$$\mathrm{Orb}(\mathbf{\Pi}) = \{\, \mathrm{Ad}^*_{\mathbf{A}^{-1}}(\mathbf{\Pi}) \mid \mathbf{A} \in \mathrm{SO}(3) \,\} = \{\, \mathbf{A}\mathbf{\Pi} \mid \mathbf{A} \in \mathrm{SO}(3) \,\}, \tag{14.1.2}$$

which is the sphere in $\mathbb{R}^3$ of radius $\|\mathbf{\Pi}\|$.   ◆

(b) Affine Group on $\mathbb{R}$.   Consider the Lie group of transformations of $\mathbb{R}$ of the form $T(x) = ax + b$ where $a \neq 0$. Identify G with the set of pairs $(a, b) \in \mathbb{R}^2$ with $a \neq 0$. Since

$$(T_1 \circ T_2)(x) = a_1(a_2 x + b_2) + b_1 = a_1 a_2 x + a_1 b_2 + b_1$$

and $T^{-1}(x) = (x - b)/a$, we take group multiplication to be

$$(a_1, b_1) \cdot (a_2, b_2) = (a_1 a_2, a_1 b_2 + b_1). \tag{14.1.3}$$

The identity element is $(1, 0)$, and the inverse of (a, b) is

$$(a, b)^{-1} = \left(\frac{1}{a}, -\frac{b}{a} \right). \tag{14.1.4}$$

Thus, G is a two-dimensional Lie group. It is an example of a *semidirect product*. (See Exercise 9.3-1.) As a set, the Lie algebra of G is $\mathfrak{g} = \mathbb{R}^2$; to compute the bracket on $\mathfrak{g}$ we shall first compute the adjoint representation. The inner automorphisms are given by

$$I_{(a,b)}(c, d) = (a, b) \cdot (c, d) \cdot (a, b)^{-1} = (ac, ad + b) \cdot \left(\frac{1}{a}, -\frac{b}{a} \right)$$
$$= (c, ad - bc + b), \tag{14.1.5}$$

---

[1] The coadjoint orbits are also embedded (but not necessarily closed) submanifolds of $\mathfrak{g}^*$ if G is an algebraic group.

and so differentiating (14.1.5) with respect to (c, d) at the identity in the direction of $(u, v) \in \mathfrak{g}$ gives

$$\mathrm{Ad}_{(a,b)}(u, v) = (u, av - bu). \tag{14.1.6}$$

Differentiating (14.1.6) with respect to (a, b) in the direction (r, s) gives the Lie bracket

$$[(r, s), (u, v)] = (0, rv - su). \tag{14.1.7}$$

An alternative approach is to realize (a, b) as the matrix

$$\begin{pmatrix} a & b \\ 0 & 1 \end{pmatrix};$$

one checks that group multiplication corresponds to matrix multiplication. Then the Lie algebra, identified with matrices

$$\begin{pmatrix} u & v \\ 0 & 0 \end{pmatrix},$$

has the bracket given by the commutator.

The adjoint orbit through (u, v) is $\{u\} \times \mathbb{R}$ if $(u, v) \neq (0, 0)$ and is $\{(0, 0)\}$ if $(u, v) = (0, 0)$. The adjoint orbit $\{u\} \times \mathbb{R}$ cannot be symplectic, as it is one-dimensional. To compute the coadjoint orbits, denote elements of $\mathfrak{g}^*$ by the column vector

$$(\alpha, \beta)^T = \begin{pmatrix} \alpha \\ \beta \end{pmatrix}$$

and use the pairing

$$\left\langle (u, v), \begin{pmatrix} \alpha \\ \beta \end{pmatrix} \right\rangle = \alpha u + \beta v \tag{14.1.8}$$

to identify $\mathfrak{g}^*$ with $\mathbb{R}^2$. Then

$$\left\langle \mathrm{Ad}^*_{(a,b)} \begin{pmatrix} \alpha \\ \beta \end{pmatrix}, (u, v) \right\rangle = \left\langle \begin{pmatrix} \alpha \\ \beta \end{pmatrix}, \mathrm{Ad}_{(a,b)}(u, v) \right\rangle = \left\langle \begin{pmatrix} \alpha \\ \beta \end{pmatrix}, (u, av - bu) \right\rangle$$

$$= \alpha u + \beta av - \beta bu. \tag{14.1.9}$$

Thus,

$$\mathrm{Ad}^*_{(a,b)} \begin{pmatrix} \alpha \\ \beta \end{pmatrix} = \begin{pmatrix} \alpha - \beta b \\ \beta a \end{pmatrix}. \tag{14.1.10}$$

If $\beta = 0$, the coadjoint orbit through $(\alpha, \beta)^T$ is a single point. If $\beta \neq 0$, the orbit through $(\alpha, \beta)^T$ is $\mathbb{R}^2$ minus the α-axis.

It is sometimes convenient to identify the dual $\mathfrak{g}^*$ with $\mathfrak{g}$, that is, with matrices

$$\begin{pmatrix} \alpha & \beta \\ 0 & 0 \end{pmatrix}$$

via the pairing of $\mathfrak{g}^*$ with $\mathfrak{g}$ that is given by the trace of matrix multiplication of an element of $\mathfrak{g}^*$ with the transpose conjugate of an element of $\mathfrak{g}$. ♦

(c) Orbits in $\mathfrak{X}^*_{\mathrm{div}}$. Let $G = \mathrm{Diff}_{\mathrm{vol}}(\Omega)$, the group of volume-preserving diffeomorphisms of a region Ω in $\mathbb{R}^n$, with Lie algebra $\mathfrak{X}_{\mathrm{div}}(\Omega)$. In Example (d) of §10.2 we identified $\mathfrak{X}^*_{\mathrm{div}}(\Omega)$ with $\mathfrak{X}_{\mathrm{div}}(\Omega)$ by using the L^2-pairing on vector fields. Here we begin by finding a different representation of the dual $\mathfrak{X}^*_{\mathrm{div}}(\Omega)$, which is more convenient for explicitly determining the coadjoint action. Then we return to the identification above and will find the expression for the coadjoint action on $\mathfrak{X}_{\mathrm{div}}(\Omega)$.

The main technical ingredient used below is the Hodge decomposition theorem for manifolds with boundary. Here we state only the relevant facts to be used below. A k-form α is called *tangent* to $\partial\Omega$ if $i^*(*\alpha) = 0$. Let $\Omega^k_t(\Omega)$ denote all k-forms on M that are tangent to $\partial\Omega$. One of the Hodge decomposition theorems states that there is an L^2-orthogonal decomposition

$$\Omega^k(\Omega) = \mathbf{d}\Omega^{k-1}(\Omega) \oplus \{\, \alpha \in \Omega^k_t(\Omega) \mid \delta\alpha = 0 \,\}.$$

This implies that the pairing

$$\langle\,,\rangle : \{\, \alpha \in \Omega^1_t(\Omega) \mid \delta\alpha = 0 \,\} \times \mathfrak{X}_{\mathrm{div}}(\Omega) \to \mathbb{R}$$

given by

$$\langle M, X\rangle = \int_\Omega M_i X^i d^n x \tag{14.1.11}$$

is weakly nondegenerate. Indeed, if $M \in \{\, \alpha \in \Omega^1_t(M) \mid \delta\alpha = 0 \,\}$ and $\langle M, X\rangle = 0$ for all $X \in \mathfrak{X}_{\mathrm{div}}(\Omega)$, then $\langle M, B\rangle = 0$ for all $B \in \{\, \Omega^1_t(\Omega) \mid \delta B = 0 \,\}$ because the index-lowering operator $\flat$ given by the metric on Ω induces an isomorphism between $\mathfrak{X}_{\mathrm{div}}(\Omega)$ and $\{\, \alpha \in \Omega^1_t(\Omega) \mid \delta B = 0 \,\}$. Therefore, by the L^2-orthogonal decomposition quoted above, $M = \mathbf{d}f$, and hence $M = 0$. Similarly, if $X \in \mathfrak{X}_{\mathrm{div}}(\Omega)$ and $\langle M, X\rangle = 0$ for all $M \in \{\, \alpha \in \Omega^1_t(M) \mid \delta\alpha = 0 \,\}$, then $\langle M, X^\flat\rangle = 0$ for all such M, and as before, $X^\flat = \mathbf{d}f$, that is, $X = \nabla f$. But this implies $X = 0$, since $\mathfrak{X}_{\mathrm{div}}(\Omega)$ and gradients are L^2-orthogonal by Stokes' theorem. Therefore, we can identify

$$\mathfrak{X}^*_{\mathrm{div}}(\Omega) = \{\, M \in \Omega^1_t(\Omega) \mid \delta M = 0 \,\}. \tag{14.1.12}$$

The coadjoint action of $\text{Diff}_{\text{vol}}(\Omega)$ on $\mathfrak{X}^*_{\text{div}}(\Omega)$ is computed in the following way. Recall from Chapter 9 that $\text{Ad}_\varphi(X) = \varphi_* X$ for $\varphi \in \text{Diff}_{\text{vol}}(\Omega)$ and $X \in \mathfrak{X}_{\text{div}}(\Omega)$. Thus,

$$\langle \text{Ad}^*_{\varphi^{-1}} M, X \rangle = \langle M, \text{Ad}_{\varphi^{-1}} X \rangle = \int_\Omega M \cdot \varphi^* X \, d^n x = \int_\Omega \varphi_* M \cdot X \, d^n x$$

by the change of variables formula. Therefore,

$$\text{Ad}^*_{\varphi^{-1}} M = \varphi_* M, \tag{14.1.13}$$

and so

$$\text{Orb} \, M = \{ \varphi_* M \mid \varphi \in \text{Diff}_{\text{vol}}(\Omega) \}.$$

Next, let us return to the identification of $\mathfrak{X}_{\text{div}}(\Omega)$ with itself by the L^2-pairing on vector fields

$$\langle X, Y \rangle = \int_\Omega X \cdot Y \, d^n x. \tag{14.1.14}$$

The Helmholtz decomposition says that any vector field on Ω can be uniquely decomposed orthogonally in a sum of a gradient of a function and a divergence-free vector field tangent to $\partial \Omega$; this decomposition is equivalent to the Hodge decomposition on one-forms quoted before. This shows that (14.1.14) is a weakly nondegenerate pairing. For $\varphi \in \text{Diff}_{\text{vol}}(\Omega)$, denote by $(T\varphi)^\dagger$ the adjoint of $T\varphi : T\Omega \to T\Omega$ relative to the metric (14.1.14). By the change of variables formula,

$$\langle \text{Ad}^*_{\varphi^{-1}} Y, X \rangle = \langle Y, \text{Ad}_{\varphi^{-1}} X \rangle = \int_\Omega Y \cdot \varphi^* X \, d^n x$$

$$= \int_\Omega Y \cdot (T\varphi^{-1} \circ X \circ \varphi) \, d^n x = \int_\Omega ((T\varphi^{-1})^\dagger \circ Y \circ \varphi) \cdot X \, d^n x,$$

that is,

$$\text{Ad}^*_{\varphi^{-1}} Y = (T\varphi^{-1})^\dagger \circ Y \circ \varphi \tag{14.1.15}$$

and

$$\text{Orb} \, Y = \{ (T\varphi^{-1})^\dagger \circ Y \circ \varphi \mid \varphi \in \text{Diff}_{\text{vol}}(\Omega) \}. \tag{14.1.16}$$

This example shows that different pairings give rise to different formulas for the coadjoint action and that the choice of dual is dictated by the specific application one has in mind. For example, the pairing (14.1.14) was convenient for the Lie–Poisson bracket on $\mathfrak{X}_{\text{div}}(\Omega)$ in Example (d) of §10.2. On the other hand, many computations involving the coadjoint action are simpler with the choice (14.1.12) of the dual corresponding to the pairing (14.1.11). ◆

(d) Orbits in $\mathfrak{X}^*_{\mathrm{can}}$.   Let $G = \mathrm{Diff}_{\mathrm{can}}(P)$ be the group of canonical transformations of a symplectic manifold P with $H^1(P) = 0$. Letting k be a function on P, and X_k the corresponding Hamiltonian vector field, and $\varphi \in G$, we have

$$\mathrm{Ad}_\varphi X_k = \varphi_* X_k = X_{k \circ \varphi^{-1}}, \tag{14.1.17}$$

so identifying $\mathfrak{g}$ with $\mathcal{F}(P)$ modulo constants, or equivalently with functions on P with zero average, we get $\mathrm{Ad}_\varphi k = \varphi_* k = k \circ \varphi^{-1}$. On the dual space, which is identified with $\mathcal{F}(P)$ (modulo constants) via the L^2-pairing, a straightforward verification shows that

$$\mathrm{Ad}^*_{\varphi^{-1}} f = \varphi_* f = f \circ \varphi^{-1}. \tag{14.1.18}$$

One sometimes says that $\mathrm{Orb}(f) = \{\, f \circ \varphi^{-1} \mid \varphi \in \mathrm{Diff}_{\mathrm{can}}(P) \,\}$ consists of *canonical rearrangements* of f.   ◆

(e) Toda Orbit.   Another interesting example is the Toda orbit, which arises in the study of completely integrable systems. Let

$\mathfrak{g}$ = Lie algebra of real $n \times n$ lower triangular matrices
   with trace zero,

G = lower triangular matrices with determinant one,

and identify $\mathfrak{g}^*$ = the upper triangular matrices, using the pairing

$$\langle \xi, \mu \rangle = \mathrm{trace}(\xi\mu),$$

where $\xi \in \mathfrak{g}$ and $\mu \in \mathfrak{g}^*$. Since $\mathrm{Ad}_A \, \xi = A\xi A^{-1}$, we get

$$\mathrm{Ad}^*_{A^{-1}} \, \mu = P(A\mu A^{-1}), \tag{14.1.19}$$

where $P : \mathfrak{sl}(n, \mathbb{R}) \to \mathfrak{g}^*$ is the projection sending any matrix to its upper triangular part. Now let

$$\mu = \begin{bmatrix} 0 & 1 & 0 & \cdots & 0 & 0 \\ 0 & 0 & 1 & \cdots & 0 & 0 \\ 0 & 0 & 0 & \cdots & 0 & 0 \\ \vdots & \vdots & \vdots & \ddots & \vdots & \vdots \\ 0 & 0 & 0 & \cdots & 0 & 1 \\ 0 & 0 & 0 & \cdots & 0 & 0 \end{bmatrix} \in \mathfrak{g}^*. \tag{14.1.20}$$

One finds that $\mathrm{Orb}(\mu) = \{\, P(A\mu A^{-1}) \mid A \in G \,\}$ consists of matrices of the form

$$
L = \begin{bmatrix} b_1 & a_1 & 0 & 0 & \cdots & 0 & 0 \\ 0 & b_2 & a_2 & 0 & \cdots & 0 & 0 \\ 0 & 0 & b_3 & a_3 & \cdots & 0 & 0 \\ 0 & 0 & 0 & b_4 & \cdots & 0 & 0 \\ \vdots & \vdots & \vdots & \vdots & \ddots & \vdots & \vdots \\ 0 & 0 & 0 & 0 & \cdots & b_{n-1} & a_{n-1} \\ 0 & 0 & 0 & 0 & \cdots & 0 & b_n \end{bmatrix}, \tag{14.1.21}
$$

where $\sum b_n = 0$. See Kostant [1979] and Symes [1982a, 1982b] for further information.   ◆

(f) Coadjoint Orbits That Are Not Submanifolds.   This example presents a Lie group G whose generic coadjoint orbits in $\mathfrak{g}^*$ are *not* submanifolds, which is due to Kirillov [1976b, p. 293]. Let α be irrational, define

$$
G = \left\{ \begin{bmatrix} e^{it} & 0 & z \\ 0 & e^{i\alpha t} & w \\ 0 & 0 & 1 \end{bmatrix} \middle| \, t \in \mathbb{R}, \ z, w \in \mathbb{C} \right\}, \tag{14.1.22}
$$

and note the G is diffeomorphic to $\mathbb{R}^5$. As a group, it is the semidirect product of

$$
H = \left\{ \begin{bmatrix} e^{it} & 0 \\ 0 & e^{i\alpha t} \end{bmatrix} \middle| \, t \in \mathbb{R} \right\}
$$

with $\mathbb{C}^2$, the action being by left multiplication of vectors in $\mathbb{C}^2$ by elements of H (see Exercise 9.3-1). Thus, the identity element of G is the 3×3 identity matrix and

$$
\begin{bmatrix} e^{it} & 0 & z \\ 0 & e^{i\alpha t} & w \\ 0 & 0 & 1 \end{bmatrix}^{-1} = \begin{bmatrix} e^{-it} & 0 & -ze^{-it} \\ 0 & e^{-i\alpha t} & -we^{-i\alpha t} \\ 0 & 0 & 1 \end{bmatrix}.
$$

The Lie algebra $\mathfrak{g}$ of G is

$$
\mathfrak{g} = \left\{ \begin{bmatrix} it & 0 & x \\ 0 & i\alpha t & y \\ 0 & 0 & 0 \end{bmatrix} \middle| \, t \in \mathbb{R}, \ x, y \in \mathbb{C} \right\} \tag{14.1.23}
$$

with the usual commutator bracket as Lie bracket. Identify $\mathfrak{g}^*$ with

$$
\mathfrak{g}^* = \left\{ \begin{bmatrix} is & 0 & 0 \\ 0 & i\alpha s & 0 \\ a & b & 0 \end{bmatrix} \middle| \, s \in \mathbb{R}, \ a, b \in \mathbb{C} \right\} \tag{14.1.24}
$$

via the nondegenerate pairing in $\mathfrak{gl}(3,\mathbb{C})$ given by

$$\langle A, B \rangle = \mathrm{Re}(\mathrm{trace}(AB)).$$

The adjoint action of

$$g = \begin{bmatrix} e^{it} & 0 & z \\ 0 & e^{i\alpha t} & w \\ 0 & 0 & 1 \end{bmatrix} \quad \text{on} \quad \xi = \begin{bmatrix} is & 0 & x \\ 0 & i\alpha s & y \\ 0 & 0 & 0 \end{bmatrix}$$

is given by

$$\mathrm{Ad}_g\,\xi = \begin{bmatrix} is & 0 & e^{it}x - isz \\ 0 & i\alpha s & e^{i\alpha t}y - i\alpha sw \\ 0 & 0 & 0 \end{bmatrix}. \tag{14.1.25}$$

The coadjoint action of the same group element g on

$$\mu = \begin{bmatrix} iu & 0 & 0 \\ 0 & i\alpha u & 0 \\ a & b & 0 \end{bmatrix}$$

is given by

$$\mathrm{Ad}^*_{g^{-1}}\,\mu = \begin{bmatrix} iu' & 0 & 0 \\ 0 & i\alpha u' & 0 \\ ae^{-it} & be^{-i\alpha t} & 0 \end{bmatrix}, \tag{14.1.26}$$

where

$$u' = u + \frac{1}{1+\alpha^2}\,\mathrm{Im}(ae^{-it}z + be^{-i\alpha t}\alpha w). \tag{14.1.27}$$

If $a, b \neq 0$, the orbit through μ is two-dimensional; it is a cylindrical surface whose generator is the u'-axis and whose base is the curve in $\mathbb{C}^2$ given parametrically by $t \mapsto (ae^{-it}, be^{-i\alpha t})$. This curve, however, is the irrational flow on the torus with radii $|a|$ and $|b|$, that is, the cylindrical surface accumulates on itself and thus is not a submanifold of $\mathbb{R}^5$. In addition, note that the closure of this orbit is the three-dimensional manifold that is the product of the u'-line with the two-dimensional torus of radii $|a|$ and $|b|$. We shall return to this example in the Internet supplement. ◆

Exercises

⋄ **14.1-1.** Show that for $\mu \in \mathfrak{g}^*$,

$$\mathrm{Orb}(\mu) = J_R\left[J_L^{-1}(\mu)\right] = J_L\left[J_R^{-1}(\mu)\right].$$

⋄ **14.1-2.** Work out (14.1.10) using matrix notation.

14.2   Tangent Vectors to Coadjoint Orbits

In general, orbits of a Lie group action, while manifolds in their own right, are not submanifolds of the ambient manifold; they are only injectively immersed manifolds. A notable exception occurs in the case of compact Lie groups: Then all their orbits are closed embedded submanifolds. Coadjoint orbits are no exception to this global problem, as we saw in the preceding examples. We shall always regard them as injectively immersed submanifolds, diffeomorphic to G/G_μ, where $G_\mu = \{ g \in G \mid \mathrm{Ad}_g^* \mu = \mu \}$ is the isotropy subgroup of the coadjoint action at a point μ in the orbit.

We now describe tangent vectors to coadjoint orbits. Let $\xi \in \mathfrak{g}$ and let $g(t)$ be a curve in G tangent to ξ at $t = 0$; for example, let $g(t) = \exp(t\xi)$. Let $\mathcal{O}$ be a coadjoint orbit and $\mu \in \mathcal{O}$. If $\eta \in \mathfrak{g}$, then

$$\mu(t) = \mathrm{Ad}^*_{g(t)^{-1}} \mu \qquad (14.2.1)$$

is a curve in $\mathcal{O}$ with $\mu(0) = \mu$. Differentiating the identity

$$\langle \mu(t), \eta \rangle = \langle \mu, \mathrm{Ad}_{g(t)^{-1}} \eta \rangle \qquad (14.2.2)$$

with respect to t at $t = 0$, we get

$$\langle \mu'(0), \eta \rangle = -\langle \mu, \mathrm{ad}_\xi \eta \rangle = -\langle \mathrm{ad}_\xi^* \mu, \eta \rangle,$$

and so

$$\mu'(0) = -\mathrm{ad}_\xi^* \mu. \qquad (14.2.3)$$

Thus,

$$T_\mu \mathcal{O} = \{ \mathrm{ad}_\xi^* \mu \mid \xi \in \mathfrak{g} \}. \qquad (14.2.4)$$

This calculation also proves that the infinitesimal generator of the coadjoint action is given by

$$\xi_{\mathfrak{g}^*}(\mu) = -\mathrm{ad}_\xi^* \mu. \qquad (14.2.5)$$

The following characterization of the tangent space to coadjoint orbits is often useful. We let $\mathfrak{g}_\mu = \{ \xi \in \mathfrak{g} \mid \mathrm{ad}_\xi^* \mu = 0 \}$ be the coadjoint isotropy algebra of μ; it is the Lie algebra of the coadjoint isotropy group

$$G_\mu = \{ g \in G \mid \mathrm{Ad}_g^* \mu = \mu \}.$$

Proposition 14.2.1. *Let* $\langle , \rangle : \mathfrak{g}^* \times \mathfrak{g} \to \mathbb{R}$ *be a weakly nondegenerate pairing and let* $\mathcal{O}$ *be the coadjoint orbit through* $\mu \in \mathfrak{g}^*$. *Let*

$$\mathfrak{g}_\mu^\circ := \{ \nu \in \mathfrak{g}^* \mid \langle \nu, \eta \rangle = 0 \text{ for all } \eta \in \mathfrak{g}_\mu \}$$

be the annihilator of $\mathfrak{g}_\mu$ *in* $\mathfrak{g}^*$. *Then* $T_\mu \mathcal{O} \subset \mathfrak{g}_\mu^\circ$. *If* $\mathfrak{g}$ *is finite-dimensional, then* $T_\mu \mathcal{O} = \mathfrak{g}_\mu^\circ$. *The same equality holds if* $\mathfrak{g}$ *and* $\mathfrak{g}^*$ *are Banach spaces,* $T_\mu \mathcal{O}$ *is closed in* $\mathfrak{g}^*$, *and the pairing is strongly nondegenerate.*

Proof.   For any $\xi \in \mathfrak{g}$ and $\eta \in \mathfrak{g}_\mu$ we have

$$\langle \mathrm{ad}^*_\xi \, \mu, \eta \rangle = \langle \mu, [\xi, \eta] \rangle = -\langle \mathrm{ad}^*_\eta \, \mu, \xi \rangle = 0,$$

which proves the inclusion $T_\mu \mathcal{O} \subset \mathfrak{g}^\circ_\mu$. If $\mathfrak{g}$ is finite-dimensional, equality holds, since $\dim T_\mu \mathcal{O} = \dim \mathfrak{g} - \dim \mathfrak{g}_\mu = \dim \mathfrak{g}^\circ_\mu$. If $\mathfrak{g}$ and $\mathfrak{g}^*$ are infinite-dimensional Banach spaces and $\langle , \rangle : \mathfrak{g}^* \times \mathfrak{g} \to \mathbb{R}$ is a strong pairing, we can assume without loss of generality that it is the natural pairing between a Banach space and its dual. If $\mathfrak{g}^\circ_\mu \neq T_\mu \mathcal{O}$, pick $\nu \in \mathfrak{g}^\circ_\mu$ such that $\nu \neq 0$ and $\nu \notin T_\mu \mathcal{O}$. By the Hahn–Banach theorem there is an $\eta \in \mathfrak{g}$ such that $\langle \nu, \eta \rangle = 1$ and $\langle \mathrm{ad}^*_\xi \, \mu, \eta \rangle = 0$ for all $\xi \in \mathfrak{g}$. The latter condition is equivalent to $\eta \in \mathfrak{g}_\mu$. On the other hand, since $\nu \in \mathfrak{g}^\circ_\mu$, we have $\langle \nu, \eta \rangle = 0$, which is a contradiction.   ∎

Examples of Tangent Vectors

(a) Rotation Group.   Identifying $(\mathfrak{so}(3), [\cdot, \cdot]) \cong (\mathbb{R}^3, \times)$ and $\mathfrak{so}(3)^* \cong \mathbb{R}^3$ via the natural pairing given by the Euclidean inner product, formula (14.2.5) reads as follows for $\mathbf{\Pi} \in \mathfrak{so}(3)^*$ and $\xi, \eta \in \mathfrak{so}(3)$:

$$\langle \xi_{\mathfrak{so}(3)^*}(\mathbf{\Pi}), \eta \rangle = -\mathbf{\Pi} \cdot (\xi \times \eta) = -(\mathbf{\Pi} \times \xi) \cdot \eta, \qquad (14.2.6)$$

so that $\xi_{\mathfrak{so}(3)^*}(\mathbf{\Pi}) = -\mathbf{\Pi} \times \xi = \xi \times \mathbf{\Pi}$. As expected, $\xi_{\mathfrak{so}(3)^*}(\mathbf{\Pi}) \in T_{\mathbf{\Pi}} \, \mathrm{Orb}(\mathbf{\Pi})$ is tangent to the sphere $\mathrm{Orb}(\mathbf{\Pi})$. Allowing ξ to vary in $\mathfrak{so}(3) \cong \mathbb{R}^3$, one obtains all of $T_{\mathbf{\Pi}} \, \mathrm{Orb}(\mathbf{\Pi})$.   ◆

(b) Affine Group on $\mathbb{R}$.   Let $(u, v) \in \mathfrak{g}$ and consider the coadjoint orbit through the point

$$\begin{pmatrix} \alpha \\ \beta \end{pmatrix} \in \mathfrak{g}^*.$$

Then (14.2.5) reads

$$(u, v)_{\mathfrak{g}^*} \begin{pmatrix} \alpha \\ \beta \end{pmatrix} = \left\langle \begin{pmatrix} \alpha \\ \beta \end{pmatrix}, [\,\cdot\,, (u, v)] \right\rangle. \qquad (14.2.7)$$

But

$$\left\langle \begin{pmatrix} \alpha \\ \beta \end{pmatrix}, [(r, s), (u, v)] \right\rangle = \left\langle \begin{pmatrix} \alpha \\ \beta \end{pmatrix}, (0, rv - su) \right\rangle = rv\beta - su\beta,$$

and so

$$(u, v)_{\mathfrak{g}^*} \begin{pmatrix} \alpha \\ \beta \end{pmatrix} = \begin{pmatrix} v\beta \\ -u\beta \end{pmatrix}. \qquad (14.2.8)$$

If $\beta \neq 0$, these vectors span $\mathfrak{g}^* = \mathbb{R}^2$, as they should.   ◆

(c) The Group $\mathrm{Diff}_{\mathrm{vol}}$. For $G = \mathrm{Diff}_{\mathrm{vol}}$ and $M \in \mathfrak{X}^*_{\mathrm{div}}$, we get the tangent vectors to $\mathrm{Orb}(M)$ by differentiating (14.1.13) with respect to φ, yielding

$$T_M \, \mathrm{Orb}(M) = \{ -\pounds_v M \mid v \text{ is divergence free and tangent to } \partial\Omega \}.$$
$$(14.2.9)$$

$\blacklozenge$

(d) The Group $\mathrm{Diff}_{\mathrm{can}}$ **(P).** For $G = \mathrm{Diff}_{\mathrm{can}}(P)$, we have

$$T_f \, \mathrm{Orb}(f) = \{ -\{f, k\} \mid k \in \mathcal{F}(P) \}.$$
$$(14.2.10)$$

$\blacklozenge$

(e) The Toda Lattice. The tangent space to the Toda orbit consists of matrices of the same form as L in (14.1.21), since those matrices form a linear space. The reader can check that (14.2.4) gives the same answer. $\blacklozenge$

Exercises

$\diamond$ **14.2-1.** Show that for the affine group on $\mathbb{R}$, the Lie–Poisson bracket is

$$\{f, g\}(\alpha, \beta) = \beta \left(\frac{\partial f}{\partial \alpha} \frac{\partial g}{\partial \beta} - \frac{\partial f}{\partial \beta} \frac{\partial g}{\partial \alpha} \right).$$

14.3 The Symplectic Structure on Coadjoint Orbits

Theorem 14.3.1 (Coadjoint Orbit Theorem). *Let G be a Lie group and let $\mathcal{O} \subset \mathfrak{g}^*$ be a coadjoint orbit. Then*

$$\omega^{\pm}(\mu)(\xi_{\mathfrak{g}^*}(\mu), \eta_{\mathfrak{g}^*}(\mu)) = \pm \langle \mu, [\xi, \eta] \rangle \qquad (14.3.1)$$

*for all $\mu \in \mathcal{O}$ and $\xi, \eta \in \mathfrak{g}$ define symplectic forms on $\mathcal{O}$. We refer to $\omega^{\pm}$ as the **coadjoint orbit symplectic structures** and, if there is danger of confusion, denote it by $\omega^{\pm}_{\mathcal{O}}$.*

Proof. We prove the result for ω^-, the argument for ω^+ being similar. First we show that formula (14.3.1) gives a well-defined form; that is, the right-hand side is independent of the particular $\xi \in \mathfrak{g}$ and $\eta \in \mathfrak{g}$ that define the tangent vectors $\xi_{\mathfrak{g}^*}(\mu)$ and $\eta_{\mathfrak{g}^*}(\mu)$. This follows by observing that $\xi_{\mathfrak{g}^*}(\mu) = \xi'_{\mathfrak{g}^*}(\mu)$ implies $-\langle \mu, [\xi, \eta] \rangle = -\langle \mu, [\xi', \eta] \rangle$ for all $\eta \in \mathfrak{g}$. Therefore,

$$\omega^-(\mu)(\xi_{\mathfrak{g}^*}(\mu), \eta_{\mathfrak{g}^*}(\mu)) = \omega^-(\xi'_{\mathfrak{g}^*}(\mu), \eta_{\mathfrak{g}^*}(\mu)),$$

so ω^- is well-defined.

Second, we show that ω^- is nondegenerate. Since the pairing $\langle\,,\rangle$ is non-degenerate, $\omega^-(\mu)(\xi_{\mathfrak{g}^*}(\mu), \eta_{\mathfrak{g}^*}(\mu)) = 0$ for all $\eta_{\mathfrak{g}^*}(\mu)$ implies $-\langle\mu, [\xi, \eta]\rangle = 0$ for all η. This means that $0 = -\langle\mu, [\xi, \cdot]\rangle = \xi_{\mathfrak{g}^*}(\mu)$.

Finally, we show that ω^- is closed, that is, $\mathbf{d}\omega^- = 0$. To do this we begin by defining, for each $\nu \in \mathfrak{g}^*$, the one-form ν_L on G by

$$\nu_L(g) = (T_g^* L_{g^{-1}})(\nu),$$

where $g \in G$. The one-form ν_L is readily checked to be left invariant; that is, $L_g^* \nu_L = \nu_L$ for all $g \in G$. For $\xi \in \mathfrak{g}$, let ξ_L be the corresponding left-invariant vector field on G, so $\nu_L(\xi_L)$ is a constant function on G (whose value at any point is $\langle\nu, \xi\rangle$). Choose $\nu \in \mathcal{O}$ and consider the surjective map $\varphi_\nu : G \to \mathcal{O}$ defined by $g \mapsto \operatorname{Ad}_{g^{-1}}^*(\nu)$ and the two-form $\sigma = \varphi_\nu^* \omega^-$ on G. We claim that

$$\sigma = \mathbf{d}\nu_L. \tag{14.3.2}$$

To prove this, notice that

$$(T_e\varphi_\nu)(\eta) = \eta_{\mathfrak{g}^*}(\nu), \tag{14.3.3}$$

so that the surjective map φ_ν is submersive at e. By definition of pull-back, $\sigma(e)(\xi, \eta)$ equals

$$(\varphi_\nu^* \omega^-)(e)(\xi, \eta) = \omega^-(\varphi_\nu(e))(T_e\varphi_\nu \cdot \xi, T_e\varphi_\nu \cdot \eta)$$
$$= \omega^-(\nu)(\xi_{\mathfrak{g}^*}(\nu), \eta_{\mathfrak{g}^*}(\nu)) = -\langle\nu, [\xi, \eta]\rangle. \tag{14.3.4}$$

Hence

$$\sigma(\xi_L, \eta_L)(e) = \sigma(e)(\xi, \eta) = -\langle\nu, [\xi, \eta]\rangle = -\langle\nu_L, [\xi_L, \eta_L]\rangle(e). \tag{14.3.5}$$

We shall need the relation $\sigma(\xi_L, \eta_L) = -\langle\nu_L, [\xi_L, \eta_L]\rangle$ at each point of G; to get it, we first prove two lemmas.

Lemma 14.3.2. *The map* $\operatorname{Ad}_{g^{-1}}^* : \mathcal{O} \to \mathcal{O}$ *preserves* ω^-, *that is,*

$$(\operatorname{Ad}_{g^{-1}}^*)^* \omega^- = \omega^-.$$

Proof. To prove this, we recall two identities from Chapter 9. First,

$$(\operatorname{Ad}_g \xi)_{\mathfrak{g}^*} = \operatorname{Ad}_{g^{-1}}^* \circ \xi_{\mathfrak{g}^*} \circ \operatorname{Ad}_g^*, \tag{14.3.6}$$

which is proved by letting ξ be tangent to a curve $h(\varepsilon)$ at $\varepsilon = 0$, recalling that

$$\operatorname{Ad}_g \xi = \left.\frac{d}{d\varepsilon} gh(\varepsilon)g^{-1}\right|_{\varepsilon=0} \tag{14.3.7}$$

and noting that

$$(\mathrm{Ad}_g\,\xi)_{\mathfrak{g}^*}(\mu) = \frac{d}{d\varepsilon}\,\mathrm{Ad}^*_{(gh(\varepsilon)g^{-1})^{-1}}\,\mu\Big|_{\varepsilon=0}$$

$$= \frac{d}{d\varepsilon}\,\mathrm{Ad}^*_{g^{-1}}\,\mathrm{Ad}^*_{h(\varepsilon)^{-1}}\,\mathrm{Ad}^*_g(\mu)\Big|_{\varepsilon=0}. \qquad (14.3.8)$$

Second, we require the identity

$$\mathrm{Ad}_g[\xi,\eta] = [\mathrm{Ad}_g\,\xi,\mathrm{Ad}_g\,\eta], \qquad (14.3.9)$$

which follows by differentiating the relation

$$I_g(I_h(k)) = I_g(h)I_g(k)I_g(h^{-1}) \qquad (14.3.10)$$

with respect to h and k and evaluating at the identity.

Evaluating (14.3.6) at $\nu = \mathrm{Ad}^*_{g^{-1}}\,\mu$, we get

$$(\mathrm{Ad}_g\,\xi)_{\mathfrak{g}^*}(\nu) = \mathrm{Ad}^*_{g^{-1}}\cdot\xi_{\mathfrak{g}^*}(\mu) = T_\mu\,\mathrm{Ad}^*_{g^{-1}}\cdot\xi_{\mathfrak{g}^*}(\mu), \qquad (14.3.11)$$

by linearity of $\mathrm{Ad}^*_{g^{-1}}$. Thus,

$$\begin{aligned}
((\mathrm{Ad}^*_{g^{-1}})^*\omega^-)(\mu)(\xi_{\mathfrak{g}^*}(\mu),\eta_{\mathfrak{g}^*}(\mu)) & \\
= \omega^-(\nu)(T_\mu\,\mathrm{Ad}^*_{g^{-1}}\cdot\xi_{\mathfrak{g}^*}(\mu),T_\mu\,\mathrm{Ad}^*_{g^{-1}}\cdot\eta_{\mathfrak{g}^*}(\mu)) & \\
= \omega^-(\nu)((\mathrm{Ad}_g\,\xi)_{\mathfrak{g}^*}(\nu),(\mathrm{Ad}_g\,\eta)_{\mathfrak{g}^*}(\nu)) & \quad \text{(by (14.3.11))} \\
= -\langle\nu,[\mathrm{Ad}_g\,\xi,\mathrm{Ad}_g\,\eta]\rangle & \quad \text{(by definition of }\omega^-) \\
= -\langle\nu,\mathrm{Ad}_g[\xi,\eta]\rangle & \quad \text{(by (14.3.9))} \\
= -\langle\mathrm{Ad}^*_g\,\nu,[\xi,\eta]\rangle = -\langle\mu,[\xi,\eta]\rangle & \\
= \omega^-(\mu)(\xi_{\mathfrak{g}^*}(\mu),\eta_{\mathfrak{g}^*}(\mu)). & \qquad (14.3.12)
\end{aligned}$$

▼

Lemma 14.3.3. *The two-form σ is left invariant, that is, $L_g^*\sigma = \sigma$ for all $g \in G$.*

Proof. Using the equivariance identity $\varphi_\nu \circ L_g = \mathrm{Ad}^*_{g^{-1}}\circ\varphi_\nu$, we compute

$$L_g^*\sigma = L_g^*\varphi_\nu^*\omega^- = (\varphi_\nu \circ L_g)^*\omega^- = (\mathrm{Ad}^*_{g^{-1}}\circ\varphi_\nu)^*\omega^-$$

$$= \varphi_\nu^*(\mathrm{Ad}^*_{g^{-1}})^*\omega^- = \varphi_\nu^*\omega^- = \sigma.$$ ▼

Lemma 14.3.4. *We have the identity $\sigma(\xi_L,\eta_L) = -\langle\nu_L,[\xi_L,\eta_L]\rangle$.*

Proof. Both sides are left invariant and are equal at the identity by (14.3.5).   ▼

The exterior derivative $d\alpha$ of a one-form α is given in terms of the Jacobi–Lie bracket by

$$(d\alpha)(X,Y) = X[\alpha(Y)] - Y[\alpha(X)] - \alpha([X,Y]). \qquad (14.3.13)$$

Since $\nu_L(\xi_L)$ is constant, $\eta_L[\nu_L(\xi_L)] = 0$ and $\xi_L[\nu_L(\eta_L)] = 0$, so Lemma 14.3.4 implies[2]

$$\sigma(\xi_L, \eta_L) = (d\nu_L)(\xi_L, \eta_L). \qquad (14.3.14)$$

Lemma 14.3.5. *We have the equality*

$$\sigma = d\nu_L. \qquad (14.3.15)$$

Proof. We shall prove that for any vector fields X and Y, $\sigma(X,Y) = (d\nu_L)(X,Y)$. Indeed, since σ is left invariant,

$$
\begin{aligned}
\sigma(X,Y)(g) &= (L_{g^{-1}}^* \sigma)(g)(X(g), Y(g)) \\
&= \sigma(e)(TL_{g^{-1}} \cdot X(g), TL_{g^{-1}} \cdot Y(g)) \\
&= \sigma(e)(\xi, \eta) \quad \text{(where } \xi = TL_{g^{-1}} \cdot X(g) \text{ and } \eta = TL_{g^{-1}} \cdot Y(g)) \\
&= \sigma(\xi_L, \eta_L)(e) = (d\nu_L)(\xi_L, \eta_L)(e) \quad\quad \text{(by (14.3.14))} \\
&= (L_g^* d\nu_L)(\xi_L, \eta_L)(e) \quad\quad\quad\quad \text{(since } \nu_L \text{ is } left \text{ invariant)} \\
&= (d\nu_L)(g)(TL_g \cdot \xi_L(e), TL_g \cdot \eta_L(e)) \\
&= (d\nu_L)(g)(TL_g \cdot \xi, TL_g \cdot \eta) = (d\nu_L)(g)(X(g), Y(g)) \\
&= (d\nu_L)(X,Y)(g). \qquad\qquad\qquad\qquad\qquad\qquad \blacktriangledown
\end{aligned}
$$

Since $\sigma = d\nu_L$ by Lemma 14.3.5, $d\sigma = dd\nu_L = 0$, and so $0 = d\varphi_\nu^* \omega^- = \varphi_\nu^* d\omega^-$. From $\varphi_\nu \circ L_g = \mathrm{Ad}_{g^{-1}}^* \circ \varphi_\nu$, it follows that submersivity of φ_ν at e is equivalent to submersivity of φ_ν at any $g \in G$, that is, φ_ν is a surjective submersion. Thus, φ_ν^* is injective, and hence $d\omega^- = 0$. $\blacksquare$

Since coadjoint orbits are symplectic, we get the following:

Corollary 14.3.6. *Coadjoint orbits of finite-dimensional Lie groups are even-dimensional.*

Corollary 14.3.7. *Let $G_\nu = \{ g \in G \mid \mathrm{Ad}_{g^{-1}}^* \nu = \nu \}$ be the isotropy subgroup of the coadjoint action of $\nu \in \mathfrak{g}^*$. Then G_ν is a closed subgroup of G, and so the quotient G/G_ν is a smooth manifold with smooth projection*

---

[2] Any Lie group carries a natural connection associated to the left (or right) action. The calculation (14.3.13) is essentially the calculation of the curvature of this connection and is closely related to the *Maurer–Cartan equations* (see §9.1).

$\pi : G \to G/G_\nu$, $g \mapsto g \cdot G_\nu$. We identify $G/G_\nu \cong \text{Orb}(\nu)$ via the diffeomorphism $\rho : g \cdot G_\nu \in G/G_\nu \mapsto \text{Ad}^*_{g^{-1}}(\nu) \in \text{Orb}(\nu)$. Thus, G/G_ν is symplectic, with symplectic form ω^- induced from $d\nu_L$, that is,

$$d\nu_L = \pi^* \rho^* \omega^-$$

(respectively, $d\nu_R = \pi^* \rho^* \omega^+$).

As we shall see in Example (a) of §14.5, ω^- is not exact in general, even though $\pi^* \rho^* \omega^-$ is.

Examples

(a) Rotation Group. Consider $\text{Orb}(\Pi)$, the coadjoint orbit through $\Pi \in \mathbb{R}^3$; then

$$\xi_{\mathbb{R}^3}(\Pi) = \xi \times \Pi \in T_\Pi(\text{Orb}(\Pi))$$

and

$$\eta_{\mathbb{R}^3}(\Pi) = \eta \times \Pi \in T_\Pi(\text{Orb}(\Pi)),$$

and so with the usual identification of $\mathfrak{so}(3)$ with $\mathbb{R}^3$, the $(-)$ coadjoint orbit symplectic structure becomes

$$\omega^-(\xi_{\mathbb{R}^3}(\Pi), \eta_{\mathbb{R}^3}(\Pi)) = -\Pi \cdot (\xi \times \eta). \qquad (14.3.16)$$

Recall that the oriented area of the (planar) parallelogram spanned by two vectors $\mathbf{v}, \mathbf{w} \in \mathbb{R}^3$ is given by $\mathbf{v} \times \mathbf{w}$ (the numerical area is $\|\mathbf{v} \times \mathbf{w}\|$). Thus, the oriented area spanned by $\xi_{\mathbb{R}^3}(\Pi)$ and $\eta_{\mathbb{R}^3}(\Pi)$ is

$$(\xi \times \Pi) \times (\eta \times \Pi) = [(\xi \times \Pi) \cdot \Pi] \eta - [(\xi \times \Pi) \cdot \eta] \Pi$$
$$= \Pi(\Pi \cdot (\xi \times \eta)).$$

The area element dA on a sphere in $\mathbb{R}^3$ assigns to each pair $(\mathbf{v}, \mathbf{w})$ of tangent vectors the number $dA(\mathbf{v}, \mathbf{w}) = \mathbf{n} \cdot (\mathbf{v} \times \mathbf{w})$, where $\mathbf{n}$ is the unit outward normal (this is the area of the parallelogram spanned by $\mathbf{v}$ and $\mathbf{w}$, taken "+" if $\mathbf{v}, \mathbf{w}, \mathbf{n}$ form a positively oriented basis and "−" otherwise). For a sphere of radius $\|\Pi\|$ and tangent vectors $\mathbf{v} = \xi \times \Pi$ and $\mathbf{w} = \eta \times \Pi$, we have

$$dA(\xi \times \Pi, \eta \times \Pi) = \frac{\Pi}{\|\Pi\|} \cdot ((\xi \times \Pi) \times (\eta \times \Pi))$$
$$= \frac{\Pi}{\|\Pi\|} \cdot ((\xi \times \Pi) \cdot \Pi)\eta - ((\xi \times \Pi) \cdot \eta)\Pi)$$
$$= \|\Pi\|\Pi \cdot (\xi \times \eta). \qquad (14.3.17)$$

Thus,

$$\omega^-(\Pi) = -\frac{1}{\|\Pi\|}dA. \tag{14.3.18}$$

The use of "dA" for the area element is, of course, a notational abuse, since this two-form cannot be exact. Likewise,

$$\omega^+(\Pi) = \frac{1}{\|\Pi\|}dA. \tag{14.3.19}$$

Notice that $\omega^+/\|\Pi\| = (dA)/\|\Pi\|^2$ is the solid angle subtended by the area element dA.   ◆

(b) Affine Group on $\mathbb{R}$**.** For

$$\beta \neq 0 \quad \text{and} \quad \mu = \begin{pmatrix} \alpha \\ \beta \end{pmatrix}$$

on the open orbit $\mathcal{O}$, formula (14.3.1) gives

$$\omega^-(\mu)\left((r,s)_{\mathfrak{g}^*}(\mu), (u,v)_{\mathfrak{g}^*}(\mu)\right) = -\left\langle \begin{pmatrix} \alpha \\ \beta \end{pmatrix}, [(r,s),(u,v)] \right\rangle$$
$$= \beta(rv - su). \tag{14.3.20}$$

Using the coordinates $(\alpha, \beta) \in \mathbb{R}^2$, this reads

$$\omega^-(\mu) = \frac{1}{\beta}d\alpha \wedge d\beta. \tag{14.3.21}$$

◆

(c) The Group Diff_{vol}**.** For a coadjoint orbit of $G = \text{Diff}_{\text{vol}}(\Omega)$ the $(+)$ coadjoint orbit symplectic structure at a point M becomes

$$\omega^+(M)(-\pounds_v M, -\pounds_w M) = -\int_\Omega M \cdot [v,w]\, d^n x, \tag{14.3.22}$$

where $[v,w]$ is the Jacobi–Lie bracket. Note that we have indeed a minus sign on the right-hand side of (14.3.22), since $[v,w]$ is *minus* the left Lie algebra bracket.   ◆

Exercises

◇ **14.3-1.** Let G be a Lie group. Find an action of G on T^*G for which the map

$$J(\xi)(\nu_L(g)) = -\langle \nu_L(g), \xi_L(g)\rangle = -\langle \nu, \xi\rangle$$

is an equivariant momentum map.

⋄ **14.3-2.** Relate the calculations of this section to the Maurer–Cartan equations.

⋄ **14.3-3.** Give another proof that $d\omega^\pm = 0$ by showing that X_H for $\omega^\pm$ coincides with that for the Lie–Poisson bracket and hence that Jacobi's identity holds.

⋄ **14.3-4 (The Group Diff$_{can}$).** For a coadjoint orbit for $G = \text{Diff}_{can}(P)$, show that the (+) coadjoint orbit symplectic structure is

$$\omega^+(L)(\{k, f\}, \{h, f\}) = \int_P f\{k, h\}\, dq\, dp.$$

⋄ **14.3-5 (The Toda Lattice).** For the Toda orbit, check that the orbit symplectic structure is

$$\omega^+(f) = \sum_{i=1}^{n-1} \frac{1}{a_i}\, db_i \wedge da_i.$$

⋄ **14.3-6.** Verify formula (14.3.21); that is,

$$\omega^-(\mu) = \frac{1}{\beta} d\alpha \wedge d\beta.$$

14.4   The Orbit Bracket via Restriction of the Lie–Poisson Bracket

Theorem 14.4.1 (Lie–Poisson and Coadjoint Orbit Compatibility).
The Lie–Poisson bracket and the coadjoint orbit symplectic structure are consistent in the following sense: For $F, H : \mathfrak{g}^* \to \mathbb{R}$ and $\mathcal{O}$ a coadjoint orbit in $\mathfrak{g}^*$,

$$\{F, H\}_+|\mathcal{O} = \{F|\mathcal{O}, H|\mathcal{O}\}^+. \qquad (14.4.1)$$

Here, the bracket $\{F, G\}_+$ is the (+) Lie–Poisson bracket, while the bracket on the right-hand side of (14.4.1) is the Poisson bracket defined by the (+) coadjoint orbit symplectic structure on $\mathcal{O}$. Similarly,

$$\{F, H\}_-|\mathcal{O} = \{F|\mathcal{O}, H|\mathcal{O}\}^-. \qquad (14.4.2)$$

The following paragraph summarizes the basic content of the theorem.

Two Approaches to the Lie–Poisson Bracket

There are two different ways to produce the same Lie–Poisson bracket $\{F, H\}_-$ (respectively, $\{F, H\}_+$) on $\mathfrak{g}^*$:

Extension Method:

1. Take $F, H : \mathfrak{g}^* \to \mathbb{R}$;

2. extend F, H to $F_L, H_L : T^*G \to \mathbb{R}$ by left (respectively, right) translation;

3. take the bracket $\{F_L, H_L\}$ with respect to the canonical symplectic structure on T^*G; and

4. restrict

$$\{F_L, H_L\}|\mathfrak{g}^* = \{F, H\}_-$$

(respectively, $\{F_R, H_R\}|\mathfrak{g}^* = \{F, H\}_+$).

Restriction Method:

1. Take $F, H : \mathfrak{g}^* \to \mathbb{R}$;

2. form the restrictions $F|\mathcal{O}, H|\mathcal{O}$ to a coadjoint orbit; and

3. take the Poisson bracket $\{F|\mathcal{O}, H|\mathcal{O}\}^-$ with respect to the $-$ (respectively, $+$) orbit symplectic structure ω^- (respectively, ω^+) on the orbit $\mathcal{O}$: For $\mu \in \mathcal{O}$ we have

$$\{F|\mathcal{O}, H|\mathcal{O}\}^-(\mu) = \{F, H\}_-(\mu)$$

(respectively, $\{F|\mathcal{O}, H|\mathcal{O}\}^+(\mu) = \{F, H\}_+(\mu)$).

Proof of Theorem 14.4.1. Let $\mu \in \mathcal{O}$. By definition,

$$\{F, H\}_-(\mu) = -\left\langle \mu, \left[\frac{\delta F}{\delta \mu}, \frac{\delta H}{\delta \mu} \right] \right\rangle. \qquad (14.4.3)$$

On the other hand,

$$\{F|\mathcal{O}, H|\mathcal{O}\}^-(\mu) = \omega^-(X_F, X_H)(\mu), \qquad (14.4.4)$$

where X_F and X_H are the Hamiltonian vector fields on $\mathcal{O}$ generated by $F|\mathcal{O}$ and $H|\mathcal{O}$, and ω^- is the minus orbit symplectic form. Recall that the Hamiltonian vector field X_F on $\mathfrak{g}^*_-$ is given by

$$X_F(\mu) = \mathrm{ad}^*_\xi(\mu), \qquad (14.4.5)$$

where $\xi = \delta F/\delta \mu \in \mathfrak{g}$.

Motivated by this, we prove the following:

Lemma 14.4.2. *Using the orbit symplectic form ω^-, for $\mu \in \mathcal{O}$ we have*

$$X_{F|\mathcal{O}}(\mu) = \mathrm{ad}^*_{\delta F/\delta \mu}(\mu). \qquad (14.4.6)$$

Proof.   Let $\xi, \eta \in \mathfrak{g}$, so (14.3.1) gives

$$\omega^-(\mu)(\mathrm{ad}^*_\xi\,\mu, \mathrm{ad}^*_\eta\,\mu) = -\langle\mu, [\xi, \eta]\rangle = \langle\mu, \mathrm{ad}_\eta(\xi)\rangle = \langle\mathrm{ad}^*_\eta(\mu), \xi\rangle. \quad (14.4.7)$$

Letting $\xi = \delta F/\delta\mu$ and η be arbitrary, we get

$$\omega^-(\mu)(\mathrm{ad}^*_{\delta F/\delta\mu}\,\mu, \mathrm{ad}^*_\eta\,\mu) = \left\langle \mathrm{ad}^*_\eta\,\mu, \frac{\delta F}{\delta\mu} \right\rangle = \mathbf{d}F(\mu) \cdot \mathrm{ad}^*_\eta\,\mu. \quad (14.4.8)$$

Thus, $X_{F|\mathcal{O}}(\mu) = \mathrm{ad}^*_{\delta F/\delta\mu}\,\mu$, as required.   ▼

To complete the proof of Theorem 14.4.1, note that

$$\begin{aligned}
\{F|\mathcal{O}, H|\mathcal{O}\}^-(\mu) &= \omega^-(\mu)(X_{F|\mathcal{O}}(\mu), X_{H|\mathcal{O}}(\mu)) \\
&= \omega^-(\mu)(\mathrm{ad}^*_{\delta F/\delta\mu}\,\mu, \mathrm{ad}^*_{\delta H/\delta\mu}\,\mu) \\
&= -\left\langle \mu, \left[\frac{\delta F}{\delta\mu}, \frac{\delta H}{\delta\mu}\right]\right\rangle = \{F, H\}_-(\mu). \quad (14.4.9)
\end{aligned}$$

■

Corollary 14.4.3.

(i) *For $H \in \mathcal{F}(\mathfrak{g}^*)$, the trajectory of X_H starting at μ stays in* $\mathrm{Orb}(\mu)$.

(ii) *A function $C \in \mathcal{F}(\mathfrak{g}^*)$ is a Casimir function iff $\delta C/\delta\mu \in \mathfrak{g}_\mu$ for all $\mu \in \mathfrak{g}^*$.*

(iii) *If $C \in \mathcal{F}(\mathfrak{g}^*)$ is* Ad^**-invariant (constant on orbits), then C is a Casimir function. The converse is also true if all coadjoint orbits are connected.*

Proof.   Part (i) follows from the fact that $X_H(\nu)$ is tangent to the coadjoint orbit $\mathcal{O}$ for $\nu \in \mathcal{O}$, since $X_H(\nu) = \mathrm{ad}^*_{\delta H/\delta\mu}(\nu)$. Part (ii) follows from the definitions and formula (14.4.5), and (iii) follows from (ii) by writing out the condition of Ad^*-invariance as $C(\mathrm{Ad}^*_{g^{-1}}\,\mu) = C(\mu)$ and differentiating in g at $g = e$.

For the converse, recall Proposition 10.4.7, which states that any Casimir function is constant on the symplectic leaves. Thus, since the connected components of the coadjoint orbits are the symplectic leaves of $\mathfrak{g}^*$, the Casimir functions are constant on them. In particular, if the coadjoint orbits are connected, the Casimir functions are constant on each coadjoint orbit, which then implies that they are all Ad^*-invariant.   ■

To illustrate part (iii), we note that for $G = SO(3)$, the function

$$C_\Phi(\Pi) = \Phi\left(\frac{1}{2}\|\Pi\|^2\right)$$

is invariant under the coadjoint action $(\mathbf{A}, \mathbf{\Pi}) \mapsto \mathbf{A}\mathbf{\Pi}$ and is therefore a Casimir function. Another example is given by $G = \mathrm{Diff}_{\mathrm{can}}(P)$ and the functional

$$C_\Phi(f) := \int_P \Phi(f)\, dq\, dp,$$

where $dq\, dp$ is the Liouville measure and Φ is any function of one variable. This is a Casimir function, since it is Ad^*-invariant by the change of variables formula.

In general, Ad^*-invariance of C is a stronger condition than C being a Casimir function. Indeed if C is Ad^*-invariant, differentiating the relation $C(\mathrm{Ad}^*_{g^{-1}}\mu) = C(\mu)$ relative to μ rather than g as we did in the proof of (iii), we get

$$\frac{\delta C}{\delta(\mathrm{Ad}^*_{g^{-1}}\mu)} = \mathrm{Ad}_g\, \frac{\delta C}{\delta \mu} \tag{14.4.10}$$

for all $g \in G$. Taking $g \in G_\mu$, this relation becomes $\delta C/\delta\mu = \mathrm{Ad}_g(\delta C/\delta\mu)$, that is, $\delta C/\delta\mu$ belongs to the centralizer of G_μ in $\mathfrak{g}$, that is, to the set

$$\mathrm{Cent}(G_\mu, \mathfrak{g}) := \{\xi \in \mathfrak{g} \mid \mathrm{Ad}_g\, \xi = \xi \text{ for all } g \in G_\mu\}.$$

Letting

$$\mathrm{Cent}(\mathfrak{g}_\mu, \mathfrak{g}) := \{\xi \in \mathfrak{g} \mid [\eta, \xi] = 0 \text{ for all } \eta \in \mathfrak{g}_\mu\}$$

denote the centralizer of $\mathfrak{g}_\mu$ in $\mathfrak{g}$, we see, by differentiating the relation defining $\mathrm{Cent}(G_\mu, \mathfrak{g})$ with respect to g at the identity, that $\mathrm{Cent}(G_\mu, \mathfrak{g}) \subset \mathrm{Cent}(\mathfrak{g}_\mu, \mathfrak{g})$. Thus, if C is Ad^*-invariant, then

$$\frac{\delta C}{\delta\mu} \in \mathfrak{g}_\mu \cap \mathrm{Cent}(\mathfrak{g}_\mu, \mathfrak{g}) = \mathrm{Cent}(\mathfrak{g}_\mu) = \text{ the center of } \mathfrak{g}_\mu.$$

Thus, we conclude the following:

Proposition 14.4.4 (Kostant [1979]). *If C is an Ad^*-invariant function on $\mathfrak{g}^*$, then $\delta C/\delta\mu$ lies in both $\mathrm{Cent}(G_\mu, \mathfrak{g})$ and $\mathrm{Cent}(\mathfrak{g}_\mu)$. If C is a Casimir function, then $\delta C/\delta\mu$ lies in the center of $\mathfrak{g}_\mu$.*

Proof. The first statement follows from the preceding considerations. The second statement is deduced in the following way. Let G_0 be the connected component of the identity in G. Since the Lie algebras of G and of G_0 coincide, a Casimir function C of $\mathfrak{g}^*$ is necessarily constant on the G_0-coadjoint orbits, since they are connected (see Corollary 14.4.3(iii)). Thus, by the first part, $\delta C/\delta\mu \in \mathrm{Cent}(\mathfrak{g}_\mu)$. ∎

By the theorem of Duflo and Vergne [1969] (see the Internet supplement for Chapter 9), for generic $\mu \in \mathfrak{g}^*$, the coadjoint isotropy $\mathfrak{g}_\mu$ is Abelian, and therefore $\text{Cent}(\mathfrak{g}_\mu) = \mathfrak{g}_\mu$ generically. The above corollary and proposition leave open, in principle, the possibility of non-Ad^*-invariant Casimir functions on $\mathfrak{g}^*$. This is not possible for Lie groups with connected coadjoint orbits, as we saw before. *If $C : \mathfrak{g}^* \to \mathbb{R}$ is a function such that $\delta C / \delta \mu \in \mathfrak{g}_\mu$ for all $\mu \in \mathfrak{g}^*$, but there is at least one $\nu \in \mathfrak{g}^*$ such that $\delta C / \delta \nu \notin \text{Cent}(\mathfrak{g}_\nu)$, then C is a Casimir function that is not Ad^*-invariant.* This element $\nu \in \mathfrak{g}^*$ must be such that its coadjoint orbit is disconnected, and it must be nongeneric. We know of no such example of a Casimir function.

On the other hand, the above statements provide easily verifiable criteria for the form of, or the nonexistence of, Casimir functions on duals of Lie algebras. For example, if $\mathfrak{g}^*$ has open orbits whose union is dense, it cannot have Casimir functions. Indeed, any such function would have to be constant on the connected components of each orbit, and thus by continuity, on $\mathfrak{g}^*$. An example of such a Lie algebra is that of the affine group on the line discussed in Example (b) of §14.1. The same argument shows that Lie algebras with at least one dense orbit have no Casimir functionals.

In the Internet supplement we find the Casimir functions for Example (f) of §14.1 and use it to show that *Casimir functions need not characterize generic coadjoint orbits.*

A mathematical reason that coadjoint orbits and the Lie–Poisson bracket are so important is that Hamiltonian systems with symmetry are sometimes a covering of a coadjoint orbit. This is proved below.

If X and Y are topological spaces, a continuous surjective map $p : X \to Y$ is called a **covering map** if every point in Y has an open neighborhood U such that $p^{-1}(U)$ is a disjoint union of open sets in X, called the **decks** over U. Note that each deck is homeomorphic to U by p. If $p : M \to N$ is a surjective proper map of smooth manifolds that is also a local diffeomorphism, then it is a covering map. For example, SU(2) (the spin group) forms a covering space of SO(3) with two decks over each point, and SU(2) is simply connected, while SO(3) is not. (See Chapter 9.)

Transitive Hamiltonian actions have been characterized by Lie, Kostant, Kirillov, and Souriau in the following manner (see Kostant [1966]):

Theorem 14.4.5 (Kostant's Coadjoint Orbit Covering Theorem). *Let P be a Poisson manifold and let $\Phi : G \times P \to P$ be a left, transitive, Hamiltonian action with equivariant momentum map $\mathbf{J} : P \to \mathfrak{g}^*$. Then:*

(i) $\mathbf{J} : P \to \mathfrak{g}_+^*$ *is a canonical submersion onto a coadjoint orbit of G in $\mathfrak{g}^*$.*

(ii) *If P is symplectic, then $\mathbf{J}$ is a symplectic local diffeomorphism onto a coadjoint orbit endowed with the "+" orbit symplectic structure. If $\mathbf{J}$ is also proper, then it is a covering map.*

Proof.   (i) That $\mathbf{J}$ is a canonical map was proved in §12.4. Since Φ is transitive, if we choose a $z_0 \in P$, then any $z \in P$ can be written as $z = \Phi_g(z_0)$ for some $g \in G$. Thus, by equivariance,

$$\mathbf{J}(P) = \{\,\mathbf{J}(z) \mid z \in P\,\} = \{\,\mathbf{J}(\Phi_g(z_0)) \mid g \in G\,\}$$
$$= \{\,\mathrm{Ad}^*_{g^{-1}} \mathbf{J}(z_0) \mid g \in G\,\} = \mathrm{Orb}(\mathbf{J}(z_0)).$$

Again by equivariance, for $z \in P$ we have $T_z \mathbf{J}(\xi_P(z)) = -\,\mathrm{ad}^*_\xi \mathbf{J}(z)$, which has the form of a general tangent vector at $\mathbf{J}(z)$ to the orbit $\mathrm{Orb}(\mathbf{J}(z_0))$; thus, $\mathbf{J}$ is a submersion.

(ii) If P is symplectic with symplectic form Ω, $\mathbf{J}$ is a symplectic map if the orbit has the "$+$" symplectic form: $\omega^+(\mu)(\mathrm{ad}^*_\xi \mu, \mathrm{ad}^*_\eta \mu) = \langle \mu, [\xi, \eta]\rangle$. This is seen in the following way. Since $T_z P = \{\,\xi_P(z) \mid \xi \in \mathfrak{g}\,\}$ by transitivity of the action,

$$(\mathbf{J}^* \omega^+)(z)(\xi_P(z), \eta_P(z)) = \omega^+(\mathbf{J}(z))(T_z\mathbf{J}(\xi_P(z)), T_z\mathbf{J}(\eta_P(z)))$$
$$= \omega^+(\mathbf{J}(z))(\mathrm{ad}^*_\xi \mathbf{J}(z), \mathrm{ad}^*_\eta \mathbf{J}(z))$$
$$= \langle \mathbf{J}(z), [\xi, \eta]\rangle = \mathbf{J}([\xi, \eta])(z)$$
$$= \{J(\xi), J(\eta)\}(z) \qquad \text{(by equivariance)}$$
$$= \Omega(z)(X_{J(\xi)}(z), X_{J(\eta)}(z))$$
$$= \Omega(z)(\xi_P(z), \eta_P(z)), \qquad\qquad (14.4.11)$$

which shows that $\mathbf{J}^* \omega^+ = \Omega$, that is, $\mathbf{J}$ is symplectic. Since any symplectic map is an immersion, $\mathbf{J}$ is a local diffeomorphism. If $\mathbf{J}$ is also proper, it is a symplectic covering map, as discussed above. ∎

If $\mathbf{J}$ is proper and the symplectic manifold P is simply connected, the covering map in (ii) is a diffeomorphism; this follows from classical theorems about covering spaces (Spanier [1966]). It is clear that if Φ is not transitive, then $\mathbf{J}(P)$ is a union of coadjoint orbits. See Guillemin and Sternberg [1984] and Grigore and Popp [1989] for more information.

Exercises

◇ **14.4-1.**   Show that if C is a Casimir function on a Poisson manifold, then $\{F, K\}_C = C\{F, K\}$ is also a Poisson structure. If X_H is a Hamiltonian vector field for $\{\,,\}$, show that it is also Hamiltonian for $\{\,,\}_C$ with the Hamiltonian function CH.

◇ **14.4-2.**   Does Kostant's coadjoint orbit covering theorem *ever* apply to group actions on cotangent bundles by cotangent lift?

14.5 The Special Linear Group of the Plane

In the Lie algebra $\mathfrak{sl}(2, \mathbb{R})$ of traceless real 2×2 matrices, introduce the basis

$$\mathbf{e} = \begin{bmatrix} 0 & 1 \\ 0 & 0 \end{bmatrix}, \quad \mathbf{f} = \begin{bmatrix} 0 & 0 \\ 1 & 0 \end{bmatrix}, \quad \mathbf{h} = \begin{bmatrix} 1 & 0 \\ 0 & -1 \end{bmatrix}.$$

Note that $[\mathbf{h}, \mathbf{e}] = 2\mathbf{e}$, $[\mathbf{h}, \mathbf{f}] = -2\mathbf{f}$, and $[\mathbf{e}, \mathbf{f}] = \mathbf{h}$. Identify $\mathfrak{sl}(2, \mathbb{R})$ with $\mathbb{R}^3$ via

$$\xi := x\mathbf{e} + y\mathbf{f} + z\mathbf{h} \in \mathfrak{sl}(2, \mathbb{R}) \mapsto (x, y, z) \in \mathbb{R}^3. \tag{14.5.1}$$

The nonzero structure constants are $c_{12}^3 = 1$, $c_{13}^1 = -2$, and $c_{23}^2 = 2$. We identify the dual space $\mathfrak{sl}(2, \mathbb{R})^*$ with $\mathfrak{sl}(2, \mathbb{R})$ via the nondegenerate pairing

$$\langle \alpha, \xi \rangle = \operatorname{trace}(\alpha\xi). \tag{14.5.2}$$

In particular, the dual basis of $\{\mathbf{e}, \mathbf{f}, \mathbf{h}\}$ is $\{\mathbf{f}, \mathbf{e}, \frac{1}{2}\mathbf{h}\}$, and we identify $\mathfrak{sl}(2, \mathbb{R})^*$ with $\mathbb{R}^3$ using this basis, that is,

$$\alpha = a\mathbf{f} + b\mathbf{e} + c\frac{1}{2}\mathbf{h} \mapsto (a, b, c) \in \mathbb{R}^3. \tag{14.5.3}$$

The $(\pm)$ Lie–Poisson bracket on $\mathfrak{sl}(2, \mathbb{R})^*$ is thus given by

$$\{F, H\}_\pm (\alpha) = \pm \operatorname{trace} \left(\alpha \left[\frac{\delta F}{\delta \alpha}, \frac{\delta H}{\delta \alpha} \right] \right),$$

where

$$\left\langle \delta\alpha, \frac{\delta F}{\delta \alpha} \right\rangle = \operatorname{trace} \left(\delta\alpha \frac{\delta F}{\delta \alpha} \right) = \mathbf{D}F(\alpha) \cdot \delta\alpha$$

$$= \frac{d}{dt}\bigg|_{t=0} F(\alpha + t\delta\alpha)$$

$$= \frac{\partial F}{\partial a} \delta a + \frac{\partial F}{\partial b} \delta b + \frac{\partial F}{\partial c} \delta c,$$

and where

$$\delta\alpha = \begin{bmatrix} \frac{1}{2}\delta c & \delta b \\ \delta a & -\frac{1}{2}\delta c \end{bmatrix} \quad \text{and} \quad \frac{\delta F}{\delta \alpha} = \begin{bmatrix} \frac{\partial F}{\partial c} & \frac{\partial F}{\partial a} \\ \frac{\partial F}{\partial b} & -\frac{\partial F}{\partial c} \end{bmatrix}.$$

The expression of the Lie–Poisson bracket in coordinates is therefore

$$\{F, G\}_\pm (a, b, c) = \mp 2a \left(\frac{\partial F}{\partial a} \frac{\partial G}{\partial c} - \frac{\partial F}{\partial c} \frac{\partial G}{\partial a} \right) \pm 2b \left(\frac{\partial F}{\partial b} \frac{\partial G}{\partial c} - \frac{\partial F}{\partial c} \frac{\partial G}{\partial b} \right)$$

$$\pm c \left(\frac{\partial F}{\partial u} \frac{\partial G}{\partial b} - \frac{\partial F}{\partial b} \frac{\partial G}{\partial a} \right). \tag{14.5.4}$$

Since SL$(2,\mathbb{R})$ is connected, the Casimir functions are the Ad*-invariant functions on $\mathfrak{sl}(2,\mathbb{R})^*$. Since Ad$_g\,\xi = g\xi g^{-1}$, if $g \in$ SL$(2,\mathbb{R})$ and $\xi \in \mathfrak{sl}(2,\mathbb{R})$, it follows that
$$\mathrm{Ad}^*_{g^{-1}}\,\alpha = g\alpha g^{-1},$$
for $\alpha \in \mathfrak{sl}(2,\mathbb{R})^*$. The determinant of
$$\begin{bmatrix} \tfrac{1}{2}c & b \\ a & -\tfrac{1}{2}c \end{bmatrix}$$
is obviously invariant under conjugation. Therefore, for $\mathbb{R}^3$ endowed with the $(\pm)$ Lie–Poisson bracket of $\mathfrak{sl}(2,\mathbb{R})^*$, any function of the form
$$C(a,b,c) = \Phi\left(ab + \frac{1}{4}c^2\right), \tag{14.5.5}$$
for a C^1 function $\Phi:\mathbb{R}\to\mathbb{R}$, is a Casimir function. The symplectic leaves are sheets of the hyperboloids
$$C_0(a,b,c) := \frac{1}{2}\left(ab + \frac{1}{4}c^2\right) = \text{constant} \neq 0, \tag{14.5.6}$$
the two nappes (without vertex) of the cone
$$ab + \frac{1}{4}c^2 = 0,$$
and the origin. One can verify this directly by using Ad$^*_{g^{-1}}\,\alpha = g\alpha g^{-1}$. The orbit symplectic structure on these hyperboloids is given by
$$\omega^-(a,b,c)(\mathrm{ad}^*_{(x,y,z)}(a,b,c), \mathrm{ad}^*_{(x',y',z')}(a,b,c))$$
$$= -a(2zx' - 2xz') - b(2yz' - 2zy') - c(xy' - yx')$$
$$= -\frac{1}{\|\nabla C_0(a,b,c)\|}\ (\text{area element of the hyperboloid}). \tag{14.5.7}$$
To prove the last equality in (14.5.7), use the formulas
$$\mathrm{ad}^*_{(x,y,z)}(a,b,c) = (2az - cy, cx - 2bz, 2by - 2zx),$$
$$\mathrm{ad}^*_{(x,y,z)}(a,b,c) \times \mathrm{ad}^*_{(x',y',z')}(a,b,c)$$
$$= \big(2bc(xy' - yx') + 4b^2(yz' - zy') + 4ab(zx' - xz'),$$
$$2ac(xy' - yx') + 4ab(yz' - zy') + 4a^2(zx' - xz'),$$
$$c^2(xy' - yx') + 2bc(yz' - zy') + 2ac(zx' - xz')\big),$$
and the fact that $\nabla(ab + \frac{1}{4}c^2) = (b,a,\frac{1}{2}c)$ is normal to the hyperboloid to get, as in (14.3.18),
$$dA(a,b,c)(\mathrm{ad}^*_{(x,y,z)}(a,b,c), \mathrm{ad}^*_{(x',y',z')}(a,b,c))$$
$$= \frac{(b,a,\frac{1}{2}c)}{\|(b,a,\frac{1}{2}c)\|} \cdot (\mathrm{ad}^*_{(x,y,z)}(a,b,c) \times \mathrm{ad}^*_{(x',y',z')}(a,b,c))$$
$$= -\|\nabla C_0(a,b,c)\| \cdot \omega^-(a,b,c)(\mathrm{ad}^*_{(x,y,z)}(a,b,c), \mathrm{ad}^*_{(x',y',z')}(a,b,c)).$$

Exercises

◇ **14.5-1.** Using traces, find a Casimir function for $\mathfrak{sl}(3,\mathbb{R})^*$.

14.6   The Euclidean Group of the Plane

We use the notation and terminology from Exercise 11.4-3. Recall that the group SE(2) consists of matrices of the form

$$(\mathbf{R}_\theta, \mathbf{a}) := \begin{bmatrix} \mathbf{R}_\theta & \mathbf{a} \\ \mathbf{0} & 1 \end{bmatrix}, \tag{14.6.1}$$

where $\mathbf{a} \in \mathbb{R}^2$ and $\mathbf{R}_\theta$ is the rotation matrix

$$\mathbf{R}_\theta = \begin{bmatrix} \cos\theta & -\sin\theta \\ \sin\theta & \cos\theta \end{bmatrix}. \tag{14.6.2}$$

The identity element is the 3×3 identity matrix, and the inverse is given by

$$\begin{bmatrix} \mathbf{R}_\theta & \mathbf{a} \\ \mathbf{0} & 1 \end{bmatrix}^{-1} = \begin{bmatrix} \mathbf{R}_{-\theta} & -\mathbf{R}_{-\theta}\mathbf{a} \\ \mathbf{0} & 1 \end{bmatrix}. \tag{14.6.3}$$

The Lie algebra $\mathfrak{se}(2)$ of SE(2) consists of 3×3 block matrices of the form

$$\begin{bmatrix} -\omega\mathbf{J} & \mathbf{v} \\ \mathbf{0} & 0 \end{bmatrix}, \tag{14.6.4}$$

where

$$\mathbf{J} = \begin{bmatrix} 0 & 1 \\ -1 & 0 \end{bmatrix} \tag{14.6.5}$$

(note, as usual, that $\mathbf{J}^T = \mathbf{J}^{-1} = -\mathbf{J}$) with the usual commutator bracket. If we identify $\mathfrak{se}(2)$ with $\mathbb{R}^3$ by the isomorphism

$$\begin{bmatrix} -\omega\mathbf{J} & \mathbf{v} \\ \mathbf{0} & 0 \end{bmatrix} \in \mathfrak{se}(2) \mapsto (\omega, \mathbf{v}) \in \mathbb{R}^3, \tag{14.6.6}$$

the expression for the Lie algebra bracket becomes

$$[(\omega, v_1, v_2), (\zeta, w_1, w_2)] = (0, \zeta v_2 - \omega w_2, \omega w_1 - \zeta v_1)$$
$$= (0, \omega\mathbf{J}^T\mathbf{w} - \zeta\mathbf{J}^T\mathbf{v}), \tag{14.6.7}$$

where $\mathbf{v} = (v_1, v_2)$ and $\mathbf{w} = (w_1, w_2)$.

The adjoint action of

$$(\mathbf{R}_\theta, \mathbf{a}) = \begin{bmatrix} \mathbf{R}_\theta & \mathbf{a} \\ \mathbf{0} & 1 \end{bmatrix} \quad \text{on} \quad (\omega, \mathbf{v}) = \begin{bmatrix} -\omega\mathbf{J} & \mathbf{v} \\ \mathbf{0} & 0 \end{bmatrix}$$

is given by conjugation,

$$\begin{bmatrix} \mathbf{R}_\theta & \mathbf{a} \\ \mathbf{0} & 1 \end{bmatrix} \begin{bmatrix} -\omega \mathbb{J} & \mathbf{v} \\ \mathbf{0} & 0 \end{bmatrix} \begin{bmatrix} \mathbf{R}_{-\theta} & -\mathbf{R}_{-\theta}\mathbf{a} \\ \mathbf{0} & 1 \end{bmatrix} = \begin{bmatrix} -\omega \mathbb{J} & \omega \mathbb{J}\mathbf{a} + \mathbf{R}_\theta \mathbf{v} \\ \mathbf{0} & 0 \end{bmatrix}, \quad (14.6.8)$$

or, in coordinates,

$$\mathrm{Ad}_{(\mathbf{R}_\theta, \mathbf{a})}(\omega, \mathbf{v}) = (\omega, \omega \mathbb{J}\mathbf{a} + \mathbf{R}_\theta \mathbf{v}). \quad (14.6.9)$$

In proving this, we used the identity $\mathbf{R}_\theta \mathbb{J} = \mathbb{J}\mathbf{R}_\theta$. Identify $\mathfrak{se}(2)^*$ with matrices of the form

$$\begin{bmatrix} \frac{\mu}{2}\mathbb{J} & \mathbf{0} \\ \alpha & 0 \end{bmatrix} \quad (14.6.10)$$

via the nondegenerate pairing given by the trace of the product. Thus, $\mathfrak{se}(2)^*$ is isomorphic to $\mathbb{R}^3$ via

$$\begin{bmatrix} \frac{\mu}{2}\mathbb{J} & \mathbf{0} \\ \alpha & 0 \end{bmatrix} \in \mathfrak{se}(2)^* \mapsto (\mu, \alpha) \in \mathbb{R}^3, \quad (14.6.11)$$

so that in these coordinates, the pairing between $\mathfrak{se}(2)^*$ and $\mathfrak{se}(2)$ becomes

$$\langle (\mu, \alpha), (\omega, \mathbf{v}) \rangle = \mu\omega + \alpha \cdot \mathbf{v}, \quad (14.6.12)$$

that is, the usual dot product in $\mathbb{R}^3$. The coadjoint action is thus given by

$$\mathrm{Ad}^*_{(\mathbf{R}_\theta, \mathbf{a})^{-1}}(\mu, \alpha) = (\mu - \mathbf{R}_\theta \alpha \cdot \mathbb{J}\mathbf{a}, \mathbf{R}_\theta \alpha). \quad (14.6.13)$$

Indeed, by (14.6.3), (14.6.5), (14.6.9), (14.6.12), and (14.6.13) we get

$$\begin{aligned}
\langle \mathrm{Ad}^*_{(\mathbf{R}_\theta, \mathbf{a})^{-1}}(\mu, \alpha), (\omega, \mathbf{v}) \rangle &= \langle (\mu, \alpha), \mathrm{Ad}_{(\mathbf{R}_{-\theta}, -\mathbf{R}_{-\theta}\mathbf{a})}(\omega, \mathbf{v}) \rangle \\
&= \langle (\mu, \alpha), (\omega, -\omega \mathbb{J}\mathbf{R}_{-\theta}\mathbf{a} + \mathbf{R}_{-\theta}\mathbf{v}) \rangle \\
&= \mu\omega - \omega\alpha \cdot \mathbb{J}\mathbf{R}_{-\theta}\mathbf{a} + \alpha \cdot \mathbf{R}_{-\theta}\mathbf{v} \\
&= (\mu - \alpha \cdot \mathbf{R}_{-\theta}\mathbb{J}\mathbf{a})\omega + \mathbf{R}_\theta \alpha \cdot \mathbf{v} \\
&= \langle (\mu - \mathbf{R}_\theta \alpha \cdot \mathbb{J}\mathbf{a}, \mathbf{R}_\theta \alpha), (\omega, \mathbf{v}) \rangle.
\end{aligned}$$

Coadjoint Orbits in $\mathfrak{se}(2)^*$. Formula (14.6.13) shows that the coadjoint orbits are the cylinders $T^*S^1_\alpha = \{ (\mu, \alpha) \mid \|\alpha\| = \text{constant} \}$ if $\alpha \neq 0$ together with the points are on the μ-axis. The canonical cotangent bundle projection, denoted by $\pi : T^*S^1_\alpha \to S^1_\alpha$, is defined by $\pi(\mu, \alpha) = \alpha$.

Connectedness of SE(2) implies by Corollary 14.4.3(iii) that the Casimir functions coincide with the functions invariant under the coadjoint action (14.6.13), that is, all Casimir functions have the form

$$C(\mu, \alpha) = \Phi\left(\tfrac{1}{2}\|\alpha\|^2\right) \quad (14.6.14)$$

for a smooth function $\Phi : [0, \infty) \to \mathbb{R}$.

The Lie–Poisson Bracket on $\mathfrak{se}(2)^*$. We next determine the ($\pm$) Lie–Poisson bracket on $\mathfrak{se}(2)^*$. If $F : \mathfrak{se}(2)^* \cong \mathbb{R} \times \mathbb{R}^2 \to \mathbb{R}$, its functional derivative is

$$\frac{\delta F}{\delta(\mu, \alpha)} = \left(\frac{\partial F}{\partial \mu}, \nabla_\alpha F \right), \tag{14.6.15}$$

where $(\mu, \alpha) \in \mathfrak{se}(2)^* \cong \mathbb{R} \times \mathbb{R}^2$ and $\nabla_\alpha F$ denotes the gradient of F with respect to α. The ($\pm$) Lie–Poisson structure on $\mathfrak{se}(2)^*$ is given by

$$\{F, G\}_\pm(\mu, \alpha) = \pm \left(\frac{\partial F}{\partial \mu} \mathbf{J}\alpha \cdot \nabla_\alpha G - \frac{\partial G}{\partial \mu} \mathbf{J}\alpha \cdot \nabla_\alpha F \right). \tag{14.6.16}$$

It can now be directly verified that the functions given by (14.6.14) are indeed Casimir functions for the bracket (14.6.16).

The Symplectic Form on Orbits. The coadjoint action of $\mathfrak{se}(2)$ on $\mathfrak{se}(2)^*$ is given by

$$\mathrm{ad}^*_{(\xi, \mathbf{u})}(\mu, \alpha) = (-\mathbf{J}\alpha \cdot \mathbf{u}, \xi \mathbf{J}\alpha). \tag{14.6.17}$$

On the coadjoint orbit representing a cylinder about the μ-axis, the orbit symplectic structure is

$$\begin{aligned}
\omega(\mu, \alpha) & \left(\mathrm{ad}^*_{(\xi, \mathbf{u})}(\mu, \alpha), \mathrm{ad}^*_{(\eta, \mathbf{v})}(\mu, \alpha) \right) \\
&= \pm(\xi \mathbf{J}\alpha \cdot \mathbf{v} - \eta \mathbf{J}\alpha \cdot \mathbf{u}) \\
&= \pm(\text{area element } dA \text{ on the cylinder})/\|\alpha\|.
\end{aligned} \tag{14.6.18}$$

The last equality is proved in the following way. Since the outward unit normal to the cylinder is $(0, \alpha)/\|\alpha\|$, by (14.6.17) the area element dA is given by

$$\begin{aligned}
dA(\mu, \alpha)&((-\mathbf{J}\alpha \cdot \mathbf{u}, \xi \mathbf{J}\alpha), (-\mathbf{J}\alpha \cdot \mathbf{v}, \eta \mathbf{J}\alpha)) \\
&= \frac{(0, \alpha)}{\|\alpha\|} \cdot [((-\mathbf{J}\alpha \cdot \mathbf{u}, \xi \mathbf{J}\alpha) \times (-\mathbf{J}\alpha \cdot \mathbf{u}, \xi \mathbf{J}\alpha)] \\
&= \|\alpha\|(\xi \mathbf{J}\alpha \cdot \mathbf{v} - \eta \mathbf{J}\alpha \cdot \mathbf{u}).
\end{aligned}$$

Let us show that on the orbit through (μ, α), the symplectic form $\|\alpha\|\omega^-$, equals the canonical symplectic form of the cotangent bundle $T^* S^1_\alpha$. Since $\pi(\mu, \alpha) = \alpha$, it follows by (14.6.17) that

$$T_{(\mu, \alpha)}\pi \left(\mathrm{ad}^*_{(\xi, \mathbf{u})}(\mu, \alpha) \right) = \xi \mathbf{J}\alpha,$$

thought of as a tangent vector to S^1 at α. The length of this vector is $|\xi| \|\alpha\|$, so we identify it with the pair $(\xi \|\alpha\|, \alpha) \in T_\alpha S^1_\alpha$. The canonical one-form is given by

$$\Theta(\mu, \alpha) \cdot \text{ad}^*_{(\xi, \mathbf{u})}(\mu, \alpha) = (\mu, \alpha) \cdot T_{(\mu, \alpha)} \pi \left(\text{ad}^*_{(\xi, \mathbf{u})}(\mu, \alpha) \right)$$

$$= (\mu, \alpha) \cdot (\xi \|\alpha\|, \alpha) = \mu \xi \|\alpha\|. \qquad (14.6.19)$$

To compute the canonical symplectic form Ω on $T^* S^1$ in this notation, we extend the tangent vectors

$$\text{ad}^*_{(\xi, \mathbf{u})}(\mu, \alpha) \quad \text{and} \quad \text{ad}^*_{(\eta, \mathbf{v})}(\mu, \alpha) \in T_{(\mu, \alpha)} \left(T^* S^1_\alpha \right)$$

to vector fields

$$X : (\mu, \alpha) \mapsto \text{ad}^*_{(\xi, \mathbf{u})}(\mu, \alpha) \quad \text{and} \quad Y : (\mu, \alpha) \mapsto \text{ad}^*_{(\eta, \mathbf{v})}(\mu, \alpha)$$

to get

$$\text{ad}^*_{(\xi, \mathbf{u})}(\mu, \alpha) \cdot [\Theta(Y)](\mu, \alpha) = \mathbf{d}\Theta(Y)(\mu, \alpha) \cdot \text{ad}^*_{(\xi, \mathbf{u})}(\mu, \alpha)$$

$$= \left. \frac{d}{dt} \right|_{t=0} \Theta(Y)(\mu(t), \alpha(t)),$$

where $(\mu(t), \alpha(t))$ is a curve in $T^* S^1_\alpha$ such that

$$(\mu(0), \alpha(0)) = (\mu, \alpha) \quad \text{and} \quad (\mu'(0), \alpha'(0)) = \text{ad}^*_{(\xi, \mathbf{u})}(\mu, \alpha).$$

Since $\|\alpha(t)\| = \|\alpha\|$, we conclude that this is equal to

$$\left. \frac{d}{dt} \right|_{t=0} \mu(t) \eta \|\alpha\| = \mu'(0) \eta \|\alpha\| = -\mathbf{J}\alpha \cdot \mathbf{u}\eta \|\alpha\|. \qquad (14.6.20)$$

Similarly,

$$\text{ad}^*_{(\eta, \mathbf{v})}(\mu, \alpha) \cdot (\Theta(X))(\mu, \alpha) = -\mathbf{J}\alpha \cdot \mathbf{v}\xi \|\alpha\|. \qquad (14.6.21)$$

Since $X = (\xi, \mathbf{u})_{\mathfrak{se}(2)^*}$ and $Y = (\eta, \mathbf{v})_{\mathfrak{se}(2)^*}$, we conclude that

$$[X, Y](\mu, \alpha) = -[(\xi, \mathbf{u}), (\eta, \mathbf{v})]_{\mathfrak{se}(2)^*}(\mu, \alpha) = -(0, \xi \mathbf{J}^T \mathbf{v} - \eta \mathbf{J}^T \mathbf{u})_{\mathfrak{se}(2)^*}(\mu, \alpha)$$

$$= -\text{ad}^*_{(0, \xi \mathbf{J}^T \mathbf{v} - \eta \mathbf{J}^T \mathbf{u})}(\mu, \alpha)$$

and by (14.6.19) that

$$\Theta([X, Y])(\mu, \alpha) = 0. \qquad (14.6.22)$$

We also recall a general formula from Chapter 4:

$$\mathbf{d}\Theta(X, Y) = X[\Theta(Y)] - Y[\Theta(X)] - \Theta([X, Y]). \qquad (14.6.23)$$

Using this, (14.6.21), and (14.6.22), we get

$$\Omega(\mu, \alpha)\left(\text{ad}^*_{(\xi,\mathbf{u})}(\mu, \alpha), \text{ad}^*_{(\eta,\mathbf{v})}(\mu, \alpha)\right)$$
$$= -\mathbf{d}\Theta(X, Y)(\mu, \alpha)$$
$$= -\text{ad}^*_{(\xi,\mathbf{u})}(\mu, \alpha) \cdot [\Theta(Y)](\mu, \alpha)$$
$$+ \text{ad}^*_{(\eta,\mathbf{v})}(\mu, \alpha) \cdot [\Theta(X)](\mu, \alpha) + \Theta([X, Y])(\mu, \alpha)$$
$$= -\|\alpha\| \left(\xi \mathbf{J}\alpha \cdot \mathbf{v} - \eta \mathbf{J}\alpha \cdot \mathbf{u}\right),$$

which shows that

$$\Omega = \|\alpha\|\omega^- = -(\text{area form on the cylinder of radius } \|\alpha\|).$$

Lie Algebra Deformations. The Poisson structures of $\mathfrak{so}(3)^*$, $\mathfrak{sl}(2, \mathbb{R})^*$, and $\mathfrak{se}(2)^*$ fit together in a larger Poisson manifold. Weinstein [1983b] considers for every $\varepsilon \in \mathbb{R}$ the Lie algebra $\mathfrak{g}_\varepsilon$ with abstract basis $\mathbf{X}_1, \mathbf{X}_2, \mathbf{X}_3$ and relations

$$[\mathbf{X}_3, \mathbf{X}_1] = \mathbf{X}_2, \quad [\mathbf{X}_2, \mathbf{X}_3] = \mathbf{X}_1, \quad [\mathbf{X}_1, \mathbf{X}_2] = \varepsilon \mathbf{X}_3. \tag{14.6.24}$$

If $\varepsilon > 0$, the map

$$\mathbf{X}_1 \mapsto \sqrt{\varepsilon}(1, 0, 0)\hat{}, \quad \mathbf{X}_2 \mapsto \sqrt{\varepsilon}(0, 1, 0)\hat{}, \quad \mathbf{X}_3 \mapsto (0, 0, 1)\hat{}, \tag{14.6.25}$$

defines an isomorphism of $\mathfrak{g}_\varepsilon$ with $\mathfrak{so}(3)$, while if $\varepsilon = 0$, the map

$$\mathbf{X}_1 \mapsto (0, 0, -1), \quad \mathbf{X}_2 \mapsto (0, -1, 0), \quad \mathbf{X}_3 \mapsto (-1, 0, 0), \tag{14.6.26}$$

defines an isomorphism of $\mathfrak{g}_0$ with $\mathfrak{se}(2)$, and if $\varepsilon < 0$, the map

$$\mathbf{X}_1 \mapsto \frac{\sqrt{-\varepsilon}}{2}\begin{bmatrix} 1 & 0 \\ 0 & -1 \end{bmatrix}, \quad \mathbf{X}_2 \mapsto \frac{\sqrt{-\varepsilon}}{2}\begin{bmatrix} 0 & 1 \\ 1 & 0 \end{bmatrix}, \quad \mathbf{X}_3 \mapsto \frac{1}{2}\begin{bmatrix} 0 & -1 \\ 1 & 0 \end{bmatrix},$$

defines an isomorphism of $\mathfrak{g}_\varepsilon$ with $\mathfrak{sl}(2, \mathbb{R})$.

The (+) Lie–Poisson structure of $\mathfrak{g}_\varepsilon^*$ is given by the bracket relations

$$\{x_3, x_1\} = x_2, \quad \{x_2, x_3\} = x_1, \quad \{x_1, x_2\} = \varepsilon x_3, \tag{14.6.27}$$

for the coordinate functions $x_i \in \mathfrak{g}_\varepsilon^* = \mathbb{R}^3$, $\langle x_i, x_j \rangle = \delta_{ij}$.

In $\mathbb{R}^4$ with coordinate functions $(x_1, x_2, x_3, \varepsilon)$ consider the above bracket relations plus $\{\varepsilon, x_1\} = \{\varepsilon, x_2\} = \{\varepsilon, x_3\} = 0$. This defines a Poisson structure on $\mathbb{R}^4$ that is not of Lie–Poisson type. The leaves of this Poisson structure are all two-dimensional in the space (x_1, x_2, x_3), and the Casimir functions are all functions of $x_1^2 + x_2^2 + \varepsilon x_3^2$ and ε. The inclusion of $\mathfrak{g}_\varepsilon^*$ in $\mathbb{R}^4$ with the above Poisson structure is a canonical map. The leaves of $\mathbb{R}^4$ with the above Poisson structure as ε passes through zero are given in Figure 14.6.1.

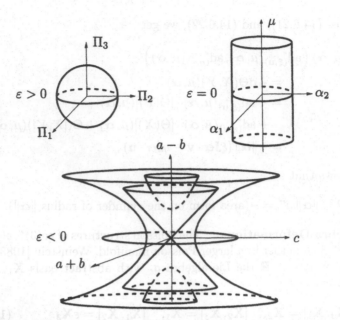

FIGURE 14.6.1. The orbit structure for $\mathfrak{so}(3)^*$, $\mathfrak{se}(2)^*$, and $\mathfrak{sl}(2,\mathbb{R})^*$.

14.7 The Euclidean Group of Three-Space

The Euclidean Group, Its Lie Algebra, and Its Dual. An element
of SE(3) is a pair $(\mathbf{A}, \mathbf{a})$ where $\mathbf{A} \in$ SO(3) and $\mathbf{a} \in \mathbb{R}^3$; the action of SE(3)
on $\mathbb{R}^3$ is the rotation $\mathbf{A}$ followed by translation by the vector $\mathbf{a}$ and has the
expression

$$(\mathbf{A}, \mathbf{a}) \cdot \mathbf{x} = \mathbf{A}\mathbf{x} + \mathbf{a}. \tag{14.7.1}$$

Using this formula, one sees that multiplication and inversion in SE(3) are
given by

$$(\mathbf{A}, \mathbf{a})(\mathbf{B}, \mathbf{b}) = (\mathbf{A}\mathbf{B}, \mathbf{A}\mathbf{b} + \mathbf{a}) \quad \text{and} \quad (\mathbf{A}, \mathbf{a})^{-1} = (\mathbf{A}^{-1}, -\mathbf{A}^{-1}\mathbf{a}),$$
$$\tag{14.7.2}$$

for $\mathbf{A}, \mathbf{B} \in$ SO(3) and $\mathbf{a}, \mathbf{b} \in \mathbb{R}^3$. The identity element is $(\mathbf{I}, \mathbf{0})$. Note that
SE(3) embeds into SL(4; $\mathbb{R}$) via the map

$$(\mathbf{A}, \mathbf{a}) \mapsto \begin{bmatrix} \mathbf{A} & \mathbf{a} \\ 0 & 1 \end{bmatrix}; \tag{14.7.3}$$

thus one can operate with SE(3) as one would with matrix Lie groups
by using this embedding. In particular, the Lie algebra $\mathfrak{se}(3)$ of SE(3) is
isomorphic to a Lie subalgebra of $\mathfrak{sl}(4; \mathbb{R})$ with elements of the form

$$\begin{bmatrix} \hat{\mathbf{x}} & \mathbf{y} \\ 0 & 0 \end{bmatrix}, \quad \text{where } \mathbf{x}, \mathbf{y} \in \mathbb{R}^3, \tag{14.7.4}$$

and whose Lie algebra bracket is equal to the commutator bracket of matrices. This shows that the Lie bracket operation on $\mathfrak{se}(3)$ is given by

$$[(\mathbf{x}, \mathbf{y}), (\mathbf{x}', \mathbf{y}')] = (\mathbf{x} \times \mathbf{x}', \mathbf{x} \times \mathbf{y}' - \mathbf{x}' \times \mathbf{y}). \tag{14.7.5}$$

Since

$$\begin{bmatrix} A & \mathbf{a} \\ 0 & 1 \end{bmatrix}^{-1} = \begin{bmatrix} A^{-1} & -A^{-1}\mathbf{a} \\ 0 & 1 \end{bmatrix}$$

and

$$\begin{bmatrix} A & \mathbf{a} \\ 0 & 1 \end{bmatrix} \begin{bmatrix} \hat{\mathbf{x}} & \mathbf{y} \\ 0 & 0 \end{bmatrix} \begin{bmatrix} A^{-1} & -A^{-1}\mathbf{a} \\ 0 & 1 \end{bmatrix} = \begin{bmatrix} A\hat{\mathbf{x}}A^{-1} & -A\hat{\mathbf{x}}A^{-1}\mathbf{a} + A\mathbf{y} \\ 0 & 0 \end{bmatrix},$$

we see that the adjoint action of SE(3) on $\mathfrak{se}(3)$ has the expression

$$\mathrm{Ad}_{(A,\mathbf{a})}(\mathbf{x}, \mathbf{y}) = (A\mathbf{x}, A\mathbf{y} - A\mathbf{x} \times \mathbf{a}). \tag{14.7.6}$$

The 6×6 matrix of $\mathrm{Ad}_{(A,\mathbf{a})}$ is given by

$$\begin{bmatrix} A & 0 \\ \hat{\mathbf{a}}A & A \end{bmatrix}. \tag{14.7.7}$$

Identifying the dual of $\mathfrak{se}(3)$ with $\mathbb{R}^3 \times \mathbb{R}^3$ by the dot product in every factor, it follows that the matrix of $\mathrm{Ad}^*_{(A,\mathbf{a})^{-1}}$ is given by the inverse transpose of the 6×6 matrix (14.7.7), that is, it equals

$$\begin{bmatrix} A & \hat{\mathbf{a}}A \\ 0 & A \end{bmatrix}. \tag{14.7.8}$$

Thus, the coadjoint action of SE(3) on $\mathfrak{se}(3)^* = \mathbb{R}^3 \times \mathbb{R}^3$ has the expression

$$\mathrm{Ad}^*_{(A,\mathbf{a})^{-1}}(\mathbf{u}, \mathbf{v}) = (A\mathbf{u} + \mathbf{a} \times A\mathbf{v}, A\mathbf{v}). \tag{14.7.9}$$

(This Lie algebra is a semidirect product, and all formulas derived here "by hand" are special cases of general ones that may be found in works on semidirect products; see, for example, Marsden, Ratiu, and Weinstein [1984a, 1984b].)

Coadjoint Orbits in $\mathfrak{se}(3)^*$. Let $\{\mathbf{e}_1, \mathbf{e}_2, \mathbf{e}_3, \mathbf{f}_1, \mathbf{f}_2, \mathbf{f}_3\}$ be an orthonormal basis of $\mathfrak{se}(3) = \mathbb{R}^3 \times \mathbb{R}^3$ such that $\mathbf{e}_i = \mathbf{f}_i$, $i = 1, 2, 3$. The dual basis of $\mathfrak{se}(3)^*$ via the dot product is again $\{\mathbf{e}_1, \mathbf{e}_2, \mathbf{e}_3, \mathbf{f}_1, \mathbf{f}_2, \mathbf{f}_3\}$. Let $\mathbf{e}$ and $\mathbf{f}$ denote arbitrary vectors satisfying $\mathbf{e} \in \mathrm{span}\{\mathbf{e}_1, \mathbf{e}_2, \mathbf{e}_3\}$ and $\mathbf{f} \in \mathrm{span}\{\mathbf{f}_1, \mathbf{f}_2, \mathbf{f}_3\}$. For the coadjoint action the only zero-dimensional orbit is the origin. Since $\mathfrak{se}(3)$ is six-dimensional, there can also be two- and four-dimensional coadjoint orbits. These in fact occur and fall into three types.

Type I: The orbit through $(\mathbf{e}, 0)$ equals

$$SE(3) \cdot (\mathbf{e}, 0) = \{ (A\mathbf{e}, 0) \mid A \in SO(3) \} = S_{\|\mathbf{e}\|}^2, \qquad (14.7.10)$$

the two-sphere of radius $\|\mathbf{e}\|$.

Type II: The orbit through $(0, \mathbf{f})$ is given by

$$
\begin{aligned}
SE(3) \cdot (0, \mathbf{f}) &= \{ (\mathbf{a} \times A\mathbf{f}, A\mathbf{f}) \mid A \in SO(3), \mathbf{a} \in \mathbb{R}^3 \} \\
&= \{ (\mathbf{u}, A\mathbf{f}) \mid A \in SO(3), \mathbf{u} \perp A\mathbf{f} \} = TS_{\|\mathbf{f}\|}^2, \qquad (14.7.11)
\end{aligned}
$$

the tangent bundle of the two-sphere of radius $\|\mathbf{f}\|$; note that the vector part is in the first slot.

Type III: The orbit through $(\mathbf{e}, \mathbf{f})$, where $\mathbf{e}, \mathbf{f} \neq 0$, equals

$$SE(3) \cdot (\mathbf{e}, \mathbf{f}) = \{ (A\mathbf{e} + \mathbf{a} \times A\mathbf{f}, A\mathbf{f}) \mid A \in SO(3), \mathbf{a} \in \mathbb{R}^3 \}. \qquad (14.7.12)$$

We will prove below that this orbit is diffeomorphic to $TS_{\|\mathbf{f}\|}^2$. Consider the smooth map

$$\varphi : (A, \mathbf{a}) \in SE(3) \mapsto \left(A\mathbf{e} + \mathbf{a} \times A\mathbf{f} - \frac{\mathbf{e} \cdot \mathbf{f}}{\|\mathbf{f}\|^2} A\mathbf{f}, A\mathbf{f} \right) \in TS_{\|\mathbf{f}\|}^2, \quad (14.7.13)$$

which is right invariant under the isotropy group

$$SE(3)_{(\mathbf{e}, \mathbf{f})} = \{ (B, \mathbf{b}) \mid B\mathbf{e} + \mathbf{b} \times \mathbf{f} = \mathbf{e}, \; B\mathbf{f} = \mathbf{f} \} \qquad (14.7.14)$$

(see (14.7.9)), that is, $\varphi((A, \mathbf{a})(B, \mathbf{b})) = \varphi(A, \mathbf{a})$ for all $(A, \mathbf{a}) \in SE(3)$ and $(B, \mathbf{b}) \in SE(3)_{(\mathbf{e}, \mathbf{f})}$. Thus, φ induces a smooth map $\bar{\varphi} : SE(3)/SE(3)_{(\mathbf{e}, \mathbf{f})} \to TS_{\|\mathbf{f}\|}^2$. The map $\bar{\varphi}$ is injective, for if $\varphi(A, \mathbf{a}) = \varphi(A', \mathbf{a}')$, then

$$(A, \mathbf{a})^{-1}(A', \mathbf{a}') = (A^{-1}A', A^{-1}(\mathbf{a}' - \mathbf{a})) \in SE(3)_{(\mathbf{e}, \mathbf{f})},$$

as is easily checked. To see that φ (and hence $\bar{\varphi}$) is surjective, let $(\mathbf{u}, \mathbf{v}) \in TS_{\|\mathbf{f}\|}^2$, that is, $\|\mathbf{v}\| = \|\mathbf{f}\|$ and $\mathbf{u} \cdot \mathbf{v} = 0$. Then choose an $A \in SO(3)$ such that $A\mathbf{f} = \mathbf{v}$ and let $\mathbf{a} = [\mathbf{v} \times (\mathbf{u} - A\mathbf{e})]/\|\mathbf{f}\|^2$. It is then straightforward to check that $\varphi(A, \mathbf{a}) = (\mathbf{u}, \mathbf{v})$ by (14.7.13). Thus, $\bar{\varphi}$ is a bijective map. Since the derivative of φ at $(A, \mathbf{a})$ in the direction $T_{(I,0)}L_{(A,\mathbf{a})}(\hat{\mathbf{x}}, \mathbf{y}) = (A\hat{\mathbf{x}}, A\mathbf{y})$ equals

$$
\begin{aligned}
T_{(A,\mathbf{a})}\varphi(A\hat{\mathbf{x}}, A\mathbf{y}) &= \left. \frac{d}{dt} \right|_{t=0} \varphi(Ae^{t\hat{\mathbf{x}}}, \mathbf{a} + tA\mathbf{y}) \\
&= (A(\mathbf{x} \times \mathbf{e} + \mathbf{y} \times \mathbf{f}) + \mathbf{a} \times A(\mathbf{x} \times \mathbf{f}) \\
&\quad - \frac{\mathbf{e} \cdot \mathbf{f}}{\|\mathbf{f}\|^2} A(\mathbf{x} \times \mathbf{f}), A(\mathbf{x} \times \mathbf{f})), \qquad (14.7.15)
\end{aligned}
$$

we see that its kernel consists of left translates by $(\mathbf{A}, \mathbf{a})$ of

$$\{(\mathbf{x}, \mathbf{y}) \in \mathfrak{se}(3) \mid \mathbf{x} \times \mathbf{e} + \mathbf{y} \times \mathbf{f} = \mathbf{0}, \; \mathbf{x} \times \mathbf{f} = \mathbf{0}\}. \tag{14.7.16}$$

However, taking the derivatives of the defining relations in (14.7.14) at $(\mathbf{B}, \mathbf{b}) = (1, \mathbf{0})$, we see that (14.7.16) coincides with $\mathfrak{se}(3)_{(\mathbf{e}, \mathbf{f})}$. This shows that $\overline{\varphi}$ is an immersion, and hence, since

$$\dim(\mathrm{SE}(3)/\mathrm{SE}(3)_{(\mathbf{e}, \mathbf{f})}) = \dim T S_{\|\mathbf{f}\|}^2 = 4,$$

it follows that $\overline{\varphi}$ is a local diffeomorphism. Therefore, $\overline{\varphi}$ is a diffeomorphism.

To compute the tangent spaces to these orbits, we use Proposition 14.2.1, which states that the annihilator of the coadjoint isotropy subalgebra at μ equals $T_\mu \mathcal{O}$. The coadjoint action of the Lie algebra $\mathfrak{se}(3)$ on its dual $\mathfrak{se}(3)^*$ is computed to be

$$\mathrm{ad}^*_{(\mathbf{x}, \mathbf{y})}(\mathbf{u}, \mathbf{v}) = (\mathbf{u} \times \mathbf{x} + \mathbf{v} \times \mathbf{y}, \mathbf{v} \times \mathbf{x}). \tag{14.7.17}$$

Thus, the isotropy subalgebra $\mathfrak{se}(3)_{(\mathbf{u}, \mathbf{v})}$ is given again by (14.7.16), that is, it equals $\{(\mathbf{x}, \mathbf{y}) \in \mathfrak{se}(3) \mid \mathbf{u} \times \mathbf{x} + \mathbf{v} \times \mathbf{y} = \mathbf{0}, \; \mathbf{v} \times \mathbf{x} = \mathbf{0}\}$. Let $\mathcal{O}$ denote a nonzero coadjoint orbit in $\mathfrak{se}(3)^*$. Then the tangent space at a point in $\mathcal{O}$ is given as follows for each of the three types of orbits:

Type I: Since

$$\mathfrak{se}(3)_{(\mathbf{e}, \mathbf{0})} = \{(\mathbf{x}, \mathbf{y}) \in \mathfrak{se}(3) \mid \mathbf{e} \times \mathbf{x} = \mathbf{0}\} = \mathrm{span}(\mathbf{e}) \times \mathbb{R}^3, \tag{14.7.18}$$

it follows that the tangent space to $\mathcal{O}$ at $(\mathbf{e}, \mathbf{0})$ is the tangent space to the sphere of radius $\|\mathbf{e}\|$ at the point $\mathbf{e}$ in the first factor.

Type II: Since

$$\mathfrak{se}(3)_{(\mathbf{0}, \mathbf{f})} = \{(\mathbf{x}, \mathbf{y}) \in \mathfrak{se}(3) \mid \mathbf{f} \times \mathbf{y} = \mathbf{0}, \; \mathbf{f} \times \mathbf{x} = \mathbf{0}\} = \mathrm{span}(\mathbf{f}) \times \mathrm{span}(\mathbf{f}), \tag{14.7.19}$$

it follows that the tangent space to $\mathcal{O}$ at $(\mathbf{0}, \mathbf{f})$ equals $\mathbf{f}^\perp \times \mathbf{f}^\perp$, where $\mathbf{f}^\perp$ denotes the plane perpendicular to $\mathbf{f}$.

Type III: Since

$$\begin{aligned} \mathfrak{se}(3)_{(\mathbf{e}, \mathbf{f})} &= \{(\mathbf{x}, \mathbf{y}) \in \mathfrak{se}(3) \mid \mathbf{e} \times \mathbf{x} + \mathbf{f} \times \mathbf{y} = \mathbf{0} \text{ and } \mathbf{f} \times \mathbf{x} = \mathbf{0}\} \\ &= \{(c_1 \mathbf{f}, c_1 \mathbf{e} + c_2 \mathbf{f}) \mid c_1, c_2 \in \mathbb{R}\}, \end{aligned} \tag{14.7.20}$$

the tangent space at $(\mathbf{e}, \mathbf{f})$ to $\mathcal{O}$ is the orthogonal complement of the space spanned by $(\mathbf{f}, \mathbf{e})$ and $(\mathbf{0}, \mathbf{f})$, that is, it equals

$$\{(\mathbf{u}, \mathbf{v}) \mid \mathbf{u} \cdot \mathbf{f} + \mathbf{v} \cdot \mathbf{e} = 0 \text{ and } \mathbf{v} \cdot \mathbf{f} = 0\}.$$

The Symplectic Form on Orbits. Let $\mathcal{O}$ denote a nonzero orbit of $\mathfrak{se}(3)^*$. We consider the different orbit types separately, as above.

Type I: If $\mathcal{O}$ contains a point of the form $(\mathbf{e}, \mathbf{0})$, the orbit $\mathcal{O}$ equals $S^2_{\|\mathbf{e}\|} \times \{\mathbf{0}\}$. The minus orbit symplectic form is

$$\omega^-(\mathbf{e}, \mathbf{0})(\mathrm{ad}^*_{(\mathbf{x},\mathbf{y})}(\mathbf{e}, \mathbf{0}), \mathrm{ad}^*_{(\mathbf{x}',\mathbf{y}')}(\mathbf{e}, \mathbf{0})) = -\mathbf{e} \cdot (\mathbf{x} \times \mathbf{x}'). \tag{14.7.21}$$

Thus, the symplectic form on $\mathcal{O}$ at $(\mathbf{e}, \mathbf{0})$ is $-1/\|\mathbf{e}\|$ times the area element of the sphere of radius $\|\mathbf{e}\|$ (see (14.3.16) and (14.3.18)).

Type II: If $\mathcal{O}$ contains a point of the form $(\mathbf{0}, \mathbf{f})$, then $\mathcal{O}$ equals $TS^2_{\|\mathbf{f}\|}$. Let $(\mathbf{u}, \mathbf{v}) \in \mathcal{O}$, that is, $\|\mathbf{v}\| = \|\mathbf{f}\|$ and $\mathbf{u} \perp \mathbf{v}$. The symplectic form in this case is

$$\omega^-(\mathbf{u}, \mathbf{v})(\mathrm{ad}^*_{(\mathbf{x},\mathbf{y})}(\mathbf{u}, \mathbf{v}), \mathrm{ad}^*_{(\mathbf{x}',\mathbf{y}')}(\mathbf{u}, \mathbf{v}))$$
$$= -\mathbf{u} \cdot (\mathbf{x} \times \mathbf{x}') - \mathbf{v} \cdot (\mathbf{x} \times \mathbf{y}' - \mathbf{x}' \times \mathbf{y}). \tag{14.7.22}$$

We shall prove below that this form is exact, namely, $\omega^- = -\mathbf{d}\theta$, where

$$\theta(\mathbf{u}, \mathbf{v}) \cdot \mathrm{ad}^*_{(\mathbf{x},\mathbf{y})}(\mathbf{u}, \mathbf{v}) = \mathbf{u} \cdot \mathbf{x}. \tag{14.7.23}$$

First, note that θ is indeed well-defined, for if

$$\mathrm{ad}^*_{(\mathbf{x},\mathbf{y})}(\mathbf{u}, \mathbf{v}) = \mathrm{ad}^*_{(\mathbf{x}',\mathbf{y}')}(\mathbf{u}, \mathbf{v}),$$

by (14.7.17) we have $(\mathbf{x} - \mathbf{x}') \times \mathbf{v} = 0$, that is, $\mathbf{x} - \mathbf{x}' = c\mathbf{v}$ for some constant $c \in \mathbb{R}$, and since $\mathbf{u} \perp \mathbf{v}$, we conclude from here that $\mathbf{u} \cdot \mathbf{x} = \mathbf{u} \cdot \mathbf{x}'$. Second, to compute $\mathbf{d}\theta$, we shall use the formula

$$\mathbf{d}\theta(X, Y) = X[\theta(Y)] - Y[\theta(X)] - \theta([X, Y])$$

for any vector fields X, Y on $\mathcal{O}$. Third, we shall choose X and Y as follows:

$$X(\mathbf{u}, \mathbf{v}) = (\mathbf{x}, \mathbf{y})_{\mathfrak{se}(3)^*}(\mathbf{u}, \mathbf{v}) = -\mathrm{ad}^*_{(\mathbf{x},\mathbf{y})}(\mathbf{u}, \mathbf{v}),$$
$$Y(\mathbf{u}, \mathbf{v}) = (\mathbf{x}', \mathbf{y}')_{\mathfrak{se}(3)^*}(\mathbf{u}, \mathbf{v}) = -\mathrm{ad}^*_{(\mathbf{x}',\mathbf{y}')}(\mathbf{u}, \mathbf{v}),$$

for fixed $\mathbf{x}, \mathbf{y}, \mathbf{x}', \mathbf{y}' \in \mathbb{R}^3$. Fourth, to compute $X[\theta(Y)](\mathbf{u}, \mathbf{v})$, consider the path $(\mathbf{u}(\epsilon), \mathbf{v}(\epsilon)) = (e^{\epsilon\hat{\mathbf{x}}}\mathbf{u} - \epsilon(\mathbf{v} \times \mathbf{y}), e^{\epsilon\hat{\mathbf{x}}}\mathbf{v})$, which satisfies $(\mathbf{u}(0), \mathbf{v}(0)) = (\mathbf{u}, \mathbf{v})$ and

$$(\mathbf{u}'(0), \mathbf{v}'(0)) = -(\mathbf{u} \times \mathbf{x} + \mathbf{v} \times \mathbf{y}, \mathbf{v} \times \mathbf{x}) = -\mathrm{ad}^*_{(\mathbf{x},\mathbf{y})}(\mathbf{u}, \mathbf{v}) = X(\mathbf{u}, \mathbf{v}).$$

Then

$$X[\theta(Y)](\mathbf{u}, \mathbf{v}) = \left.\frac{d}{d\epsilon}\right|_{\epsilon=0} \theta(Y)(\mathbf{u}(\epsilon), \mathbf{v}(\epsilon))$$
$$= \left.\frac{d}{d\epsilon}\right|_{\epsilon=0} -\mathbf{u}(\epsilon) \cdot \mathbf{x}' = (\mathbf{u} \times \mathbf{x} + \mathbf{v} \times \mathbf{y}) \cdot \mathbf{x}'.$$

Similarly, $Y[\theta(X)](\mathbf{u}, \mathbf{v}) = (\mathbf{u} \times \mathbf{x}' + \mathbf{v} \times \mathbf{y}') \cdot \mathbf{x}$. Finally,

$$
\begin{aligned}
[X, Y](\mathbf{u}, \mathbf{v}) &= [(\mathbf{x}, \mathbf{y})_{se(3)^*}, (\mathbf{x}', \mathbf{y}')_{se(3)^*}](\mathbf{u}, \mathbf{v}) \\
&= -[(\mathbf{x}, \mathbf{y}), (\mathbf{x}', \mathbf{y}')]_{se(3)^*}(\mathbf{u}, \mathbf{v}) \\
&= -(\mathbf{x} \times \mathbf{x}', \mathbf{x} \times \mathbf{y}' - \mathbf{x}' \times \mathbf{y})_{se(3)^*}(\mathbf{u}, \mathbf{v}) \\
&= \mathrm{ad}^*_{(\mathbf{x} \times \mathbf{x}', \mathbf{x} \times \mathbf{y}' - \mathbf{x}' \times \mathbf{y})}(\mathbf{u}, \mathbf{v}).
\end{aligned}
$$

Therefore,

$$
\begin{aligned}
&- d\theta(\mathbf{u}, \mathbf{v})(\mathrm{ad}^*_{(\mathbf{x}, \mathbf{y})}(\mathbf{u}, \mathbf{v}), \mathrm{ad}^*_{(\mathbf{x}', \mathbf{y}')}(\mathbf{u}, \mathbf{v})) \\
&= -X[\theta(Y)](\mathbf{u}, \mathbf{v}) + Y[\theta(X)](\mathbf{u}, \mathbf{v}) + \theta([X, Y])(\mathbf{u}, \mathbf{v}) \\
&= -(\mathbf{u} \times \mathbf{x} + \mathbf{v} \times \mathbf{y}) \cdot \mathbf{x}' + (\mathbf{u} \times \mathbf{x}' + \mathbf{v} \times \mathbf{y}') \cdot \mathbf{x} + \mathbf{u} \cdot (\mathbf{x} \times \mathbf{x}') \\
&= -\mathbf{u} \cdot (\mathbf{x} \times \mathbf{x}') - \mathbf{v} \cdot (\mathbf{x} \times \mathbf{y}' - \mathbf{x}' \times \mathbf{y}),
\end{aligned}
$$

which coincides with (14.7.22).

The form θ given by (14.7.23) is the canonical symplectic structure when we identify $TS^2_{\|\mathbf{f}\|}$ with $T^*S^2_{\|\mathbf{f}\|}$ using the Euclidean metric.

Type III: If $\mathcal{O}$ contains $(\mathbf{e}, \mathbf{f})$, where $\mathbf{e} \neq 0$ and $\mathbf{f} \neq 0$, then $\mathcal{O}$ is diffeomorphic to $T^*S^2_{\|\mathbf{f}\|}$ in the following way. The map $\varphi : \mathrm{SE}(3) \to T^*S^2_{\|\mathbf{f}\|}$ given by (14.7.13) induces a diffeomorphism $\bar{\varphi} : \mathrm{SE}(3)/\mathrm{SE}(3)_{(\mathbf{e}, \mathbf{f})} \to T^*S^2_{\|\mathbf{f}\|}$. However, the orbit $\mathcal{O}$ through $(\mathbf{e}, \mathbf{f})$ is diffeomorphic to $\mathrm{SE}(3)/\mathrm{SE}(3)_{(\mathbf{e}, \mathbf{f})}$ by the diffeomorphism

$$
(\mathbf{A}, \mathbf{a}) \mapsto \mathrm{Ad}^*_{(\mathbf{A}, \mathbf{a})^{-1}}(\mathbf{e}, \mathbf{f}). \tag{14.7.24}
$$

Therefore, the diffeomorphism $\Phi : \mathcal{O} \to T^*S^2_{\|\mathbf{f}\|}$ is given by

$$
\Phi(\mathrm{Ad}^*_{(\mathbf{A}, \mathbf{a})^{-1}}(\mathbf{e}, \mathbf{f})) = \Phi(\mathbf{A}\mathbf{e} + \mathbf{a} \times \mathbf{A}\mathbf{f}, \mathbf{A}\mathbf{f}) \tag{14.7.25}
$$

$$
= (\mathbf{A}\mathbf{e} + \mathbf{a} \times \mathbf{A}\mathbf{f} - \frac{\mathbf{e} \cdot \mathbf{f}}{\|\mathbf{f}\|^2}\mathbf{A}\mathbf{f}, \mathbf{A}\mathbf{f}).
$$

If $(\bar{\mathbf{u}}, \bar{\mathbf{v}}) \in \mathcal{O}$, the orbit symplectic structure is given by formula (14.7.22), where $\bar{\mathbf{u}} = \mathbf{A}\mathbf{e} + \mathbf{a} \times \mathbf{A}\mathbf{f}$, $\bar{\mathbf{v}} = \mathbf{A}\mathbf{f}$ for some $\mathbf{A} \in \mathrm{SO}(3)$, $\mathbf{a} \in \mathbb{R}^3$. Let

$$
\mathbf{u} = \mathbf{A}\mathbf{e} + \mathbf{a} \times \mathbf{A}\mathbf{f} - \frac{\mathbf{e} \cdot \mathbf{f}}{\|\mathbf{f}\|^2}\mathbf{A}\mathbf{f} = \bar{\mathbf{u}} - \frac{\mathbf{e} \cdot \mathbf{f}}{\|\mathbf{f}\|^2}\bar{\mathbf{v}},
$$

$$
\mathbf{v} = \mathbf{A}\mathbf{f} = \bar{\mathbf{v}}, \tag{14.7.26}
$$

the pair of vectors $(\mathbf{u}, \mathbf{v})$ representing an element of TS^2. Note that $\|\mathbf{v}\| = \|\mathbf{f}\|$ and $\mathbf{u} \cdot \mathbf{v} = 0$. Then a tangent vector to $TS^2_{\|\mathbf{f}\|}$ at $(\mathbf{u}, \mathbf{v})$ can be represented as $\mathrm{ad}^*_{(\mathbf{x}, \mathbf{y})}(\mathbf{u}, \mathbf{v}) = (\mathbf{u} \times \mathbf{x} + \mathbf{v} \times \mathbf{y}, \mathbf{v} \times \mathbf{x})$, so that by (14.7.25) we get

$$T_{(\mathbf{u},\mathbf{v})}\Phi^{-1}(\mathrm{ad}^*_{(\mathbf{x},\mathbf{y})}(\mathbf{u},\mathbf{v})) = \left.\frac{d}{d\epsilon}\right|_{\epsilon=0} \Phi^{-1}(e^{-\epsilon\hat{\mathbf{x}}}\mathbf{u} + \epsilon(\mathbf{v}\times\mathbf{y}), e^{\epsilon\hat{\mathbf{x}}}\mathbf{v})$$

$$= \left.\frac{d}{d\epsilon}\right|_{\epsilon=0}\left(e^{-\epsilon\hat{\mathbf{x}}}\mathbf{u} + \epsilon(\mathbf{v}\times\mathbf{y}) + \frac{\mathbf{e}\cdot\mathbf{f}}{\|f\|^2}e^{-\epsilon\hat{\mathbf{x}}}\mathbf{v}, e^{-\epsilon\hat{\mathbf{x}}}\mathbf{v}\right)$$

$$= \left(\mathbf{u}\times\mathbf{x} + \mathbf{v}\times\mathbf{y} + \frac{\mathbf{e}\cdot\mathbf{f}}{\|f\|^2}(\mathbf{v}\times\mathbf{x}), \mathbf{v}\times\mathbf{x}\right)$$

$$= (\bar{\mathbf{u}}\times\mathbf{x} + \bar{\mathbf{v}}\times\mathbf{y}, \bar{\mathbf{v}}\times\mathbf{x})$$

$$= \mathrm{ad}^*_{(\mathbf{x},\mathbf{y})}(\bar{\mathbf{u}},\bar{\mathbf{v}}).$$

Therefore, the push-forward of the orbit symplectic form ω^- to $TS^2_{\|\mathbf{f}\|}$ is

$$(\Phi_*\omega^-)(\mathbf{u},\mathbf{v})(\mathrm{ad}^*_{(\mathbf{x},\mathbf{y})}(\mathbf{u},\mathbf{v}), \mathrm{ad}^*_{(\mathbf{x}',\mathbf{y}')}(\mathbf{u},\mathbf{v}))$$

$$= \omega^-(\bar{\mathbf{u}},\bar{\mathbf{v}})(T_{(\mathbf{u},\mathbf{v})}\Phi^{-1}(\mathrm{ad}^*_{(\mathbf{x},\mathbf{y})}(\mathbf{u},\mathbf{v})), T_{(\mathbf{u},\mathbf{v})}\Phi^{-1}(\mathrm{ad}^*_{(\mathbf{x}',\mathbf{y}')}(\mathbf{u},\mathbf{v}))$$

$$= \omega^-(\bar{\mathbf{u}},\bar{\mathbf{v}})(\mathrm{ad}^*_{(\mathbf{x},\mathbf{y})}(\bar{\mathbf{u}},\bar{\mathbf{v}}), \mathrm{ad}^*_{(\mathbf{x}',\mathbf{y}')}(\bar{\mathbf{u}},\bar{\mathbf{v}}))$$

$$= -\bar{\mathbf{u}}\cdot(\mathbf{x}\times\mathbf{x}') - \bar{\mathbf{v}}\cdot(\mathbf{x}\times\mathbf{y}' - \mathbf{x}'\times\mathbf{y})$$

$$= -\mathbf{u}\cdot(\mathbf{x}\times\mathbf{x}') - \mathbf{v}\cdot(\mathbf{x}\times\mathbf{y}' - \mathbf{x}'\times\mathbf{y}) - \frac{\mathbf{e}\cdot\mathbf{f}}{\|f\|^2}\mathbf{v}\cdot(\mathbf{x}\times\mathbf{x}').$$

$$(14.7.27)$$

The first two terms represent the canonical symplectic structure on $TS^2_{\|\mathbf{f}\|}$ (identified via the Euclidean metric with $T^*S^2_{\|\mathbf{f}\|}$), as we have seen in the analysis of type II orbits. The third term is the following two-form on $TS^2_{\|\mathbf{f}\|}$:

$$\beta(\mathbf{u},\mathbf{v})\left(\mathrm{ad}^*_{(\mathbf{x},\mathbf{y})}(\mathbf{u},\mathbf{v}), \mathrm{ad}^*_{(\mathbf{x}',\mathbf{y}')}(\mathbf{u},\mathbf{v})\right) = -\frac{\mathbf{e}\cdot\mathbf{f}}{\|f\|^2}\mathbf{v}\cdot(\mathbf{x}\times\mathbf{x}'). \quad (14.7.28)$$

As in the case of θ for type II orbits, it is easily seen that (14.7.27) correctly defines a two-form on $TS^2_{\|\mathbf{f}\|}$. It is necessarily closed, since it is the difference between $\Phi_*\omega^-$ and the canonical two-form on $TS^2_{\|\mathbf{f}\|}$. The two-form β is a magnetic term in the sense of §6.6.

We remark that the semidirect product theory of Marsden, Ratiu, and Weinstein [1984a, 1984b], combined with cotangent bundle reduction theory (see, for example, Marsden [1992]), can be used to give an alternative approach to the computation of the orbit symplectic forms. We refer to Marsden, Misiolek, Perlmutter, and Ratiu [1998] for details.

Exercises

⋄ **14.7-1.** Let K be a quadratic form on $\mathbb{R}^3$ and let $\mathbf{K}$ be the associated symmetric 3×3 matrix. Let

$$\{F,L\}_K = -\nabla K\cdot(\nabla F\times\nabla L).$$

Show that this is the Lie–Poisson bracket for the Lie algebra structure

$$[\mathbf{u}, \mathbf{v}]_K = \mathbf{K}(\mathbf{u} \times \mathbf{v}).$$

What is the underlying Lie group?

◇ **14.7-2.**   Determine the coadjoint orbits for the Lie algebra in the preceding exercise and calculate the orbit symplectic structure. Specialize to the case SO(2, 1).

◇ **14.7-3.**   Classify the coadjoint orbits of SU(1, 1), namely, the group of complex 2 × 2 matrices of determinant 1, of the form

$$g = \begin{pmatrix} a & b \\ \bar{b} & \bar{a} \end{pmatrix}$$

where $|a|^2 - |b|^2 = 1$.

◇ **14.7-4.**   The *Heisenberg group* is defined as follows. Start with the commutative group $\mathbb{R}^2$ with its standard symplectic form ω, the usual area form on the plane. Form the group $H = \mathbb{R}^2 \oplus \mathbb{R}$ with multiplication

$$(u, \alpha)(v, \beta) = (u + v, \alpha + \beta + \omega(u, v)).$$

Note that the identity element is $(0, 0)$ and the inverse of (u, α) is given by $(u, \alpha)^{-1} = (-u, -\alpha)$. Compute the coadjoint orbits of this group.

15
The Free Rigid Body

As an application of the theory developed so far, we discuss the motion of a free rigid body about a fixed point. We begin with a discussion of the kinematics of rigid-body motion. Our description of the kinematics of rigid bodies follows some of the notation and conventions of continuum mechanics, as in Marsden and Hughes [1983].

15.1 Material, Spatial, and Body Coordinates

Consider a rigid body, free to move in $\mathbb{R}^3$. A *reference configuration* $\mathcal{B}$ of the body is the closure of an open set in $\mathbb{R}^3$ with a piecewise smooth boundary. Points in $\mathcal{B}$, denoted by $X = (X^1, X^2, X^3) \in \mathcal{B}$, relative to an orthonormal basis $(\mathbf{E}_1, \mathbf{E}_2, \mathbf{E}_3)$ are called *material points*, and X^i, $i = 1, 2, 3$, are called *material coordinates*. A *configuration* of $\mathcal{B}$ is a mapping $\varphi : \mathcal{B} \to \mathbb{R}^3$ that is (for our purposes) C^1, orientation preserving, and invertible on its image. Points in the image of φ are called *spatial points* and are denoted by lowercase letters. Let $(\mathbf{e}_1, \mathbf{e}_2, \mathbf{e}_3)$ be a right-handed orthonormal basis of $\mathbb{R}^3$. Coordinates for spatial points, such as $x = (x^1, x^2, x^3) \in \mathbb{R}^3$, $i = 1, 2, 3$, relative to the basis $(\mathbf{e}_1, \mathbf{e}_2, \mathbf{e}_3)$ are called *spatial coordinates*. See Figure 15.1-1. Dually, one can consider material quantities such as maps defined on $\mathcal{B}$, say $Z : \mathcal{B} \to \mathbb{R}$. Then we can form spatial quantities by composition: $z_t = Z_t \circ \varphi_t^{-1}$. Spatial quantities are also called *Eulerian quantities*, and material quantities are often called

Lagrangian quantities. A *motion* of B is a time-dependent family of

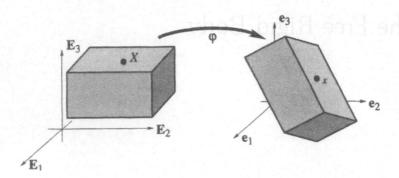

FIGURE 15.1.1. Configurations, spatial and material points.

configurations, written $x = \varphi(X,t) = \varphi_t(X)$ or simply $x(X,t)$ or $x_t(X)$. Spatial quantities are functions of x and are typically written as lowercase letters. By composition with φ_t, spatial quantities become functions of the material points X.

Rigidity of the body means that the distances between points of the body are fixed as the body moves. We shall assume that no external forces act on the body and that the center of mass is fixed at the origin (see Exercise 15.1-1). Since any isometry of $\mathbb{R}^3$ that leaves the origin fixed is a rotation (a 1932 theorem of Mazur and Ulam), we can write

$$x(X,t) = \mathbf{R}(t)X, \quad \text{i.e.,} \quad x^i = \mathbf{R}^i_j(t)X^j, \quad i,j = 1,2,3, \text{ sum on } j,$$

where x^i are the components of x relative to the basis $\mathbf{e}_1, \mathbf{e}_2, \mathbf{e}_3$ fixed in space, and $[\mathbf{R}^i_j]$ is the matrix of $\mathbf{R}$ relative to the basis $(\mathbf{E}_1, \mathbf{E}_2, \mathbf{E}_3)$ and $(\mathbf{e}_1, \mathbf{e}_2, \mathbf{e}_3)$. The motion is assumed to be continuous, and $\mathbf{R}(0)$ is the identity, so $\det(\mathbf{R}(t)) = 1$ and thus $\mathbf{R}(t) \in \mathrm{SO}(3)$, the proper orthogonal group. Thus, *the configuration space for the rotational motion of a rigid body may be identified with* $\mathrm{SO}(3)$. *Consequently, the velocity phase-space of the free rigid body is* $T\,\mathrm{SO}(3)$, *and the momentum phase space is the cotangent bundle* $T^*\,\mathrm{SO}(3)$. Euler angles, *discussed shortly, are the traditional way to parametrize* $\mathrm{SO}(3)$.

In addition to the material and spatial coordinates, there is a third set, the *convected*, or *body*, *coordinates*. These are the coordinates associated with the moving basis, and the description of the rigid-body motion in these coordinates, due to Euler, becomes very simple. As before, let $\mathbf{E}_1, \mathbf{E}_2, \mathbf{E}_3$ be an orthonormal basis fixed in the reference configuration. Let the time-dependent basis $\boldsymbol{\xi}_1, \boldsymbol{\xi}_2, \boldsymbol{\xi}_3$ be defined by $\boldsymbol{\xi}_i = \mathbf{R}(t)\mathbf{E}_i$, $i = 1,2,3$, so $\boldsymbol{\xi}_1, \boldsymbol{\xi}_2, \boldsymbol{\xi}_3$ move attached to the body. The *body coordinates* of a vector in $\mathbb{R}^3$ are its components relative to $\boldsymbol{\xi}_i$. For the rigid body anchored at the origin and rotating in space, $(\mathbf{e}_1, \mathbf{e}_2, \mathbf{e}_3)$ is thought of as a basis fixed in space, whereas

(ξ_1, ξ_2, ξ_3) is a basis fixed in the body and moving with it. For this reason (e_1, e_2, e_3) is called the *spatial coordinate system* and (ξ_1, ξ_2, ξ_3) the *body coordinate system*. See Figure 15.1-2.

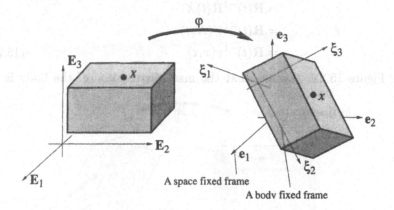

FIGURE 15.1.2. Spatial and body frames.

Exercises

◇ **15.1-1.** Start with SE(3) as the configuration space for the rigid body and "reduce out" (see §10.7, and the Euler–Poincaré and Lie–Poisson reduction theorems) translations to arrive at SO(3) as the configuration space.

15.2   The Lagrangian of the Free Rigid Body

If $X \in \mathcal{B}$ is a material point of the body, the corresponding trajectory followed by X in space is $x(t) = \mathbf{R}(t)X$, where $\mathbf{R}(t) \in$ SO(3). The *material* or *Lagrangian velocity* $V(X, t)$ is defined by

$$V(X, t) = \frac{\partial x(X, t)}{\partial t} = \dot{\mathbf{R}}(t)X, \qquad (15.2.1)$$

while the *spatial*, or *Eulerian, velocity* $v(x, t)$ is defined by

$$v(x, t) = V(X, t) = \dot{\mathbf{R}}(t)\mathbf{R}(t)^{-1}x. \qquad (15.2.2)$$

Finally, the *body*, or *convective, velocity* $\mathcal{V}(X, t)$ is defined by taking the velocity regarding X as time-dependent and x fixed, that is, we write

$X(x,t) = \mathbf{R}(t)^{-1}x$, and define

$$\begin{aligned}
\mathcal{V}(X,t) &= -\frac{\partial X(x,t)}{\partial t} = \mathbf{R}(t)^{-1}\dot{\mathbf{R}}(t)\mathbf{R}(t)^{-1}x \\
&= \mathbf{R}(t)^{-1}\dot{\mathbf{R}}(t)X \\
&= \mathbf{R}(t)^{-1}V(X,t) \\
&= \mathbf{R}(t)^{-1}v(x,t).
\end{aligned} \tag{15.2.3}$$

See Figure 15.2.1. Assume that the mass distribution of the body is de-

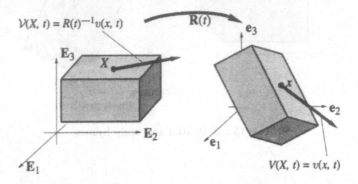

FIGURE 15.2.1. Material velocity V, spatial velocity v, and body velocity $\mathcal{V}$.

scribed by a compactly supported density measure $\rho_0 d^3 X$ in the reference configuration, which is zero at points outside the body. The Lagrangian, taken to be the kinetic energy, is given by any of the following expressions, which are related to one another by a change of variables and the identities $\|\mathcal{V}\| = \|V\| = \|v\|$:

$$L = \frac{1}{2}\int_B \rho_0(X)\|V(X,t)\|^2 \, d^3X \qquad \text{(material)} \tag{15.2.4}$$

$$= \frac{1}{2}\int_{\mathbf{R}(t)B} \rho_0(\mathbf{R}(t)^{-1}x)\|v(x,t)\|^2 \, d^3x \quad \text{(spatial)} \tag{15.2.5}$$

$$= \frac{1}{2}\int_B \rho_0(X)\|\mathcal{V}(X,t)\|^2 \, d^3X \qquad \text{(convective or body).} \tag{15.2.6}$$

Differentiating $\mathbf{R}(t)^T\mathbf{R}(t) = $ Identity and $\mathbf{R}(t)\mathbf{R}(t)^T = $ Identity with respect to t, it follows that both $\mathbf{R}(t)^{-1}\dot{\mathbf{R}}(t)$ and $\dot{\mathbf{R}}(t)\mathbf{R}(t)^{-1}$ are skew-symmetric. Moreover, by (15.2.2), (15.2.3), and the classical definition

$$\mathbf{v} = \boldsymbol{\omega} \times \mathbf{r} = \hat{\boldsymbol{\omega}}\mathbf{r}$$

of *angular velocity*, it follows that the vectors $\boldsymbol{\omega}(t)$ and $\boldsymbol{\Omega}(t)$ in $\mathbb{R}^3$ defined by

$$\hat{\boldsymbol{\omega}}(t) = \dot{\mathbf{R}}(t)\mathbf{R}(t)^{-1} \tag{15.2.7}$$

and

$$\hat{\Omega}(t) = \mathbf{R}(t)^{-1}\dot{\mathbf{R}}(t) \tag{15.2.8}$$

represent the *spatial* and *convective angular velocities* of the body. Note that $\boldsymbol{\omega}(t) = \mathbf{R}(t)\boldsymbol{\Omega}(t)$, or as matrices,

$$\hat{\omega} = \mathrm{Ad}_{\mathbf{R}}\,\hat{\Omega} = \mathbf{R}\hat{\Omega}\mathbf{R}^{-1}.$$

Let us show that $L : T\,\mathrm{SO}(3) \to \mathbb{R}$ given by (15.2.4) is left invariant. Indeed, if $\mathbf{B} \in \mathrm{SO}(3)$, then left translation by $\mathbf{B}$ is

$$L_{\mathbf{B}}\mathbf{R} = \mathbf{B}\mathbf{R} \quad \text{and} \quad TL_{\mathbf{B}}(\mathbf{R}, \dot{\mathbf{R}}) = (\mathbf{B}\mathbf{R}, \mathbf{B}\dot{\mathbf{R}}),$$

so

$$L(TL_{\mathbf{B}}(\mathbf{R}, \dot{\mathbf{R}})) = \frac{1}{2}\int_{\mathcal{B}} \rho_0(X)\|\mathbf{B}\dot{\mathbf{R}}X\|^2\,d^3X$$

$$= \frac{1}{2}\int_{\mathcal{B}} \rho_0(X)\|\dot{\mathbf{R}}X\|^2\,d^3X = L(\mathbf{R}, \dot{\mathbf{R}}), \tag{15.2.9}$$

since $\mathbf{R}$ is orthogonal.

By Lie–Poisson reduction of dynamics (Chapter 13), the corresponding Hamiltonian system on $T^*\,\mathrm{SO}(3)$, which is necessarily also left invariant, induces a Lie–Poisson system on $\mathfrak{so}(3)^*$, and this system leaves invariant the coadjoint orbits $\|\mathbf{\Pi}\| = \text{constant}$. Alternatively, by Euler–Poincaré reduction of dynamics, we get a system of equations in terms of body angular velocity on $\mathfrak{so}(3)$.

Reconstruction of the dynamics on $T\,\mathrm{SO}(3)$ is simply this: Given $\hat{\Omega}(t)$, determine $\mathbf{R}(t) \in \mathrm{SO}(3)$ from (15.2.8),

$$\dot{\mathbf{R}}(t) = \mathbf{R}(t)\hat{\Omega}(t), \tag{15.2.10}$$

which is a time-dependent linear differential equation for $\mathbf{R}(t)$.

15.3 The Lagrangian and Hamiltonian for the Rigid Body in Body Representation

From (15.2.6), (15.2.3), and (15.2.8) of the previous section, the rigid-body Lagrangian is

$$L = \frac{1}{2}\int_{\mathcal{B}} \rho_0(X)\|\Omega \times X\|^2\,d^3X. \tag{15.3.1}$$

Introducing the new inner product

$$\langle\!\langle \mathbf{a}, \mathbf{b} \rangle\!\rangle := \int_B \rho_0(X)(\mathbf{a} \times X) \cdot (\mathbf{b} \times X) \, d^3 X,$$

which is determined by the density $\rho_0(X)$ of the body, (15.3.1) becomes

$$L(\mathbf{\Omega}) = \frac{1}{2}\langle\!\langle \mathbf{\Omega}, \mathbf{\Omega} \rangle\!\rangle. \tag{15.3.2}$$

In what follows, it is useful to keep in mind the vector identity

$$(\mathbf{a} \times X) \cdot (\mathbf{b} \times X) = (\mathbf{a} \cdot \mathbf{b})\|X\|^2 - (\mathbf{a} \cdot X)(\mathbf{b} \cdot X).$$

Define the linear isomorphism $l : \mathbb{R}^3 \to \mathbb{R}^3$ by $l\mathbf{a} \cdot \mathbf{b} = \langle\!\langle \mathbf{a}, \mathbf{b} \rangle\!\rangle$ for all $\mathbf{a}, \mathbf{b} \in \mathbb{R}^3$; this is possible and uniquely determines l, since both the dot product and $\langle\!\langle\,,\rangle\!\rangle$ are nondegenerate bilinear forms (assuming that the rigid body is not concentrated on a line). It is clear that l is symmetric with respect to the dot product and is positive definite. Let $(\mathbf{E}_1, \mathbf{E}_2, \mathbf{E}_3)$ be an orthonormal basis for material coordinates. The matrix of l is

$$l_{ij} = \mathbf{E}_i \cdot l\mathbf{E}_j = \langle\!\langle \mathbf{E}_i, \mathbf{E}_j \rangle\!\rangle = \begin{cases} -\displaystyle\int_B \rho_0(X) X^i X^j \, d^3 X, & i \neq j, \\ \displaystyle\int_B \rho_0(X)(\|X\|^2 - (X^i)^2) \, d^3 X, & i = j, \end{cases}$$

which are the classical expressions of the matrix of the *inertia tensor*.

If $\mathbf{c}$ is a unit vector, then $\langle\!\langle \mathbf{c}, \mathbf{c} \rangle\!\rangle$ is the (classical) *moment of inertia* about the axis $\mathbf{c}$. Since l is symmetric, it can be diagonalized; an orthonormal basis in which it is diagonal is a *principal axis body frame*, and the diagonal elements I_1, I_2, I_3 are the *principal moments of inertia* of the rigid body. In what follows we work in a principal axis reference and body frame, $(\mathbf{E}_1, \mathbf{E}_2, \mathbf{E}_3)$.

Since $\mathfrak{so}(3)^*$ and $\mathbb{R}^3$ are identified by the dot product (not by $\langle\!\langle\,,\rangle\!\rangle$), the linear functional $\langle\!\langle \mathbf{\Omega}, \cdot \rangle\!\rangle$—the Legendre transformation of $\mathbf{\Omega}$—on $\mathfrak{so}(3) \cong \mathbb{R}^3$ is identified with $l\mathbf{\Omega} := \mathbf{\Pi} \in \mathfrak{so}(3)^* \cong \mathbb{R}^3$ because $\mathbf{\Pi} \cdot \mathbf{a} = \langle\!\langle \mathbf{\Omega}, \mathbf{a} \rangle\!\rangle$ for all $\mathbf{a} \in \mathbb{R}^3$. With $l = \mathrm{diag}(I_1, I_2, I_3)$, (15.3.2) defines a function

$$K(\mathbf{\Pi}) = \frac{1}{2}\left(\frac{\Pi_1^2}{I_1} + \frac{\Pi_2^2}{I_2} + \frac{\Pi_3^2}{I_3}\right) \tag{15.3.3}$$

that represents the expression for the kinetic energy on $\mathfrak{so}(3)^*$; note that $\mathbf{\Pi} = l\mathbf{\Omega}$ is the *angular momentum in the body frame*. Indeed, for any $\mathbf{a} \in \mathbb{R}^3$, the identity $(X \times (\mathbf{\Omega} \times X)) \cdot \mathbf{a} = (\mathbf{\Omega} \times X) \cdot (\mathbf{a} \times X)$ and the classical expression of the angular momentum in the body frame, namely,

$$\int_B (X \times \mathcal{V})\rho_0(X) \, d^3 X, \tag{15.3.4}$$

give

$$\left(\int_{\mathcal{B}} (X \times V) \rho_0(X) \, d^3 X \right) \cdot \mathbf{a} = \int_{\mathcal{B}} (X \times (\mathbf{\Omega} \times X)) \cdot \mathbf{a} \rho_0(X) \, d^3 X$$

$$= \int_{\mathcal{B}} (\mathbf{\Omega} \times X) \cdot (\mathbf{a} \times X) \rho_0(X) \, d^3 X$$

$$= \langle\!\langle \mathbf{\Omega}, \mathbf{a} \rangle\!\rangle = \mathbb{I} \mathbf{\Omega} \cdot \mathbf{a} = \mathbf{\Pi} \cdot \mathbf{a},$$

that is, the expression (15.3.4) equals $\mathbf{\Pi}$.

The *angular momentum in space* has the expression

$$\boldsymbol{\pi} = \int_{R(\mathcal{B})} (x \times v) \rho(x) \, d^3 x, \tag{15.3.5}$$

where $\rho(x) = \rho_0(X)$ is the *spatial mass density* and $v = \boldsymbol{\omega} \times x$ is the spatial velocity (see (15.2.2) and (15.2.7)). For any $\mathbf{a} \in \mathbb{R}^3$,

$$\boldsymbol{\pi} \cdot \mathbf{a} = \int_{R(\mathcal{B})} (x \times (\boldsymbol{\omega} \times x)) \cdot \mathbf{a} \rho(x) \, d^3 X$$

$$= \int_{R(\mathcal{B})} (\boldsymbol{\omega} \times x) \cdot (\mathbf{a} \times x) \rho(x) \, d^3 X. \tag{15.3.6}$$

Changing variables $x = \mathbf{R}X$, (15.3.6) becomes

$$\int_{\mathcal{B}} (\boldsymbol{\omega} \times \mathbf{R}X) \cdot (\mathbf{a} \times \mathbf{R}X) \rho_0(X) \, d^3 X$$

$$= \int_{\mathcal{B}} (\mathbf{R}^T \boldsymbol{\omega} \times X) \cdot (\mathbf{R}^T \mathbf{a} \times X) \rho_0(X) \, d^3 X$$

$$= \langle\!\langle \mathbf{\Omega}, \mathbf{R}^T \mathbf{a} \rangle\!\rangle = \mathbf{\Pi} \cdot \mathbf{R}^T \mathbf{a} = \mathbf{R}\mathbf{\Pi} \cdot \mathbf{a},$$

that is,

$$\boldsymbol{\pi} = \mathbf{R}\mathbf{\Pi}. \tag{15.3.7}$$

Since L given by (15.3.2) is left invariant on $T\,\mathrm{SO}(3)$, the function K defined on $\mathfrak{so}(3)^*$ by (15.3.3) defines the Lie–Poisson equations of motion on $\mathfrak{so}(3)^*$ relative to the rigid-body bracket

$$\{F, H\}(\mathbf{\Pi}) = -\mathbf{\Pi} \cdot (\nabla F(\mathbf{\Pi}) \times \nabla H(\mathbf{\Pi})). \tag{15.3.8}$$

Since $\nabla K(\mathbf{\Pi}) = \mathbb{I}^{-1}\mathbf{\Pi}$, we get from (15.3.8) the rigid-body equations

$$\dot{\mathbf{\Pi}} = -\nabla K(\mathbf{\Pi}) \times \mathbf{\Pi} = \mathbb{I} \times \mathbb{I}^{-1}\mathbf{\Pi}, \tag{15.3.9}$$

that is, they are the classical *Euler equations*:

$$\dot{\Pi}_1 = \frac{I_2 - I_3}{I_2 I_3} \Pi_2 \Pi_3,$$

$$\dot{\Pi}_2 = \frac{I_3 - I_1}{I_1 I_3} \Pi_1 \Pi_3, \qquad (15.3.10)$$

$$\dot{\Pi}_3 = \frac{I_1 - I_2}{I_1 I_2} \Pi_1 \Pi_2.$$

The fact that these equations preserve coadjoint orbits amounts, in this case, to the easily verified fact that

$$\Pi^2 := \|\mathbf{\Pi}\|^2 \qquad (15.3.11)$$

is a constant of the motion. In terms of coadjoint orbits, these equations are Hamiltonian on each sphere in $\mathbb{R}^3$ with Hamiltonian function K. The functions

$$C_\Phi(\mathbf{\Pi}) = \Phi\left(\frac{1}{2}\|\mathbf{\Pi}\|^2\right), \qquad (15.3.12)$$

for any $\Phi : \mathbb{R} \to \mathbb{R}$, are easily seen to be Casimir functions.

The conserved momentum resulting from left invariance is the *spatial angular momentum*:

$$\pi = \mathbf{R}\mathbf{\Pi}. \qquad (15.3.13)$$

Using left invariance, or a direct calculation, one finds that π is constant in time. Indeed,

$$\dot{\pi} = (\mathbf{R}\mathbf{\Pi})^{\cdot} = \dot{\mathbf{R}}\mathbf{\Pi} + \mathbf{R}\dot{\mathbf{\Pi}} = \omega \times \mathbf{R}\mathbf{\Pi} + \mathbf{R}\dot{\mathbf{\Pi}}$$

$$= \mathbf{R}\Omega \times \mathbf{R}\mathbf{\Pi} + \mathbf{R}\dot{\mathbf{\Pi}} = \mathbf{R}(-\mathbf{\Pi} \times \mathbf{I}^{-1}\mathbf{\Pi} + \dot{\mathbf{\Pi}}) = 0.$$

The flow lines are given by intersecting the ellipsoids $K = $ constant with the coadjoint orbits, which are two-spheres. For distinct moments of inertia $I_1 > I_2 > I_3$ or $I_1 < I_2 < I_3$, the flow on the sphere has saddle points at $(0, \pm\Pi, 0)$ and centers at $(\pm\Pi, 0, 0), (0, 0, \pm\Pi)$. The saddles are connected by four heteroclinic orbits, as indicated in Figure 15.3.1. In §15.10 we prove the following theorem.

Theorem 15.3.1 (Rigid-Body Stability Theorem). *In the motion of a free rigid body, rotation around the long and short axes are (Liapunov), stable and rotation about the middle axis is unstable.*

Even though we completely solved the rigid-body equations in body representation, the actual configuration of the body, that is, its attitude in space, has not been determined yet. This will be done in §15.8. Also, one

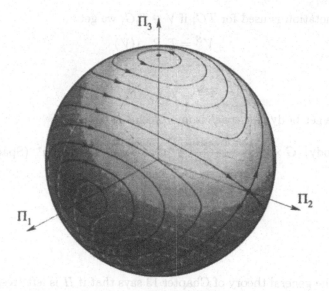

FIGURE 15.3.1. Rigid-body flow on the angular momentum spheres for the case $I_1 < I_2 < I_3$.

has to be careful about the meaning of stability in space versus material versus body representation.

Euler's equations are very general. The n-dimensional case has been treated by Mishchenko and Fomenko [1976, 1978a], Adler and van Moerbeke [1980a, 1980b], and Ratiu [1980, 1981, 1982] in connection with Lie algebras and algebraic geometry. The Russian school has generalized these equations further to a large class of Lie algebras and proved their complete integrability in a long series of papers starting in 1978; see the treatise of Fomenko and Trofimov [1989] and references therein.

15.4 Kinematics on Lie Groups

We now generalize the notation used for the rigid body to any Lie group. This abstraction unifies ideas common to rigid bodies, fluids, and plasmas in a consistent way. If G is a Lie group and $H : T^*G \to \mathbb{R}$ is a Hamiltonian for a mechanical system, we say that the system is described in the *material picture*. If $\alpha \in T_g^*G$, its *spatial representation* is defined by

$$\alpha^S = T_e^* R_g(\alpha), \tag{15.4.1}$$

while its *body representation* is

$$\alpha^B = T_e^* L_g(\alpha). \tag{15.4.2}$$

Similar notation is used for TG; if $V \in T_gG$, we get

$$V^S = T_g R_{g^{-1}}(V) \tag{15.4.3}$$

and

$$V^B = T_g L_{g^{-1}}(V). \tag{15.4.4}$$

Thus, we get body and space isomorphisms as follows:

$$\text{(Body)} \quad G \times \mathfrak{g}^* \xleftarrow{\text{Left Translate}} T^*G \xrightarrow{\text{Right Translate}} G \times \mathfrak{g}^* \quad \text{(Space)}.$$

Thus,

$$\alpha^S = \mathrm{Ad}^*_{g^{-1}}\, \alpha^B \tag{15.4.5}$$

and

$$V^S = \mathrm{Ad}_g\, V^B. \tag{15.4.6}$$

Part of the general theory of Chapter 13 says that if H is left (respectively, right) invariant on T^*G, it induces a Lie–Poisson system on $\mathfrak{g}^*_-$ (respectively, $\mathfrak{g}^*_+$).

Exercises

15.4-1 (Cayley–Klein parameters). Recall that the Lie algebras of SO(3) and SU(2) are the same. Recall also that SU(2) acts symplectically on $\mathbb{C}^2$ by multiplication of (complex) matrices. Use this to produce a momentum map $\mathbf{J} : \mathbb{C}^2 \to \mathfrak{su}(2)^* \cong \mathbb{R}^3$.

  (a) Write down $\mathbf{J}$ explicitly.

  (b) Verify by hand that $\mathbf{J}$ is a Poisson map.

  (c) If H is the rigid-body Hamiltonian, compute $H_{\mathrm{CK}} = H \circ \mathbf{J}$.

  (d) Write down Hamilton's equations for H_{CK} and discuss the collective Hamiltonian theorem in this context.

  (e) Find this material and relate it to the present context in one of the standard books (Whittaker, Pars, Hamel, or Goldstein, for example).

15.5 Poinsot's Theorem

Recall from §15.3 that the spatial angular momentum vector $\boldsymbol{\pi}$ is constant under the flow of the free rigid body. Also, if $\boldsymbol{\omega}$ is the angular velocity in space, then

$$\boldsymbol{\omega} \cdot \boldsymbol{\pi} = \boldsymbol{\Omega} \cdot \boldsymbol{\Pi} = 2K \tag{15.5.1}$$

is a constant. From this, it follows that ω moves in an (affine) plane perpendicular to the fixed vector π, called the *invariable plane*. The distance from the origin to this plane is $2K/\|\pi\|$; hence the equation of this plane is $\mathbf{u} \cdot \pi = 2K$. See Figure 15.5.1. The *ellipsoid of inertia in the body*

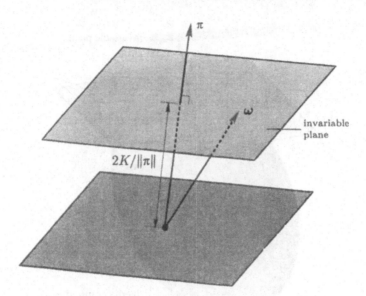

FIGURE 15.5.1. The invariable plane is orthogonal to π.

is defined by

$$\mathfrak{E} = \{\, \Omega \in \mathbb{R}^3 \mid \Omega \cdot \mathbb{I}\Omega = 2K \,\}.$$

The *ellipsoid of inertia in space* is

$$\mathbf{R}(\mathfrak{E}) = \{\, \mathbf{u} \in \mathbb{R}^3 \mid \mathbf{u} \cdot \mathbf{R}\mathbb{I}\mathbf{R}^{-1}\mathbf{u} = 2K \,\},$$

where $\mathbf{R} = \mathbf{R}(t) \in \mathrm{SO}(3)$ denotes the configuration of the body at time t.

Theorem 15.5.1 (Poinsot's Theorem). *The moment of inertia ellipsoid in space rolls without slipping on the invariable plane.*

Proof. First, note that $\omega \in \mathbf{R}(\mathfrak{E})$ if ω has energy K. Next, we determine the planes perpendicular to the fixed vector π and tangent to $\mathbf{R}(\mathfrak{E})$. See Figure 15.5.2. To do this, note that $\mathbf{R}(\mathfrak{E})$ is the level set of

$$\varphi(u) = \frac{1}{2}u \cdot \mathbf{R}\mathbb{I}\mathbf{R}^{-1}u,$$

so that at ω,

$$\nabla\varphi(\omega) = \mathbf{R}\mathbb{I}\mathbf{R}^{-1}\omega = \mathbf{R}\mathbb{I}\Omega = \mathbf{R}\Pi = \pi.$$

Thus, the tangent plane to $\mathbf{R}(\mathfrak{E})$ at ω is the invariable plane.

Since the point of tangency is ω, which is the instantaneous axis of rotation, its velocity is zero, that is, the rolling of the inertia ellipsoid on the invariable plane takes place without slipping.   ∎

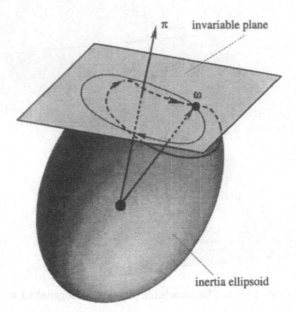

FIGURE 15.5.2. The geometry of Poinsot's theorem.

Exercises

◇ **15.5-1.** Prove a generalization of Poinsot's theorem to any Lie algebra $\mathfrak{g}$ as follows. Assume that $l : \mathfrak{g} \to \mathbb{R}$ is a quadratic Lagrangian, that is, a map of the form

$$l(\xi) = \frac{1}{2} \langle \xi, A\xi \rangle$$

where $A : \mathfrak{g} \to \mathfrak{g}^*$ is a (symmetric) isomorphism.

Define the *energy ellipsoid with value* E_0 to be

$$\mathcal{E}_0 = \{ \xi \in \mathfrak{g} \mid l(\xi) = E_0 \}.$$

If $\xi(t)$ is a solution of the Euler–Poincaré equations and

$$g(t)^{-1}\dot{g}(t) = \xi(t),$$

with $g(0) = e$, call the set

$$\mathcal{E}_t = g(t)(\mathcal{E}_0)$$

the *energy ellipsoid* at time t. Let $\mu = A\xi$ be the body momentum and

$$\mu^S = \mathrm{Ad}^*_{g^{-1}} \mu$$

the conserved spatial momentum. Define the *invariable plane* to be the affine plane

$$\mathcal{I} = \xi(0) + \{\, \xi \in \mathfrak{g} \mid \langle \mu^S, \xi \rangle = 0 \,\},$$

where $\xi(0)$ is the initial condition.

(a) Show that $\xi^S(t) = \mathrm{Ad}_{g(t)} \xi(t)$, the spatial velocity, lies in $\mathcal{I}$ for all t; that is, $\mathcal{I}$ is *invariant*.

(b) Show that $\xi^S(t) \in \mathcal{E}_t$ and that the surface $\mathcal{E}_t$ is tangent to $\mathcal{I}$ at this point.

(c) Show in a precise sense that $\mathcal{E}_t$ rolls without slipping on the invariable plane.

15.6   Euler Angles

In what follows, we adopt the conventions of Arnold [1989], Cabannes [1962], Goldstein [1980], and Hamel [1949]; these are different from the ones used by the British school (Whittaker [1927] and Pars [1965]).

Let (x^1, x^2, x^3) and (χ^1, χ^2, χ^3) denote the components of a vector written in the bases $(\mathbf{e}_1, \mathbf{e}_2, \mathbf{e}_3)$ and $(\boldsymbol{\xi}_1, \boldsymbol{\xi}_2, \boldsymbol{\xi}_3)$, respectively. We pass from the basis $(\mathbf{e}_1, \mathbf{e}_2, \mathbf{e}_3)$ to the basis $(\boldsymbol{\xi}_1, \boldsymbol{\xi}_2, \boldsymbol{\xi}_3)$ by means of three consecutive counterclockwise rotations (see Figure 15.6.1). First rotate $(\mathbf{e}_1, \mathbf{e}_2, \mathbf{e}_3)$ by an angle φ around $\mathbf{e}_3$ and denote the resulting basis and coordinates by $(\mathbf{e}'_1, \mathbf{e}'_2, \mathbf{e}'_3)$ and (x'_1, x'_2, x'_3), respectively. The new coordinates (x'^1, x'^2, x'^3) are expressed in terms of the old coordinates (x^1, x^2, x^3) of the *same point* by

$$\begin{bmatrix} x'^1 \\ x'^2 \\ x'^3 \end{bmatrix} = \begin{bmatrix} \cos\varphi & \sin\varphi & 0 \\ -\sin\varphi & \cos\varphi & 0 \\ 0 & 0 & 1 \end{bmatrix} \begin{bmatrix} x^1 \\ x^2 \\ x^3 \end{bmatrix}. \qquad (15.6.1)$$

Denote the change of basis matrix (15.6.1) in $\mathbb{R}^3$ by $\mathbf{R}_1$. Second, rotate $(\mathbf{e}'_1, \mathbf{e}'_2, \mathbf{e}'_3)$ by the angle θ around $\mathbf{e}'_1$ and denote the resulting basis and coordinate system by $(\mathbf{e}''_1, \mathbf{e}''_2, \mathbf{e}''_3)$ and (x''^1, x''^2, x''^3), respectively. The new coordinates (x''^1, x''^2, x''^3) are expressed in terms of the old coordinates (x'^1, x'^2, x'^3) by

$$\begin{bmatrix} x''^1 \\ x''^2 \\ x''^3 \end{bmatrix} = \begin{bmatrix} 1 & 0 & 0 \\ 0 & \cos\theta & \sin\theta \\ 0 & -\sin\theta & \cos\theta \end{bmatrix} \begin{bmatrix} x'^1 \\ x'^2 \\ x'^3 \end{bmatrix}. \qquad (15.6.2)$$

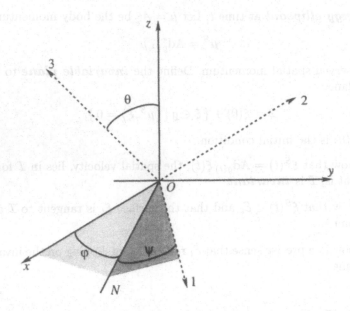

FIGURE 15.6.1. Euler angles.

Denote the change of basis matrix in (15.6.2) by $\mathbf{R}_2$. The $\mathbf{e}_1'$-axis, that is, the intersection of the $(\mathbf{e}_1, \mathbf{e}_2)$-plane with the $(\mathbf{e}_1'', \mathbf{e}_2'')$-plane, is called the *line of nodes* and is denoted by ON. Finally, rotate by the angle ψ around $\mathbf{e}_3''$. The resulting basis is $(\boldsymbol{\xi}_1, \boldsymbol{\xi}_2, \boldsymbol{\xi}_3)$, and the new coordinates (χ^1, χ^2, χ^3) are expressed in terms of the old coordinates (x''^1, x''^2, x''^3) by

$$\begin{bmatrix} \chi^1 \\ \chi^2 \\ \chi^3 \end{bmatrix} = \begin{bmatrix} \cos\psi & \sin\psi & 0 \\ -\sin\psi & \cos\psi & 0 \\ 0 & 0 & 1 \end{bmatrix} \begin{bmatrix} x''^1 \\ x''^2 \\ x''^3 \end{bmatrix}. \qquad (15.6.3)$$

Let $\mathbf{R}_3$ denote the change of basis matrix in (15.6.3). The rotation $\mathbf{R}$ sending (x^1, x^2, x^3) to (χ^1, χ^2, χ^3) is described by the matrix $\mathbf{P} = \mathbf{R}_3\mathbf{R}_2\mathbf{R}_1$ given by

$$\begin{bmatrix} \cos\psi\cos\varphi - \cos\theta\sin\varphi\sin\psi & \cos\psi\sin\varphi + \cos\theta\cos\varphi\sin\psi & \sin\theta\sin\psi \\ -\sin\psi\cos\varphi - \cos\theta\sin\varphi\cos\psi & -\sin\psi\sin\varphi + \cos\theta\cos\varphi\cos\psi & \sin\theta\cos\psi \\ \sin\theta\sin\varphi & -\sin\theta\cos\varphi & \cos\theta \end{bmatrix}.$$

Thus, $\chi = \mathbf{P}x$; equivalently, since the *same* point is expressed in two ways as $\sum_{i=1}^3 \chi^i\boldsymbol{\xi}_i = \sum_{j=1}^3 x^i\mathbf{e}_j$, we get

$$\sum_{j=1}^3 x^j\mathbf{e}_j = \sum_{i=1}^3 \chi^i\boldsymbol{\xi}_i = \sum_{i=1}^3 \left(\sum_{j=1}^3 P_{ij}x^j\right)\boldsymbol{\xi}_i = \sum_{j=1}^3 x^j\sum_{i=1}^3 P_{ij}\boldsymbol{\xi}_i,$$

that is,

$$e_j = \sum_{i=1}^{3} P_{ij}\xi_i, \qquad (15.6.4)$$

and hence $\mathbf{P}$ is the change of basis matrix between the rotated basis (ξ_1, ξ_2, ξ_3) and the fixed spatial basis (e_1, e_2, e_3). On the other hand, (15.6.4) represents the matrix expression of the rotation $\mathbf{R}^T$ sending ξ_j to e_j, that is, the matrix $[\mathbf{R}]_\xi$ of $\mathbf{R}$ in the basis (ξ_1, ξ_2, ξ_3) is P^T:

$$[\mathbf{R}]_\xi = \mathbf{P}^T, \quad \text{i.e.,} \quad \mathbf{R}\xi_i = \sum_{i=1}^{3} P_{ij}\xi_j. \qquad (15.6.5)$$

Consequently, the matrix $[\mathbf{R}]_e$ of $\mathbf{R}$ in the basis (e_1, e_2, e_3) is given by P:

$$[\mathbf{R}]_e = \mathbf{P}, \quad \text{i.e.,} \quad \mathbf{R}e_j = \sum_{i=1}^{3} P_{ij}e_i. \qquad (15.6.6)$$

It is straightforward to check that if

$$0 \le \varphi < 2\pi, \quad 0 \le \psi < 2\pi, \quad 0 \le \theta < \pi,$$

then there is a bijective map between the (φ, ψ, θ) variables and SO(3). However, this bijective map does not define a chart, since its differential vanishes, for example at $\varphi = \psi = \theta = 0$. The differential is nonzero for

$$0 < \varphi < 2\pi, \quad 0 < \psi < 2\pi, \quad 0 < \theta < \pi,$$

and on this domain the Euler angles do form a chart.

15.7   The Hamiltonian of the Free Rigid Body in the Material Description via Euler Angles

To express the kinetic energy in terms of Euler angles, we choose the basis E_1, E_2, E_3 of $\mathbb{R}^3$ in the reference configuration to equal the basis (e_1, e_2, e_3) of $\mathbb{R}^3$ in the spatial coordinate system. Thus, the matrix representation of $\mathbf{R}(t)$ in the basis ξ_1, ξ_2, ξ_3 equals $\mathbf{P}^T$, where $\mathbf{P}$ is given by (15.6). In this way, $\boldsymbol{\omega}$ and $\boldsymbol{\Omega}$ have the following expressions in the basis ξ_1, ξ_2, ξ_3:

$$\boldsymbol{\omega} = \begin{bmatrix} \dot{\theta}\cos\varphi + \dot{\psi}\sin\varphi\sin\theta \\ \dot{\theta}\sin\varphi - \dot{\psi}\cos\varphi\sin\theta \\ \dot{\varphi} + \dot{\psi}\cos\theta \end{bmatrix}, \quad \boldsymbol{\Omega} = \begin{bmatrix} \dot{\theta}\cos\psi + \dot{\varphi}\sin\psi\sin\theta \\ -\dot{\theta}\sin\psi + \dot{\varphi}\cos\psi\sin\theta \\ \dot{\varphi}\cos\theta + \dot{\psi} \end{bmatrix}.$$

$$(15.7.1)$$

By definition of $\boldsymbol{\Pi}$, it follows that

$$\boldsymbol{\Pi} = \left[\begin{array}{c} I_1(\dot{\varphi}\sin\theta\sin\psi + \dot{\theta}\cos\psi) \\ I_2(\dot{\varphi}\sin\theta\cos\psi - \dot{\theta}\sin\psi) \\ I_3(\dot{\varphi}\cos\theta + \dot{\psi}) \end{array} \right]. \tag{15.7.2}$$

This expresses $\boldsymbol{\Pi}$ in terms of coordinates on $T\,\mathrm{SO}(3)$. Since $T\,\mathrm{SO}(3)$ and $T^*\,\mathrm{SO}(3)$ are to be identified by the metric defined as the left-invariant metric given at the identity by $\langle\!\langle\,,\rangle\!\rangle$, the variables $(p_\varphi, p_\psi, p_\theta)$ canonically conjugate to (φ, ψ, θ) are given by the Legendre transformation

$$p_\varphi = \partial K/\partial\dot{\varphi}, \quad p_\psi = \partial K/\partial\dot{\psi}, \quad p_\theta = \partial K/\partial\dot{\theta},$$

where the expression of the kinetic energy on $T\,\mathrm{SO}(3)$ is obtained by plugging (15.7.2) into (15.3.3). We get

$$\begin{aligned} p_\varphi &= I_1(\dot{\varphi}\sin\theta\sin\psi + \dot{\theta}\cos\psi)\sin\theta\sin\psi \\ &\quad + I_2(\dot{\varphi}\sin\theta\cos\varphi - \dot{\theta}\sin\psi)\sin\theta\cos\psi + I_3(\dot{\varphi}\cos\theta + \dot{\psi})\cos\theta, \\ p_\psi &= I_3(\dot{\varphi}\cos\theta + \dot{\psi}), \\ p_\theta &= I_1(\dot{\varphi}\sin\theta\sin\psi + \dot{\theta}\cos\psi)\cos\psi \\ &\quad - I_2(\dot{\varphi}\sin\theta\cos\psi - \dot{\theta}\sin\psi)\sin\psi, \end{aligned} \tag{15.7.3}$$

whence, by (15.7.2),

$$\boldsymbol{\Pi} = \left[\begin{array}{c} ((p_\varphi - p_\psi\cos\theta)\sin\psi + p_\theta\sin\theta\cos\psi)/\sin\theta \\ ((p_\varphi - p_\psi\cos\theta)\cos\psi - p_\theta\sin\theta\sin\psi)/\sin\theta \\ p_\psi \end{array} \right], \tag{15.7.4}$$

and so by (15.3.3) we get the coordinate expression of the kinetic energy in the material picture to be

$$\begin{aligned} K(\varphi, &\psi, \theta, p_\varphi, p_\psi, p_\theta) \\ &= \frac{1}{2}\left\{ \frac{[(p_\varphi - p_\psi\cos\theta)\sin\psi + p_\theta\sin\theta\cos\psi]^2}{I_1\sin^2\theta} \right. \\ &\quad + \left. \frac{[(p_\varphi - p_\psi\cos\theta)\cos\psi - p_\theta\sin\theta\sin\psi]^2}{I_1\sin^2\theta} + \frac{p_\psi^2}{I_3} \right\}. \end{aligned} \tag{15.7.5}$$

This expression for the kinetic energy has an invariant expression on the cotangent bundle $T^*\,\mathrm{SO}(3)$. In fact,

$$K(\alpha_R) = \frac{1}{2}\langle\!\langle \boldsymbol{\Omega}, \boldsymbol{\Omega} \rangle\!\rangle = \frac{1}{4}\,\mathrm{trace}\,(\mathbb{I}R^{-1}\dot{R}R^{-1}\dot{R}), \tag{15.7.6}$$

where $\alpha_R \in T_R^*\,\mathrm{SO}(3)$ is defined by $\langle \alpha, R\hat{\mathbf{v}} \rangle = \langle\!\langle \boldsymbol{\Omega}, \mathbf{v} \rangle\!\rangle$ for all $\mathbf{v} \in \mathbb{R}^3$.

The equation of motion (15.3.9) can also be derived "by hand," without appeal to Lie–Poisson or Euler–Poincaré reduction, as follows. Hamilton's canonical equations

$$\dot{\varphi} = \frac{\partial K}{\partial p_\varphi}, \quad \dot{\psi} = \frac{\partial K}{\partial p_\psi}, \quad \dot{\theta} = \frac{\partial K}{\partial p_\theta},$$

$$\dot{p}_\varphi = -\frac{\partial K}{\partial \varphi}, \quad \dot{p}_\psi = -\frac{\partial K}{\partial \psi}, \quad \dot{p}_\theta = -\frac{\partial K}{\partial \theta},$$

in a chart given by the Euler angles become after direct substitution and a somewhat lengthy calculation,

$$\dot{\mathbf{\Pi}} = \mathbf{\Pi} \times \mathbf{\Omega}.$$

For $F, G : T^* \mathrm{SO}(3) \to \mathbb{R}$, that is, F, G are functions of $(\varphi, \psi, \theta, p_\varphi, p_\psi, p_\theta)$ in a chart given by Euler angles, the standard canonical Poisson bracket is

$$\{F, G\} = \frac{\partial F}{\partial \varphi}\frac{\partial G}{\partial p_\varphi} - \frac{\partial F}{\partial p_\varphi}\frac{\partial G}{\partial \varphi} + \frac{\partial F}{\partial \psi}\frac{\partial G}{\partial p_\psi} - \frac{\partial F}{\partial p_\psi}\frac{\partial G}{\partial \psi} + \frac{\partial F}{\partial \theta}\frac{\partial G}{\partial p_\theta} - \frac{\partial F}{\partial p_\theta}\frac{\partial G}{\partial \theta}.$$
$$(15.7.7)$$

A computation shows that after the substitution

$$(\varphi, \psi, \theta, p_\varphi, p_\psi, p_\theta) \mapsto (\Pi_1, \Pi_2, \Pi_3),$$

this becomes

$$\{F, G\}(\mathbf{\Pi}) = -\mathbf{\Pi} \cdot (\nabla F(\mathbf{\Pi}) \times \nabla G(\mathbf{\Pi})), \qquad (15.7.8)$$

which is the $(-)$ Lie–Poisson bracket. This provides a direct check on the Lie–Poisson reduction theorem in Chapter 13. Thus (15.7.4) defines a canonical map between Poisson manifolds. The apparently "miraculous" groupings and cancellations of terms that occur in this calculation should make the reader appreciate the general theory.

Exercises

$\diamond$ **15.7-1.** Verify that (15.7.8), namely,

$$\{F, G\}(\mathbf{\Pi}) = -\mathbf{\Pi} \cdot (\nabla F(\mathbf{\Pi}) \times \nabla G(\mathbf{\Pi})),$$

holds by a **direct** calculation using substitution and the chain rule starting with the canonical brackets in Euler angle representation.

15.8   The Analytical Solution of the Free Rigid-Body Problem

We now give the analytical solution of the Euler equations. These formulas are useful when, for example, one is dealing with perturbations leading to chaos via the Poincaré–Melnikov method, as in Ziglin [1980a, 1980b], Holmes and Marsden [1983], and Koiller [1985]. For the last part of this section, the reader is assumed to be familiar with Jacobi's elementary elliptic functions; see, for example, Lawden [1989]. Let us make the following simplifying notation:

$$a_1 = \frac{I_2 - I_3}{I_2 I_3} \geq 0, \quad a_2 = \frac{I_3 - I_1}{I_1 I_3} \leq 0, \quad \text{and} \quad a_3 = \frac{I_1 - I_2}{I_1 I_2} \geq 0,$$

where we assume $I_1 \geq I_2 \geq I_3 > 0$. Then Euler's equations $\dot{\Pi} = \Pi \times 1^{-1}\Pi$ can be written as

$$\dot{\Pi}_1 = a_1 \Pi_2 \Pi_3,$$
$$\dot{\Pi}_2 = a_2 \Pi_3 \Pi_1, \qquad (15.8.1)$$
$$\dot{\Pi}_3 = a_3 \Pi_1 \Pi_2.$$

For the analysis that follows it is important to recall that *the angular momentum in space is fixed* and that the instantaneous axis of rotation of the body in body coordinates is given by the angular velocity vector $\mathbf{\Omega}$.

Case 1. $I_1 = I_2 = I_3$. Then $a_1 = a_2 = a_3 = 0$, and we conclude that Π, and thus $\mathbf{\Omega}$, are both constant. Hence the body rotates with constant angular velocity about a fixed axis. In Figure 15.3.1 all points on the sphere become fixed points.

Case 2. $I_1 = I_2 > I_3$. Then $a_3 = 0$ and $a_2 = -a_1$. Since $a_3 = 0$, it follows from (15.8.1) that $\Pi_3 = $ constant, and thus setting $\lambda = -a_1 \Pi_3$, we get $a_2 \Pi_3 = \lambda$. Thus, (15.8.1) becomes

$$\dot{\Pi}_1 + \lambda \Pi_2 = 0,$$
$$\dot{\Pi}_2 - \lambda \Pi_1 = 0,$$

which has a solution for initial data given at time $t = 0$ given by

$$\Pi_1 = \Pi_1(0) \cos \lambda t - \Pi_2(0) \sin \lambda t,$$
$$\Pi_2 = \Pi_2(0) \cos \lambda t + \Pi_1(0) \sin \lambda t.$$

These formulas say that the axis of symmetry OZ of the body rotates *relative to the body* with angular velocity λ. It is straightforward to check that OZ, $\mathbf{\Omega}$, and Π are in the same plane and that Π and $\mathbf{\Omega}$ make constant angles with OZ and thus among themselves. In addition, since $I_1 = I_2$, we have

$$
\begin{aligned}
\|\mathbf{\Omega}\|^2 &= \frac{\Pi_1^2}{I_1^2} + \frac{\Pi_2^2}{I_2^2} + \frac{\Pi_3^2}{I_3^2} \\
&= \left(\frac{\Pi_1^2}{I_1} + \frac{\Pi_2^2}{I_2} + \frac{\Pi_3^2}{I_3} \right) \frac{1}{I_1} - \frac{\Pi_3^2}{I_3} \left(\frac{1}{I_1} - \frac{1}{I_3} \right) \\
&= \frac{2K}{I_1} - \frac{a_2 \Pi_3^2}{I_3} = \text{constant.}
\end{aligned}
$$

Therefore, the corresponding spatial objects Oz (the symmetry axis of the inertia ellipsoid in space), $\boldsymbol{\omega}$, and $\boldsymbol{\pi}$ enjoy the same properties, and hence the axis of rotation in the body (given by $\mathbf{\Omega}$) makes a constant angle with the angular momentum vector that is fixed in space, and thus the axis of rotation describes a right circular cone of constant angle in space. At the same time, the axis of rotation in the body (given by $\mathbf{\Omega}$) makes a constant angle with Oz, thus tracing a second cone in the body. See Figure 15.8.1.

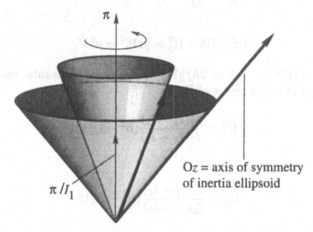

FIGURE 15.8.1. The geometry for integrating Euler's equations.

Consequently, the motion can be described by the rolling of a cone of constant angle in the body on a second cone of constant angle fixed in space. Whether the cone in the body rolls outside or inside the cone in space is determined by the sign of λ. Since Oz, $\boldsymbol{\omega}$, and $\boldsymbol{\pi}$ remain coplanar during the motion, $\boldsymbol{\omega}$ and Oz rotate about the fixed vector $\boldsymbol{\pi}$ with the same angular velocity, namely, the component of $\boldsymbol{\omega}$ along $\boldsymbol{\pi}$ in the decomposition of $\boldsymbol{\omega}$ relative to $\boldsymbol{\pi}$ and the Oz-axis. This angular velocity is called the **angular velocity of precession**. Let **e** denote the unit vector along Oz and write

$\omega = \alpha\pi + \beta e$. Therefore,

$$2K = \omega \cdot \pi = \alpha\|\pi\|^2 + \beta e \cdot \pi = \alpha\|\pi\|^2 + \beta\Pi_3,$$

$$\frac{\Pi_3}{I_3} = \Omega^3 = \omega \cdot e = \alpha\pi \cdot e + \beta = \alpha\Pi_3 + \beta,$$

and

$$\beta = -a_2\Pi_3,$$

so that $\alpha = 1/I_1$ and $\beta = -a_2\Pi_3$. Therefore, the *angular velocity of precession equals* Π_S/I_1.

On the Π-sphere, the dynamics reduce to two fixed points surrounded by oppositely oriented periodic lines of latitude and separated by an equator of fixed points. A similar analysis applies if $I_1 > I_2 = I_3$.

Case 3. $I_1 > I_2 > I_3$. The two integrals of energy and angular momentum,

$$\frac{\Pi_1^2}{I_1} + \frac{\Pi_2^2}{I_2} + \frac{\Pi_3^3}{I_3} = 2h = ab^2, \qquad (15.8.2)$$

$$\Pi_1^2 + \Pi_2^2 + \Pi_3^2 = \|\Pi\|^2 = a^2b^2, \qquad (15.8.3)$$

where $a = \|\Pi\|^2/(2h)$, $b = 2h/\|\Pi\|$ are positive constants, enable us to express Π_1 and Π_3 in terms of Π_2 as

$$\Pi_1^2 = \frac{I_1(I_2 - I_3)}{I_2(I_1 - I_3)}(\alpha^2 - \Pi_2^2) \qquad (15.8.4)$$

and

$$\Pi_3^2 = \frac{I_3(I_1 - I_2)}{I_2(I_1 - I_3)}(\beta^2 - \Pi_2^2), \qquad (15.8.5)$$

where α and β are positive constants given by

$$\alpha^2 = \frac{aI_2(a - I_3)b^2}{I_2 - I_3} \quad \text{and} \quad \beta^2 = \frac{aI_2(I_1 - a)b^2}{I_1 - I_2}. \qquad (15.8.6)$$

By the definition of a, note that $I_1 \geq a \geq I_3$. The endpoints of the interval $[I_1, I_3]$ are easy to deal with. If $a = I_1$, then $\Pi_2 = \Pi_3 = 0$, and the motion is a steady rotation about the Π-axis with body angular velocity $\pm b$. Similarly, if $a = I_3$, then $\Pi_1 = \Pi_2 = 0$. So we can assume that $I_1 > a > I_3$. With these expressions, the square of (15.8.1) becomes

$$(\dot{\Pi}_2)^2 = a_1a_3(\alpha^2 - \Pi_2^2)(\beta^2 - \Pi_2^2), \qquad (15.8.7)$$

that is,

$$t = \int_{\Pi_2(0)}^{\Pi_2} \frac{du}{\sqrt{a_1 a_3 (\alpha^2 - u^2)(\beta^2 - u^2)}}, \tag{15.8.8}$$

which shows that Π_2, and hence Π_1, Π_3, are elliptic functions of time.

In case the quartic under the square root has double roots, that is, $\alpha = \beta$, (15.8.8) can be integrated explicitly by means of elementary functions. By (15.8.6) it follows that

$$\beta^2 - \alpha^2 = \frac{ab^2 I_2 (I_1 - I_3)(I_2 - a)}{(I_1 - I_2)(I_2 - I_3)}.$$

Thus $\alpha = \beta$ if and only if $a = I_2$, which in turn forces $\alpha = \beta = ab = \|\Pi\|$ and $\|\Pi\|^2 = 2hI_2$. Thus (15.8.7) becomes

$$(\dot{\Pi}_2)^2 = a_1 a_3 (\|\Pi\|^2 - \Pi_2^2)^2. \tag{15.8.9}$$

If $\|\Pi\|^2 = 2hI_2$ is satisfied, the intersection of the sphere of constant angular momentum $\|\Pi\|$ with the elliptical energy surface corresponding to the value $2h$ consists of two great circles on the sphere going through the Π_2-axis in the planes

$$\Pi_3 = \pm\Pi_1 \sqrt{\frac{a_3}{a_1}}.$$

In other words, the solution of (15.8.9) consists of four heteroclinic orbits and the values $\Pi_2 = \pm\|\Pi\|$. Equation (15.8.9) is solved by putting $\Pi_2 = \|\Pi\| \tanh \theta$. Setting $\Pi_2(0) = 0$ for simplicity, we get the four heteroclinic orbits

$$\Pi_1^+(t) = \pm\|\Pi\| \sqrt{\frac{a_1}{-a_2}} \operatorname{sech}\left(-\sqrt{a_1 a_3} \|\Pi\| t\right),$$

$$\Pi_2^+(t) = \pm\|\Pi\| \tanh\left(-\sqrt{a_1 a_3} \|\Pi\| t\right), \tag{15.8.10}$$

$$\Pi_3^+(t) = \pm\|\Pi\| \sqrt{\frac{a_3}{-a_2}} \operatorname{sech}\left(-\sqrt{a_1 a_3} \|\Pi\| t\right),$$

when

$$\Pi_3 = \Pi_1 \sqrt{\frac{a_3}{a_1}},$$

and

$$\Pi_1^-(t) = \Pi_1^+(-t), \quad \Pi_2^-(t) = \Pi_2^+(-t), \quad \Pi_3^-(t) = \Pi_3^+(-t),$$

when

$$\Pi_3 = -\Pi_1 \sqrt{\frac{a_3}{a_1}}.$$

If $\alpha \neq \beta$, then $a \neq I_2$, and the integration is performed with the aid of Jacobi's elliptic functions (see Whittaker and Watson [1940, Chapter 22], or Lawden [1989]). For example, the elliptic function sn u with modulus k is given by

$$\operatorname{sn} u = u - \frac{1}{3!}(1+k^2)u^3 + \frac{1}{5!}(1+14k^2+k^4)u^5 - \cdots,$$

and its inverse is

$$\operatorname{sn}^{-1} x = \int_0^x \frac{1}{\sqrt{(1-t^2)(1-k^2t^2)}}\, dt, \quad 0 \le x \le 1.$$

Assuming $I_1 > I_2 > a > I_3$ or, equivalently, $\alpha < \beta$, the substitution of the elliptic function $\Pi_2 = \alpha \operatorname{sn} u$ in (15.8.8) with the modulus

$$k = \alpha/\beta = \left[\frac{(I_1 - I_2)(a - I_3)}{(I_1 - a)(I_2 - I_3)}\right]^{1/2}$$

gives $\dot{u}^2 = ab^2(I_1 - a)(I_2 - I_3)/(I_1 I_2 I_3) = \mu^2$. We will need the following identities that define the functions cn u and dn u

$$\operatorname{cn}^2 u = 1 - \operatorname{sn}^2 u, \quad \operatorname{dn}^2 u = 1 - k^2 \operatorname{sn}^2 u, \quad \text{and} \quad \frac{d}{dx}\operatorname{sn} u = \operatorname{cn} u \operatorname{dn} u.$$

With initial condition $\Pi_2(0) = 0$, this gives

$$\Pi_2 = \alpha \operatorname{sn}(\mu t). \tag{15.8.11}$$

Thus, Π_2 varies between α and $-\alpha$. Choosing the time direction appropriately, we can assume without loss of generality that $\dot{\Pi}_2(0) > 0$. Note that Π_1 vanishes when Π_2 equals $\pm\alpha$ by (15.8.4) but that Π_3^2 attains its maximal value

$$\frac{I_3(I_1 - I_2)}{I_2(I_1 - I_3)}(\beta^2 - \alpha^2) = \frac{I_3(I_2 - a)ab^2}{(I_2 - I_3)} \tag{15.8.12}$$

by (15.8.5). The minimal value of Π_3^2 occurs when $\Pi_2 = 0$. This minimal value is

$$\frac{I_3(I_1 - I_2)}{I_2(I_1 - I_3)}\beta^2 = \frac{I_3(I_1 - a)ab^2}{(I_1 - I_3)} =: \delta^2, \tag{15.8.13}$$

again by (15.8.5). Thus the sign of Π_3 is constant throughout the motion. Let us assume that it is positive. This hypothesis together with $\dot{\Pi}_2(0) > 0$ and $a_2 < 0$ imply that $\Pi_1(0) < 0$.

Solving for Π_1 and Π_3 from (15.8.2) and (15.8.3) and remembering that $\Pi_1(0) < 0$ gives $\Pi_1(t) = -\gamma \operatorname{cn}(\mu t)$, $\Pi_3(t) = \delta \operatorname{dn}(\mu t)$, where δ is given by (15.8.13) and

$$\gamma^2 = \frac{I_1(a - I_3)ab^2}{(I_1 - I_3)}. \tag{15.8.14}$$

Note that $\beta > \alpha > \gamma$, and as usual, the values of γ and δ are taken to be positive. The solution of the Euler equations is therefore

$$\Pi_1(t) = -\gamma\,\mathrm{cn}(\mu t), \quad \Pi_2(t) = \alpha\,\mathrm{sn}(\mu t), \quad \Pi_3(t) = \delta\,\mathrm{dn}(\mu t), \qquad (15.8.15)$$

with α, γ, δ given by (15.8.6), (15.8.13), (15.8.14). If κ denotes the period invariant of Jacobi's elliptic functions, then Π_1 and Π_2 have period $4\kappa/\mu$, whereas Π_3 has period $2\kappa/\mu$.

Exercises

$\diamond$ **15.8-1.** Continue this integration process and find formulas for the attitude matrix $A(t)$ as functions of time with $A(0) = $ Identity and with given
body angular momentum (or velocity).

15.9 Rigid-Body Stability

Following the energy–Casimir method step by step (see the Introduction),
we begin with the equations

$$\dot{\Pi} = \frac{d\Pi}{dt} = \Pi \times \Omega, \qquad (15.9.1)$$

where $\Pi, \Omega \in \mathbb{R}^3, \Omega$ is the angular velocity, and Π is the angular momentum, both viewed in the body; the relation between Π and Ω is given by
$\Pi_j = I_j\Omega^j$, $j = 1, 2, 3$, where $I = (I_1, I_2, I_3)$ is the diagonalized moment
of inertia tensor, $I_1, I_2, I_3 > 0$. This system is Hamiltonian in the Lie–
Poisson structure of $\mathbb{R}^3$ given by (15.3.8) and relative to the kinetic energy
Hamiltonian

$$H(\Pi) = \frac{1}{2}\Pi \cdot \Omega = \frac{1}{2}\sum_{i=1}^{3}\frac{\Pi_i^2}{I_i}. \qquad (15.9.2)$$

Recall from (15.3.12) that for a smooth function $\Phi : \mathbb{R} \to \mathbb{R}$,

$$C_\Phi(\Pi) = \Phi\left(\frac{1}{2}\|\Pi\|^2\right) \qquad (15.9.3)$$

is a Casimir function.

1. First Variation. We find a Casimir function C_Φ such that $H_{C_\Phi} :=
H + C_\Phi$ has a critical point at a given equilibrium point of (15.9.1). Such
points occur when Π is parallel to Ω. We can assume without loss of generality that Π and Ω point in the Ox-direction. After normalizing if necessary, we can assume that the equilibrium solution is $\Pi_e = (1, 0, 0)$. The

derivative of

$$H_{C_\Phi}(\Pi) = \frac{1}{2}\sum_{i=1}^{3}\frac{\Pi_i^2}{I_i} + \Phi\left(\frac{1}{2}\|\Pi\|^2\right)$$

is

$$DH_{C_\Phi}(\Pi) \cdot \delta\Pi = \left(\Omega + \Phi'\left(\frac{1}{2}\|\Pi\|^2\right)\Pi\right) \cdot \delta\Pi. \tag{15.9.4}$$

This equals zero at $\Pi_e = (1,0,0)$, provided that

$$\Phi'\left(\frac{1}{2}\right) = -\frac{1}{I_1}. \tag{15.9.5}$$

2. Second Variation.   Using (15.9.4), the second derivative of H_{C_Φ} at the equilibrium $\Pi_e = (1,0,0)$ is

$$D^2 H_{C_\Phi}(\Pi_e) \cdot (\delta\Pi, \delta\Pi)$$

$$= \delta\Omega \cdot \delta\Pi + \Phi'\left(\frac{1}{2}\|\Pi_e\|^2\right)\|\delta\Pi\|^2 + (\Pi_e \cdot \delta\Pi)^2\Phi''\left(\frac{1}{2}\|\Pi_e\|^2\right)$$

$$= \sum_{i=1}^{3}\frac{(\delta\Pi_i)^2}{I_i} - \frac{\|\delta\Pi\|^2}{I_1} + \Phi''\left(\frac{1}{2}\right)(\delta\Pi_1)^2$$

$$= \left(\frac{1}{I_2} - \frac{1}{I_1}\right)(\delta\Pi_2)^2 + \left(\frac{1}{I_3} - \frac{1}{I_1}\right)(\delta\Pi_3)^2 + \Phi''\left(\frac{1}{2}\right)(\delta\Pi_1)^2. \tag{15.9.6}$$

3. Definiteness.   This quadratic form is positive definite if and only if

$$\Phi''\left(\frac{1}{2}\right) > 0 \tag{15.9.7}$$

and

$$I_1 > I_2, \quad I_1 > I_3. \tag{15.9.8}$$

Consequently,

$$\Phi(x) = -\frac{1}{I_1}x + \left(x - \frac{1}{2}\right)^2$$

satisfies (15.9.5) and makes the second derivative of H_{C_Φ} at $(1,0,0)$ positive definite, so *stationary rotation around the shortest axis is (Liapunov) stable.*

The quadratic form is negative definite, provided that

$$\Phi''\left(\frac{1}{2}\right) < 0 \tag{15.9.9}$$

and

$$I_1 < I_2, \quad I_1 < I_3. \tag{15.9.10}$$

It is obvious that we may find a function Φ satisfying the requirements (15.9.5) and (15.9.9); for example, $\Phi(x) = -(1/I_1)x - (x - \frac{1}{2})^2$. This proves that *rotation around the long axis is (Liapunov) stable*.

Finally, the quadratic form (15.9.6) is indefinite if

$$I_1 > I_2, \quad I_3 > I_1, \tag{15.9.11}$$

or with the inequalities reversed. We cannot show by this method that rotation around the middle axis is *unstable*. We shall prove, by using a spectral analysis, that rotation about the middle axis is, in fact, unstable. Linearizing (15.9.1) at $\mathbf{\Pi}_e = (1, 0, 0)$ yields the linear constant-coefficient system

$$
(\delta\dot{\Pi}) = \delta\Pi \times \Omega_e + \Pi_e \times \delta\Omega
$$
$$
= \left(0, \frac{I_3 - I_1}{I_3 I_1} \delta\Pi_3, \frac{I_1 - I_2}{I_1 I_2} \delta\Pi_2 \right)
$$
$$
= \begin{bmatrix} 0 & 0 & 0 \\ 0 & 0 & \dfrac{I_3 - I_1}{I_3 I_1} \\ 0 & \dfrac{I_1 - I_2}{I_1 I_2} & 0 \end{bmatrix} \delta\Pi. \tag{15.9.12}
$$

On the tangent space at $\mathbf{\Pi}_e$ to the sphere of radius $\|\mathbf{\Pi}_e\| = 1$, the linear operator given by this linearized vector field has a matrix given by the lower right 2×2 block whose eigenvalues are

$$\pm \frac{1}{I_1\sqrt{I_2 I_3}} \sqrt{(I_1 - I_2)(I_3 - I_1)}.$$

Both of them are real by (15.9.11), and one is strictly positive. Thus $\mathbf{\Pi}_e$ is spectrally unstable and thus is unstable.

We summarize the results in the following theorem.

Theorem 15.9.1 (Rigid-Body Stability Theorem). *In the motion of a free rigid body, rotation around the long and short axes is (Liapunov) stable and around the middle axis is unstable.*

It is important to keep the Casimir functions as general as possible, because otherwise (15.9.5) and (15.9.9) could be contradictory. Had we simply chosen

$$\Phi(x) = -\frac{1}{I_1}x + \left(x - \frac{1}{2}\right)^2,$$

(15.9.5) would be satisfied, but (15.9.9) would not. It is only the choice of *two different* Casimirs that enables us to prove the two stability results, even though the level surfaces of these Casimirs are the same.

Remarks.

1. As we have seen, rotations about the intermediate axis are unstable, and this is even for the linearized equations. The unstable homoclinic orbits that connect the two unstable points have interesting features. Not only are they interesting because of the chaotic solutions via the Poincaré–Melnikov method that can be obtained in various perturbed systems (see Holmes and Marsden [1983], Wiggins [1988], and references therein), but already the orbit itself is interesting, since a rigid body tossed about its middle axis will undergo an interesting half twist when the opposite saddle point is reached, even though the rotation axis has returned to where it was. The reader can easily perform the experiment; see Ashbaugh, Chicone, and Cushman [1990] and Montgomery [1991a] for more information.

2. The same stability theorem can also be proved by working with the second derivative along a coadjoint orbit in $\mathbb{R}^3$, that is, a two-sphere; see Arnold [1966a]. This coadjoint orbit method also *suggests* instability of rotation around the intermediate axis.

3. Dynamic stability on the Π-sphere has been shown. What about the stability of the dynamical rigid body we "see"? This can be deduced from what we have done. Probably the best approach, though, is to use the relation between the reduced and unreduced dynamics; see Simo, Lewis, and Marsden [1991] and Lewis [1992] for more information.

4. When the body angular momentum undergoes a periodic motion, the actual motion of the rigid body in space is not periodic. In the Introduction we described the associated geometric phase.

5. See Lewis and Simo [1990] and Simo, Lewis, and Marsden [1991] for related work on deformable elastic bodies (pseudo-rigid bodies).   ♦

Exercises

◊ **15.9-1.** Let **B** be a given fixed vector in $\mathbb{R}^3$ and let M evolve by $\dot{M} = M \times B$. Show that this evolution is Hamiltonian. Determine the equilibria and their stability.

◊ **15.9-2** (Double bracket dissipation).   Consider the following modification of the Euler equations:

$$\dot{\Pi} = \Pi \times \Omega + \alpha \Pi \times (\Pi \times \Omega),$$

where α is a positive constant. Show that:

(a) The spheres $\|\Pi\|^2$ are preserved.

(b) Energy is strictly decreasing except at equilibria.

(c) The equations can be written in the form

$$\dot{F} = \{F, H\}_{\mathrm{rb}} + \{F, H\}_{\mathrm{sym}},$$

where the first bracket is the usual rigid-body bracket and the second is the *symmetric* bracket

$$\{F, K\}_{\mathrm{sym}} = \alpha(\Pi \times \nabla F) \cdot (\Pi \times \nabla K).$$

15.10 Heavy Top Stability

The heavy top equations are

$$\frac{d\Pi}{dt} = \Pi \times \Omega + Mgl\Gamma \times \chi, \qquad (15.10.1)$$

$$\frac{d\Gamma}{dt} = \Gamma \times \Omega, \qquad (15.10.2)$$

where $\Pi, \Gamma, \chi \in \mathbb{R}^3$. Here Π and Ω are the angular momentum and angular velocity in the body, $\Pi_i = I_i\Omega^i$, $I_i > 0$, $i = 1, 2, 3$, with $I = (I_1, I_2, I_3)$ the moment of inertia tensor. The vector Γ represents the motion of the unit vector along the Oz-axis as seen from the body, and the constant vector χ is the unit vector along the line segment of length l connecting the fixed point to the center of mass of the body; M is the total mass of the body, and g is the strength of the gravitational acceleration, which is along Oz pointing down.

This system is Hamiltonian in the Lie–Poisson structure of $\mathbb{R}^3 \times \mathbb{R}^3$ given in the Introduction relative to the heavy top Hamiltonian

$$H(\Pi, \Gamma) = \frac{1}{2}\Pi \cdot \Omega + Mgl\Gamma \cdot \chi. \qquad (15.10.3)$$

The Poisson structure (with $\|\Pi\| = 1$ imposed) foreshadows that of

$$T^* SO(3)/S^1,$$

where S^1 acts by rotation about the axis of gravity. The fact that one gets the Lie–Poisson bracket for a semidirect product Lie algebra is a special case of the general theory of reduction and semidirect products (Marsden, Ratiu, and Weinstein [1984a, 1984b]).

The functions $\Pi \cdot \Gamma$ and $\|\Gamma\|^2$ are Casimir functions, as is

$$C(\Pi, \Gamma) = \Phi(\Pi \cdot \Gamma, \|\Gamma\|^2), \qquad (15.10.4)$$

where Φ is any smooth function from $\mathbb{R}^2$ to $\mathbb{R}$.

We shall be concerned here with the Lagrange top. This is a heavy top for which $I_1 = I_2$, that is, it is symmetric, and the center of mass lies on the axis of symmetry in the body, that is, $\boldsymbol{\chi} = (0, 0, 1)$. This assumption simplifies the equations of motion (15.10.1) to

$$\dot{\Pi}_1 = \frac{I_2 - I_3}{I_2 I_3} \Pi_2 \Pi_3 + Mgl\Gamma_2,$$

$$\dot{\Pi}_2 = \frac{I_3 - I_1}{I_1 I_3} \Pi_1 \Pi_3 - Mgl\Gamma_1,$$

$$\dot{\Pi}_3 = \frac{I_1 - I_2}{I_1 I_2} \Pi_1 \Pi_2.$$

Since $I_1 = I_2$, we have $\dot{\Pi}_3 = 0$; thus Π_3, and hence any function $\varphi(\Pi_3)$ of Π_3 is conserved.

1. First Variation. We shall study the equilibrium solution

$$\boldsymbol{\Pi}_e = (0, 0, \Pi_3^0), \quad \boldsymbol{\Gamma}_e = (0, 0, 1),$$

where $\Pi_3^0 \neq 0$, which represents the spinning of a symmetric top in its upright position. To begin, we consider conserved quantities of the form $H_{\Phi,\varphi} = H + \Phi(\boldsymbol{\Pi} \cdot \boldsymbol{\Gamma}, \|\boldsymbol{\Gamma}\|^2) + \varphi(\Pi_3)$ that have a critical point at the equilibrium. The first derivative of $H_{\Phi,\varphi}$ is given by

$$\begin{aligned}
\mathbf{D}H_{\Phi,\varphi}(\boldsymbol{\Pi}, \boldsymbol{\Gamma}) \cdot (\delta\boldsymbol{\Pi}, \delta\boldsymbol{\Gamma}) = &(\boldsymbol{\Omega} + \dot{\Phi}(\boldsymbol{\Pi} \cdot \boldsymbol{\Gamma}, \|\boldsymbol{\Gamma}\|^2)\boldsymbol{\Gamma}) \cdot \delta\boldsymbol{\Pi} \\
&+ [Mgl\boldsymbol{\chi} + \dot{\Phi}(\boldsymbol{\Pi} \cdot \boldsymbol{\Gamma}, \|\boldsymbol{\Gamma}\|^2)\boldsymbol{\Pi} \\
&+ 2\Phi'(\boldsymbol{\Pi} \cdot \boldsymbol{\Gamma}, \|\boldsymbol{\Gamma}\|^2)\boldsymbol{\Gamma}] \cdot \delta\boldsymbol{\Gamma} + \varphi'(\Pi_3)\delta\Pi_3,
\end{aligned}$$

where $\dot{\Phi} = \partial\Phi/\partial(\boldsymbol{\Pi} \cdot \boldsymbol{\Gamma})$ and $\Phi' = \partial\Phi/\partial(\|\boldsymbol{\Gamma}\|^2)$. At the equilibrium solution $(\boldsymbol{\Pi}_e, \boldsymbol{\Gamma}_e)$ the first derivative of $H_{\Phi,\varphi}$ vanishes, provided that

$$\frac{\Pi_3^0}{I_3} + \dot{\Phi}(\Pi_3^0, 1) + \varphi'(\Pi_3^0) = 0$$

and that

$$Mgl + \dot{\Phi}(\Pi_3^0, 1)\Pi_3^0 + 2\Phi'(\Pi_3^0, 1) = 0;$$

the remaining equations, involving indices 1 and 2, are trivially satisfied. Solving for $\dot{\Phi}(\Pi_3^0, 1)$ and $\Phi'(\Pi_3^0, 1)$ we get the conditions

$$\dot{\Phi}(\Pi_3^0, 1) = -\left(\frac{1}{I_3} + \frac{\varphi'(\Pi_3^0)}{\Pi_3^0}\right) \Pi_3^0, \tag{15.10.5}$$

$$\Phi'(\Pi_3^0, 1) = \frac{1}{2}\left(\frac{1}{I_3} + \frac{\varphi'(\Pi_3^0)}{\Pi_3^0}\right)(\Pi_3^0)^2 - \frac{1}{2}Mgl. \tag{15.10.6}$$

2. Second Variation. We shall check for definiteness of the second variation of $H_{\Phi,\varphi}$ at the equilibrium point $(\mathbf{\Pi}_e, \mathbf{\Gamma}_e)$. To simplify the notation we shall set

$$a = \varphi''(\Pi_3^0), \qquad b = 4\Phi''(\Pi_3^0, 1),$$
$$c = \ddot{\Phi}(\Pi_3^0, 1), \qquad d = 2\dot{\Phi}'(\Pi_3^0, 1).$$

With this notation, the matrix of the second derivative at $(\mathbf{\Pi}_e, \mathbf{\Gamma}_e)$ is

$$
\begin{bmatrix}
1/I_1 & 0 & 0 & \dot{\Phi}(\Pi_3^0, 1) & 0 & 0 \\
0 & 1/I_1 & 0 & 0 & \dot{\Phi}(\Pi_3^0, 1) & 0 \\
0 & 0 & (1/I_3) + a + c & 0 & 0 & a_{36} \\
\dot{\Phi}(\Pi_3^0, 1) & 0 & 0 & 2\Phi'(\Pi_3^0, 1) & 0 & 0 \\
0 & \dot{\Phi}(\Pi_3^0, 1) & 0 & 0 & 2\Phi'(\Pi_3^0, 1) & 0 \\
0 & 0 & a_{36} & 0 & 0 & a_{66}
\end{bmatrix},
$$

$$(15.10.7)$$

where

$$a_{36} = \dot{\Phi}(\Pi_3^0, 1) + \Pi_3^0 c + d,$$
$$a_{66} = 2\Phi'(\Pi_3^0, 1) + b + (\Pi_3^0)^2 c + \Pi_3^0 d.$$

3. Definiteness. The computations for this part will be done using the following formula from linear algebra. If

$$
M = \begin{bmatrix} A & B \\ C & D \end{bmatrix}
$$

is a $(p+q) \times (p+q)$ matrix and if the $p \times p$ matrix A is invertible, then

$$\det M = \det A \, \det(D - CA^{-1}B).$$

If the quadratic form given by (15.10.7) is definite, it must be positive definite, since the $(1,1)$-entry is positive. Recalling that $I_1 = I_2$, the six principal determinants have the following values:

$$\frac{1}{I_1}, \quad \frac{1}{I_1^2}, \quad \frac{1}{I_1^2}\left(\frac{1}{I_3} + a + c\right),$$

$$\frac{1}{I_1}\left(\frac{1}{I_3} + a + c\right)\left(\frac{2}{I_1}\Phi'(\Pi_3^0, 1) - \dot{\Phi}(\Pi_3^0, 1)^2\right),$$

$$\left(\frac{2}{I_1}\Phi'(\Pi_3^0, 1) - \dot{\Phi}(\Pi_3^0, 1)^2\right)^2\left(\frac{1}{I_3} + a + c\right),$$

and

$$\left(\frac{2}{I_1}\Phi'(\Pi_3^0, 1) - \dot{\Phi}(\Pi_3^0, 1)^2\right)^2\left[a_{66}\left(\frac{1}{I_3} + a + c\right) - a_{36}^2\right].$$

Consequently, the quadratic form given by (15.10.7) is positive definite if and only if

$$\frac{1}{I_3} + a + c > 0, \tag{15.10.8}$$

$$\frac{2}{I_1} \Phi'(\Pi_3^0, 1) - \dot{\Phi}(\Pi_3^0, 1)^2 > 0, \tag{15.10.9}$$

and

$$a_{66} \left(\frac{1}{I_3} + a + c \right) - \left(\dot{\Phi}(\Pi_3^0, 1) + \Pi_3^0 c + d \right)^2 > 0. \tag{15.10.10}$$

Conditions (15.10.8) and (15.10.10) can always be satisfied if we choose the numbers a, b, c, and d appropriately; for example, $a = c = d = 0$ and b sufficiently large and positive. Thus, the determining condition for stability is (15.10.9). By (15.10.5) and (15.10.6), this becomes

$$\frac{1}{I_1} \left[\left(\frac{1}{I_3} + \frac{\varphi'(\Pi_3^0)}{\Pi_3^0} \right) (\Pi_3^0)^2 - Mgl \right] - \left(\frac{1}{I_3} + \frac{\varphi'(\Pi_3^0)}{\Pi_3^0} \right)^2 (\Pi_3^0)^2 > 0. \tag{15.10.11}$$

We can choose $\varphi'(\Pi_3^0)$ such that

$$\frac{1}{I_3} + \frac{\varphi'(\Pi_3^0)}{\Pi_3^0} = e$$

has any value we wish. The left side of (15.10.11) is a quadratic polynomial in e, whose leading coefficient is negative. In order for this to be positive for some e, it is necessary and sufficient for the discriminant

$$\frac{(\Pi_3^0)^4}{I_1^2} - \frac{4(\Pi_3^0)^2 Mgl}{I_1}$$

to be positive; that is, $(\Pi_3^0)^2 > 4MglI_1$, which is the classical stability condition for a fast top. We have proved the first part of the following:

Theorem 15.10.1 (Heavy Lagrange Top Stability Theorem). *An upright spinning Lagrange top is stable, provided that the angular velocity is strictly larger than $2\sqrt{MglI_1}/I_3$. It is unstable if the angular velocity is smaller than this value.*

The second part of the theorem is proved, as in §15.9, by a spectral analysis of the linearized equations, namely

$$(\delta \dot{\Pi}) = \delta \Pi \times \Omega + \Pi_e \times \delta \Omega + Mgl\delta \Gamma \times \chi, \tag{15.10.12}$$

$$(\delta \dot{\Gamma}) = \delta \Gamma \times \Omega + \Gamma_e \times \delta \Omega, \tag{15.10.13}$$

on the tangent space to the coadjoint orbit in $\mathfrak{se}(3)^*$ through $(\mathbf{\Pi}_e, \mathbf{\Gamma}_e)$ given by

$$\{ (\delta\mathbf{\Pi}, \delta\mathbf{\Gamma}) \in \mathbb{R}^3 \times \mathbb{R}^3 \mid \delta\mathbf{\Pi} \cdot \mathbf{\Gamma}_e + \mathbf{\Pi}_e \cdot \delta\mathbf{\Gamma} = 0 \text{ and } \delta\mathbf{\Gamma} \cdot \mathbf{\Gamma}_e = 0 \}$$
$$\cong \{ (\delta\Pi_1, \delta\Pi_2, \delta\Gamma_1, \delta\Gamma_2) \} = \mathbb{R}^4. \tag{15.10.14}$$

The matrix of the linearized system of equations on this space is computed to be

$$\begin{bmatrix} 0 & \dfrac{\Pi_3^0}{I_3} \dfrac{I_1 - I_3}{I_1} & 0 & Mgl \\[2ex] -\dfrac{\Pi_3^0}{I_3} \dfrac{I_1 - I_3}{I_1} & 0 & -Mgl & 0 \\[2ex] 0 & -\dfrac{1}{I_1} & 0 & \dfrac{\Pi_3^0}{I_3} \\[2ex] \dfrac{1}{I_1} & 0 & -\dfrac{\Pi_3^0}{I_3} & 0 \end{bmatrix}. \tag{15.10.15}$$

The matrix (15.10.15) has characteristic polynomial

$$\lambda^4 + \frac{1}{I_1^2} \left[(I_1^2 + (I_1 - I_3)^2) \left(\frac{\Pi_3^0}{I_3} \right)^2 - 2MglI_1 \right] \lambda^2$$
$$+ \frac{1}{I_1^2} \left[(I_1 - I_3) \left(\frac{\Pi_3^0}{I_3} \right)^2 + Mgl \right]^2, \tag{15.10.16}$$

whose discriminant as a quadratic polynomial in λ^2 is

$$\frac{1}{I_1^4} (2I_1 - I_3)^2 \left(\frac{\Pi_3^0}{I_3} \right)^2 \left(I_3^2 \left(\frac{\Pi_3^0}{I_3} \right)^2 - 4MglI_1 \right).$$

This discriminant is negative if and only if $\Pi_3^0 < 2\sqrt{MglI_1}$. Under this condition the four roots of the characteristic polynomial are all distinct and equal to

$$\lambda_0, \quad \bar{\lambda}_0, \quad -\lambda_0, \quad -\bar{\lambda}_0 \quad \text{for some } \lambda_0 \in \mathbb{C},$$

where

$$\operatorname{Re} \lambda_0 \neq 0 \quad \text{and} \quad \operatorname{Im} \lambda_0 \neq 0.$$

Thus, at least two of these roots have real part strictly larger than zero, thereby showing that $(\mathbf{\Pi}_e, \mathbf{\Gamma}_e)$ is spectrally unstable and hence unstable.

When $I_2 = I_1 + \epsilon$ for small ϵ, the conserved quantity $\varphi(\Pi_3)$ is no longer available. In this case, a sufficiently fast top is still linearly stable, and nonlinear stability can be assessed by KAM theory. Other regions of phase-space are known to possess chaotic dynamics in this case (Holmes and Marsden [1983]). For more information on stability and bifurcation in the heavy top, we refer to Lewis, Ratiu, Simo, and Marsden [1992].

Exercises

◇ **15.10-1.**

(a) Show that $\tilde{H}(\mathbf{\Pi}, \mathbf{\Gamma}) = H(\mathbf{\Pi}, \mathbf{\Gamma}) + \|\mathbf{\Gamma}\|^2/2$, where H is given by (15.10.3), generates the same equations of motion (15.10.1) and (15.10.2).

(b) Taking the Legendre transform of $\tilde{H}$, show that the equations can be written in Euler–Poincaré form.

15.11 The Rigid Body and the Pendulum

This section, following Holm and Marsden [1991], shows how the rigid body and the pendulum are linked.

Euler's equations are expressible in vector form as

$$\frac{d}{dt}\mathbf{\Pi} = \nabla L \times \nabla H, \tag{15.11.1}$$

where H is the energy

$$H = \frac{\Pi_1^2}{2I_1} + \frac{\Pi_2^2}{2I_2} + \frac{\Pi_3^2}{2I_3}, \tag{15.11.2}$$

$$\nabla H = \left(\frac{\partial H}{\partial \Pi_1}, \frac{\partial H}{\partial \Pi_2}, \frac{\partial H}{\partial \Pi_3} \right) = \left(\frac{\Pi_1}{I_1}, \frac{\Pi_2}{I_2}, \frac{\Pi_3}{I_3} \right) \tag{15.11.3}$$

is the gradient of H, and L is the square of the body angular momentum

$$L = \frac{1}{2} \left(\Pi_1^2 + \Pi_2^2 + \Pi_3^2 \right). \tag{15.11.4}$$

Since both H and L are conserved, the rigid-body motion itself takes place, as we know, along the intersections of the level surfaces of the energy (ellipsoids) and the angular momentum (spheres) in $\mathbb{R}^3$. The centers of the energy ellipsoids and the angular momentum spheres coincide. This, along with the $(\mathbb{Z}_2)^3$ symmetry of the energy ellipsoids, implies that the two sets of level surfaces in $\mathbb{R}^3$ develop collinear gradients (for example, tangencies) at pairs of points that are diametrically opposite on an angular momentum sphere. At these points, collinearity of the gradients of H and L implies stationary rotations, that is, equilibria.

Euler's equations for the rigid body may also be written as

$$\frac{d}{dt}\mathbf{\Pi} = \nabla N \times \nabla K, \tag{15.11.5}$$

where K and N are linear combinations of energy and angular momentum of the form

$$\begin{pmatrix} N \\ K \end{pmatrix} = \begin{bmatrix} a & b \\ c & d \end{bmatrix} \begin{pmatrix} H \\ L \end{pmatrix}, \qquad (15.11.6)$$

with real constants a, b, c, and d satisfying $ad - bc = 1$. To see this, recall that

$$K = \frac{1}{2}\left(\frac{c}{I_1} + d\right)\Pi_1^2 + \frac{1}{2}\left(\frac{c}{I_2} + d\right)\Pi_2^2 + \frac{1}{2}\left(\frac{c}{I_3} + d\right)\Pi_3^2.$$

Thus, if $I_1 = I_2 = I_3$, the choice $c = -dI_1$ yields $K = 0$, and so equation (15.11.5) becomes $\dot{\Pi} = 0$ for any choice of N, which is precisely the equation $\dot{\Pi} = \Pi \times \Omega$, for $I_1 = I_2 = I_3$. If $I_1 \neq I_2 = I_3$, the choice $c = -dI_2$, $d \neq 0$, yields

$$K = \frac{d}{2}\left(1 - \frac{I_2}{I_1}\right)\Pi_1^2.$$

If one now takes

$$N = \frac{I_1}{2I_2 d}\left(\Pi_2^2 + \Pi_3^2\right),$$

then equation (15.11.5) becomes the rigid-body equation $\dot{\Pi} = \Pi \times \Omega$. Finally, if $I_1 < I_2 < I_3$, the choice

$$c = 1, \quad d = -\frac{1}{I_3}, \quad a = -\frac{I_1 I_3}{I_3 - I_1} < 0, \quad \text{and} \quad b = \frac{I_3}{I_3 - I_1} < 0 \quad (15.11.7)$$

gives

$$K = \frac{1}{2}\left(\frac{1}{I_1} - \frac{1}{I_3}\right)\Pi_1^2 + \frac{1}{2}\left(\frac{1}{I_2} - \frac{1}{I_3}\right)\Pi_2^2 \qquad (15.11.8)$$

and

$$N = \frac{I_3(I_2 - I_1)}{2I_2(I_3 - I_1)}\Pi_2^2 + \frac{1}{2}\Pi_3^2. \qquad (15.11.9)$$

Then equations (15.11.5) are the rigid-body equation $\dot{\Pi} = \Pi \times \Omega$.

With this choice, the orbits for Euler's equations for rigid-body dynamics are realized as motion along the intersections of two, orthogonally oriented, *elliptic cylinders*, one elliptic cylinder being a level surface of K, with its translation axis along Π_3 (where $K = 0$), and the other a level surface of N, with its translation axis along Π_1 (where $N = 0$).

For a general choice of K and N, equilibria occur at points where the gradients of K and N are collinear. This can occur at points where the level sets are tangent (and the gradients both are nonzero) or at points where one of the gradients vanishes. In the elliptic cylinder case above, these two cases are points where the elliptic cylinders are tangent, and points where the axis of one cylinder punctures normally through the surface of the other. The elliptic cylinders are tangent at one $\mathbb{Z}_2$-symmetric pair of points along the Π_2 axis, and the elliptic cylinders have normal axial punctures at two other $\mathbb{Z}_2$-symmetric pairs of points along the Π_1 and Π_3 axes.

Let us pursue the elliptic cylinders point of view further. We now change variables in the rigid-body equations within a level surface of K. To simplify notation, we first define the three positive constants k_i^2, $i = 1, 2, 3$, by setting in (15.11.8) and (15.11.9)

$$K = \frac{\Pi_1^2}{2k_1^2} + \frac{\Pi_2^2}{2k_2^2} \quad \text{and} \quad N = \frac{\Pi_2^2}{2k_3^2} + \frac{1}{2}\Pi_3^2 \tag{15.11.10}$$

for

$$\frac{1}{k_1^2} = \frac{1}{I_1} - \frac{1}{I_3}, \quad \frac{1}{k_2^2} = \frac{1}{I_2} - \frac{1}{I_3}, \quad \frac{1}{k_3^2} = \frac{I_3(I_2 - I_1)}{I_2(I_3 - I_1)}. \tag{15.11.11}$$

On the surface $K = $ constant, and setting $r = \sqrt{2K} = $ constant, define new variables θ and p by

$$\Pi_1 = k_1 r \cos\theta, \quad \Pi_2 = k_2 r \sin\theta, \quad \Pi_3 = p. \tag{15.11.12}$$

In terms of these variables, the constants of the motion become

$$K = \frac{1}{2}r^2 \quad \text{and} \quad N = \frac{1}{2}p^2 + \left(\frac{k_2^2}{2k_3^2}r^2 \right) \sin^2\theta. \tag{15.11.13}$$

From Exercise 1.3-2 it follows that

$$\{F_1, F_2\}_K = -\nabla K \cdot (\nabla F_1 \times \nabla F_2) \tag{15.11.14}$$

is a Poisson bracket on $\mathbb{R}^3$ having K as a Casimir function. One can now verify directly that the symplectic structure on the leaf $K = $ constant is given by the following Poisson bracket on this elliptic cylinder (see Exercise 15.11-1):

$$\{F, G\}_{\text{EllipCyl}} = \frac{1}{k_1 k_2} \left(\frac{\partial F}{\partial p}\frac{\partial G}{\partial \theta} - \frac{\partial F}{\partial \theta}\frac{\partial G}{\partial p} \right). \tag{15.11.15}$$

In particular,

$$\{p, \theta\}_{\text{EllipCyl}} = \frac{1}{k_1 k_2}. \tag{15.11.16}$$

The restriction of the Hamiltonian H to the elliptic cylinder $K = $ constant is by (15.11.3)

$$H = \frac{k_1^2 K}{I_1} + \frac{1}{I_3}\left[\frac{1}{2}p^2 + \frac{I_3^2(I_2 - I_1)}{2(I_3 - I_2)(I_3 - I_1)}r^2 \sin^2\theta \right] = \frac{k_1^2 K}{I_1} + \frac{1}{I_3}N,$$

that is, N/I_3 can be taken as the Hamiltonian on this symplectic leaf. Note that N/I_3 has the form of kinetic plus potential energy. The equations of motion are thus given by

$$\frac{d}{dt}\theta = \left\{\theta, \frac{N}{I_3}\right\}_{\text{EllipCyl}} = \frac{1}{k_1 k_2 I_3}\frac{\partial N}{\partial p} = -\frac{1}{k_1 k_2 I_3}p, \qquad (15.11.17)$$

$$\frac{d}{dt}p = \left\{p, \frac{N}{I_3}\right\}_{\text{EllipCyl}} = \frac{1}{k_1 k_2 I_3}\frac{\partial N}{\partial \theta} = \frac{1}{k_1 k_2 I_3}\frac{k_2^2}{k_3^2}r^2 \sin\theta\cos\theta. \quad (15.11.18)$$

Combining these equations of motion gives

$$\frac{d^2\theta}{dt^2} = -\frac{r^2}{2k_1^2 k_3^2 I_3^2}\sin 2\theta, \qquad (15.11.19)$$

or, in terms of the original rigid-body parameters,

$$\frac{d^2}{dt^2}\theta = -\frac{K}{I_3^2}\left(\frac{1}{I_1} - \frac{1}{I_2}\right)\sin 2\theta. \qquad (15.11.20)$$

Thus, we have proved the following result:

Proposition 15.11.1. *Rigid-body motion reduces to pendulum motion on level surfaces of K.*

Another way of saying this is as follows: Regard rigid-body angular momentum space as the union of the level surfaces of K, so the dynamics of the rigid body are recovered by looking at the dynamics on each of these level surfaces. On each level surface, the dynamics are equivalent to a simple pendulum. In this sense, we have proved the following:

Corollary 15.11.2. *The dynamics of a rigid body in three-dimensional body angular momentum space are a union of two-dimensional simple pendula phase portraits.*

By restricting to a nonzero level surface of K, the pair of rigid-body equilibria along the Π_3 axis are excluded. (This pair of equilibria can be included by permuting the indices of the moments of inertia.) The other two pairs of equilibria, along the Π_1 and Π_2 axes, lie in the $p = 0$ plane at $\theta = 0$, $\pi/2$, π, and $3\pi/2$. Since K is positive, the stability of each equilibrium point is determined by the relative sizes of the principal moments of inertia, which affect the overall sign of the right-hand side of the pendulum equation. The well-known results about stability of equilibrium rotations along the least and greatest principal axes, and instability around the intermediate axis, are immediately recovered from this overall sign, combined with the stability properties of the pendulum equilibria. For $K > 0$ and $I_1 < I_2 < I_3$, this overall sign is negative, so the equilibria at $\theta = 0$ and π (along the Π_1 axis) are stable, while those at $\theta = \pi/2$ and $3\pi/2$ (along the Π_2 axis) are unstable. The factor of 2 in the argument of the sine in the pendulum equation is explained by the $\mathbb{Z}_2$ symmetry of the level surfaces of K (or, just as well, by their invariance under $\theta \mapsto \theta + \pi$). Under this discrete symmetry operation, the equilibria at $\theta = 0$ and $\pi/2$ exchange with their counterparts at $\theta = \pi$ and $3\pi/2$, respectively, while the elliptical level surface of K is left invariant. By construction, the Hamiltonian N/I_3 in the reduced variables θ and p is also invariant under this discrete symmetry.

The rigid body can, correspondingly, be regarded as a left-invariant system on the group $O(K)$ or SE(2). The special case of SE(2) is the one in which the orbits are cotangent bundles. The fact that one gets a cotangent bundle in this situation is a special case of the cotangent bundle reduction theorem using the semidirect product reduction theorem; see Marsden, Ratiu, and Weinstein [1984a, 1984b]. For the Euclidean group this theorem says that the coadjoint orbits of the Euclidean group of the plane are given by reducing the cotangent bundle of the rotation group of the plane by the trivial group, giving the cotangent bundle of a circle with its canonical symplectic structure up to a factor. This is the abstract explanation of why, in the elliptic cylinder case above, the variables θ and p were, up to a factor, canonically conjugate. This general theory is also consistent with the fact that the Hamiltonian N/I_3 is of the form kinetic plus potential energy. In fact, in the cotangent bundle reduction theorem, one always gets a Hamiltonian of this form, with the potential being changed by the addition of an amendment to give the *amended potential*. In the case of the pendulum equation, the original Hamiltonian is purely kinetic energy, and so the potential term in N/I_3, namely $(k_2^2 r^2/(2k_3^2 I_3)) \sin^2 \theta$, is entirely amendment.

Putting the above discussion together with Exercises 14.7-1 and 14.7-2, one gets the following theorem.

Theorem 15.11.3. *Euler's equations for a free rigid body are Lie–Poisson with the Hamiltonian N for the Lie algebra $\mathbb{R}_K^3$ where the underlying Lie group is the orthogonal group of K if the quadratic form is nondegenerate, and is the Euclidean group of the plane if K has signature $(+, +, 0)$. In particular, all the groups SO(3), SO(2, 1), and SE(2) occur as the parameters a, b, c, and d are varied. (If the body is a Lagrange body, then the Heisenberg group occurs as well.)*

The same richness of Hamiltonian structure was found in the Maxwell–Bloch system in David and Holm [1992] (see also David, Holm, and Tratnik [1990]). As in the case of the rigid body, the $\mathbb{R}^3$ motion for the Maxwell–Bloch system may also be realized as motion along the intersections of two orthogonally oriented cylinders. However, in this case, one cylinder is parabolic in cross section, while the other is circular. Upon passing to parabolic cylindrical coordinates, the Maxwell–Bloch system reduces to the ideal Duffing equation, while in circular cylindrical coordinates, the pendulum equation results. The SL(2, ℝ) matrix transformation in the Maxwell–

Bloch case provides a parametrized array of (offset) ellipsoids, hyperboloids, and cylinders, along whose intersections the $\mathbb{R}^3$ motion takes place.

Exercises

⋄ **15.11-1.**   Consider the Poisson bracket on $\mathbb{R}^3$ given by

$$\{F_1, F_2\}_K(\mathbf{\Pi}) = -\nabla K(\mathbf{\Pi}) \cdot (\nabla F_1(\mathbf{\Pi}) \times (\nabla F_2(\mathbf{\Pi}))$$

with

$$K(\mathbf{\Pi}) = \frac{\Pi_1^2}{2k_1^2} + \frac{\Pi_2^2}{2k_2^2}.$$

Verify that the Poisson bracket on the two-dimensional leaves of this bracket given by $K = $ constant has the expression

$$\{\theta, p\}_{\text{Ellip Cyl}} = -\frac{1}{k_1 k_2},$$

where $p = \Pi_3$ and $\theta = \tan^{-1}(k_1\Pi_2/(k_2\Pi_1))$. What is the symplectic form on these leaves?

Bloch case provides a parametrized array of (offset) ellipsoids, hyperboloids and cylinders, along whose intersections the R^3 motion takes place.

Exercises

15.11-1.   Consider the Poisson bracket on R^3 given by

$$\{F, K\}k(\Pi) = -\nabla K(\Pi) \cdot (\nabla H(\Pi) \times \nabla H(\Pi))$$

with

$$K(\Pi) = \frac{\Pi_1^2}{2k_1} + \frac{\Pi_2^2}{2k_2}$$

Verify that the Poisson bracket on the two-dimensional leaves of this bracket given by K = constant has the expression

$$\{f, g\}(\phi, p) = \frac{1}{k_1 k_2}$$

where $p = \Pi_3$ and $\theta = \tan^{-1}(k_1 \Pi_2 / (k_2 \Pi_1))$. What is the symplectic form on these leaves?

References

Abarbanel, H. D. I. and D. D. Holm [1987] Nonlinear stability analysis of inviscid flows in three dimensions: incompressible fluids and barotropic fluids. *Phys. Fluids* **30**, 3369–3382.

Abarbanel, H. D. I., D. D. Holm, J. E. Marsden, and T. S. Ratiu [1986] Nonlinear stability analysis of stratified fluid equilibria. *Phil. Trans. Roy. Soc. London A* **318**, 349–409; also Richardson number criterion for the nonlinear stability of three-dimensional stratified flow. *Phys. Rev. Lett.* **52** [1984], 2552–2555.

Abraham, R. and J. E. Marsden [1978] *Foundations of Mechanics*. Second Edition, Addison-Wesley.

Abraham, R., J. E. Marsden, and T. S. Ratiu [1988] *Manifolds, Tensor Analysis, and Applications*. Second Edition, Applied Mathematical Sciences **75**, Springer-Verlag.

Adams, J. F. [1969] *Lectures on Lie groups*. Benjamin-Cummings, Reading, Mass.

Adams, J. F. [1996] *Lectures on Exceptional Lie groups*. University of Chicago Press.

Adams, M. R., J. Harnad, and E. Previato [1988] Isospectral Hamiltonian flows in finite and infinite dimensions I. Generalized Moser systems and moment maps into loop algebras. *Comm. Math. Phys.* **117**, 451–500.

Adams, M. R., T. S. Ratiu, and R. Schmid [1986a] A Lie group structure for pseudo-differential operators. *Math. Ann.* **273**, 529–551.

Adams, M. R., T. S. Ratiu, and R. Schmid [1986b] A Lie group structure for Fourier integral operators. *Math. Ann.* **276**, 19–41.

Adler, M. and P. van Moerbeke [1980a] Completely integrable systems, Euclidean Lie algebras and curves. *Adv. in Math.* **38**, 267–317.

Adler, M. and P. van Moerbeke [1980b] Linearization of Hamiltonian systems, Jacobi varieties and representation theory. *Adv. in Math.* **38**, 318–379.

Aeyels, D. and M. Szafranski [1988] Comments on the stabilizability of the angular velocity of a rigid body. *Systems Control Lett.* **10**, 35–39.

Aharonov, Y. and J. Anandan [1987] Phase change during acyclic quantum evolution. *Phys. Rev. Lett.* **58**, 1593–1596.

Alber, M. and J. E. Marsden [1992] On geometric phases for soliton equations. *Comm. Math. Phys.* **149**, 217–240.

Alber, M. S., R. Camassa, D. D. Holm, and J. E. Marsden [1994] The geometry of peaked solitons and billiard solutions of a class of integrable PDEs. *Lett. Math. Phys.* **32**, 137–151.

Alber, M. S., R. Camassa, D. D. Holm, and J. E. Marsden [1995] On the link between umbilic geodesics and soliton solutions of nonlinear PDEs. *Proc. Roy. Soc.* **450**, 677–692.

Alber, M. S., G. G. Luther, and J. E. Marsden [1997a] Energy Dependent Schrödinger Operators and Complex Hamiltonian Systems on Riemann Surfaces. *Nonlinearity* **10**, 223–242.

Alber, M. S., G. G. Luther, and J. E. Marsden [1997b] Complex billiard Hamiltonian systems and nonlinear waves, in: A.S. Fokas and I.M. Gelfand, eds., *Algebraic Aspects of Integrable Systems: In Memory of Irene Dorfman, Progress in Nonlinear Differential Equations* **26**, Birkhäuser, 1–15.

Alber, M. S., G. G. Luther, J. E. Marsden, and J. W. Robbins [1998] Geometric phases, reduction and Lie–Poisson structure for the resonant three-wave interaction. *Physica D* **123**, 271–290.

Altmann, S. L. [1986] *Rotations, Quaternions, and Double Groups*. Oxford University Press.

Anandan, J. [1988] Geometric angles in quantum and classical physics. *Phys. Lett.* **A 129**, 201–207.

Arens, R. [1970] A quantum dynamical, relativistically invariant rigid-body system. *Trans. Amer. Math. Soc.* **147**, 153–201.

Armero, F. and J. C. Simo [1996a] Long-Term Dissipativity of Time-Stepping Algorithms for an Abstract Evolution Equation with Applications to the Incompressible MHD and Navier–Stokes Equations. *Comp. Meth. Appl. Mech. Eng.* **131**, 41–90.

Armero, F. and J. C. Simo [1996b] Formulation of a new class of fraction-step methods for the incompressible MHD equations that retains the long-term dissipativity of the continuum dynamical system. *Fields Institute Comm.* **10**, 1–23.

Arms, J. M. [1981] The structure of the solution set for the Yang–Mills equations. *Math. Proc. Camb. Philos. Soc.* **90**, 361–372.

Arms, J. M., R. H. Cushman, and M. Gotay [1991] A universal reduction procedure for Hamiltonian group actions. *The Geometry of Hamiltonian systems*, T. Ratiu, ed., MSRI Series **22**, Springer-Verlag, 33–52.

Arms, J. M., A. Fischer, and J. E. Marsden [1975] Une approche symplectique pour des théorèmes de décomposition en géométrie ou relativité générale. *C. R. Acad. Sci. Paris* **281**, 517–520.

Arms, J. M., J. E. Marsden, and V. Moncrief [1981] Symmetry and bifurcations of momentum mappings. *Comm. Math. Phys.* **78**, 455–478.

Arms, J. M., J. E. Marsden, and V. Moncrief [1982] The structure of the space solutions of Einstein's equations: II Several Killings fields and the Einstein–Yang–Mills equations. *Ann. of Phys.* **144**, 81–106.

Arnold, V. I. [1964] Instability of dynamical systems with several degrees of freedom. *Dokl. Akad. Nauk SSSR* **156**, 9–12.

Arnold, V. I. [1965a] Sur une propriété topologique des applications globalement canoniques de la mécanique classique. *C.R. Acad. Sci. Paris* **26**, 3719–3722.

Arnold, V. I. [1965b] Conditions for nonlinear stability of the stationary plane curvilinear flows of an ideal fluid. *Dokl. Mat. Nauk SSSR* **162**, 773–777.

Arnold, V. I. [1965c] Variational principle for three-dimensional steady-state flows of an ideal fluid. *J. Appl. Math. Mech.* **29**, 1002–1008.

Arnold, V. I. [1966a] Sur la géometrie differentielle des groupes de Lie de dimension infinie et ses applications à l'hydrodynamique des fluids parfaits. *Ann. Inst. Fourier, Grenoble* **16**, 319–361.

Arnold, V. I. [1966b] On an a priori estimate in the theory of hydrodynamical stability. *Izv. Vyssh. Uchebn. Zaved. Mat. Nauk* **54**, 3–5; English Translation: *Amer. Math. Soc. Transl.* **79** [1969], 267–269.

Arnold, V. I. [1966c] Sur un principe variationnel pour les découlements stationaires des liquides parfaits et ses applications aux problèmes de stabilité non linéaires. *J. Mécanique* **5**, 29–43.

Arnold, V. I. [1967] Characteristic class entering in conditions of quantization. *Funct. Anal. Appl.* **1**, 1–13.

Arnold, V. I. [1968] Singularities of differential mappings. *Russian Math. Surveys* **23**, 1–43.

Arnold, V. I. [1969] Hamiltonian character of the Euler equations of the dynamics of solids and of an ideal fluid. *Uspekhi Mat. Nauk* **24**, 225–226.

Arnold, V. I. [1972] Note on the behavior of flows of a three dimensional ideal fluid under a small perturbation of the initial velocity field. *Appl. Math. Mech.* **36**, 255–262.

Arnold, V. I. [1984] *Catastrophe Theory.* Springer-Verlag.

Arnold, V. I. [1988] *Dynamical Systems III.* Encyclopedia of Mathematics **3**, Springer-Verlag.

Arnold, V. I. [1989] *Mathematical Methods of Classical Mechanics.* Second Edition, Graduate Texts in Mathematics **60**, Springer-Verlag.

Arnold, V. I. and B. Khesin [1992] Topological methods in hydrodynamics. *Ann. Rev. Fluid Mech.* **24**, 145–166.

Arnold, V. I. and B. Khesin [1998] *Topological Methods in Hydrodynamics.* Appl. Math. Sciences **125**, Springer-Verlag.

Arnold, V. I., V. V. Kozlov, and A. I. Neishtadt [1988] Mathematical aspects of classical and celestial mechanics, in: *Dynamical Systems III*, V.I. Arnold, ed. Springer-Verlag.

Arnold, V. I. and S. P. Novikov [1994] *Dynamical systems VII*, Encyclopedia of Math. Sci. **16**, Springer-Verlag.

Ashbaugh, M. S., C. C. Chicone, and R. H. Cushman [1990] The twisting tennis racket. *Dyn. Diff. Eqns.* **3**, 67–85.

Atiyah, M. [1982] Convexity and commuting Hamiltonians. *Bull. London Math. Soc.* **14**, 1–15.

Atiyah, M. [1983] Angular momentum, convex polyhedra and algebraic geometry. *Proc. Edinburgh Math. Soc.* **26**, 121–138.

Atiyah, M. and R. Bott [1984] The moment map and equivariant cohomology. *Topology* **23**, 1–28.

Audin, M. [1991] *The Topology of Torus Actions on Symplectic Manifolds.* Progress in Math **93**, Birkhäuser.

Austin, M. and P. S. Krishnaprasad [1993] Almost Poisson Integration of Rigid-Body Systems. *J. Comp. Phys.* **106**.

Baider, A., R. C. Churchill, and D. L. Rod [1990] Monodromy and nonintegrability in complex Hamiltonian systems. *J. Dyn. Diff. Eqns.* **2**, 451–481.

Baillieul, J. [1987] Equilibrium mechanics of rotating systems. *Proc. CDC* **26**, 1429–1434.

Baillieul, J. and M. Levi [1987] Rotational elastic dynamics. *Physica D* **27**, 43–62.

Baillieul, J. and M. Levi [1991] Constrained relative motions in rotational mechanics. *Arch. Rat. Mech. Anal.* **115**, 101–135.

Ball, J. M. and J. E. Marsden [1984] Quasiconvexity at the boundary, positivity of the second variation and elastic stability. *Arch. Rat. Mech. Anal.* **86**, 251–277.

Bambusi, D. [1998] A necessary and sufficient condition for Darboux' theorem in weak symplectic manifolds. *Proc. Am. Math. Soc.* (to appear).

Bao, D., J. E. Marsden, and R. Walton [1984] The Hamiltonian structure of general relativistic perfect fluids. *Comm. Math. Phys.* **99**, 319–345.

Bao, D. and V. P. Nair [1985] A note on the covariant anomaly as an equivariant momentum mapping. *Comm. Math. Phys.* **101**, 437–448.

Bates, L. and R. Cushman [1997] Global Aspects of Classical Integrable Systems, Birkhäuser, Boston.

Bates, L. and E. Lerman [1997] Proper group actions and symplectic stratified spaces. *Pacific J. Math.* **181**, 201–229.

Bates, L. and J. Sniatycki [1993] Nonholonomic reduction. *Reports on Math. Phys.* **32**, 99–115.

Bates, S. and A. Weinstein [1997] *Lectures on the Geometry of Quantization*, CPA-MUCB, Am. Math. Soc.

Batt, J. and G. Rein [1993] A rigorous stability result for the Vlasov–Poisson system in three dimensions. *Ann. Mat. Pura Appl.* **164**, 133–154.

Benjamin, T. B. [1972] The stability of solitary waves. *Proc. Roy. Soc. London* **328A**, 153–183.

Benjamin, T. B. [1984] Impulse, flow force and variational principles. *IMA J. Appl. Math.* **32**, 3–68.

Benjamin, T. B. and P. J. Olver [1982] Hamiltonian structure, symmetrics and conservation laws for water waves. *J. Fluid Mech.* **125**, 137–185.

Berezin, F. A. [1967] Some remarks about the associated envelope of a Lie algebra. *Funct. Anal. Appl.* **1**, 91–102.

Bernstein, B. [1958] Waves in a plasma in a magnetic field. *Phys. Rev.* **109**, 10–21.

Berry, M. [1984] Quantal phase factors accompanying adiabatic changes. *Proc. Roy. Soc. London A* **392**, 45–57.

Berry, M. [1985] Classical adiabatic angles and quantal adiabatic phase. *J. Phys. A. Math. Gen.* **18**, 15–27.

Berry, M. [1990] Anticipations of the geometric phase. *Physics Today*, December, 1990, 34–40.

Berry, M. and J. Hannay [1988] Classical non-adiabatic angles. *J. Phys. A. Math. Gen.* **21**, 325–333.

Besse, A. L. [1987] *Einstein Manifolds.* Springer-Verlag.

Bhaskara, K. H. and K. Viswanath [1988] *Poisson Algebras and Poisson Manifolds.* Longman (UK) and Wiley (US).

Bialynicki-Birula, I., J. C. Hubbard, and L. A. Turski [1984] Gauge-independent canonical formulation of relativistic plasma theory. *Physica A* **128**, 509–519.

Birnir, B. [1986] Chaotic perturbations of KdV. *Physica D* **19**, 238–254.

Birnir, B. and R. Grauer [1994] An explicit description of the global attractor of the damped and driven sine–Gordon equation. *Comm. Math. Phys.* **162**, 539–590.

Bloch, A. M., R. W. Brockett, and T. S. Ratiu [1990] A new formulation of the generalized Toda lattice equations and their fixed point analysis via the momentum map. *Bull. Amer. Math. Soc.* **23**, 477–485.

Bloch, A. M., R. W. Brockett, and T. S. Ratiu [1992] Completely integrable gradient flows. *Comm. Math. Phys.* **147**, 57–74.

Bloch, A. M., H. Flaschka, and T. S. Ratiu [1990] A convexity theorem for isospectral manifolds of Jacobi matrices in a compact Lie algebra. *Duke Math. J.* **61**, 41–65.

Bloch, A. M., H. Flaschka, and T. S. Ratiu [1993] A Schur–Horn–Kostant convexity theorem for the diffeomorphism group of the annulus. *Inv. Math.* **113**, 511–529.

Bloch, A. M., P. S. Krishnaprasad, J. E. Marsden, and R. Murray [1996] Nonholonomic mechanical systems with symmetry. *Arch. Rat. Mech. An.* **136**, 21–99.

Bloch, A. M., P. S. Krishnaprasad, J. E. Marsden, and T. S. Ratiu [1991] Asymptotic stability, instability, and stabilization of relative equilibria. *Proc. ACC., Boston IEEE*, 1120–1125.

Bloch, A. M., P. S. Krishnaprasad, J. E. Marsden, and T. S. Ratiu [1994] Dissipation Induced Instabilities. *Ann. Inst. H. Poincaré, Analyse Nonlinéaire* **11**, 37–90.

Bloch, A. M., P. S. Krishnaprasad, J. E. Marsden, and T. S. Ratiu [1996] The Euler–Poincaré equations and double bracket dissipation. *Comm. Math. Phys.* **175**, 1–42.

Bloch, A. M., P. S. Krishnaprasad, J. E. Marsden, and G. Sánchez de Alvarez [1992] Stabilization of rigid-body dynamics by internal and external torques. *Automatica* **28**, 745–756.

Bloch, A. M., N. Leonard, and J. E. Marsden [1997] Stabilization of Mechanical Systems Using Controlled Lagrangians. *Proc CDC* **36**, 2356–2361.

Bloch, A. M., N. Leonard, and J. E. Marsden [1998] Matching and Stabilization by the Method of Controlled Lagrangians. *Proc CDC* **37** (to appear).

Bloch, A. M. and J. E. Marsden [1989] Controlling homoclinic orbits. *Theoretical and Computational Fluid Mechanics* **1**, 179–190.

Bloch, A. M. and J. E. Marsden [1990] Stabilization of rigid-body dynamics by the energy–Casimir method. *Systems Control Lett.* **14**, 341–346.

Bobenko, A. I., A. G. Reyman, and M. A. Semenov-Tian-Shansky [1989] The Kowalewski top 99 years later: A Lax pair, generalizations and explicit solutions. *Comm. Math. Phys.* **122**, 321–354.

Bogoyavlensky, O. I. [1985] *Methods in the Qualitative Theory of Dynamical Systems in Astrophysics and Gas Dynamics.* Springer-Verlag.

Bolza, O. [1973] *Lectures on the Calculus of Variations.* Chicago University Press (1904). Reprinted by Chelsea, (1973).

Bona, J. [1975] On the stability theory of solitary waves. *Proc. Roy. Soc. London* **344A**, 363–374.

Born, M. and L. Infeld [1935] On the quantization of the new field theory. *Proc. Roy. Soc. London A* **150**, 141.

Bortolotti, F. [1926] *Rend. R. Naz. Lincei* **6a**, 552.

Bourbaki, N. [1971] *Variétés differentielles et analytiqes. Fascicule de résultats* **33**, Hermann.

Bourguignon, J. P. and H. Brezis [1974] Remarks on the Euler equation. *J. Funct. Anal.* **15**, 341–363.

Boya, L. J., J. F. Carinena, and J. M. Gracia-Bondia [1991] Symplectic structure of the Aharonov–Anandan geometric phase. *Phys. Lett. A* **161**, 30–34.

Bretherton, F. P. [1970] A note on Hamilton's principle for perfect fluids. *J. Fluid Mech.* **44**, 19-31.

Bridges, T. [1990] Bifurcation of periodic solutions near a collision of eigenvalues of opposite signature. *Math. Proc. Camb. Philos. Soc.* **108**, 575-601.

Bridges, T. [1994] Hamiltonian spatial structure for 3-D water waves relative to a moving frame of reference. *J. Nonlinear Sci.* **4**, 221-251.

Bridges, T. [1997] Multi-symplectic structures and wave propagation. *Math. Proc. Camb. Phil. Soc.* **121**, 147-190.

Brizard, A. [1992] Hermitian structure for linearized ideal MHD equations with equilibrium flow. *Phys. Lett. A* **168**, 357-362.

Brockett, R. W. [1973] Lie algebras and Lie groups in control theory. *Geometric Methods in Systems Theory*, Proc. NATO Advanced Study Institute, R. W. Brockett and D.Q. Mayne (eds.), Reidel, 43–82.

Brockett, R. W. [1976] Nonlinear systems and differential geometry. *Proc. IEEE* **64**, No. 1, 61–72.

Brockett, R. W. [1981] Control theory and singular Riemannian geometry. *New Directions in Applied Mathematics*, P. J. Hilton and G.S. Young (eds.), Springer-Verlag.

Brockett, R. W. [1983] Asymptotic stability and feedback stabilization. *Differential Geometric Control Theory*, R. W. Brockett, R. S. Millman, and H. Sussman (eds.), Birkhäuser.

Brockett, R. W. [1987] On the control of vibratory actuators. *Proc. 1987 IEEE Conf. Decision and Control*, 1418–1422.

Brockett, R. W. [1989] On the rectification of vibratory motion. *Sensors and Actuators* **20**, 91–96.

Broer, H., S. N. Chow, Y. Kim, and G. Vegter [1993] A normally elliptic Hamiltonian bifurcation. *Z. Angew Math. Phys.* **44**, 389-432.

Burov, A. A. [1986] On the non-existence of a supplementary integral in the problem of a heavy two-link pendulum. *PMM USSR* **50**, 123-125.

Busse, F. H. [1984] Oscillations of a rotating liquid drop. *J. Fluid Mech.* **142**, 1-8.

Cabannes, H. [1962] *Cours de Mécanique Générale.* Dunod.

Calabi, E. [1970] On the group of automorphisms of a symplectic manifold. In *Problems in Analysis*, Princeton University Press, 1–26.

Camassa, R. and D. D. Holm [1992] Dispersive barotropic equations for stratified mesoscale ocean dynamics. *Physica D* **60**, 1-15.

Carinena, J. F., E. Martinez, and J. Fernandez-Nunez [1992] Noether's theorem in time-dependent Lagrangian mechanics. *Rep. Math. Phys.* **31**, 189-203.

Carr, J. [1981] *Applications of center manifold theory.* Springer-Verlag: New York, Heidelberg, Berlin.

Cartan, E. [1922] Sur les petites oscillations d'une masse fluide. *Bull. Sci. Math.* **46**, 317-352 and 356-369.

Cartan, E. [1923] Sur les variétés a connexion affine et théorie de relativité généralisée. *Ann. Ecole Norm. Sup.* **40**, 325-412; **41**, 1-25.

Cartan, E. [1928a] Sur la représentation géométrique des systèmes matériels non holonomes. *Atti. Cong. Int. Matem.* **4**, 253-261.

Cartan, E. [1928b] Sur la stabilité ordinaire des ellipsoides de Jacobi. *Proc. Int. Math. Cong. Toronto* **2**, 9-17.

Casati, P., G. Falqui, F. Magri, and M. Pedroni, M. [1998] Bihamiltonian reductions and w_n-algebras. *J. Geom. and Phys.* **26**, 291–310.

Casimir, H. B. G. [1931] *Rotation of a Rigid Body in Quantum Mechanics*. Thesis, J.B. Wolters' Uitgevers-Maatschappij, N. V. Groningen, den Haag, Batavia.

Cendra, H., A. Ibort, and J. E. Marsden [1987] Variational principal fiber bundles: a geometric theory of Clebsch potentials and Lin constraints. *J. Geom. Phys.* **4**, 183–206.

Cendra, H., D. D. Holm, M. J. W. Hoyle, and J. E. Marsden [1998] The Maxwell–Vlasov equations in Euler–Poincaré form. *J. Math. Phys.* **39**, 3138–3157.

Cendra, H. and J. E. Marsden [1987] Lin constraints, Clebsch potentials and variational principles. *Physica D* **27**, 63–89.

Cendra, H., D. D. Holm, J. E. Marsden and T. S. Ratiu [1998] Lagrangian Reduction, the Euler–Poincaré Equations, and Semidirect Products. *AMS Transl.* **186**, 1–25.

Cendra, H., J. E. Marsden, and T. S. Ratiu [1999] Lagrangian reduction by stages. *Preprint*.

Chandrasekhar, S. [1961] *Hydrodynamic and Hydromagnetic Instabilities*. Oxford University Press.

Chandrasekhar, S. [1977] *Ellipsoidal Figures of Equilibrium*. Dover.

Channell, P. [1983] Symplectic integration algorithms. *Los Alamos National Laboratory Report AT-6:ATN-83-9*.

Channell, P. and C. Scovel [1990] Symplectic integration of Hamiltonian Systems. *Nonlinearity* **3**, 231–259.

Chen, F. F. [1974] *Introduction to Plasma Physics*. Plenum.

Chern, S. J. [1997] Stability Theory for Lumped Parameter Electromechanical Systems. *Preprint*.

Chern, S. J. and J. E. Marsden [1990] A note on symmetry and stability for fluid flows. *Geo. Astro. Fluid. Dyn.* **51**, 1–4.

Chernoff, P. R. and J. E. Marsden [1974] *Properties of Infinite Dimensional Hamiltonian systems*. Springer Lect. Notes in Math. **425**.

Cherry, T. M. [1959] The pathology of differential equations. *J. Austral. Math. Soc.* **1**, 1–16.

Cherry, T. M. [1968] Asymptotic solutions of analytic Hamiltonian systems. *J. Differential Equations* **4**, 142–149.

Chetaev, N. G. [1961] *The Stability of Motion*. Pergamon.

Chetaev, N. G. [1989] *Theoretical Mechanics*. Springer-Verlag.

Chirikov, B. V. [1979] A universal instability of many dimensional oscillator systems. *Phys. Rep.* **52**, 263–379.

Chorin, A. J., T. J. R. Hughes, J. E. Marsden, and M. McCracken [1978] Product formulas and numerical algorithms. *Comm. Pure Appl. Math.* **31**, 205–256.

Chorin, A. J. and J. E. Marsden [1993] *A Mathematical Introduction to Fluid Mechanics*. Third Edition, Texts in Applied Mathematical Sciences 4, Springer-Verlag.

Chow, S. N. and J. K. Hale [1982] *Methods of Bifurcation Theory*. Springer-Verlag.

Chow, S. N., J. K. Hale, and J. Mallet-Paret [1980] An example of bifurcation to homoclinic orbits. *J. Diff. Eqns.* **37**, 351–373.

Clebsch, A. [1857] Über eine allgemeine Transformation der hydrodynamischen Gleichungen. *Z. Reine Angew. Math.* **54**, 293–312.

Clebsch, A. [1859] Über die Integration der hydrodynamischen Gleichungen. *Z. Reine Angew. Math.* **56**, 1–10.

Clemmow, P. C. and J. P. Dougherty [1959] *Electrodynamics of Particles and Plasmas.* Addison-Wesley.

Conn, J. F. [1984] Normal forms for Poisson structures. *Ann. of Math.* **119**, 576–601, **121**, 565–593.

Cordani, B. [1986] Kepler problem with a magnetic monopole. *J. Math. Phys.* **27**, 2920–2921.

Corson, E. M. [1953] *Introduction to Tensors, Spinors and Relativistic Wave Equations.* Hafner.

Crouch, P. E. [1986] Spacecraft attitude control and stabilization: application of geometric control to rigid-body models. *IEEE Trans. Auto. Cont.* **29**, 321–331.

Cushman, R. and D. Rod [1982] Reduction of the semi-simple 1:1 resonance. *Physica D* **6**, 105–112.

Cushman, R. and R. Sjamaar [1991] On singular reduction of Hamiltonian spaces. *Symplectic Geometry and Mathematical Physics*, ed. by P. Donato, C. Duval, J. El-hadad, and G.M. Tuynman, Birkhäuser, 114–128.

Dashen, R. F. and D. H. Sharp [1968] Currents as coordinates for hadrons. *Phys. Rev.* **165**, 1857–1866.

David, D. and D. D. Holm [1992] Multiple Lie–Poisson structures. Reductions, and geometric phases for the Maxwell–Bloch travelling wave equations. *J. Nonlinear Sci.* **2**, 241–262.

David, D., D. D. Holm, and M. Tratnik [1990] Hamiltonian chaos in nonlinear optical polarization dynamics. *Phys. Rep.* **187**, 281–370.

Davidson, R. C. [1972] *Methods in Nonlinear Plasma Theory.* Academic Press.

de Leon, M., M. H. Mello, and P. R. Rodrigues [1992] Reduction of nondegenerate nonautonomous Lagrangians. *Cont. Math. AMS* **132**, 275–306.

Deift, P. A. and L. C. Li [1989] Generalized affine lie algebras and the solution of a class of flows associated with the QR eigenvalue algorithm. *Comm. Pure Appl. Math.* **42**, 963–991.

Dellnitz, M. and I. Melbourne [1993] The equivariant Darboux theorem. *Lect. Appl. Math.* **29**, 163–169.

Dellnitz, M., J. E. Marsden, I. Melbourne, and J. Scheurle [1992] Generic bifurcations of pendula. *Int. Series on Num. Math.* **104**, 111-122. ed. by G. Allgower, K. Böhmer, and M. Golubitsky, Birkhäuser.

Dellnitz, M., I. Melbourne, and J. E. Marsden [1992] Generic bifurcation of Hamiltonian vector fields with symmetry. *Nonlinearity* **5**, 979–996.

Delshams, A. and T. M. Seara [1991] An asymptotic expression for the splitting of separatrices of the rapidly forced pendulum. *Comm. Math. Phys.* **150**, 433–463.

Delzant, T. [1988] Hamiltoniens périodiques et images convexes de l'application moment. *Bull. Soc. Math. France* **116**, 315–339.

Delzant, T. [1990] Classification des actions hamiltoniennes complètement intégrables de rang deux. *Ann. Global Anal. Geom.* **8**, 87–112.

Deprit, A. [1983] Elimination of the nodes in problems of N bodies. *Celestial Mech.* **30**, 181–195.

de Vogelaére, R. [1956] Methods of integration which preserve the contact transformation property of the Hamiltonian equations. Department of Mathematics, *University of Notre Dame Report*, **4**.

Diacu, F. and P. Holmes [1996] *Celestial encounters. The origins of chaos and stability.* Princeton Univ. Press, Princeton, NJ.

Dirac, P. A. M. [1930] *The Principles of Quantum Mechanics.* Oxford University Press.

Dirac, P. A. M. [1950] Generalized Hamiltonian mechanics. *Canad. J. Math.* **2**, 129–148.

Dirac, P. A. M. [1964] *Lectures on Quantum Mechanics.* Belfer Graduate School of Science, Monograph Series **2**, Yeshiva University.

Duflo, M. and M. Vergne [1969] Une propriété de la représentation coadjointe d'une algèbre de Lie. *C.R. Acad. Sci. Paris* **268**, 583–585.

Duistermaat, H. [1974] Oscillatory integrals, Lagrange immersions and unfolding of singularities. *Comm. Pure and Appl. Math.* **27**, 207–281.

Duistermaat, H. [1983] *Bifurcations of periodic solutions near equilibrium points of Hamiltonian systems.* Springer Lect. Notes in Math. **1057**, 57–104.

Duistermaat, H. [1984] Non-integrability of 1:2:2 resonance. *Ergodic Theory Dynamical Systems* **4**, 553.

Duistermaat, J. J. [1980] On global action angle coordinates. *Comm. Pure Appl. Math.* **33**, 687–706.

Duistermaat, J. J. and G. J. Heckman [1982] On the variation in the cohomology of the symplectic form of the reduced phase space. *Inv. Math* **69**, 259–269, **72**, 153–158.

Dzyaloshinskii, I. E. and G. E. Volovick [1980] Poisson brackets in condensed matter physics. *Ann. of Phys.* **125**, 67–97.

Ebin, D. G. [1970] On the space of Riemannian metrics. *Symp. Pure Math., Am. Math. Soc.* **15**, 11–40.

Ebin, D. G. and J. E. Marsden [1970] Groups of diffeomorphisms and the motion of an incompressible fluid. *Ann. Math.* **92**, 102–163.

Ebin, D. G. [1982] Motion of slightly compressible fluids in a bounded domain I. *Comm. Pure Appl. Math.* **35**, 452–485.

Eckard, C. [1960] Variational principles of hydrodynamics. *Phys. Fluids* **3**, 421–427.

Eckmann J.-P. and R. Seneor [1976] The Maslov–WKB method for the (an-)harmonic oscillator. *Arch. Rat. Mech. Anal.* **61** 153–173.

Emmrich, C. and H. Römer [1990] Orbifolds as configuration spaces of systems with gauge symmetries. *Comm. Math. Phys.* **129**, 69–94.

Enos, M. J. [1993] On an optimal control problem on SO(3) × SO(3) and the falling cat. *Fields Inst. Comm.* **1**, 75–112.

Ercolani, N., M. G. Forest, and D. W. McLaughlin [1990] Geometry of the modulational instability, III. Homoclinic orbits for the periodic sine–Gordon equation. *Physica D* **43**, 349.

Ercolani, N., M. G. Forest, D. W. McLaughlin, and R. Montgomery [1987] Hamiltonian structure of modulation equation for the sine–Gordon equation. *Duke Math. J.* **55**, 949–983.

Fedorov, Y. N. [1994] Generalized Poinsot interpretation of the motion of a multidimensional rigid body. (Russian) *Trudy Mat. Inst. Steklov.* **205** Novye Rezult. v Teor. Topol. Klassif. Integr. Sistem, 200–206.

Fedorov, Y. N. and V. V. Kozlov [1995] Various aspects of n-dimensional rigid-body dynamics. Dynamical systems in classical mechanics, 141–171, *Amer. Math. Soc. Transl. Ser. 2*, **168**.

Feng, K. [1986] Difference schemes for Hamiltonian formalism and symplectic geometry. *J. Comp. Math.* **4**, 279–289.

Feng, K. and Z. Ge [1988] On approximations of Hamiltonian systems. *J. Comp. Math.* **6**, 88–97.

Finn, J. M. and G. Sun [1987] Nonlinear stability and the energy–Casimir method. *Comm. on Plasma Phys. and Controlled Fusion* **XI**, 7–25.

Fischer, A. E. and J. E. Marsden [1972] The Einstein equations of evolution—a geometric approach. *J. Math. Phys.* **13**, 546–568.

Fischer, A. E. and J. E. Marsden [1979] Topics in the dynamics of general relativity. *Isolated Gravitating Systems in General Relativity*, J. Ehlers (ed.), Italian Physical Society, 322–395.

Fischer, A. E., J. E. Marsden, and V. Moncrief [1980] The structure of the space of solutions of Einstein's equations, I: One Killing field. *Ann. Inst. H. Poincaré* **33**, 147–194.

Flaschka, H. [1976] The Toda lattice. *Phys. Rev. B* **9**, 1924–1925.

Flaschka, H., A. Newell, and T. S. Ratiu [1983a] Kac–Moody Lie algebras and soliton equations II. Lax equations associated with $A_1^{(1)}$. *Physica D* **9**, 300–323.

Flaschka, H., A. Newell, and T. S. Ratiu [1983b] Kac–Moody Lie algebras and soliton equations III. Stationary equations associated with $A_1^{(1)}$. *Physica D* **9**, 324–332.

Fomenko, A. T. [1988a] *Symplectic Geometry.* Gordon and Breach.

Fomenko, A. T. [1988b] *Integrability and Nonintegrability in Geometry and Mechanics.* Kluwer Academic.

Fomenko, A. T. and V. V. Trofimov [1989] *Integrable Systems on Lie Algebras and Symmetric Spaces.* Gordon and Breach.

Fong, U. and K. R. Meyer [1975] Algebras of integrals. *Rev. Colombiana Mat.* **9**, 75–90.

Fontich, E. and C. Simo [1990] The splitting of separatrices for analytic diffeomorphisms. *Erg. Thy. Dyn. Syst.* **10**, 295–318.

Fowler, T. K. [1963] Liapunov's stability criteria for plasmas. *J. Math. Phys.* **4**, 559–569.

Friedlander, S. and M. M. Vishik [1990] Nonlinear stability for stratified magnetohydrodynamics. *Geophys. Astrophys. Fluid Dyn.* **55**, 19–45.

Fukumoto, Y. [1997] Stationary configurations of a vortex filament in background flows. *Proc. R. Soc. Lon. A* **453**, 1205–1232.

Fukumoto, Y. and Miyajima, M. [1996] The localized induction hierarchy and the Lund–Regge equation. *J. Phys. A: Math. Gen.* **29**, 8025–8034.

Glgani, L. Giorgilli, A. and Strelcyn, J.-M. [1981] Chaotic motions and transition to stochasticity in the classical problem of the heavy rigid body with a fixed point. *Nuovo Cimento* **61**, 1–20.

Galin, D. M. [1982] Versal deformations of linear Hamiltonian systems. *AMS Transl.* **118**, 1–12 (1975 *Trudy Sem. Petrovsk.* **1**, 63–74).

Gallavotti, G. [1983] *The Elements of Mechanics.* Springer-Verlag.

Gantmacher, F. R. [1959] *Theory of Matrices.* Chelsea.

Gardner, C. S. [1971] Korteweg–de Vries equation and generalizations IV. The Korteweg–de Vries equation as a Hamiltonian system. *J. Math. Phys.* **12**, 1548–1551.

Ge, Z. [1990] Generating functions, Hamilton–Jacobi equation and symplectic groupoids over Poisson manifolds. *Indiana Univ. Math. J.* **39**, 859–876.

Ge, Z. [1991a] Equivariant symplectic difference schemes and generating functions. *Physica D* **49**, 376–386.

Ge, Z. [1991b] A constrained variational problem and the space of horizontal paths. *Pacific J. Math.* **149**, 61–94.

Ge, Z., H. P. Kruse, and J. E. Marsden [1996] The limits of Hamiltonian structures in three-dimensional elasticity, shells and rods. *J. Nonlin. Sci.* **6**, 19–57.

Ge, Z. and J. E. Marsden [1988] Lie–Poisson integrators and Lie–Poisson Hamilton–Jacobi theory. *Phys. Lett. A* **133**, 134–139.

Gelfand, I. M. and I. Y. Dorfman [1979] Hamiltonian operators and the algebraic structures connected with them. *Funct. Anal. Appl.* **13**, 13–30.

Gelfand, I. M. and S. V. Fomin [1963] *Calculus of Variations*. Prentice-Hall.

Gibbons, J. [1981] Collisionless Boltzmann equations and integrable moment equations. *Physica A* **3**, 503–511.

Gibbons, J., D. D. Holm, and B. A. Kuperschmidt [1982] Gauge-invariance Poisson brackets for chromohydrodynamics. *Phys. Lett.* **90A**.

Godbillon, C. [1969] *Géométrie Différentielle et Mécanique Analytique*. Hermann.

Goldin, G. A. [1971] Nonrelativistic current algebras as unitary representations of groups. *J. Math. Phys.* **12**, 462–487.

Goldman, W. M. and J. J. Millson [1990] Differential graded Lie algebras and singularities of level sets of momentum mappings. *Comm. Math. Phys.* **131**, 495–515.

Goldreich, P. and A. Toomre [1969] Some remarks on polar wandering, *J. Geophys. Res.* **10**, 2555–2567.

Goldstein, H. [1980] *Classical Mechanics*. Second Edition, Addison-Wesley.

Golin, S., A. Knauf, and S. Marmi [1989] The Hannay angles: geometry, adiabaticity, and an example. *Comm. Math. Phys.* **123**, 95–122.

Golin, S. and S. Marmi [1990] A class of systems with measurable Hannay angles. *Nonlinearity* **3**, 507–518.

Golubitsky, M., M. Krupa, and C. Lim [1991] Time reversibility and particle sedimentation. *SIAM J. Appl. Math.* **51**, 49–72.

Golubitsky, M., J. E. Marsden, I. Stewart, and M. Dellnitz [1994] The constrained Liapunov–Schmidt procedure and periodic orbits. *Fields Inst. Comm.* (to appear).

Golubitsky, M., and D. Schaeffer [1985] *Singularities and Groups in Bifurcation Theory*. Vol. 1, Applied Mathematical Sciences **69**, Springer-Verlag.

Golubitsky, M. and I. Stewart [1987] Generic bifurcation of Hamiltonian systems with symmetry. *Physica D* **24**, 391–405.

Golubitsky, M., I. Stewart, and D. Schaeffer [1988] *Singularities and Groups in Bifurcation Theory*. Vol. 2, Applied Mathematical Sciences **69**, Springer-Verlag.

Goodman, L. E. and A. R. Robinson [1958] Effects of finite rotations on gyroscopic sensing devices. *J. of Appl. Mech.* **28**, 210–213. (See also *Trans. ASME* **80**, 210–213.)

Gotay, M. J. [1988] A multisymplectic approach to the KdV equation: in *Differential Geometric Methods in Theoretical Physics*. Kluwer, 295–305.

Gotay, M. J., J. A. Isenberg, J. E. Marsden, and R. Montgomery [1997] Momentum Maps and Classical Relativistic Fields. *Preprint*.

Gotay, M. J., R. Lashof, J. Sniatycki, and A. Weinstein [1980] Closed forms on symplectic fiber bundles. *Comm. Math. Helv.* **58**, 617–621.

Gotay, M. J. and J. E. Marsden [1992] Stress–energy–momentum tensors and the Belifante–Resenfeld formula. *Cont. Math. AMS* **132**, 367–392.

Gotay, M. J., J. M. Nester, and G. Hinds [1979] Presymplectic manifolds and the Dirac–Bergmann theory of constraints. *J. Math. Phys.* **19**, 2388–2399.

Gozzi, E. and W. D. Thacker [1987] Classical adiabatic holonomy in a Grassmannian system. *Phys. Rev. D* **35**, 2388–2396.

Greenspan, B. D. and P. J. Holmes [1983] Repeated resonance and homoclinic bifurcations in a periodically forced family of oscillators. *SIAM J. Math. Anal.* **15**, 69–97.

Griffa, A. [1984] Canonical transformations and variational principles for fluid dynamics. *Physica A* **127**, 265–281.

Grigore, D. R. and O. T. Popp [1989] The complete classification of generalized homogeneous symplectic manifolds. *J. Math. Phys.* **30**, 2476–2483.

Grillakis, M., J. Shatah, and W. Strauss [1987] Stability theory of solitary waves in the presence of symmetry, I & II. *J. Funct. Anal.* **74**, 160–197 and **94** (1990), 308–348.

Grossman, R., P. S. Krishnaprasad, and J. E. Marsden [1988] The dynamics of two coupled rigid bodies. *Dynamical Systems Approaches to Nonlinear Problems in Systems and Circuits*, Salam and Levi (eds.). SIAM, 373–378.

Gruendler, J. [1985] The existence of homoclinic orbits and the methods of Melnikov for systems. *SIAM J. Math. Anal.* **16**, 907–940.

Guckenheimer, J. and P. Holmes [1983] *Nonlinear Oscillations, Dynamical Systems and Bifurcations of Vector Fields*. Applied Mathematical Sciences **43**, Springer-Verlag.

Guckenheimer, J. and P. Holmes [1988] Structurally stable heteroclinic cycles. *Math. Proc. Camb. Philos. Soc.* **103**, 189–192.

Guckenheimer, J. and A. Mahalov [1992a] Resonant triad interactions in symmetric systems. *Physica D* **54**, 267–310.

Guckenheimer, J. and A. Mahalov [1992b] Instability induced by symmetry reduction. *Phys. Rev. Lett.* **68**, 2257–2260.

Guichardet, A. [1984] On rotation and vibration motions of molecules. *Ann. Inst. H. Poincaré* **40**, 329–342.

Guillemin, V. and A. Pollack [1974] *Differential Topology*. Prentice-Hall.

Guillemin, V. and E. Prato [1990] Heckman, Kostant, and Steinberg formulas for symplectic manifolds. *Adv. in Math.* **82**, 160–179.

Guillemin, V. and S. Sternberg [1977] *Geometric Asymptotics*. Amer. Math. Soc. Surveys **14**. (Revised edition, 1990.)

Guillemin, V. and S. Sternberg [1980] The moment map and collective motion. *Ann. of Phys.* **1278**, 220–253.

Guillemin, V. and S. Sternberg [1982] Convexity properties of the moment map. *Inv. Math.* **67**, 491–513; **77** (1984) 533–546.

Guillemin, V. and S. Sternberg [1983] On the method of Symes for integrating systems of the Toda type. *Lett. Math. Phys.* **7**, 113–115.

Guillemin, V. and S. Sternberg [1984] *Symplectic Techniques in Physics*. Cambridge University Press.

Guillemin, V. and A. Uribe [1987] Reduction, the trace formula, and semiclassical asymptotics. *Proc. Nat. Acad. Sci.* **84**, 7799–7801.

Hahn, W. [1967] *Stability of Motion*. Springer-Verlag.

Hale, J. K. [1963] *Oscillations in Nonlinear Systems*. McGraw-Hill.

Haller, G. [1992] Gyroscopic stability and its loss in systems with two essential coordinates. *Int. J. Nonlinear Mech.* **27**, 113–127.

Haller, G. and I. Mezić [1998] Reduction of three-dimensional, volume-preserving flows with symmetry. *Nonlinearity* **11**, 319–339.

Haller, G. and S. Wiggins [1993] Orbit homoclinic to resonances: the Hamiltonian case. *Physica D* **66**, 293–346.

Hamel, G. [1904] Die Lagrange-Eulerschen Gleichungen der Mechanik. *Z. Mathematik u. Physik* **50**, 1–57.

Hamel, G. [1949] *Theoretische Mechanik.* Springer-Verlag.

Hamilton, W. R. [1834] On a general method in dynamics. *Phil. Trans. Roy. Soc. Lon.* 95–144, 247–308.

Hannay, J. [1985] Angle variable holonomy in adiabatic excursion of an itegrable Hamiltonian. *J. Phys. A: Math. Gen.* **18**, 221–230.

Hanson, A., T. Regge, and C. Teitelboim [1976] Constrained Hamiltonian systems. *Accademia Nazionale Dei Lincei, Rome*, 1–135.

Helgason, S. [1978] *Differential Geometry, Lie Groups and Symmetric Spaces*, Academic Press.

Henon, M. [1982] Vlasov Equation. *Astron. Astrophys.* **114**, 211–212.

Herivel, J. W. [1955] The derivation of the equation of motion of an ideal fluid by Hamilton's principle. *Proc. Camb. Phil. Soc.* **51**, 344–349.

Hermann, R. [1962] The differential geometry of foliations. *J. Math. Mech.* **11**, 303–315.

Hermann, R. [1964] An incomplete compact homogeneous Lorentz metric *J. Math. Mech.* **13**, 497–501.

Hermann, R. [1968] *Differential Geometry and the Calculus of Variations.* Math. Science Press.

Hermann, R. [1973] *Geometry, Physics, and Systems.* Marcel Dekker.

Hirsch, M. and S. Smale [1974] *Differential Equations, Dynamical Systems and Linear Algebra.* Academic Press.

Hirschfelder, J. O. and J. S. Dahler [1956] The kinetic energy of relative motion. *Proc. Nat. Acad. Sci.* **42**, 363–365.

Holm, D.D. [1996] Hamiltonian balance equations. *Physica D* **98**, 379–414.

Holm, D. D. and B. A. Kuperschmidt [1983] Poisson brackets and Clebsch representations for magnetohydrodynamics, multifluid plasmas, and elasticity. *Physica D* **6**, 347–363.

Holm, D. D., B. A. Kupershmidt, and C. D. Levermore [1985] Hamiltonian differencing of fluid dynamics. *Adv. in Appl. Math.* **6**, 52–84.

Holm, D. D. and J. E. Marsden [1991] The rotor and the pendulum. *Symplectic Geometry and Mathematical Physics*, P. Donato, C. Duval, J. Elhadad, and G.M. Tuynman (eds.), Birkhäuser, pp. 189–203.

Holm, D. D., J. E. Marsden, and T. S. Ratiu [1986] The Hamiltonian structure of continuum mechanics in material, spatial and convective representations. *Séminaire de Mathématiques Supérieurs, Les Presses de l'Univ. de Montréal* **100**, 11–122.

Holm, D. D., J. E. Marsden and T. S. Ratiu [1998] The Euler–Poincaré equations and semidirect products with applications to continuum theories. *Adv. in Math.* **137**, 1–81.

Holm, D. D., J. E. Marsden, and T. Ratiu [1998b] The Euler–Poincaré equations in geophysical fluid dynamics, in *Proceedings of the Isaac Newton Institute Programme on the Mathematics of Atmospheric and Ocean Dynamics*, Cambridge University Press (to appear).

Holm, D. D., J. E. Marsden, and T. S. Ratiu [1998c] Euler–Poincaré models of ideal fluids with nonlinear dispersion. *Phys. Rev. Lett.* **349**, 4173–4177.

Holm, D. D., J. E. Marsden, T. S. Ratiu, and A. Weinstein [1985] Nonlinear stability of fluid and plasma equilibria. *Phys. Rep.* **123**, 1–116.

Holmes, P. J. [1980a] *New Approaches to Nonlinear Problems in Dynamics*. SIAM.

Holmes, P. J. [1980b] Averaging and chaotic motions in forced oscillations. *SIAM J. Appl. Math.* **38**, 68–80 and **40**, 167–168.

Holmes, P. J., J. R. Jenkins, and N. Leonard [1998] Dynamics of the Kirchhoff equations I: coincident centers of gravity and buoyancy. *Physica D* **118**, 311–342.

Holmes, P. J. and J. E. Marsden [1981] A partial differential equation with infinitely many periodic orbits: chaotic oscillations of a forced beam. *Arch. Rat. Mech. Anal.* **76**, 135–166.

Holmes, P. J. and J. E. Marsden [1982a] Horseshoes in perturbations of Hamiltonian systems with two-degrees-of-freedom. *Comm. Math. Phys.* **82**, 523–544.

Holmes, P. J. and J. E. Marsden [1982b] Melnikov's method and Arnold diffusion for perturbations of integrable Hamiltonian systems. *J. Math. Phys.* **23**, 669–675.

Holmes, P. J. and J. E. Marsden [1983] Horseshoes and Arnold diffusion for Hamiltonian systems on Lie groups. *Indiana Univ. Math. J.* **32**, 273–310.

Holmes, P. J., J. E. Marsden, and J. Scheurle [1988] Exponentially small splittings of separatrices with applications to KAM theory and degenerate bifurcations. *Contemp. Math.* **81**, 213–244.

Hörmander, L. [1995] Symplectic classification of quadratic forms, and general Mehler formulas. *Math. Zeit.* **219**, 413–449.

Horn, A. [1954] Doubly stochastic matrices and the diagonal of a rotation matrix. *Amer. J. Math.* **76**, 620–630.

Howard, J. E. and R. S. MacKay [1987a] Linear stability of symplectic maps. *J. Math. Phys.* **28**, 1036–1051.

Howard, J. E. and R. S. MacKay [1987b] Calculation of linear stability boundaries for equilibria of Hamiltonian systems. *Phys. Lett. A* **122**, 331–334.

Hughes, T. J. R., T. Kato, and J. E. Marsden [1977] Well-posed quasi-linear second-order hyperbolic systems with applications to nonlinear elastodynamics and general relativity. *Arch. Rat. Mech. Anal.* **90**, 545–561.

Iacob, A. [1971] Invariant manifolds in the motion of a rigid body about a fixed point. *Rev. Roumaine Math. Pures Appl.* **16**, 1497–1521.

Ichimaru, S. [1973] *Basic Principles of Plasma Physics*. Addison-Wesley.

Isenberg, J. and J. E. Marsden [1982] A slice theorem for the space of solutions of Einstein's equations. *Phys. Rep.* **89**, 179–222.

Ishlinskii, A. [1952] *Mechanics of Special Gyroscopic Systems* (*in Russian*). National Acad. Ukrainian SSR, Kiev.

Ishlinskii, A. [1963] *Mechanics of Gyroscopic Systems* (*in English*). Israel Program for Scientific Translations, Jerusalem, 1965 (also available as a NASA technical translation).

Ishlinskii, A. [1976] *Orientation, Gyroscopes and Inertial Navigation* (*in Russian*). Nauka, Moscow.

Iwai, T. [1982] The symmetry group of the harmonic oscillator and its reduction. *J. Math. Phys.* **23**, 1088–1092.

Iwai, T. [1985] On reduction of two-degrees-of-freedom Hamiltonian systems by an S^1 action, and $SO_0(1,2)$ as a dynamical group. *J. Math. Phys.* **26**, 885–893.

Iwai, T. [1987a] A gauge theory for the quantum planar three-body system. *J. Math. Phys.* **28**, 1315–1326.

Iwai, T. [1987b] A geometric setting for internal motions of the quantum three-body system. *J. Math. Phys.* **28**, 1315–1326.

Iwai, T. [1987c] A geometric setting for classical molecular dynamics. *Ann. Inst. Henri Poincaré, Phys. Théor.* **47**, 199–219.

Iwai, T. [1990a] On the Guichardet/Berry connection. *Phys. Lett. A* **149**, 341–344.

Iwai, T. [1990b] The geometry of the SU(2) Kepler problem. *J. Geom. Phys.* **7**, 507–535.

Iwiński, Z. R. and L. A. Turski [1976] Canonical theories of systems interacting electromagnetically. *Letters in Applied and Engineering Sciences* **4**, 179–191.

Jacobi, C. G. K. [1837] Note sur l'intégration des équations différentielles de la dynamique. *C.R. Acad. Sci., Paris* **5**, 61.

Jacobi, C. G. K. [1843] *J. Math.* **26**, 115.

Jacobi, C. G. K. [1866] *Vorlesungen über Dynamik.* (Based on lectures given in 1842–3) Verlag G. Reimer; Reprinted by Chelsea, 1969.

Jacobson, N. [1962] *Lie Algebras.* Interscience, reprinted by Dover.

Jeans, J. [1919] *Problems of Cosmogony and Stellar Dynamics.* Cambridge University Press.

Jellinek, J. and D. H. Li [1989] Separation of the energy of overall rotations in an N-body system. *Phys. Rev. Lett.* **62**, 241–244.

Jepson, D. W. and J. O. Hirschfelder [1958] Set of coordinate systems which diagonalize the kinetic energy of relative motion. *Proc. Nat. Acad. Sci.* **45**, 249–256.

Jost, R. [1964] Poisson brackets (An unpedagogical lecture). *Rev. Mod. Phys.* **36**, 572–579.

Kammer, D. C. and G. L. Gray [1992] A nonlinear control design for energy sink simulation in the Euler–Poinsot problem. *J. Astr. Sci.* **41**, 53–72.

Kane, T. R. and M. Scher [1969] A dynamical explanation of the falling cat phenomenon. *Int. J. Solids Structures* **5**, 663–670.

Kaplansky, I. [1971] *Lie Algebras and Locally Compact Groups*, University of Chicago Press.

Karasev, M. V. and V. P. Maslov [1993] *Nonlinear Poisson Brackets. Geometry and Quantization.* Transl. of Math. Monographs. **119**. Amer. Math. Soc.

Kato, T. [1950] On the adiabatic theorem of quantum mechanics. *J. Phys. Soc. Japan* **5**, 435–439.

Kato, T. [1967] On classical solutions of the two-dimensional non-stationary Euler equation. *Arch. Rat. Mech. Anal.* **25**, 188–200.

Kato, T. [1972] Nonstationary flows of viscous and ideal fluids in $\mathbb{R}^3$. *J. Funct. Anal.* **9**, 296–305.

Kato, T. [1975] On the initial value problem for quasi-linear symmetric hyperbolic systems. *Arch. Rat. Mech. Anal.* **58**, 181–206.

Kato, T. [1984] *Perturbation Theory for Linear Operators.* Springer-Verlag.

Kato, T. [1985] *Abstract Differential Equations and Nonlinear Mixed Problems.* Lezioni Fermiane, Scuola Normale Superiore, Accademia Nazionale dei Lincei.

Katz, S. [1961] Lagrangian density for an inviscid, perfect, compressible plasma. *Phys. Fluids* **4**, 345–348.

Kaufman, A. [1982] Elementary derivation of Poisson structures for fluid dynamics and electrodynamics. *Phys. Fluids* **25**, 1993–1994.

Kaufman, A. and R. L. Dewar [1984] Canonical derivation of the Vlasov-Coulomb noncanonical Poisson structure. *Contemp. Math.* **28**, 51–54.

Kazhdan, D., B. Kostant, and S. Sternberg [1978] Hamiltonian group actions and dynamical systems of Calogero type. *Comm. Pure Appl. Math.* **31**, 481–508.

Khesin, B. A. [1992] Ergodic interpretation of integral hydrodynamic invariants. *J. Geom. Phys.* **9**, 101–110.

Khesin, B. A. and Y. Chekanov [1989] Invariants of the Euler equations for ideal or barotropic hydrodynamics and superconductivity in *D* dimensions. *Physica D* **40**, 119–131.

Kijowski, J. and W. Tulczyjew [1979] *A Symplectic Framework for Field Theories.* Springer Lect. Notes in Phys. **107**.

Kirillov, A. A. [1962] Unitary representations of nilpotent Lie groups. *Russian Math. Surveys* **17**, 53–104.

Kirillov, A. A. [1976a] Local Lie Algebras. *Russian Math. Surveys* **31**, 55–75.

Kirillov, A. A. [1976b] *Elements of the Theory of Representations.* Grundlehren Math. Wiss., Springer-Verlag.

Kirillov, A. A. [1981] The orbits of the group of diffeomorphisms of the circle and local Lie superalgebras. (Russian) *Funktsional. Anal. i Prilozhen.* **15**, 75–76.

Kirillov, A. A. [1993] The orbit method. II. Infinite-dimensional Lie groups and Lie algebras. Representation theory of groups and algebras. *Contemp. Math.* **145**, 33–63.

Kirk, V., J. E. Marsden, and M. Silber [1996] Branches of stable three-tori using Hamiltonian methods in Hopf bifurcation on a rhombic lattice. *Dyn. and Stab. of Systems* **11**, 267–302.

Kirwan, F. C. [1984] *Cohomology Quotients in Symplectic and Algebraic Geometry.* Princeton Math. Notes **31**, Princeton University Press.

Kirwan, F. C. [1985] Partial desingularization of quotients of nonsingular varieties and their Betti numbers. *Ann. of Math.* **122**, 41–85.

Kirwan, F. C. [1988] The topology of reduced phase spaces of the motion of vortices on a sphere. *Physica D* **30**, 99–123.

Kirwan, F. C. [1998] Momentum maps and reduction in algebraic geometry. *Diff. Geom. and Appl.* **9**, 135–171.

Klein, F. [1897] *The Mathematical Theory of the Top.* Scribner.

Klein, M. [1970] *Paul Ehrenfest.* North-Holland.

Klingenberg, W. [1978] *Lectures on Closed Geodesics.* Grundlehren Math. Wiss. **230**, Springer-Verlag.

Knapp, A. W. [1996] *Lie Groups: Beyond an Introduction.* Progress in Mathematics **140**, Birkhäuser, Boston.

Knobloch, E. and J. D. Gibbon [1991] Coupled NLS equations for counterpropagating waves in systems with reflection symmetry. *Phys. Lett. A* **154**, 353–356.

Knobloch, E., Mahalov, and J. E. Marsden [1994] Normal forms for three-dimensional parametric instabilities in ideal hydrodynamics. *Physica D* **73**, 49–81.

Knobloch, E. and M. Silber [1992] Hopf bifurcation with $Z_4 \times T^2$ symmetry: in *Bifurcation and Symmetry*, K. Boehmer (ed.). Birkhäuser.

Kobayashi and Nomizu [1963] *Foundations of Differential Geometry.* Wiley.

Kocak, H., F. Bisshopp, T. Banchoff, and D. Laidlaw [1986] Topology and Mechanics with Computer Graphics. *Adv. in Appl. Math.* **7**, 282–308.

Koiller, J. [1985] On Aref's vortex motions with a symmetry center. *Physica D* **16**, 27–61.

Koiller, J. [1992] Reduction of some classical nonholonomic systems with symmetry. *Arch. Rat. Mech. Anal.* **118**, 113–148.

Koiller, J., J. M. Balthazar, and T. Yokoyama [1987] Relaxation–Chaos phenomena in celestial mechanics. *Physica D* **26**, 85–122.

Koiller, J., I. D. Soares, and J. R. T. Melo Neto [1985] Homoclinic phenomena in gravitational collapse. *Phys. Lett.* **110A**, 260–264.

Koon, W. S. and J. E. Marsden [1997] Optimal control for holonomic and nonholonomic mechanical systems with symmetry and Lagrangian reduction. *SIAM J. Control and Optim.* **35**, 901–929.

Koon, W. S. and J. E. Marsden [1997b] The Hamiltonian and Lagrangian approaches to the dynamics of nonholonomic systems. *Rep. Math. Phys.* **40**, 21–62.

Koon, W. S. and J. E. Marsden [1998] The Poisson reduction of nonholonomic mechanical systems. *Reports on Math. Phys.* (to appear).

Kopell, N. and R. B. Washburn Jr. [1982] Chaotic motions in the two degree-of-freedom swing equations. *IEEE Trans. Circuits and Systems* **29**, 738–746.

Korteweg, D. J. and G. de Vries [1895] On the change of form of long waves advancing in a rectangular canal and on a new type of long stationary wave. *Phil. Mag.* **39**, 422–433.

Kozlov, V. V. [1996] *Symmetries, Topology and Resonances in Hamiltonian Mechanics.* Translated from the Russian manuscript by S. V. Bolotin, D. Treshchev, and Y. Fedorov. Ergebnisse der Mathematik und ihrer Grenzgebiete **31**. Springer-Verlag, Berlin.

Kosmann-Schwarzbach, Y. and F. Magri [1990] Poisson–Nijenhuis structures. *Ann. Inst. H. Poincaré* **53**, 35–81.

Kostant, B. [1966] Orbits, symplectic structures and representation theory. *Proc. US-Japan Seminar on Diff. Geom., Kyoto. Nippon Hyronsha, Tokyo* **77**.

Kostant, B. [1970] *Quantization and unitary representations.* Springer Lect. Notes in Math. **570**, 177–306.

Kostant, B. [1973] On convexity, the Weyl group and the Iwasawa decomposition. *Ann. Sci. École Norm. Sup.* **6**, 413–455.

Kostant, B. [1978] On Whittaker vectors and representation theory. *Inv. Math.* **48**, 101–184.

Kostant, B. [1979] The solution to a generalized Toda lattice and representation theory. *Adv. in Math.* **34**, 195–338.

Koszul, J. L. [1985] Crochet de Schouten–Nijenhuis et cohomologie. *É. Cartan et les mathématiques d'Aujourdhui, Astérisque hors série, Soc. Math. France*, 257–271.

Kovačič, G. and S. Wiggins [1992] Orbits homoclinic to resonances, with an application to chaos in the damped and forced sine–Gordon equation. *Physica D* **57**, 185.

Krall, N. A. and A. W. Trivelpiece [1973] *Principles of Plasma Physics.* McGraw-Hill.

Krein, M. G. [1950] A generalization of several investigations of A. M. Liapunov on linear differential equations with periodic coefficients. *Dokl. Akad. Nauk. SSSR* **73**, 445–448.

Krishnaprasad, P. S. [1985] Lie–Poisson structures, dual-spin spacecraft and asymptotic stability. *Nonlinear Anal. TMA* **9**, 1011–1035.

Krishnaprasad, P. S. [1989] Eulerian many-body problems. *Cont. Math. AMS* **97**, 187–208.

Krishnaprasad, P. S. and J. E. Marsden [1987] Hamiltonian structure and stability for rigid bodies with flexible attachments. *Arch. Rat. Mech. Anal.* **98**, 137–158.

Krupa, M. [1990] Bifurcations of relative equilibria. *SIAM J. Math. Anal.* **21**, 1453–1486.

Kruse, H. P. [1993] The dynamics of a liquid drop between two plates. Thesis, University of Hamburg.

Kruse, H. P., J. E. Marsden, and J. Scheurle [1993] On uniformly rotating field drops trapped between two parallel plates. *Lect. in Appl. Math. AMS* **29**, 307–317.

Kummer, M. [1975] An interaction of three resonant modes in a nonlinear lattice. *J. Math. Anal. App.* **52**, 64.

Kummer, M. [1979] On resonant classical Hamiltonians with two-degrees-of-freedom near an equilibrium point. *Stochastic Behavior in Classical and Quantum Hamiltonian Systems.* Springer Lect. Notes in Phys. **93**.

Kummer, M. [1981] On the construction of the reduced phase space of a Hamiltonian system with symmetry. *Indiana Univ. Math. J.* **30**, 281–291.

Kummer, M. [1986] On resonant Hamiltonian systems with finitely many degrees of freedom. In *Local and Global Methods of Nonlinear Dynamics*, Lect. Notes in Phys. **252**, Springer-Verlag.

Kummer, M. [1990] On resonant classical Hamiltonians with n frequencies. *J. Diff. Eqns.* **83**, 220–243.

Kunzle, H. P. [1969] Degenerate Lagrangian systems. *Ann. Inst. H. Poincaré* **11**, 393–414.

Kunzle, H. P. [1972] Canonical dynamics of spinning particles in gravitational and electromagnetic fields. *J. Math. Phys.* **13**, 739–744.

Kuperschmidt, B. A. and T. Ratiu [1983] Canonical maps between semidirect products with applications to elasticity and superfluids. *Comm. Math. Phys.* **90**, 235–250.

Lagrange, J. L. [1788] *Mécanique Analytique.* Chez la Veuve Desaint.

Lanczos, C. [1949] *The Variational Principles of Mechanics.* University of Toronto Press.

Larsson, J. [1992] An action principle for the Vlasov equation and associated Lie perturbation equations. *J. Plasma Phys.* **48**, 13–35; **49**, 255–270.

Laub, A. J. and K. R. Meyer [1974] Canonical forms for symplectic and Hamiltonian matrices. *Celestial Mech.* **9**, 213–238.

Lawden, D. F. [1989] *Elliptic Functions and Applications.* Applied Mathematical Sciences **80**, Springer-Verlag.

Leimkuhler, B. and G. Patrick [1996] Symplectic integration on Riemannian manifolds. *J. of Nonl. Sci.* **6**, 367–384.

Leimkuhler, B. and R. Skeel [1994] Symplectic numerical integrators in constrained Hamiltonian systems. *Journal of Computational Physics* **112**, 117–125.

Leonard, N. E. [1997] Stability of a bottom-heavy underwater vehicle. *Automatica J. IFAC* **33**, 331–346.

Leonard, N. E. and P. S. Krishnaprasad [1995] Motion control of drift-free, left-invariant systems on Lie groups. *IEEE Trans. Automat. Control* **40**, 1539–1554.

Leonard, N. E. and W. S. Levine [1995] *Using Matlab to Analyze and Design Control Systems*, Second Edition, Benjamin-Cummings Publishing Co.

Leonard, N. E. and J. E. Marsden [1997] Stability and Drift of Underwater Vehicle Dynamics: Mechanical Systems with Rigid Motion Symmetry. *Physica D* **105**, 130–162.

Lerman, E. [1989] On the centralizer of invariant functions on a Hamiltonian G-space. *J. Diff. Geom.* **30**, 805–815.

Levi, M. [1989] Morse theory for a model space structure. *Cont. Math. AMS* **97**, 209–216.

Levi, M. [1993] Geometric phases in the motion of rigid bodies. *Arch. Rat. Mech. Anal.* **122**, 213–229.

Lewis, A. [1996] The geometry of the Gibbs–Appell equations and Gauss' principle of least constraint. *Reports on Math. Phys.* **38**, 11–28.

Lewis, A. and R. M. Murray [1995] Variational principles in constrained systems: theory and experiments. *Int. J. Nonl. Mech.* **30**, 793–815.

Lewis, D. [1989] Nonlinear stability of a rotating planar liquid drop. *Arch. Rat. Mech. Anal.* **106**, 287–333.

Lewis, D. [1992] Lagrangian block diagonalization. *Dyn. Diff. Eqns.* **4**, 1–42.

Lewis, D., J. E. Marsden, R. Montgomery, and T. S. Ratiu [1986] The Hamiltonian structure for dynamic free boundary problems. *Physica D* **18**, 391–404.

Lewis, D., J. E. Marsden, and T. S. Ratiu [1987] Stability and bifurcation of a rotating liquid drop. *J. Math. Phys.* **28**, 2508–2515.

Lewis, D., T. S. Ratiu, J. C. Simo, and J. E. Marsden [1992] The heavy top, a geometric treatment. *Nonlinearity* **5**, 1–48.

Lewis, D. and J. C. Simo [1990] Nonlinear stability of rotating pseudo-rigid bodies. *Proc. Roy. Soc. Lon. A* **427**, 281–319.

Lewis, D. and J. C. Simo [1995] Conserving algorithms for the dynamics of Hamiltonian systems on Lie groups. *J. Nonlinear Sci.* **4**, 253–299.

Li, C. W. and M. Z. Qin [1988] A symplectic difference scheme for the infinite-dimensional Hamilton system. *J. Comp. Math.* **6**, 164–174.

Liapunov, A. M. [1892] *Problème Générale de la Stabilité du Mouvement.* Kharkov. French translation in *Ann. Fac. Sci. Univ.* Toulouse, **9**, 1907; reproduced in *Ann. Math. Studies* **17**, Princeton University Press, 1949.

Liapunov, A. M. [1897] Sur l'instabilité de l'équilibre dans certains cas où la fonction de forces n'est pas maximum. *J. Math. Appl.* **3**, 81–84.

Libermann, P. and C. M. Marle [1987] *Symplectic Geometry and Analytical Mechanics.* Kluwer Academic.

Lichnerowicz, A. [1951] Sur les variétés symplectiques. *C.R. Acad. Sci. Paris* **233**, 723–726.

Lichnerowicz, A. [1973] Algèbre de Lie des automorphismes infinitésimaux d'une structure de contact. *J. Math. Pures Appl.* **52**, 473–508.

Lichnerowicz, A. [1975a] Variété symplectique et dynamique associée à une sous-variété. *C.R. Acad. Sci. Paris, Sér. A* **280**, 523–527.

Lichnerowicz, A. [1975b] Structures de contact et formalisme Hamiltonien invariant. *C.R. Acad. Sci. Paris, Sér. A* **281**, 171–175.

Lichnerowicz, A. [1976] Variétés symplectiques, variétés canoniques, et systèmes dynamiques, in *Topics in Differential Geometry*, H. Rund and W. Forbes (eds.). Academic Press.

Lichnerowicz, A. [1977] Les variétés de Poisson et leurs algèbres de Lie associées. *J. Diff. Geom.* **12**, 253–300.

Lichnerowicz, A. [1978] Deformation theory and quantization. *Group Theoretical Methods in Physics*, Springer Lect. Notes in Phys. **94**, 280–289.

Lichtenberg, A. J. and M. A. Liebermann [1983] *Regular and Stochastic Motion.* Applied Mathematical Sciences **38**. Springer-Verlag, 2nd edition [1991].

Lie, S. [1890] *Theorie der Transformationsgruppen. Zweiter Abschnitt.* Teubner.

Lin, C. C. [1963] Hydrodynamics of helium II. *Proc. Int. Sch. Phys.* **21**, 93–146.

Littlejohn, R. G. [1983] Variational principles of guiding center motion. *J. Plasma Physics* **29**, 111-125.

Littlejohn, R. G. [1984] Geometry and guiding center motion. *Cont. Math. AMS* **28**, 151-168.

Littlejohn, R. G. [1988] Cyclic evolution in quantum mechanics and the phases of Bohr–Sommerfeld and Maslov. *Phys. Rev. Lett.* **61**, 2159-2162.

Littlejohn, R. and M. Reinch [1997] Gauge fields in the separation of rotations and internal motions in the *n*-body problem. *Rev. Mod. Phys.* **69**, 213-275.

Love, A. E. H. [1944] *A Treatise on the Mathematical Theory of Elasticity*. Dover.

Low, F. E. [1958] A Lagrangian formulation of the Boltzmann–Vlasov equation for plasmas. *Proc. Roy. Soc. London A* **248**, 282-287.

Lu, J. H. and T. S. Ratiu [1991] On Kostant's convexity theorem. *J. AMS* **4**, 349-364.

Lundgren, T. S. [1963] Hamilton's variational principle for a perfectly conducing plasma continuum. *Phys. Fluids* **6**, 898-904.

MacKay, R. S. [1991] Movement of eigenvalues of Hamiltonian equilibria under non-Hamiltonian perturbation. *Phys. Lett. A* **155**, 266-268.

MacKay, R. S. and J. D. Meiss [1987] *Hamiltonian Dynamical Systems*. Adam Higler, IOP Publishing.

Maddocks, J. [1991] On the stability of relative equilibria. *IMA J. Appl. Math.* **46**, 71-99.

Maddocks, J. and R. L. Sachs [1993] On the stability of KdV multi-solitons. *Comm. Pure and Applied Math.* **46**, 867-901.

Manin, Y. I. [1979] Algebraic aspects of nonlinear differential equations. *J. Soviet Math.* **11**, 1-122.

Marle, C. M. [1976] Symplectic manifolds, dynamical groups and Hamiltonian mechanics; in *Differential Geometry and Relativity*, M. Cahen and M. Flato (eds.), Reidel.

Marsden, J. E. [1981] *Lectures on Geometric Methods in Mathematical Physics*. SIAM.

Marsden, J. E. [1982] A group theoretic approach to the equations of plasma physics. *Canad. Math. Bull.* **25**, 129-142.

Marsden, J. E. [1987] *Appendix to* Golubitsky and Stewart [1987].

Marsden, J. E. [1992] *Lectures on Mechanics*. London Mathematical Society Lecture Note Series, **174**, Cambridge University Press.

Marsden, J. E., D. G. Ebin, and A. Fischer [1972] Diffeomorphism groups, hydrodynamics and relativity. *Proceedings of the 13th Biennial Seminar on Canadian Mathematics Congress*, pp. 135-279.

Marsden, J. E. and T. J. R. Hughes [1983] *Mathematical Foundations of Elasticity*. Prentice-Hall. Dover edition [1994].

Marsden, J. E. and M. McCracken [1976] *The Hopf Bifurcation and its Applications*. Springer Applied Mathematics Series **19**.

Marsden, J. E., G. Misiolek, M. Perlmutter, and T. S. Ratiu [1998] Symplectic reduction for semidirect products and central extensions. *Diff. Geometry and Appl.* **9**, 173-212.

Marsden, J. E., R. Montgomery, P. Morrison, and W. B. Thompson [1986] Covariant Poisson brackets for classical fields. *Ann. of Phys.* **169**, 29-48.

Marsden, J. E., R. Montgomery, and T. Ratiu [1989] Cartan–Hannay–Berry phases and symmetry. *Cont. Math. AMS* **97**, 279-295.

Marsden, J. E., R. Montgomery, and T. S. Ratiu [1990] *Reduction, Symmetry, and Phases in Mechanics*. Memoirs AMS **436**.

Marsden, J. E. and P. J. Morrison [1984] Noncanonical Hamiltonian field theory and reduced MHD. *Cont. Math. AMS* **28**, 133–150.

Marsden, J. E., P. J. Morrison, and A. Weinstein [1984] The Hamiltonian structure of the BBGKY hierarchy equations. *Cont. Math. AMS* **28**, 115–124.

Marsden, J. E., O. M. O'Reilly, F. J. Wicklin, and B. W. Zombro [1991] Symmetry, stability, geometric phases, and mechanical integrators. *Nonlinear Science Today* **1**, 4–11; **1**, 14–21.

Marsden, J.E. and J. Ostrowski [1998] Symmetries in Motion: Geometric Foundations of Motion Control. *Nonlinear Sci. Today* (http://link.springer-ny.com).

Marsden, J. E., G. W. Patrick, and W. F. Shadwick [1996] *Integration Algorithms and Classical Mechanics*. Fields Institute Communications **10**, Am. Math. Society.

Marsden, J. E., G. W. Patrick, and S. Shkoller [1998] Multisymplectic Geometry, Variational Integrators, and Nonlinear PDEs. *Comm. Math. Phys.* **199**, 351–395.

Marsden, J. E. and T. S. Ratiu [1986] Reduction of Poisson manifolds. *Lett. Math. Phys.* **11**, 161–170.

Marsden, J. E., T. S. Ratiu, and G. Raugel [1991] Symplectic connections and the linearization of Hamiltonian systems. *Proc. Roy. Soc. Edinburgh A* **117**, 329–380.

Marsden, J. E., T. S. Ratiu, and A. Weinstein [1984a] Semi-direct products and reduction in mechanics. *Trans. Amer. Math. Soc.* **281**, 147–177.

Marsden, J. E., T. S. Ratiu, and A. Weinstein [1984b] Reduction and Hamiltonian structures on duals of semidirect product Lie Algebras. *Cont. Math. AMS* **28**, 55–100.

Marsden, J. E. and J. Scheurle [1987] The Construction and Smoothness of Invariant Manifolds by the Deformation Method. *SIAM J. Math. Anal.* **18**, 1261–1274.

Marsden, J. E. and J. Scheurle [1993a] Lagrangian reduction and the double spherical pendulum. *ZAMP* **44**, 17–43.

Marsden, J. E. and J. Scheurle [1993b] The reduced Euler–Lagrange equations. *Fields Institute Comm.* **1**, 139–164.

Marsden, J. E. and J. Scheurle [1995] Pattern evocation and geometric phases in mechanical systems with symmetry. *Dyn. and Stab. of Systems* **10**, 315–338.

Marsden, J. E., J. Scheurle, and J. Wendlandt [1996] Visualization of orbits and pattern evocation for the double spherical pendulum. *ICIAM 95: Mathematical Research, Academie Verlag, Ed. by K. Kirchgässner, O. Mahrenholtz, and R. Mennicken* **87**, 213–232.

Marsden, J. E. and S. Shkoller [1997] Multisymplectic geometry, covariant Hamiltonians and water waves. *Math. Proc. Camb. Phil. Soc.* (to appear).

Marsden, J. E. and A. J. Tromba [1996] *Vector Calculus*. Fourth Edition, W.H. Freeman.

Marsden, J. E. and A. Weinstein [1974] Reduction of symplectic manifolds with symmetry. *Rep. Math. Phys.* **5**, 121–130.

Marsden, J. E. and A. Weinstein [1979] Review of *Geometric Asymptotics* and *Symplectic Geometry and Fourier Analysis*, *Bull. Amer. Math. Soc.* **1**, 545–553.

Marsden, J. E. and A. Weinstein [1982] The Hamiltonian structure of the Maxwell–Vlasov equations. *Physica D* **4**, 394–406.

Marsden, J. E. and A. Weinstein [1983] Coadjoint orbits, vortices and Clebsch variables for incompressible fluids. *Physica D* **7**, 305–323.

Marsden, J. E., A. Weinstein, T. S. Ratiu, R. Schmid, and R. G. Spencer [1983] Hamiltonian systems with symmetry, coadjoint orbits and plasma physics. Proc. IUTAM-IS1MM Symposium on *Modern Developments in Analytical Mechanics*, Torino 1982, *Atti della Acad. della Sc. di Torino* **117**, 289–340.

Marsden, J. E. and J. M. Wendlandt [1997] Mechanical systems with symmetry, variational principles and integration algorithms. *Current and Future Directions in Applied Mathematics*, Edited by M. Alber, B. Hu, and J. Rosenthal, Birkhäuser, 219–261.

Martin, J. L. [1959] Generalized classical dynamics and the "classical analogue" of a Fermi oscillation. *Proc. Roy. Soc. London A* **251**, 536.

Maslov, V. P. [1965] *Theory of Perturbations and Asymptotic Methods*. Moscow State University.

Mazer, A. and T. S. Ratiu [1989] Hamiltonian formulation of adiabatic free boundary Euler flows. *J. Geom. Phys.* **6**, 271–291.

Melbourne, I., P. Chossat, and M. Golubitsky [1989] Heteroclinic cycles involving periodic solutions in mode interactions with $O(2)$ symmetry. *Proc. Roy. Soc. Edinburgh* **133A**, 315–345.

Melbourne, I. and M. Dellnitz [1993] Normal forms for linear Hamiltonian vector fields commuting with the action of a compact Lie group. *Proc. Camb. Phil. Soc.* **114**, 235–268.

Melnikov, V. K. [1963] On the stability of the center for time periodic perturbations. *Trans. Moscow Math. Soc.* **12**, 1–57.

Meyer, K. R. [1973] Symmetries and integrals in mechanics. In *Dynamical Systems*, M. Peixoto (ed.). Academic Press, pp. 259–273.

Meyer, K. R. [1981] Hamiltonian systems with a discrete symmetry. *J. Diff. Eqns.* **41**, 228–238.

Meyer, K. R. and R. Hall [1992] *Hamiltonian Mechanics and the n-body Problem*. Applied Mathematical Sciences **90**, Springer-Verlag.

Meyer, K. R. and D. G. Saari [1988] *Hamiltonian Dynamical Systems*. Cont. Math. AMS **81**.

Mielke, A. [1992] *Hamiltonian and Lagrangian Flows on Center Manifolds, with Applications to Elliptic Variational Problems*. Springer Lect. Notes in Math. **1489**.

Mikhailov, G. K. and V. Z. Parton [1990] *Stability and Analytical Mechanics. Applied Mechanics, Soviet Reviews.* **1**, Hemisphere.

Miller, S. C. and R. H. Good [1953] A WKB-type approximation to the Schrödinger equation. *Phys. Rev.* **91**, 174–179.

Milnor, J. [1963] *Morse Theory*. Princeton University Press.

Milnor, J. [1965] *Topology from the Differential Viewpoint*. University of Virginia Press.

Mishchenko, A. S. and A. T. Fomenko [1976] On the integration of the Euler equations on semisimple Lie algebras. *Sov. Math. Dokl.* **17**, 1591–1593.

Mishchenko, A. S. and A. T. Fomenko [1978a] Euler equations on finite-dimensional Lie groups. *Math. USSR, Izvestija* **12**, 371–389.

Mishchenko, A. S. and A. T. Fomenko [1978b] Generalized Liouville method of integration of Hamiltonian systems. *Funct. Anal. Appl.* **12**, 113–121.

Mishchenko, A. S. and A. T. Fomenko [1979] Symplectic Lie group actions. Springer Lecture Notes in Mathematics **763**, 504–539.

Mishchenko, A. S., V. E. Shatalov, and B. Y. Sternin [1990] *Lagrangian Manifolds and the Maslov Operator*. Springer-Verlag.

Misiolek, G. [1998] A shallow water equation as a geodesic flow on the Bott-Virasoro group. *J. Geom. Phys.* **24**, 203–208.

Misner, C., K. Thorne, and J. A. Wheeler [1973] *Gravitation*. W.H. Freeman, San Francisco.

Mobbs, S. D. [1982] Variational principles for perfect and dissipative fluid flows. *Proc. Roy. Soc. London A* **381**, 457–468.

Montaldi, J. A., R. M. Roberts, and I. N. Stewart [1988] Periodic solutions near equilibria of symmetric Hamiltonian systems. *Phil. Trans. Roy. Soc. London A* **325**, 237–293.

Montaldi, J. A., R. M. Roberts, and I. N. Stewart [1990] Existence of nonlinear normal modes of symmetric Hamiltonian systems. *Nonlinearity* **3**, 695–730, 731–772.

Montgomery, R. [1984] Canonical formulations of a particle in a Yang–Mills field. *Lett. Math. Phys.* **8**, 59–67.

Montgomery, R. [1985] Analytic proof of chaos in the Leggett equations for superfluid ^{3}He. *J. Low Temp. Phys.* **58**, 417–453.

Montgomery, R. [1988] The connection whose holonomy is the classical adiabatic angles of Hannay and Berry and its generalization to the non-integrable case. *Comm. Math. Phys.* **120**, 269–294.

Montgomery, R. [1990] Isoholonomic problems and some applications. *Comm. Math. Phys.* **128**, 565–592.

Montgomery, R. [1991a] How much does a rigid body rotate? A Berry's phase from the eighteenth century. *Amer. J. Phys.* **59**, 394–398.

Montgomery, R. [1991b] *The Geometry of Hamiltonian Systems*, T. Ratiu ed., MSRI Series **22**, Springer-Verlag.

Montgomery, R., J. E. Marsden, and T. S. Ratiu [1984] Gauged Lie–Poisson structures. *Cont. Math. AMS* **28**, 101–114.

Moon, F. C. [1987] *Chaotic Vibrations*, Wiley-Interscience.

Moon, F. C. [1998] *Applied Dynamics*, Wiley-Interscience.

Morozov, V. M., V. N. Rubanovskii, V. V. Rumiantsev, and V. A. Samsonov [1973] On the bifurcation and stability of the steady state motions of complex mechanical systems. *PMM* **37**, 387–399.

Morrison, P. J. [1980] The Maxwell–Vlasov equations as a continuous Hamiltonian system. *Phys. Lett. A* **80**, 383–386.

Morrison, P. J. [1982] Poisson brackets for fluids and plasmas, in *Mathematical Methods in Hydrodynamics and Integrability in Related Dynamical Systems*. M. Tabor and Y. M. Treve (eds.) AIP Conf. Proc. **88**.

Morrison, P. J. [1986] A paradigm for joined Hamiltonian and dissipative systems. *Physica D* **18**, 410–419.

Morrison, P. J. [1987] Variational principle and stability of nonmonotone Vlasov–Poisson equilibria. *Z. Naturforsch.* **42a**, 1115–1123.

Morrison, P. J. and S. Eliezer [1986] Spontaneous symmetry breaking and neutral stability on the noncanonical Hamiltonian formalism. *Phys. Rev. A* **33**, 4205.

Morrison, P. J. and D. Pfirsch [1990] The free energy of Maxwell–Vlasov equilibria. *Phys. Fluids B* **2**, 1105–1113.

Morrison, P. J. and D. Pfirsch [1992] Dielectric energy versus plasma energy, and Hamiltonian action-angle variables for the Vlasov equation. *Phys. Fluids B* **4**, 3038–3057.

Morrison, P. J. and J. M. Greene [1980] Noncanonical Hamiltonian density formulation of hydrodynamics and ideal magnetohydrodynamics. *Phys. Rev. Lett.* **45**, 790–794, errata **48** (1982), 569.

Morrison, P. J. and R. D. Hazeltine [1984] Hamiltonian formulation of reduced magnetohydrodynamics. *Phys. Fluids* **27**, 886–897.

Moser, J. [1958] New aspects in the theory of stability of Hamiltonian systems. *Comm. Pure Appl. Math.* **XI**, 81–114.

Moser, J. [1965] On the volume elements on a manifold. *Trans. Amer. Math. Soc.* **120**, 286–294.

Moser, J. [1973] *Stable and Random Motions in Dynamical Systems with Special Emphasis on Celestial Mechanics*. Princeton University Press.

Moser, J. [1974] *Finitely Many Mass Points on the Line Under the Influence of an Exponential Potential*. Springer Lect. Notes in Phys. **38**, 417–497.

Moser, J. [1975] Three integrable Hamiltonian systems connected with isospectral deformations. *Adv. in Math.* **16**, 197–220.

Moser, J. [1976] Periodic orbits near equilibrium and a theorem by Alan Weinstein. *Comm. Pure Appl. Math.* **29**, 727–747.

Moser, J. [1980] Various aspects of integrable Hamiltonian systems. *Dynamical Systems*, Progress in Math. **8**, Birkhäuser.

Moser, J. and A. P. Veselov [1991] Discrete versions of some classical integrable systems and factorization of matrix polynomials. *Comm. Math. Phys.* **139**, 217–243.

Murray, R. M. and S. S. Sastry [1993] Nonholonomic motion planning: steering using sinusoids. *IEEE Trans. on Automatic Control* **38**, 700–716.

Naimark, J. I. and N. A. Fufaev [1972] *Dynamics of Nonholonomic Systems*. Translations of Mathematical Monographs, Amer. Math. Soc., vol. **33**.

Nambu, Y. [1973] Generalized Hamiltonian dynamics. *Phys. Rev. D* **7**, 2405–2412.

Nekhoroshev, N. M. [1971a] Behavior of Hamiltonian systems close to integrable. *Funct. Anal. Appl.* **5**, 338–339.

Nekhoroshev, N. M. [1971b] Action angle variables and their generalizations. *Trans. Moscow Math. Soc.* **26**, 180–198.

Nekhoroshev, N. M. [1977] An exponential estimate of the time of stability of nearly integrable Hamiltonian systems. *Russ. Math. Surveys* **32**, 1–65.

Neishtadt, A. [1984] The separation of motions in systems with rapidly rotating phase. *P.M.M. USSR* **48**, 133–139.

Nelson, E. [1959] Analytic vectors. *Ann. Math.* **70**, 572–615.

Neumann, C. [1859] de problemate quodam mechanico, quod ad primam integralium ultra-ellipticorum clossem revocatur. *J. Reine u. Angew. Math.* **56**, 54–66.

Newcomb, W. A. [1958] Appendix in Bernstein [1958].

Newcomb, W. A. [1962] Lagrangian and Hamiltonian methods in Magnetohydrodynamics. *Nuc. Fusion* **Suppl.**, part **2**, 451–463.

Newell, A. C. [1985] *Solitons in Mathematics and Physics*. SIAM.

Newton, P. [1994] Hannay–Berry phase and the restricted three-vortex problem. *Physica D* **79**, 416–423.

Nijenhuis, A. [1953] On the holonomy ogroup of linear connections. *Indag. Math.* **15**, 233–249; **16** (1954), 17–25.

Nijenhuis, A. [1955] Jacobi-type identities for bilinear differential concomitants of certain tensor fields. *Indag. Math.* **17**, 390–403.

Nill, F. [1983] An effective potential for classical Yang–Mills fields as outline for bifurcation on gauge orbit space. *Ann. Phys.* **149**, 179–202.

Nirenberg, L. [1959] On elliptic partial differential equations. *Ann. Scuola. Norm. Sup. Pisa* 13(3), 115–162.

Noether, E. [1918] Invariante Variationsprobleme. *Kgl. Ges. Wiss. Nachr. Göttingen. Math. Physik.* 2, 235–257.

Oh, Y. G. [1987] A stability criterion for Hamiltonian systems with symmetry. *J. Geom. Phys.* 4, 163–182.

Oh, Y. G., N. Sreenath, P. S. Krishnaprasad, and J. E. Marsden [1989] The dynamics of coupled planar rigid bodies. Part 2: bifurcations, periodic solutions, and chaos. *Dynamics Diff. Eqns.* 1, 269–298.

Olver, P. J. [1980] On the Hamiltonian structure of evolution equations. *Math. Proc. Camb. Philps. Soc.* 88, 71–88.

Olver, P. J. [1984] Hamiltonian perturbation theory and water waves. *Cont. Math. AMS* 28, 231–250.

Olver, P. J. [1986] *Applications of Lie Groups to Differential Equations*. Graduate Texts in Mathematics 107, Springer-Verlag.

Olver, P. J. [1988] Darboux' theorem for Hamiltonian differential operators. *J. Diff. Eqns.* 71, 10–33.

Ortega, J.-P., and Ratiu, T. S. [1997] Persistence and smoothness of critical relative elements in Hamiltonian systems with symmetry. *C. R. Acad. Sci. Paris Sér. I Math.* 325, 1107–1111.

Ortega, J.-P., and Ratiu, T. S. [1998] Symmetry, Reduction, and Stability in Hamiltonian Systems. In preparation.

O'Reilly, O. M. [1996] The dynamics of rolling disks and sliding disks. *Nonlinear Dynamics* 10, 287–305.

O'Reilly, O. M. [1997] On the computation of relative rotations and geometric phases in the motions of rigid bodies. *Preprint.*

O'Reilly, O., N. K. Malhotra, and N. S. Namamchchivaya [1996] Some aspects of destabilization in reversible dynamical systems with application to follower forces. *Nonlinear Dynamics* 10, 63–87.

Otto, M. [1987] A reduction scheme for phase spaces with almost Kähler symmetry regularity results for momentum level sets. *J. Geom. Phys.* 4, 101–118.

Ovsienko, V. Y. and B. A. Khesin [1987] Korteweg–de Vries superequations as an Euler equation. *Funct. Anal. Appl.* 21, 329–331.

Palais, R. S. [1968] *Foundations of Global Non-Linear Analysis*. Benjamin.

Paneitz, S. M. [1981] Unitarization of symplectics and stability for causal differential equations in Hilbert space. *J. Funct. Anal.* 41, 315–326.

Pars, L. A. [1965] *A Treatise on Analytical Dynamics*. Wiley.

Patrick, G. [1989] The dynamics of two coupled rigid bodies in three-space. *Cont. Math. AMS* 97, 315–336.

Patrick, G. [1992] Relative equilibria in Hamiltonian systems: The dynamic interpretation of nonlinear stability on a reduced phase space. *J. Geom. and Phys.* 9, 111–119.

Patrick, G. [1995] Relative equilibria of Hamiltonian systems with symmetry: linearization, smoothness and drift. *J. Nonlinear Sci.* 5, 373–418.

Pauli, W. [1933] *General Principles of Quantum Mechanics*. Reprinted in English translation by Springer-Verlag (1981).

Pauli, W. [1953] On the Hamiltonian structure of non local field theories. *Il Nuovo Cimento* 10, 648–667.

Pekarsky, S. and J. E. Marsden [1998] Point Vortices on a Sphere: Stability of Relative Equilibria. *J. of Math. Phys.* **39**, 5894–5907.

Penrose, O. [1960] Electrostatic instabilities of a uniform non-Maxwellian plasma. *Phys. Fluids* **3**, 258–265.

Percival, I. and D. Richards [1982] *Introduction to Dynamics*. Cambridge Univ. Press.

Perelomov, A. M. [1990] *Integrable Systems of Classical Mechanics and Lie Algebras*. Birkhäuser.

Poincaré, H. [1885] Sur l'équilibre d'une masse fluide animée d'un mouvement de rotation. *Acta Math.* **7**, 259.

Poincaré, H. [1890] Sur la problème des trois corps et les équations de la dynamique. *Acta Math.* **13**, 1–271.

Poincaré, H. [1892] Les formes d'équilibre d'une masse fluide en rotation. *Revue Générale des Sciences* **3**, 809–815.

Poincaré, H. [1901a] Sur la stabilité de l'équilibre des figures piriformes affectées par une masse fluide en rotation. *Philos. Trans.* A **198**, 333–373.

Poincaré, H. [1901b] Sur une forme nouvelle des équations de la mécanique. *C.R. Acad. Sci.* **132**, 369–371.

Poincaré, H. [1910] Sur la precession des corps déformables. *Bull. Astron.* **27**, 321–356.

Potier-Ferry, M. [1982] On the mathematical foundations of elastic stability theory. *Arch. Rat. Mech. Anal.* **78**, 55–72.

Pressley, A. and G. Segal [1988] *Loop Groups*. Oxford Univ. Press.

Pullin, D. I. and P. G. Saffman [1991] Long time symplectic integration: the example of four-vortex motion. *Proc. Roy. Soc. London A* **432**, 481–494.

Puta, M. [1993] *Hamiltonian Mechanical Systems and Geometric Quantization*. Kluwer.

Rais, M. [1972] Orbites de la représentation coadjointe d'un groupe de Lie, *Représentations des Groupes de Lie Résolubles*. P. Bernat, N. Conze, M. Duflo, M. Lévy-Nahas, M. Rais, P. Renoreard, M. Vergne, eds. *Monographies de la Société Mathématique de France, Dunod, Paris* **4**, 15–27.

Ratiu, T. S. [1980] *Thesis*. University of California at Berkeley.

Ratiu, T. S. [1981a] Euler–Poisson equations on Lie algebras and the N-dimensional heavy rigid body. *Proc. Nat. Acad. Sci. USA* **78**, 1327–1328.

Ratiu, T. S. [1981b] The C. Neumann problem as a completely integrable system on an adjoint orbit. *Trans. Amer. Math. Soc.* **264**, 321–329.

Ratiu, T. S. [1982] Euler–Poisson equations on Lie algebras and the N-dimensional heavy rigid body. *Amer. J. Math.* **104**, 409–448, 1337.

Rayleigh, J. W. S. [1880] On the stability or instability of certain fluid motions. *Proc. London. Math. Soc.* **11**, 57–70.

Rayleigh, L. [1916] On the dynamics of revolving fluids. *Proc. Roy. Soc. London A* **93**, 148–154.

Reeb, G. [1949] Sur les solutions périodiques de certains systèmes différentiels canoniques. *C.R. Acad. Sci. Paris* **228**, 1196–1198.

Reeb, G. [1952] Variétés symplectiques, variétés presque-complexes et systèmes dynamiques. *C.R. Acad. Sci. Paris* **235**, 776–778.

Reed, M. and B. Simon [1974] *Methods on Modern Mathematical Physics*. Vol. 1: *Functional Analysis*. Vol. 2: *Self-adjointness and Fourier Analysis*. Academic Press.

Reyman, A. G. and M. A. Semenov-Tian-Shansky [1990] Group theoretical methods in the theory of integrable systems. *Encyclopedia of Mathematical Sciences* **16**, Springer-Verlag.

Riemann, B. [1860] Untersuchungen über die Bewegung eines flüssigen gleich-artigen Ellipsoides. *Abh. d. Königl. Gesell. der Wiss. zu Göttingen* **9**, 3–36.

Riemann, B. [1861] Ein Beitrag zu den Untersuchungen über die Bewegung eines flüssigen gleichartigen Ellipsoides. *Abh. d. Königl. Gesell. der Wiss. zu Göttingen*.

Robbins, J. M. and M. V. Berry [1992] The geometric phase for chaotic systems. *Proc. Roy. Soc. London A* **436**, 631–661.

Robinson, C. [1970] Generic properties of conservative systems, I, II. *Amer. J. Math.* **92**, 562–603.

Robinson, C. [1975] Fixing the center of mass in the *n*-body problem by means of a group action. *Colloq. Intern. CNRS* **237**.

Robinson, C. [1988] Horseshoes for autonomous Hamiltonian systems using the Melnikov integral. *Ergodic Theory Dynamical Systems* **8***, 395–409.

Rosenbluth, M. N. [1964] Topics in microinstabilities. *Adv. Plasma Phys.* **137**, 248.

Routh, E. J. [1877] *Stability of a given state of motion*. Macmillan. Reprinted in *Stability of Motion*, A. T. Fuller (ed.), Halsted Press, 1975.

Routh, E. J. [1884] *Advanced Rigid Dynamics*. Macmillian.

Rubin, H. and P. Ungar [1957] Motion under a strong constraining force. *Comm. Pure Appl. Math.* **10**, 65–87.

Rumjantsev, V. V. [1982] On stability problem of a top. *Rend. Sem. Mat. Univ. Padova* **68**, 119–128.

Ruth, R. [1983] A canonical integration technique. *IEEE Trans. Nucl. Sci.* **30**, 2669–2671.

Rytov, S. M. [1938] Sur la transition de l'optique ondulatoire à l'optique géométrique. *Dokl. Akad. Nauk SSSR* **18**, 263–267.

Salam, F. M. A., J. E. Marsden, and P. P. Varaiya [1983] Arnold diffusion in the swing equations of a power system. *IEEE Trans. CAS* **30**, 697–708, **31** 673–688.

Salam, F. A. and S. Sastry [1985] Complete Dynamics of the forced Josephson junction; regions of chaos. *IEEE Trans. CAS* **32** 784–796.

Salmon, R. [1988] Hamiltonian fluid mechanics. *Ann. Rev. Fluid Mech.* **20**, 225–256.

Sánchez de Alvarez, G. [1986] Thesis. University of California at Berkeley.

Sánchez de Alvarez, G. [1989] Controllability of Poisson control systems with symmetry. *Cont. Math. AMS* **97**, 399–412.

Sanders, J. A. [1982] Melnikov's method and averaging. *Celestial Mech.* **28**, 171–181.

Sanders, J. A. and F. Verhulst [1985] *Averaging Methods in Nonlinear Dynamical Systems*. Applied Mathematical Sciences **59**, Springer-Verlag.

Sanz-Serna, J. M. and M. Calvo [1994] *Numerical Hamiltonian Problems*. Chapman and Hall, London.

Sattinger, D. H. and D. L. Weaver [1986] *Lie Groups and Lie Algebras in Physics, Geometry, and Mechanics*. Applied Mathematical Sciences **61**, Springer-Verlag.

Satzer, W. J. [1977] Canonical reduction of mechanical systems invariant under abelian group actions with an application to celestial mechanics. *Indiana Univ. Math. J.* **26**, 951–976.

Scheurle, J. [1989] Chaos in a rapidly forced pendulum equation. *Cont. Math. AMS* **97**, 411–419.

Scheurle, J., J. E. Marsden, and P. J. Holmes [1991] Exponentially small estimates for separatrix splittings. *Proc. Conf. Beyond all Orders*, II. Segur and S. Tanveer (eds.), Birkhäuser.

Schouten, J. A. [1940] *Ricci Calculus* (2nd Edition 1954). Springer-Verlag.

Schur, I. [1923] Über eine Klasse von Mittelbildungen mit Anwendungen auf Determinantentheorie. *Sitzungsberichte der Berliner Math. Gessellshaft* **22**, 9–20.

Scovel, C. [1991] Symplectic numerical integration of Hamiltonian systems. *Geometry of Hamiltonian systems*, ed. T. Ratiu, MSRI Series **22**, Springer-Verlag, pp. 463–496.

Segal, G. [1991] The geometry of the KdV equation. *Int. J. Mod. Phys. A* **6**, 2859–2869.

Segal, I. [1962] Nonlinear semigroups. *Ann. Math.* **78**, 339–364.

Seliger, R. L. and G. B. Whitham [1968] Variational principles in continuum mechanics. *Proc. Roy. Soc. London* **305**, 1–25.

Serrin, J. [1959] Mathematical principles of classical fluid mechanics. *Handbuch der Physik* **VIII-I**, 125–263, Springer-Verlag.

Shahshahani, S. [1972] Dissipative systems on manifolds. *Inv. Math.* **16**, 177–190.

Shapere, A. and F. Wilczek [1987] Self-propulsion at low Reynolds number. *Phys. Rev. Lett.* **58**, 2051–2054.

Shapere, A. and F. Wilczeck [1989] Geometry of self-propulsion at low Reynolds number. *J. Fluid Mech.* **198**, 557–585.

Shinbrot, T., C. Grebogi, J. Wisdom, and J. A. Yorke [1992] Chaos in a double pendulum. *Amer. J. Phys.* **60**, 491–499.

Simo, J. C. and D. D. Fox [1989] On a stress resultant, geometrically exact shell model. Part I: Formulation and optimal parametrization. *Comp. Meth. Appl. Mech. Engr.* **72**, 267–304.

Simo, J. C., D. D. Fox, and M. S. Rifai [1990] On a stress resultant geometrically exact shell model. Part III: Computational aspects of the nonlinear theory. *Comput. Methods Applied Mech. and Engr.* **79**, 21–70.

Simo, J. C., D. R. Lewis, and J. E. Marsden [1991] Stability of relative equilibria I: The reduced energy momentum method. *Arch. Rat. Mech. Anal.* **115**, 15–59.

Simo, J. C. and J. E. Marsden [1984] On the rotated stress tensor and a material version of the Doyle Ericksen formula. *Arch. Rat. Mech. Anal.* **86**, 213–231.

Simo, J. C., J. E. Marsden, and P. S. Krishnaprasad [1988] The Hamiltonian structure of nonlinear elasticity: The material, spatial, and convective representations of solids, rods, and plates. *Arch. Rat. Mech. Anal.* **104**, 125–183.

Simo, J. C., T. A. Posbergh, and J. E. Marsden [1990] Stability of coupled rigid body and geometrically exact rods: block diagonalization and the energy–momentum method. *Phys. Rep.* **193**, 280–360.

Simo, J. C., T. A. Posbergh, and J. E. Marsden [1991] Stability of relative equilibria II: Three dimensional elasticity. *Arch. Rat. Mech. Anal.* **115**, 61–100.

Simo, J. C., M. S. Rifai, and D. D Fox [1992] On a stress resultant geometrically exact shell model. Part VI: Conserving algorithms for nonlinear dynamics. *Comp. Meth. Appl. Mech. Engr.* **34**, 117–164.

Simo, J. C. and N. Tarnow [1992] The discrete energy–momentum method. Conserving algorithms for nonlinear elastodynamics. *ZAMP* **43**, 757–792.

Simo, J. C., N. Tarnow, and K. K. Wong [1992] Exact energy–momentum conserving algorithms and symplectic schemes for nonlinear dynamics. *Comput. Methods Appl. Mech. Engr.* **1**, 63–116.

Simo, J. C. and L. VuQuoc [1985] Three-dimensional finite strain rod model. Part II. Computational Aspects. *Comput. Methods Appl. Mech. Engr.* **58**, 79–116.

Simo, J. C. and L. VuQuoc [1988a] On the dynamics in space of rods undergoing large overall motions—a geometrically exact approach. *Comput. Methods Appl. Mech. Engr.* **66**, 125–161.

Simo, J. C. and L. VuQuoc [1988b] The role of nonlinear theories in the dynamics of fast rotating flexible structures. *J. Sound Vibration* **119**, 487–508.

Simo, J. C. and K. K. Wong [1989] Unconditionally stable algorithms for the orthogonal group that exactly preserve energy and momentum. *Int. J. Num. Meth. Engr.* **31**, 19–52.

Simon, B. [1983] Holonomy, the Quantum Adiabatic Theorem, and Berry's Phase. *Phys. Rev. Letters* **51**, 2167–2170.

Sjamaar, R. and E. Lerman [1991] Stratified symplectic spaces and reduction. *Ann. of Math.* **134**, 375–422.

Slawianowski, J. J. [1971] Quantum relations remaining valid on the classical level. *Rep. Math. Phys.* **2**, 11–34.

Slebodzinski, W. [1931] Sur les équations de Hamilton. *Bull. Acad. Roy. de Belg.* **17**, 864–870.

Slebodzinski, W. [1970] *Exterior Forms and Their Applications.* Polish Scientific.

Slemrod, M. and J. E. Marsden [1985] Temporal and spatial chaos in a van der Waals fluid due to periodic thermal fluctuations. *Adv. Appl. Math.* **6**, 135–158.

Smale, S. [1964] Morse theory and a nonlinear generalization of the Dirichlet problem. *Ann. Math.* **80**, 382–396.

Smale, S. [1967] Differentiable dynamical systems. *Bull. Amer. Math. Soc.* **73**, 747–817. (Reprinted in *The Mathematics of Time.* Springer-Verlag, by S. Smale [1980].)

Smale, S. [1970] Topology and Mechanics. *Inv. Math.* **10**, 305–331; **11**, 45–64.

Sniatycki, J. [1974] Dirac brackets in geometric dynamics. *Ann. Inst. H. Poincaré* **20**, 365–372.

Sniatycki, J. and W. Tulczyjew [1971] Canonical dynamics of relativistic charged particles. *Ann. Inst. H. Poincaré* **15**, 177–187.

Sontag, E. D. and H. J. Sussman [1988] Further comments on the stabilization of the angular velocity of a rigid body. *Systems Control Lett.* **12**, 213–217.

Souriau, J. M. [1966] Quantification géométrique. *Comm. Math. Phys.* **1**, 374–398.

Souriau, J. M. [1967] Quantification géométrique. Applications. *Ann. Inst. H. Poincaré* **6**, 311–341.

Souriau, J. M. [1970] (©1969) *Structure des Systèmes Dynamiques.* Dunod, Paris. English translation by R. H. Cushman and G. M. Tuynman. Progress in Mathematics, **149**. Birkhäuser Boston, 1997.

Spanier, E. H. [1966] *Algebraic Topology.* McGraw-Hill (Reprinted by Springer-Verlag).

Spencer, R. G. and A. N. Kaufman [1982] Hamiltonian structure of two-fluid plasma dynamics. *Phys. Rev. A* **25**, 2437–2439.

Spivak, M. [1976] *A Comprehensive Introduction to Differential Geometry.* Publish or Perish.

Sreenath, N., Y. G. Oh, P. S. Krishnaprasad, and J. E. Marsden [1988] The dynamics of coupled planar rigid bodies. Part 1: Reduction, equilibria and stability. *Dyn. Stab. Systems* **3**, 25–49.

Stefan, P. [1974] Accessible sets, orbits and foliations with singularities. *Proc. Lond. Math. Soc.* **29**, 699–713.

Sternberg, S. [1963] *Lectures on Differential Geometry.* Prentice-Hall. (Reprinted by Chelsea.)

Sternberg, S. [1969] *Celestial Mechanics,* Vols. I, II. Benjamin-Cummings.

Sternberg, S. [1975] Symplectic homogeneous spaces. *Trans. Amer. Math. Soc.* **212**, 113–130.

Sternberg, S. [1977] Minimal coupling and the symplectic mechanics of a classical particle in the presence of a Yang–Mills field. *Proc. Nat. Acad. Sci.* **74**, 5253–5254.

Su, C. A. [1961] Variational principles in plasma dynamics. *Phys. Fluids* **4**, 1376–1378.

Sudarshan, E. C. G. and N. Mukunda [1974] *Classical Mechanics: A Modern Perspective.* Wiley, 1974; Second Edition, Krieber, 1983.

Sussman, H. [1973] Orbits of families of vector fields and integrability of distributions. *Trans. Amer. Math. Soc.* **180**, 171–188.

Symes, W. W. [1980] Hamiltonian group actions and integrable systems. *Physica D* **1**, 339–374.

Symes, W. W. [1982a] Systems of Toda type, inverse spectral problems and representation theory. *Inv. Math.* **59**, 13–51.

Symes, W. W. [1982b] The QR algorithm and scattering for the nonperiodic Toda lattice. *Physica D* **4**, 275–280.

Szeri, A. J. and P. J. Holmes [1988] Nonlinear stability of axisymmetric swirling flow. *Phil. Trans. Roy. Soc. London A* **326**, 327–354.

Temam, R. [1975] On the Euler equations of incompressible perfect fluids. *J. Funct. Anal.* **20**, 32–43.

Thirring, W. E. [1978] *A Course in Mathematical Physics.* Springer-Verlag.

Thomson, W. (Lord Kelvin) and P. G. Tait [1879] *Treatise on Natural Philosophy.* Cambridge University Press.

Toda, M. [1975] Studies of a non-linear lattice. *Phys. Rep. Phys. Lett.* **8**, 1–125.

Tulczyjew, W. M. [1977] The Legendre transformation. *Ann. Inst. Poincaré* **27**, 101–114.

Vaisman, I. [1987] *Symplectic Geometry and Secondary Characteristic Classes.* Progress in Mathematics **72**, Birkhäuser.

Vaisman, I. [1994] *Lectures on the Geometry of Poisson Manifolds.* Progress in Mathematics **118**, Birkhäuser.

Vaisman, I. [1996] Reduction of Poisson–Nijenhuis manifolds. *J. Geom. and Physics* **19**, 90–98.

van der Meer, J. C. [1985] *The Hamiltonian Hopf Bifurcation.* Springer Lect. Notes in Math. **1160**.

van der Meer, J. C. [1990] Hamiltonian Hopf bifurcation with symmetry. *Nonlinearity* **3**, 1041–1056.

van der Schaft, A. J. [1982] Hamiltonian dynamics with external forces and observations. *Math. Systems Theory* **15**, 145–168.

van der Schaft, A. J. [1986] Stabilization of Hamiltonian systems. *Nonlinear Amer. TMA* **10**, 1021–1035.

van der Schaft, A. J. and D. E. Crouch [1987] Hamiltonian and self-adjoint control systems. *Systems Control Lett.* **8**, 289–295.

van Kampen, N. G. and B. U. Felderhof [1967] *Theoretical Methods in Plasma Physics.* North-Holland.

van Kampen, N. G. and J. J. Lodder [1984] Constraints. *Am. J. Phys.* **52**(5), 419–424.

van Saarloos, W. [1981] A canonical transformation relating the Lagrangian and Eulerian descriptions of ideal hydrodynamics. *Physica A* **108**, 557–566.

Varadarajan, V. S. [1974] *Lie Groups, Lie Algebras and Their Representations.* Prentice Hall. (Reprinted in Graduate Texts in Mathematics, Springer-Verlag.)

Vershik, A. M. and L. Faddeev [1981] Lagrangian mechanics in invariant form. *Sel. Math. Sov.* **1**, 339–350.

Vershik, A. M. and V. Ya Gershkovich [1988] Non-holonomic Riemannian manifolds. Encyclopedia of Math. *Dynamical Systems* **7**, Springer-Verlag.

Veselov, A. P. [1988] Integrable discrete-time systems and difference operators. *Funct. An. and Appl.* **22**, 83–94.

Veselov, A. P. [1991] Integrable Lagrangian correspondences and the factorization of matrix polynomials. *Funct. An. and Appl.* **25**, 112–123.

Vinogradov, A. M. and I. S. Krasilshchik [1975] What is the Hamiltonian formalism? *Russ. Math. Surveys* **30**, 177–202.

Vinogradov, A. M. and B. A. Kuperschmidt [1977] The structures of Hamiltonian mechanics. *Russ. Math. Surveys* **32**, 177–243.

Vladimirskii, V. V. [1941] Über die Drehung der Polarisationsebene im gekrümmten Lichtstrahl *Dokl. Akad. Nauk USSR* **21**, 222–225.

Wald, R.M. [1993] Variational principles, local symmetries and black hole entropy. *Proc. Lanczos Centenary Volume* SIAM, 231–237.

Wan, Y. H. [1986] The stability of rotating vortex patches. *Comm. Math. Phys.* **107**, 1–20.

Wan, Y. H. [1988a] Instability of vortex streets with small cores. *Phys. Lett. A* **127**, 27–32.

Wan, Y. H. [1988b] Desingularizations of systems of point vortices. *Physica D* **32**, 277–295.

Wan, Y. H. [1988c] Variational principles for Hill's spherical vortex and nearly spherical vortices. *Trans. Amer. Math. Soc.* **308**, 299–312.

Wan, Y. H. and M. Pulvirente [1984] Nonlinear stability of circular vortex patches. *Comm. Math. Phys.* **99**, 435–450.

Wang, L. S. and P. S. Krishnaprasad [1992] Gyroscopic control and stabilization. *J. Nonlinear Sci.* **2**, 367–415.

Wang, L. S., P. S. Krishnaprasad, and J. H. Maddocks [1991] Hamiltonian dynamics of a rigid body in a central gravitational field. *Cel. Mech. Dyn. Astr.* **50**, 349–386.

Weber, R. W. [1986] Hamiltonian systems with constraints and their meaning in mechanics. *Arch. Rat. Mech. Anal.* **91**, 309–335.

Weinstein, A. [1971] Symplectic manifolds and their Lagrangian submanifolds. *Adv. in Math.* **6**, 329–346 (see also *Bull. Amer. Math. Soc.* **75** (1969), 1040–1041).

Weinstein, A. [1973] Normal modes for nonlinear Hamiltonian systems. *Inv. Math.* **20**, 47–57.

Weinstein, A. [1977] *Lectures on Symplectic Manifolds.* CBMS Regional Conf. Ser. in Math. **29**, Amer. Math. Soc.

Weinstein, A. [1978a] A universal phase space for particles in Yang–Mills fields. *Lett. Math. Phys.* **2**, 417–420.

Weinstein, A. [1978b] Bifurcations and Hamilton's principle. *Math. Z.* **159**, 235–248.

Weinstein, A. [1981] Neighborhood classification of isotropic embeddings. *J. Diff. Geom.* **16**, 125–128.

Weinstein, A. [1083a] Sophus Lie and symplectic geometry. *Exposition Math.* **1**, 95–96.

Weinstein, A. [1983b] The local structure of Poisson manifolds. *J. Diff. Geom.* **18**, 523–557.

Weinstein, A. [1984] Stability of Poisson–Hamilton equilibria. *Cont. Math. AMS* **28**, 3–14.

Weinstein, A. [1990] Connections of Berry and Hannay type for moving Lagrangian submanifolds. *Adv. in Math.* **82**, 133–159.

Weinstein, A. [1996] Lagrangian Mechanics and Groupoids. *Fields Inst. Comm.* **7**, 207–231.

Wendlandt, J. M. and J. E. Marsden [1997] Mechanical integrators derived from a discrete variational principle. *Physica D* **106**, 223–246.

Whittaker, E. T. [1927] *A Treatise on the Analytical Dynamics of Particles and Rigidbodies.* Cambridge University Press.

Whittaker, E. T. and G. N. Watson [1940] *A Course of Modern Analysis, 4th ed.* Cambridge University Press.

Wiggins, S. [1988] *Global Bifurcations and Chaos.* Texts in Applied Mathematical Sciences **73**, Springer-Verlag.

Wiggins, S. [1990] *Introduction to Applied Nonlinear Dynamical Systems and Chaos.* Texts in Applied Mathematical Sciences **2**, Springer-Verlag.

Wiggins, S. [1992] *Chaotic Transport in Dynamical Systems.* Interdisciplinary Mathematical Sciences, Springer-Verlag.

Wiggins, S. [1993] *Global Dynamics, Phase Space Transport, Orbits Homoclinic to Resonances and Applications.* Fields Institute Monographs. **1**, Amer. Math. Soc.

Wilczek, F. and A. Shapere [1989] Geometry of self-propulsion at low Reynold's number. Efficiencies of self-propulsion at low Reynold's number. *J. Fluid Mech.* **198**, 587–599.

Williamson, J. [1936] On an algebraic problem concerning the normal forms of linear dynamical systems. *Amer. J. Math.* **58**, 141–163; **59**, 599–617.

Wintner, A. [1941] *The Analytical Foundations of Celestial Mechanics.* Princeton University Press.

Wisdom, J., S. J. Peale, and F. Mignard [1984] The chaotic rotation of Hyperion. *Icarus* **58**, 137–152.

Wong, S. K. [1970] Field and particle equations for the classical Yang–Mills field and particles with isotopic spin. *Il Nuovo Cimento* **65**, 689–694.

Woodhouse, N. M. J. [1992] *Geometric Quantization.* Clarendon Press, Oxford University Press, 1980, Second Edition, 1992.

Xiao, L. and M. E. Kellman [1989] Unified semiclassical dynamics for molecular resonance spectra. *J. Chem. Phys.* **90**, 6086–6097.

Yang, C. N. [1985] Fiber bundles and the physics of the magnetic monopole. *The Chern Symposium.* Springer-Verlag, pp. 247–254.

Yang, R. and P. S. Krishnaprasad [1990] On the dynamics of floating four bar linkages. *Proc. 28th IEEE Conf. on Decision and Control.*

Zakharov, V. E. [1971] Hamiltonian formalism for hydrodynamic plasma models. *Sov. Phys. JETP* **33**, 927–932.

Zakharov, V. E. [1974] The Hamiltonian formalism for waves in nonlinear media with dispersion. *Izvestia Vuzov, Radiofizika* **17**.

Zakharov, V. E. and L. D. Faddeev [1972] Korteweg–de Vries equation: a completely integrable Hamiltonian system. *Funct. Anal. Appl.* **5**, 280–287.

Zakharov, V. E. and E. A. Kuznetsov [1971] Variational principle and canonical variables in magnetohydrodynamics. *Sov. Phys. Dokl.* **15**, 913–914.

Zakharov, V. E. and E. A. Kuznetsov [1974] Three-dimensional solitons. *Sov. Phys. JETP* **39**, 285–286.

Zakharov, V. E. and E. A. Kuznetsov [1984] Hamiltonian formalism for systems of hydrodynamic type. *Math. Phys. Rev.* **4**, 167–220.

Zenkov, D. V. [1995] The Geometry of the Routh Problem. *J. Nonlinear Sci.* **5**, 503–519.

Zenkov, D. V., A. M. Bloch, and J. E. Marsden [1998] The Energy Momentum Method for the Stability of Nonholonomic Systems. *Dyn. Stab. of Systems.* **13**, 123–166.

Zhuravlev, V. F. [1996] The solid angle theorem in rigid-body dynamics. *J. Appl. Math. and Mech. (PMM)* **60**, 319–322.

Ziglin, S. L. [1980a] Decomposition of separatrices, branching of solutions and nonexistena of an integral in the dynamics of a rigid body. *Trans. Moscow Math. Soc.* **41**, 287.

Ziglin, S. L. [1980b] Nonintegrability of a problem on the motion of four point vortices. *Sov. Math. Dokl.* **21**, 296–299.

Ziglin, S. L. [1981] Branching of solutions and nonexistence of integrals in Hamiltonian systems. *Dokl. Akad. Nauk SSSR* **257**, 26–29; *Funct. Anal. Appl.* **16**, 30–41, **17**, 8–23.

Index